Gas-Liquid-Solid Fluidization Engineering

BUTTERWORTHS SERIES IN CHEMICAL ENGINEERING

SERIES TITLES

Chemical Process Equipment: Selection and Design *Stanley M. Walas*

Constitutive Equations for Polymer Melts and Solutions *Ronald G. Larson*

Fundamental Process Control *David M. Prett and Carlos E. García*

Gas-Liquid-Solid Fluidization Engineering *Liang-Shih Fan*

Gas Separation by Adsorption Process *Ralph T. Yang*

Granular Filtration of Aerosols and Hydrosols *Chi Tien*

Heterogeneous Reactor Design *Hong H. Lee*

Molecular Thermodynamics of Nonideal Fluids *Lloyd L. Lee*

Phase Equilibria in Chemical Engineering *Stanley M. Walas*

Physicochemical Hydrodynamics: An Introduction *Ronald F. Probstein*

Transport Processes in Chemically Reacting Flow Systems *Daniel E. Rosner*

Viscous Flows: The Practical Use of Theory *Stuart Winston Churchill*

Gas-Liquid-Solid Fluidization Engineering

Liang-Shih Fan
Professor of Chemical Engineering
The Ohio State University

Butterworths
Boston London Singapore Sydney Toronto Wellington

Library of Congress Cataloging-in-Publication Data

Fan, Liang-Shih.
Gas-liquid-solid fluidization engineering.

(Butterworths series in chemical engineering)
Includes bibliographies and index.
1. Fluidization. I. Title. II. Series.
TP156.F65F35 1989 660'.284292 89-7081
ISBN 0-409-95179-X

British Library Cataloguing in Publication Data
Fan, Liang - Shih
Gas - liquid - solid fluidization engineering.
1. Chemical engineering. Fluidisation
I. Title
660.2'842

ISBN 0-409-95179-X

Butterworth Publishers
80 Montvale Avenue
Stoneham, MA 02180

10 9 8 7 6 5 4 3 2 1

Printed in the United States of America

In memory of my father
and to my mother

Contents

* by L.-S. Fan and K. Muroyama

Preface

The field of gas-liquid-solid fluidization, in the broad sense, encompasses any gas-liquid-solid system where solids are in a nonstationary state. Such systems have gained considerable importance as evidenced in their commercial or demonstrated applications in physical, chemical, petrochemical, electrochemical, and biochemical processing. Successful application of gas-liquid-solid fluidization systems lies in comprehensive understanding of the transport phenomena, including hydrodynamics, heat and mass transfer, and mixing. The transport behavior in these systems is, however, very complex. This book attempts to provide a unified treatment of the principles underlying transfer phenomena of various fluidization systems and to reflect their impact on reactor applications.

The book consists of three parts. Part I is an introduction which describes a unique method of classification of a wide range of gas-liquid-solid fluidization systems varying from fluidized beds, slurry bubble columns, and turbulent bed contactors, to three-phase transport. Numerous application examples using these systems are also given. Part II discusses transport fundamentals of gas-liquid-solid fluidization, described according to the classifications given in Part I. Here, mechanistic interpretations based on theoretical and semi-theoretical approaches to transport phenomena systems are described. Moreover, a comprehensive account of the empirical approaches in terms of correlations is given. In Part III, numerous industrial applications of three-phase fluidization systems including results obtained from large scale systems are presented in the context of the transport phenomena approaches given in Part II. Pertinent reactor engineering problems associated with these applications are also addressed. A classical approach to catalytic reactor analysis utilizing transport properties described in Part II is exemplified in Appendix A. Appendix B summarizes the SI units on which all the empirical correlations given in this book are based.

The book can be used as a graduate textbook in a three-phase fluidization engineering course or as a supplemental textbook to a general fluidization engineering course. It is desirable that students who use this book have already had some background in two-phase systems and are familiar with the classical material presented in, for example, Wallis (1969) (One-Dimensional Two-Phase Flow, McGraw Hill) and Kunii and Levenspiel (1969) (Fluidization Engineering, John Wiley). The book can also be used as a supplemental textbook or reference for multiphase reaction engineering courses covering areas including biochemical engineering, environmental engineering, petroleum engineering and energy engineering. The book complements well books on general three-phase reactors by Shah (1979) (Gas-Liquid-Solid Reactor Design, McGraw Hill) and Ramachandran and Chaudhari (1983) (Three-Phase Catalytic Reactors, Gordon and Breach) which are more reaction engineering oriented. The book will also be useful to practicing engineers or industrial researchers who require readily available transport properties information for system design, operation or simulation.

Completion of this book would not have been timely without numerous very dedicated and capable members of my research group who have helped with the project over the years. I am especially grateful to Dr. K. Muroyama for providing an extensive literature survey, particularly on empirical correlations which are included in the tables and figures of Chapters 3, 4, 5, and 6. I owe much to Mr. D. C. Arters, Dr. Y.-M. Chen, Dr. K. Fujie, Mr. R. L. Gorowara, Mr. R.-H. Jean, Dr. K. Kitano, Mr. B. E. Kreischer, Dr. T. Miyahara, Mr. P. Murray, Mr. G. H. Song, Mr. W.-T. Tang, Dr. K. Tsuchiya, Professor K. D. Wisecarver, and Dr. Y.-S. Yang for providing literature material, and for making valuable original contributions which are directly incorporated in the text. Other research group members have read the entire text and made useful technical and editorial suggestions. My inquisitive students in the Chemical Engineering 815.15 course, entitled "Fluidization Engineering", have provided valuable feedback about some portions of the text.

I am deeply grateful to several colleagues who have reviewed a specific chapter in the application part in the area of their expertise and made significant comments: Mr. R. D. Buttke, Professor J. J. Chalmers, Dr. C. S. Chang, Mr. J. F. Mosby, Dr. L. E. Peck, Professor S. T. Yang and Ms. L.-P. Zhang. I express my special thanks to Professor M. A. Bergougnou, Professor N. Epstein, Professor J. P. Hartnett, Professor R. Leyva-Ramos and Mr. D. N. Smith, who have provided valuable suggestions and overviews on the text, and to many others who graciously responded to my requests for specific information for inclusion in this book. I want to especially acknowledge my colleagues in the Department of Chemical Engineering at The Ohio State University: Professor R. S. Brodkey, Professor T. L. Sweeney, and

Professor J. L. Zakin, for providing consistent encouragement over the years, the late Professor C. Y. Wen at West Virginia University for first introducing me to this field some fifteen years ago, and Professor L. T. Fan at Kansas State University and Mr. L. F. Fan at Texas Instruments for their inspiration and personal efforts. Thanks are due to Mrs. P. J. Osborn and Mrs. S. E. McDonald for their superb use of the word processor in typing numerous editions of the manuscript, to Mr. P. Cai and Mr. B. E. Kreischer, as well as Mr. D. C. Arters, Mr. C.-S. Jang, Mr. K. M. Russ, Dr. K. Raghunathan, Mr. G. H. Song and Dr. K. Tsuchiya, for excellent editorial work on the text, tables, and figures, and to the students at Tottori University for outstanding efforts on figure drawing. Several agencies including Amoco Oil Company, the National Science Foundation, The Ohio State University Water Resources Center, and U.S. Geological Survey, provided financial support for the research on which much of this book is based. Their assistance is gratefully acknowledged.

Finally, I am most indebted to my wife, Shu-Huan, for her support and enduring sacrifices throughout my academic career and to my children, Jonathan and Joline, for brightening the days.

L.-S. Fan
Columbus, Ohio

Special Acknowledgments

Chapters 3, 4, 5, and 6 of this book were written, in part, based on a literature survey and pertinent comments provided by Dr. Katsuhiko Muroyama. He is duly credited as a coauthor of these chapters for his contribution. Dr. Muroyama is presently with the Department of Environmental Chemistry and Technology at Tottori University, Japan.

L.-S. Fan

Part I

Introduction

Chapter 1

Classification and Significance

Gas-liquid-solid systems for physical, chemical, petrochemical, electrochemical, and biochemical operations can refer to a variety of operational schemes. Processing alternatives include the choice of continuous phase, relative flow directions, and continuous flow or batch. For example, both the liquid and solids can be either in continuous flow or batch while the gas is in continuous flow; the flow of gas and liquid can be cocurrent upward, cocurrent downward, countercurrent, or crosscurrent; the flow of solids can be upward or downward independent of or dependent on the gas or liquid flow; the gas may either be a continuous phase or discrete bubbles; the liquid may be a continuous phase, a film, or droplets; and the solids are a discrete phase in either a packed or suspended state (Ostergaard, 1968; Shah, 1979; Epstein, 1981; Shah et al., 1982; Muroyama and Fan, 1985; Smith et al., 1986; Fan et al., 1988). Because of this diversity, establishing a unified classification scheme for gas-liquid-solid systems is not an easy task. This chapter describes the classification of gas-liquid-solid systems, operating modes of fluidization, and applications of gas-liquid-solid fluidization and compares fluidization systems with fixed bed systems.

1.1 CLASSIFICATION

It is desirable to classify gas-liquid-solid systems through appropriate differences in hydrodynamic behavior. The hydrodynamic behavior is characterized by the type of operation, relative direction of flows, and continuity of the phases and is inherently related to the state of motion of the solid phase. Classification of three-phase systems can readily be extended from that of gas-liquid, gas-solid, and liquid-solid systems. It is, however, convenient and practical to classify them according to the state of particle motion in a manner similar to that for gas-solid or liquid-solid systems[1]. The state of particle motion can be subdivided into three basic operating regimes: the fixed bed regime, the expanded bed regime, and the transport regime. The fixed bed regime exists when the drag force on the particles induced by the flow of a gas-liquid mixture is smaller than the effective weight of the particles in the system. When, with an increase in gas and/or liquid velocity, the drag force counterbalances the effective weight of the particles, the bed is in the state of minimum fluidization and marks the onset of the expanded bed regime. With a further increase in gas and/or liquid velocity beyond the minimum fluidization velocity, operation is in the expanded bed regime until the gas or liquid velocity[2] reaches the terminal velocity of the particle in the gas-liquid medium (U_t'). At a gas or liquid velocity above U_t', operation is in the transport regime. An example of an operating regime map for an air-water-solid system with cocurrent upward flow of gas and liquid with liquid as the continuous phase is shown in Fig. 1.1. In this figure, the variations of the minimum fluidization velocity ($U_{\ell mf}$) and U_t' with the particle terminal velocity in the liquid medium (U_t) and the gas velocity[3] (U_g) are given. As shown, $U_{\ell mf}$ increases with increasing U_t but decreases with increasing gas velocity. For particles with a small U_t, U_t' is unaffected by gas velocity; but for particles with a large U_t, U_t' decreases with increasing gas velocity. Note that Fig. 1.1 was obtained for particles with uniform properties in the bed.

For gas-liquid upward flow with liquid as the continuous phase, fluidized bed and slurry bubble column systems are the two most extensively investigated three-phase systems involving a solid phase in suspension. A diverse range of operating conditions have been labeled slurry bubble column operation; indeed, the operating conditions for "slurry" systems are loosely defined. Typical operating ranges for three-phase fluidized beds and slurry bubble columns for air-water-solid systems are shown in Fig. 1.2 (Fan et al., 1987). Note that three-phase fluidized beds operate in the expanded bed regime (covering U_t from 3 to 50 cm/s in the figure) and slurry bubble columns may operate in both the expanded bed regime and the transport regime (covering U_t from 0.03 to 7 cm/s in the figure). Thus, overlap for U_t from 3 to 7 cm/s exists between fluidized bed

---

1: The classification described here is based on a vertical system.
2: It depends upon which phase is the continuous phase (see Section 2.3.5).
3: Throughout this book, the operating variables for the gas phase are expressed based on the prevailing operating conditions of the system unless otherwise indicated.

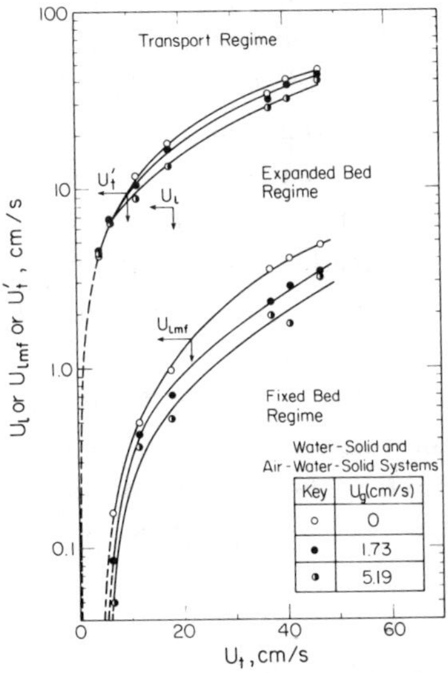

Figure 1.1 Operating regimes for cocurrent upward gas-liquid-solid systems with liquid as the continuous phase (from Fan et al., 1987).

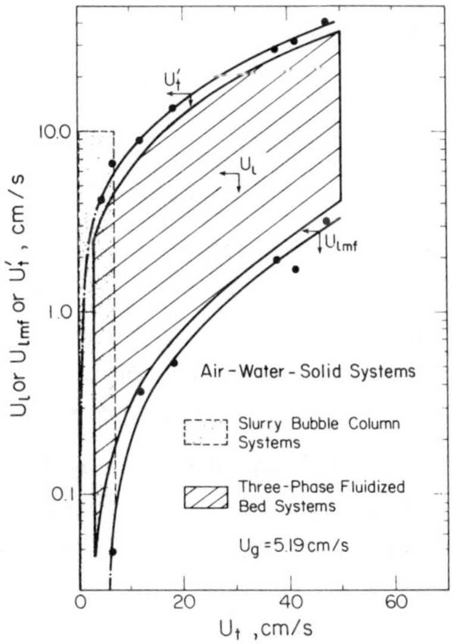

Figure 1.2 Common operating ranges for three-phase fluidized bed and slurry bubble column systems (from Fan et al., 1987).

and slurry bubble column operations in the expanded bed regime. For convenience, the demarcation between the fluidized bed and slurry bubble column operated in the expanded bed regime is roughly set at 5 cm/s. In slurry bubble columns, the liquid velocity upper limit varies significantly with the application intended. In reactor applications, the upper limit for U_ℓ is usually set as high as 10 cm/s.

Generalization of the concepts leading to the operating regime map given in Fig. 1.1 is needed. Naturally, $U_{\ell mf}$ and U_t' vary with the relative flow direction and phase continuity of the systems. A regime map such as that given in Fig. 1.1 can provide a unified and rigorous classification scheme. Such a classification scheme permits appropriate bounds to be established for correlations to be employed without the undue extrapolation which has occurred with gas-liquid mass transfer correlations as noted by Kim and Kim (1986).

1.2 OPERATING MODES FOR FLUIDIZATION SYSTEMS

In a broad sense, gas-liquid-solid "fluidization" encompasses both the expanded bed and transport regimes with U_t ranging from 0.03 to 50 cm/s. Figure 1.3 gives the basic operational schemes, or modes, for each operating regime.

In the expanded bed regime with a continuous liquid phase flowing cocurrently upward with the gas, solids can be introduced either batchwise or continuously. For large or dense particles (U_t > 5 cm/s, see Section 1.1), the solids can be charged and discharged independent of the liquid flow (Mode E-I-a-1); in such a system, the particles are uniformly distributed, in a global sense, in the bed and the bed height is well defined within the system[4]. For small or light particles (U_t < 5 cm/s, see Section 1.1), the charge and discharge of solids usually depend on the liquid flow (Mode E-I-a-2); in such a system, the particle concentration is high at the bottom and progressively decreases axially upward and the bed surface is ill-defined. For both Modes E-I-a-1 and E-I-a-2, bed expansion can be supported by the liquid phase, the gas bubbles, or both. In the expanded bed regime with a continuous gas phase flowing cocurrently upward with the liquid (Mode E-I-b), the solids can be continuously charged and discharged independent of the gas flow; the particle bed is mainly supported by the flow of gas.

4: This behavior generally refers to systems without internals.

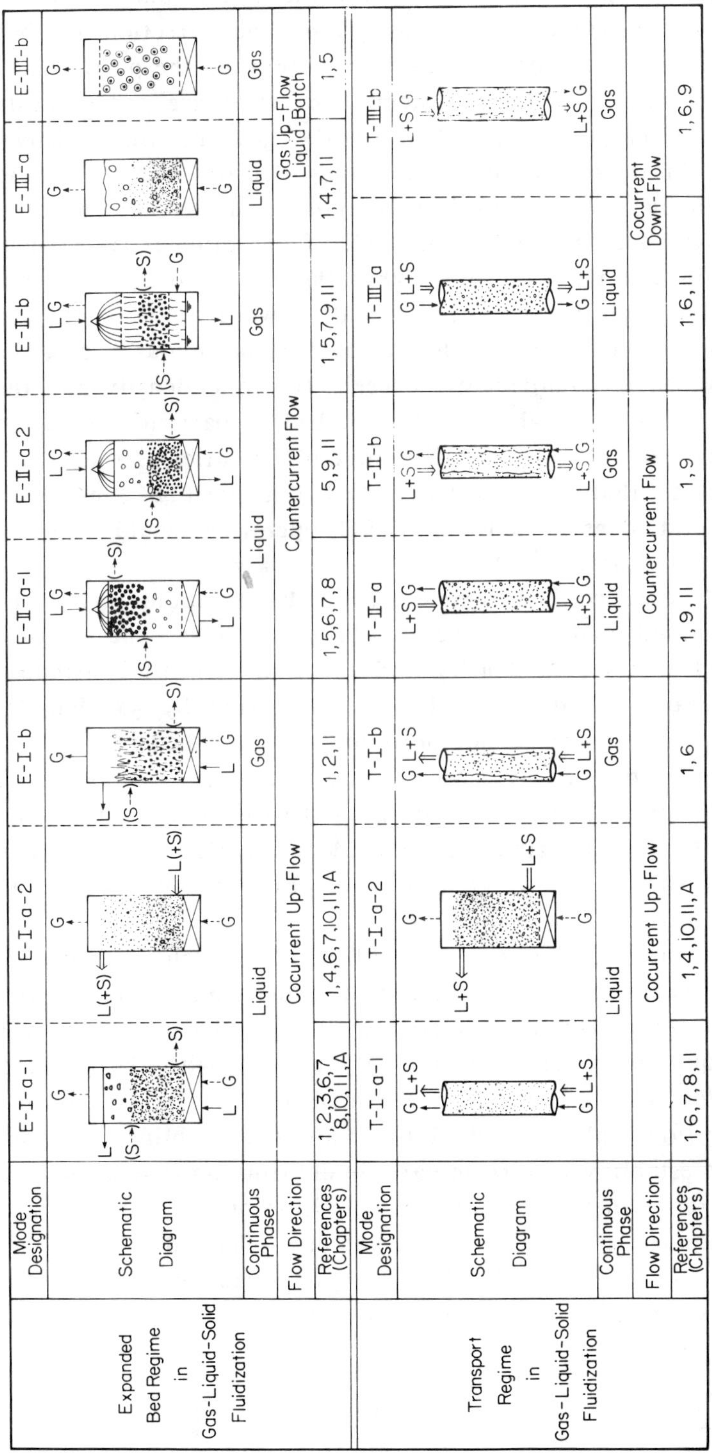

(S--→) Independent Charge of Solids from Fluid (--→S) Independent Discharge of Solids from Fluid (+S) Charge or Discharge of Solids with Fluid
A : Appendix A

Figure 1.3 Basic classification of gas-liquid-solid fluidization systems.

In the expanded bed regime with the liquid phase flowing downward countercurrent to the gas phase, the direction of bed expansion depends on the density difference between the particle and the continuous phase; in these systems the solids are continuously charged and discharged independent of the gas and liquid flows. When the liquid density exceeds the solids density and the liquid phase is continuous, the particle bed expands downward supported by the liquid flow (Mode E-II-a-1); this mode of operation is referred to as that of an inverse fluidized bed. When the particle density exceeds the liquid density, the particle bed expands upward supported by gas bubbles (Mode E-II-a-2). When the gas is the continuous phase, the liquid density usually significantly exceeds the solid density and the liquid trickles down through an expanded bed of particles supported by the gas phase (Mode E-II-b). This mode of operation is typical of, but not limited to, that of the turbulent bed contactor (TBC); however, in a high solid density situation TBC operation could also be conducted under Mode E-II-a-2.

In the expanded bed regime with a batch liquid phase, the solids are charged and discharged batchwise. When the particle density exceeds the liquid density and the liquid is the continuous phase, the expanded bed of particles is supported by gas bubbles (Mode E-III-a). When gas is the continuous phase, a bed of particles expands upward supported by the gas phase (E-III-b); the liquid phase exists as a film or droplets. Depending on the gas velocity, the axial solids concentration distribution for Modes E-III-a and E-III-b operation may vary from an exponential decrease at low velocities to near uniformity at high gas velocities[5].

In the transport regime, gas and/or liquid velocities exceed those in the expanded bed regime and continuous charging and discharging of solids usually accompanies the liquid flow. When the continuous liquid phase flows cocurrently upward with the gas and solids, particles with either large or small U_t can be transported mainly by the liquid phase (Mode T-I-a-1) and the bubbling gas phase can be used to establish a uniform particle distribution (Mode T-I-a-2). Mode T-I-a-2 represents an operational condition of the slurry bubble column where particles are usually charged or discharged continuously as slurries (a liquid-solid mixture). The system can also be operated with gas as the continuous phase (Mode T-I-b).

5: See footnote 4.

In the transport regime with a countercurrent flow situation, slurries flow downward relative to the upward flow of gas with either liquid as the continuous phase (Mode T-II-a) or gas as the continuous phase (Mode T-II-b). In cocurrent downward systems, the continuous phase can be liquid (Mode T-III-a) or gas (Mode T-III-b). Note that the flow with gas as the continuous phase shown in Fig. 1.3 is exemplified by an annular flow.

The presence of internals within a system, such as draft tubes, heat exchangers and baffles, or geometric irregularities such as a tapered geometry, greatly increases the complexity of the transport phenomena; however, the classification scheme given above is still valid for such systems. For the transport regime, the same mode designations also apply to situations where solids are initially present in the gas phase forming a gas-solid mixture; they then progressively transfer to the liquid phase as the mixture flows through the system. Although the general classifications are based on a vertical system, similar considerations for classification can also be made for a horizontal system involving gas-liquid-solid transport. Note that for all fluidization systems described above or generally in this book, solids suspension is considered solely induced by flows of gas and/or liquid phases without the aid of external means such as mechanical agitation (e.g., Nienow, 1984) or magnetic fields (e.g., Hu and Wu, 1987; Wu et al., 1988), unless otherwise noted. Information sources in reference to modes of operation and chapters are provided in Fig. 1.3.

1.3 FIXED BED SYSTEMS VERSUS FLUIDIZATION SYSTEMS

The modes of operation for three-phase fixed bed systems can be classified in a manner similar to those for fluidization systems. Fixed bed systems can be operated with either gas or liquid as the continuous phase in cocurrent upward flow, cocurrent downward flow, or countercurrent with downward liquid flow and upward gas flow.

In cocurrent downward flow systems, four types of flow can be distinguished depending on the relative magnitude of the gas-to-liquid flow rate: mist or spray flow, trickle flow, pulse or slug flow, and dispersed bubble flow. In both mist and trickle flow, the gas is the continuous phase and the liquid is a discrete phase. In slug flow, the liquid rich part of the slug, which is followed by the gas rich part, appears longer in length and greater in liquid content as the liquid velocity increases. As the liquid velocity increases at a moderate gas velocity the gas rich part of the slug disappears and dispersed bubble flow occurs where the liquid is the continuous phase and the gas is a discrete phase. Similar flow patterns can also be distinguished for cocurrent upward flow systems including mist or spray flow, pulse or slug flow, and dispersed bubble flow.

Fixed bed operation is of considerable industrial interest. Common fixed bed operations are in cocurrent downward and countercurrent trickle flows. Fixed beds in cocurrent downward trickle flow are generally used as reactors for such reactions as catalytic desulfurization and hydrocracking of petroleum residues and catalytic hydrogenation and oxidation of chemical feed stocks. Fixed beds in countercurrent trickle flow, on the other hand, are generally used as mass transfer devices with or without chemical reactions such as wastewater aeration and carbon dioxide absorption.

There are inherent differences in transport properties between fixed bed and fluidization systems. The major differences lie in that fluidization systems yield considerably higher macromixing and heat transfer in gas, liquid and solid phases than do fixed bed systems. For process applications, comparisons of these systems are complex and difficult. It requires consideration of not only the transport properties but also reaction kinetics for the process application intended. However, commonly acknowledged general advantages of fixed bed systems over fluidization systems, and vice versa, used as catalytic reactors (e.g., Germain et al., 1979; Shah 1979) are described in the following.

The advantages of fixed bed systems over fluidized bed systems are low macromixing, yielding small axial dispersion of phases and high reactant conversions for reaction kinetics favoring plug flow patterns; high controllability over product selectivity for complex reactions; low solids attrition and consumption, hence permitting precious metal catalysts to be used for the reaction. The major advantages of fluidization systems over fixed bed systems are high macromixing, yielding large axial dispersion of phases and high reactant conversions for reaction kinetics favoring completely mixed flow patterns; ability in achieving significant temperature uniformity without the aid of external means; ease in heat supply and removal and hence temperature controllability; low intraparticle diffusion resistance and external liquid-solid mass transfer resistance; ease in catalyst replacement and hence high controllability of catalyst activity and minimum flow maldistribution. Specific comparisons of advantages and disadvantages between fixed bed and fluidized bed systems based on two individual process applications, namely, aerobic wastewater treatment and hydrotreating and conversion of residues are given in Chapters 8 and 10, respectively.

Extensive information regarding the transport properties and reaction engineering of gas-liquid-solid fixed bed systems is available in the literature: hydrodynamics and reaction engineering, especially relating to hydrogenation - e.g., Satterfield (1975), Hofmann (1978, 1986), L'Homme (1979), Shah (1979), Van Landeghem (1980), Ramachandran and Chaudhari (1983); reaction engineering relating to oxidation - e.g., Goto et al. (1977); design and modeling - e.g., Herskowitz and Smith (1983); mass transfer - e.g., Goto and Smith (1975); Charpentier (1976, 1981, 1986). Readers are referred to these references for details about three-phase fixed bed operation.

1.4 APPLICATIONS OF FLUIDIZATION SYSTEMS

Numerous applications of gas-liquid-solid fluidization systems exist which vary in size from bench scale to commercial scale and include all of the basic modes of operation given in Fig. 1.3. In these applications, the individual phases can be reactants, products, catalysts, or inert (Shah, 1979; L'Homme, 1979; Ramachandran and Chaudhari, 1983). For example, all three phases are either reactants or products as in coal liquefaction (Ashland Synthetic Fuels, Inc., 1984); the gas and liquid phases can be either reactants or products and the solid can be a catalyst as in hydrogenation of α-methyl styrene in a slurry of alumina-supported palladium catalyst (Satterfield et al., 1969); two phases can be reactants or products and the third phase can be inert as in ammonia scrubbing by a boric acid solution (Douglas, 1964); and the phases can be neither reactants nor products as in purely physical processes such as air humidification (Douglas, 1964). In the following, historical highlights of commercial or large scale operation of three-phase fluidization systems are described. In addition, numerous processing examples of these systems conducted on a relatively small scale are given to reflect the importance of three-phase fluidization applications.

1.4.1 Historical Highlights of Commercial/Large Scale Operation

The use of gas-liquid-solid fluidization systems in reactor applications can be traced back to the work of Bergius from 1912 through 1926 on direct coal liquefaction or coal hydrogenation (Storch, 1945; Probstein and Hicks, 1982). In his studies, pulverized coal-oil slurries were reacted with hydrogen at high temperature and pressure using slurry bubble column systems (Mode E-I-a-2). The study led to the development of a direct coal liquefaction process at a mini-commercial scale, i.e., Bergius-Pier process, for the production of motor fuels from lignite in Germany in 1927 with a capacity of 2,500 barrels per day (bbl/day) (Wenzel et al., 1986). It is interesting to note that the timing of commercialization of this process approximately parallels that of the first commercialization of gas-solid fluidization systems, i.e., Winkler process, for coal gasification developed also in Germany (Kunii and Levenspiel, 1969). In 1935, a mini-commercial process for direct coal liquefaction was also operated in England. During World War II, the commercial production using this process in Germany peaked at an average annual production of 4.2 million tons of mostly aviation gasoline (Donath, 1963). The direct production of liquid fuels from coal was phased out, however, following the end of the war.

Liquefaction of coal to produce liquid fuels could also be achieved indirectly from liquid phase reactions, involving hydrogen and synthesis gas derived from coal gasification, in the presence of a catalyst. The reactions known as Fischer-Tropsch (F-T) synthesis were discovered in 1926 (Frohning et al., 1977). F-T synthesis has been conducted in various types of reactors including a slurry bubble column reactor (Mode E-III-a) developed by Kolbel and Ackermann (1956) during the period 1938-1953 in Germany (see Section 11.9). The demonstration reactor developed by the companies Rheinpreussen and Koppers for the liquid phase F-T synthesis was operated from 1952-1953 at a daily production rate of 11.5 tons of liquid fuels (Kolbel and Ralek, 1980). Production was then discontinued due to the availability of cheap petroleum crudes and natural gas.

Though a bench scale study was first performed in the 1940's, it was not until 1987 that a slurry bubble column reactor (Mode T-I-a-2) was successfully proven at a mini-commercial scale in Canada for hydrotreating of tar sand and petroleum resids (see Chapter 10). The reactor is a part of the Canmet hydrocracking process which has a processing capacity of 5,000 barrels per day (bbl/day) (Lunin et al., 1985).

Slurry bubble column or transport reactors (Modes E-I-a-2, E-III-a, T-I-a-1, T-I-a-2, or T-III-a) have been successfully utilized as an alternative to gas-liquid-solid trickle bed reactors or gas-solid fluidized bed reactors for catalytic production of organic chemicals or polyolefins since the 1950's. Slurry bubble column or transport reactors appeared in various designs and were operated over a wide range of flow conditions for catalytic reaction applications. The earlier production came mostly from Europe. Major reaction applications in commercial production (e.g., Germain et al., 1979) included hydrogenation of glucose to sorbitol, benzene to cyclohexane, benzoic acid to cyclohexanecarboxylic acid, butynediol to butenediol, adiponitrile to hexamethylene diamine, esters to fatty alcohols, 3,5 nitrotoluene to 3,5 toluene diamine, and aluminum and ethylene to Ziegler alcohol (ALFOL process), and ethylene polymerization (Solvay process using Ziegler-Natta catalysts). Clearly, hydrogenation reactions are typical of these applications.

The three-phase fluidized bed reactor (Mode E-I-a-1) was first used commercially in 1968 for hydrotreating petroleum resids (see Chapter 10). The process, known as the H-Oil process, was developed in the U.S. following research and development beginning in the late 1950's. Based on similar technology, the LC-Fining resid hydrocracking process was later developed in the U.S. Several commercial reactors based on either process are currently in operation in refineries in Kuwait and Mexico as well. The processing capacities in these refineries range from 18,500 to 60,000 bbl/day. The reactor is commonly referred to in industry as an "ebullated bed" where particles are in "ebullation" induced by gas-liquid phases. It is noted that the technical term "ebullated bed" or "ebullation" was first defined by P.W. Garbo in the patent of Johanson (1961) to describe a gas-liquid contacting process in contrast to the common industrial term "fluidized bed" where particles are in "fluidization" induced by the gas phase alone.

The energy crisis in 1973 brought renewed interest in the synthetic fuel area. Demonstration or semi-demonstration process units for direct coal liquefaction were developed in the U.S. These processes included the Exxon Donor Solvent (EDS) process and Solvent Refined Coal (SRC) process, and H-Coal process. The EDS and SRC processes use slurry bubble column (Mode E-I-a-2) reactors operated at a coal feed rate of 250 ton/day and 50 ton/day, respectively. The H-Coal process was developed based on the same technology as the H-Oil process using a fluidized bed reactor (Mode E-I-a-1) and was operated at a coal feed rate of 500 ton/day. All of these processes were, however, phased out in the early 1980's, again due to the low price of petroleum crudes.

The energy crisis also led to innovative development in flue gas cleaning systems for coal combustion processes. A three-phase fluidized bed system (wet scrubbing system) known as the Turbulent Bed Contactor (TBC) (Mode E-II-b, mostly), first introduced in the late 1950's (Chem. Eng., 1959), was used for this purpose (see Chapter 9). In the mid-1970's, TBC scrubbers were operated at a commercial scale to remove sulfur dioxide and particulates from flue gas generated from coal combustion. Typically, each TBC had the operational capacity to treat 100 MW equivalent of flue gas containing sulfur dioxide and particulates. The flue gas scrubbing application of the TBC was terminated in the early 1980's, partly due to maldistribution and inconsistent durability problems associated with the low density particles used and the excessive pressure drop over a slurry spray tower (Mode T-II-b) (see Chapter 9). Since the late 1970's, commercial flue gas desulfurization has been conducted mainly using slurry spray towers.

Three-phase fluidization systems have made inroads into biotechnology covering biological wastewater treatment and fermentation. The fluidized bed bioreactor using immobilized living cells for wastewater treatment was conceived as early as the 1930's. Commercial application of three-phase fluidized bed bioreactors with immobilized living cells, however, was not demonstrated until the late 1970's and early 1980's, when the Biolex process bioreactors (Mode T-I-a-1[6]) with sizes ranging from 34 to 342 m^3 were put in operation to treat night soil, tobacco mill and paper-board mill wastewater (Suzuki et al., 1981) (see Chapter 8). Development of immobilized living cell technology for fluidized bed or slurry bubble column bioreactor applications in fermentation significantly lags that in wastewater treatment. Although the initial success in cell immobilization for fermentation was demonstrated in the early 1960's (Linko and Linko, 1983), large scale three-phase fluidized bed or slurry bubble column fermentors were not demonstrated until the early 1980's, as exemplified by ethanol production in a 10,000 liter fermentor (Mode E-I-a-2) (Samejima et al., 1984) and penicillin production in a 200 liter fermentor (Mode E-III-a) (Endo et al., 1986), both in Japan (see Chapter 7). In the U.K. around the same time, a slurry bubble column bioreactor (Mode T-I-a-1[6]) of 1000 liter working volume was put in

---

6: The Biolex reactor is a draft-tube three-phase bioreactor. The mode designation given here is based on the behavior in the draft tube region of the bioreactor.

operation for the production of several kilograms per year monoclonal antibodies using suspended mammalian cells[7] (Birch et al., 1985).

It is noted that in each of the large scale reactor operations described above, three-phase transport systems (operated in the transport regime) were naturally also in place on a large scale for reactant feeding, product discharging, internal/external fluid/solid recycling, etc.

1.4.2 Application Examples

Beyond those commercial or large scale applications given above, there are numerous examples of applications of three-phase fluidization systems demonstrated at relatively small scales. These examples are grouped according to physical, chemical (petrochemical or electrochemical) and biochemical applications as given in Tables 1.1 through 1.3, respectively. Tables 1.1 through 1.3 also show modes of operation for each application. It is seen that for all applications, modes in both the expanded bed regime and transport regime are used[8]. However, Modes E-I-a-1, E-I-a-2, E-II-b, E-III-a and T-I-a-2 are the most common modes for reactor application. Note that a given application may possibly be conducted in more than one mode of operation (e.g., hydrotreating of heavy petroleum residue using Modes E-I-a-1 and T-I-a-2). In some instances, a part of the system can be operated in a three-phase condition while the remaining part can be operated in a two-phase condition (e.g., urea granulation and production of polyethylene and polyolefins). In most aerobic fermentation systems (e.g., penicillin production), the liquid flow, a nutrient solution flow, is usually very small. Consequently, these systems are regarded as being batch with respect to the liquid flow. For some anaerobic fermentation systems with no inlet flow of gas, but with significant outlet flow of gas generated from metabolic reactions (e.g., ethanol production), the systems are regarded as being a special case of three-phase fluidization operation where the gas velocity in the feed is zero.

7: It is noted that for suspended cell systems which include those of microbial, plant and mammalian cells, only the plant and mammalian cells are treated fluid mechanically as being heterogeneous to the liquid phase as they are substantially larger than microbial cells. Hence, for suspended cell systems, only those involving plant and mammalian cells are considered in the context of three-phase operation.

8: For systems with non-uniform particle properties, the mode designation is based on the behavior of the average particle properties.

Table 1.1 Examples of Physical Processing

Applications	Modes of Operation	Examples of References (Chapters or Sections)
Sand filter cleaning	E-I-a-1	Brownell (1951) (also see Sec. 11.4)
Crystallization	E-I-a-1 E-I-a-2	Thompson (1955) Okochi et al. (1978) (also see Sec. 11.5)
Air flotation	E-I-a-2	Fuerstenau (1980) Foot et al. (1986) (also see Sec. 11.6)
Urea granulation, simultaneous granulation and drying	E-I-b	Kono (1980) Morl and Kunne (1986) (also see Sec. 11.2)
Dust collection	E-I-b	Nienow and Killick (1983) (also see Sec. 9.2)
Air cooling, humidification or dehumidification	E-II-b	Douglas (1964) Barile and Meyer (1971) Barile et al. (1974) Adamiec et al. (1976)
Particulate collection	E-II-b	Mlodzinski and Warych (1975) Uysal (1978) (also see Sec. 9.2)
Lactose granulation	E-II-b	Waldie et al. (1987) (also see Sec. 11.2)
Drying of granular material	E-III-b	Palancz (1983)
Three-phase transport	T-I-a-1	Toda et al. (1982) Kim and Choi (1984) Kato et al. (1986) Hatate et al. (1986)
Lifting of solids in shaft, well, and tunnel drilling or lifting of coal with underground water; Pumping manganese nodules from below ocean surface (air-lift pump applications)	T-I-a-1	Weber and Dedegil (1976) Giot (1982)
Three-phase transport	T-I-b	Kato et al. (1986) Hatate et al. (1986)
Drying of calcium carbonate and polyvinylchloride particles	T-I-b	Martin and Saleh (1984)
Three-phase transport	T-III-a	Shah et al. (1983) Hatate et al. (1986)
Three-phase transport	T-III-b	Hatate et al. (1986)

Table 1.2 Examples of Chemical, Petrochemical, or Electrochemical Processing

Applications	Modes of Operation	Examples for References (Chapters or Sections)
Hydrotreating and conversion of heavy petroleum and synthetic crude (H-Oil, LC-Fining or ebullated bed reactor)	E-I-a-1 (with internal recycle tube)	Johnson et al. (1985) Beaton et al. (1986) (also see Chap. 10)
Coal liquefaction (H-Coal or ebullated bed reactor)	E-I-a-1 (with internal recycle tube)	Li and Liu (1981) Ashland Synthetic Fuels Inc. (1984)
Electrodes	E-I-a-1	Oloman and Watkinson (1975) Kusakabe et al. (1981) (also see Sec. 11.1)
Methanation	E-I-a-1	Blum and Toman (1977) (also see Sec. 11.8)
Production of zinc hydrosulfite	E-I-a-1	Sastri et al. (1983)
Hydrogenation of heptane	E-I-a-1	Ermakova et al. (1973)
Hydrogenation of 1-heptene	E-I-a-1	Gartsman et al. (1977)
Calcium bisulfite production	E-I-a-1 and E-I-b	Volpicelli and Massimilla (1965, 1970) (also see Sec. 11.3)
Methanol production	E-I-a-1 E-III-a T-I-a-2	Sherwin and Frank (1976) Studer et al. (1987) (also see Sec. 11.10)
Coal gasification	E-I-a-2	Cover et al. (1973) Kohl et al. (1978) Yosim and Barclay (1981) (also see Sec. 11.8)
Coal liquefaction (Exxon Donor Solvent Reactor)	E-I-a-2	Tarmy et al. (1984) Mueller (1985)
Oxydesulfurization of coal	E-I-a-2	Joshi et al. (1981)
Production of polyethylene and polyolefins	E-I-b	Jenkins et al. (1985, 1986) (also see Sec. 11.7)
Adiponitrile production	E-I-b	Kono (1980) (also see Sec. 11.2)
Flue gas desulfurization using limestone slurries	E-I-b	Beaver et al. (1982)
Flue gas desulfurization using limestone slurries	E-II-b	McMichael et al. (1976) Wang and Burbank (1982) (also see Chap. 9)
Absorption of bromine from a bromide-water mixture by sodium hydroxide and iron bromide solutions	E-II-b	Gel'perin et al. (1965,1966)
Rectification of a binary ethyl alcohol-water mixture	E-II-b	Gel'perin et al. (1968)

(Table 1.2 continued)

Applications	Modes of Operation	Examples for References (Chapters or Sections)
Ammonia, carbon dioxide and fluorine gases scrubbing	E-II-b	Douglas (1964) Wozniak and Ostergaard (1973) Adamiec et al. (1976)
Hydrogen sulfite desorption	E-II-b	Kudra et al. (1978)
Fischer-Tropsch synthesis	E-III-a	Schlesinger et al. (1951, 1954) Kolbel and Ackermann (1956) Kolbel and Ralek (1980) Deckwer et al. (1980, 1981, 1982) Satterfield and Huff (1981) (also see Sec 11.9)
Oxidation of sodium sulfite	E-III-a	Kato et al. (1973)
Oxidation of ethylene	E-III-a	Fujimoto et al. (1976)
Hydrogenation of cyclohexane	E-III-a	Sherwood and Farkas (1966)
Hydrogenation of acetylene	E-III-a	Heck and Smith (1970)
Oxidation of sulfur dioxide	E-III-a	Goto and Kojima (1985)
Hydrogenation of α-methylstyrene	E-III-a	Govindarao and Chidambaram (1986)
Methanation	E-III-a	Hammer (1979) (also see Sec. 11.8)
Absorption of carbon dioxide in sodium hydroxide solution	E-III-b	Furusaki et al. (1984)
Coal slurry and hydrogen transport in preheater of coal liquefaction	T-I-a-1	Shah (1981) Kornosky et al. (1982)
Leaching of copper ores	T-I-a-2	Reilly et al. (1982)
Hydrogenation of glucose	T-I-a-2	Germain et al. (1979) Brahme et al. (1984)
Hydrotreating of tar sand and petroleum resids	T-I-a-2	de'Bruijn et al. (1982) Lunin et al. (1985)
Flue gas desulfurization using magnesia slurries	T-I-b	Egan and Felker (1986)
Flue gas desulfurization using limestone slurries	T-II-a	Idemura et al. (1978) (also see Chap. 9)
Flue gas desulfurization using limestone slurries	T-II-b	EPRI (1987) (also see Chap. 9)
Catalyst slurry and hydrogen transport in ebullated bed downcomer for coal liquefaction and resid hydrotreating	T-III-a	Beaton et al. (1986) Ashland Synthetic Fuels, Inc. (1984) (also see Chap. 10)

Table 1.3 Examples of Biochemical Processing (also see Chapters 7 and 8)

Applications	Modes of Operation	Examples for References
Treatment of lactose wastewater	E-I-a-1* (tapered bio-reactor)	Scott and Hancher (1976)
Treatment of phenol wastewater	E-I-a-1* (tapered bio-reactor)	Holladay et al. (1978) Lee et al. (1979)
Treatment of thiocyanate and coal hydrocarbonization waste-water	E-I-a-1 (tapered bio-reactor)	Lee et al. (1979)
Treatment of slaughter house wastewater	E-I-a-1	Qian (1980)
Treatment of municipal sewage wastewater	E-I-a-1	Shen and Wang (1980)
Treatment of phenol wastewater	E-I-a-1	Hirata et al. (1982) Hosaka et al. (1985)
Treatment of glucose-BOD wastewater	E-I-a-1	Takase et al. (1983)
Treatment of gasification wastewater	E-I-a-1	Donaldson et al. (1984)
Treatment of synthetic milk wastewater sewage	E-I-a-1* (tapered and non-tapered bio-reactor	Forster et al. (1986)
Treatment of s-triazine wastewater	E-I-a-1* (tapered bioreactor)	Hogrefe et al. (1986)
Treatment of isobutanol or acetic acid wastewater	E-I-a-2	Chatib et al. (1981) Grasmick et al. (1981)
Ethanol production by yeast cells	E-I-a-2	Samejima et al. (1984)
Conversion of sucrose to glucose by plant cells <u>D. carota</u>	E-I-a-2	Prenosil and Pedersen (1985)
Bioprocessing of coal for liquefaction and removal of sulfur, nitrogen and trace metals	E-I-a-2 E-III-a	Wilson et al. (1986) Narayan (1987)
Simultaneous selective floccu-lation and microbial desulfuri-zation of high sulfur coal	E-I-a-2	Fan and Attia (1988)
Treatment of phenol wastewater	E-II-a-1	Edwards (1981)
Treatment of oil refinery or kitchen wastewater	E-II-a-1	Hasegawa (1982) Shimodaira and Yushina (1983)
Ferrous iron oxidation by <u>Thiobacillus ferrooxidans</u>	E-II-a-1	Karamanev and Nikolov (1988)
Ethanol production by <u>Saccharomyces cerevisiae</u>	E-II-b	Moebus and Teuber (1982) Bauer (1986)
Bioleaching of metals from ores	E-III-a	Karaivko et al. (1977)
L-isoleucine production by <u>Serratia marcescens</u>	E-III-a	Wada et al. (1980)

(Table 1.3 continued)

Applications	Modes of Operation	Examples for References
Patulin production by Penicillium urticae	E-III-a	Berk et al. (1984a, 1984b)
Penicillin production by Penicillium chrysogenum	E-III-a	Endo et al. (1986) Kalogerakis et al. (1986) Jones et al. (1986)
Pre-oxidation and bacterial pre-conditioning for coal desulfurization	E-III-a	Zeky and Attia (1986)
Secondary metabolite production by plant cells	E-III-a	Kargi and Rosenberg (1987)
Enzyme production by Aspergillus sojae	E-III-b	Akao and Okamoto (1984)
Betamines production by plant cells B. vulgaris	T-I-a-1** (internal draft tube bioreactor)	Wagner and Vogelmann (1977)
Treatment of domestic wastewater	T-I-a-1** (internal draft tube bioreactor)	Sawada (1978) Kataoka et al. (1979) Suzuki et al. (1981)
Treatment of night soil waste-water	T-I-a-1** (internal draft tube bioreactor)	Kataoka et al. (1979)
Production of animal cells (Namalwa and BHK21); production of monoclonal ABO blood-typing reagent production by hybridoma cells	T-I-a-1** (internal draft tube bioreactor	Katinger et al. (1979) Birch et al. (1985)
Treatment of effluent of septic tank, photo processing, food processing or kraft pulp waste-water	T-I-a-1** (internal draft tube bioreactor)	Suzuki et al. (1981)
Treatment of municipal sewage	T-I-a-1** (internal draft tube bioreactor)	Zhang et al. (1981)
Treatment of fish processing wastewater	T-I-a-1** (internal draft tube bioreactor)	Sawai et al. (1982)
Treatment of phenol wastewater	T-I-a-1 (external circulating bioreactor)	Suzuki et al. (1983)
Treatment of phenol wastewater	T-I-a-1** (internal draft tube bioreactor)	Fan et al. (1984) Tang and Fan (1987) Tang et al. (1987)
Treatment of Naphthalene-2-sulfonate wastewater	T-I-a-2 (external circulating bioreactor)	Wagner and Hempel (1988)

* The mode designation given here is based on the criterion of $U_t > 5$ cm/s.

** The mode designation given here is based on the behavior in the draft tube region of the bioreactor.

1.5 SCOPE OF PARTS II AND III

The remaining parts of this book are devoted to the fundamental understanding of gas-liquid-solid fluidization engineering. They are divided into two parts, i.e., Parts II and III. Part II considers the fundamentals and Part III considers applications.

In Part II, system behavior including hydrodynamics, mixing, and heat and mass transfer are covered. Information regarding this behavior is essential to modeling and design of fluidization systems for process applications. In this part, hydrodynamics of cocurrent fluidized bed systems represented by operating Modes E-I-a-1 and E-I-b are covered in Chapter 2. Mass transfer, mixing and heat transfer for Mode E-I-a-1 are given in Chapter 3. Slurry bubble column systems encompassing operating Modes E-I-a-2, E-III-a, and T-I-a-2 are described in Chapter 4. Chapter 5 addresses countercurrent and liquid-batch fluidized bed systems (i.e., Modes E-II-a-1, E-II-a-2, E-II-b and E-III-b). Fundamentals of miscellaneous systems including draft tube, semifluidization and annular fluidization systems and vertical and horizontal three-phase transport are given in Chapter 6.

In Part III, biological applications in fermentation and aerobic wastewater treatment are described in Chapters 7 and 8, respectively. Chapter 9 elucidates application to flue gas desulfurization and particulate removal. Hydrotreating and conversion of resids are illustrated in Chapter 10. In Chapter 11, numerous miscellaneous applications are covered including electrodes, granulation, adiponitrile production, calcium bisulfite acid production, sand filter cleaning, crystallization, flotation, polymerization, coal gasification, methanation, Fischer-Tropsch synthesis and methanol synthesis. In Appendix A, an example is provided which illustrates the interplay between the transport properties and reaction kinetics in evaluating the reactant conversion in a three-phase fluidization system where a catalytic reaction takes place. The example also illustrates the impact of this interplay on the optimum selection of a mode of operation (among Modes E-I-a-1, E-I-a-2 and T-I-a-2) for the reaction. The SI units are consistently used for the empirical correlation equations throughout the entire text. They are summarized in Appendix B.

1.6 CONCLUDING REMARKS

It is of practical significance to classify complex gas-liquid-solid systems according to the solid phase state of motion in a manner similar to that for gas-solid or liquid-solid systems. Accordingly, the solid phase state of motion can be classified into three basic operating regimes: the fixed bed regime, the expanded bed regime, and the transport regime. The scope of gas-liquid-solid fluidization, in a broad sense, encompasses operation in both the expanded bed and transport regimes. For each of these regimes, various operating modes can be further classified based on the relative direction and continuity of the phases. Considerable work is needed in generalizing the transition velocities between these regimes for each operating mode.

1.7 NOMENCLATURE

U_g : superficial gas velocity

U_ℓ : superficial liquid velocity

$U_{\ell mf}$: minimum fluidization velocity in a three-phase system

U_t : particle terminal velocity in a liquid medium

U_t' : particle terminal velocity in a gas-liquid medium

1.8 REFERENCES

Adamiec, J., T. Kudra, and C. Strumillo, paper presented at the 2nd International Conference on "The Control of Gaseous Sulfur and Nitrogen Compound Emission," and 170th Event of the European Federation of Chemical Engineering, Salford, England (1976).

Akao, T., and Y. Okamoto, in "Fluidization," ed. by D. Kunii and R. Toei, p. 631, Engineering Foundation Press (1984).

Ashland Synthetic Fuels Inc., Final Report to U.S. Department of Energy, Contract No. DE-AC05-76ET10143, April (1984).

Barile, R.G., and D.W. Meyer, Chem. Eng. Prog. Symposium Series, 67, No. 119, 134 (1971).

Barile, R.G., J.L. Dengler, and T.A. Hertwig, AIChE Symposium Series, 70, No. 138, 154 (1974).

Bauer, W.J., in "Fluidization," ed. by K. Ostergaard and A. Sorensen, p. 619, Engineering Foundation Press (1986).

Beaton, W.I., N.K. McDaniel, W.E. McWhirter, R.D. Petersen, and R.P. Van Driesen, Technology (Oil and Gas J.), p. 47, July 7 (1986).

Beaver, L.E., L.-S. Fan, and K. Muroyama, paper presented at the AIChE National Meeting, Aug. 29 - Sep. 3 (1982).

Berk, D., L.A. Behie, A. Jones, B.H. Lesser, and G.M. Gaucher, Can. J. Chem. Eng., 62, 112 (1984a).

Berk, D., L.A. Behie, A. Jones, B.H. Lesser, and G.M. Gaucher, Can. J. Chem. Eng., 62, 120 (1984b).

Birch, J.R., R. Boraston, and L. Wood, Trends in Biotech., 3(7), 1 (1985).

Blum, D.B., and J.J. Toman, AIChE Symposium Series, 73, No. 161, 115 (1977).

Brahme, P.H., R.V. Chaudhari, and P.A. Ramachandran, I&EC Process Des. & Dev., 23, 857 (1984).

Brownell, L.E., in "Encyclopedia of Chemical Technology," ed. by R.E. Kirk and D.F. Othmer, Vol. 6, p. 516, Interscience Encyclopedia, New York (1951).

Chatib, B., A. Grasmick, S. Elmaleh, and R. Ben Aim, in "Biological Fluidized Bed Treatment of Water and Wastewater," ed. by P.F. Cooper and B. Atkinson, p. 192, Ellis Horwood (1981).

Charpentier, J.C., Chem. Eng. J., 11, 161 (1976).

Charpentier, J.C., in "Advances in Chemical Engineering," Vol. 11, ed. by T.B. Drew and T. Vermeulen, Academic Press (1981).

Charpentier, J.C., in "Multiphase Chemical Reactors - Theory, Design, Scale-up," ed. by A. Gianetto and P.L. Silveston, p. 289, Hemisphere (1986).

Cover, A.E., W.C. Schreiner, and G.T. Skaperdas, Chem. Eng. Prog., 69(3), 31 (1973).

de'Bruijn, T.J.W., D.J. Patmore, G. Flaherty, and L.W. Chambers, paper presented at the Synfuels 2nd Worldwide Symposium, Brussels, Belgium, Oct. 11-13 (1982).

Deckwer, W.-D., Y. Louisi, A. Zaidi, and M. Ralek, I&EC Process Des. & Dev., 19(4), 699 (1980).

Deckwer, W.-D., Y. Serpemen, M. Ralek, and B. Schmidt, Chem. Eng. Sci., 36, 765 (1981).

Deckwer, W.-D., Y. Serpemen, M. Ralek, and B. Schmidt, I&EC Process Des. & Dev., 21, 222 (1982).

Donaldson, T.L., G.W. Strandberg, J.D. Hewitt, and G.S. Shields, Environ. Prog., 3, 248 (1984).

Donath, E.E., in "Chemistry of Coal Utilization," Supplementary Volume, ed. by H.H. Lowry, p. 1041, John Wiley (1963).

Douglas, W.J.M., Chem. Eng. Prog., 60(7), 66 (1964).

Edwards, A.M., M.S. Thesis, The Ohio State University (1981).

Egan, B.Z., and L.K. Felker, I&EC Process Des. & Dev., 25, 558 (1986).

Endo, I., T. Nagamune, N. Kato, T. Kobayashi, and M. Nishimura, paper 25c presented at the AIChE Annual Meeting, Miami Beach, Nov. 2-7 (1986).

EPRI (Electric Power Research Institute), "Introduction to Limestone Flue Gas Desulfurization," Tech Tape CC87-01 (1987).

Epstein, N., Can. J. Chem. Eng., 59, 649 (1981).

Ermakova, A., G.K. Ziganshin, E.F. Stefoglo, and M.G. Slin'ko, Theor. Osnovy Khim. Tekhnol., 7(1), 35 (1973).

Fan, L.-S., K. Fujie, and T.R. Long, AIChE Symposium Series, 80, No. 241, 102 (1984).

Fan, L.-S., R.-H. Jean, and K. Kitano, Chem. Eng. Sci., 42, 1853 (1987).

Fan, L.-S., B.K. Kreischer, and K. Tsuchiya, in "Transport Processes in Fluidized Beds," ed. by A. Mujumdar and L.K. Doraiswamy, Elsevier (in press, 1988).

Fan, L.-S., and Y.A. Attia, Interim Report to Ohio Coal Development Office, Research Project No. RN1-87-090 (1988).

Foot, D.G., Jr., J.D. McKay, and J.L. Huiatt, Can. Metal. Q., 25(1), 15 (1986).

Forster, C.F., A.P. Boyes, B.A. Hay, and J.A. Butt, Chem. Eng. Res. Des., 64, 425 (1986).

Frohning, C.D., H. Kolbel, M. Ralek, W. Rottig, F. Schnur, and H. Schultz, in "Chemie-rohstoffe aus Kohle," ("Chemical Raw Materials from Coal,") ed. by J. Falbe, p. 219, Thieme, Stuttgart (1977).

Fuerstenau, D.W., in "Fine Particle Flotation," ed. by P. Somasundaran, p. 669, Am. Inst. Mining, Metallurgical and Petroleum Engrs. (1980).

Fujimoto, K., O. Kuchi-ishi, and T. Kunugi, I&EC Product Res. & Dev., 15(4), 259 (1976).

Furusaki, S., Y. Nozaki, and K. Nakagiri, Can. J. Chem. Eng., 62, 610 (1984).

Gartsman, A.N., A. Ermakova, V.P. Bakhbalova, and N.I. Rassadnikova, Int. Chem. Eng., 17, 697 (1977).

Gel'perin, N.I., V.I. Savchenko, V.I. Ksenzenko, V.Z. Grishko, and E.A. Dianov, Khim. Prom., No 11, 832 (1965).

Gel'perin, N.I., V.Z. Grishko, V.I. Savchenko, and V.M. Shchedro, Khim. i Neft. Mash., No. 1, 22 (1966).

Gel'perin, N.I., Yu.M. Latyshev, and L.I. Blyakhman, Int. Chem. Eng., 8(4), 691 (1968).

Germain, A., G.A. L'Homme, and A. Lefebvre, in "Chemical Engineering of Gas-Liquid-Solid Catalyst Reactions," ed. by G.A. L'Homme, p. 265, CEBEDOC, Liege (1979).

Giot, M., in "Handbook of Multiphase Systems," ed. by G. Hetsroni, p. 7-29, McGraw-Hill (1982).

Goto, S., and J.M. Smith, AIChE J., 21, 706 (1975).

Goto, S., J. Levec, and J.M. Smith, Catal. Rev. - Sci. Eng., 15(2), 187 (1977).

Goto, S., and Y. Kojima, Chem. Eng. Commun., 34, 213 (1985).

Govindarao, V.M.H., and M. Chidambaram, J. Chem. Eng. Japan, 19, 243 (1986).

Grasmick, A., B. Chatib, S. Elmaleh, and R. Ben Aim, Water Research, 15, 719 (1981).

Hammer, H., in "Chemical Engineering of Gas-Liquid-Solid Catalyst Reactions," ed. by G.A. L'Homme, p. 334, CEBEDOC, Liege (1979).

Hasegawa, J., in Proceedings of the Technical Seminar on Wastewater Treatment and Environment Purification, Hakko Kogyo Kyokai, Japan, Oct. 15 (1982).

Hatate, Y., H. Nomura, T. Fujita, S. Tajiri, and A. Ikari, J. Chem. Eng. Japan, 19, 330 (1986).

Heck, R.M., and T.G. Smith, I&EC Process Des. & Dev., 9(4), 537 (1970).

Herskowitz, M., and J.M. Smith, AIChE J., 29, 1 (1983).

Hirata, A., Y. Hosaka, H. Mukai, and M. Kaiho, Water Purification and Liquid Wastes Treatment, 23, 15 (1982).

Hofmann, H.P., Catal. Rev. - Sci. Eng., 17, 71 (1978).

Hofmann, H.P., in "Multiphase Chemical Reactors - Theory, Design, Scale-up," ed. by A. Gianetto and P.L. Silveston, p. 257, Hemisphere (1986).

Hogrefe, W., H. Grossenbacher, A.M. Cook, and R. Hutter, Biotech. Bioeng., 28, 1577 (1986).

Holladay, D.W., C.W. Hancher, C.D. Scott, and D.D. Chilcote, J. Water Pollution Control Fed., 50, 2573 (1978).

Hosaka, Y., M. Kaihou, and A. Hirata, Water Sci. and Technol., 17, 1437 (1985).

Hu, T.-T., and J.Y. Wu, Chem. Eng. Res. Des., 65, 238 (1987).

Idemura, H., T. Kanai, and H. Yanagioka, Chem. Eng. Prog., 74 (2), 46 (1978).

Jenkins, J.M. III, R.L. Jones, and T.M. Jones, U.S. Patent 4,543,399 (1985).

Jenkins, J.M. III, R.L. Jones, T.M. Jones, and S. Beret, U.S. Patent 4,588,790 (1986).

Johanson, E.S., U.S. Patent 2,987,465 (1961).

Johnson, T.E., J.R. Murphy, and K.G. Tasker, Technology (Oil and Gas J.), p. 50, July 1 (1985).

Jones, A., D.N. Wood, T. Razniewska, G.M. Gaucher, and L.A. Behie, Can. J. Chem. Eng., 64, 547 (1986).

Joshi, J.B., J.S. Abichandani, Y.T. Shah, J.A. Ruether, and H.J. Ritz, AIChE J., 27, 937 (1981).

Kalogerakis, N., L.A. Behie, and G.M. Gaucher, in "Fluidization," ed. by K. Ostergaard and A. Sorensen, p. 611, Engineering Foundation Press (1986).

Karaivko, G.I., S.I. Kuznetsov, and A.I. Golonizik, "The Bacterial Leaching of Metals from Ores," Technicopy Limited (1977).

Karamanev, D.G., and L.N. Nikolov, Biotech. Bioeng., 31, 295 (1988).

Kargi, F., and M.Z. Rosenberg, Biotech. Prog., 3, 1 (1987).

Kataoka, K., S. Ito, and S. Tsunoda, Kankyo Souzou, Japan, p. 59, Sep. (1979).

Katinger, H.W.D., W. Scheirer, and E. Kromer, Ger. Chem. Eng., 2, 31 (1979).

Kato, Y., A. Nishiwaki, T. Kago, T. Fukuda, and S. Tanaka, Int. Chem. Eng., 13, 562 (1973).

Kato, Y., N. Hidaka, and H. Kamimura, Kagaku Kogaku Ronbunshu, 12, 726 (1986).

Kim, S.D., and J.H. Choi, Can. J. Chem. Eng., 62, 85 (1984).

Kim, S.D., and J.O. Kim, J. Chem. Eng. Japan, 19, 86 (1986).

Kohl, A.L., R.B. Harty, and J.G. Johanson, Chem. Eng. Prog., 74, 73, Aug. (1978).

Kolbel, H., and P. Ackermann, Chem.-Ing.-Tech., 28, 381 (1956).

Kolbel, H., and M. Ralek, Catal. Rev. - Sci. Eng., 21(2), 225 (1980).

Kono, H., Hydrocarbon Processing, 59, 123, Jan. (1980).

Kornosky, R.M., M. Perlmutter, W. Fuchs, and J.A. Ruether, U.S. Department of Energy Report DOE/PETC/TR-82/8 (DE82012013) (1982).

Kudra, T., G. Bednarek, J. Adamiec, and C. Strumillo, Inst. Chem. Engrs. Symposium Series, No. 57 (1978).

Kunii, D., and O. Levenspiel, "Fluidization Engineering," John Wiley (1969).

Kusakabe, K., S. Morooka, and Y. Kato, J. Chem. Eng. Japan, 14(3), 208 (1981).

Lee, D.D., C.D. Scott, and C.W. Hancher, J. Water Pollution Control Fed., 51, 974 (1979).

L'Homme, G.A., "Chemical Engineering of Gas-Liquid-Solid Catalyst Reactions," CEBEDOC, Liege (1979).

Li, A., and D. Liu, in Proceedings of the 2nd World Congress of Chemical Engineering, IV, p. 170 (1981).

Linko, P., and Y.-Y. Linko, in "Applied Biochemistry and Bioengineering," ed. by I. Chibata and L.B. Wingard, Jr., 4, 53 (1983).

Lunin, G., J. Lamarre, and K. Whitham, paper presented at the UNITAR 3rd International Conference on "Heavy Crude and Tar Sands," Long Beach, CA, July 22-31 (1985).

Martin, H., and A.H. Saleh, Int. Chem. Eng., 24, 13 (1984).

McMichael, W.J., L.-S. Fan, and C.Y. Wen, I&EC Process Des. & Dev., 15, 459 (1976).

Mlodzinski, B., and J. Warych, Chem.-Ing.-Tech., 47, 813 (1975).

Moebus, O., and M. Teuber, Eur. J. Appl. Microbiol. Biotech., 15, 194 (1982).

Morl, L., and L.-J. Kunne, Int. Chem. Eng., 26(3), 423 (1986).

Mueller, W.H., Interim Report (DOE/ET/10069-T113) to U.S. Department of Energy, Contract No. DE-FC05-77ET10069 (1985).

Muroyama, K., and L.-S. Fan, AIChE J., 31, 1 (1985).

Narayan, R., paper presented at the 1st Annual Workshop on Biological Processing of Coal, sponsored by Electric Power Research Inst., Monterey, CA (1987).

Nienow, A.W., and R.C. Killick, AIChE Symposium Series, 79, No. 222, 1 (1983).

Nienow, A.W., paper presented at the AIChE Annual Meeting, San Francisco, Nov. 25-30 (1984).

Okochi, I., H. Yamazaki, K. Izumi, and S. Takahashi, Japan Kokai Tokkyo Koho, 78,119,782, Oct. 19 (1978).

Oloman, C., and A.P. Watkinson, Can. J. Chem. Eng., 53, 268 (1975).

Ostergaard, K., in "Advances in Chemical Engineering," Vol. 7, ed. by T.B. Drew, G.R. Cokelet, J.W. Hoopes, Jr., and T. Vermeulen, p. 71, Academic Press (1968).

Palancz, B., Chem. Eng. Sci., 38(7), 1045 (1983).

Prenosil, J.E., and H. Pedersen, Enzyme Microb. Technol., 5, 323 (1985).

Probstein, R.F., and R.E. Hicks, "Synthetic Fuels," McGraw-Hill (1982).

Qian, W.X., Research Report, No. 5 (Jan.), Wuhan Institute of Municipal Engineering, China (1980).

Ramachandran, P.A., and R.V. Chaudhari, "Three-Phase Catalytic Reactors," Gordon and Breach Science (1983).

Reilly, I.G., D.S. Scott, and M. Abou-el-Hassan, Can. J. Chem. Eng., 60, 399 (1982).

Samejima, H., M. Nagashima, M. Azuma, S. Noguchi, and K. Inuzuka, Annals New York Academy of Sciences, 434, 394 (1984).

Sastri, N.V.S., N. Epstein, A. Hirata, I. Koshijima, and M. Izumi, Can. J. Chem. Eng., 61, 635 (1983).

Satterfield, C.N., A.A. Pelossof, and T.K. Sherwood, AIChE J., 15, 226 (1969).

Satterfield, C.N., AIChE J., 21, 209 (1975).

Satterfield, C.N., and G.A. Huff, Jr., Chem. Eng. Sci., 36, 790 (1981).

Sawada, S., Yosui to Haisui, Japan, 20, 1051 (1978).

Sawai, T., K. Sugada, and T. Miki, The Mitsubishi Heavy Industry Technical Report, 19(5), 510, Sep. (1982).

Schlesinger, M.D., J.H. Crowell, M. Leva, and H.H. Storch, I&EC, 43, 1474 (1951).

Schlesinger, M.D., H.E. Benson, E.M. Murphy, and H.H. Storch, I&EC, 46, 1322 (1954).

Scott, C.D., and C.W. Hancher, Biotech. Bioeng., 18, 1393 (1976).

Shah, Y.T., "Gas-Liquid-Solid Reactor Design," McGraw-Hill (1979).

Shah, Y.T., "Reaction Engineering in Direct Coal Liquefaction," Addison-Wesley (1981).

Shah, Y.T., B.G. Kelkar, S.P. Godbole, and W.-D. Deckwer, AIChE J., 28, 353 (1982).

Shah, Y.T., A.A. Kulkarni, J.H. Wieland, and N.L. Carr, Chem. Eng. J., 26, 95 (1983).

Shen, Y.M., and B.L. Wang, Research Report, No. 2, Chen-du Institute of Municipal Engineering, China (1980).

Sherwin, M.B., and M.E. Frank, Hydrocarbon Processing, 55(11), 122 (1976).

Sherwood, T.K., and E.J. Farkas, Chem. Eng. Sci., 21, 573 (1966).

Shimodaira, C., and Y. Yushina, in Proceedings of the 3rd Pacific Chemical Engineering Congress, IV, p. 239, Seoul (1983).

Smith, D.N., G.T. Stiegel, and J.A. Ruether, in "Encyclopedia of Fluid Mechanics," Vol. 6, ed. by N.P. Cheremisinoff, Chap. 15, p. 535, Gulf Publishing (1986).

Storch, H.H., in "Chemistry of Coal Utilization," II, ed. by H.H. Lowry, p. 1750, John Wiley (1945).

Studer, D.W., E.P. Holley, T.H. Hsiung, and R.L. Mednick, paper presented at the U.S. DOE/FE Indirect Liquefaction Contractors' Review Meeting, Pittsburgh, PA, Dec. 8-9 (1987).

Suzuki, K., K. Kataoka, and S. Tsunoda, Ebara Infilco Report, No. 85 (1981).

Suzuki, M., S. Momonoi, and H. Harada, in Proceedings of the Symposium on Biological Wastewater Treatment, p. 34, Soc. Chem. Engrs. Japan, Sendai, Japan (1983).

Takase, H., D. Ishibashi, and M. Sugimoto, in Proceedings of the Symposium on Biological Wastewater Treatment, p. 17, Soc. Chem. Engrs. Japan, Sendai, Japan (1983).

Tang, W.-T., and L.-S. Fan, AIChE J., 33, 239 (1987).

Tang, W.-T., K.D. Wisecarver, and L.-S. Fan, Chem. Eng. Sci., 42, 2123 (1987).

Tarmy, B.L., M. Chang, C.A. Coulaloglou, and P.R. Ponzi, paper presented at the 8th International Symposium on Chemical Reaction Engineering, Edinburgh (1984).

Thompson, A.R., in "Encyclopedia of Chemical Technology," ed. by R.E. Kirk and D.F. Othmer, Vol. 4, p. 619, Interscience Encyclopedia, New York (1955).

Toda, M., E. Harada, M. Kuriyama, S. Saruta, and H. Konno, Kagaku Kogaku Ronbunshu, 8, 380 (1982).

Uysal, B.Z., Ph.D. Thesis, McGill University (1978).

Van Landeghem, H., Chem. Eng. Sci., 35, 1912 (1980).

Volpicelli, G., and L. Massimilla, Pulp Paper Mag. Canada, 66(10), T512, Oct. (1965).

Volpicelli, G., and L. Massimilla, Chem. Eng. Sci., 25, 1361 (1970).

Wada, M., T. Uchida, J. Kato, and I. Chibata, Biotech. Bioeng., 22, 1175 (1980).

Wagner, F., and H. Vogelmann, in "Plant Tissue Culture and Its Biotechnological Application," ed. by W. Barz, E. Reinhard, and M.H. Zenk, p. 245, Springer-Velag, Berlin (1977).

Wagner, K., and D.C. Hempel, Biotech. Bioeng., 31, 559 (1988).

Waldie, B., D. Wilkinson, and L. Zachra, Chem. Eng. Sci., 42, 653 (1987).

Wang, S.C., and D. Burbank, EPA Final Report EPA-600/7-82-010a (1982).

Weber, M., and Y. Dedegil, paper presented at the 4th International Conference on Hydraulic Transport Solids Pipes, p. 1, BHRA Fluid Engineering, Cranfield, Bedford, U.K. (1976).

Wenzel, F.W., U. Graeser, and K. Niemann, paper presented at the Japan Petroleum Institute, Petroleum Refining Conference, Tokyo, Oct. 27-28 (1986).

Wilson, B.W., R.M. Bean, J. Pyne, D.L. Stewart, and J. Fredickson, in Proceedings of Biological Treatment of Coals Workshop, p. 114, sponsored by U.S. Department of Energy, Herndon, VA, June 23-25 (1986).

Wozniak, M., and K. Ostergaard, Chem. Eng. Sci., 28, 167 (1973).

Wu, Y., O. Fan, and M. Kwauk, in "Fluidization '88 Science and Technology -- Conference Papers 3rd China-Japan Symposium," ed. by M. Kwauk and D. Kunii, p. 268, Science Press (Beijing, China) (1988).

Yosim, S.J., and K.M. Barclay, in Proceedings of the 2nd International Symposium on Molten Salts, ed. by J.R. Selman and J. Braunstein, p. 313, Electrochemical Society (1981).

Zeky, M.EL., and Y.A. Attia, paper presented at the 8th International Symposium on Coal Slurry Fuels Preparation and Utilization, Orlando, FL, May 27-30 (1986).

Zhang, Z.J., B.N. Lu, and X. Gao, Research Report, No. 4, Harbin Institute of Architecture and Engineering, China, April (1981).

——, Chem. Eng., 66, 106, Dec. 14 (1959).

Part II

Fundamentals

Chapter 2

Hydrodynamics of Cocurrent Upward Fluidized Bed Systems (Modes E-I-a-1 and E-I-b; Figure 1.3)

This chapter deals with hydrodynamics of fluidized bed systems with cocurrent upward gas and liquid flows. Operation is Mode E-I-a-1 when the liquid is the continuous phase and Mode E-I-b when gas is the continuous phase. Mode E-I-a-1 is one of the most important industrial applications of gas-liquid-solid fluidized bed technology and considerable fundamental work has been performed regarding this mode. Little is known, however, about Mode E-I-b. The bulk of this chapter pertains to Mode E-I-a-1 operation. The last part concerns Mode E-I-b operation.

The hydrodynamic behavior in Mode E-I-a-1 operation is a complex subject and is discussed in three areas, i.e., the general, or overall, bed behavior, the mechanics of bubbles and bubble wakes, and flow models. The description of the general bed behavior includes observations about and empirical relations for pressure drop, flow regimes, incipient fluidization, pressure fluctuations, phase holdups and solids entrainment, solids wettability, surfactant and liquid surface tension effects, and overall bed rheology. Subsequently, the mechanics of bubbles and bubble wakes are illustrated. The fundamental understanding of bubble and bubble wake mechanics provides a basis for the development of flow models to account for individual phase holdups in the bed and freeboard regions and particle terminal velocities in a gas-liquid medium. The discussion presented is limited to systems without internals.

2.1 GENERAL BED BEHAVIOR IN MODE E-I-a-1

The overall fluid mechanic or hydrodynamic behavior of three-phase fluidized beds reflects the complex interactions between the individual phases. The most prominent interaction occurs between the rising gas bubbles and the surrounding liquid-solid medium. Three distinct regions above the gas-liquid distributor are identifiable based on the prevailing physical phenomena: the distributor region, the bulk fluidized bed region, and the freeboard region. A schematic diagram for Mode E-I-a-1 operation is shown in Fig. 2.1.

The distributor region refers to the region immediately above the gas-liquid distributor where gas spouts may occur. It includes the region from initial bubble formation to the establishment of the final bubble shape. The hydrodynamic behavior in the distributor region inherently depends on the gas-liquid distributor design and the physical properties of the liquid-solid medium. The bulk fluidized bed region includes the main portion of the fluidized bed. The hydrodynamic behavior in the bulk fluidized bed region varies drastically over large ranges of operating conditions. However, for a given operating condition, there is a minimum axial transport property variation in the region. The freeboard region mainly contains entrained particles from the bulk fluidized bed region. Particle entrainment leads to a solids holdup profile above the fluidized bed surface which decreases axially in a manner similar to that in a gas-solid fluidized bed. The demarcation between the freeboard region and the bulk fluidized bed region is much more distinct for large/heavy[1] particles than for small/light[1] particles. Understanding the hydrodynamic behavior in the freeboard region is instrumental for a proper reactor design.

2.1.1 Pressure Drop

The pressure drop through the bed is strongly related to the individual phase holdups in the bed. In the fluidized bed with low solids entrainment rates (see Section 2.1.5.5), the volume fraction of the solids, or solids holdup, can be expressed in terms of the total mass of solids (W_S), the solid density (ρ_S), the cross section of the column (A), and the effective bed height (H) as

$$\epsilon_S = 1 - \epsilon = \frac{W_S}{\rho_S AH} \tag{2.1}$$

1: Throughout this chapter, small or small/light particles refer to fluidized bed particles with a terminal velocity in water of less than about 30 cm/s, whereas large or large/heavy particles refer to fluidized bed particles with a terminal velocity in water of larger than about 30 cm/s, unless otherwise noted.

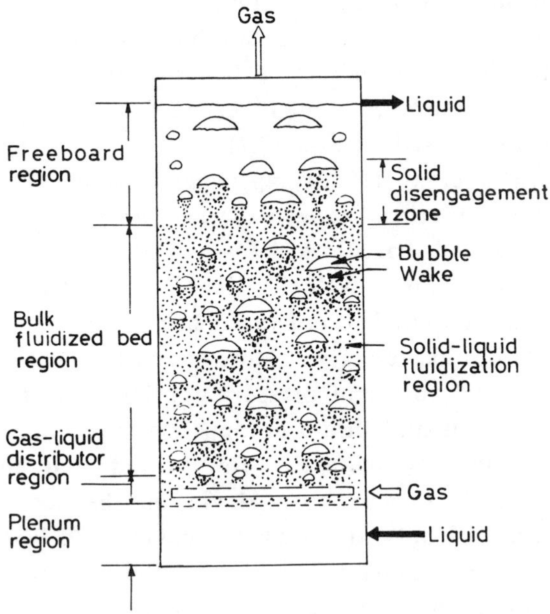

Figure 2.1 Schematic representation of the Mode E-I-a-1 fluidized bed.

where ϵ_s is the solids holdup and ϵ is the bed porosity. A necessary requirement of the phase volume fractions, or individual phase holdups, is

$$\epsilon_g + \epsilon_\ell + \epsilon_s = 1 \tag{2.2}$$

where ϵ_g and ϵ_ℓ are the gas and liquid holdups, respectively.

At steady state conditions with low or moderate liquid and gas velocities, the total vertical pressure gradient (static pressure gradient) can be readily obtained from (e.g., Ostergaard, 1969)

$$-\frac{dP}{dz} = (\epsilon_s\rho_s + \epsilon_\ell\rho_\ell + \epsilon_g\rho_g)g \tag{2.3}$$

In this equation the frictional drag on the column wall and the acceleration terms for the gas and liquid flows are neglected. Since the contribution of the gas phase is usually negligible compared to the other terms, Eqn. (2.3) can be simplified to

$$-\frac{dP}{dz} = (\epsilon_s\rho_s + \epsilon_\ell\rho_\ell)g \tag{2.4}$$

Equation (2.4) permits evaluation of individual phase holdups from the pressure gradient. ϵ_s can be directly obtained from Eqn. (2.1) with the measured effective bed height. ϵ_ℓ can be calculated from the total pressure gradient based on Eqn. (2.4) and ϵ_g is obtained by the difference based on Eqn. (2.2).

Ermakova et al. (1970) experimentally verified Eqn. (2.3) and found that the deviation between the measured total pressure gradient and that calculated based on known values of the phase holdups decreased as the liquid velocity increased. The improvement in agreement was attributed to enhanced bed homogeneity resulting from the increased liquid velocity. More recently, Dhanuka (1978) compared experimental values of pressure gradient with those predicted by Eqn. (2.3), using gas and liquid holdups measured independently by an electrical conductivity method. He showed that the prediction is satisfactory with a maximum deviation between predicted and measured values of only about 5%. Begovich and Watson (1978b) also showed that holdup values obtained from the pressure gradient method agree with those measured by the conductivity method within 5% deviation.

It is interesting to note that Eqn. (2.3) is also valid for countercurrent fluidization systems (see Chapter 5) such as turbulent bed contactor systems (Barile and Meyer, 1971; Tichy and Douglas, 1972; Kuroda and Tabei, 1979) and inverse three-phase fluidized bed systems (Fan et al. 1982). In these systems, the bed is in a state of uniform expansion; the contributions of fluid acceleration and wall friction are negligible.

When liquid is the continuous phase, the dynamic pressure gradient for the liquid ($-dP_d/dz$), defined as the total pressure gradient corrected for the hydrostatic head of the liquid (Epstein, 1981), is

$$-\frac{dP_d}{dz} = \left(-\frac{dP}{dz}\right) - \rho_\ell g \tag{2.5}$$

Combining Eqns. (2.2), (2.3) and (2.5) yields

$$-\frac{dP_d}{dz} = \epsilon_s(\rho_s - \rho_\ell)g - \epsilon_g(\rho_\ell - \rho_g)g \tag{2.6}$$

Equation (2.6) shows that the dynamic pressure gradient as measured by differential manometry corresponds to a force balancing the weight of particles (downward direction) and the buoyancy of bubbles (upward direction).

The frictional pressure gradient ($-dP_f/dz$), defined as the total pressure gradient corrected for the hydrostatic head of the gas-liquid two-phase mixture (Epstein, 1981), is

$$-\frac{dP_f}{dz} = \left(-\frac{dP}{dz}\right) - \rho_f g \tag{2.7}$$

where ρ_f, the average density of the two-phase mixture, is

$$\rho_f = \frac{\epsilon_g \rho_g + \epsilon_\ell \rho_\ell}{\epsilon_g + \epsilon_\ell} = \frac{\epsilon_g \rho_g + \epsilon_\ell \rho_\ell}{1 - \epsilon_s} \tag{2.8}$$

Combining Eqns. (2.2), (2.3), (2.7), and (2.8), and simplifying gives (Epstein, 1981)

$$-\frac{dP_f}{dz} = \epsilon_s(\rho_s - \rho_f)g = (1 - \epsilon)(\rho_s - \rho_f)g \tag{2.9}$$

Examining this equation reveals that the frictional pressure gradient balances the effective weight of solid particles in the two-phase medium. In the absence of gas, Eqns. (2.6) and (2.9) reduce to

$$-\frac{dP_d}{dz} = -\frac{dP_f}{dz} = (1 - \epsilon_\ell)(\rho_s - \rho_\ell)g \tag{2.10}$$

which is the characteristic equation for liquid-solid fluidization.

Figure 2.2 shows typical examples of axial dynamic pressure variations for a three-phase fluidized bed containing glass beads of uniform size. The dynamic pressure distribution shows linear behavior in both the bulk fluidized bed region and the freeboard region above the bed. Both the dynamic pressure and the pressure gradient decrease with increasing gas velocity because of increased gas holdup. The location of the expanded bed height can be determined from the intersection of the linear pressure profiles in the bulk fluidized bed and freeboard regions.

2.1.2 Flow Regimes

Three flow regimes can be identified based on the bubble flow behavior in three-phase fluidized beds: the coalesced bubble, the dispersed bubble, and the slugging regimes. In the coalesced bubble regime, bubbles tend to coalesce and both the bubble size and velocity become large and show a wide distribution. Coalesced bubbles rise

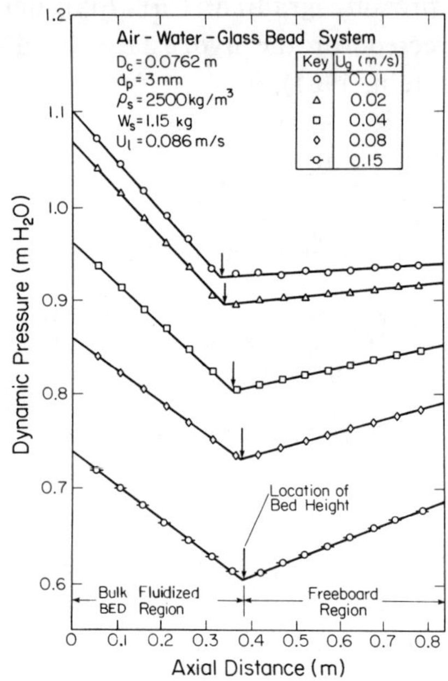

Figure 2.2 Variation of dynamic pressure with axial distance as a function of U_g in a gas-liquid-solid fluidized bed (from Jean, 1988).

near the column center with high velocity and stir the bed violently. The coalesced bubble regime predominates at low liquid and high gas velocities. In the dispersed bubble regime, no bubble coalescence occurs and the bubbles are of uniform, small size. The dispersed bubble regime predominates at high liquid velocities and at low and intermediate gas velocities. In a small diameter column (e.g., $D_c < 15$ cm), the gas bubbles can easily grow to the size of the column diameter at high gas flow rates creating "slug" bubbles which occupy nearly the whole cross section. In columns of large diameter, however, slugging may not occur.

The flow regimes can be mapped out on a plot of U_ℓ vs. U_g as shown in Fig. 2.3. It is seen that for air-water-particle systems, the flow regimes vary significantly with the column diameter. Particle properties also profoundly affect the prevailing flow regime at given gas and liquid velocities (Fan et al. 1986a). The terminal velocity of the fluidized particles affects the liquid velocity of transition from the coalesced to the dispersed bubble regime (U_{tr}) as shown in Fig 2.4. For a low U_t, an increase in particle size increases the transition

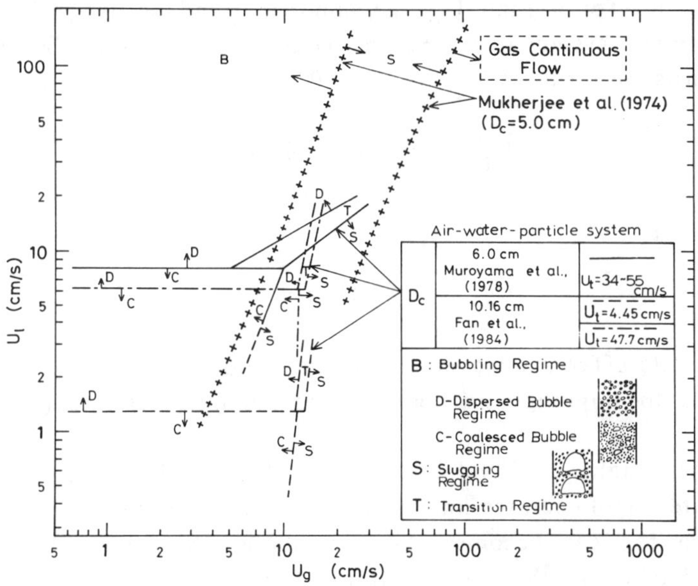

Figure 2.3 Flow regime diagram for cocurrent gas-liquid-solid fluidized beds (from Muroyama and Fan, 1985).

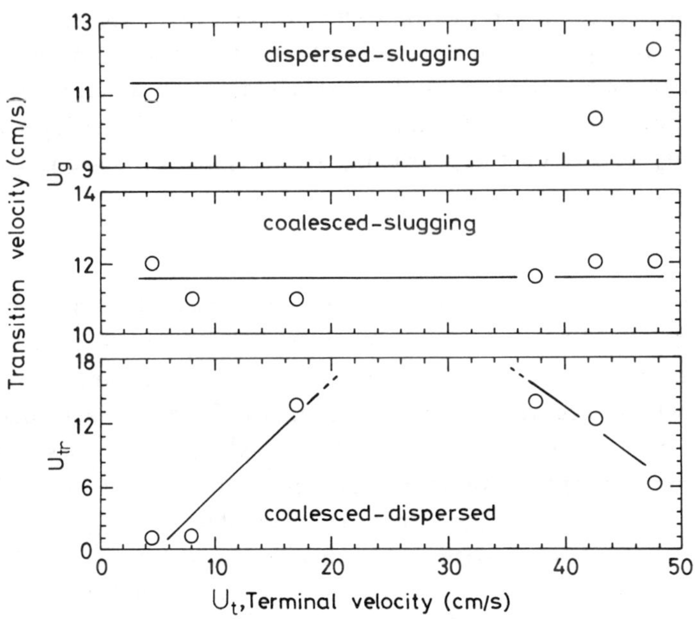

Figure 2.4 Variation in flow regime transition velocity with particle terminal velocity for air-water-particle systems in a 7.62 cm ID column (from Fan et al., 1986a).

velocity, but for a high U_t, the opposite trend occurs. For an air-water-particle system in a 7.62 cm ID column, U_{tr} could be expressed (in SI units) in terms of U_t, for $U_t < 0.273$ m/s, as

$$U_{tr} = 1.08\ U_t - 0.0538 \tag{2.11}$$

and, for $U_t > 0.273$ m/s, as

$$U_{tr} = -\ 0.81\ U_t + 0.461 \tag{2.12}$$

U_t has little effect, however, on the gas velocities for transition from dispersed to slugging and coalesced to slugging regimes (Fan et al., 1986a).

At high gas velocities, the liquid-solid mixture forms fragmented agglomerates and eventually an operating state is reached in which the gas becomes the continuous phase and the liquid a dispersed phase in the form of liquid films or droplets (Mukherjee et al., 1974). The transition from a continuous liquid phase to a continuous gas phase has not been clearly investigated.

It is noted that the bed pressure or presence of surfactants significantly affect the prevailing flow regime (also see Sections 2.1.7 and 10.5). At high pressures or in surfactant systems, the bubble size can be very small (less than 1 mm) yielding an operation in the dispersed bubble regime. Furthermore, the effect of particles greater than the bubble size on the prevailing flow regime becomes minimal.

2.1.3 Incipient Fluidization

For a given gas velocity, the minimum liquid flow rate required to fluidize a bed of particles ($U_{\ell mf}$) may generally be determined from the change in the bed dynamic pressure drop ($-\Delta P_d$) behavior which occurs as the bed changes from a fixed bed to a fluidized bed. As shown in Fig. 2.5, for a bed of 3 mm glass beads, ($-\Delta P_d$) increases sharply with an increase in liquid velocity in the fixed bed state. Whereas in the fluidized bed state, ($-\Delta P_d$) shows only relatively small changes with increasing liquid velocity. In fact, for the case of no gas flow, ($-\Delta P_d$) remains independent of liquid velocity.

There are considerable variations on minimum fluidization phenomena among small/light, large/heavy and mixed particle systems. However, $U_{\ell mf}$ in general can be evaluated mechanistically by considering an intrinsic condition for minimum fluidization where the

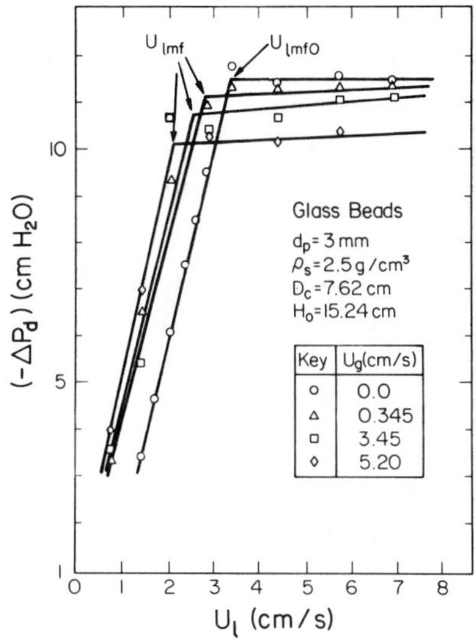

Figure 2.5 Variation of dynamic pressure drop across the bed with U_ℓ as a function of U_g in an air-water-glass bead fluidized bed (from Jean, 1988).

total pressure drop over a bed of particles at the fixed state is equal to the total bed weight per unit bed cross-sectional area as formulated by Song et al. (1987). In this evaluation, the pressure drop in the fixed bed can be described by a flow model developed by Chern et al. (1983, 1984) (see the account of the pressure drop behavior in the packed section of a semifluidized bed in Section 6.2.1).

2.1.3.1 Small/Light Particle Systems

Determination of the incipient fluidization velocity for beds of small/light particles is more difficult than that for beds of large/heavy particles and somewhat subjective since the transition from a fixed bed to a fluidized bed is not distinct. For three-phase fluidization utilizing small particles, Ermakova et al. (1970) defined three hydrodynamic states for the transition from the fixed bed state to the fully fluidized bed state: fixed bed, heterogeneous fluidized bed, and homogeneous fluidized bed. The heterogeneous fluidization state is characterized by two zones with one consisting of a relatively dense fixed bed at the bottom and the other consisting of a fluidized bed at the top. With increasing liquid velocity, the transition from a heterogeneous fluidized bed to a homogeneous fluidized bed or a fully fluidized bed takes place. The onset velocity for heterogeneous fluidization decreases with increasing gas velocity. On the other hand, the onset velocity of homogeneous fluidization increases with increasing gas velocity and decreasing particle size. El-Temtamy and Epstein (1986) studied the effects of pulsations and vibrations superimposed over the continuous liquid flow on the minimum fluidization velocity for beds of small particles. They found that pulsations or vibrations made a fixed bed more compact, increasing the interstitial liquid velocity and pressure drop due to friction for a given superficial liquid velocity, so that minimum fluidization was obtained at a lower liquid velocity.

2.1.3.2 Large/Heavy Particle Systems

In the case of large/heavy particle fluidization, the bed expands uniformly with an increase of either the liquid velocity or the gas velocity beyond the onset of fluidization so that the transition from a fixed bed to a fluidized bed is distinct. The liquid velocity required for the onset of fluidization decreases with an increase in the gas velocity (Begovich and Watson, 1978a; Lee and Al-Dabbagh, 1978; Fan et al., 1985a; Costa et al., 1986; Song et al., 1987). Begovich and Watson (1978a) found that the initial bed height and column diameter have no effect on the onset of minimum fluidization. They developed empirical equations for the liquid velocity at the onset of fluidization based on a wide range of data. Lee and Al-Dabbagh (1978) confirmed that three-phase fluidization occurs when the dynamic pressure drop over the bed is equal to the effective weight of the particles. Fan et al. (1985a) reported that $U_{\ell mf}$ is proportional to $U_t^{1.64}$ under otherwise constant operating conditions.

Recently, Song et al. (1987) studied the effect of particle shape on the minimum fluidization velocity. Minimum fluidization velocities of cylindrical catalysts and data for spherical particles reported in the literature were used to develop a general empirical correlation equation (in SI units) as given below:

$$\frac{U_{\ell mf}}{U_{\ell mf0}} = 1 - 376\, U_g^{0.327}\, \mu_\ell^{0.227}\, d_e^{0.213}\, (\rho_s - \rho_\ell)^{-0.423} \tag{2.13}$$

where d_e is the equivalent particle diameter. $U_{\ell mf0}$ can be calculated from Wen and Yu's correlation (1966) for liquid-solid fluidized beds:

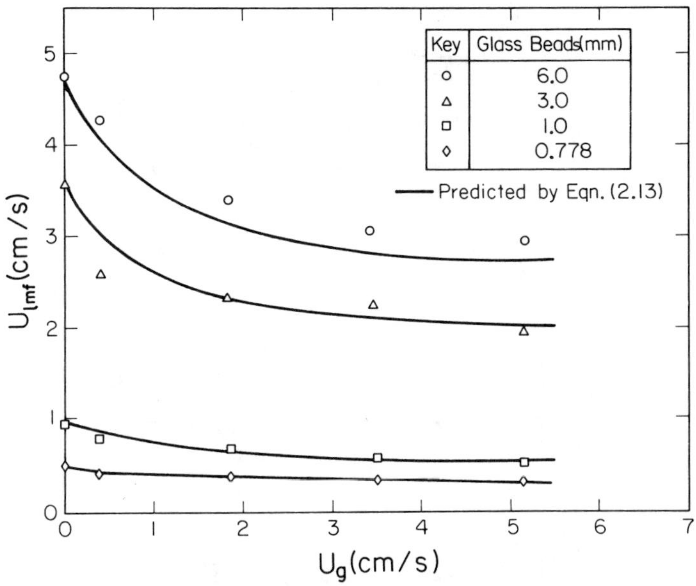

Figure 2.6 Variation of $U_{\ell mf}$ with U_g for glass beads of various sizes in an air-water-glass bead fluidized bed (from Jean, 1988).

$$Re_{mf0} = \sqrt{(33.7)^2 + 0.0408\ Ar} - 33.7 \tag{2.14}$$

where

$$Re_{mf0} = \frac{d_e\ U_{\ell mf0}\ \rho_\ell}{\mu_\ell}$$

Equation (2.13) is given in a similar form as that of Begovich and Watson's (1978a), but has the widest range of applicability among correlations for $U_{\ell mf}$ proposed to date, i.e., $0 \le U_g \le 17$ cm/s, 0.9 (or 9×10^{-3}) $\le \mu_\ell \le 11.4$ (or 0.114) cP (or g/cm•s), $0.046 \le d_e \le 0.63$ cm, and $1.8 \le \rho_s \le 2.5$ g/cm^3. Predictions of the minimum fluidization velocity based on Eqn. (2.13) are compared with data of various sized glass beads in water and air as given in Fig. 2.6. It can be seen that Eqn. (2.13) yields good predictions for beds of either small or large particles.

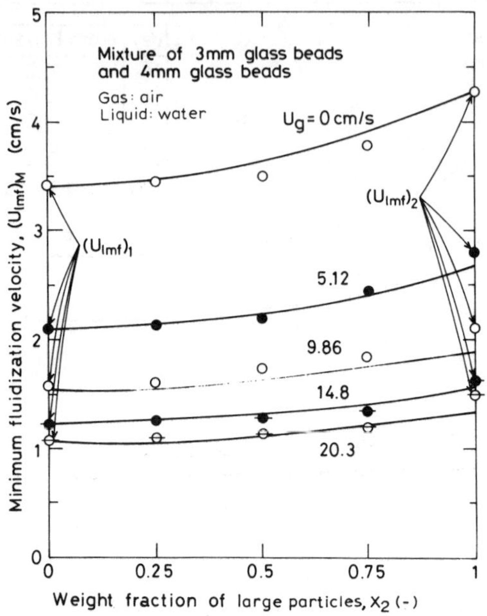

Figure 2.7 Variation of minimum fluidization velocity, as a function of U_g, with the weight fraction of larger particles in a bed containing a binary mixture of particles (from Fan et al., 1985a).

2.1.3.3 Mixed Particle Systems

Three-phase fluidized beds generally contain particles of nonuniform size. Studies of incipient fluidization for mixed particle beds in three-phase systems are, however, scarce in the literature. Oh and Kim (1980) observed the incipient fluidization behavior for beds of a 50 wt% mixture of 1.63 and 7.85 mm glass beads. They indicated that the minimum fluidization velocity in beds of mixed particles was lower than that in beds of large particles alone and the addition of small particles both reduced the minimum fluidization velocity and improved bed homogeneity.

Fan et al. (1985a) studied the incipient fluidization characteristics of a bed containing a binary mixture of particles of various sizes and/or densities. Figure 2.7 shows the minimum fluidization velocities at various compositions for binary mixtures of 3 and 4 mm glass beads. For all mixtures of 3 and 4 mm beads, the minimum fluidization velocity of the binary mixture, $(U_{\ell mf})_M$, occurs between that of the small particles (3 mm glass beads) and that of the large particles (4 mm glass beads). The effect of the weight fraction of large particles (x_2) on $(U_{\ell mf})_M$ decreases as U_g increases. $(U_{\ell mf})_M$ depends on both the incipient fluidization velocity of the small particles, $(U_{\ell mf})_1$, and that of the large particles, $(U_{\ell mf})_2$, in addition to x_2. Fan et al. (1985a) correlated $(U_{\ell mf})_M$ with the equation

$$(U_{\ell mf})_M/(U_{\ell mf})_1 = [(U_{\ell mf})_2/(U_{\ell mf})_1]^{x_2^{1.69}} \qquad (2.15)$$

where $(U_{\ell mf})_i$ can be predicted by Eqn. (2.13) for particle i.

2.1.4 Pressure Fluctuation

Utilizing a 0.102 m ID column, Fan et al. (1986a) investigated the pressure fluctuation behavior and flow regime transition in a three-phase fluidized bed. Figure 2.8 shows typical power spectral density functions of the pressure signals obtained in the dispersed bubble, coalesced bubble, and slugging regimes, respectively. The power spectral density functions appear distinctly different for the three regimes. For each regime, a distinct peak exists between 0 and 2 Hz and a broad peak exists between 16 and 25 Hz, although the relative magnitude of these peaks differs for each regime. There appears to be almost no contribution to the power spectrum of frequencies between 2 and 16 Hz in the dispersed bubble regime, while the contribution of these frequencies is fairly large in the coalesced bubble regime and quite large in the slugging regime.

The power spectral density function, at very low frequencies, was much larger for the dispersed bubble regime than for the coalesced bubble or slugging regime. This behavior suggests a potential objective method for determining the transition from the dispersed bubble regime to the coalesced bubble regime and from the dispersed bubble regime to the slugging regime. The cumulative power near zero frequency (the sum of the discrete power spectral density function for frequency values below approximately 0.6 Hz) showed a sharp rise at the observed transition from the coalesced bubble regime to the dispersed bubble regime and drops sharply at the transition from the dispersed bubble regime to the slugging regime.

The root mean square (RMS) of the pressure fluctuations showed a distinct dependency on the superficial gas and liquid velocities in each flow regime (Fan et al., 1986a). This behavior indicates that the variation in the RMS of the fluctuations with flow

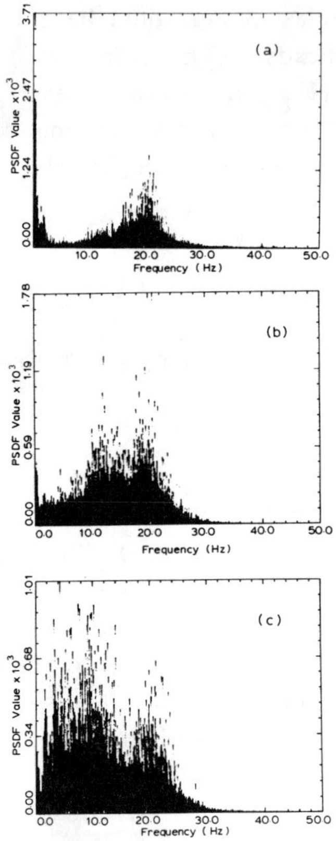

Figure 2.8 Power spectral density function for an air-water-3.04 mm glass bead fluidized bed (from Fan et al., 1986c; reproduced by permission of the American Institute of Chemical Engineers).
(a) Dispersed bubble regime: U_ℓ = 15.8 cm/s, U_g = 3.3 cm/s.
(b) Coalesced bubble regime: U_ℓ = 9.03 cm/s, U_g = 3.3 cm/s.
(c) Slugging regime: U_ℓ = 6.77 cm/s, U_g = 13.2 cm/s.

rate may also provide an objective method for characterizing the transitions from the coalesced bubble regime to the dispersed bubble regime and to the slugging regime.

2.1.5 Phase Holdups

Phase holdups are important transport properties of three-phase fluidized beds. In the following, measurement methods, and behavior and correlations of gas, liquid and solids holdups are described.

2.1.5.1 Measurement Methods

When the expanded bed height is known, the solid holdup can easily be obtained from Eqn. (2.1). The bed height can be determined either from visual observation or from the pressure distribution curves as reported by Kim et al. (1972) and Bhatia and Epstein (1974). The determination of bed height based on visual observation is reasonably accurate for large/heavy particle beds due to existence of a distinct bed surface. In contrast, visual observation becomes quite subjective and unreliable for small/light particle beds due to an indistinct bed surface caused by the considerable entrainment of particles into the freeboard region. Particle entrainment is of particular significance at high gas flow rates. For moderately small particles, the bed height can be determined, as noted earlier, by the interception of the axial pressure distribution curves for the bulk fluidized bed region and the freeboard region due to the different slopes between the two regions. For very small particles with a U_t close to 5 cm/s (see Section 1.2), however, the determination of bed height from the pressure distribution curves also presents difficulty. Beds of very small particles tend to form a rather deep disengagement zone with solids concentration continuously decreasing in the upward axial direction. In such a case, the local value of either liquid or solid holdup should be known independently from the static pressure distribution, from which the rest of the holdups can be calculated. The electrical conductivity probe method has been commonly employed to measure the liquid holdup as described below. Direct measurement of the concentration of metallic catalysts in three-phase systems may be possible using induction transducers mounted on the column wall (Kafarov et al., 1973). Sampling methods may also be adopted in the case of dilute operations.

Both the overall and local liquid holdup can be evaluated independently using an electrical conductivity measurement method. In this approach, a quantitative relationship between the effective conductivity of the mixture and the volume fraction of the dispersed or continuous phase is necessary. Maxwell (1881), based upon a theoretical analysis, suggested the relationship between the effective conductivity and the dispersed phase volume fraction in a heterogeneous two-phase system containing a homogeneously distributed spherical dispersed phase to be

$$\frac{\Lambda_e}{\Lambda_{ec}} = \frac{1+2\beta\epsilon_d}{1-\beta\epsilon_d} \tag{2.16}$$

where Λ_e is the effective conductivity of the heterogeneous two-phase material, Λ_{ec} is the conductivity of the pure continuous phase, and ϵ_d is the volume fraction of the dispersed phase. β is given by

$$\beta = \frac{\Lambda_{ed}-\Lambda_{ec}}{\Lambda_{ed}+2\Lambda_{ec}} = \frac{\alpha_e-1}{\alpha_e+2}, \quad \alpha_e = \frac{\Lambda_{ed}}{\Lambda_{ec}}$$

where Λ_{ed} is the conductivity of the dispersed phase. Although only applicable for low ϵ_d (< 0.1) in principle, Maxwell's equation has been found to hold well even at high ϵ_d (up to fixed bed values) and low α_e (< 10) in a liquid-solid fluidized bed of uniform, spherical particles with a size smaller than 1.0 mm (Turner, 1976). Meredith and Tobias (1962), on the other hand, found that as the volume fraction of the nonconductive dispersed phase was increased, data on spherical particles scattered toward the curve described by Bruggeman's equation: Bruggeman's equation (Bruggeman, 1935) for a nonconductive dispersed phase is

$$\frac{\Lambda_e}{\Lambda_{ec}} = (1-\epsilon_d)^{3/2} \tag{2.17}$$

The same behavior was reported for a fluidized bed of 0.6 mm sand particles of irregular shape (Nasr-El-Din et al., 1987).

For fluidized beds of uniform, spherical, large (greater than 4.0 mm) nonconductive particles, Begovich and Watson (1978b) reported that the empirical relationship

$$\frac{\Lambda_e}{\Lambda_{ec}} = 1-\epsilon_d \tag{2.18}$$

related the effective conductivity to the solids holdup on their data better than the Maxwell equation. This equation was also found to be applicable in determining the liquid holdup in a liquid-solid fluidized bed (Dhanuka and Stepanek, 1978). Kato et al. (1981) found that for fluidized beds of uniform glass beads of 0.42, 0.66, 1.2, and 2.2 mm, the following empirical equation fit their data:

$$\frac{\Lambda_e}{\Lambda_{ec}} = (1-\epsilon_d)^{1.2} \tag{2.19}$$

Nasr-El-Din et al. (1987) also observed that the measured Λ_e/Λ_{ec} was higher than that predicted by the Maxwell equation for fluidized beds of uniform, spherical glass beads of 1.5, 2.8, and 5.5 mm. Similar behavior was found in fluidized beds of uniform, low density (1.05 to 1.3 g/cm^3) spherical particles with a size greater than 1.0 mm by Tang and Fan (1987); in addition, they found that such behavior can be very closely described by the following equation derived by Buyevich (1974) for $\Lambda_{ed}/\Lambda_{ec} << 1$:

$$\frac{\Lambda_e}{\Lambda_{ec}} = \frac{24-17\epsilon_d}{24+19\epsilon_d} \tag{2.20}$$

Equation (2.20) was originally derived for evaluation of the effective thermal conductivity of granular materials and took into account the effects of paired interaction between two dispersed particles and the size variation of the dispersed particles. This equation was also found to fit closely the data of Nasr-El-Din et al. (1987) and Kato et al. (1981).

When extending the same technique to the measurement of local phase holdup of a three-phase fluidized bed, caution should be exercised to ensure the equation used still applies. This is because in three-phase fluidized beds bubbles have sizes and shapes distinctly different from the solid particles. As mentioned above, the size variation and shape of the dispersed phase may render some theoretically derived equations inadequate for use, not to mention the empirical equations which are usually not appropriate for extrapolation. Tang and Fan (1987) reported that for a gas-liquid system, the Maxwell and Buyevich equations very accurately depict the relationship between the effective conductivity and the gas volume fraction up to 0.13 and above 0.25, respectively. The applicability of the Buyevich equation to both the liquid-solid and liquid-gas systems provides the foundation for its use in the calculation of local liquid holdup from measured conductivity of gas-liquid-solid system. The applicability of the electrical conductivity probe and the Buyevich equation to local liquid holdup measurement in three-phase fluidized beds of low density particles with size greater than 1 mm was tested and verified by Tang and Fan (1987).

The overall phase holdups can be determined by measuring the volume fraction of each phase isolated in a test section after simultaneously closing valves located at the inlet and outlet of the fluidized bed, as conducted by Bhatia and Epstein (1974). Kim et al.

(1972) showed that the holdups can be measured by simultaneously shutting off the liquid and gas flows; however, when this method is used careful correction for the gas-liquid two-phase region above the bulk fluidized bed region is necessary.

The γ-ray absorption method may be used to measure the gas holdup just above the surface of a three-phase fluidized bed (Lee and Worthington, 1974). The gas holdup in the freeboard region, however, is greater than that in the bulk fluidized bed region because of decreased liquid velocity in the freeboard region (especially for beds of large particles). Using a radioactive gaseous tracer permits measurement of the residence time distribution of the gas phase from which the mean gas holdup can be obtained together with information on gas mixing. Vasalos et al. (1977, 1979, 1980, 1982) used a radioactive tracer technique to measure the gas holdup and mixing properties in a gas-slurry-solid fluidized bed system. Measurement of the average bubble velocity by means of a motion analyzer and cinematographic films is an indirect method of evaluating gas holdup and was conducted by Baker et al. (1977) and Kim et al. (1977).

An impedance or conductivity double probe can be used to measure the in-bed behavior of gas holdup and bubble properties such as bubble size (vertical chord length) and bubble velocity in three-phase fluidized beds (Rigby et al., 1970; Darton and Harrison, 1974a; Fan et al., 1985b). Rigby et al. (1970) reported that the local gas holdup was higher in the core region than in the near-wall region in a column containing small particles. Fan et al. (1985b) noted that the higher gas holdup in the core region occurred in the coalesced bubble regime whereas in the dispersed bubble regime the radial gas holdup distribution was essentially uniform.

Optical fiber techniques have been reported for measuring local phase holdups and bubble properties in three-phase fluidized beds. Optical fiber probes have been developed that detect either reflected (e.g., Miller and Mitchie, 1970; Calderbank and Pereira, 1977; Abuaf et al., 1978; Delhaye, 1981; Ishida and Tanaka, 1982; de Lasa et al., 1984; Hu et al., 1985; and Frijlink, 1987) or transmitted light (Kitano and Fan, 1988) to measure local bubble properties. Unlike the conductance probe technique, the optical fiber technique is not limited to use with conductive fluids only and a small layer of fluid covering the probe surface does not reduce its effectiveness. Ishida and Tanaka (1982) modified the dual optical fiber probe shown in Fig. 2.9(a) used by Abuaf et al. (1978) in gas-liquid systems (Fig. 2.9(b-1)). They utilized a single quartz fiber to detect characteristic signals of reflected light

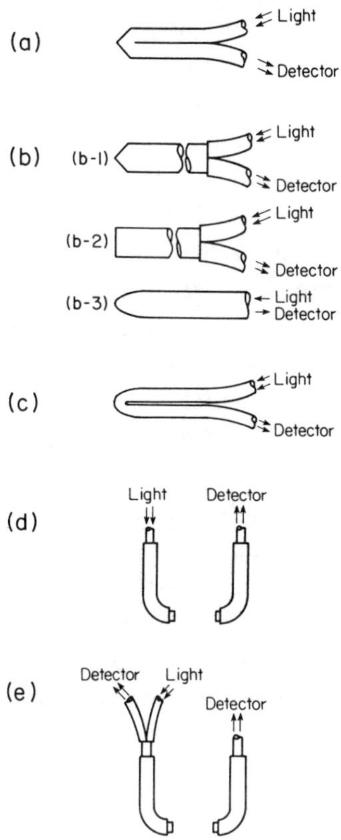

Figure 2.9 Optical fiber probes: (a) dual fiber reflectance probe (from Abuaf et al., 1978), (b) single fiber reflectance probe (b-1 - from Ishida and Tanaka, 1982; b-2 - from Hu et al., 1985; b-3 - from Frijlink, 1987), (c) U-shaped reflectance probe (from de Lasa et al., 1984), (d) transmittance probe (from Kitano and Fan, 1988), (e) transmittance-reflectance probe (from Kitano and Fan, 1987).

for both bubbles and particles in a three-phase fluidized bed. A flat-topped single fiber probe, shown in Fig. 2.9(b-2), could distinguish between the reflected light characteristics of bubbles and particles in three-phase systems. Such an optical probe system was used by Hu et al. (1985) to measure the local behavior of gas bubbles and solids in three-phase fluidized beds. Frijlink (1987) examined various shapes of the single fiber tip, including flat-topped and triangular tips, and concluded that the rounded tip shown in Fig. 2.9(b-3) yielded the highest reflected light intensity, signifying an improved tip design. de Lasa et al. (1984), Lee et al. (1984), and Lee and de Lasa (1986) used silica optical fiber probes, which have a U-shaped tip of 0.5 mm radius as shown in Fig. 2.9(c), to sense the difference between light reflected by a bubble and that reflected by the liquid. The probes successfully detected the movement of bubbles and could be used to measure the vertical bubble length, the rise velocity, and their distributions. Kitano and Fan (1988) developed a light transmittance optical probe, shown in Fig. 2.9(d), to investigate local solids holdup behavior in a three-phase fluidized bed. The probe is based on the principle that the intensity of the transmitted light depends on the local solids holdup. They (Kitano and Fan, 1987) also developed a probe, shown in Fig. 2.9(e), which combines light transmittance and reflectance techniques to simultaneously determine gas and solids holdups.

Note that a multiple fiber or conductivity probe assembly can be used to detect bubble shape or to improve measurement signal selection. In addition, probes must be developed that are able to detect bubbles with size less than 1 mm; these probes must rely on techniques which do not intercept or pierce the bubble. Light transmission methods or, possibly, optical microscopy may fill this need.

2.1.5.2 Overall Solids Holdup and Empirical Correlations

Solids holdup, bed expansion, or bed porosity behavior has been extensively investigated. Three-phase fluidized beds using small particles display unique bed expansion characteristics: i.e., upon initial introduction of the gas into the liquid-solid fluidized bed, contraction, instead of expansion, of the bed occurs. An increasing gas flow rate causes further contraction up to a critical gas flow rate beyond which the bed expands. Massimilla et al. (1959), who first reported this contraction phenomenon, observed that the bed contraction is larger when the liquid velocity is greater. Turner (1964) and Ostergaard (1964) also reported this phenomenon and presented some data on bed expansion. A quantitative elucidation of bed contraction was reported by Stewart and Davidson (1964) and El-Temtamy and Epstein (1979), among others. Basically, bed contraction is attributed to the behavior of the bubble wake. Details of the bubble wake structure and its relevance to bed contraction are given in Sections 2.2.2 and 2.3.1.3.

Various correlations for bed porosity are given in Table 2.1. Kim et al. (1975) proposed separate correlations for expanding and contracting beds. The correlations of Dakshinamurty et al. (1971) and

Kim et al. (1975) include effects of liquid surface tension and the correlations of Oh and Kim (1980) and Sinha et al. (1984) are based on systems using binary particle mixtures and polydispersed particle size distributions, respectively. Note that several correlations in Table 2.1 fail to apply as the gas velocity approaches zero, i.e., in the liquid-solid fluidization state. Bruce and Revel-Chion (1974) compared their bed porosity data with the predictions of Dakshinamurty et al.'s (1971, 1972) correlations and showed that the predicted values were higher than the experimental ones, especially at high gas rates. Kim et al. (1975) also pointed out that, in general, the correlations of Dakshinamurty et al. (1971 and 1972) overestimated their experimental porosities; Kim et al.'s (1975) data, however, were from a rectangular column of narrow width while Dakshinamurty's correlation was based on data from a cylindrical column. Baker et al. (1978) indicated that bed porosities obtained in a cylindrical column were about 10% higher than those of Kim et al. (1975) in the two-dimensional column. These results suggest that the effect of column geometry is slight but still not fully certain.

In the limit as the gas velocity approaches zero, bed porosity correlations should reduce to those for liquid-solid fluidized beds. Bed expansion in liquid-solid fluidized beds often follows the Richardson-Zaki equation (Richardson and Zaki, 1954) given by

$$\epsilon = \left(\frac{U_\ell}{U_i} \right)^{1/n} \tag{2.21}$$

where U_i is the extrapolated liquid velocity as the bed voidage approaches 1. n is the Richardson-Zaki index (see Table 2.2 for estimation). The exponent on the liquid velocity for most correlations of Table 2.1 falls within the range of 1/n obtained from Table 2.2.

Begovich and Watson (1978a) suggested Eqn. (T-2.1.1) in Table 2.1 for bed porosity based on a large quantity of experimental data taken both from the literature and their own work. This correlation does not accurately reflect bed porosity behavior under low gas velocities where bed contraction may occur; note that at no gas flow the bed porosity becomes zero. In addition, Eqn. (T-2.1.1) does not reflect the correct general interdependency of bed porosity with the column diameter; experimental evidence shows that $(\epsilon_g+\epsilon_\ell)$ increases with increasing D_c. Despite these inadequacies, Eqn. (T-2.1.1) remains the most widely accepted empirical correlation for the prediction of bed expansion.

Table 2.1 Empirical Correlations of Phase Holdups for Three-Phase Fluidization (after Muroyama and Fan, 1985).

Investigators	Correlation Equations (in SI Units)	Gas-Liquid	Solid	Column Dimension	Remarks
Ostergaard and Michelsen (1968)	$\varepsilon_\ell = K_1 U_\ell^{K_2} 10^{-K_3 U_g}$ $d_p \times 10^3$(m) / $U_\ell \times 10^2$(m/s) / $U_g \times 10^2$(m/s) / K_1 / K_2 / K_3 0.25 / 0.5 ~ 1.25 / 0.35 ~ 1.7 / 1.86 / 0.211 / 1.8 1 / 2 ~ 10.5 / 0 ~ 2.2 / 2.16 / 0.397 / 2.3 6 / 9 ~ 18 / 0 ~ 2.2 / 1.414 / 0.415 / 1.1	air/water	glass beads (0.25,1 and 6 mm)	216 mm I.D.	purely emirical Data were taken also for axial mixings of gas and liquid.
Dakshinamurty et al. (1971)	$\varepsilon = \varepsilon_g + \varepsilon_\ell = K_1' (\frac{U_\ell}{U_t})^{K_2'} (\frac{U_g \mu_\ell}{\sigma})^{0.08}$ Re_t / K_1' / K_2' < 500 / 2.12 / 0.41 > 500 / 2.65 / 0.60	air/water air/kerosene	Rockwool shot (1.3 mm) sand (1.06 and 2.235 mm) glass beads (3.35, 4.89 and 6.84 mm) iron shot (3 mm)	56 mm I.D.	bed expansion only purely empirical
Dakshinamurty et al. (1972)	$(\varepsilon_g + \varepsilon_\ell) = 2.85 (\frac{U_\ell}{U_t})^{0.60} (\frac{U_g \mu_\ell}{\sigma})^{0.08}$ for $Re_t > 500$	air/water N_2 /electrolyte solution	lead shot (2.13 mm) Rockwool shot (1.3 mm) glass beads (3.35, 4.18, 6.03 and 6.84 mm)	circular and annular column	bed expansion only purely empirical
Kim et al. (1972)	$(\varepsilon_\ell)_{U_g=0} - \varepsilon_\ell = 0.0025 (Fr_\ell \frac{\rho_s}{\rho_\ell})^{0.149} (Fr_g \frac{\rho_s}{\rho_g})^{0.161} (Re_\ell Re_g)^{0.259}$ $(\varepsilon_\ell)_{U_g=0} = 0.409 (Fr_\ell \frac{\rho_s}{\rho_\ell})^{0.193} Re_\ell^{0.074}$ $(\frac{H}{H_0}) - (\frac{H}{H_0})_{U_g=0} = 0.0026 (Fr_\ell \frac{\rho_s}{\rho_\ell})^{0.178} (Fr_g \frac{\rho_s}{\rho_\ell})^{0.345} (Re_\ell Re_g)^{0.252}$ $(\frac{H}{H_0})_{U_g=0} = 1.80 (Fr_\ell \frac{\rho_s}{\rho_\ell})^{0.202} Re_\ell^{-0.012}$	air/water	glass beads (6 mm) irregular gravel (2.6 mm)	660.4 x 25.4 mm rectangular cross section column	two-dimensional column Data were taken also for liquid mixing purely empirical.

Razumov et al. (1973)	$\varepsilon_s = 0.578 - 3.198\ U_\ell - 0.538\ U_g$ $\varepsilon_\ell = 0.422 + 0.135\ U_\ell/d_p^{0.562} - 1.82\ U_g$	air/water	sand particles and slag beads (0.49 to 1.27 mm)	300 mm I.D.	purely empirical limited liquid and gas velocities
Kim et al. (1975)	$\varepsilon_g + \varepsilon_\ell = 1.40\ Fr_\ell^{0.170} \left(\frac{U_g \mu_\ell}{\sigma}\right)^{0.078}$ for expanded beds $\varepsilon_g + \varepsilon_\ell = 1.301\ Fr_\ell^{0.128} \left(\frac{U_g \mu_\ell}{\sigma}\right)^{0.073} \exp[0.031 \left(\frac{U_\ell}{U_g}\right)(\varepsilon_\ell)_{U_g=0}]$ for contracted beds $(\varepsilon_\ell)_{U_g=0} = 1.353\ Fr_\ell^{0.206}\ Re_\ell'^{-0.100}$ $\varepsilon_\ell = 1.504\ Fr_\ell^{0.234}\ Fr_g^{-0.86}\ Re_\ell'^{-0.082}\ Re_\ell'^{-0.082} \left(\frac{U_g \mu_\ell}{\sigma}\right)^{0.092}$ $Fr_\ell = \frac{U_\ell^2}{d_p g}$, $Fr_g = \frac{U_g^2}{d_p g}$, $\mu_\ell = K\gamma^{\bar{n}-1}$ $Re_\ell' = \frac{U_\ell^{2-\bar{n}} d_p^{\bar{n}}}{\nu_\ell}$	air/sugar solution air/carboxymethyl cellulose air/water-acetone	glass beads (6 mm) irregular gravel (2.6 mm)	660.4 x 25.4 mm rectangular cross section column	two-dimentional column purely empirical
Blum and Toman (1977)	$\frac{(\varepsilon_g + \varepsilon_\ell) - (\varepsilon_\ell)_{U_g=0}}{1 - (\varepsilon_\ell)_{U_g=0}} = f(U_g)$	N_2/light mineral oil	cylindrical catalyst Dia. (mm) / Length (mm) 3.18 / 3.18 4.76 / 2.38 4.76 / 4.76	101.6 mm I.D.	graphical description for $f(U_g)$
Soung (1978)	$\left(\frac{H_O}{H}\right)/\left(\frac{H_O}{H}\right)_{U_g=0} = 1.50 + 0.16 \ln\left(\frac{U_\ell}{U_g}\right) - 0.065 \ln(\phi_s Re_t)$ for $0.06 \leq \frac{U_\ell}{U_g} < 0.6$ $\left(\frac{H_O}{H}\right)/\left(\frac{H_O}{H}\right)_{U_g=0} = 2.09 - 0.17 \ln(\phi_s Re_t)$ for $0.6 \leq \frac{U_\ell}{U_g} < 5$ $\left(\frac{H_O}{H}\right)_{U_g=0} = 1 - 4.50 \left(\frac{U_\ell}{U_t}\right)^{2.15}$, for $\frac{U_\ell}{U_t} \leq 0.25$ $\left(\frac{H_O}{H}\right)_{U_g=0} = 1 - 1.22 \left(\frac{U_\ell}{U_t}\right)^{1.20}$, for $\frac{U_\ell}{U_t} > 0.25$	N_2/n-heptane	Co-Mo extrudate catalyst (0.635, 1.27 and 1.6 mm Dia.; 4.76 mm in length)	127 and 152.4 mm I.D. columns	purely empirical

(Table 2.1 continued)

Investigators	Correlation Equations (in SI Units)	Gas-Liquid	Solid	Column Dimension	Remarks
Begovich and Watson (1978a)	$\varepsilon_g + \varepsilon_\ell = (3.93 \pm 0.18)\, U_\ell^{(0.271 \pm 0.011)}\, U_g^{(0.041 \pm 0.005)}\, (\rho_s - \rho_\ell)^{(-0.316 \pm 0.011)}\, d_p^{(-0.268 \pm 0.010)}\, \mu_\ell^{(0.055 \pm 0.008)}\, D_c^{(-0.033 \pm 0.013)}$ (T-2.1.1) $\varepsilon_g = (1.61 \pm 0.336)\, U_g^{(0.72 \pm 0.028)}\, d_p^{(0.168 \pm 0.061)}\, D_c^{(-0.125 \pm 0.088)}$ (T-2.1.2)	air/water	alumina beads (6.2 mm) glass beads (4.6 and 6.2 mm) alumino-silicate (1.9 mm) Plexiglas (6.3 mm)	76.2 and 152 mm I.D.	purely empirical 2381 points for $\varepsilon_g + \varepsilon_\ell$ 913 points for ε_g
Oh and Kim (1980)	$\varepsilon = 0.46\, Fr_\ell^{0.19}\, Fr_g^{0.01}\, Re_\ell^{0.026}$ $\varepsilon_\ell = 0.26\, Fr_\ell^{0.44}\, Fr_g^{-0.032}\, Re_\ell^{0.006}$	air/water	glass beads (1.63, 2.92, 7.85 mm and their binary mixtures)	15 cm I.D.	holdup behaviors in beds of mixed particles d_p = harmonic mean diameter
Kato et al. (1981)	$\left(\frac{U_\ell}{U_t}\right) = (\varepsilon_\ell/\varepsilon_\ell^*)^n$ (T-2.1.3) $\varepsilon_\ell^* = 1 - 9.7\,(350 + Re_t^{1.1})^{-0.5}\,(\rho_\ell U_g^4/g\sigma)^{0.092}$ (T-2.1.4) $\frac{5.1 + 86.2(\rho_\ell U_g^4/g\sigma)^{0.285} - n}{n - 1.7} = [0.1 + 0.443(\rho_\ell U_g^4/g\sigma)^{0.165}]\, Re_t^{0.9}$ (T-2.1.5) $\mu_\ell = 9.0 \times 10^{-4} \sim 5.2 \times 10^{-2}$ Pa·s	air/water air/carboxymethyl cellulose soln.	glass sphere (0.42, 0.66, 1.2 and 2.2 mm)	52 and 120 mm I.D.	modified from the correlation of Garside and Al-Dibouni (1977)

Saberian-Broudjenni et al. (1984)	$\varepsilon=(14\phi_s)^{-1/3}\left(\frac{U_\ell}{U_{\ell mf0}}\right)^{0.27}\left[1+0.070\left(\frac{d_pU_g\rho_\ell}{\mu_\ell}\right)^{0.34}\right]$ (T-2.1.6) $\frac{U_{\ell mf0}\rho_\ell d_p}{\mu_\ell}=\sqrt{(25.25)^2+0.0651\left(\frac{gd_p^3\rho_\ell(\rho_s-\rho_\ell)}{\mu_\ell^2}\right)}-25.25$ $\varepsilon_g=\varepsilon(U_g-j'_{g\ell})/(U_g+U_\ell)$ (T-2.1.7) where $j'_{g\ell}=0.017(\rho_\ell U_g^2)^{0.45}$ (T-2.1.7a) $\varepsilon_\ell=\varepsilon\frac{j'_{g\ell}+U_\ell}{U_g+U_\ell}$ (T-2.1.8)	gas: air, He, CO_2 liquid: cyclohexane, kerosene, water, gas oil, C_2Cl_4	glass beads (1.37, 3.18 mm) alumina beads (1.74, 1.87, 2.55 mm) alumina cylinder (2.0, 1.8 mm)	5.0 and 15 cm I.D.	Correlation is not valid for foaming liquid.
Sinha et al. (1984)	$\varepsilon-\varepsilon_o=0.125(Re_\ell)^{0.115}\left(\frac{\mu_\ell U_\ell}{\sigma}\right)^{-0.40}\left(\frac{\mu_\ell U_g}{\sigma}\right)^{0.431}$ $-0.201\,(Fr_g)^{0.167}$ where $\varepsilon_o=1.714(Fr_\ell)^{0.205}(Re_\ell)^{-0.133}$	N_2/kerosene N_2/heptane	alumina catalyst (0.5, 1.3 and 2 mm)	7.6 cm I.D.	holdup behavior in beds of polydispersed particles
Kato et al. (1985)	Dense region: $\varepsilon_s=\varepsilon_s^*[1-1.09(1+0.00124\,Re_t^{0.8})(U_\ell/U_t)^{0.85}]$ $\varepsilon_g=0.3\,W^{1.3}/[(1+1.1\,W^{1.15})(1+0.01\,Re_t^{0.62})]$ where $W=(gD_c^2\rho_\ell/\sigma)^{0.198}(gD_c^3/\nu_\ell^2)^{0.035}(U_g/\sqrt{gD_c})$ $\varepsilon_s^*=0.63$ for spherical particles	air/water	glass beads (0.52, 2.2 and 5.2 mm) porous alumina (1.5mm)	12 cm I.D.	gas-liquid-solid system; gas-slurry-solid system Axial distributions of phase holdups were correlated.
Chern et al. (1984) Fan et al. (1985a)	$\frac{U_g}{\varepsilon_g}=\frac{U_g}{1-\varepsilon_s}+\frac{U_\ell}{1-\varepsilon_s}+0.1016+1.488\left(\frac{U_g}{1-\varepsilon_s}\right)^{1/2}$ (T-2.1.9) for dispersed bubble flow regime $\frac{U_g}{\varepsilon_g}=1.783\left(\frac{U_g}{1-\varepsilon_s}+\frac{U_\ell}{1-\varepsilon_s}\right)+0.35\,(gD_c)^{1/2}$ (T-2.1.10) for slug flow $\varepsilon_g=C\,U_\ell^{-0.98}\,U_g^{0.70}$ (T-2.1.11) for coalesced bubble flow regime $U_\ell\times10^2$(m/s) — C 2.2-6.0 — 0.0270 > 6.0 — 0.0491	air/water	nylon beads (2.5 mm) activated carbon beads (1.74 mm) glass beads (1, 3.04, 3.99 and 6.11 mm) alumina beads (2.27 and 5.50 mm)	7.62 cm I.D.	relationship extendible to a binary mixture of particles

(Table 2.1 continued)

Investigators	Correlation Equations (in SI Units)	Gas-Liquid	Solid	Column Dimension	Remarks
Jean and Fan (1986)	$U_\ell' = U_i\,(\varepsilon)^n$ (T-2.1.12) $U_\ell' = U_\ell\,\left(\frac{U_{\ell mf}}{U_{\ell mf0}}\right)^y$ (T-2.1.13) $\frac{U_{\ell mf}}{U_{\ell mf0}}$ predicted by Eqn. (2.13) $y = -0.0869\left[\frac{d_p}{D_c}\left(\frac{\rho_s}{\rho_s-\rho_\ell}\right)^{3.0}\right]^{-0.896}$ (T-2.1.14) $0 < U_g < 0.15$ m/s ; $0.06 < U_\ell < 0.25$ m/s	air/water	glass beads (3.04, 3.99 and 6.11 mm) alumina beads (2.27, 2.53, 5.50 and 6.69 mm) lead particle (3.33 mm)	7.62 cm I.D.	bed expansion correlation extendable to a binary mixture of particles

Table 2.2 Index for the Richardson-Zaki Equation (from Richardson and Zaki, 1954).

$n = 4.65 + 20\, d_p/D_c$;	$Re_t < 0.2$
$n = (4.4 + 18 d_p/D_c) Re_t^{-0.03}$;	$0.2 < Re_t < 1$
$n = (4.4 + 18 d_p/D_c) Re_t^{-0.1}$;	$1 < Re_t < 200$
$n = 4.4\, Re_t^{-0.1}$;	$200 < Re_t < 500$
$n = 2.4$;	$500 < Re_t$

where $Re_t = U_t\, d_p\, \rho_\ell/\mu_\ell$

Several investigators considered the effect of particle shape on the solids holdup behavior (Soung, 1978; Saberian-Broudjenni et al. 1984; Song and Fan, 1986b; Sinha et al., 1986; Song et al., 1987). For systems using nonspherical particles, such as extrudates in cylindrical shape, Song and Fan (1986b) reported that Begovich and Watson's (1978a) correlation can be modified to account for the bed expansion characteristics by replacing the particle diameter d_p by a product of the equivalent particle diameter and sphericity, $(d_p\phi_s)$. Saberian-Broudjenni et al. (1984) proposed Eqn. (T-2.1.6) for nonspherical particles which appears valid as U_g goes to zero. This correlation was based on a study using various cylindrical alumina catalysts fluidized by cyclohexane, kerosene, gas oil, C_2Cl_4, or water as the liquid phase and air, helium, or carbon dioxide as the gas phase. The exponent on the liquid velocity falls within the range predicted by the Richardson-Zaki relationship.

Jean and Fan (1986) proposed a unified empirical relationship for predicting solids holdup in both liquid-solid and gas-liquid-solid fluidized beds. They modified the Richardson-Zaki equation (Richardson and Zaki, 1954) to include effects of the gas phase. The two-phase (liquid-solid) equivalent expression of the Richardson-Zaki equation for a three-phase fluidized bed is given by Eqn. (T-2.1.12). They found that U'_ℓ can be simply related to U_ℓ, $U_{\ell mf}$, and $U_{\ell mf0}$ by Eqn. (T-2.1.13). Note that $U_{\ell mf}$ in Eqn. (T-2.1.13) is a function of the gas velocity, particle properties, and liquid properties (see Eqn. (2.13)). The applicability of Eqns. (T-2.1.12) through (T-2.1.14) covers the dispersed bubble, coalesced bubble, and slugging regimes. The above analysis was also found to be applicable for a three-phase fluidized bed involving a binary mixture of particles.

Most bed expansion studies, and consequently the correlations derived from them, used solids of nearly uniform size; however, this only represents an ideal situation. Oh and Kim (1980) found that bed porosity behavior for mixed particle beds was intermediate between that for beds of mono-sized particles. For beds of a 50 wt% mixture of 1.63 and 2.92 mm particles, the bed porosity showed an initial reduction on introducing the gas at low liquid rates in a manner similar to that in beds of 1.63 mm beads alone; the reduction, however, was much smaller in the mixed particle bed. In the case of a 50 wt% mixture of 1.63 and 7.85 mm particles, the bed slightly expanded with gas velocity, but the expansion was much less than that for a bed consisting solely of 7.85 mm beads.

Schaefer et al. (1983) showed that in three-phase fluidization operations the addition of up to 7 vol% of fines did not affect bed expansion behavior, but addition of a larger quantity of fines significantly increased bed expansion because of an increased apparent viscosity of the liquid-solid mixture or slurry. According to their observations the bed expanded with gas velocity at low gas velocities when the slurry concentration was low and the slurry velocity was high. The bed, however, eventually contracted at high gas velocities (U_g > 3 cm/s) when the slurry concentration was high or the slurry velocity was low; this phenomenon can be attributed to bubble coalescence and its interaction with bubble wake at low slurry velocities or high fines concentration. Addition of a large quantity of large particles, on the other hand, reduced the bed porosity as reported by Sinha et al. (1984).

In summary, for the independent estimation of porosity of a bed containing relatively large particles fluidized by a water-like liquid, the correlations of Dakshinamurty et al. (1971) and Begovich and Watson (1978a) can be employed, since these correlations are based on a large number of data obtained with varying solid and liquid physical properties. For estimation of bed porosity in a bed of particles fluidized with a non-foaming organic liquid the correlations by Soung (1978) and Saberian-Broudjenni et al. (1984) can be used.

2.1.5.3 Overall Gas Holdup and Empirical Correlations

Gas holdup behavior strongly depends on the flow regime, as shown by Michelsen and Ostergaard (1970) and hence, on both particle and liquid properties. For example, a decrease in liquid surface tension can dramatically increase gas holdup (Blum and Toman, 1977). Gas holdup in three-phase fluidized beds can be lower than that in a corresponding bubble column (Viswanathan et al., 1964; Adlington and Thompson, 1965; Vasalos et al., 1980) when the particles promote bubble coalescence, and it can be higher than that in a corresponding bubble column when the particles break up gas bubbles. The gas holdup and the bubble rise velocity are interrelated. At any given axial distance and at steady state, a material balance on the gas phase, assuming no reaction and constant density, yields a relationship between $\epsilon_g(r)$ and $U_b(r)$ as

$$U_g = \frac{1}{A} \int_A \epsilon_g(r)\, U_b(r)\, dA \tag{2.22}$$

For systems in the dispersed bubble regime where both the gas holdup and bubble rise velocity are uniformly distributed in the bed, Eqn. (2.22) reduces to

$$U_g = \epsilon_g U_b \tag{2.23}$$

Correlations of gas holdup are based either on a strictly empirical approach or on a semi-empirical description of the relative gas-liquid flow. Begovich and Watson (1978a) suggested the empirical relationship (Eqn. (T-2.1.2)) based on data collected from a large number of literature sources. This correlation is, however, developed from data obtained for aqueous liquid systems. The general interdependency of ϵ_g on D_c is not correctly reflected by this correlation.

The complex nature of the three-phase fluidized bed has led to several different semi-empirical approaches for describing the relative motion of gas bubbles with respect to the liquid phase. Darton and Harrison (1975) modified Wallis' drift flux theory (also see Section 2.3.6) for two-phase flow systems (Wallis, 1969) by simply substituting the relative velocity between the gas and liquid phases $(U_g/\epsilon_g - U_\ell/\epsilon_\ell)$ into the relationship for a two-phase system and obtained

$$j_{g\ell} = [(1-\epsilon_g)/\epsilon_\ell]\,(U_g\epsilon_\ell - U_\ell\epsilon_g) \tag{2.24}$$

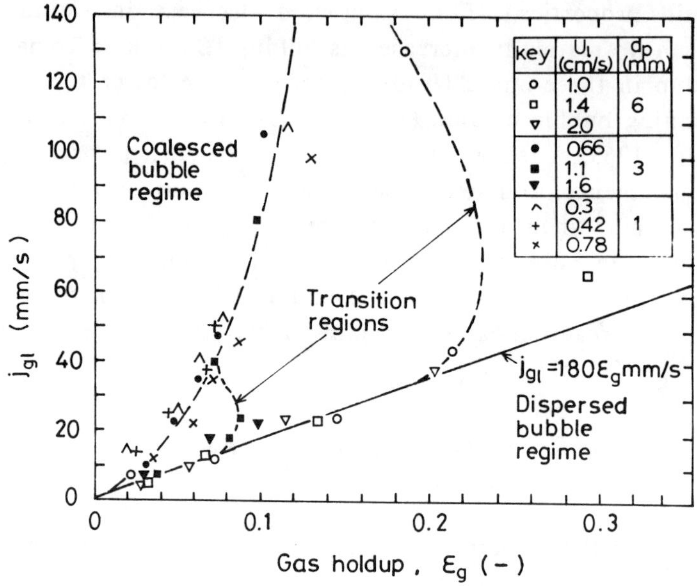

Figure 2.10 Variation of the drift flux of gas relative to liquid flow with the gas holdup in an air-water-glass bead fluidized bed based on the data of Michelsen and Ostergaard (1970).

The drift flux, $j_{g\ell}$, defined by Eqn. (2.24) is a function of ϵ_g as shown in Fig. 2.10. In the dispersed bubble regime, the drift flux could be correlated (in SI units) by

$$j_{g\ell} = 0.18\ \epsilon_g \tag{2.25}$$

However, in the coalesced bubble regime it is difficult to correlate $j_{g\ell}$ as an explicit function of ϵ_g. The gas holdup can be estimated from Eqn. (2.25) for the dispersed bubble regime and from Fig. 2.10 for the transition and coalesced bubble regimes with an independent estimation of ϵ_ℓ. Schaefer et al. (1983) used Eqn. (2.24) to correlate gas holdup behavior in a gas-slurry-solid fluidized bed. Figure 2.11 shows the relationship between the gas drift flux and ϵ_g for the gas-slurry-solid fluidized bed systems using kerosene with coal char fines as the slurry phase. As expected, the linearly correlated $j_{g\ell}$ values for the dispersed bubble regime in kerosene with fines are significantly lower than those in the water system obtained by Darton and Harrison. The drift flux in the coal char slurry system lies mainly in the transition regime and shows wide variation depending on the liquid velocity.

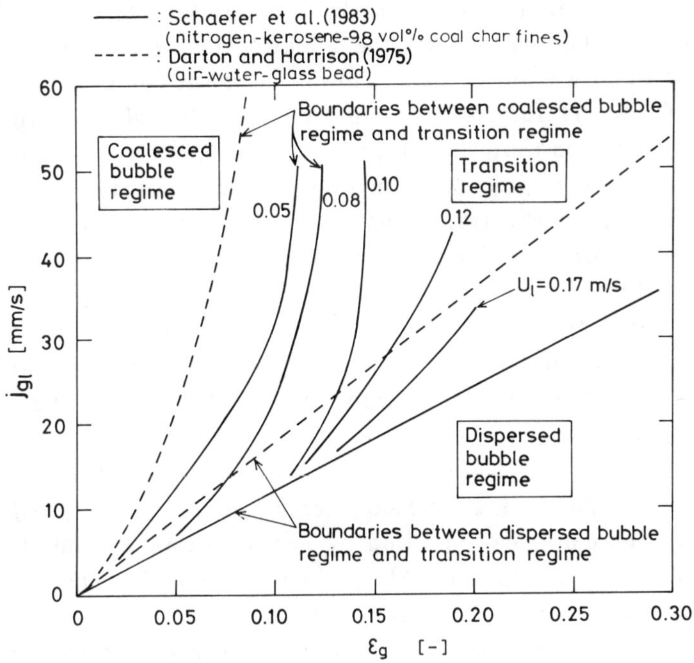

Figure 2.11 Relationship between $j_{g\ell}$ and ϵ_g as a function of U_ℓ based on the data of Schaefer et al. (1983) and Darton and Harrison (1975) (from Muroyama, 1987).

Saberian-Broudjenni et al. (1984) defined the gas phase drift flux for the void channel in a three-phase system as

$$U_g = (U_g + U_\ell)\delta_g + j'_{g\ell} \tag{2.26}$$

where

$$\delta_g = \epsilon_g/(\epsilon_g + \epsilon_\ell) \tag{2.27}$$

Thus the gas drift flux is

$$j'_{g\ell} = (1/\epsilon)(U_g\epsilon_\ell - U_\ell\epsilon_g) \tag{2.28}$$

They proposed a correlation for the gas drift flux given by Eqn. (T-2.1.7a). The gas holdup can be calculated from Eqn. (T-2.1.7). Equation (T-2.1.7) is good for various liquids such as cyclohexane, gas oil, C_2Cl_4, and water, but is poor for kerosene.

Chern et al. (1984) correlated gas holdup separately for each flow regime. They modified the Nicklin theory (Nicklin, 1962) for two-phase bubbly flow and slug flow to account for gas holdup in the gas-liquid-solid fluidized bed. In line with the Nicklin theory, the flow of gas can be represented by Eqns. (T-2.1.9) and (T-2.1.10) for the dispersed bubble regime and slugging regime, respectively. Note that the gas holdup in the transition regime can be approximated by the average of those calculated from Eqns. (T-2.1.9) and (T-2.1.10). In order to estimate ϵ_g using the above correlation, a correlation for ϵ_s based on the generalized wake model (see Section 2.3.1) was developed by the same authors. Chern et al. (1984) and Fan et al. (1985a) proposed an empirical correlation for gas holdup in the coalesced bubble regime as Eqn. (T-2.1.11).

In summary, the behavior of the gas holdup varies significantly with flow regime and a unified correlation equation is difficult to establish. The empirical equation by Begovich and Watson (1978a) can be used for the coalesced bubble regime in beds of particles fluidized by a water-like liquid, but it yields a conservative value for the dispersed bubble regime. Note that for a foaming system involving low surface tension liquids the correlations proposed so far considerably underestimate the experimental gas holdup. Even though the semi-empirical approaches employed by Darton and Harrison (1975), Chern et al. (1984) and Saberian-Broudjenni et al. (1984), are all based on extensions of Wallis' two-phase flow theory, their expressions are different from one another. A more fundamental definition of the drift flux in three-phase systems is given in Section 2.3.6. In general, the analysis based on such bubble flow models should be developed further.

2.1.5.4 Overall Liquid Holdup and Empirical Correlations

Liquid holdup increases with liquid velocity and decreases monotonically with increasing gas velocity (Michelsen and Ostergaard, 1970). The liquid holdup increases with liquid viscosity in the range from 1 to 70 cP; the trend is more pronounced for small particles (Kim et al., 1975). Liquid holdup decreases slightly with surface tension (σ varied from 40 to 73 dyne/cm) for 1 mm glass beads, but for 2.6 mm gravel particles liquid holdup increases with increasing surface tension (Kim et al., 1975).

Kato et al. (1981) developed correlations (Eqns. (T-2.1.3) through (T-2.1.5)) for liquid holdup in three-phase fluidized beds by modifying an empirical equation proposed by Garside and Al-Dibouni (1977) for liquid-solid fluidized beds. Kato et al.'s experiments used liquid viscosities varying over a wide range. Equations (T-2.1.3) through (T-2.1.5) predict the behavior of liquid holdup in three-phase fluidized beds with reasonable accuracy, as verified by recent studies of Muroyama et al. (1984) and Chiu and Ziegler (1985).

According to Saberian-Broudjenni et al. (1984), liquid holdup can be predicted by a correlation, Eqn. (T-2.1.8), similar to that for gas holdup. The correlations for ϵ and $j'_{g\ell}$ are in Eqns. (T-2.1.6) and (T-2.1.7a), respectively.

In summary, for liquid holdup estimation, the correlation of Kato et al. (1981) is recommended because of the wide range of particle diameters and liquid viscosities covered.

2.1.5.5 Solids Holdup in the Freeboard Region

Behavior of solids holdup in the freeboard region has significant effects on the sizing of the region for design purposes; the effects are greater for the small/light particles than for the large/heavy particles. As noted previously, solids holdup in the disengagement zone of the bed (see Fig. 2.1) decreases sharply with the axial distance and is bound by two near constant phase holdups in the bulk fluidized bed region and the region beyond the disengagement zone in the freeboard. Figure 2.12 shows a typical example of such a holdup profile; the solids holdup in the region beyond the disengagement zone is zero signifying no solids elutriation from the bed.

According to Page and Harrison (1974) the fundamental mechanisms for particle entrainment and de-entrainment in the disengagement zone of the freeboard is as follows. Particles are drawn from the upper surface of the fluidized bed into the freeboard in the wake behind a bubble (also see Section 2.2.2). Vortices containing particles are shed from the wake in the freeboard. The particles in the shed vortices settle back into the bed when the liquid velocity in the freeboard is less than the particle terminal velocity. They indicated that particle entrainment decreased with a decrease in both bubble size and bubble frequency and with an increase in both liquid velocity and particle size. In addition, inserting a wire mesh baffle into the freeboard section significantly reduces particle entrainment.

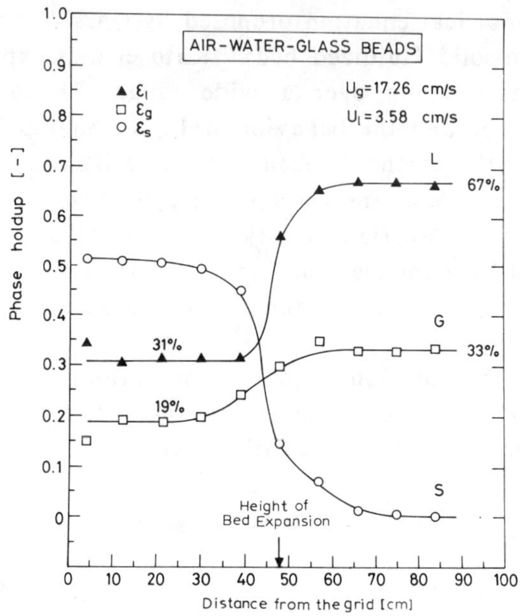

Figure 2.12 Axial variation of phase holdups in a bed containing 4.6 mm glass beads (from Begovich and Watson, 1978b).

2.1.5.6 Gas Holdup in the Freeboard Region

Catros et al. (1985) measured the gas holdup in the freeboard region of a bed containing 3 mm glass beads. In their apparatus, the four independent nozzles (2.54 cm ID) used as the gas distributor produced a spouted behavior of the gas phase at the bottom of the bed. They found that gas holdup decreased significantly with increasing liquid velocity. They proposed a correlation for ϵ_g in the freeboard region for $U_\ell/(U_\ell + U_g) > 0.08$ as

$$\epsilon_g = 0.066\ [U_\ell/(U_\ell + U_g)]^{-0.424} \tag{2.29}$$

Kato et al. (1985), using a porous plate gas distributor with 2 mm holes to ensure a uniform gas distribution, found that the gas holdup in the freeboard region was greater than that in the bulk fluidized bed region but smaller than that in a bubble column. They proposed the following correlation for estimating gas holdup in the freeboard region:

$$\epsilon_g = 0.3\ W^{1.3}/(1 + 1.1W^{1.15}) \tag{2.30}$$

where the parameter W is defined as

$$W = (gD_c^{\,2}\rho_\ell/\sigma)^{0.198}\,(gD_c^{\,3}/\nu_\ell^{\,2})^{0.035}(U_g/\sqrt{gD_c}) \tag{2.31}$$

According to Eqn. (2.31), W is inversely proportional to the liquid viscosity to the 0.07 power, but is independent of column diameter. Note that no significant axial variation of gas holdup in the freeboard region is evident for a given operation.

2.1.6 Effect of Solids Wettability

The discussion on the hydrodynamic behavior to this point has been mainly concerned with systems using wettable particles. However, the wettability of the solid surface has some influence on holdup behavior.

Bhatia et al. (1972) observed that a bed of non-wettable particles always expands upon introducing gas in contrast to a bed of clean wettable beads which contracts upon the initial introduction of gas. Armstrong et al. (1976) found that in beds of wettable 6 mm beads bubble disintegration occurred at low gas velocities. Beds of non-wettable 6 mm beads exhibit greater expansion compared to those of wettable ones and the gas holdup is much lower than that for the wettable ones. These results may be due to adherence of bubbles to the solids for non-wettable particles. When bubbles adhere to non-wettable particles, the apparent density of the particle-bubble aggregates becomes less than that of the particle alone. The lesser apparent density leads to greater bed expansion and less bubble breakage, and hence lower gas holdup.

The work of adhesion (W_{SLV}), the energy required to separate a liquid from a solid per unit interfacial area, is (Adamson, 1982)

$$W_{SLV} = \sigma\,(1 + \cos\theta_c) \tag{2.32}$$

where σ is the liquid surface tension and θ_c is the contact angle, which is the angle formed by the liquid phase as it contacts the gas and solid phases (see Eqn. (11.3)). Figure 2.13 shows three examples of three-phase contact and the location of the contact angle in each case. Figure 2.13(a) shows a gas bubble contacting a wetted (or wettable) solid surface (θ_c < 90°). Figure 2.13(b) shows a liquid drop on a non-wetted (or non-wettable) solid surface (θ_c > 90°) and Fig. 2.13(c) shows a liquid drop wetting a (wettable) solid surface (θ_c < 90°). W_{SLV}

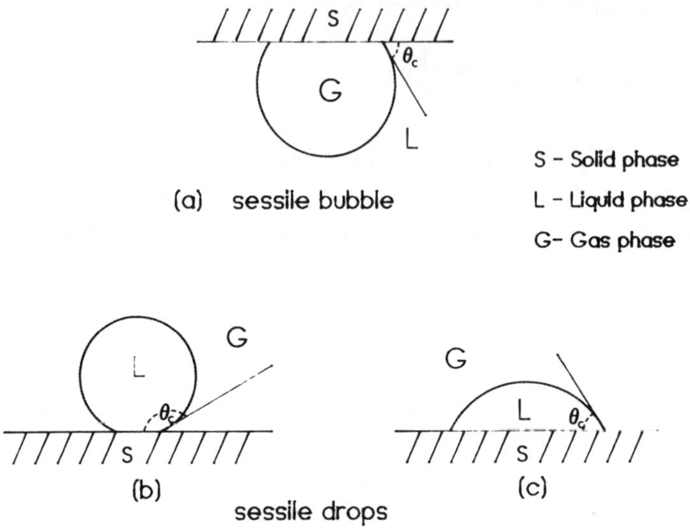

Figure 2.13 Contact angle for three different cases (from Adamson, 1982).

decreases as the solids become less wettable or as θ_c increases, which favors the contact between bubbles and solids.

The contact angle, θ_c, defined in Eqn. (2.32) is the static contact angle which accounts for situations when the point of three-phase contact is not in motion or is at an equilibrium condition, shown in Fig. 2.14(a), as occurs when the particle-bubble aggregates are formed after their collision. When the point of three-phase contact is in motion, the dynamic contact angle may depend on the direction of the relative motion, as well as the speed of the motion. At sufficiently slow motion, up to about 1 mm/min (Huh and Scriven, 1971), the dynamic contact angle remains constant, depending only on the direction of movement, as shown in Fig. 2.14(b). The receding contact angle refers to the contact angle established when the solid surface is pulling out of the liquid phase, and the advancing contact angle refers to that when the solid surface is pushing into the liquid phase. As the speed increases beyond 1 mm/min, the dynamic contact angle becomes velocity dependent until a critical speed (on the order of 10 mm/min) is reached, beyond which the dynamic contact angle yields another constant value, i.e., 0° for the receding contact angle or 180° for the advancing contact angle, shown in Fig. 2.14(c). The effect of the dynamic contact angle on the bubble breakage by a penetrating particle is discussed in Section 2.2.1.2.

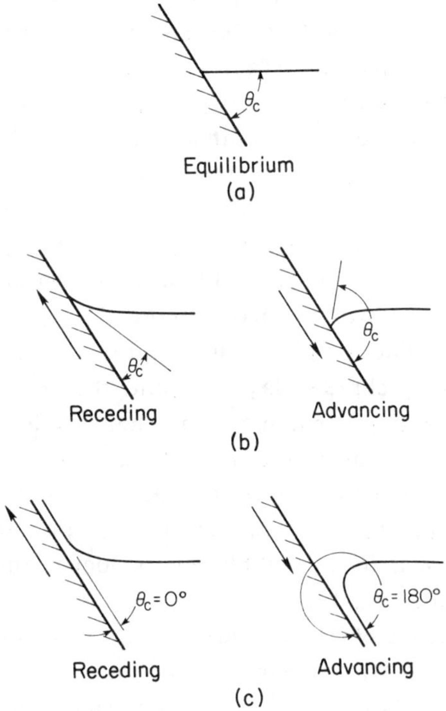

Figure 2.14 Illustrations of dynamic contact angles (from Huh and Scriven, 1971).
(a) Static contact angle.
(b) Dynamic contact angle below the critical speed.
(c) Dynamic contact angle beyond the critical speed.

2.1.7 Surfactant and Liquid Surface Tension Effects

A pure fluid which has a low surface tension (compared to water) initially forms small bubbles that may later undergo coalescence and increase in size. The equilibrium bubble size is still much less than that in an air-water system. In surfactant systems, the low surface tension liquid phase facilitates the formation of small bubbles; however, there is a reported non-coalescing tendency of gas bubbles in the liquid phase (Keitel and Onken, 1982). Consequently, the formation of small non-coalescing bubbles leads to higher gas holdup.

Surfactants are classified into three general classes (anionic, cationic, and non-ionic) and contribute two major effects (lower surface tension and coalescence inhibition). A surfactant can reduce the surface tension and inhibit formation of large bubbles. A surfactant molecule reduces interfacial tension since each end of the molecule shows different solubility characteristics (amphipatic character). One end of the surfactant molecule is polar and hence, is attracted to other polar molecules; the other end of the molecule is non-polar. Since air is a non-polar fluid while water is a polar liquid, the surfactant molecules are able to coat the surface of the bubble with the polar ends of the molecules dissolved in water. The resulting reduced surface tension leads to smaller bubble sizes and consequently, larger gas holdup. The aqueous surfactant, in air-water systems, produces an electrically charged layer around the gas bubble which in turn reduces the coalescing tendencies of the bubble. Such behavior does not occur in pure liquid systems. In addition, the presence of surfactants can also increase drag on the gas bubble due to the formation of a surface tension gradient on the bubble surface. The higher drag results in a lower bubble rise velocity and hence, higher gas holdup (e.g., Shah et al., 1985).

Aqueous solutions of alcohol surfactants were shown to produce higher gas holdup than water in bubble columns (Kelkar et al., 1983; Shah et al., 1985). Kelkar et al. (1983) reported that n-butanol was more effective in increasing gas holdup in a bubble column than methanol, ethanol, n-propanol, and iso-propanol. Kelkar et al. also found that varying the alcohol concentration from 0.5 to 2.4 wt% did not significantly affect the gas holdup behavior. Although longer carbon chain alcohols were not studied by Kelkar et al. (1983), Kelkar et al.'s correlation suggests a further 6% increase in gas phase holdup is expected for a five carbon chain alcohol. Fan et al. (1986c) found that a 0.5 wt% t-pentanol solution had a significantly higher gas holdup in a bubble column, up to 70% at high gas velocity, than that predicted by Kelkar et al.'s correlation. This was attributed to different distributor design and limitations of Kelkar et al.'s correlation. Fan et al. (1986c) also reported that t-pentanol solutions had a higher gas holdup than n-butanol solutions in a three-phase fluidized bed but found that concentration was not a significant factor. t-Pentanol is more soluble in water than both n-pentanol and n-butanol, and is a more efficient surfactant than n-butanol at low concentrations.

Keitel and Onken (1982) attributed the coalescence inhibition property of surfactants to the formation of an electrically charged monolayer at the gas-liquid interface. They concluded that the stability of the surfactant monolayer increased with the hydrophobic chain length, and that mono-carboxylic acids were the most efficient of the surfactants they studied for inhibiting bubble coalescence. Zieminski et al. (1967) noted similar findings, commenting that n-alcohols seemed to be comparable to mono-carboxylic acids in reducing bubble size. Keitel and Onken (1982) noted that inhibition of bubble coalescence does not directly correlate to an increase in gas holdup.

In three-phase fluidization, the use of a low surface tension liquid, such as kerosene, may eventually lead to a foaming system where bed expansion, rather than contraction, takes place when gas is introduced. Such bed expansion behavior was reported by Dakshinamurty et al. (1971) and Saberian-Broudjenni et al. (1984). Saberian-Broudjenni et al. (1984) claimed that correlations for holdup developed from data for non-foaming liquids do not apply in the case of foaming liquids. The effect of surface tension on phase holdup was also measured by Kim et al. (1975).

Fan et al. (1986c) found that gas holdup in surfactant systems (0.5 wt% t-pentanol aqueous solution) at ambient conditions approached that reported by Tarmy et al. (1984) for pilot plant scale coal liquefaction systems. Note, however, that significant differences exist between the operating conditions of Tarmy's pilot units and those of Fan et al.'s work: the pilot units experiments used high operating temperatures (450°C) and pressures (17,100 kPa) (also see Chapter 10). Figure 2.15 shows a comparison between the gas holdup in 1 mm glass bead systems and the reported data by Tarmy et al. The experimental results reasonably simulate Tarmy et al.'s data. However, a comparison of solids holdup shows a discrepancy possibly due to high pressure effects. Gas holdup data for larger particles (3 ~ 6 mm diameter) are, in general, lower than Tarmy et al.'s data. In a similar approach, Morooka et al. (1986) used a Triton X-405 (Rohm and Haas product) surfactant in an air-water slurry bubble column system to simulate the hydrodynamics of a coal dissolver.

In summary, gas holdup in surfactant solutions is dependent upon molecular composition, liquid velocity, and distributor design, but is not dependent upon equilibrium surface tension, nor surfactant concentration above a certain value (e.g., 0.5 wt% for alcohol). Surfactants reduce bubble size, inhibit coalescence and reduce the bubble rise velocity. More work is recommended to further characterize the effects of this phenomenon as well as in mixture systems. Much of the current work has been limited to short-chain alcohols (up to C_8) with a few exceptions. Characterization of the

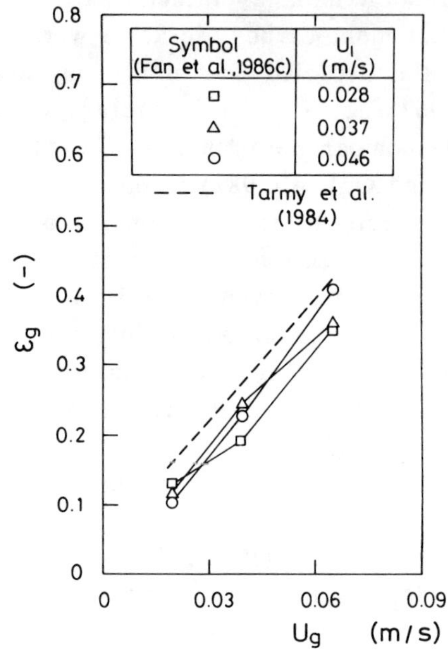

Figure 2.15 Comparison of the experimental data for gas (air) holdup obtained in 0.5 wt% of t-pentanol aqueous solution with 1 mm glass beads with those reported by Tarmy et al. (1984) (from Fan et al., 1986c).

behavior of long-chain surfactants, such as those produced in biological systems, needs to be determined. Caution must be exercised in interpreting the results of studies on surfactant systems due to the complex interfacial phenomena and possible foaming tendencies inherent in these systems.

2.1.8 Overall Rheological Behavior

Analogous to the viscosity of liquids, an apparent viscosity can be defined for a three-phase fluidized bed. The overall rheological behavior can be characterized by a simple rheological equation of state given by

$$\bar{\mu}_3 = \tau/\gamma \tag{2.33}$$

The overall viscosity obtained is an "apparent" quantity which comprises factors due to the viscosities of the gas and the liquid phases, particle-particle interactions, and bed expansion. The rheological behavior of three-phase fluidized beds reflects the fluid dynamics of the bed such as the bubbling behavior and bed contraction phenomena.

In three-phase fluidized beds, bed voidage is predominantly affected by the liquid flow rate for a liquid supported bed, but the extent of turbulence, which is associated with the kinetic energy of the particles, is influenced predominantly by the gas flow rate. The apparent viscosity ($\bar{\mu}_3$) in a three-phase fluidized bed reflects a combined effect of "frictional viscosity" and "kinetic viscosity" of the system (Song and Fan, 1986a). The frictional viscosity can be defined as a shearing viscosity which depends on the bed voidage. The kinetic viscosity is contributed by the fluid turbulence and the particle kinetic energies which inherently reflect particle accelerations and interactions. Consequently, the apparent viscosity varies with the bed voidage and the degree of turbulence in the bed and hence, liquid and gas flow rates. In addition, any factors, such as particle size and density, which affect bed expansion behavior necessarily affect bed rheology.

For a bed of glass beads of less than 1.5 mm where bed contraction is observed, the bed voidage variation with gas velocity is reflected in an increasing and then decreasing $\bar{\mu}_3$ with increasing gas velocity as shown in Fig. 2.16(a).

Figure 2.16(b) shows the variation of apparent viscosity with gas flow rate for 4.0 mm glass beads as a function of liquid flow rate. At a liquid velocity as low as 3.5 cm/s, $\bar{\mu}_3$ decreases sharply with an increase in gas flow rate. The results signify a sharp variation in apparent viscosity due to transition from a fixed to a fluidized bed state. At high liquid flow rates the apparent viscosity decreases moderately with increasing gas flow rate.

Figure 2.16(c) shows the variation in $\bar{\mu}_3$ with gas velocity as a function of liquid velocity for 2.27 mm alumina beads. For alumina beads, the apparent viscosity increases with an increase in gas velocity at a constant liquid velocity indicating the kinetic energy of the particles dominates and hence the kinetic viscosity of the system predominates. An increase in the liquid velocity at a constant gas velocity, however, decreases $\bar{\mu}_3$.

Song and Fan (1986a) developed a correlation for three-phase fluidized beds where the frictional viscosity predominates based on the liquid-solid fluidized bed correlation. The correlation, a function of a liquid Reynolds number and a gas Froude number, is

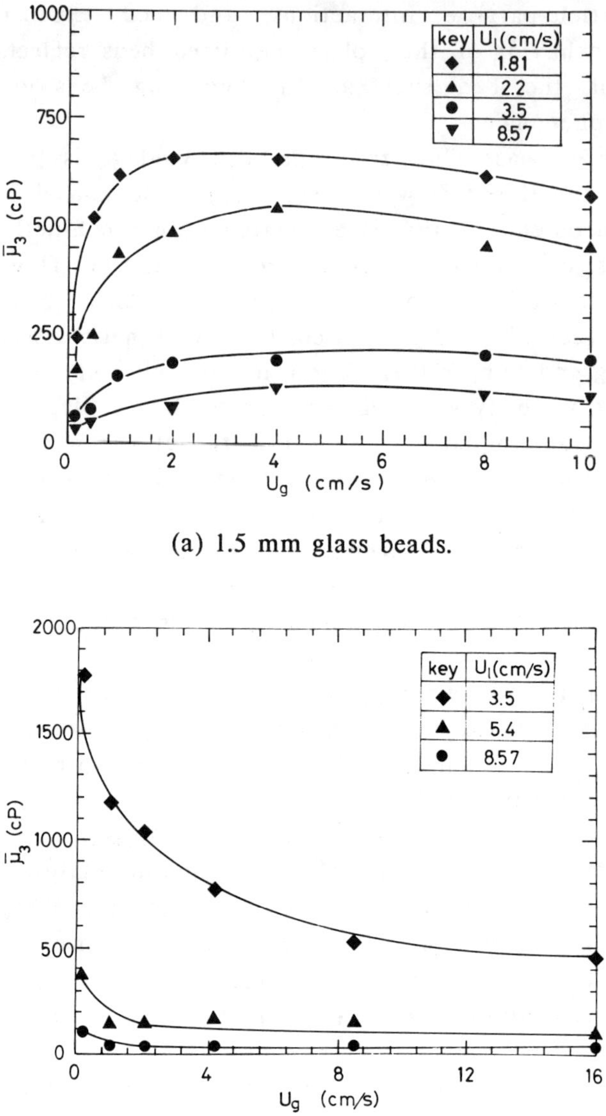

(b) 4.0 mm glass beads.

Figure 2.16 Effect of gas velocity on apparent viscosity of an air-water-particle fluidized bed using different particles at different liquid velocities (from Song and Fan, 1986a).

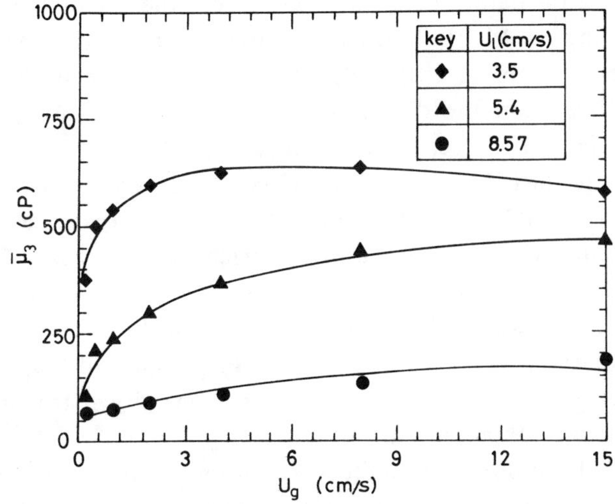

(Figure 2.16 continued) (c) 2.27 mm alumina beads.

$$\bar{\mu}_3 / \bar{\mu}_2 = 1 + 0.057\, \mathrm{Re}_\ell^{0.9458}\, \mathrm{Fr}_g^{0.5153} \tag{2.34}$$

$$\bar{\mu}_2 / \mu_\ell = \mathrm{Fr}_\ell^{0.1226} \exp(2.466/\epsilon) \tag{2.35}$$

and is valid for $0.4 < \epsilon < 0.85$, $79.45 < \mathrm{Re}_\ell < 195.0$, and $0 < \mathrm{Fr}_g < 1.011$.

2.2 MECHANICS OF BUBBLES AND BUBBLE WAKES IN MODE E-I-a-1

As described previously, the complex behavior of phase holdups and flow patterns in a three-phase fluidized bed is closely related to bubble and bubble wake flow characteristics. Consequently, a fundamental understanding of bubble and bubble wake behavior can provide the basis for a mechanistic approach to the analysis of the various hydrodynamic phenomena. This section examines bubble motion behavior and bubble wake structure in relation to the present understanding of the bed hydrodynamics.

2.2.1 Bubble Geometry and Motion

The interaction between a rising gas bubble and the surrounding liquid or liquid-solid medium results in the observed bubble shape and determines the extent of the disturbance in the surrounding flow field. Bubbles in motion are generally classified by shape as spherical, oblate ellipsoidal, and spherical-cap with the actual shape depending upon the dominating forces, such as surface tension, viscous, or inertial force, acting on a bubble. A more detailed classification is possible. In low viscosity liquids such as water, bubbles are very irregular and oscillate unsteadily in a transition region between the ellipsoidal and spherical-cap shapes. In viscous liquids under very quiescent conditions the formation of "skirts," whose thickness is about 100 μm (Guthrie and Bradshaw, 1969), has been experimentally observed (Hnat and Buckmaster, 1976; Bhaga and Weber, 1981). The skirt is a thin annular film of gas trailing behind the rim of a large spherical-cap bubble.

For small bubbles (diameter of volume-equivalent sphere (D_b) less than 0.1 cm in water), the bubble shape is approximately a sphere, the surface tension force dominates, and the bubble rises steadily in a rectilinear path. For intermediate size bubbles, effects of both surface tension and inertia of liquid flowing around the bubble are important and liquid viscosity and the presence of surface-active contaminants influence the bubble dynamics. The motion of intermediate size bubbles is extremely complex; ellipsoidal bubbles exhibit secondary motion characteristics, such as zig-zag or spiral trajectories and shape dilations or oscillations, as they rise. A systematic summary of these complex motions was given by Clift et al. (1978). The inertial effects of flows surrounding the bubble predominate for large bubbles (volumes > 3 cm^3 or D_b > 1.8 cm in general; Clift et al., 1978) and the effects of surface tension, viscosity, and cleanliness of the liquid medium are negligible. Large bubbles have an approximately spherical-cap shape with an included angle of about 100° and a relatively flat base. The bubbles follow a rectilinear path with some rocking and/or base oscillations.

The differences in the motion of these bubbles inherently depend on the differences in the flow fields immediately behind the bubble, or the so-called bubble wake. For small bubbles, the flow around the bubble can be described by the creeping flow approximation, or Stoke's law, and the wake size is negligible. If the internal circulation within the bubble is important, as in pure liquids, the Hadamard-Rybczynski theory provides a better prediction (Clift et al., 1978). The secondary motion associated with intermediate size bubbles probably results from periodic shedding of vortices from the bubble wake. The bubble oscillations start with the onset of vortex shedding from the wake (Edge and Grant, 1971).

2.2.1.1 Bubble Size

Basically, bubble sizes are measured by photographic or probe techniques. In two-dimensional beds, photographic techniques or, in three-dimensional beds, off-axis holographic techniques (Peterson et al., 1987) can provide a cross-sectional view of the rising bubble; from this view several possible characteristic bubble diameters may be defined. The size and shape of the rising bubble can be determined, in general, by specifying the width (maximum horizontal length) and height (maximum vertical length). In two-dimensional beds, based on these two dimensions and assuming an elliptical bubble shape, the bubble area, and consequently an area-equivalent circular bubble diameter (D_b), can be reasonably estimated for most bubble size ranges. For extremely large bubbles, the bubble has a circular-cap shape with a flat bubble base and information such as the included angle and the frontal radius of curvature will help estimate D_b. In three-dimensional beds, a volume-equivalent spherical bubble diameter can be obtained. The equivalent bubble diameter thus defined is the most common representation of the bubble size. In some cases, however, a characteristic length is a more suitable representation of the bubble size. For example, Gorring and Katz (1962) preferred, on the basis of theoretical arguments, the bubble height (minor diameter) in the estimation of drag coefficients for a rising bubble and the bubble width (major diameter) to evaluate the Reynolds number. Tsuchiya and Fan (1986) used the bubble width (breadth) in characterizing the wake flow behind a rising bubble based on the idea that the distance between the separation points around the bubble edge controls the wake phenomena.

The second measurement technique involves the use of a probe (conductivity probe, optical fiber probe, etc., see Section 2.1.5.1) and is applicable to three-dimensional beds. In this case, the probe measures a chord length of the bubble as it passes; the actual chord length measured may not, and probably does not, pass through the bubble center, hence leading to difficulties in the data analysis. Tsutsui and Miyauchi (1980) converted the measured chord length distribution (chord length probability density function, $Z(\ell_b)$, and the mean chord length, $\bar{\ell}_b$) to a bubble size distribution by the following method. They defined the vertical chord length passing through the center of the bubble as the bubble diameter (d_b: minor diameter). The bubble size distribution and the chord length distribution can be related by

$$Z(\ell_b) = \int_{\ell_b}^{\infty} p(d_b, \ell_b)\, r(d_b)\, dd_b \tag{2.36}$$

where $p(d_b, \ell_b)$ is the probability function. They indicated that for a spherical-cap bubble or a nonspherical bubble which can be approximated by a pair of concentric spheroids, $p(d_b, \ell_b)$ can be expressed, for $\ell_b \le d_b$, as

$$p(d_b, \ell_b) = 2\,\frac{\ell_b}{d_b^2} \tag{2.37}$$

The measured chord length distribution can readily be converted to the bubble size distribution with Eqns. (2.36) and (2.37). Matsuura and Fan (1984) showed that the mean bubble size can be correlated (in SI units) with the mean chord length as follows:

$$\bar{d}_b = 1.34\,\bar{\ell}_b + 4.2 \times 10^{-4} \tag{2.38}$$

The values of $\bar{d}_b$ from the correlation of Eqn. (2.38) lie in between those from the correlation for uniform sized spherical bubbles by Ueyama and Miyauchi (1976) ($\bar{d}_b = 1.5\,\bar{\ell}_b$) and those from the correlation for spheroidal bubbles of uniform size ($\bar{d}_b = 1.33\,\bar{\ell}_b$).

In multi-bubble systems, a mean bubble size is usually used to describe the system. When the probe measurement technique is used, the mean size is derived directly from the chord length distribution; when the photographic technique, or any technique which provides individual bubble sizes directly, is used, proper averaging must be made. In such a case, the mean size is commonly expressed through the Sauter, or volume-surface, mean. For a group of bubbles with measured diameters, the Sauter mean is

$$d_{vs} = \frac{\Sigma n_i\, d_{bi}^{\;3}}{\Sigma n_i\, d_{bi}^{\;2}} \tag{2.39}$$

where n_i is the number of bubbles with size d_{bi}. Note that d_{bi} can be any equivalent diameter (based on equivalent volume, etc.); the exact choice will naturally affect the magnitude of the Sauter mean size.

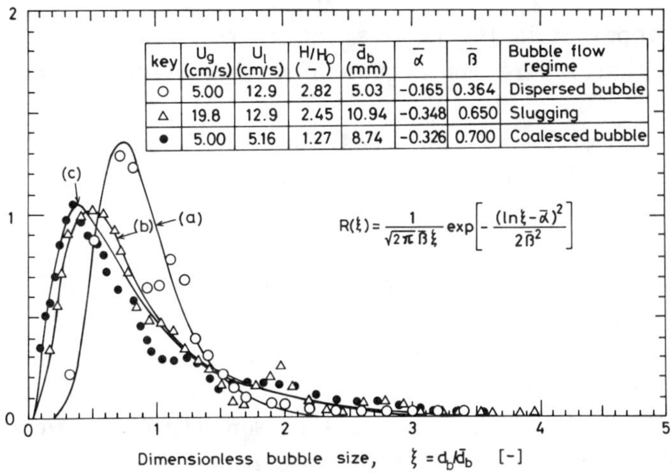

Figure 2.17 Bubble size distribution in an air-water-particle fluidized bed containing a binary mixture of 3 and 6 mm glass beads for various flow regimes (from Matsuura and Fan, 1984).

Several investigators have measured bubble size/chord length distributions in three-phase fluidized beds. At low liquid velocities, Lee et al. (1984) found a normal bubble chord length distribution; at higher liquid velocities, they found a log-normal bubble chord length distribution. The log-normal bubble size distribution was also observed by Darton and Harrison (1974a) and Matsuura and Fan (1984). Figure 2.17 shows bubble size distributions for a binary mixture of large particles in the dispersed bubble, slugging, and coalesced bubble regimes. All size distributions (based on the number frequency of the bubbles) were shown to follow a log-normal distribution, R(ξ), given below in a dimensionless form:

$$R(\xi) = \frac{1}{\sqrt{2\pi}\ \xi\bar{\beta}} \exp\left[-\frac{(\ln\xi - \bar{\alpha})^2}{2\bar{\beta}^2}\right] \tag{2.40}$$

where $\bar{\alpha}$ and $\bar{\beta}$ are a scale parameter and a shape parameter, respectively. The mean, $\underline{\mu}$, and variance, $\underline{\sigma}^2$, of the log-normal distribution are given as

$$\underline{\mu} = \exp(\bar{\alpha} + \bar{\beta}^2/2) \tag{2.41}$$

$$\underline{\sigma}^2 = \exp(2\bar{\alpha} + \bar{\beta}^2)\{\exp(\bar{\beta}^2) - 1\} \tag{2.42}$$

As is apparent in the figures, the distributions are similar in shape except that a large mean bubble size is observed for the slugging regime.

Matsuura and Fan (1984) found a simple relationship between the mean and variance of the bubble size distribution as given (in SI units) by

$$\underline{\sigma}^2 = 4.1 \times 10^3 \ \bar{d}_b^{\,2.0} \tag{2.43}$$

This relation indicates that a small bubble size is associated with a small variance, while a large mean bubble size is associated with a large variance. Darton and Harrison (1974a) indicated that the variance increased with an increase in gas flow rate; however, they found no consistent trend in the variance with changes in the liquid flow rate.

Ostergaard (1964) and Rigby et al. (1970) reported that an increase in the liquid velocity decreased the bubble size; Lee et al. (1974) found similar results for low liquid velocities, but at velocities above a certain value no further effect was noticed. In the dispersed bubble regime, Peterson et al. (1987) noted that the mean bubble size reaches a minimum near minimum fluidization conditions, then increases with an increase in liquid velocity.

Matsuura and Fan (1984) found that the mean bubble size increased with an increase in the bed expansion parameter (H/H_0) for various flow regimes, as shown in Fig. 2.18. In a two-dimensional bed, Kim et al. (1977) studied liquid surface tension and viscosity effects on the average bubble diameter; but their study involved large bubbles ($3 < \bar{D}_b < 8$ cm) on which inertial effects of the surrounding flow predominate and hence, surface tension and viscosity effects are not important. Kim et al. (1977) proposed

$$\bar{D}_b = 0.142\, U_\ell^{\,0.052} U_g^{\,0.248}\, \gamma_\ell^{\,0.008} \sigma^{0.034} \tag{2.44}$$

as a correlation (in SI units) for the equivalent spherical diameter where γ_ℓ is the generalized liquid viscosity constant defined by $(K \cdot 8^{\bar{n}-1})$. Note the near independency of Eqn. (2.44) on surface tension and viscosity.

The bubble size decreases with an increase in bed pressure. At high pressure, when the particle size is greater than the bubble size, the particle size effects on the prevailing bubble size become minimal, yielding bubble size characteristics in the bed similar to those in the

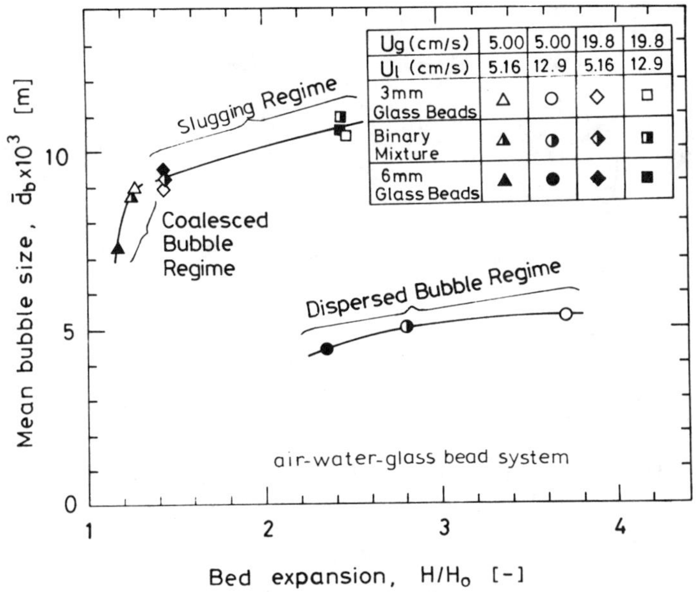

Figure 2.18 Relationship between mean bubble size based on bubble size number distribution and bed expansion parameter (from Matsuura and Fan, 1984).

gas-liquid bubble column under the same operating conditions. Idogawa et al. (1986), in a study of high pressure gas-liquid bubble columns, reported that above 5 MPa, the bubble size distribution became narrower and approached a near uniform size. In addition, the effect of gas distribution on bubble behavior became less noticeable and vanished completely above 10 MPa.

2.2.1.2 Bubble Coalescence and Breakup

The bubble size at any axial location within a three-phase fluidized bed reflects a number of complex interactions. In the distributor region bubbles are formed and their initial sizes depend on the distributor design as well as forces acting on them such as buoyancy, viscous drag, surface tension and inertia (Azbel, 1981). As the bubble rises, coalescence or breakup may occur due to interactions between the rising bubble and other bubbles, individual particles, and the surrounding liquid-solid medium.

Several investigators have reported the importance of the gas distributor region to the bubble size. Page and Harrison (1972)

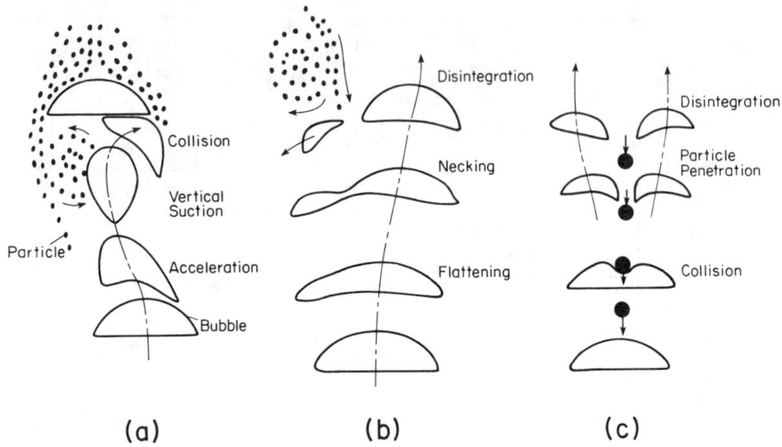

Figure 2.19 Mechanisms of coalescence and disintegration of bubbles in gas-liquid-solid fluidized beds.

reported that the initial bubble size depends strongly on the type of gas distributor. Both Massimilla et al. (1961) and Ostergaard (1966) found the bubble size to increase within a short distance from the distributor and Rigby et al. (1970) noted that the mean bubble size increased and the bubble size distribution broadened with an increase in the axial distance. The increase in the bubble size with the axial distance from a distributor is due to bubble coalescence. Ostergaard (1966) indicated that bubble coalescence resulted in a decrease in bubble frequency within a short distance from the distributor. Ostergaard (1964) found that as the bed porosity increased, the coalescence rate decreased.

In a three-phase fluidized bed bubble coalescence may occur according to a mechanism similar to that in a gas-liquid system or in a gas-solid fluidized system. When two successive bubbles rise in a liquid-solid suspension, the later bubble accelerates and eventually overtakes the preceding bubble (Fig. 2.19(a)). Fan et al. (1988b) attributed the acceleration of the trailing bubble to the suction caused by a low pressure region immediately below the preceding bubble. Bubble coalescence commonly occurs in fluidized beds of small particles and it has been attributed to the liquid-solid suspension acting as a pseudo-homogeneous medium of higher apparent viscosity and density than those of the liquid medium alone.

A limit exists as to how large a bubble can form in a fluidized bed. Large bubbles may be deformed and ruptured to form smaller bubbles by the turbulent eddies generated in the liquid-solid suspension (Fig. 2.19(b)). Bubble breakup may also be attributed to horizontal entrainment, flattening, necking, and disintegration of the later bubble affected by the turbulent shear and vortical flow near the edges of the preceding bubble (Fan et al., 1988b). Interaction between a bubble and individual particles can also lead to bubble breakup (Fig. 2.19(c)); several theories have been proposed to date on bubble splitting in beds with large or heavy particles as given below.

Lee (1965) and Lee et al. (1974) proposed a bubble breakup mechanism by considering collision between a single particle and a single bubble. The bubble was assumed to disintegrate when a particle with adequate inertia induced a hemispherical indentation on the bubble roof. This approach led to a critical Weber number (We = $\rho_s U_b^2 d_p/\sigma$) of three beyond which the bubble will break. The physical mechanism underlying the theory proposed by Lee et al. was, however, refuted by Henriksen and Ostergaard (1974b) who found that the theory prediction contradicts an observation in which a single air bubble could hardly be disintegrated by single glass spheres of 3 to 6 mm diameter or a 5 mm diameter steel sphere, even though the particle penetrated all the way through the bubble.

Henriksen and Ostergaard (1974b) suggested bubble disintegration was caused by Rayleigh-Taylor instability on the bubble roof. Based on Bellman and Pennington's theory (1954) for a plane surface with a two-dimensional sinusoidal disturbance, they suggested that a particle with a diameter greater than half of the critical wavelength will generate enough disturbance to break the bubble. However, the Bellman and Pennington theory is not directly applicable to the case of bubble breakage, as indicated by Chen and Fan (1988a), since neither is a bubble roof flat nor is the disturbance by a particle two-dimensional.

Chen and Fan (1988a) employed a local force balance to study the Rayleigh-Taylor instability on two- and three-dimensional curved surfaces. It was found that an interface is more stable when it is subject to a disturbance in three dimensions than a comparable disturbance in two dimensions. If bubble breakup in three-phase fluidized beds is assumed to be induced merely by the Rayleigh-Taylor instability and the particle diameter is taken as half of the critical wavelength of a three-dimensional disturbance, it would require a particle of at least 1.2 cm to penetrate the bubble in an air-water system. This prediction contradicts experimental observations (e.g., Henriksen and Ostergaard, 1974b; and Chen and Fan, 1988b) in which bubble breakage occurs at particle sizes considerably smaller than 1.2

cm. Thus it can be concluded that the bubble breakage is not due to merely the Rayleigh-Taylor instability.

Chen and Fan (1988b) proposed a mechanism for the breakage of a spherical-cap bubble by the collision of a spherical particle. In this mechanism, they considered separately the conditions for the full penetration of a particle through the bubble, and the actual breakage of the bubble. Accordingly, when a bubble collides with a particle with a relative velocity (U_r), the particle is subject to an upward acceleration (A_c) expressed by

$$\frac{\pi}{6} d_p^3 \rho_s A_c = \frac{\pi}{6} d_p^3 \rho_\ell g - \frac{\pi}{6} d_p^3 \rho_s g + \sigma \pi d_p + \left(\frac{2\sigma}{r_o} - \rho_\ell g \ell\right) \frac{\pi}{4} d_p^2 \tag{2.45}$$

They obtained three conditions for particle penetration:

$$A_{co} < 0 \tag{2.46}$$

$$\frac{U_r \omega}{A_{co}} > 1 \tag{2.47}$$

and

$$1 > \frac{h\omega^2}{A_{co} - (A_{co}^2 - U_r^2 \omega^2)^{1/2}} \tag{2.48}$$

where

$$A_{co} = g \frac{\rho_\ell - \rho_s}{\rho_s} + \frac{6\sigma}{\rho_s d_p^2} \left(1 + \frac{d_p}{2r_o}\right) \tag{2.49a}$$

and

$$\omega = \left(\frac{3}{2} \frac{g}{d_p} \frac{\rho_\ell}{\rho_s}\right)^{1/2} \tag{2.49b}$$

A particle will penetrate through the bubble if any of the three conditions given by Eqns. (2.46) through (2.48), is satisfied. However, this criterion constitutes only a necessary condition for bubble breakup with a single particle.

Chen and Fan (1988b) also indicated that the penetrated bubble should deform into a doughnut shape, with the central hole being the size of the particle, d_p, and the bubble height being h_d, which can be estimated by requiring that the deformed bubble have the same volume as the original bubble. In order for a single particle to break a bubble, the particle must not only be able to penetrate through the bubble, but also the particle diameter must be greater than the height of the deformed bubble; that is

$$d_p > h_d \tag{2.50}$$

The subsequent bubble breakup of the doughnut-shaped bubble is by the necking mechanism due to Boys' (1890) instability.

Based on the criterion described above (Eqns. (2.46) through (2.48) and Eqn. (2.50)), three markedly different collision phenomena between a single particle and a single bubble were predicted: a small/light particle is unable to penetrate and is ejected after collision; a medium size particle is able to penetrate but is unable to break the bubble; a large and heavy particle penetrates and subsequently, breaks the bubble. These predictions agreed very well with the experimental results of Chen and Fan (1988b) and those of Henriksen and Ostergaard (1974b). Indeed, in Chen and Fan's experiments only an 11.5 mm iron oxide particle was observed to break a bubble with an equivalent diameter of about 18 mm. In a multi-particle (fluidized bed) system, the necessary conditions for particle penetration alone could be adequate to account for bubble breakage. This is because when particle penetration takes place in a multi-particle system, it often involves more than one particle penetrating, simultaneously or consecutively, through a bubble. Multi-particle penetration induces an unstable bubble and hence, causes bubble breakage. If particle penetration is taken as a sufficient condition for bubble breakage in a three-phase fluidized bed, the critical particle size for an air-water-glass bead system is predicted as 2.2 mm by Chen and Fan's criterion given above, close to the value of 2.5 mm experimentally measured by Kim et al. (1977).

The particles dealt with above have been wettable. For non-wettable particles such as Teflon coated glass beads, Chen and Fan (1988d) found that the contact surface between the particle and a bubble remains hemispherical, similar to that of a wettable particle. This is due to the fact that the relative velocity between a penetrating particle and a bubble is far beyond the critical speed (see Section 2.1.6) and the dynamic receding angle is always 0° regardless of the surface condition. They also found that a 2 mm Teflon coated glass sphere can penetrate a spherical-cap bubble whereas a clean 2 mm glass sphere cannot. A non-wettable particle is more capable of penetrating because the hemispherical contact surface is no longer surrounded by a liquid film and consequently, the buoyant force effect becomes smaller. Nevertheless, the basic phenomena of a non-wettable particle collision are very similar to those of a wettable particle.

The concept of bubble breakage by large/heavy particles has been applied to three-phase fluidized bed operation. For instance, gas holdups can be enhanced by the use of floating bubble breakers in the bed. Chang and Kim (1979) and Kim and Kim (1987) employed floating bubble breakers with a density of 2.28 g/cm^3, made of a mixture of lead particle and paraffin placed in an acryl capsule, fluidized in an air-water bed containing glass beads with sizes varying from 1 mm to 6 mm. The optimum ratio of breaker volume to total particle volume in the bed yielding minimum bubble size and hence maximum gas holdup is about 0.2.

The competing tendencies of bubble coalescence and disintegration may result in an equilibrium bubble size distribution. Therefore, the bubble size distribution for a deep bed is virtually independent of the initial size of bubbles immediately above the distributor (Page and Harrison, 1972).

2.2.1.3 Single Bubble Rise Velocity

The rise velocity of a single gas bubble inherently depends on its size: for small bubbles, the rise velocity strongly depends on liquid properties such as surface tension and viscosity; for large bubbles, the rise velocity is insensitive to liquid properties as predicted by the well-known Davies-Taylor relationship for spherical-cap bubbles in inviscid liquids. Haberman and Morton (1953) and Tadaki and Maeda (1961) found experimentally that, for bubbles with equivalent spherical diameters greater than 1.8 cm, no appreciable difference occurred in the bubble rise velocity for various liquids. However, for bubbles smaller than 1.0 cm in pure liquids, the rise velocity is higher, in general, in liquids with higher surface tension or lower viscosity for a given bubble size.

As mentioned previously, the rise of single bubbles in liquid-solid fluidized beds with small/light particles is similar to that in a liquid alone but with a higher viscosity. Massimilla et al. (1961) found that the rise velocity of a single bubble in a water fluidized bed was significantly smaller than that in a solid-free system for bubbles (D_b) smaller than 1 cm. The rise velocity increases with an increase in bed expansion, but the size and density of the particles apparently have little effect on the rise velocity.

For large single bubbles (equivalent circular bubble diameter ranging from 2 to 7 cm) in a two-dimensional column, Henriksen and Ostergaard (1974a) found that, except for bed voidages less than 0.5, the bubble rise velocity behavior could be successfully correlated with a Davies-Taylor type relationship; however, they correlated the absolute rise velocity (U_b) rather than the rise velocity relative to the liquid phase (U_B). The absolute bubble rise velocity and that relative to the interstitial liquid velocity are related by

$$U_B = U_b - U_\ell/\epsilon_\ell \tag{2.51}$$

As pointed out by Darton and Harrison (1974b) and El-Temtamy and Epstein (1980a), the relative rise velocity should be used in liquid-solid fluidized beds since the intrinsic bubble rise characteristics depend on the relative motion between the bubble and the surrounding liquid-solid medium. In this sense, Mendes and Qassim (1984) suggested a rise velocity relative to an average velocity of the homogenous liquid-solid medium rather than relative to the interstitial liquid velocity. In general, the difference between the two relative velocities is not significant. El-Temtamy and Epstein (1980a) re-correlated Henriksen and Ostergaard's (1974a) data in terms of the relative bubble rise velocity. They found that the two-dimensional Davies-Taylor equation, after accounting for increases in the velocity due either to three-dimensional effects or nonspherical-cap bubble shapes, successfully predicted the data provided that the bed voidage was greater than 0.5.

As is apparent from the work of Henriksen and Ostergaard (1974a) and El-Temtamy and Epstein (1980a), the bed voidage plays a prominent role in the bubble rise characteristics. Darton and Harrison (1974b) systematically studied the effect of bed porosity on the relative bubble rise velocity of intermediate size bubbles ($0.5 < D_b < 2.5$ cm) in three-dimensional water-sand ($d_p = 0.5$ or 1.0 mm) fluidized beds. Their results for 0.5 mm sand particles, given in Fig. 2.20, indicate a significant effect of bed voidage, especially for small bubbles: for a fixed bubble size, the rise velocity increases as the extent of bed expansion increases up to a bed voidage of 0.55; beyond a bed voidage of 0.55 no appreciable changes occur. Massimilla et al.

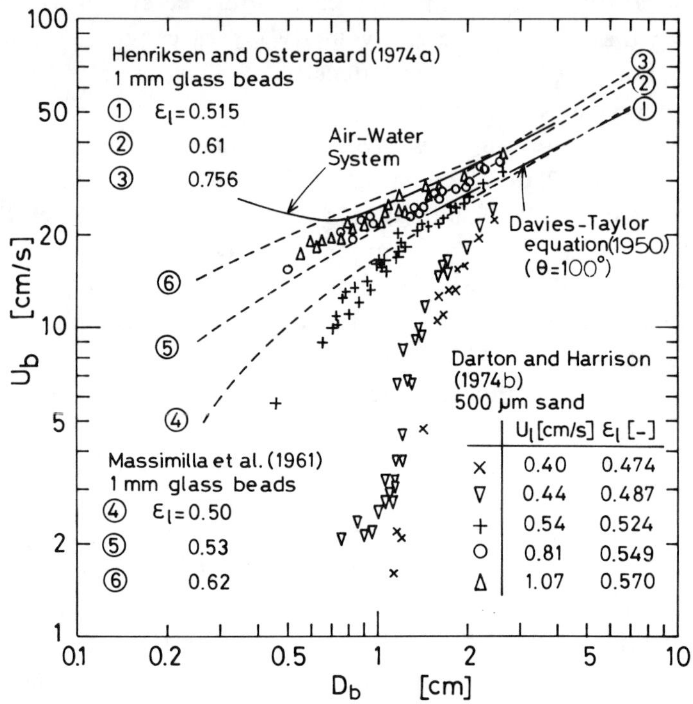

Figure 2.20 Effect of ϵ_ℓ on the rise velocity of a single air bubble in liquid (water)-solid fluidized beds (after Darton, 1985).

(1961) and Henriksen and Ostergaard (1974a) found similar trends as shown in Fig. 2.20; the dashed lines in Fig. 2.20 are their predicted values.

The rise velocity of single bubbles can also be interpreted in terms of a drag coefficient. Darton and Harrison (1974b) analyzed drag coefficients on rising bubbles for two special cases: spherical (low Reynolds numbers) bubbles and spherical-cap (high Reynolds numbers) bubbles. Darton and Harrison suggested that spherical bubbles move slowly through the bed under the action of viscous drag and that the rise velocity is governed by the flow field around a sphere. They found that the drag coefficient for small bubbles in beds of 500 μm particles at low bed expansions (ϵ_ℓ = 0.474 and ϵ_ℓ = 0.487, see Fig. 2.21) could be correlated by

$$C_D = 38\ Re_d^{-1.5} \tag{2.52}$$

where

$$C_D = \frac{4}{3} g \frac{D_b}{U_B^2} \quad , \quad Re_d = \frac{\rho_2 U_B D_b}{\bar{\mu}_2}$$

Here ρ_2 and $\bar{\mu}_2$ are the average (bulk) density and the apparent viscosity of the liquid-solid fluidized bed, respectively. For high Reynolds numbers ($Re_d > 2$), the drag coefficient could be correlated by

$$C_D = 2.7 + \frac{24}{Re_d} \tag{2.53}$$

Figure 2.21 shows a plot of C_D against Reynolds number (Darton and Harrison, 1974b).

The apparent bed viscosity used in Eqns. (2.52) and (2.53) was obtained by Darton and Harrison (1974b) by assuming that air bubbles

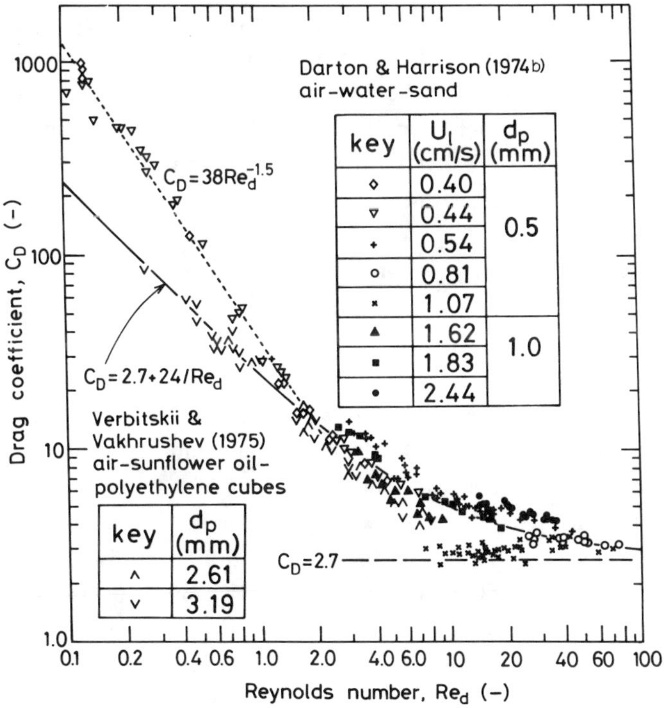

Figure 2.21 C_D-Re_d relationship for gas bubbles in liquid-solid fluidized beds and some suspensions (after Darton and Harrison, 1974b).

in water fluidized beds have rigid surfaces. Henriksen and Ostergaard (1974a) found a relationship between the included angle of a bubble rising in liquid-solid fluidized beds and the apparent bed kinematic viscosity. The apparent bed viscosity was found to vary from 0.12 to 3.10 P as the bed voidage decreased from 0.756 to 0.515 for 1 mm glass bead systems. Darton (1985) correlated the data for bed viscosity with high solids concentration obtained by Henriksen and Ostergaard (1974a) and Darton and Harrison (1974b) by

$$\bar{\mu}_2/\mu_\ell = \exp\,(36.15\ \epsilon_s^{\,2.5}) \tag{2.54}$$

The bubble shape, bubble rise velocity, and drag coefficient of single bubbles in liquid-solid fluidized beds can be predicted based on theories available for gas-liquid systems provided the bed is treated as a homogenous liquid-solid mixture of known viscosity. The apparent viscosity of the liquid-solid mixture depends on the bed porosity and significantly affects the bubble characteristics. The effect of bed voidage becomes insignificant above 0.55 where the relative rise velocity for large bubbles follows the Davies-Taylor equation; and in general, the higher the bed voidage, the larger the bubble rise velocity. The discrepancy between the predicted and experimental values at low bed voidage may be attributed to the non-Newtonian behavior of the concentrated suspension.

2.2.1.4 Rise Velocity in Multi-Bubble Systems

Bubble-bubble interactions play a prominent role in the resulting observed bubble rise velocity behavior. In multi-bubble systems, the bubble size follows a log-normal distribution as indicated previously and hence, the bubble rise velocity must be expected to be nonuniform within the bed. Figure 2.22 shows bubble rise velocity distributions for three flow regimes in a bed where a binary mixture of particles are present (Matsuura and Fan, 1984). Note that $Q(U_b)$ is based on the number frequency of the bubble rise velocities. In the dispersed bubble and coalesced bubble regimes, the bubble rise velocity distribution is narrow; the variations in the mean and variance of the bubble rise velocity distribution are insensitive to the particle size used. In the slugging regime, the bubble rise velocity distribution shows a broad multi-modal distribution.

Several investigators have related the bubble rise velocity in multi-bubble systems to the bubble size; Table 2.3 lists these

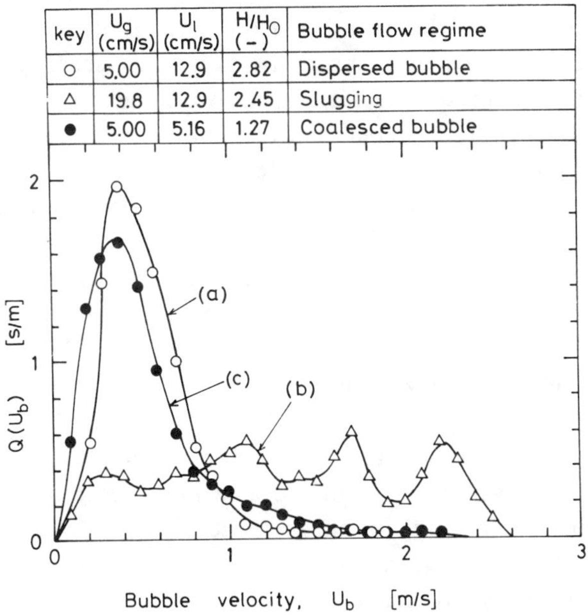

Figure 2.22 Bubble rise velocity distribution in an air-water-particle fluidized bed containing a binary mixture of 3 and 6 mm glass beads for various flow regimes (from Matsuura and Fan, 1984).

Table 2.3 Bubble Rise Velocities in Multi-Bubble Systems.

Investigators	Correlations (in SI Units)	Ranges
Rigby et al. (1970)	$[U_b - (U_g + U_\ell)][\frac{1-\varepsilon}{\varepsilon}]^2 = 373\ \ell_b^{1.53}$	$0.4 < \ell_b < 4$ cm
Kim et al. (1977)[1]	$\bar{U}_B = 17.1\ \bar{D}_b^{0.989}$	$3 < \bar{D}_b < 8$ cm
Matsuura and Fan (1984)	$\bar{U}_b = 10.6\ \bar{d}_b^{0.5}$ (slug flow)	$0.86 < \bar{d}_b < 1.1$ cm
Lee and de Lasa (1986)	$U_B = 1.565\ \ell_b^{0.293}$	$0.2 < \ell_b < 1.01$ cm
	$U_B = 6.357\ \ell_b^{0.598}$	$1.01 < \ell_b < 4$ cm

1: 2-Dimensional Bed

relationships. Some points are noted about these studies and correlations: first, the two-dimensional beds were used to investigate large bubbles and hence, large rise velocities; second, Lee and de Lasa (1986) divided their correlation based on the ranges of bubble length; and third, no unique dependency on bubble size is apparent. Lee and de Lasa (1986) arrived at a correlation for relative bubble rise velocity although the data (aslso see de Lasa et al., 1984; and Lee et al., 1984) showed a large degree of scatter. Their interpretation of the relative rise velocity in terms of the average linear liquid velocity instead of the actual local linear liquid velocity may contribute to this scatter.

2.2.2 Bubble Wake Structure

In addition to its role in the motion of rising bubbles, the bubble wake has been recognized as a key factor in explaining various phenomena occurring in three-phase fluidized beds such as solids mixing, particle entrainment into the freeboard (Page and Harrison, 1974), and bed contraction upon the initial introduction of gas bubbles into liquid-solid fluidized beds (Stewart and Davidson, 1964; Ostergaard, 1965; Rigby and Capes, 1970).

Stewart and Davidson (1964) observed the behavior of gas bubbles ascending in a two-dimensional liquid-solid fluidized bed and identified two different types of solid-free regions behind the bubbles: (1) a void of water, or so-called water bubble, stabilized by a thin layer of gas or small gas bubbles collecting on its roof in a bed containing dense particles such as iron shot or lead shot, and (2) a particle-depleted wake behind a single gas bubble in a bed containing glass beads. In addition, Stewart and Davidson (1964) observed that shed vortices, which are largely free of particles, were left on the zig-zag trail of the bubble and these vortices moved upward with a velocity somewhat lower than that of the bubble itself.

2.2.2.1 Wake Nature

In gas-liquid systems the wake geometry has long been disputed and historically grouped into three types: (1) closed laminar/toroidal wake, (2) closed turbulent wake, and (3) open turbulent wake (Coppus et al., 1977). Bubble wakes in gas-liquid-solid fluidized beds may be somewhat different from those in gas-liquid systems due to the presence of solid particles and because of the external disturbances prevailing in the bulk flow. There should, however, exist a universal resemblance in the nature of the wakes between these two systems, especially when small/light particles are used in three-phase fluidization. The following discussion is thus mainly based on the wake behavior in gas-liquid systems unless otherwise noted.

Closed laminar/toroidal wakes have been observed behind large spherical-cap bubbles rising in viscous liquids or behind the corresponding circular-cap bubbles in two-dimensional vessels. The closed laminar wake is hydrodynamically stable, consists of a well-defined boundary and a toroidal vortex ring inside, and exchanges no liquid with the external flow (Coppus et al., 1977; Bhaga and Weber, 1981). The stability of the closed laminar wake is due to viscous and/or wall effects. Keeping the bubble volume constant and decreasing viscosity, or increasing the Reynolds number, Bhaga and Weber (1981) demonstrated that the flow in the wake becomes less stable and starts shedding vortices (transition to the open wake). The wake behind circular-cap bubbles rising between two flat plates appears to be more stable than that behind three-dimensional bubbles. Using a two-dimensional system with a 6.4 mm gap, Collins (1965) showed the wake to be closed and to consist of a large stable vortex pair. The Reynolds number based on the bubble radius of curvature was as high as 2.5×10^4. Over the same Reynolds number range, on the other hand, Crabtree and Bridgwater (1967) found considerable wake shedding. In the experiments, they used a larger vessel gap (12.7 mm). Accordingly, the smaller the gap between the plates of a two-dimensional system, the more stabilized the wake flow.

The circulatory flow within a closed wake and the flow field external to the wake are symmetric along the bubble centerline. These flow fields can be reasonably described by applying the modified Hill's spherical vortex model (proposed by Harper and Moore (1968)) to the spherical region approximately containing the spherical-cap and its closed wake. The actual shape of the closed wake is somewhat elongated downward, but the dimensionless distance from the axis of symmetry to the stagnation ring is essentially the same as the value of 0.707 for Hill's spherical vortex (Bhaga and Weber, 1981).

Turbulent wakes have been observed behind large bubbles rising in less viscous liquids. Considerable disagreement exists among researchers (e.g., Maxworthy, 1967; Wegener and Parlange, 1973; Hills, 1975; Coppus et al., 1977) as to whether turbulent wakes have an open or closed structure.

A closed turbulent wake was first visualized by Davies and Taylor (1950) applying flash shadow photography to a large spherical-cap bubble rising in nitrobenzene. Owing to optical anisotropy of the liquid under shear, a closed region of high shear (high turbulence) was clearly shown immediately behind the bubble. The wake boundary approximately completed the sphere but no detailed flow pattern inside the wake was visible. Care should be exercised, however, when interpreting this result. As pointed out by Maxworthy (1967) and supported by Wegener and Parlange (1973), a turbulent wake cannot be confined to a completely closed region due to the momentum defect in the liquid following the passage of the bubble; the bubble must experience a finite drag.

Accordingly, an open geometry appears to be more accurate in describing turbulent wakes. Employing Schlieren photography, Wegener and Parlange (1973) showed that the wakes extended far behind the bubbles at Reynolds numbers (Re_e) greater than 3000. In their photographs the flow inside the wake appears to be chaotic and no circulatory flow patterns were recognized; nevertheless, they indicated that, based on the motion of small satellite bubbles, there should exist a vortical motion within the near wake. The fine detail of the Schlieren technique may obscure large scale flow patterns, as pointed out by Hills (1975).

Note that the word "turbulent" may be misleading as an expression of the wake nature at higher Reynolds numbers if it indicates totally "chaotic" motion inside the wake. Increasing the Reynolds number up to 1.5×10^4, Coppus et al. (1977) still found an enclosed region immediately behind the bubble in which a broad circulatory flow pattern was identified, although less stable and rather irregular. This near wake region was followed by a vortex street extending far behind the bubble. Even at a lower Reynolds number (Re_e = 56), Slaughter and Wraith (1968) observed a streaming tail below a toroidal vortex travelling with the bubble; the tail extended along the bubble rise path. This structure was revealed by intensifying the optical effect with a shadow-graph technique in an 87 wt% aqueous glycerol solution.

Figure 2.23 shows a photograph (Fig. 2.23(a)) of a relatively large two-dimensional nitrogen bubble rising through a water-774 μm glass bead fluidized bed and a schematic interpretation (Fig. 2.23(b)) of the wake flow. It can be seen that the internal structure of the bubble wake is characterized by the dynamic behavior of vortices. Note, in the figure, the region that includes two growing vortices: one

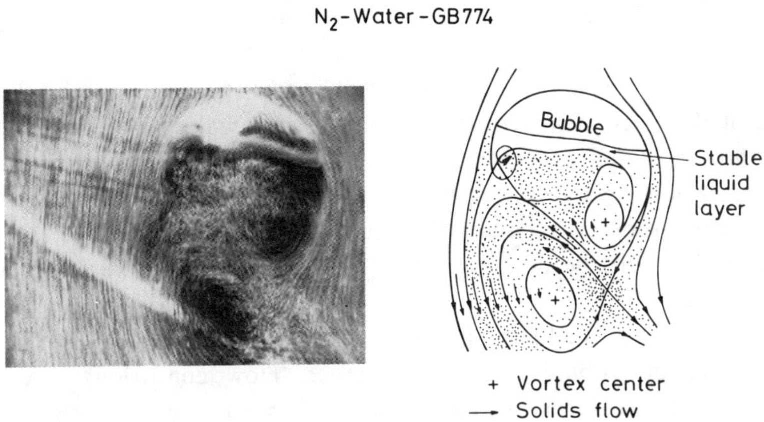

(a) (b)

Figure 2.23 (a) Photograph of a circular-cap nitrogen bubble and its wake rising through a water - 774 μm glass bead fluidized bed. (b) Schematic interpretation of the wake flow (from Tsuchiya and Fan, 1986; reprinted with permission from Chemical Engineering Science, 43(5), K.Tsuchiya and L.-S. Fan, Near-wake structure of a single gas bubble in a two-dimensional liquid-solid fluidized bed: vortex shedding and wake size variation, copyright (1988), Pergamon Press plc.)

is represented by a well-established circulation (right-hand side) and the other is just being formed (left-hand side). This region, denoted as the primary wake (Tsuchiya and Fan, 1986), has nearly the same rise velocity as the bubble, and thus plays a primary role in the various hydrodynamic/transport phenomena mentioned earlier. The region beneath the primary wake, demarcated by streams of the external flow across the wake from right to left, includes shed vortices (note the slightly deformed vortex seen in the figure) and has secondary importance. In this sense, the region which extends far behind the bubble below the primary wake is denoted as the secondary wake. Also seen in the figure is the solids concentration gradient in the wake: lower concentration regions are observed immediately beneath the bubble base and around a vortex center; higher concentration regions occurs around the vortices, especially in regions where the two vortices interact.

2.2.2.2 Wake Instability - Formation and Shedding

The most important fluid dynamic characteristic of the wake is probably its instability. Initially, the wake consists of a closed laminar region of a toroidal vortex ring (main portion) and a narrow tail following it. As the bubble accelerates and the wake grows in size by continuously accumulating material from outside, the symmetry of this flow will be perturbed and eventually some wake material will be discharged.

Wake shedding phenomena have been observed for a wide variety of bubble shapes and hence, flow conditions. Any bubble which experiences secondary motion is suspected to shed its wake. The bubble shape may vary from an ellipsoid to a spherical-cap. The critical Reynolds number for the onset of wake shedding is reported as low as 100-110 based on D_b for gas-liquid systems (Wegener and Parlange, 1973; Bhaga and Weber, 1981).

Wake shedding phenomena are closely related to the flow path a bubble follows. For instance, the periodic nature of asymmetric wake shedding causes a zig-zag path in bubble motion or bubble rocking. Asymmetric shedding of a vortex from one side takes place as the inherently symmetric nature of the flow around small bubbles or bubbles in a viscous medium is disturbed by increasing Reynolds number. Subsequently, vortices shed periodically from alternating sides. Two-dimensional wakes are considered to at least approximate the projection, or cross section, of three-dimensional wakes despite having different degrees of wake instability.

<u>Formation-Shedding Mechanisms</u>: Mechanisms of wake formation and shedding can be described based on localized fluid dynamic properties of the flow field around a rising bubble. In experiments performed in a two-dimensional gas-liquid-solid fluidized bed system, Tsuchiya and Fan (1986) observed that wake shedding often occurs asymmetrically (alternate shedding) and that bubbles of oblate shape such as ellipsoidal and/or spherical-cap have fixed separation regions along the edge - the portion of the bubble surface with the sharpest curvature. As shown in Fig. 2.24, the formation of vortices in the near wake region originates from the separation of the external flow at the bubble edge. The separation of the flow induces the generation of vorticity at the separation points, which is followed by the conveyance of vorticity along the free shear layer. This free shear layer, or vortex sheet, tends to roll up into a spiral form due to differences in the characteristic velocities between the outer and inner

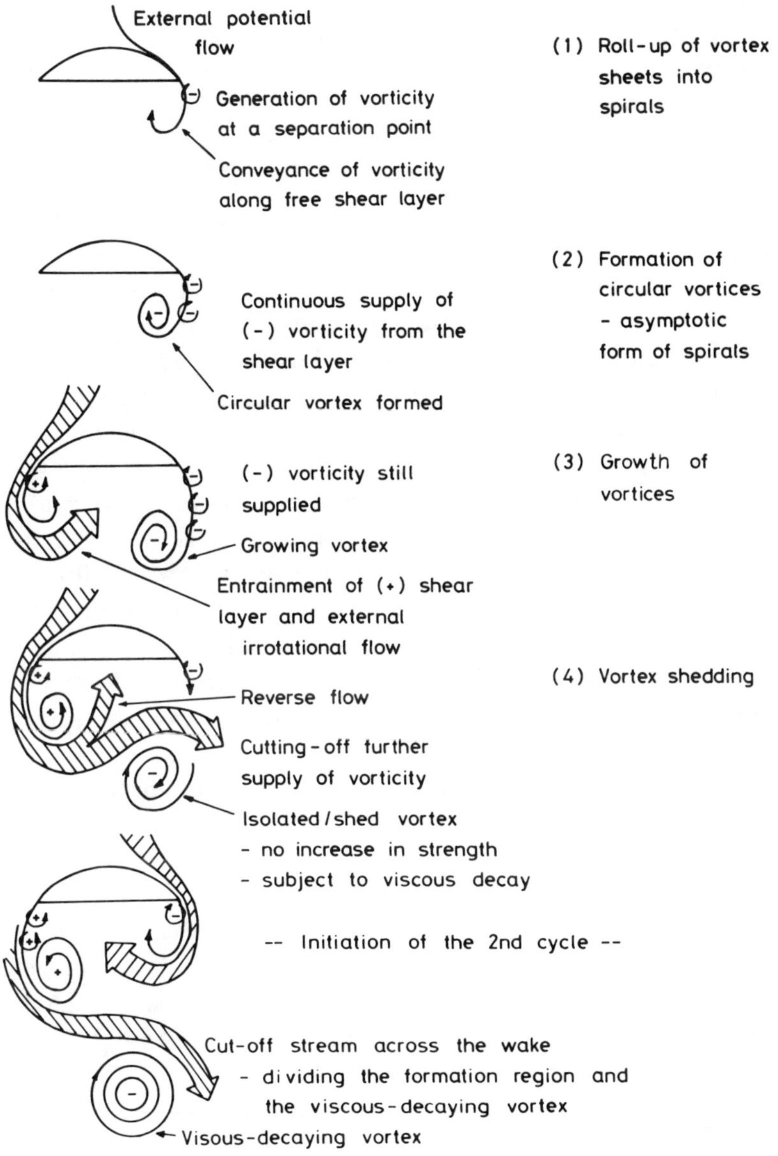

Figure 2.24 Mechanistic description of vortex formation and shedding - alternate shedding (from Tsuchiya and Fan, 1986).

boundaries of the shear layer. Thus the rate of increase in the strength of the vorticity induced at the bubble edge can be approximated (Fage and Johansen, 1927; Sarpkaya, 1975; Kiya and Arie, 1977) by

$$\frac{\partial \Gamma}{\partial t} = \frac{1}{2}(V_{out}^2 - V_{in}^2) \approx \frac{1}{2} V_{out}^2 \qquad (2.55)$$

The rolled up vortex sheet will eventually form a circular-cross-sectioned vortex. At this stage, the vortex is still attached to the bubble edge via the shear layer and grows as vorticity is continuously supplied through the shear layer. The vortex continues to grow until it becomes strong enough to draw the external flow and opposite shear layer across the wake. When this happens, oppositely-signed vorticity of sufficient concentration cuts off further supply of vorticity to the vortex and the vortex ceases to increase in strength and starts detaching from the bubble. At this moment the vortex is said to be shed. A very similar description was given by Gerrard (1966) on the basic mechanism controlling the frequency of vortex shedding from bluff bodies. The cut-off stream divides the near wake, or what Gerrard (1966) called the formation region, from the viscous-decaying shed vortex. The roll-up of the vortex sheet through the cut-off of the vorticity supply constitutes one cycle of the vortex formation-shedding process.

Parallel shedding may take place if the above process proceeds almost simultaneously from both edges. Most likely, however, parallel shedding occurs when the vortex sheets become unstable before forming well-established spiral/circulation flow patterns or even before rolling up. This is the case for large circular-cap bubbles.

The form of instability induced along a shear layer, or velocity discontinuity in a homogeneous fluid, is known as Helmholtz or Rayleigh instability. Employing a simple sinusoidal wave form at the interface along the shear layer, $y' = \eta(x',t)$:

$$\eta(x',t) = A_0 \exp\left[\frac{2\pi i}{\lambda}(x' - \kappa t)\right] \qquad (2.56)$$

and knowing (Currie, 1974)

$$\kappa = \frac{V_{out} + V_{in}}{2} \pm i\,\frac{V_{out} - V_{in}}{2} \qquad (2.57)$$

it can be easily deduced that the shear layer is unstable; the interfacial wave will exponentially grow with time. Here A_0 is the amplitude, λ is the wavelength, x' is the flow direction, t is the time, and κ is a measure of the sheet strength. As argued by Saffman and Baker (1979), the vortex sheet in an unsteady flow stretches/shrinks in length resulting in its strength being decreased/increased according to Kelvin's theorem (the conservation of circulation). In addition, this local variation in length corresponds to the same rate of increase/decrease in the wavelength of a disturbance. Now, based on Eqns. (2.56) and (2.57), the more compressed the sheet (i.e., the stronger the sheet and the shorter the wave), the more unstable the sheet tends to be. External disturbances due to the turbulence in the bulk flow also promote wake instability.

The local stretching/shrinking of the vortex sheet as well as diffusion and dissipation of vorticity by the action of viscosity eventually breaks up the deforming sheet, and blobs of vorticity are shed from the bubble edge. A formation-shedding process of this kind is depicted schematically in Fig. 2.25. The occurrence of this vorticity-blob discharge on one side should affect neither the other side discharge nor the bubble motion appreciably because the shed vorticity blob is weak in "hydrodynamic" forces, or of small momentum, and does not invade into the interior of the near wake. Vorticity-blob discharge only causes the dilations of the bubble edge and not systematic alternative shedding. The discharged vorticity blobs from both edges may later organize themselves into a large-scale vortical motion since each blob possesses a unique sign of vorticity. The "organized" vorticity blobs at both sides now draw the external flows into the wake: some fluid elements are spouted towards the bubble base while others from each side collide around the wake centerline leaving a chaotic turbulent region.

Generality of the above descriptions is supported by Perry et al. (1982) in their study on the vortex shedding process behind two-dimensional bluff bodies. They believe that "a deep insight into the mechanism of vortex shedding can be obtained by studying the instantaneous streamline patterns at various phases of the vortex-shedding cycle.... when the flow is non-turbulent, the instantaneous streamline patterns for each part of the vortex-shedding cycle will have the same qualitative features as the fully turbulent phase-averaged results."

Vortex Shedding Frequency: Shedding frequency has been measured in gas-liquid-solid fluidized beds. In a two-dimensional (7 mm gap) three-phase fluidized bed of 775 μm sand particles, Rigby and Capes (1970) found that the vortex shedding rate (for an

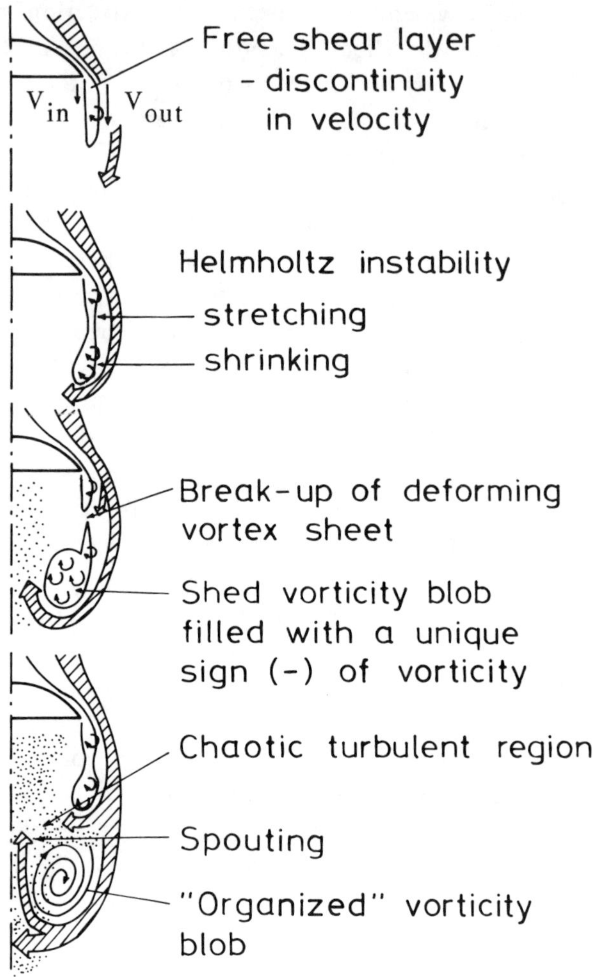

Figure 2.25 Mechanistic description of vortex formation and shedding - parallel shedding (from Tsuchiya and Fan, 1986).

individual vortex), f_v, slightly increased from 10.3 to 12.3 s^{-1} as the Reynolds number increased from 1800 (51) to 11200 (320). These Reynolds numbers were estimated based on the breadth of bubbles measured from the photographs, the reported bubble rise velocities, and the viscosity of water; for reference, the Re_b based on the apparent bed viscosity of 0.35 P is given in parentheses. Over this range of Reynolds numbers the bubble breadth varied from 0.86 to 3.9 cm and the Strouhal number Sr_b (= $f_v b/U_B$) from 0.43 to 1.65. For

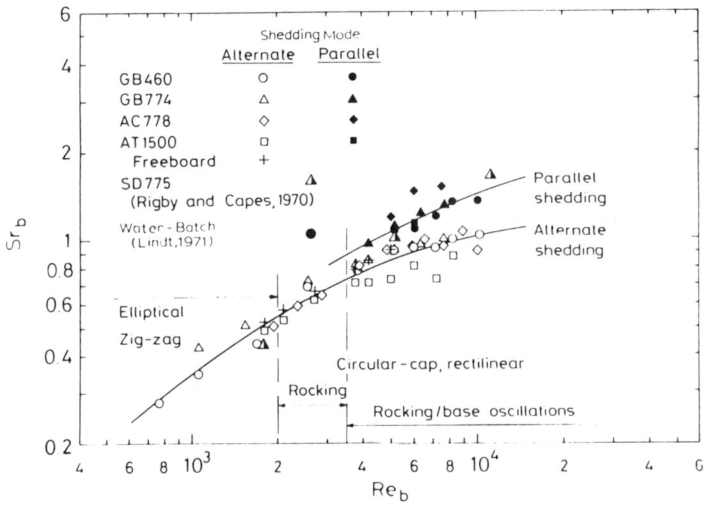

Figure 2.26 Relationship between Strouhal number and Reynolds number for vortex shedding from single nitrogen bubbles in a two-dimensional liquid (water)-solid fluidized bed (from Tsuchiya and Fan, 1986).

bubbles rising in quiescent water which was confined in a two-dimensional bed with 1 cm gap, on the other hand, Lindt (1971) reported the vortex shedding frequency (for a pair of vortices) decreased slightly from 4.4 to 3.5 s^{-1} as the Re_b increased from 2000 to 5000. The corresponding bubble breadth ranged from 1.0 to 2.0 cm and the Sr_b increased monotonically from 0.4 to 0.6 over this range.

In their recent study on the bubble wake structure in a two-dimensional liquid-solid fluidized bed, Tsuchiya and Fan (1986) reported the Sr_b as a function of the Re_b (based on water viscosity) for four different spherical particles. Figure 2.26 shows the Sr_b as a function of the Re_b for all four particles and the results of Rigby and Capes (1970) and Lindt (1971). Each data point represents the average for bubbles of matching size range. The range of data encompasses elliptical bubbles with a zig-zag motion to circular-cap bubbles in a rectilinear path with rocking as well as base oscillations. Periodic, alternate shedding of vortices was observed at low Re_b (< 3500) and the Sr_b was uniquely defined and monotonically (linearly in log-log scale) increased for each particle. At higher Re_b, shedding occurred rather irregularly. Careful observations, however, revealed the co-existence of alternate and parallel shedding modes. The parallel shedding mode gives higher shedding frequencies. Accordingly, bubble motion could not be clearly defined; both rocking and base oscillations occurred for a single bubble during its rise.

Essentially no appreciable effects of particles on the Sr_b were observed; however, the largest particles (1.5 mm acetate) gave slightly smaller values at high Re_b. Furthermore, Sr_b evaluated indirectly from the bubble rocking frequency in the freeboard region over a water-acetate particle fluidized bed equaled those evaluated in the bed. Since the freeboard can be considered as a gas-liquid system, this suggests that, in this case, the presence of particles does not significantly alter the shedding frequency. For single bubbles rising in stationary water, Lindt (1971) found the same trends in the Sr_b but the values were smaller by a factor of approximately 1/3. This behavior could be attributed to the nature of the surroundings: the systems other than that of Lindt (1971) were all disturbed by the liquid flows which may have destabilized the wake flow appreciably.

For a two-dimensional bed, Tsuchiya and Fan (1986) suggested the following correlation for the Sr_b: for alternate shedding

$$Sr_b = \frac{1}{(2420Re_b^{-1.02} + 0.776)} \tag{2.58a}$$

and for parallel shedding

$$Sr_b = \frac{1}{(981Re_b^{-0.888} + 0.425)} \tag{2.58b}$$

where the empirical parameters were evaluated employing a nonlinear regression. For a given confidence level in predicting Sr_b from Eqns. (2.58a) and (2.58b), the parameter can vary over wide ranges. The range of applicability is: $800 < Re_b < 10^4$; $0.46 < d_p < 1.5$ mm; $1.25 < \rho_s < 2.50$ g/cm^3; and $1.4 < H/H_0 < 2.0$. Here H/H_0 gives the bed expansion ratio.

2.2.2.3 Wake Boundaries and Sizes

Because of the significance of the primary wake to various hydrodynamic/transport phenomena, determination of its size is of particular importance. Except for the case of a steady laminar wake, the wake flow is unsteady and its internal structure is constantly changing. Consequently, such complicated hydrodynamic flow situations often prevent the location of a clear boundary between the primary and secondary wakes.

Tsuchiya and Fan (1986) considered the periodic nature of the primary wake and identified boundaries qualitatively between the primary and secondary wakes using a two-dimensional bed for several important representative wake structures. For steady flows, the primary wake observed for large bubbles in a viscous medium or at low Reynolds numbers ($Re_e < 110$) is laminar and closed as shown in Fig. 2.27. This type of primary wake, as shown enclosed by dotted lines and shaded inside, consists mostly of a well-developed toroidal vortex ring (three-dimensional) and a stagnation region near the rear stagnation point. Depending upon the Reynolds number and/or wake stability, this stagnant region may have negligible volume (Fig. 2.27(a)) as reported by Collins (1965) or a non-negligible volume in a cusp shape (Fig. 2.27(b)) as claimed by Batchelor (1956) and Bessler (1984). The shape and size of the boundary is invariant with time. Note that in both gas-liquid and gas-liquid-solid systems such a primary wake has also been observed during the initial stable growth period of the wake before the symmetry of the wake flow is disturbed; consequently, this type of structure is preliminary to the non-steady structure.

At higher Reynolds numbers for intermediate to large bubbles in liquids and/or liquid suspensions with low inertia solid particles, the wake flow is unsteady but has steady (cyclic) vortex shedding; Fig. 2.27(c) depicts a representative primary wake boundary for this case. The vortices generated in the near wake region are well developed before leaving the primary wake. The shedding scheme is the asymmetric alternate mode discussed previously. Thus, the primary wake boundary is marked by the cut-off stream which crosses the wake center axis from one side of the free shear layer to the other. The shape and size of the boundary vary periodically due to vortex shedding; the minimum size occurs at the moment of vortex shedding from one side, and the maximum size occurs just before shedding.

Over this same Reynolds number range, but for larger inertia particles, the wake flow is still asymmetric shedding and the shed material has large-scale vortical motion. However, there is a region of chaotic turbulence about the wake center axis in addition to the stagnant turbulent layer directly behind the bubble as shown in Fig. 2.27(d). This region may result from either the high inertia solid particles deviating from the liquid flow pattern and being expelled

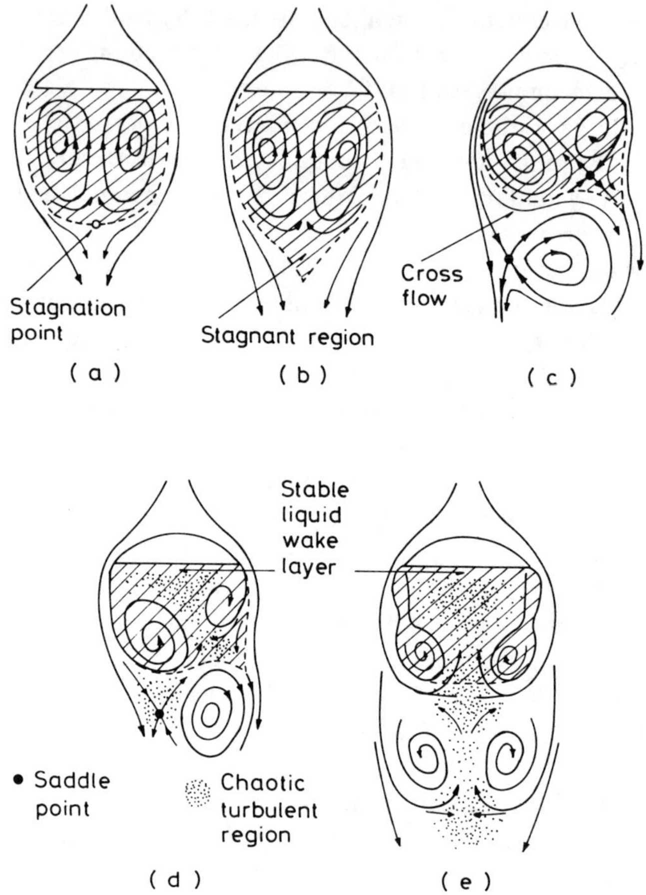

Figure 2.27 Schematic definition of primary wake boundary (from Tsuchiya and Fan, 1986).

from the vortex region to the central wake region or the turbulent shear layers being dragged into the interior. Consequently, the primary wake is said to be "open" since it cannot be distinguished from the adjacent secondary wake region due to this turbulent region. Nevertheless, the boundary can be defined, as shown in the figure, by connecting the outermost streamlines of the vortices at each side (dotted line).

For very large bubbles rising at high Reynolds numbers the wake flow is unsteady and has parallel wake shedding; Fig. 2.27(e) shows a phase (or time) averaged primary wake boundary for this case. As explained previously, the shedding possibly takes place due to the instability induced along the free shear layer. Thus, the primary wake

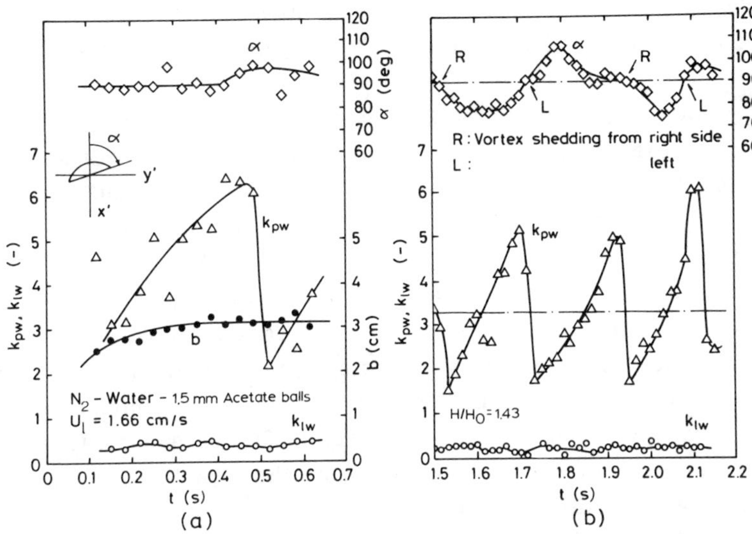

Figure 2.28 Dynamic variation in primary wake area (k_{pw}), liquid wake area ($k_{\ell w}$) and angle of attack (α) during (a) steady growing period and (b) steady shedding period (from Tsuchiya and Fan, 1986).

boundary can be designated by connecting the end points of two vortex sheets hanging from the bubble edges. The primary wake contains the vortex sheets or blobs, which still may not have established a well-defined vortical motion, and rather a wide region of chaotic turbulence around the wake center axis.

After making a suitable boundary definition based on the flow field, the primary wake size can be determined. Quantitative results on the size of the primary wake region in unsteady wakes are scarce. Tsuchiya and Fan (1986) measured the dynamic variation in the primary wake size in a two-dimensional bed containing water-particle mixtures using a video camera which tracked a vertically rising bubble. Figure 2.28 shows the dynamic variation in the primary wake area, normalized with respect to the bubble area at each instant (k_{pw}), and the angle of attack (α), which is defined in the figure. A circular-cap bubble of size b ≈ 3 cm was injected into a water-1.5 mm acetate particle fluidized bed. Although the bubble reached its steady-state geometry quickly (less than 0.2 s), the bubble wake steadily grew in size until about 0.5 s had elapsed before it shed any wake material. Figure 2.28(a) shows the increase in the primary wake area in the steady growth period. The primary wake grew to about 6 times the bubble area then dropped in size to about twice the bubble size. Note that during this period the angle of attack remained almost constant at around 90°.

The dynamic variations in the steady shedding period are shown in Fig. 2.28(b). The figure includes three cycles of vortex shedding with frequency of about 5 s^{-1} in this particular case. The primary wake size periodically varied in the form of a saw-tooth wave function from about 1.5 to 5.5 times the bubble size. Also clearly demonstrated in this example is that the shedding frequency can be estimated by the corresponding bubble rocking frequency if vortex shedding occurs alternately. Note the one-to-one correspondence between the frequency in the primary wake size variation and the angle of attack variation as shown in the figure: a decrease in α from 90° corresponds to vortex shedding from the right side and an increase in α indicates shedding from the left side.

Although the amplitude and frequency of the primary wake size fluctuations in the steady shedding period differ from condition to condition, Tsuchiya and Fan (1986) found that its periodic nature can be universally expressed by a saw-tooth wave function. In addition, they found that the mean primary wake size, obtained over at least three cycles of wake size variations, was a weak function of Re_b and reported that the ratio of the primary wake area to the bubble area was fairly constant at 3.3 ± 1.2 over a wide range of Re_b (1500 - 8150) with or without the presence of particles. The near constant wake size with respect to the Reynolds number was also found by other investigators (Henriksen and Ostergaard, 1974a; Kojima et al., 1975; Coppus et al., 1977; and Kitano and Fan, 1988). All investigators except Coppus et al. (1977) found the ratio of the primary wake volume to the bubble volume to fall between 3 and 6.

The primary wake usually includes a stable liquid layer with negligible solids concentration located immediately beneath the bubble base (see Fig. 2.23). Its existence, from a hydrodynamic point of view, has been inherently shown by Lazarek and Littman (1974) who measured the pressure field around a rising bubble.

A theoretical estimation of the height of the liquid-wake layer was made by Darton and Harrison (1976) for three-phase fluidization. Although it may be affected by wake shedding, the overall (time-averaged) height can be estimated based on steady wake configurations; as shown in Fig. 2.28, no appreciable cyclic variation occurred for the liquid layer size ($k_{\ell w}$). They assumed two extreme cases of wake flow: (1) Hill's spherical vortex (see Fig. 2.29(a)) and (2)

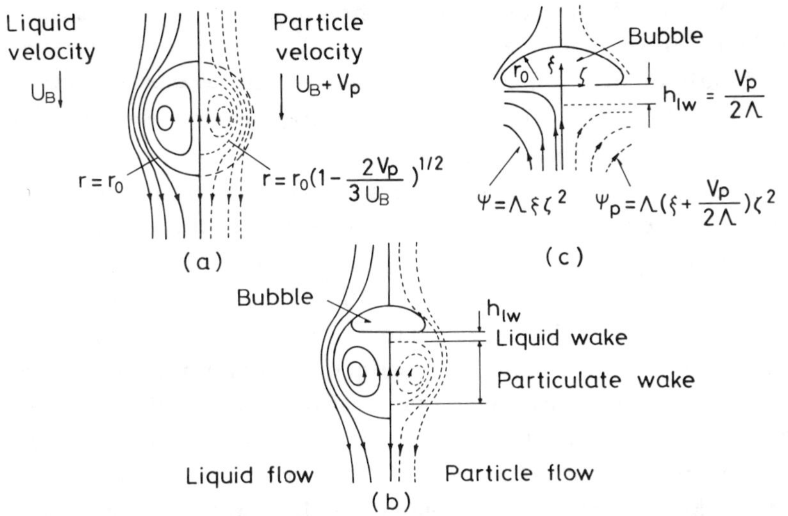

Figure 2.29 Liquid and particle steady flow patterns near a bubble in a three-phase fluidized bed based on: (a) Hill's spherical vortex; (b) actual flow; and (c) flat-topped wake (from Darton and Harrison, 1976).

a flat-topped wake represented by an axisymmetric flow towards a stagnation point on the flat base (Fig. 2.29(c)). The actual flow should be distorted due to the presence of the bubble and take an intermediate flow pattern as shown in Fig. 2.29(b). The difference in the streamlines between liquid and particles was accounted for by introducing a downward velocity of the particles relative to the liquid, V_p, equal to the interstitial liquid velocity. This approach predicted the liquid vortex size to exceed the particle vortex size. Darton and Harrison (1976) postulated that this difference in the radius of circulation region should give an estimate of the height of the liquid wake ($h_{\ell w}$). For both cases (1) and (2), they obtained the following expression

$$\frac{h_{\ell w}}{r_0} = \frac{U_\ell}{3\epsilon_\ell U_B} \quad \text{(three-dimensional)} \tag{2.59}$$

where r_0 is the radius of the sphere completing wake.

Darton and Harrison (1976) compared the above theoretical estimation in a corresponding two-dimensional case with data from photographs of bubbles and wakes in two-dimensional beds. They found that the theory satisfactorily predicted the liquid wake size for spherical-cap bubbles greater than about 0.5 cm^3, but for small non-cap shaped bubbles it underestimated the size. However, the increase in $k_{\ell w}$ with decreasing bubble size and increasing particle terminal velocity was reasonably demonstrated.

Direct measurements of the size of the stable liquid wake have been conducted only for single bubbles rising in two-dimensional liquid-solid fluidized beds. Rigby and Capes (1970) calculated the liquid wake volume normalized with respect to the bubble volume ($\underline{V}_{\ell w}/\underline{V}_b$) based on photographs from two-dimensional beds of 550 and 775 μm sand particles fluidized with different liquid velocities. For the calculation they assumed that a two-dimensional bubble with its wake has the same cross-sectional shape as that in a three-dimensional bed. The reported results are given in Fig. 2.30 as a plot of $\underline{V}_{\ell w}/\underline{V}_b$ versus bubble volume (or estimated bubble breadth). As shown, the relative proportion of liquid wake decreased with increasing bubble size. For large bubbles (~3 cm^3) in beds of fine particles (~160-290 μm) Rigby and Capes (1970) observed from photographs almost no liquid wake.

2.2.2.4 Wake Solids Concentration and Particle Trajectories

The variation in solids concentrations between the wake and the surrounding medium and within the wake region itself results from the inability of the particles to follow exactly the flow paths of the liquid elements. When the liquid elements undergo rapid changes in flow direction as, for example, in the circulatory flow patterns within the vortices of the primary wake the larger inertial particles deviate from the liquid flow path. Note that the extent of deviation of the particle flow pattern from that of the liquid elements strongly depends on the particle size and density. The circulatory flow pattern brings about centrifugal forces which in turn tend to expel larger inertia particles out of the vortex toward the surrounding medium. Some particles, however, remain within the vortex and keep circulating because the surroundings are fully occupied with a higher concentration of particles and because the tangential velocity is much greater than the radial velocity of the particles. The particle (also liquid) flow direction is downward near the free shear layer and upward around the central axis of the wake. The latter upward flow of particles tends to penetrate very close to the bubble base. Nevertheless, there is a local pressure maximum along the plane

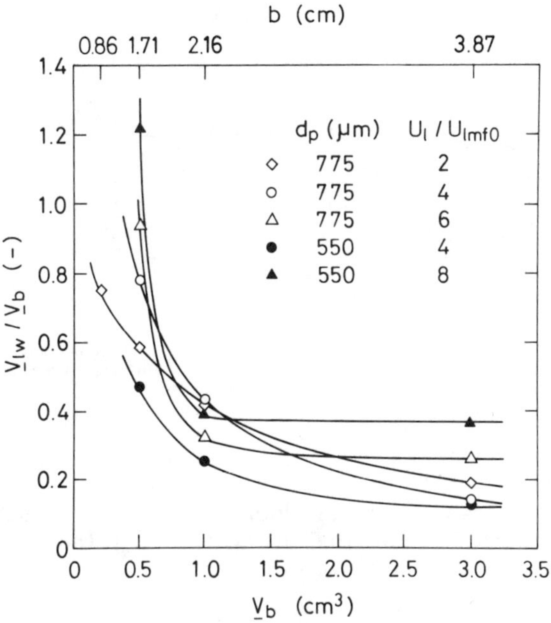

Figure 2.30 Variation of ratio of stable liquid wake volume to bubble volume for bubbles rising in water-sand particle fluidized beds (from Rigby and Capes, 1970).

parallel to but somewhat below the bubble base established by the liquid flow (Lazarek and Littman, 1974), which resists this particle penetration. The competing mechanism is schematically shown in Fig. 2.31. The upward particle inertia must overcome the pressure gradient, drag force, and gravitational force downward in order to invade into the stagnant layer of low particle concentration. This is a possible mechanism to account for the formation and maintenance of the stable liquid wake.

Based on the above physical description of local force fields around the near wake region, information about the wake structure can be revealed by analyzing particle concentration profiles and particle trajectories.

Wake Solids Concentration: Tsuchiya and Fan (1986) observed regions of low particle concentration under bubbles rising in a two-dimensional water-774 μm glass beads fluidized bed (see Fig. 2.23). Immediately beneath the bubble base, a particle-free region (stable liquid layer) existed whose size and shape were almost invariant with respect to time for large bubbles (b $\approx$ 3.5 cm). Low solids

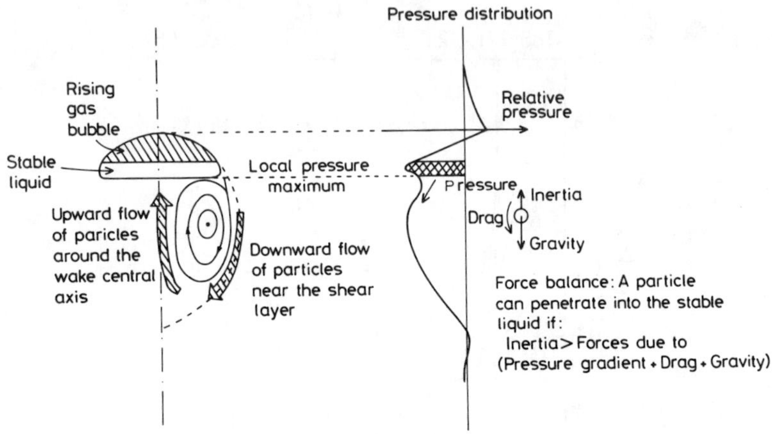

Figure 2.31 A competing mechanism for the penetration of the particles flowing upwards along the wake central axis into the stable liquid layer.

concentration regions were also observed along the vortex sheets generated from the edges of the bubble. Note that this part of the particle-free region seems to stay near the vortex centers and continues descending. The size of the low concentration region keeps expanding since the vortex core size increases with time due to viscous action and the solids concentration in this region keeps increasing since the centrifugal force field weakens (viscous decay); eventually this region becomes indistinguishable from the liquid-solid fluidized region.

Quantitative measurements of the local solids concentration/holdup in the bubble wake area have been recently conducted by Kitano and Fan (1988) using an optical fiber probe system for single bubbles in a two-dimensional liquid-solid fluidized bed. Based on the axial solids concentration distributions and the information obtained from the visual observation, Kitano and Fan (1988) subdivided the near-wake region into four regions: a stable liquid wake region (I), a stable solids wake region (II), a vortex sheet region (III), and a fluctuating solids wake region (IV). The stable liquid wake region occurs immediately beneath the bubble base and has negligible solids concentration; note this region may not be observed under certain flow and particle conditions. The stable solids wake region (II) follows the liquid wake and includes part of the growing vortices and the turbulent region around the wake centerline; this region is made up of a liquid-solid emulsion and is partitioned from region IV by a relatively low solids concentration layer. The fluctuating solids wake region is distinguished by streamwise flow from the far wake which interacts with the external flow field. The vortex sheet region (III), which has a low particle concentration, resides outside the stable and fluctuating solids wake regions. The size of each wake region varies periodically due to vortex shedding; the fluctuating solids wake region varies its size remarkably.

Kitano and Fan (1988) computed the average solids holdup within the primary wake region based on measurement of the local wake areas and solids concentrations in different locations using an optical fiber probe system. Based on comprehensive analysis of the local solids holdup distribution in the wake region, the average solids holdup in this region was found to be accurately represented by the solids holdup averaged over the wake center axis. The average solids holdups in the wake region (ϵ_{sw}), expressed in terms of relative value to that in the surrounding liquid-solid fluidized region (ϵ_{sf}), i.e., $(\epsilon_{sw}-\epsilon_{sf})/\epsilon_{sf}$ are presented in Fig. 2.32 for various particles. Denoting x as

$$x = \frac{\epsilon_{sw}}{\epsilon_{sf}} \tag{2.60}$$

the figure shows that, at small ϵ_{sf}, x can be significantly greater than unity. This indicates that the assumption made in all the other existing literature work regarding the wake models (see Section 2.3) which states the solids holdup in the wake is always equal to or less than that in the liquid-solid fluidized region, may not always be valid. Also given in the figure is the state of bubble wake observed for each of the particles employed. The figure reveals that x was not appreciably influenced by the bubble size but was strongly affected by the solids holdup in the liquid-solid fluidized region. It is seen that x decreases sharply as ϵ_{sf} increases.

Kitano and Fan (1988) found that ϵ_{sw} was nearly independent of Re_b with an average value of 0.42 and almost unaffected by the type of particles used. This behavior yields a simple correlation expression for x of

$$x = 0.42/\epsilon_{sf} \quad , \quad 0.25 < \epsilon_{sf} < 0.52 \tag{2.61}$$

Comparisons of the calculated (x-1) based on Eqn. (2.61) (solid curve) with the experimental data are shown in Fig. 2.32. Considering the

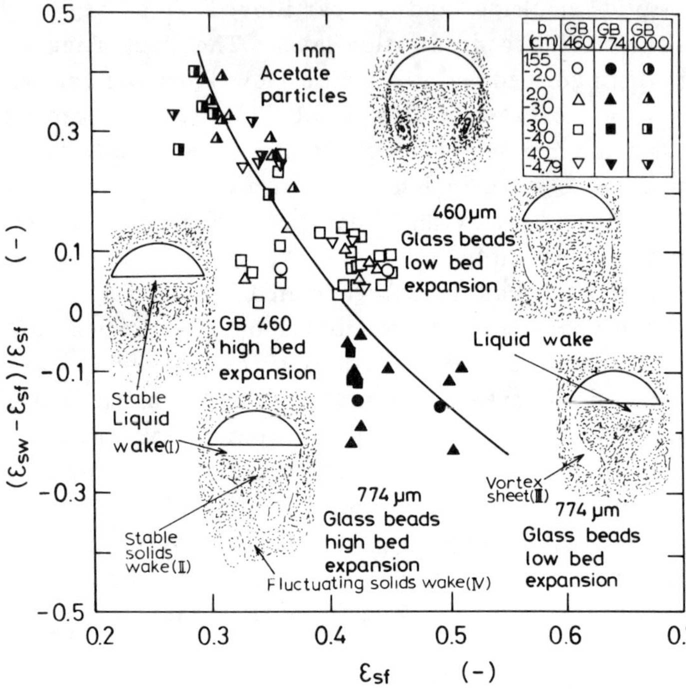

Figure 2.32 Relative deviation in local average solids holdup in the wake of a single nitrogen bubble from that in the liquid (water)-solid fluidized region (from Kitano and Fan, 1988).

non-steady nature of the wake, the agreement of Eqn. (2.61) with the data is quite satisfactory.

In three-dimensional systems, the stability of the wake region is lower than in a two-dimensional system, and consequently, the solids holdup profile in the wake shows greater variability. Kreischer et al. (1988), using the same optical fiber probe technique as Kitano and Fan (1988), found that the shape of the solids holdup profile behind a single large bubble depends significantly on the average solids holdup in the liquid-solid fluidized region (ϵ_{sf}). At high ϵ_{sf}, the wake behavior seems to be damped and the concentration differences between the wake and the liquid-solid fluidized region are limited to only a small region beneath the rising bubble. However, at low ϵ_{sf}, the wake effect can extend to a very great depth beneath the bubble; both local regions of high and low solids holdup can be found. In multi-bubble systems, the depth beneath the bubble to which the wake

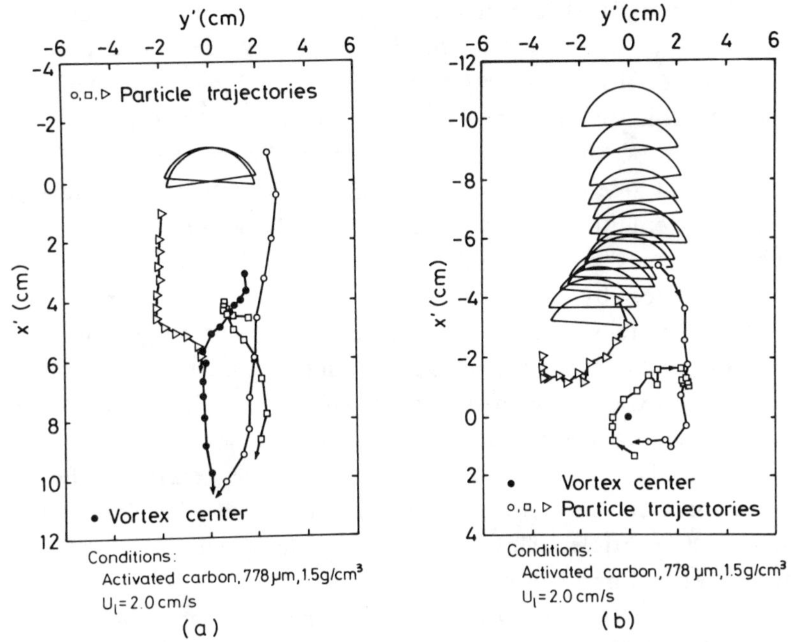

Figure 2.33 Particle trajectories around the near wake region in a water-778 μm activated carbon particle fluidized bed: observer moving with (a) bubble and (b) vortex center (from Tsuchiya and Fan, 1986).

effect extends is expected to be reduced drastically. Quantitatively, the average relative solids holdup (x) is near one for a high ϵ_{sf} and, in general, decreases with a decrease in ϵ_{sf}. Locally, however, the relative holdup can exceed unity to a significant degree which was also evident in the two-dimensional studies of Kitano and Fan (1988). For small bubble systems, the concentration difference in the wake should be less pronounced.

Particle Trajectories: Particle trajectories can provide an estimate of the average residence time of solids in the primary wake region and consequently, insight into solids mixing and other hydrodynamic phenomena.

Tsuchiya and Fan (1986) studied the particle trajectories around a single bubble rising through a two-dimensional liquid-solid fluidized bed. They used colored particles as the tracers of the fluidized solids. Figure 2.33 gives an example of a set of particle trajectories. Spherical activated carbon particles of 778 μm were used as the bed material. Three types of trajectories were viewed from an observer moving with the center point of the bubble base as shown in Fig. 2.33(a). In this example, the bubble had a circular-cap shape and rose in a rectilinear path with rocking. Since vortices play an influential role in the particle flow patterns, the trajectory of the vortex center of interest is also shown (solid circles). The particle, represented by open circles, was originally located in the external flow region and kept descending almost vertically without any appreciable influence from the vortex. The open-squared particle was, however, greatly affected by the presence of the vortex in its neighborhood: first the particle was trapped by the vortex and stayed momentarily stationary with respect to the bubble motion and then it moved down with the vortex. The open-triangled particle, originally located on the opposite side, continued descending until it was influenced by the induced circular flow due to the vortex. Here details of the particle flow patterns in the very near wake region were masked due to the presence of particles flowing between the front wall of the two-dimensional bed and the bubble.

Figure 2.33(b) shows the trajectories of the same particles as those in Fig. 2.33(a) when the observer is moving with the center of the vortex of interest. The vortex center, represented by a black dot, is fixed at the origin of the coordinates. In this frame of reference, the bubble appears to ascend relative to the vortex center. The trajectories of the three particles are as shown in the figure. Note the circulating pattern of the squared particle around the vortex center and that this particle is being expelled out of the vortical region partly due to the centrifugal force and partly because of the expanding nature of the vortex core.

2.2.2.5 Wake Velocity Profiles

Unlike the case of steady wakes, the velocity distribution for unsteady wakes has to be obtained at each instant during the course of its variation. Tsuchiya and Fan (1986) obtained instantaneous flow fields of solid particles flowing with liquid around single bubbles. Figure 2.34 shows a typical flow field represented by particle streaks which are the resultant of 1/60 s of exposure time in the video tape. A typical circulation pattern about Vortex 2 represented by a solid circle was seen. Also seen was the cross flow of the external streams from right to left between Vortex 1 and Vortex 2. This cross flow divides the near wake region from the far wake region. The regions of horizontal bands, designated by A and B, are specified to

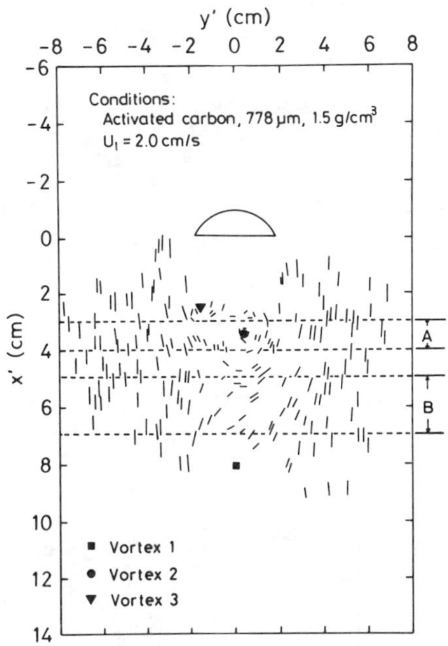

Figure 2.34 Instantaneous particle flow field around a circular-cap bubble rising in a water-778 μm activated carbon particle fluidized bed, represented by particle streaks during an exposure time of 1/60 s (from Tsuchiya and Fan, 1986).

obtain the instantaneous velocity distributions of particles along the horizontal coordinate. Region A covers the vortex center while Region B covers the cross flow region.

Figure 2.35(a) shows a plot of the vertical components of the instantaneous velocities versus the horizontal distance. The relative location of the bubble is also given in the figure. The constant velocity region corresponds to the external flow field and has about the same velocity as the absolute rise velocity of the bubble. The distinct feature in the distribution occurs between Regions A and B. The velocity distribution in Region A characterizes the rotational vortical motion, which is a typical distribution expected around the vortex center. On the other hand, the distribution in Region B is symmetric along the bubble center line, which characterizes the cross flow.

Figure 2.35(b) gives a plot of the angles of the velocity vectors whose tangent is defined as the ratio of horizontal velocity component

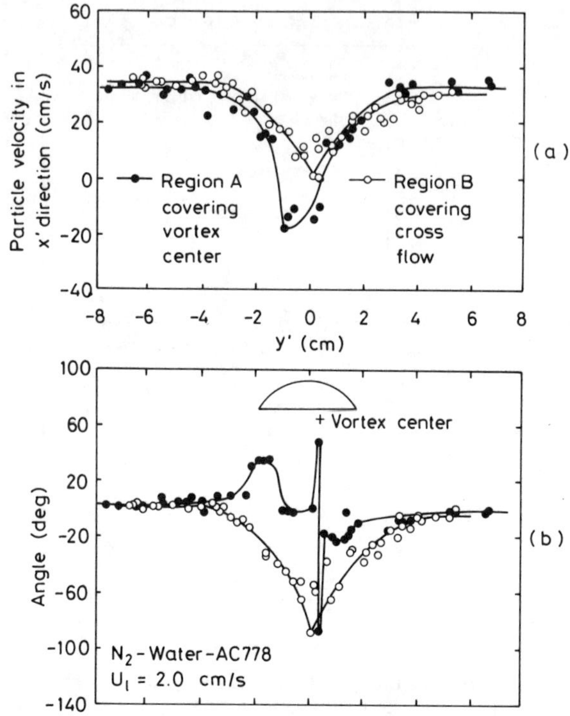

Figure 2.35 Instantaneous particle velocity profiles in the regions across the wake in terms of (a) vertical velocity component and (b) direction of velocity vector (from Tsuchiya and Fan, 1986).

to the vertical velocity component. For a no bubble situation, the angle is around zero degrees. The disturbance caused by the rising bubble changes the direction of the local velocity vectors leading to the angle profiles shown in the figure. Notice the symmetric nature of the angle distribution in Region B again showing the characteristic feature of the cross flow.

2.3 FLOW MODELS FOR MODE E-I-a-1

The development of flow models is essential to the quantitative description of the transport properties in three-phase fluidized beds. The wake concept discussed thus far has provided a fundamental framework for a global treatment of the bed behavior. It is important to note, however, that in the wake concept adopted by most of the flow models, e.g., generalized wake model (Bhatia and Epstein, 1974), indirect approaches are used in determining the average size of the bubble wake. In these approaches, a dimensionless bubble wake size parameter, k, is employed, which is defined as the ratio of the wake volume fraction to the bubble volume fraction, i.e.,

$$k = \frac{\epsilon_w}{\epsilon_g} \tag{2.62}$$

When solid particles are considered not to be present in the bubble wake, the models are referred to as the solid-free wake models. In the solid-free wake models, the parameter k_0 is employed in place of k. From the physical viewpoint, k corresponds to k_{pw} in three dimensions whereas k_0 corresponds to $k_{\ell w}$ (see Section 2.2.2.3). k or k_0 and x (see Eqn. (2.60)) are two key parameters in the wake models. Since k or k_0 as well as x are evaluated in conjunction with the flow models employed, they should be treated, in a strict sense, as empirical or, at best, semi-empirical model parameters and should only be applied jointly with the specific flow models that evaluate them. In the following, details of the wake models including k, k_0 and x along with other important models are given to account for such bed behavior as bed contraction, overall phase holdup, axial solids holdup distribution in the freeboard, liquid velocity profiles and particle terminal velocity.

2.3.1 Generalized Wake Model

The influence of gas bubble wakes on bed expansion (or solids holdup) and liquid holdup behavior has been satisfactorily explained by wake models. In models based on the wake concept, or simple wake models, a common assumption is that the bed can be subdivided into a gas bubble region, a wake region, and a liquid-solid (fluidized) region, as illustrated in Fig. 2.36. The sum of the volume of each of these regions must equal the total volume, or in terms of volume fractions:

$$\epsilon_g + \epsilon_{\ell\text{-}s} + \epsilon_w = 1 \tag{2.63}$$

Basic equations for the generalized wake model developed by Bhatia and Epstein (1974) have proved widely applicable for design purposes. Several basic assumptions are inherent in the model development: first, the solids content in the wake can be an arbitrary

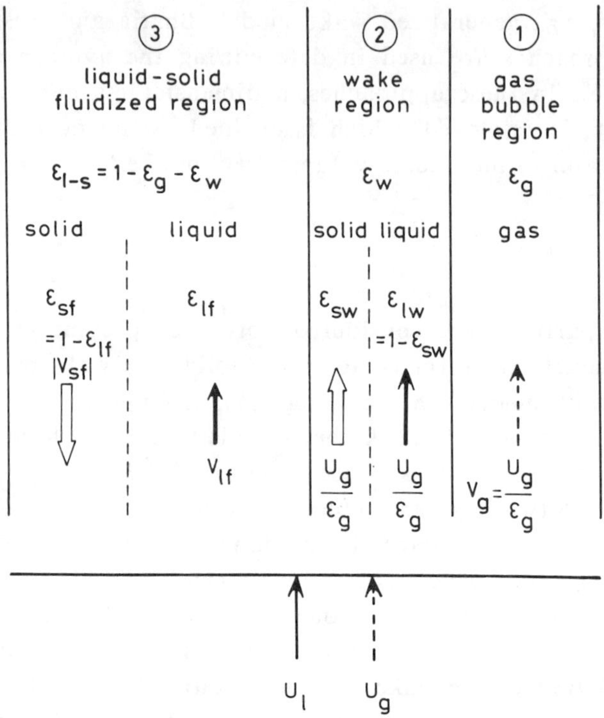

Figure 2.36 The generalized wake model for the Mode E-I-a-1 fluidized bed (from Bhatia and Epstein, 1974).

value different from that in the liquid-solid fluidized region; second, the wake rises at the same velocity as that of the bubble; and third, the Richardson-Zaki correlation (see Eqn. (2.21)) between solids holdup and liquid velocity applies in the liquid-solid fluidized region. The model describes the overall liquid holdup and bed porosity in terms of two key parameters.

The key parameters in this model are x (Eqn. (2.60)) and k (Eqn. (2.62)). The bubble velocity (U_b or V_g) can be related to the superficial gas velocity (see Eqn. (2.23)). Referring to Fig. 2.36, a mass balance on the solids yield

$$V_g \epsilon_w \epsilon_{sw} + V_{sf}(1 - \epsilon_g - \epsilon_w)\epsilon_{sf} = 0 \tag{2.64}$$

where V_{sf} is the linear solids velocity in the liquid-solid fluidized region. Solving Eqn. (2.64) for the mean velocity of solids in the liquid-solid fluidized region gives

$$V_{sf} = - \frac{V_g \ \epsilon_w x}{1 - \epsilon_g - \epsilon_w} \tag{2.65}$$

Likewise, an overall mass balance on the liquid yields

$$U_\ell = \epsilon_w V_g (1 - \epsilon_{sw}) + \epsilon_{\ell f} V_{\ell f} (1 - \epsilon_g - \epsilon_w) \tag{2.66}$$

and solving for the linear liquid velocity in the liquid-solid fluidized region gives

$$V_{\ell f} = \frac{U_\ell - \epsilon_w V_g (1 - x(1 - \epsilon_{\ell f}))}{\epsilon_{\ell f} (1 - \epsilon_g - \epsilon_w)} \tag{2.67}$$

The linear liquid and solid velocities in the liquid-solid fluidized region can then be related in terms of a Richardson-Zaki (1954) type expression by

$$V_{\ell s} = V_{\ell f} - V_{sf} = U_i (\epsilon_{\ell f})^{n-1} \tag{2.68}$$

where U_i is the extrapolated superficial liquid velocity at $\epsilon_{\ell f} = 1$ and the values for n are given in Table 2.2.

The desired relationship for the overall liquid holdup can be determined from the liquid holdup in the wake and liquid-solid fluidized regions by

$$\epsilon_\ell = \epsilon_w (1 - \epsilon_{sw}) + \epsilon_{\ell f} (1 - \epsilon_g - \epsilon_w) \tag{2.69}$$

Combining Eqns. (2.23), (2.60), (2.62), (2.65), (2.67), (2.68), and (2.69) gives analytical expressions for the liquid holdup:

$$\epsilon_\ell = \left[\frac{U_\ell - U_g \ k(1-x)}{U_i (1 - \epsilon_g - k\epsilon_g)}\right]^{1/n} [1 - \epsilon_g (1+k(1-x))] + \epsilon_g k(1-x) \tag{2.70}$$

and for the bed porosity:

$$1 - \epsilon_s = (\epsilon_g + \epsilon_\ell) = \left[\frac{U_\ell - U_g k(1-x)}{U_i \ (1 - \epsilon_g - k\epsilon_g)}\right]^{1/n} [1 - \epsilon_g (1+k(1-x))] + \epsilon_g (1+k(1-x)) \tag{2.71}$$

in terms of the the gas holdup, x, and k.

The solid-free wake model proposed by Efremov and Vakhrushev (1970) may be readily derived from the generalized wake model by setting x equal to zero and replacing the general k with k_0. In this model, the key parameter is the ratio of the solid-free wake holdup to the gas holdup, k_0. The resulting equations for the liquid holdup and bed porosity are

$$\epsilon_\ell = \left[\frac{U_\ell - U_g k_0}{U_i}\right]^{1/n} (1 - \epsilon_g - k_0 \epsilon_g)^{1-1/n} + k_0 \epsilon_g \tag{2.72}$$

and

$$1 - \epsilon_s = \left[\frac{U_\ell - U_g k_0}{U_i}\right]^{1/n} (1-\epsilon_g - k_0 \epsilon_g)^{1-1/n} + (1+k_0)\epsilon_g \tag{2.73}$$

The two key parameters defined within the model are evaluated empirically or, in some cases, through theoretical developments based on several physically justifiable arguments. In order to estimate the liquid holdup or the bed expansion by the generalized wake model k and x need to be known; however, one is unable to evaluate k and x independently from the holdup data. In the solid-free wake model x is set to zero and k_0 must be known to estimate the holdup.

2.3.1.1 Ratio of Wake Volume to Bubble Volume

Some correlations for k or k_0 proposed are summarized in Table 2.4. Efremov and Vakhrushev (1970), who showed that k_0 decreased with increasing U_g/U_ℓ, presented the first correlation for k_0. The correlation, however, produced a large deviation in the prediction of bed porosity except at low gas velocities (Muroyama, 1976). Darton and Harrison (1975) found that k_0 becomes negligible at $U_\ell/U_g < 0.4$, under which the flow pattern changes to a highly turbulent mode in the bubbling regime, as suggested by Schuman et al. (1965). El-Temtamy (1974), Muroyama (1976), and recently Chern et al. (1984) also proposed empirical correlations for k_0. According to these correlations, k_0 increases with U_ℓ, decreases with U_g, and is a weak function of particle properties. Baker et al. (1977), investigating the effects of gas and liquid velocities, liquid viscosity and surface tension on k_0, found that k_0 decreased with gas velocity steeply at low gas velocities, but gradually at high gas velocities; k_0 increased with liquid velocity; k_0 decreased with liquid viscosity for low viscosity liquids, but went through a minimum with continued increases in liquid viscosity; and k_0 was relatively insensitive to surface tension.

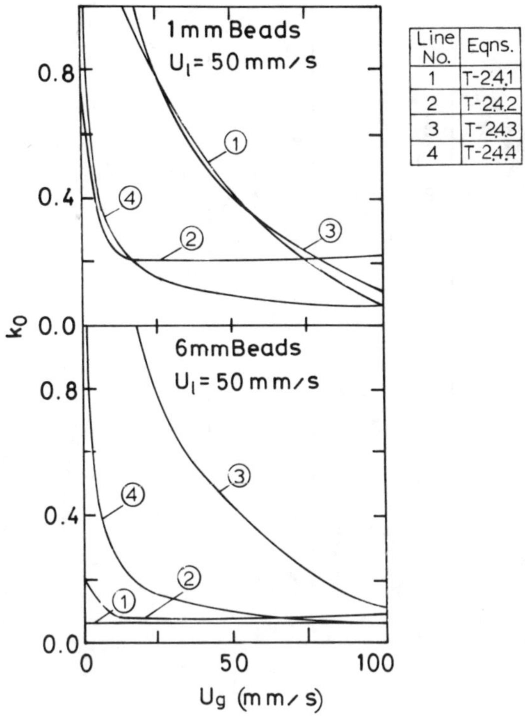

Figure 2.37 Comparison of the different correlations for k_0 (from Baker et al., 1977).

They proposed an empirical correlation for k_0 and also for the bed porosity relationship in the liquid-solid fluidized bed in place of the Richardson-Zaki equation commonly employed in wake models. Figure 2.37 compares the values of k_0 calculated from various correlations for the solid-free wake model. For reference, the values of k used by the generalized wake model (Curve 2, Bhatia and Epstein, 1974) are also shown in the figure. Note a wide discrepancy among different correlations. The discrepancy reflects the difference in indirect evaluation of the wake volume through the wake models. The correlations proposed by Bhatia and Epstein (1974) (Eqn. (T-2.4.2)), and by Baker et al. (1977) (Eqn. (T-2.4.4)), however, show general agreement.

Darton and Harrison (1975) showed that k_0, for single bubble systems observed in a two-dimensional column by Rigby and Capes (1970), decreased with an increase in the bubble volume. Recently, Darton (1985) found that k_0 increased with a decrease in mean bubble diameter.

Table 2.4 Correlations of Phase Holdups Based on the Wake Models (after Muroyama and Fan, 1985)

Investigators	Correlation Equation for k (in SI Units)	Assumption for x	Liquid-Solid Interrelation and Gas-Liquid Interrelation (in SI Units)	Remarks
Ostergaard (1965)	$k = 14.0\,\varepsilon_g^{-0.5}[U_\ell - U_{\ell mf0}]$	x = 1 without solid circulation	Richardson-Zaki equation $\frac{U_g}{\varepsilon_g} = 0.0052 - 0.046\ \ln U_g + U_\ell$	air/water system $U_g = 0 \sim 2$ cm/s $U_\ell = 1 \sim 3$ cm/s
Efremov and Vakhrushev (1970)	$k_0 = 5.1(\varepsilon_\ell)^{4.85}_{U_g=0}\{1-\tanh[\frac{40U_g(\varepsilon_\ell)^{10}_{U_g=0}}{U_\ell} - 3.32(\varepsilon_\ell)^{5.45}_{U_g=0}]\}$ (T-2.4.1)	x = 0	Richardson-Zaki equation $\varepsilon_g = (\varepsilon_g)_{g-\ell}\{\frac{\rho_\ell}{(\varepsilon_\ell)_{U_g=0}\rho_\ell + [1-(\varepsilon_\ell)_{U_g=0}]\rho_s}\}^{1.22}$ $(\varepsilon_g)_{g-\ell}$ = gas holdup in gas-solid system	air/water system $d_p = 0.32 \sim 2.15$ mm $U_g = 0 \sim 7$ cm/s U_ℓ = around 11 cm/s
Rigby and Capes (1970)	Graphical description	x = 0 or x = 1	Richardson-Zaki equation Bubble velocity was estimated by Rigby et al. (1970)	air/water system d_p = 0.29, 0.47 and 0.775 mm U_g = around 2.0 cm/s $U_\ell/U_{\ell mf0} = 2 \sim 8$
Bhatia and Epstein (1974)	$k = k_0'(1 - \varepsilon_s)^3$ According to El-Temtamy and Epstein (1978) $k_0' = 0.61 + \frac{0.037}{\varepsilon_g + 0.013}$ (T-2.4.2)	$0 \leq x \leq 1$ adjustable parameter	Richardson-Zaki equation large columns: $\frac{U_g}{\varepsilon_g} - \frac{U_\ell}{\varepsilon_\ell} = U_{b\infty} + 2U_g$ small columns: $\frac{U_g}{\varepsilon_g} - \frac{U_\ell}{\varepsilon_\ell} = \frac{U_{b\infty}\sqrt{\tanh(0.25\,\varepsilon_g^{-1/3})}}{\varepsilon_\ell}$ (bubble flows) $\frac{U_g}{\varepsilon_g} - \frac{U_\ell}{\varepsilon_\ell} = \frac{0.2(U_g+U_\ell)+0.35\sqrt{gD_c}}{\varepsilon_\ell}$ (slug flows)	air/water air/aqueous glycerol air/aqueous polyethylene glycol $d_p = 0.25 \sim 3$ mm $\rho_p = 2.5 \sim 11$ g/cm^3 $\mu_\ell = 1 \sim 63$ cP $U_g = 0 \sim 20$ cm/s $U_\ell = 0 \sim 39$ cm/s
Darton and Harrison (1975)	$k_0 = 1.4(\frac{U_\ell}{U_g})^{0.33} - 1,$ for $\frac{U_\ell}{U_g} \geq 0.4$ $k_0 = 0$ for $\frac{U_\ell}{U_g} < 0.4$ (T-2.4.3)	x = 0	Richardson-Zaki equation $U_g(1-\varepsilon_g) - U_\ell\varepsilon_g\frac{(1-\varepsilon_g)}{\varepsilon_\ell} = f(\varepsilon_g, U_{b\infty})$ $f(\varepsilon_g, U_{b\infty}) = 0.18\,\varepsilon_g$ (uniform bubbling regime) Graphical presentation (churn-turbulent regime)	air/water $d_p = 0.32 \sim 6$ mm $U_\ell/U_g = 0.4 \sim 20$

Muroyama (1976)	$k_0=1.72\ U_g^{-0.66}U_\ell$	$x = 0$	Richardson-Zaki equation $\varepsilon_g=0.3405\ U_g^{2/3}\ U_\ell^{-0.2}$ (for coalesced bubble regime)	$d_p=0.215 \sim 6.9$ mm $\rho_s = 2.475 \sim 3.96$ g/cm^3
Baker et al. (1977)	$k_0= 0.01765\ (\frac{U_\ell}{U_g})^{0.61}\ \sigma^{-0.654}$ (T-2.4.4)	$x = 0$	$\varepsilon_{\ell f} = 0.631\ U_{\ell f}^{0.38} d_p^{-0.285} \gamma_\ell^{0.115}$ γ_ℓ = generalized viscosity constant Measured bubble velocity	air/water air/acetone aq. soln. air/sugar aq. soln. air/carboxymethyl cellulose soln. $d_p = 1 \sim 6$ mm $U_g = 0 \sim 85$ cm/s U_ℓ = around 10 cm/s
El-Temtamy and Epstein (1978)	$k = k_1\ \exp(-5.08\ \varepsilon_g)$ k_1: relative wake holdup for a single bubble system	$x = 1-0.877y$ for $0 \leq y \leq 1.14$ $x = 0$ for $y > 1.14$ where $y = \frac{U_i}{\frac{U_g}{\varepsilon_g} - \frac{U_\ell}{\varepsilon_\ell}}$	Richardson-Zaki equation Measured ε_g	$\rho_s = 2.5 \sim 3.0$ g/cm^3 $d_p = 0.273 \sim 2.00$ mm $\mu_\ell = 1 \sim 6.3$ cP $0.07 \leq y \leq 1.08$
Khang et al. (1983)	$k=(1-\varepsilon_s)^3\{\exp(-1.2\ \varepsilon_g) + 2.5\ \exp(-32.8\ \varepsilon_g)\}$	$x \doteqdot 0.1$	A modified Richardson-Zaki equation New assumption for linear particle velocity in the wake region as $V_{sw} = [U_g/\varepsilon_g - U_t]$ $(1-\varepsilon_g)U_g+\varepsilon_g U_\ell = \varepsilon_g(0.2267\ \varepsilon_g+0.1044)$	Modification of generalized wake model
Chern et al. (1984)	$k_0 = 0.0631\ U_g^{-0.646} U_\ell^{0.246}$ (T-2.4.5)	virtually $x = 0$	See table 2.1.	See Table 2.1.

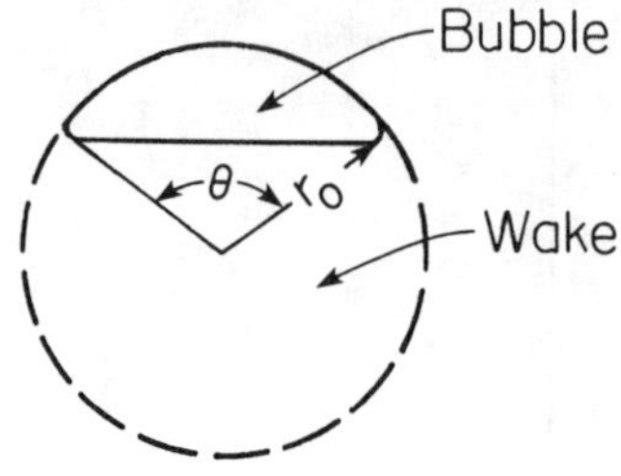

Figure 2.38 Geometry of sphere-completing bubble wake.

In terms of the generalized wake model k takes on a more theoretical look in terms of bubble geometry. Bhatia and Epstein (1974) presented a heuristic equation for k assuming that the wake behavior in the three-phase system is analogous to that in a liquid-liquid dispersion system (Letan and Kehat, 1968). El-Temtamy and Epstein (1977) obtained k by assuming that the effective volume of the bubble wake equals the sphere-completing volume of spherical-cap bubbles. Henriksen and Ostergaard (1974a) found the included angle θ of the spherical-cap bubble (see the definition in Fig. 2.38) to be a unique function of the kinematic viscosity of the medium surrounding the bubble. The apparent viscosity $\bar{\mu}_2$ of a liquid-solid fluidized bed can be estimated from correlation equations presented by several investigators, e.g., Hetzler and Williams (1969), who proposed the following equation (Henriksen and Ostergaard, 1974a):

$$\bar{\mu}_2 = 454\ \mu_\ell \nu_\ell^{\ 4/3}\ \frac{\sqrt{\rho_s}\,U_\ell}{d_p \epsilon_{\ell 2}}\ \exp(3.75/\epsilon_{\ell 2}) \qquad (2.74)$$

The average density of the liquid-solid fluidized bed, ρ_2, may be represented as the volume average of the mixture. Knowing $\bar{\mu}_2/\rho_2$, the included angle θ can be read from the figure presented by Henriksen and Ostergaard (1974a). k in this model can be calculated from the following correlation which was taken, by analogy, from the liquid-liquid study of Yeheskel and Kehat (1973):

$$k = k_1 \exp(-5.08\,\epsilon_g) \qquad (2.75)$$

The wake-bubble volume ratio k_1 of a single bubble can be determined from the geometry of the sphere-completing wake and is

$$k_1 = \frac{8 - \cos(3\theta/2) + 9\cos(\theta/2)}{8 + \cos(3\theta/2) - 9\cos(\theta/2)} \tag{2.76}$$

In summary, most correlations available in the literature to estimate the relative holdup in the wake region are purely empirical. The values of k or k_0 used in obtaining those correlations are not directly measured based on local fluid mechanic behavior of the bubble wake (see Section 2.2.2.3). Precautions should be taken whenever using these correlations. The trends of k_0 with respect to the operating conditions among various studies, however, are generally in agreement: k_0 increases with U_ℓ to the power of 0.2 ~ 1.0 and decreases with U_g to the power of 0.6 ~ 0.7 (see Table 2.4). Among the correlations presented in Table 2.4, Eqn. (T-2.4.2) (Bhatia and Epstein, 1974) or Eqn. (T-2.4.4) (Baker et al., 1977) is recommended.

2.3.1.2 Ratio of Solids Holdup in the Wake and the Liquid-Solid Region

In most cases, x is restricted to values between 0 and 1 representing only lower solid concentrations in the wake. This difference in concentration (lower concentrations in the wake region) between the two regions allows for the occurrence of the bed contraction phenomena.

x can be evaluated from holdup data. For solid-containing wakes, an empirical correlation for x was presented by El-Temtamy and Epstein (1977) as

$$x = 1 - 0.877\,\frac{U_i}{\dfrac{U_g}{\epsilon_g} - \dfrac{U_\ell}{\epsilon_\ell}} \quad \text{for} \quad \frac{U_i}{\dfrac{U_g}{\epsilon_g} - \dfrac{U_\ell}{\epsilon_\ell}} < 1.14 \tag{2.77}$$

$$x = 0 \quad \text{for} \quad \frac{U_i}{\dfrac{U_g}{\epsilon_g} - \dfrac{U_\ell}{\epsilon_\ell}} > 1.14 \tag{2.78}$$

The equation shows that x increases as the particle size or superficial liquid velocity decreases, and as the liquid viscosity or superficial gas velocity increases. Meaningful values of x can be obtained for particles smaller than about 1 mm with specific gravities less than 3 in water. For larger or heavier particles, U_ℓ/ϵ_ℓ in the denominator dominates at low gas velocities and x becomes zero except for viscous liquids. For water systems, Dhanuka and Stepanek (1978) indicated x is virtually zero for glass particles larger than 2 mm.

2.3.1.3 Extensions of the Generalized Wake Model

Stewart and Davidson (1964) suggested that bed contraction phenomena could be explained by the presence of solid-free wakes which allows some liquid flow to by-pass the liquid-solid fluidized region at a higher velocity. This by-pass of liquid reduces the liquid velocity in the liquid-solid fluidized region, and hence the bed contracts. According to Darton and Harrison (1975) bed contraction can be expressed in terms of the phase holdups as

$$\frac{\partial}{\partial U_g}(\epsilon_g + \epsilon_\ell)_{U_g \to 0} < 0 \tag{2.79}$$

On the basis of the solid-free wake model, mathematical descriptions for the criterion of contraction-expansion behavior were presented by Darton and Harrison (1975), Epstein and Nicks (1976), and Epstein (1976). El-Temtamy and Epstein (1979), based on the generalized wake model of Bhatia and Epstein (1974), developed a criterion for bed contraction which was corrected by Jean and Fan (1987a). Bed contraction will occur when Ψ, defined as

$$\Psi = \left(\frac{n}{n-1} + k\right)\frac{U_\ell}{\epsilon_\ell} + \frac{xk(U_g/\epsilon_g)}{n-1} - \left[(1+k)U_\ell + \frac{k}{n-1}\left(\frac{U_g}{\epsilon_g} - \frac{U_\ell}{\epsilon_\ell}\right)\right] + xk\left(U_\ell - \frac{U_\ell}{\epsilon_\ell}\right)\left(\frac{n}{n-1}\right) \tag{2.80}$$

is negative. Furthermore, based on the generalized wake model, a critical value of x (= x_c) exists above which expansion and below which contraction would be expected when gas is introduced (Jean and Fan, 1987a):

$$x_c = 1 - \frac{\dfrac{U_\ell}{\epsilon_\ell}\left\{\epsilon_\ell + \frac{1}{k}[n-(n-1)\epsilon_\ell]\right\}}{nU_\ell(1-1/\epsilon_\ell) + \dfrac{U_g}{\epsilon_g}} \tag{2.81}$$

The results of these investigations indicate that a large k, a high bubble velocity, or a high liquid viscosity favors bed contraction while either large particle size or high particle density favors bed expansion. These indications are qualitatively consistent with the experimental observation of holdup behavior by Kim et al. (1975).

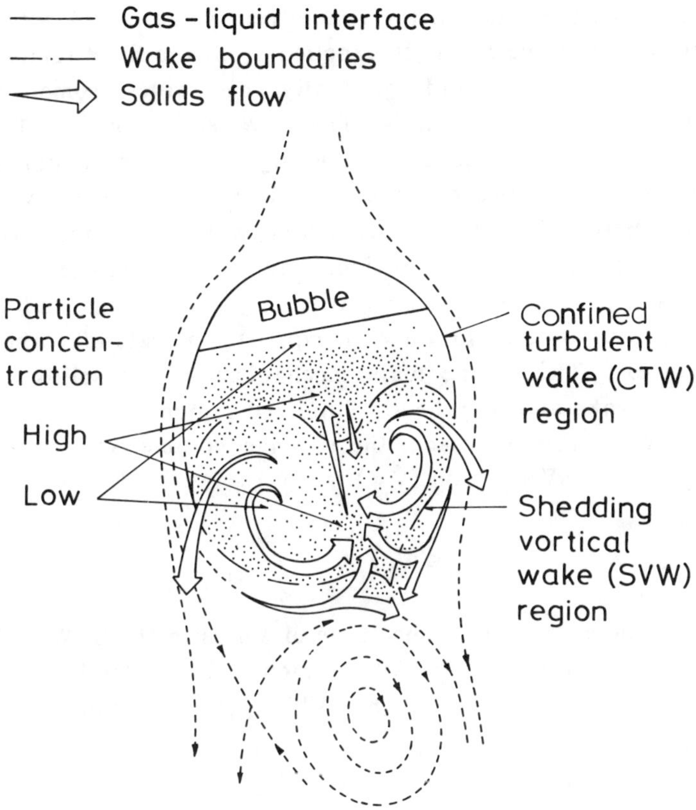

Figure 2.39 Schematic representation of postulated primary wake structure (from Tsuchiya and Fan, 1986).

2.3.2 Structural Wake Model

Based on the physical description of the specific details of wake structure presented in Section 2.2.2, the generalized wake model (Bhatia and Epstein, 1974) can be modified to account for the non-isolated nature of the solids in the wake in a three-phase fluidized bed. Fan et al. (1986b), in proposing the structural wake model, assumed that the primary wake is responsible for hydrodynamic phenomena such as bed contraction and solids mixing. They subdivided the primary wake into (1) the confined turbulent wake region (CTW) and (2) the shedding vortical wake region (SVW) as shown in Fig. 2.39. The CTW region includes the stable liquid layer (see Section 2.2.2.3) and the chaotic turbulent region about the wake central axis. It was assumed an isolated region and treated in the same manner as the wake in the generalized wake model, although in reality it collects some particles expelled from the SVW region due to the centrifugal force and gradually discharges some particles downward along the central axis. The SVW region consists of vortices or vortical blobs on diametrically opposing sides of the bubble and the spouting region in between. It constantly sheds a certain volume of solids asymmetrically into the secondary wake and thus, eventually into the liquid-solid fluidized region. Fan et al. (1986b) presumed that this shedding process was dynamically steady, or cyclic, and thus the rate of shedding of the old vortex is equal to the rate of formation of a new vortex. This is the main mechanism of the solids exchange between the wake and liquid-solid fluidized region.

Based on this approach, the average volumetric solids upward flow rate via the SVW region ($\dot{\underline{V}}_{sv}$) is

$$\dot{\underline{V}}_{sv} = V_g \bar{t}_r A \, \delta_{sv} f \, \epsilon_{ssv} \tag{2.82}$$

where $\bar{t}_r$ is the average residence time of solids in the SVW region, δ_{sv} is the volume fraction of the SVW region, f is the bubble frequency, and ϵ_{ssv} is the solids holdup in the SVW region. The volumetric solids upward flow rate via the CTW region ($\dot{\underline{V}}_{ct}$) is

$$\dot{\underline{V}}_{ct} = \underline{V}_{ct} \, f \, \epsilon_{sct} \tag{2.83}$$

where $\underline{V}_{ct}$ is the volume of the CTW region and ϵ_{sct} is the solids holdup in the CTW region. The bubble frequency is

$$f = \frac{V_g \, A \, \epsilon_g}{\underline{V}_b} \tag{2.84}$$

where $\underline{V}_b$ is the volume of a bubble. Combining Eqns. (2.83) and (2.84) gives

$$\dot{\underline{V}}_{ct} = V_g \, A \, \delta_{ct} \, \epsilon_{sct} \tag{2.85}$$

The total solids upward flow rate becomes

$$\dot{\underline{V}}_t = \dot{\underline{V}}_{ct} + \dot{\underline{V}}_{sv} \tag{2.86}$$

The upflow of solids in the bubble wake is offset by a downward flow in the liquid-solid region given by

$$\underline{\dot{V}}_t = - V_{sf} A(1 - \epsilon_g - \epsilon_w)\, \epsilon_{sf} \tag{2.87}$$

where V_{sf} is the average linear solids velocity in the liquid-solid region. Note that when

$$\bar{t}_r \geq t^* = \frac{\underline{V}_{sv}}{V_g\, \delta_{sv}\, A} \tag{2.88}$$

where $\underline{V}_{sv}$ is the volume of the SVW region, the upward flow of solids in the SVW region (Eqn. (2.82)) can be reduced to a form similar to that for solids flow in the CTW region. That is, the solids flow rate is no longer affected by the vortex shedding frequency and the solid flow rate via the wake can be treated by assuming isolated wake behavior.

Similarly, the mass balance on the liquid phase is

$$U_\ell = \epsilon_{\ell f} V_{\ell f}(1 - \epsilon_g - \epsilon_w) + \delta_{ct} \epsilon_{\ell ct} V_g + V_g\, \bar{t}_r\, \delta_{sv} f\, \epsilon_{\ell sv} \tag{2.89}$$

The following relations hold:

$$\epsilon_w = \delta_{ct} + \delta_{sv} \tag{2.90}$$

$$\epsilon_{\ell f} + \epsilon_{sf} = 1 \tag{2.91}$$

$$\epsilon_{\ell ct} + \epsilon_{sct} = 1 \tag{2.92}$$

$$\epsilon_{\ell sv} + \epsilon_{ssv} = 1 \tag{2.93}$$

Combining Eqns. (2.82), (2.85) - (2.87) and (2.89) - (2.93) with the Richardson-Zaki relationship yields

$$\epsilon_\ell = \left[\frac{U_\ell - U_g \left[\frac{k_4(1 - \bar{k}_1)}{(1 + k_3)} + \frac{k_3 k_4 (1 - k_2)}{(1 + k_3)} \bar{t}_r\, f \right]}{U_i\,(1 - \epsilon_g - k_4 \epsilon_g)} \right]^{1/n} (1 - \epsilon_g - k_4 \epsilon_g + \frac{\bar{k}_1 k_4 + k_2 k_3 k_4}{(1 + k_3)}\, \epsilon_g) + k_4 \left(\frac{1 - \bar{k}_1}{1 + k_3}\right) \epsilon_g + k_3 k_4 \left(\frac{1 - k_2}{1 + k_3}\right) \epsilon_g \tag{2.94}$$

where

$$\bar{k}_1 = \frac{\epsilon_{sct}}{\epsilon_{sf}}, \quad k_2 = \frac{\epsilon_{ssv}}{\epsilon_{sf}}, \quad k_3 = \frac{\delta_{sv}}{\delta_{ct}}, \quad k_4 = \frac{\delta_{ct} + \delta_{sv}}{\epsilon_g}$$

When Eqn. (2.88) is satisfied and ϵ_{ssv} is equal to ϵ_{sct}, the following relationships hold:

$$\bar{k}_1 = k_2 = x \tag{2.95a}$$

$$k_3 = 0 \tag{2.95b}$$

$$k_4 = k \tag{2.95c}$$

$$\bar{t}_r f = 1 \tag{2.95d}$$

Equation (2.94) can then be reduced to Eqn. (2.70) of the generalized wake model.

Examination of $\bar{t}_r$ and t^* for a single rising bubble indicates that they are of the same order of magnitude. More often than not Eqn. (2.88) is satisfied indicating that the general wake model can be used in spite of inherent solids exchange due to vortex shedding taking place in the wake. Moreover, quantitative information such as $\underline{V}_{ct}$, $\underline{V}_{sv}$, ϵ_{sct}, ϵ_{ssv}, and $\bar{t}_r$ is required to allow general correlations to be developed. The structural wake model developed by Fan et al. (1986b) can be refined by considering other solids mixing mechanisms such as time variant wake behavior. An earlier attempt to account for the non-isolated behavior of the bubble wake was made by Khang et al. (1983) by assuming that the slip velocity between the solid particles and the liquid in the wake region was equal to the particle terminal velocity. Experimental findings (e.g., Fig. 2.35), however, show that the slip velocity distributes widely and cannot be solely represented by the particle terminal velocity.

As with the generalized wake model, the structural wake model can be used to determine when bed contraction occurs. Bed contraction will occur when Ψ', defined as

$$\Psi' = -\frac{a + \bar{b}\,\bar{t}_r f}{n} V_g + (1 + a + \bar{b})V_\ell + \left[\frac{1 + k_4}{n} - (1 + k_4 - c)\right]U_\ell \tag{2.96}$$

where

$$a = \frac{k_4(1 - \bar{k}_1)}{(1 + k_3)}, \quad \bar{b} = \frac{k_3 k_4(1 - k_2)}{(1 + k_3)}, \quad c = \frac{\bar{k}_1 k_4 + k_2 k_3 k_4}{(1 + k_3)}$$

is negative.

2.3.3 Stagewise Partition Process Model

Consideration of the bubble wake structure has led to a mechanistic approach to the determination of the axial solids holdup in the freeboard. As indicated earlier, Page and Harrison (1974) suggested that wake formation and shedding were the main mechanisms for particle entrainment and de-entrainment in the transitional region of the freeboard. Consequently, the wake shedding frequency plays a prominent role in the axial variation of solids holdup.

El-Temtamy and Epstein (1980b), in developing a stagewise partition process model for solids entrainment in the freeboard region, postulated that in each stage the solid particles are carried upward in the bubble wake, transferred to the surrounding liquid-solid region, and settled in the liquid-solid region. The height of a stage in the model is assumed to be equivalent to a wake-shedding unit, defined as the distance travelled by a bubble during the time interval between the shedding of two successive wake fragments. The model agrees well with the local solids holdup data for the freeboard region. The stages are numbered from the top of the freeboard region to the surface of the fluidized bed, so that stage N is located between stage N-1 above and stage N + 1 below. A solid balance around stage N gives

$$(1 - \epsilon_g - k\epsilon_g)(-V_{sf}\epsilon_{sf})_{N-1} + U_g kx(\epsilon_{sf})_{N+1}$$
$$= U_g\, kx(\epsilon_{sf})_N + (1-\epsilon_g - k\epsilon_g)(-V_{sf}\epsilon_{sf})_N \tag{2.97}$$

Noting that the solid velocity in the liquid-solid fluidized region, V_{sf}, is assumed, after Lapidus and Elgin (1957), to be the vectorial difference between the liquid velocity in this region and the velocity of the liquid relative to the solids in the same region, $(V_{sf})_N$ can be expressed by

$$(V_{sf})_N = \frac{U_\ell - U_g k[1 - x(\epsilon_{sf})_N]}{(1 - \epsilon_g - k\epsilon_g)[1-(\epsilon_{sf})_N]} - U_i[1-(\epsilon_{sf})_N]^{n-1} \tag{2.98}$$

The solid holdup for stage N, obtained by summing up the solids contained in both the wake and liquid-solid fluidized regions, is

$$(\epsilon_s)_N = k\epsilon_g x(\epsilon_{sf})_N + (1-\epsilon_g-k\epsilon_g)(\epsilon_{sf})_N \tag{2.99}$$

Solving Eqns. (2.97), (2.98) and (2.99) gives solutions for ϵ_s in each of the stages. The height of a stage, or wake shedding unit, is

$$HWU = \frac{U_g/\epsilon_g}{f_{v2}} = \left(\frac{U_g/\epsilon_g}{U_g/\epsilon_g - U_\ell/\epsilon_\ell}\right)\left(\frac{D_b}{Sr}\right) \tag{2.100}$$

where f_{v2} is the wake shedding frequency, D_b is the effective bubble diameter, and the Strouhal number is defined based on the effective bubble diameter as

$$Sr = \left(\frac{D_b f_{v2}}{U_g/\epsilon_g - U_\ell/\epsilon_\ell}\right) \tag{2.101}$$

El-Temtamy and Epstein recommend using a Sr of 0.27 for air bubbles in water for bubble Reynolds numbers between 2200 and 6000. In the calculation of stage 1, it is necessary to assume a small, but non-zero, solids holdup for $(\epsilon_s)_1$ and zero for the term $(-V_{sf}\epsilon_{sf})_0$. Proceeding downwards, $(\epsilon_s)_N$ is calculated for each stage until it equals or exceeds the solids holdup in the main fluidized bed region.

2.3.4 Circulating Flow Model

The radial variation of gas holdup and/or liquid velocity was noted by Rigby et al. (1970), Morooka et al. (1982) and Hu et al. (1986) among others. Morooka et al. (1982) found that the liquid phase flow pattern in a three-phase fluidized bed is similar to that in a bubble column where the liquid flows upward in the central region and downward in the peripheral region. They proposed a circulating flow model, which is similar to that of Ueyama and Miyauchi (1979) for a bubble column, to simulate the lateral distributions of gas holdup and interstitial liquid velocity. The steady state equation of motion for a gas-liquid-solid fluidized bed can be given by

$$-\frac{1}{r}\frac{d}{dr}(r\tau) = \frac{dP}{dz} + (1-\epsilon_g)\rho_e g \tag{2.102}$$

where ρ_e is the effective density of the liquid-solid fluidized mixture. The shear stress, τ, is given by

$$\tau = -(\nu_\ell + \nu_T)\,\rho_e \frac{dV_\ell}{dr} \tag{2.103}$$

where ν_ℓ and ν_T are the liquid kinematic viscosity and the turbulent kinematic viscosity, respectively. Several assumptions are made to solve the above equations: $(\nu_\ell + \nu_T)$ is constant in the turbulent region, the lateral distribution of gas holdup can be approximated by

$$\epsilon_g(\phi) = 2\bar{\epsilon}_g(1 - \phi^2) \tag{2.104}$$

where ϕ is the normalized radial coordinate, the composition of the liquid-solid fluidized mixture is radially uniform, and the bed properties can be represented by

$$\epsilon_\ell(\phi)/(\epsilon_\ell(\phi) + \epsilon_s(\phi)) = \bar{\epsilon}_\ell/(\bar{\epsilon}_\ell + \bar{\epsilon}_s) \tag{2.105}$$

$$\rho_e = (\rho_\ell\bar{\epsilon}_\ell + \rho_s\bar{\epsilon}_s)/(\bar{\epsilon}_\ell + \bar{\epsilon}_s) \tag{2.106}$$

The solution of Eqn. (2.102) with Eqns. (2.103) and (2.104) was obtained by Ueyama and Miyauchi (1979) for a bubble column (ϵ_s = 0). From the solution, the recirculating liquid velocity in the gas-liquid-solid fluidized bed, derived with the aid of Eqns. (2.105) and (2.106), is

$$V_\ell - V_{\ell w} = \frac{1}{32}\frac{gD_c^2\,\bar{\epsilon}_g}{\nu_\ell + \nu_T}(1 - \phi^2)^2 \tag{2.107}$$

$$V_{\ell w} = -\frac{1}{192}\frac{gD_c^2\,\bar{\epsilon}_g}{\nu_\ell + \nu_T}\frac{2 - 3\bar{\epsilon}_g}{1 - \bar{\epsilon}_g} + \frac{U_\ell}{\bar{\epsilon}_\ell} \tag{2.108}$$

where ν_ℓ is usually negligible compared to ν_T in the three-phase fluidized bed and ν_T can be estimated by the following correlation (in SI units):

$$\nu_T = 3.1\times10^{-3}(1 + 2\times10^5 d_p^{\,2.0})\,U_g^{\,0.42} \tag{2.109}$$

for $d_p \le 2.2$ mm and $4 < U_g < 15$ cm/s (Morooka et al., 1982).

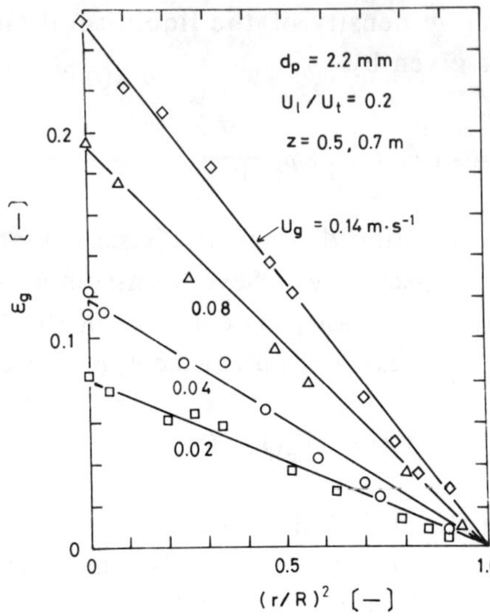

Figure 2.40 Radial distributions of gas holdup in an air-water-glass bead fluidized bed (from Morooka et al., 1982).

The gas holdup radial distributions are well correlated with Eqn. (2.104), as shown in Fig. 2.40. The liquid velocity radial distributions for various particles are illustrated in Fig. 2.41. The liquid velocity is a maximum at the axis and a minimum near the wall. The liquid velocity at the wall, $V_{\ell w}$, is not necessarily negative in a three-phase fluidized bed in contrast to the case of a bubble column under liquid-batch conditions (see Chapter 4).

2.3.5 Particle Terminal Velocity Model

The prevailing bubble behavior also implicitly affects the particle terminal velocity in a gas-liquid medium. Jean and Fan (1987b) suggested that, in addition to the usual buoyancy and drag forces on the particle, an interaction force must be considered which accounts for the effects of the bubble-particle interactions.

A force balance around a single particle in a gas-liquid medium at terminal velocity conditions can be expressed as

$$F_G - F_B = F_{DL} + F_{BWP} \tag{2.110}$$

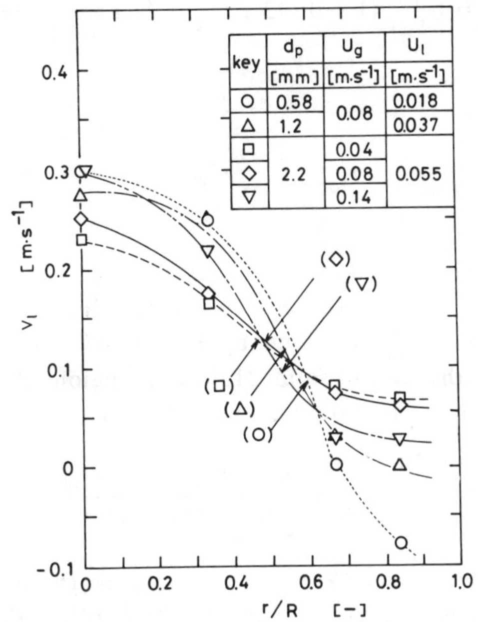

Figure 2.41 Radial distributions of liquid velocity in an air-water-glass bead fluidized bed (from Morooka et al., 1982).

where F_G is the gravity force, F_B is the buoyancy force, F_{DL} is the drag force exerted by the liquid phase, and F_{BWP} is the bubble-wake and particle interaction force.

The effective weight of the particle can be expressed as

$$F_G - F_B = \underline{V}_p(\rho_s - \rho_\ell)g \tag{2.111}$$

where $\underline{V}_p$ is the particle volume.

The drag force acting on a particle exerted by the liquid (F_{DL}) can be expressed in terms of the relative velocity between the liquid and solid as

$$F_{DL} = \frac{3}{4} C_{DL} \frac{\rho_\ell}{d_p} V_\ell^2 \underline{V}_p \tag{2.112}$$

where V_ℓ is the relative linear velocity between the liquid and solid and C_{DL} is the drag coefficient of the particle in a liquid medium. C_{DL} can be determined from the following correlation of Turton and Levenspiel (1986):

$$C_{DL} = \frac{24}{Re_{\ell t}} (1 + 0.173\, Re_{\ell t}^{0.657}) + \frac{0.413}{1+16300 Re_{\ell t}^{-1.09}} \tag{2.113}$$

where

$$Re_{\ell t} = \frac{V_\ell d_p \rho_\ell}{\mu_\ell}$$

Based on the generalized wake model concept of Bhatia and Epstein (1974), the linear velocity of the liquid in the liquid-solid fluidized region ($V_{\ell f}$) is given by Eqn. (2.67). For the limiting case where $\epsilon_{\ell f}$ in the liquid-solid fluidized region of a gas-liquid-solid system approaches one, Eqn. (2.67) reduces to

$$V_{\ell f} = V_\ell = \frac{U_\ell - V_g\, \epsilon_W}{1 - \epsilon_g - \epsilon_W} \tag{2.114}$$

The linear gas velocity (V_g) is equivalent to the absolute bubble rise velocity (U_b) and, assuming Eqn. (2.23) applies, Eqn. (2.114) can be expressed as

$$V_\ell = \frac{U_\ell - U_g\, k}{1 - (1+k)\, \epsilon_g} \tag{2.115}$$

where k is given by Eqn. (2.62).

Following the expression for the impaction force of particles with the stationary surface of a spherical shape developed by Soo (1967), the interaction force between the bubble-wake and a particle which is governed by impaction can be described by

$$F_{BWP} = \frac{3}{2} C_{DW} \frac{k \epsilon_g \rho_\ell}{d_p} V_g^{\,2}\, \underline{V}_p \tag{2.116}$$

Combining Eqns. (2.110), (2.111), (2.112), (2.115), and (2.116) yields

$$\frac{3}{4} C_{DL} \frac{\rho_\ell}{d_p} \left[\frac{U_\ell - U_g k}{1 - (1+k)\, \epsilon_g}\right]^2 + \frac{3}{2} C_{DW} \frac{k \epsilon_g \rho_\ell}{d_p} \left(\frac{U_g}{\epsilon_g}\right)^2 - (\rho_s - \rho_\ell) g = 0 \tag{2.117}$$

Note that U_ℓ in Eqn. (2.117) corresponds to the terminal velocity of a particle (U_t') in a gas-liquid medium. In addition, as U_g approaches zero, U_ℓ in Eqn. (2.117) approaches U_t.

Correlations (in SI units) for C_{DW} and ϵ_g proposed by Jean and Fan (1987b) for a 7.62 cm ID column are given in the following:

$$C_{DW} = \frac{1}{1+1.147\ Re_t^{-2.876}\left(\frac{d_p}{D_c}\right)^{-4.0}} \tag{2.118}$$

for $0 < U_g \leq 5.19$ cm/s and $3.27 \leq U_\ell \leq 32.18$ cm/s

$$\epsilon_g = 0.1552\ U_g^{0.988}(-0.0221\ U_\ell^2 - 56.4\ U_\ell + 28.821) \tag{2.119a}$$

and for $8.63 \leq U_g \leq 12.08$ cm/s and $3.27 \leq U_\ell \leq 32.18$ cm/s

$$\epsilon_g = \frac{1.404\ U_g^{0.34}}{\exp(-0.248\ R_{LG}) + \exp(0.243\ R_{LG})} \tag{2.119b}$$

where

$$R_{LG} = \frac{5.058\ U_\ell}{U_g^{0.648}} - 2.03$$

k (= k_0) in Eqn. (2.117) can be estimated by the correlation given by Eqn. (T-2.4.5).

Figure 2.42 shows the variation in U_t' with U_g over a wide range of particle sizes. The terminal velocity decreases with an increase in the superficial gas velocity for particles larger than 460 μm. For particles smaller than or equal to 460 μm, on the other hand, the terminal velocity is fairly constant with respect to the gas velocity. The prediction of the model based on Eqn. (2.117) with C_{DL}, C_{DW}, and ϵ_g, evaluated by Eqns. (2.113), (2.118), and (2.119a,b), respectively, agrees well with experimental data.

2.3.6 Drift Flux Model

In accounting for the interaction between two phases in gas-liquid or liquid-solid systems, the interaction is usually assumed to depend upon the relative motion between the phases rather than upon the absolute velocity of each phase. Under this premise, two approaches have been used to describe the phase interaction: one correlates the relative, or slip, velocity as an unique function of the phase holdups (e.g., Richardson and Zaki, 1954) and the other correlates the drift velocity or drift flux as a function of phase

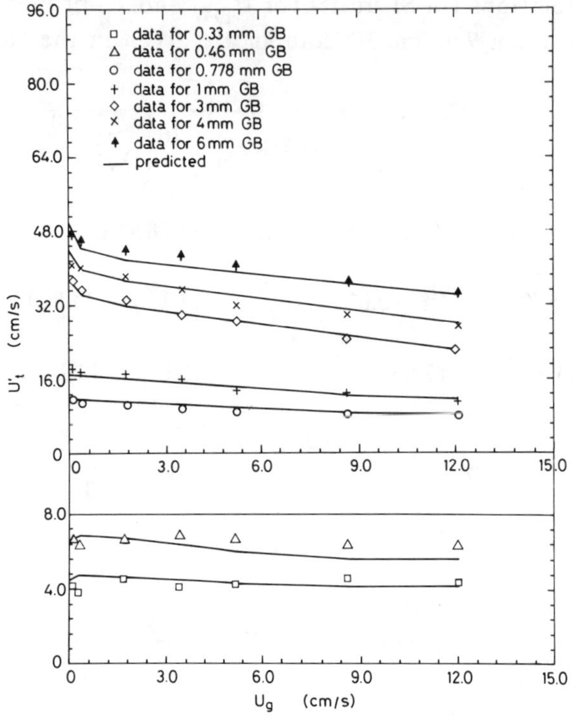

Figure 2.42 Effect of gas velocity on the particle terminal velocity in an air-water medium (from Jean and Fan, 1987b).

holdups (Wallis, 1969). While these two approaches have similarities and are interconvertible, a major conceptual difference exists: the relative velocity refers to the slower moving phase velocity and the drift velocity refers to a volumetric mean velocity of all phases.

Chen and Fan (1988c) reported that the drift velocity concept for two-phase systems is directly extendable to the analysis of bubble motion in a three-phase fluidized bed in the dispersed bubble regime. Following the analysis of gas-liquid systems proposed by Nicklin (1962), they considered a three-phase fluidized bed system, as shown in Fig. 2.43(a), with continuous flows of gas and liquid in which bubbles rise with a velocity of U_b. At a given instant, the gas and liquid flows are stopped simultaneously. The system changes to the configuration given in Fig. 2.43(b) with the lower front of the bubble swarm rising with a velocity of U_{bd}. Without continuous flows of gas and liquid the solids below the bubble front are in a denser condition with a mean solids holdup of ϵ_{s0}. Since the overall volumetric mean

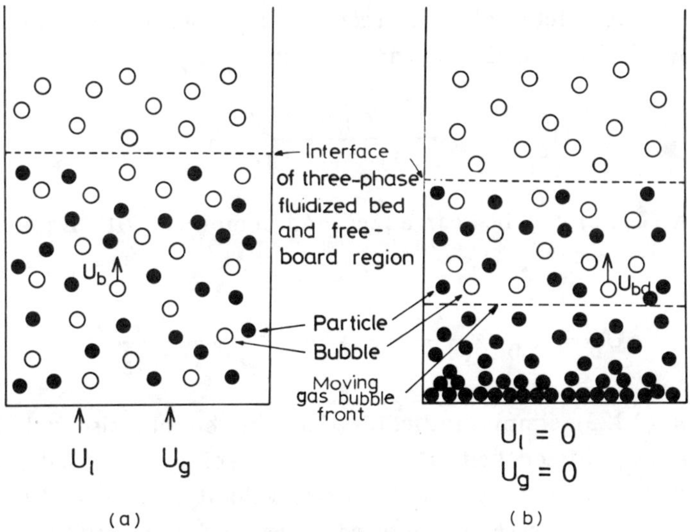

Figure 2.43 (a) Configuration of bubbles rising with continuous gas and liquid flows. (b) Configuration of bubbles rising after the gas and liquid flows are terminated (from Chen and Fan, 1988c).

velocity in Fig. 2.43(b) is zero, by definition, U_{bd} becomes the drift velocity of the bubbles in the three-phase region with a counterflow of liquid-solid mixture.

By superimposing a downward velocity U_b on Fig. 2.43(a) and U_{bd} on Fig. 2.43(b), the bubbles in both configurations become stationary. Thus the superficial liquid velocity in the case of Fig. 2.43(a) is

$$U_\ell^* = - U_b \epsilon_\ell + U_\ell \tag{2.120}$$

and in the case of Fig. 2.43(b) is

$$U_\ell^* = - U_{bd}(1 - \epsilon_{s0}) \tag{2.121}$$

where the negative sign indicates a downward flow. It is assumed that the phase holdups in the three-phase region in Fig. 2.43(b) are identical to those in Fig. 2.43(a). The assumption is reasonable in view of experimental evidence. It is assumed that the bubble motion depends on the holdups and the relative motion of the liquid phase and is independent of the relative motion of the solid phase. Equations (2.120) and (2.121) can be equated to yield

$$U_b = U_{bd} \frac{(1 - \epsilon_{s0})}{\epsilon_\ell} + \frac{U_\ell}{\epsilon_\ell} \tag{2.122}$$

Equation (2.122) can be rearranged and combined with Eqns. (2.2) and (2.23) to give

$$U_b = U_{bd} \frac{(1 - \epsilon_{s0})}{(1 - \epsilon_s)} + \frac{U_g}{(1 - \epsilon_s)} + \frac{U_\ell}{(1 - \epsilon_s)} \tag{2.123}$$

Equation (2.123) simply indicates that the bubble rise velocity in a three-phase fluidized bed is a sum of two solid-free based superficial velocities and a modified bubble drift velocity. The modification of the drift velocity is required to compensate for the difference in the solids holdup above and below the moving bubble front (see Fig. 2.43(b)). Similar to a two-phase system, U_{bd} is expected to depend on the phase holdups, but not on the bulk flows of the liquid and gas.

Equation (2.123) can also be deduced from the slip velocity approach mentioned earlier. Referring to Figs. 2.43(a) and 2.43(b), the slip velocity between the bubble and the liquid in the two systems are given, respectively, by Eqn. (2.51) and the following equation:

$$U_B = U_{bd} + \frac{U_\ell^{**}}{\epsilon_\ell} \tag{2.124}$$

where U_ℓ^{**} is the downward superficial liquid velocity in the three-phase region in Fig. 2.43(b).

A volumetric flow balance across the moving bubble front yields

$$U_\ell^{**} + U_s^{**} = U_{bd}\, \epsilon_g \tag{2.125}$$

where U_s^{**} is the downward superficial solids velocity. The ratio of the liquid to solids holdups below the bubble front equals the ratio of the volumetric flux of liquid to solid through the front; i.e.,

$$\frac{(1 - \epsilon_{s0})}{\epsilon_{s0}} = \frac{\left(\frac{U_\ell^{**}}{\epsilon_\ell} + U_{bd}\right)}{\left(\frac{U_s^{**}}{\epsilon_s} + U_{bd}\right)} \frac{\epsilon_\ell}{\epsilon_s} \tag{2.126}$$

Combining Eqns. (2.51) and (2.124) through (2.126) gives

$$U_b = U_{bd} \frac{(1 - \epsilon_{s0})}{\epsilon_\ell} + \frac{U_\ell}{\epsilon_\ell} \tag{2.127}$$

which is identical to Eqn. (2.122).

It is important to note that empirical equations with an expression very similar to Eqn. (2.123) have been proposed to account for phase holdup behavior by Chern et al. (1984) and Tarmy et al. (1984). Chern et al.'s (1984) correlation for air-water-particle systems expresses $U_{bd}(1 - \epsilon_{s0})$ as a function of U_g. Tarmy et al.'s (1984) correlation for coal liquefaction expresses $U_{bd}(1 - \epsilon_{s0})$ as a function of the holdups for three-phase fluidized bed systems, i.e.,

$$U_{bd} = U_{b\infty} \frac{(1-\delta_g)^m}{(1-\epsilon_{s0})} \tag{2.128}$$

where m is a correlation constant. Tarmy et al. (1984) operated the three-phase fluidized bed under high temperature and pressure conditions at which small bubbles are expected. Note that the drift flux, $j'_{g\ell}$, defined by Eqn. (2.28) was employed by them, which related to U_{bd} by

$$j'_{g\ell} = U_{bd}(1-\epsilon_{s0})\delta_g \tag{2.129}$$

Thus $U_{b\infty}$ in Eqn. (2.128) can be determined by

$$\left.\frac{dj'_{g\ell}}{d\delta_g}\right|_{\delta_g=0} = U_{b\infty} \tag{2.130}$$

Apparently, the drift flux described by Eqn. (2.123) can best be applied to the dispersed bubble regime.

In Nicklin's experiments (1962) for gas-liquid bubble columns, it was indicated that the drift flux approach could not be applied to high gas flow conditions where bubble coalescence becomes significant. This is due to the fact that the assumption of a uniform one-dimensional flow in the drift flux approach can no longer apply. At high gas flow conditions, bubbles tend to rise in the center core of the column yielding nonuniform flow distribution in the radial direction. In order to account for the nonuniformity under these conditions, Eqn. (2.123) must be modified by including a distribution coefficient, C_0, similar to that for gas-liquid bubble columns (Zuber and Findlay, 1965), as given by

$$U_b = C_o \frac{(U_g + U_\ell)}{(1 - \epsilon_s)} + U_{bd} \frac{(1 - \epsilon_{s0})}{(1 - \epsilon_s)} \tag{2.131}$$

The values for C_o in Eqn. (3.131) vary with prevailing flow regimes. For the dispersed bubble regime, C_o is taken as 1 whereas for coalesced bubble and slugging flow regimes, C_o is greater than 1. The values of C_o must be determined experimentally and the variation of C_o can be correlated with the flow regime transition.

2.3.7 Other Flow Models

Bhaga and Weber (1972a and 1972b) used the concept of relative velocity to develop equations describing the interrelationships among the velocities of the dispersed phases (gas and solid) and the continuous phase (liquid). This approach is in contrast to the concept of drift flux used by Wallis (1969). Bhaga and Weber defined, for each dispersed phase, a local relative velocity as the local phase velocity with respect to the local velocity of the continuous phase. Linear relationships between the individual phase velocities were formulated by assuming that the local relative velocity for the gas and solid can be expressed as a function of the terminal velocity of a single bubble or particle and the concentration of the dispersed phases. The theory was successful in explaining experimental data for individual phase holdups taken from a 3.8 cm ID column under relatively low gas flow rates (varying up to 2.5 cm/s) and low solids concentrations (varying up to only 6%). In their experiments, glass beads with diameters of 0.3 and 0.15 mm were suspended in water as a liquid-solid mixture flowing with relatively high velocity. Holdups were considered to be uniform axially. Their one-dimensional flow model, however, does not take into account the presence of the wake behind bubbles and the accompanying distribution of the liquid flow; hence it cannot predict initial bed contraction upon gas injection in a three-phase fluidized bed operation.

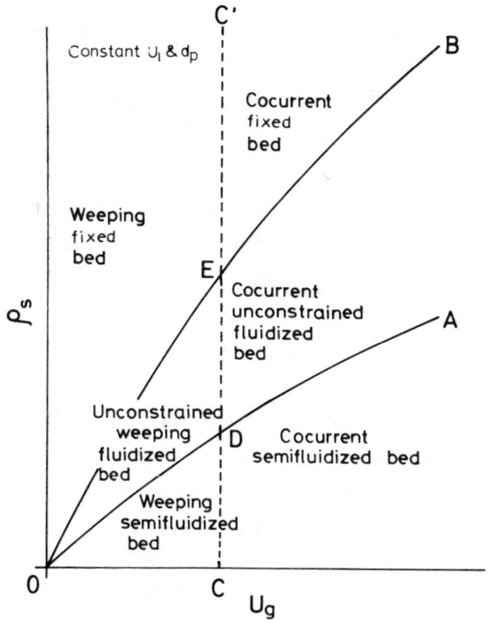

Figure 2.44 Flow regime map for the Mode E-I-b fluidized bed (from Beaver and Fan, 1984).

2.4 HYDRODYNAMIC BEHAVIOR OF MODE E-I-b

The operation of three-phase fluidization with a gas as the continuous phase is important in applications which require gas to liquid velocity ratios exceeding 100, as in the production of calcium bisulphite acid liquors (Volpicelli and Massimilla, 1970) or for polymerization (Jenkins et al., 1985) and granulation (Kono, 1980). Little fundamental information is available on this mode of fluidization (e.g., Beaver and Fan, 1984).

Beaver (1983) and Beaver and Fan (1984) studied the flow regimes, fluidized bed height, and solids holdup in a column of 10.2 cm I.D. and 40.6 cm height. (The top and the bottom grids were made of a wire mesh with an opening ratio of 0.398.) As the superficial gas velocity increased, at a constant liquid flow rate and particle size, six distinct flow regimes were observed. A qualitative flow regime map is presented as shown in Fig. 2.44. "Weeping" is characterized by a liquid weeping through the lower retaining grid into the inlet gas distributor section. This phenomenon occurs when the gas does not have a high enough velocity to prevent liquid from weeping through the lower retaining grid. With increasing gas velocity, weeping will eventually stop as designated by line CC′ in the figure. The weeping rate is also a function of the open area of the grid. When particle density is very high, fixed bed weeping occurs at low gas velocities. On top of the fixed bed a layer of frothy liquid builds up. As the gas velocity increases, the weeping stops and the flow pattern shifts to either a cocurrent fixed bed or a constrained weeping fluidized bed depending on the particle density and size. When the particle density is light to moderate, a weeping fixed bed, an unconstrained weeping fluidized bed, and a cocurrent unconstrained fluidized bed are encountered in order as the gas velocity increases. The unconstrained cocurrent fluidized bed consists of a fluidized bed immediately above the lower retaining grid and a slugging bed at the top of the entire fluidized bed. Intermediate between the two a void, or lean region, exists where particles jump up and fall down intermittently. As the gas velocity increases beyond a certain value, a packed bed will form underneath the top grid. Below the packed bed, a mobile bed exists where a great deal of churning and intermittent slugging occurs. Weeping semi-fluidized beds (also see Section 6.2) occur only for low density or small particles. A fluidized bed exists in the lower portion of the column while a packed bed is formed against the upper retaining grid, but the gas does not have a high enough velocity to prevent liquid from weeping through the lower retaining grid. The gas velocity at which weeping occurs is slightly dependent on liquid velocity, and independent of the particle diameter and density. The boundaries between the different fluidization regimes vary with respect to particle properties at a given liquid velocity. As the particle density or diameter increases, boundaries between the different flow regimes occur at lower superficial gas velocities.

2.5 CONCLUDING REMARKS

For Mode E-I-a-1, extensive information is available on bed properties such as pressure drop, pressure fluctuation, incipient fluidization, individual phase holdups, particle entrainment, effects of solids wettability, and bed rheology. The bed properties are predominantly dictated by the interactions between the rising gas bubble (and bubble wake) and the surrounding liquid-solid medium. Various correlations reflect the distinct differences in bed properties due to three identifiable flow regimes; these flow regimes are the dispersed bubble, coalesced bubble, and slugging.

For systems with large bubbles, the gas holdup is low and the bubble wake plays a prominent role in the bed hydrodynamics. The role of the bubble wake has been well recognized in the wake model analysis of bed contraction, phase holdups, and particle terminal velocities in a gas-liquid medium. The majority of the studies available in the literature treat wake properties as being adjustable parameters or as being an approximation of those for gas-liquid systems. Recent investigations into the wake structure reveals that the primary wake size, relative to that of the bubble, is a weak function of the bubble Reynolds number and the solids holdup in the primary wake, under limited conditions in a two-dimensional bed, is constant at about 0.42 independent of the bubble Reynolds number. Furthermore, the Strouhal number, and hence the wake shedding frequency, is independent of particle properties and is a unique function of the bubble Reynolds number only. Additional work is needed in fluid and particle mechanics to account for bubbles, bubble wakes, and their interaction.

For systems with small bubbles, the gas holdup is high and the effects of the bubble wake diminish while liquid drift effects due to rising bubbles become prominent. For modeling purposes, an assumption of a pseudo-homogeneous gas-liquid phase is appropriate. In general, studies performed have been limited to particles with high density (2.0 g/cm^3 or higher). Comprehensive studies are needed for systems with low density particles such as those used in biochemical operations (see Chapters 7 and 8) and systems with small bubbles such as those encountered in the hydrotreating of heavy residues (see Chapter 10).

For Mode E-I-b, the relative velocity between the gas and liquid must be high to prevent liquid weeping. As this mode of operation has gained increasing importance (see Chapter 11), more work is needed in understanding the role of the liquid phase in the already complicated gas-solid fluidized bed.

2.6 NOMENCLATURE

A : cross-sectional area of the column

A_0 : amplitude of wave form

A_c : upward acceleration of a particle

A_{co} : initial value of A_c

Ar : Archimedes number, $d_p^3\rho_\ell(\rho_s-\rho_\ell)g/\mu_\ell^2$

b : bubble breadth

C_D : drag coefficient for bubble-liquid interaction

C_{DL} : drag coefficient for the liquid-particle interaction

C_{DW} : bubble wake-particle interaction coefficient

C_o : distribution coefficient defined in Eqn. (2.131)

d_b : bubble size defined based on the vertical chord length passing through the bubble center

$\bar{d}_b$: mean size of d_b number distribution

d_e : diameter of a spherical particle of volume equivalent to that of the non-spherical particle

d_p : particle diameter

d_{vs} : Sauter, or volume-surface, mean bubble diameter

D_b : diameter of a sphere having the same volume as the bubble in three dimensions (3D) or diameter of a circle having the same area as the bubble in two dimensions (2D); $\bar{D}_b$, average bubble diameter in the bed

D_c : column diameter

f : bubble frequency in the structural wake model

f_v : individual vortex shedding frequency

f_{v2} : shedding frequency for two successive vortices

F_B : buoyancy force

F_{BWP} : bubble wake-particle interaction force

F_{DL} : drag force on a particle due to the liquid phase

F_G : gravitational force

Fr_g : gas Froude number, $U_g{}^2/gd_p$

Fr_ℓ : liquid Froude number, $U_\ell{}^2/gd_p$

g : gravitational acceleration

h : bubble height

h_d : height of the doughnut-shape bubble

$h_{\ell w}$: liquid wake height

H : effective height of bed expansion

H_0 : static bed height

HWU : height of wake-shedding unit

$j_{g\ell}$: drift flux of gas relative to the liquid flow

$j'_{g\ell}$: gas drift flux defined by Eqn. (2.28)

k : ratio of wake size (or volume) to bubble size (or volume) in the generalized wake model

k_0 : ratio of solid-free wake size to bubble size in the solid-free wake model

k'_0 : ratio of wake size to bubble size in a bubble column

k_1 : ratio of wake size to bubble size for a single bubble system

$k_{\ell w}$: ratio of liquid wake size to bubble size

k_{pw} : ratio of primary wake size to bubble size

K : fluid consistency index for power-law fluids

ℓ : depth at which particle penetrates into the bubble

ℓ_b : bubble chord length

$\bar{\ell}_b$: mean bubble chord length

m : empirical exponent

n : Richardson-Zaki index

$\bar{n}$: flow behavior index for power-law fluids

N : number of stages

$p(d_b,\ell_b)$: probability function relating $Z(\ell_b)$ to $r(d_b)$

P : pressure

$-\frac{dP}{dz}$: total axial pressure gradient

$-\Delta P_d$: dynamic pressure drop

$-\frac{dP_d}{dz}$: dynamic axial pressure gradient

$-\frac{dP_f}{dz}$: frictional axial pressure gradient

$Q(U_b)$: bubble rise velocity distribution

r : radial distance

$r(d_b)$: density function for bubble size distribution

r_o : radius of a sphere-completing wake or radius of curvature of a bubble frontal surface

R : column radius

R_{LG} : defined in Eqn. (2.119b)

$R(\xi)$: dimensionless bubble size distribution based on the number frequency of bubbles

Re_b : bubble Reynolds number based on bubble breadth, bU_B/ν_ℓ

Re_d : bubble Reynolds number, $D_bU_B\rho_2/\bar{\mu}_2$

Re_e : bubble Reynolds number based on bubble equivalent diameter, D_bU_B/ν_ℓ

Re_g : gas Reynolds number, $U_gd_p\rho_g/\mu_g$

Re_ℓ : liquid Reynolds number, $U_\ell d_p\rho_\ell/\mu_\ell$

$Re_{\ell t}$: Reynolds number, $V_\ell d_p\rho_\ell/\mu_\ell$

Re_{mf0} : Reynolds number at minimum fluidization, $d_eU_{\ell mf0}\rho_\ell/\mu_\ell$

Re_t : particle Reynolds number, $d_pU_t\rho_\ell/\mu_\ell$

Sr : Strouhal number for vortex shedding based on the effective bubble diameter, $f_{v2}D_b/U_B$

Sr_b : Strouhal number for vortex shedding based on bubble breadth, f_vb/U_B

t : time

t^* : critical residence time of solids in the SVW region

$\bar{t}_r$: average residence time of solids in the SVW region

U_b : bubble rise velocity relative to column, or absolute rise velocity

$\bar{U}_b$: mean bubble rise velocity relative to column

U_{bd} : bubble drift velocity defined in Fig. 2.43(b)

$U_{b\infty}$: single bubble rise velocity in an infinite liquid medium

U_B : bubble rise velocity relative to liquid, $U_b - U_\ell/\epsilon_\ell$

$\bar{U}_B$: mean bubble rise velocity relative to liquid

U_g : superficial gas velocity

U_i : extrapolated superficial liquid velocity as $\epsilon \rightarrow 1$ for a liquid-solid fluidized bed

U_ℓ : superficial liquid velocity

U_ℓ^* : superficial liquid velocity in a moving coordinate systems in which the bubble position is fixed

U_ℓ^{**} : downward superficial liquid velocity defined in the drift flux model

$U_{\ell mf}$: minimum fluidization velocity in a three-phase system

$(U_{\ell mf})_1$: minimum fluidization velocity of flotsam particles in a binary mixture

$(U_{\ell mf})_2$: minimum fluidization velocity of jetsam particles in a binary mixture

$(U_{\ell mf})_M$: minimum fluidization velocity of a binary mixture

$U_{\ell mf0}$: minimum fluidization velocity for a liquid-solid system

U'_ℓ : equivalent superficial liquid velocity in a three-phase fluidized bed

U_r : relative velocity between the colliding bubble and particle

U_s^{**} : downward superficial solids velocity defined in the drift flux model

U_t : particle terminal velocity in a liquid medium

U'_t : particle terminal velocity in a gas-liquid medium

U_{tr} : liquid velocity for transition from the coalesced to dispersed bubble regimes

$U_{\ell f}$: superficial liquid velocity in the liquid-solid fluidized region

V_g : linear gas velocity

V_{in} : velocity along inner boundary of a free shear layer

V_ℓ : linear liquid velocity

$V_{\ell f}$: linear liquid velocity in the liquid-solid fluidized region

$V_{\ell s}$: linear relative velocity between the liquid and solid phases in the liquid-solid fluidized region

$V_{\ell w}$: linear liquid velocity at the column wall

V_{out} : velocity along outer boundary of a free shear layer

V_p : slip velocity of particle relative to liquid

V_{sf} : average linear solids velocity in the liquid-solid fluidized region

$\underline{V}_b$: bubble volume for a single bubble or multiple bubbles

$\underline{V}_{ct}$: volume of the CTW region

$\underline{V}_{\ell w}$: liquid wake volume

$\underline{V}_p$: particle volume

$\underline{V}_{sv}$: volume of the SVW region

$\dot{\underline{V}}_{ct}$: volumetric solids flow rate in the CTW region in the structural wake model

$\underline{\dot{V}}_{sv}$: volumetric solids flow rate in the SVW region in the structural wake model

$\underline{\dot{V}}_{t}$: total upwards volumetric solids flow rate in the structural wake model

W : parameter defined by Eqn. (2.31)

We : Weber number, $\rho_s U_b^2 d_p / \sigma$

W_s : mass of solids in a fluidized bed

W_{SLV} : work of adhesion per unit interfacial area

x : ratio of solids concentration in the bubble wake region to that in the liquid-solid fluidized region

x′ : vertical downward distance from bubble base

x_2 : weight fraction of larger particles in a bed containing a binary mixture of particles

x_c : critical relative solids concentration in the bubble wake region for bed contraction

y : correlation parameter defined in Eqn. (T-2.1.13)

y′ : horizontal right-hand distance from the center of bubble base

z : axial distance

$Z(\ell_b)$: density function for bubble chord length distribution

<u>Greek Letters</u>

α : angle of attack

$\overline{\alpha}$: scale parameter in the log-normal distribution

α_e : ratio of electrical conductivity of the dispersed phase to that of the pure continuous phase in a two-phase system

β : $= (\alpha_e - 1)/(\alpha_e + 2)$

$\overline{\beta}$: shape parameter in the log-normal distribution

γ : shear rate

γ_ℓ : generalized liquid viscosity constant

Γ : circulation

δ_{ct} : volume fraction of the CTW region in the structural wake model

δ_g : solid-free gas volume fraction

δ_{sv}	: volume fraction of the SVW region in the structural wake model
ϵ	: bed porosity in a three-phase or two-phase fluidized bed
ϵ_d	: volume fraction of the dispersed phase in a two-phase system
ϵ_g	: gas holdup
$\overline{\epsilon}_g$	: cross-sectionally averaged value of gas holdup
ϵ_ℓ	: liquid holdup
$\overline{\epsilon}_\ell$	: cross-sectionally averaged value of liquid holdup
$\epsilon_{\ell 2}$	: liquid holdup in a liquid-solid fluidized bed
$\epsilon_{\ell ct}$	: liquid holdup in the CTW region in the structural wake model
$\epsilon_{\ell f}$	: liquid holdup in the liquid-solid fluidized region
$\epsilon_{\ell\text{-}s}$	: volume fraction of the liquid-solid fluidized region
$\epsilon_{\ell sv}$	: liquid holdup in the SVW region in the structural wake model
ϵ_s	: solids holdup
$\overline{\epsilon}_s$	: cross-sectionally averaged value of solid holdup
ϵ_s^*	: solids holdup in a packed bed
ϵ_{s0}	: solids holdup below the bubble front defined in Eqn. (2.121)
ϵ_{sct}	: solids holdup in the CTW region in the structural wake model
ϵ_{sf}	: solids holdup in the liquid-solid fluidized region
ϵ_{ssv}	: solids holdup in the SVW region in the structural wake model
ϵ_{sw}	: average solids holdup in primary wake region
ϵ_w	: wake volume fraction
ζ	: horizontal distance from the center of bubble base
η	: sinusoidal wave form at the shear layer
θ	: included angle of a spherical-cap bubble
θ_c	: contact angle
κ	: measure of strength of a free shear layer
λ	: wavelength
Λ	: strength of wake flow

Λ_e : effective electrical conductivity of a heterogeneous two-phase material

Λ_{ec} : electrical conductivity of the continuous phase in a two-phase system

Λ_{ed} : electrical conductivity of the dispersed phase in a two-phase system

μ_g : viscosity of gas

μ_ℓ : viscosity of liquid

$\underline{\mu}$: mean of the log-normal distribution

$\bar{\mu}_2$: apparent viscosity of a liquid-solid fluidized bed

$\bar{\mu}_3$: apparent viscosity of a gas-liquid-solid fluidized bed

ν_ℓ : kinematic viscosity of liquid

ν_T : turbulent kinematic viscosity

ξ : dimensionless bubble size $(d_b/\bar{d}_b)$ or vertical upward distance from the center of bubble base

ρ_2 : average (bulk) density of a liquid-solid fluidized bed

ρ_e : effective density of the liquid-solid mixture in a three-phase fluidized bed

ρ_f : effective density of the gas-liquid mixture in a three-phase fluidized bed

ρ_g : gas density

ρ_ℓ : liquid density

ρ_s : solid density

σ : surface tension of liquid

$\underline{\sigma}^2$: variance of the log-normal distribution

τ : shear stress

ϕ : normalized radial coordinate

ϕ_s : particle sphericity

ψ : stream function for liquid flow

ψ_p : stream function for particle flow

Ψ : function defined by Eqn. (2.80) for the bed contraction criteria

Ψ' : function defined by Eqn. (2.96) for the bed contraction criteria

ω : angular velocity, $(3g\rho_\ell/2d_p\rho_s)^{1/2}$

2.7 REFERENCES

Abuaf, N., O.C. Jones, Jr., and G.A. Zimmer, Rev. Sci. Instrum., 49, 1090 (1978).

Adamson, A.W., "Physical Chemistry of Surface," 4th ed., John Wiley (1982).

Adlington, D., and E. Thompson, in Proceedings of the 3rd European Symposium on Chemical Reaction Engineering, p. 203, Pergamon Press, Oxford (1965).

Armstrong, E.R., C.G.J. Baker, and M.A. Bergougnou, in "Fluidization Technology," ed. by D.L. Keairns, I, p. 405, Hemisphere (1976).

Azbel, D., "Two-Phase Flows in Chemical Engineering," Cambridge University Press (1981).

Baker, C.G.J., S.D. Kim, and M.A. Bergougnou, Powder Technol., 18, 201 (1977).

Baker, C.G.J., E.R. Armstrong, and M.A. Bergougnou, Powder Technol., 21, 195 (1978).

Barile, R.G., and D.W. Meyer, Chem. Eng. Prog. Symposium Series, 67, No. 119, 134 (1971).

Batchelor, G.K., J. Fluid Mech., 1, 388 (1956).

Beaver, L.E., Ph.D. Dissertation, The Ohio State University (1983).

Beaver, L.E., and L.-S. Fan, J. Chinese Inst. Chem. Engrs., 15(2), 159 (1984).

Begovich, J.M., and J.S. Watson, in "Fluidization," ed. by J.F. Davidson and D.L. Keairns, p. 190, Cambridge University Press (1978a).

Begovich, J.M., and J.S. Watson, AIChE J., 24(2), 351 (1978b).

Bellman, R., and R.H. Pennington, Quart. Appl. Math., 12, 151 (1954).

Bessler, W.F., Ph.D. Dissertation, Rensselaer Polytechnic Institute (1984).

Bhaga, D., and M.E. Weber, Can. J. Chem. Eng., 50, 323 (1972a).

Bhaga, D., and M.E. Weber, Can. J. Chem. Eng., 50, 329 (1972b).

Bhaga, D., and M.E. Weber, J. Fluid Mech., 105, 61 (1981).

Bhatia, V.K., K.A. Evans, and N. Epstein, I&EC Process Des. & Dev., 11(1), 151 (1972).

Bhatia, V.K., and N. Epstein, in "Fluidization and Its Applications," ed. by H. Angelino, J.P. Couderc, H. Gibert, and C. Laguerie, p. 380, Cepadues-Editions, Toulouse (1974).

Blum, D.B., and J.J. Toman, AIChE Symposium Series, 73, No. 161, 115 (1977).

Boys, C.V., Society for Promoting Christian Knowledge, London (1890).

Bruce, P.N., and L. Revel-Chion, Powder Technol., 10, 243 (1974).

Bruggeman, D.A.G., Ann. Physik., 24, 636 (1935).

Buyevich, Y.A., Chem. Eng. Sci., 29, 37 (1974).

Calderbank, P.H., and J. Pereira, Chem. Eng. Sci., 32, 1427 (1977).

Catros, A., J.R. Bernard, C. Briens, and M.A. Bergougnou, Can. J. Chem. Eng., 63, 754 (1985).

Chang, H.S., and S.D. Kim, Hwahak Konghak (J. Korean Inst. Chem. Engrs.), 17, 407 (1979).

Chen, Y.M., and L.-S. Fan, unpublished results (1988a).

Chen, Y.M., and L.-S. Fan, Chem. Eng. Sci. (in press, 1988b).

Chen, Y.M., and L.-S. Fan, Chem. Eng. Sci. (in review, 1988c).

Chen, Y.M., and L.-S. Fan, Chem. Eng. Sci. (in review, 1988d).

Chern, S.H., K. Muroyama, and L.-S. Fan, Chem. Eng. Sci., 38, 1167 (1983).

Chern, S.H., L.-S. Fan, and K. Muroyama, AIChE J., 30(2), 288 (1984).

Chiu, T.M., and E.N. Ziegler, AIChE J., 31(9), 1504 (1985).

Clift, R., J.R. Grace, and M.E. Weber, "Bubbles, Drops, and Particles," Academic Press, New York (1978).

Collins, R., Chem. Eng. Sci., 20, 851 (1965).

Coppus, J.H.C., K. Rietema, and S.P.P. Ottengraf, Trans. Inst. Chem. Engrs., 55, 122 (1977).

Costa, E., A. de Lucas, and P. Carcla, I&EC Process Des. & Dev., 25, 849 (1986).

Crabtree, J.R., and J. Bridgwater, Chem. Eng. Sci., 22, 1517 (1967).

Currie, I.G., "Fundamental Mechanics of Fluids," McGraw-Hill, New York (1974).

Dakshinamurty, P., V. Subrahmanyam, and J.N. Rao, I&EC Process Des. & Dev., 10(3), 322 (1971).

Dakshinamurty, P., K.V. Rao, R.V. Subbaraju, and V. Subrahmanyam, I&EC Process Des. & Dev., 11(2), 318 (1972).

Darton, R.C., and D. Harrison, Inst. Chem. Engrs. Symposium Series, No. 38, Multi-Phase Flow Systems, I, Inst. Chem. Engrs., London (1974a).

Darton, R.C., and D. Harrison, Trans. Inst. Chem. Engrs., 52, 301 (1974b).

Darton, R.C., and D. Harrison, Chem. Eng. Sci., 30, 581 (1975).

Darton, R.C., and D. Harrison, in "Fluidization Technology," ed. by D.L. Keairns, I, p. 399, Hemisphere (1976).

Darton, R.C., in "Fluidization," ed. by J.F. Davidson, R. Clift, and D. Harrison, Chapter 15, p. 495, Academic Press (1985).

Davies, R.M., and G.I. Taylor, Proc. Roy. Soc., A200, 375 (1950).

de Lasa, H., S.L.P. Lee, and M.A. Bergougnou, Can. J. Chem. Eng., 62, 165 (1984).

Delhaye, J.M., Entropie, No. 99, 3 (1981).

Dhanuka, V.R., Ph.D. Thesis, University of Salford, England (1978).

Dhanuka, V.R., and J.B. Stepanek, in "Fluidization," ed. by J.F. Davidson and D.L. Keairns, p. 179, Cambridge University Press (1978).

Edge, R.M., and C.D. Grant, Chem. Eng. Sci., 26, 1001 (1971).

Efremov, G.I., and I.A. Vakhrushev, Int. Chem. Eng., 10(1), 37 (1970).

El-Temtamy, S.A., Ph.D. Thesis, Cairo University, Egypt (1974).

El-Temtamy, S.A., and N. Epstein, in Proceedings of the 2nd Pacific Area Chemical Engineering Congress (PACHEC), p. 1397, Denver (1977); Int. J. Multiphase Flow, 4, 19 (1978).

El-Temtamy, S.A., and N. Epstein, Can. J. Chem. Eng., 57, 520 (1979).

El-Temtamy, S.A., and N. Epstein, Chem. Eng. J., 19, 153 (1980a).

El-Temtamy, S.A., and N. Epstein, in "Fluidization," ed. by J.R. Grace and J.M. Matsen, p. 519, Plenum Press (1980b).

El-Temtamy, S.A., and N. Epstein, AIChE J., 32(3), 509 (1986).

Epstein, N., Can. J. Chem. Eng., 54, 259 (1976).

Epstein, N., and D. Nicks, in "Fluidization Technology," ed. by D.L. Keairns, p. 389, Hemisphere (1976).

Epstein, N., Can. J. Chem. Eng., 59, 649 (1981).

Ermakova, A., G.K. Ziganshin, and M.G. Slinko, Theor. Found. Chem. Eng., 4, 84 (1970).

Fage, A., and F.C. Johansen, Proc. Roy. Soc., A116, 170 (1927).

Fan, L.-S., K. Muroyama, and S.H. Chern, Chem. Eng. J., 24, 143 (1982).

Fan, L.-S., A. Matsuura, and S.H. Chern, AIChE J., 31(11), 1801 (1985a).

Fan, L.-S., K. Kitano, and B.E. Kreischer, paper presented at the AIChE Annual Meeting, Chicago, Nov. 11-16 (1985b); AIChE J., 33(2), 225 (1987a).

Fan, L.-S., S. Satija, and K. Wisecarver, AIChE J., 32(2), 338 (1986a).

Fan, L.-S., B.E. Kreischer, and K. Tsuchiya, paper presented at the AIChE Annual Meeting, Miami Beach, Nov. 2-7 (1986b); in "Transport

Processes in Fluidized Beds," ed. by A.S. Mujumdar and L.K. Doraiswamy, Elsevier (in press, 1988a).

Fan, L.-S., F. Bavarian, R. Gorowara, B.E. Kreischer, R.D. Buttke, and L.E. Peck, paper presented at the AIChE Annual Meeting, Miami Beach, Nov. 2-7 (1986c); Powder Technol., 53, 285 (1987b).

Fan, L.-S., K. Tsuchiya, and T. Miyahara, in "Fluidization '88 Science and Technology - Conference Papers 3rd China-Japan Symposium," ed. by M. Kwauk and D. Kunii, p. 204, Science Press (Beijing, China) (1988).

Frijlink, J.J., Ph.D. Thesis, Technische Universitat, Delft (1987).

Garside, J., and M.R. Al-Dibouni, I&EC Process Des. & Dev., 16(2), 206 (1977).

Gerrard, J.H., J. Fluid Mech., 25, 401 (1966).

Gorring, R.L., and D.L. Katz, AIChE J., 8(1), 123 (1962).

Guthrie, R.I.L., and A.V. Bradshaw, Chem. Eng. Sci., 24, 913 (1969).

Haberman, W.L., and R.K. Morton, David W. Taylor Model Basin, Report 802, Navy Department, Washington, DC (1953).

Harper, J.F., and D.W. Moore, J. Fluid Mech., 32, 367 (1968).

Henriksen, H.K., and K. Ostergaard, Chem. Eng. J., 7, 141 (1974a).

Henriksen, H.K., and K. Ostergaard, Chem. Eng. Sci., 29, 626 (1974b).

Hetzler, R., and M.C. Williams, I&EC Fundam., 8, 668 (1969).

Hills, J.H., J. Fluid Mech., 68, 503 (1975).

Hnat, J.G., and J.D. Buckmaster, Phys. Fluids, 19(2), 182 (1976).

Hu, T., Y. Zhang, H. Chao, B. Yu, and Y. Wang, in "Fluidization '85 Science and Technology - Conference Papers 2nd China-Japan Symposium," ed. by M. Kwauk, D. Kunii, J. Zheng, and M. Hasatani, p. 262, Elsevier (1985).

Hu, T., B. Yu, and Y. Wang, in "Fluidization V," ed. by K. Ostergaard and A. Sorensen, p. 353, Engineering Foundation Press (1986).

Huh, C., and L.E. Scriven, J. Colloid Interface Sci., 35, 85 (1971).

Idogawa, K., K. Ikeda, T. Fukuda, and S. Morooka, Int. Chem. Eng., 26(3), 468 (1986).

Ishida, M., and H. Tanaka, J. Chem. Eng. Japan, 15(5), 398 (1982).

Jean, R.H., and L.-S. Fan, Chem. Eng. Sci., 41(11), 2823 (1986).

Jean, R.H., and L.-S. Fan, Can. J. Chem. Eng., 65, 351 (1987a).

Jean, R.H., and L.-S. Fan, Can. J. Chem. Eng., 65, 881 (1987b).

Jean, R.H., Ph.D. Dissertation, The Ohio State University (1988).

Jenkins, III, J.M., R.L. Jones, and T.M. Jones, U.S. Patent 4,543,399 (1985).

Kafarov, V.V., V.A. Klipinitser, and O.I. Saks, Theor. Found. Chem. Eng., 7, 742 (1973).

Kato, Y., K. Uchida, T. Kago, and S. Morooka, Powder Technol., 28, 173 (1981).

Kato, Y., S. Morooka, T. Kago, T. Saruwatari, and S.Z. Yang, J. Chem. Eng. Japan, 18(4), 308 (1985).

Keitel, G., and U. Onken, Chem. Eng. Sci., 37(11), 1635 (1982).

Kelkar, B.G., S.P. Godbole, M.F. Honath, and Y.T. Shah, AIChE J., 29(3), 361 (1983).

Khang, S.J., J.G. Schwartz, and R.D. Buttke, AIChE Symposium Series, 79, No. 222, 47 (1983).

Kim, S.D., C.G.J. Baker, and M.A. Bergougnou, Can. J. Chem. Eng., 50, 695 (1972).

Kim, S.D., C.G.J. Baker, and M.A. Bergougnou, Can. J. Chem. Eng., 53, 134 (1975).

Kim, S.D., C.G.J. Baker, and M.A. Bergougnou, Chem. Eng. Sci., 32, 1299 (1977).

Kim, J.D., and S.D. Kim, Particulate Sci. and Technol., 5, 309 (1987).

Kitano, K., and L.-S. Fan, unpublished results (1987).

Kitano, K., and L.-S. Fan, Chem. Eng. Sci., 43, 1355 (1988).

Kiya, M., and M. Arie, J. Fluid Mech., 82, 223 (1977).

Kojima, E., T. Akehata, and T. Shirai, J. Chem. Eng. Japan, 8, 108 (1975).

Kono, H., Hydrocarbon Processing, 59, 123, Jan. (1980).

Kreischer, B.E., H. Moritomi, and L.-S. Fan, paper presented at the AIChE Annual Meeting, Washington, DC, Nov. 27 - Dec. 2 (1988).

Kuroda, M., and K. Tabei, Kagaku Kogaku Ronbunshu, 5, 449 (1979); Int. Chem. Eng., 21, 219 (1981).

Lapidus, L., and J.C. Elgin, AIChE J., 3, 63 (1957).

Lazarek, G.M., and H. Littman, J. Fluid Mech., 66, 673 (1974).

Lee, J.C., in Proceedings of the 3rd European Symposium on Chemical Reaction Engineering, p. 211, Pergamon Press, Oxford (1965).

Lee, J.C., A.J. Sherrard, and P.S. Buckley, in "Fluidization and Its Applications," ed. by H. Angelino, J.P. Couderc, H. Gibert, and C. Laguerie, p. 407, Cepadues-Editions, Toulouse (1974).

Lee, J.C., and H. Worthington, Inst. Chem. Engrs. Symposium Series, No. 38, Multi-Phase Flow Systems, I, Inst. Chem. Engrs., London (1974).

Lee, J.C., and N.J. Al-Dabbagh, in "Fluidization," ed. by J.F. Davidson and D.L. Keairns, p. 184, Cambridge University Press (1978).

Lee, S.L.P., H.I. de Lasa, and M.A. Bergougnou, AIChE Symposium Series, 80, No. 241, 110 (1984).

Lee, S.L.P., and H.I. de Lasa, in "Fluidization V," ed. by K. Ostergaard and A. Sorensen, p. 361, Engineering Foundation Press (1986).

Letan, R., and E. Kehat, AIChE J., 14(3), 398 (1968).

Lindt, J.T., Chem. Eng. Sci., 26, 1776 (1971).

Massimilla, L., N. Majuri, and P. Signorini, La Ricerca Scientifica, 29, 1934 (1959).

Massimilla, L., A. Solimando, and E. Squillace, Brit. Chem. Eng., 6, 232 (1961).

Matsuura, A., and L.-S. Fan, AIChE J., 30(6), 894 (1984).

Maxwell, J.C., "A Treatise on Electricity and Magnetism," 2nd ed., I, p. 398, Clareden Press, Oxford (1881).

Maxworthy, T., J. Fluid Mech., 27, 367 (1967).

Mendes, C.L.D.O., and R.Y. Qassim, Chem. Eng. J., 28, 21 (1984).

Meredith, R.E., and C.W. Tobias, in "Advances in Electrochemistry and Electrochemical Engineering," 2, p. 15, Interscience, New York (1962).

Michelsen, M.L., and K. Ostergaard, Chem. Eng. J., 1, 37 (1970).

Miller, N., and R.E. Mitchie, J. Brit. Nucl. Energy Soc., 9, 94 (1970).

Morooka, S., K. Uchida, and Y. Kato, J. Chem. Eng. Japan, 15, 29 (1982).

Morooka, S., T. Mizoguchi, T. Kago, Y. Kato, and N. Hidaka, J. Chem. Eng. Japan, 19, 507 (1986).

Mukherjee, R.N., P. Bhattacharya, and D.K. Taraphdar, in "Fluidization and Its Applications," ed. by H. Angelino, J.P. Couderc, H. Gibert, and C. Laguerie, p. 372, Cepadues-Editions, Toulouse (1974).

Muroyama, K., Dr. of Eng. Thesis, Kyoto University, Japan (1976).

Muroyama, K., M. Fukuma, and A. Yasunishi, Can. J. Chem. Eng., 62, 199 (1984).

Muroyama, K., and L.-S. Fan, AIChE J., 31, 1 (1985).

Muroyama, K., personal communication (1987).

Nasr-El-Din, H., C.A. Shook, and J. Colwell, Int. J. Multiphase Flow, 13, 365 (1987).

Nicklin, D.J., Chem. Eng. Sci., 17, 693 (1962).

Oh, J.S., and S.D. Kim, Hwahak Konghak (J. Korean Inst. Chem. Engrs.), 18(5), 375 (1980).

Ostergaard, K., in "Fluidization," p. 58, Soc. Chem. Ind., London (1964).

Ostergaard, K., Chem. Eng. Sci., 20, 165 (1965).

Ostergaard, K., Chem. Eng. Sci., 21, 470 (1966).

Ostergaard, K., and M.L. Michelsen, Joint AIChE-IIQPR Meeting, Tampa, FL (1968).

Ostergaard, K., "Studies of Gas-Liquidization," Danish Technical Press, Copenhagen (1969).

Page, R.E., and D. Harrison, Powder Technol., 6, 245 (1972).

Page, R.E., and D. Harrison, in "Fluidization and Its Applications," ed. by H. Angelino, J.P. Couderc, H. Gibert, and C. Laguerie, p. 393, Cepadues-Editions, Toulouse (1974).

Perry, A.E., M.S. Chong, and T.T. Lim, J. Fluid Mech., 116, 77 (1982).

Peterson, D.A., R.S. Tankin, and S.G. Bankoff, Int. J. Multiphase Flow, 13(4), 477 (1987).

Razumov, I.M., V.V. Manshilin, and L.L. Nemets, Int. Chem. Eng., 13, 57 (1973).

Richardson, J.F., and W.N. Zaki, Trans. Inst. Chem. Engrs., 32, 35 (1954).

Rigby, G.R., and C.E. Capes, Can. J. Chem. Eng., 48, 343 (1970).

Rigby, G.R., G.P. Van Blockland, W.H. Park, and C.E. Capes, Chem. Eng. Sci., 25, 1729 (1970).

Saberian-Broudjenni, M., G. Wild, J.-C. Charpentier, Y. Fortin, J.-P. Euzen, and R. Patoux, Entropie, No. 120, 30 (1984).

Saffman, P.G., and G.R. Baker, Ann. Rev. Fluid Mech., 11, 95 (1979).

Sarpkaya, T., J. Fluid Mech., 68, 109 (1975).

Schaefer, R.J., D.N. Rundell, and J.K. Shou, Final Progress Report to U.S. Department of Energy, Contract No. DE-AC22-80PC30026 (1983).

Schuman, S.C., R.H. Wolk, and M.C. Chrvenak, U.S. Patent 3,183,180 (1965).

Shah, Y.T., S. Joseph, D.N. Smith, and J.A. Ruether, I&EC Process Des. & Dev., 24(4), 1140 (1985).

Sinha, V.T., M.S. Butensky, and D. Hyman, AIChE Symposium Series, 80, No. 241, 176 (1984).

Sinha, V.T., M.S. Butensky, and D. Hyman, I&EC Process Des. & Dev., 25(1), 321 (1986).

Slaughter, I., and A.E. Wraith, Chem. Eng. Sci., 23, 932 (1968).

Song, G.H., and L.-S. Fan, in Proceedings of the 3rd World Congress of Chemical Engineering, III, p. 504, Tokyo, Sep. (1986a).

Song, G.H., and L.-S. Fan, Contractual Report to Amoco Oil Co., Naperville, IL, Oct. (1986b).

Song, G.H., F. Bavarian, L.-S. Fan, R.D. Buttke, and L.B. Peck, paper presented at AIChE Annual Meeting, New York, Nov. 15-20 (1987); Can. J. Chem. Eng. (in press, 1988).

Soo, S.L., "Fluid Dynamics of Multiphase Systems," Blaisdell Publishing, Waltham, MA (1967).

Soung, W.Y., I&EC Process Des. & Dev., 17(1), 33 (1978).

Stewart, P.S.B., and J.F. Davidson, Chem. Eng. Sci., 19, 319 (1964).

Tadaki, T., and S. Maeda, Kagaku Kogaku, 25, 254 (1961).

Tang, W.-T., and L.-S. Fan, paper presented at the AIChE Annual Meeting, New York, Nov. 15-20 (1987); AIChE J. (in press, 1988).

Tarmy, B.L., M. Chang, C.A. Coulaloglou, and P.R. Ponzi, paper presented at the International Symposium on Chemical Reaction Engineering, Edinburgh, Sep. 10-13 (1984).

Tichy, J., and W.J.M. Douglas, Can. J. Chem. Eng., 50, 702 (1972).

Tsuchiya, K., and L.-S. Fan, paper presented at the AIChE Annual Meeting, Miami Beach, Nov. 2-7 (1986); Chem. Eng. Sci., 43(5), 1167 (1988).

Tsutsui, T., and T. Miyauchi, Int. Chem. Eng., 20(3), 386 (1980).

Turner, R., in "Fluidization," p. 47, Soc. Chem. Ind., London (1964).

Turner, J.C.R., Chem. Eng. Sci., 31, 487 (1976).

Turton, R., and O. Levenspiel, Powder Technol., 47(1), 83 (1986).

Ueyama, K., and T. Miyauchi, Kagaku Kogaku Ronbunshu, 2, 430 (1976).

Ueyama, K., and T. Miyauchi, AIChE J., 25, 258 (1979).

Vasalos, I.A., E.M. Bild, and D.N. Rundell, paper presented at the AIChE Annual Meeting, San Francisco, Nov. 25-29 (1977).

Vasalos, I.A., E.M. Bild, D.N. Rundell, and J.W. Gorman, Quarterly Progress Report to U.S. Department of Energy, No. 6, Contract No. EF-77-C-01-2588, June (1979).

Vasalos, I.A., E.M. Bild, D.N. Rundell, and D.F. Tatterson, Final Report to U.S. Department of Energy, Contract No. DE-AC05-10149, Feb. (1980).

Vasalos, I.A., D.N. Rundell, K.E. Megiris, and G.T. Tjatjopoulos, AIChE J., 28(2), 346 (1982).

Viswanathan, S., A.S. Kakar, and P.S. Murti, Chem. Eng. Sci., 20, 903 (1964).

Volpicelli, G., and L. Massimilla, Chem. Eng. Sci., 25, 1361 (1970).

Wallis, G.B., "One-Dimensional Two-Phase Flow," McGraw-Hill (1969).

Wegener, P.P., and J.-Y. Parlange, Ann. Rev. Fluid Mech., 5, 79 (1973).

Wen, C.Y., and Y.H. Yu, Chem. Eng. Prog. Symposium Series, 62, No. 62, 100 (1966).

Yeheskel, J., and E. Kehat, AIChE J., 19, 720 (1973).

Zieminski, S.A., M.M. Caron, and R.B. Blackmore, I&EC Fundam., 6(2), 233 (1967).

Zuber, N., and J.A. Findlay, Trans. ASME J. Heat Transfer, p. 453, Nov. (1965).

Chapter 3*

Mass Transfer, Mixing and Heat Transfer of Cocurrent Upward Fluidized Bed Systems (Mode E-I-a-1; Figure 1.3)

In continuation of describing the transport properties for Mode E-I-a-1, this chapter focuses on mass transfer, mixing behavior and heat transfer. The hydrodynamic behavior inherently relates to these properties. The interplay among various transport properties will also be elucidated.

More specifically, the coverage of mass transfer will include the mass transfer coefficient and interfacial area for gas-liquid mass transfer, the mass transfer coefficient for liquid-solid mass transfer, and wall-to-bed mass transfer. On mixing, the liquid phase mixing including radial and axial dispersion of liquid is discussed along with gas phase mixing. In addition, solids phase mixing and bed segregation are described with emphasis on the behavior involving beds containing dissimilar particles. For heat transfer, the focus is on column wall-to-bed heat transfer, immersed vertical cylinder-to-bed heat transfer and immersed horizontal cylinder-to-bed heat transfer.

3.1 MASS TRANSFER

In general, mass transfer in gas-liquid-solid fluidized bed operation for physical, chemical, catalytic or noncatalytic, petrochemical or biochemical reactions occurs through both gas-liquid and/or solid (particle)-liquid interfaces. For electrochemical reactions, mass transfer may occur at the wall-to-bed interface. Intraparticle mass transfer (diffusion) resistance plays an important role in determining the overall rate of reactions. An example which

*: by L.-S. Fan and K. Muroyama

illustrates this role and its interrelationship with other mass transfer steps is given in Appendix A.

3.1.1 Gas-Liquid Mass Transfer

The mass transfer rate across the gas-liquid interface can be described by the product of three terms: the overall mass transfer coefficient, the interfacial area, and the concentration difference. In a three-phase fluidized bed both the mass transfer coefficient and the gas-liquid interfacial area inherently depend on the bed hydrodynamics. In practice, the gas-side mass transfer resistance can usually be neglected in evaluating gas-liquid mass transfer, thus leading to the simplification of the overall mass transfer coefficient to the liquid-side mass transfer coefficient. The liquid-side mass transfer coefficient incorporates the effects of the liquid flow field surrounding the rising gas bubbles. The interfacial area reflects the system bubble behavior. Consequently, the dependency of the bubble behavior on system properties such as gas and liquid velocities and particle size and density must carry over to the interfacial area. The most common approach in treating gas-liquid mass transfer is to combine the mass transfer coefficient and interfacial area terms into a single volumetric mass transfer coefficient ($k_\ell a$) averaged over the entire column height; however, to understand the fundamental mechanisms involved in gas-liquid mass transfer, it is necessary to examine the mass transfer behavior of a single bubble.

3.1.1.1 Gas-Liquid Mass Transfer for a Single Bubble in a Liquid-Solid Fluidized Bed

Very little work (Vanderschuren et al., 1974; Jean and Fan, 1988) has been attempted toward gaining a fundamental understanding of mass transfer between a bubble and a liquid-solid suspension. In this section, gas-liquid mass transfer involving a single bubble of spherical and spherical-cap geometries is described. These two geometries represent bubbles of two extreme sizes commonly encountered in a three-phase fluidized bed. The mass transfer behavior of a two-dimensional circular-cap bubble is also described.

1. Spherical Bubble

Spherical bubbles of size considerably smaller than the particles rise through the fluidized particle matrix without colliding with the large particles or undergoing shape distortion. Thus, the mass transfer between a small bubble and a liquid-solid medium can practically be treated as that between a small bubble and a liquid medium alone.

Friedlander (1957) studied the mass transfer of a solid sphere at low Reynolds numbers, i.e., Re_b $(= \frac{2r_e U_B \rho_\ell}{\mu_\ell}) < 5$. There is no distinct difference in mass transfer behavior between a solid sphere, a liquid droplet or a gas bubble under this Reynolds number range. Based on the boundary layer integral approach, the Sherwood number, Sh_b $(= 2k_{Lt}r_e/D_M)$, was found to be related to the Peclet number, Pe $(= 2r_eU_B/D_M)$, by

$$Sh_b = 0.89\ Pe^{1/3} \qquad \text{for } Pe > 10^3 \tag{3.1a}$$

and

$$Sh_b = \frac{4}{Pe} \ln \left(\frac{1}{1-Pe/2}\right), \qquad \text{for } Pe << 1 \tag{3.1b}$$

Note that in Eqn. (3.1b), as Pe approaches zero, Sh_b approaches 2 which is the result for molecular diffusion from a stagnant spherical bubble to an infinite liquid medium. For intermediate values of Pe, Friedlander (1957) presented numerical values for the dependency of Sh_b on Pe. Additional information on mass transfer from a single spherical bubble is given by Azbel (1981).

2. Circular-Cap Bubble

Jean and Fan (1988) examined the intrinsic mass transfer behavior around a large single bubble characterized by a circular-cap (in a two-dimensional bed) or a single spherical-cap bubble (in a three-dimensional bed) and presented the following fundamental framework for the analysis of mass transfer between a bubble and a liquid-solid suspension.

Consider a bubble and its wake having a structure as described in Section 2.2, where the flow field in the dynamic (non-steady) wake beneath the bubble base significantly deviates from the flow around the bubble roof. The dynamic wake, depending on particle properties and operating conditions, may contain a stable liquid or liquid-solid layer directly beneath the bubble base. The mass transfer behavior around the bubble can be separated into that through the bubble roof and through the bubble floor; the distinction of these two mass transfer mechanisms resembles that for liquid systems considered by

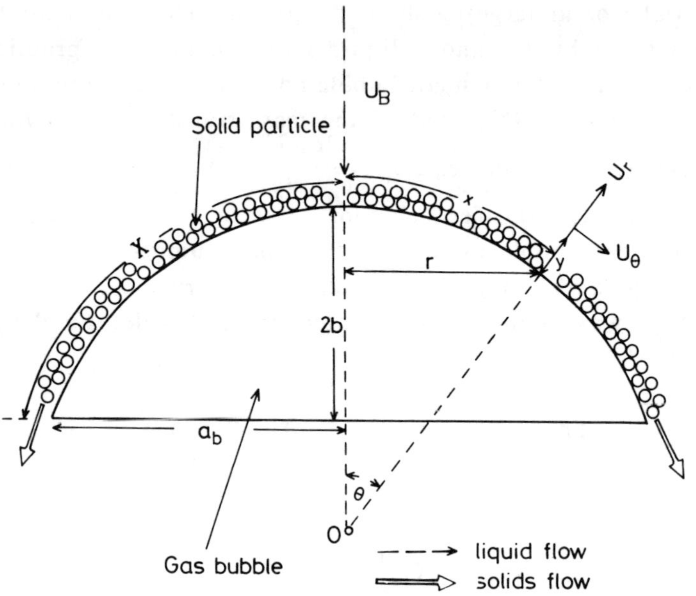

Figure 3.1 Schematic diagram for mass transfer from the roof of a single bubble (from Jean and Fan, 1988).

Calderbank et al. (1970). In the subsequent analysis of mass transfer from a bubble, the following assumptions are made:

i) The size and pressure of a bubble during the course of its rise can be represented by an axially averaged value;
ii) The bubble motion is considered to be at steady state; and
iii) A fairly thin layer of liquid exists around the bubble roof where the potential flow theory and penetration mass transfer model are applicable.

(1) Mass Transfer Through the Bubble Roof

Consider a bubble held stationary by a downward liquid flow with velocity U_B $(=U_b-U_\ell/\epsilon_\ell)$ at an infinite distance from the bubble. Referring to the coordinates defined in Fig. 3.1, the continuity equation for the liquid near the bubble roof is

$$\frac{\partial U_\theta}{\partial x} + \frac{\partial U_r}{\partial y} = 0 \tag{3.2}$$

and for the dissolved gas or solute near the bubble roof is

$$U_\theta \frac{\partial C}{\partial x} + U_r \frac{\partial C}{\partial y} = D_M \frac{\partial^2 C}{\partial y^2} \tag{3.3}$$

with the boundary conditions:

$$x = 0 , \qquad C = C_b$$

$$y = 0 , \qquad C = C^*$$

$$y = \infty , \qquad C = C_b$$

In Eqn. (3.3), D_M is the diffusivity of the solute in the liquid medium.

The mass flux of solute through the bubble roof is

$$dN_r = k_{Lr}(C^* - C_b)dA_r = - D_M \left(\frac{\partial C}{\partial y}\bigg|_{y=0}\right) dA_r \tag{3.4}$$

where C^* is the solute saturation concentration and C_b is the bulk solute concentration. Integrating Eqn. (3.4) over the entire roof gives

$$N_r = k_{Lr}\, A_r(C^* - C_b) = - D_M \int_{A_r}\!\!\int \left(\frac{\partial C}{\partial y}\bigg|_{y=0}\right)dA_r = -2D_M \int_0^X \left(\frac{\partial C}{\partial y}\bigg|_{y=0}\right)w\,dx \tag{3.5}$$

where w is the width of a circular-cap bubble. Following a similar treatment by Lochiel and Calderbank (1964) for a gas-liquid system, Eqns. (3.2) and (3.3) along with

$$U_{\theta s} \gg \frac{\partial U_\theta}{\partial y}\bigg|_{y=0} y \tag{3.6}$$

under the assumptions of high Peclet and Schmidt numbers result in a solution for the concentration distribution near the bubble roof. In Eqn. (3.6), $U_{\theta s}$ is the interfacial tangential velocity on the bubble roof. Combining the resulting concentration distribution with Eqn. (3.5) yields an expression for k_{Lr} in terms of the Sherwood number as

$$Sh_r = \left\{\frac{2}{\pi}\right\}^{1/2} \left\{\frac{4wr_e}{A_r}\right\} \left[\int_0^{X_1} U_{\theta 1}dx_1\right]^{1/2} Pe^{1/2} \tag{3.7}$$

where

$$Sh_r = \frac{2k_{Lr}\, r_e}{D_M}, \quad Pe = \frac{2r_e\, U_B}{D_M}, \quad U_{\theta 1} = \frac{U_{\theta s}}{U_B}, \quad x_1 = \frac{x}{r_e}, \quad X_1 = \frac{X}{r_e}$$

The potential flow theory (e.g., Lamb, 1945) for flow over a cylinder leads to an expression for $U_{\theta 1}$ of

$$U_{\theta 1} = 2 \sin\theta \tag{3.8}$$

Combining Eqns. (3.7) and (3.8) yields

$$Sh_r = 4.79\, \frac{\{\sin^{-1}[4E/(E^2+4)] - 4E(E^2-4)/(E^2+4)^2\}^{1/4}}{(E^2+4)^{1/2}\sin^{-1}[4E/(E^2+4)]}\, Pe^{1/2} \tag{3.9}$$

where E, equal to $(\frac{a_b}{b})$, is a function of the included angle in the circular-cap bubble and consequently, a function of the bubble Reynolds number. Note that k_{Lr} is proportional to $D_M^{1/2}$ and that the bubble diameter influences k_{Lr} both directly (through Pe and E) and indirectly through the bubble rise velocity (also through Pe).

(2) Mass Transfer Across the Bubble Base

The region immediately beneath the bubble base has been observed to be a stable liquid layer either solid-free (Case I) or containing solids (Case II) (see Chapter 2). The stable liquid layer is motionless relative to the moving bubble; diffusion from the bubble interface through the liquid layer controls the mass transfer process. A separate mass transfer analysis for each case is given below.

Case I - Stable Solid-Free Liquid Layer

The observed stable liquid layer in stagnant conditions varies in height from 0.1 cm to 0.2 cm over a wide range of bubble Reynolds number (3000 ~ 12000) (Rigby and Capes, 1970; Kitano and Fan, 1988). The penetration depth for mass transfer (δ), as derived from the diffusion equation for a semi-infinite slab, equals $4\sqrt{D_M \tau_b}$ where τ_b is the residence time of a bubble in the bed. For a single nitrogen bubble with a rise velocity of 30 cm/s in water in a column 1 m high, δ is approximately 0.032 cm and is smaller than the stable liquid layer thickness. Thus, the diffusion controlled assumption is valid under this situation and the mass transfer across the bubble base (N_f) can be described as that across a semi-infinite slab (e.g., Cussler, 1984).

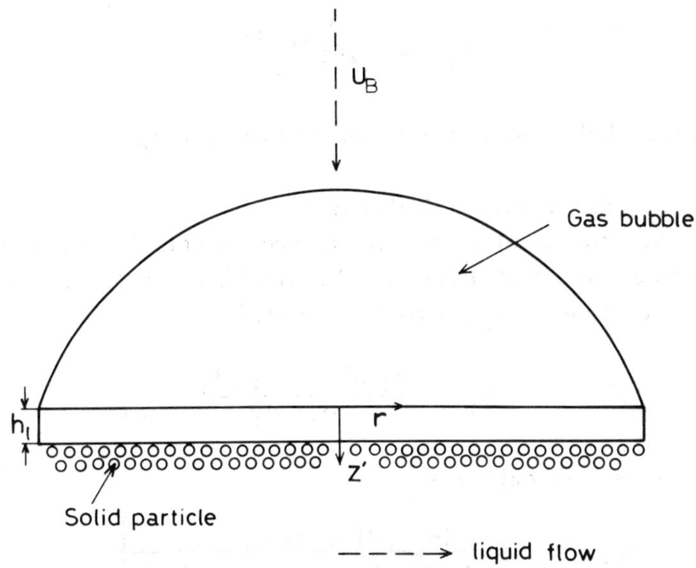

Figure 3.2 Schematic diagram for mass transfer from the base of a single bubble (from Jean and Fan, 1988).

Referring to the coordinates around the bubble base given in Fig. 3.2, N_f can be expressed as

$$N_f = k_{Lf}A_f(C^* - C_b) = A_f(-D_M \frac{\partial C}{\partial z'}) = A_f(\frac{D_M}{\pi \tau_b})^{1/2} (C^* - C_b) \tag{3.10}$$

Thus, k_{Lf} can be expressed as

$$k_{Lf} = (\frac{D_M}{\pi \tau_b})^{1/2} \tag{3.11}$$

where τ_b, which is equal to $\frac{z}{(U_B + U_\ell/\epsilon_\ell)}$, reflects bubble size effects through the relative bubble rise velocity.

Case II - Stable Liquid-Solid Layer

Assuming the thickness of the stable liquid-solid layer is also larger than the penetration depth and the solids are uniformly distributed within the stable liquid-solid layer, Eqn. (3.11) for Case I can be suitably modified to give the mass transfer coefficient (k_{Lf}) for Case II:

$$k_{Lf} = \epsilon_f \left(\frac{D_M}{\pi \tau_b}\right)^{1/2} \tag{3.12}$$

Once again, bubble size effects are contained in τ_b.

(3) Overall Mass Transfer Coefficient

The individual mass transfer coefficients for mass transfer at the bubble roof and base can be combined into an overall mass transfer coefficient (k_{Lt}) with the expression:

$$k_{Lt} = \frac{k_{Lr} A_r + k_{Lf} A_f}{A_r + A_f} \tag{3.13}$$

For a circular-cap bubble, k_{Lt} is

$$k_{Lt} = k_{Lr} \left\{ \frac{\sin^{-1}[4E/(E^2+4)]}{4E/(E^2+4) + \sin^{-1}[4E/(E^2+4)]} \right\} + k_{Lf} \left\{ \frac{4E}{4E + (E^2+4)\sin^{-1}[4E/(E^2+4)]} \right\} \tag{3.14}$$

k_{Lr} can be expressed by Eqn. (3.9) and k_{Lf} by either Eqn. (3.11) for Case I or Eqn. (3.12) for Case II.

3. Spherical-Cap Bubble

Similar to the treatment of the circular-cap bubble, the corresponding equations for the spherical-cap bubble can be summarized as follows (Jean and Fan, 1988):

(1) Mass Transfer at the Bubble Roof (Lochiel and Calderbank, 1964)

$$Sh_r = 1.79 \frac{(3E^2 + 4)^{2/3}}{E^2 + 4} Pe^{1/2} \tag{3.15}$$

(2) Mass Transfer Across the Bubble Base

Case I - Stable Solid-Free Liquid Layer

k_{Lf} can be expressed by Eqn. (3.11).

Case II - Stable Liquid-Solid Layer

k_{Lf} can be expressed by Eqn. (3.12).

(3) Overall Mass Transfer Coefficient

$$k_{Lt} = k_{Lr} \left(\frac{E^2+4}{2E^2+4}\right) + k_{Lf} \left(\frac{E^2}{2E^2+4}\right) \tag{3.16}$$

where k_{Lr} can be expressed by Eqn. (3.15) and k_{Lf} by Eqn. (3.11) for Case I or Eqn. (3.12) for Case II.

Vanderschuren et al. (1974) measured the volumetric mass transfer coefficient of a large, single spherical-cap gas bubble (carbon dioxide, benzene or chloroethane) in a three-dimensional liquid-solid fluidized bed containing 400 μm ~ 500 μm glass beads. They reported values of the mass transfer coefficient averaged over the distance between the bubble injection and the measurement points. Their results indicate the volumetric mass transfer coefficient ($k_{Lt}A_t$) initially decreases sharply with axial distance and then levels off. This behavior can be characterized by the equations described by Eqn. (3.16) with conditions for either Case I or Case II since both k_{Lf} and k_{Lr} are significant within a short distance above the distributor, but beyond this distance k_{Lf} becomes insignificant compared with k_{Lr}. The equations for either case predict the correct trend that $k_{Lt}A_t$ increases with increasing liquid velocity. Furthermore, the 0.5 power dependency of $k_{Lt}A_t$ on D_M for both cases is fairly consistent with the 0.59 power dependency empirically obtained by Vanderschuren et al. (1974).

3.1.1.2 Volumetric Gas-Liquid Mass Transfer Coefficient

Studies of the volumetric gas-liquid mass transfer coefficient involve the absorption of a gas phase component into the liquid phase. This absorption may or may not be accompanied by chemical reactions. Table 3.1 summarizes the various studies of gas-liquid mass transfer for Mode E-I-a-1 operation and indicates the methods used to obtain $k_\ell a$.

As indicated in Chapter 2 and displayed by the bubble behavior, both gas velocity and particle size have a major impact on bubble sizes prevalent in a bed; the same effects are observed in terms of $k_\ell a$. Figure 3.3 shows $k_\ell a$ increases significantly with an increase in U_g for several different particle sizes based on the data of Ostergaard and Suchozebrski (1971), Dhanuka and Stepanek (1980a,b) and Nguyen-Tien et al. (1985). The relative increase in $k_\ell a$, however, becomes smaller at high gas velocities. Also, as shown in Fig. 3.3, for particles with d_p greater than about 4mm, $k_\ell a$ is significantly greater than that in a solid-free bubble column and for particles with d_p less than about 2 mm, $k_\ell a$ is comparable to or smaller than that in a solid-free bubble column. Figure 3.4 quantifies the $k_\ell a$ dependency on d_p for various gas flow rates and U_ℓ of 0.1 m/s; $k_\ell a$ decreases sharply,

Table 3.1 Studies on Gas-Liquid Mass Transfer in Three-Phase Fluidized Beds

Investigators	Methods	Operating Conditions	Correlations (in SI Units) or Remarks
Massimilla et al. (1959)	Physical absorption; CO_2-N_2/water/silica particle, or glass beads	d_p = 0.22, 0.50 and 0.80 mm U_g = 0.7 ~ 1.3 cm/s U_ℓ = 1.1 ~ 3.0 cm/s	Graphical description absorption efficiency
Ostergaard and Suchozebrski (1971)	Physical absorption; CO_2-N_2/water/glass beads (CO_2% = 50.80 and 100)	d_p = 0.125 cm, ρ_s = 2.67 g/cm^3 d_p = 0.593 cm, ρ_s = 2.63 g/cm^3 U_g = 0 ~ 3 cm/s U_ℓ = 6.8 ~ 17 cm/s	Graphical description a very high mass transfer rate for the bed of 6 mm particles
Ostergaard and Fosbol (1972)	Physical absorption; pure O_2/water/glass beads	d_p = 1 and 6 mm U_g = 0 ~ 5.22 cm/s U_ℓ = 1.95 ~ 16.3 cm/s	Graphical description (the variation of $k_\ell a$ with the distance from the gas distributor was discussed)
Lee and Worthington (1974)	Physical absorption; CO_2/water or aqueous sodium sulphate soln./glass beads	d_p = 6 mm U_g = 1.5 ~ 3.5 cm/s U_ℓ = 5.75 ~ 16.4 cm/s	$\frac{k_\ell a}{\varepsilon_g} = 1.53$
Dakshinamurty et al. (1974a,b) Dakshinamurty and Rao (1976)	Physical absorption; CO_2/water/porcelain beads, sand particles, iron shots, or lead shots (CO_2% = 50 and 100)	d_p = 1.06 ~ 6.84 mm ρ_s = 2.4 ~ 11.18 g/cm^3 (U_t = 12 ~ 81.1 cm/s) U_g = 0 ~ 10 cm/s U_ℓ = 5.0 ~ 17.2 cm/s	Part I (1974): graphical description Part II (1976): correlations, $\frac{k_\ell a d_p^2}{D_M} = K''\left(\frac{d_p U_t \rho_\ell}{\mu_\ell}\right)^{3.3}\left(\frac{\mu_\ell U_g}{\sigma}\right)^{0.7}$ (T-3.1.1) $K'' = 3.946 \times 10^{-7}$ for $Re_t > 2000$ $K'' = 2.33 \times 10^{-5}$ for $Re_t < 2000$

Nishikawa et al. (1977)	Chemical absorption; air/0.25 N sodium sulphite soln. catalyzed by copper iron/glass beads	d_p = 1.01, 2.59 and 4.87 mm ρ_s = 2.6 g/cm^3 U_g = 0 ~ 10 cm/s U_ℓ = 0 ~ 6 cm/s	Three-phase spouted bed Graphical description
Ostergaard (1978)	Chemical absorption; measurement of $k_\ell a$ for CO_2/carbonate-bicarbonate buffer/glass beads; measurement of k_ℓ and a for CO_2/sodium hydroxide/glass beads	d_p = 1.1, 3.0 and 6.0 mm	Empirical correlations for $k_\ell a$ and a specific to each particle size
Fukushima (1979)	Using data of Lee and Worthington (1974)		$\frac{k_\ell a d_p^2 \varepsilon}{D_M \varepsilon_g} = 1.4 \times 10^4 \; (Re_\ell/\varepsilon_\ell)^{0.38} (Re_g/\varepsilon_g)^{-0.2} Sc^{0.5} (\frac{d_p}{D_c})^{2.1}$ $\frac{k_\ell d_p}{D_M} = 8.9 \times 10^2 \; (Re_\ell/\varepsilon_\ell)^{1/3} (Re_g/\varepsilon_g)^{0.2} Sc^{0.5} (\frac{d_p}{D_c})^{2.1}$ $\frac{a\, d_p}{(\varepsilon_g/\varepsilon)} = 16 \; (Re_\ell/\varepsilon_\ell)^{0.05} (Re_g/\varepsilon_g)^{-0.4}$
Dhanuka and Stepanek (1980a,b)	Chemical absorption-physical desorption; $CO_2/N_2/Na_2CO_3$-$NaHCO_3$ buffer dissociated with O_2/glass beads	d_p = 1.98, 4.08 and 5.86 mm U_g = 2 ~ 8 cm/s U_ℓ = 6.1 ~ 13.8 cm/s	Empirical correlations for $k_\ell a$ and a specific to each particle size
Alvarez-Cuenca and Nerenberg (1981) Alvarez-Cuenca et al. (1984)	Physical absorption; air/water/glass beads	d_p = 1, 3 and 5 mm u_g = 4 ~ 28 cm/s U_ℓ = 2.5 ~ 10 cm/s	Local values of $k_\ell a$ were analyzed based on the plug flow model and two zone model.

(Table 3.1 continued)

Investigators	Methods	Operating Conditions	Correlations (in SI Units) or Remarks
Lee and Buckley (1981)	physical absorption; air (O_2)/aqueous octanol soln. or distilled water/ glass beads	d_p = 6 mm, ρ_s = 2.45 g/cm^3 d_p = 4 mm, ρ_s =2.56 g/cm^3 d_p = 2 mm, U_g = 0 ~ 8 cm/s U_ℓ = 6 ~ 16 cm/s	Local values of $k_\ell a$ and k_ℓ were measured.
Nguyen-Tien et al. (1985)	physical absorption; air (O_2)/water/glass beads	d_p = 0.05 ~ 8 mm ρ_s = 2.51 to 2.95g/cm^3 U_g = 0 ~ 16 cm/s U_ℓ = 4.6 ~ 11.6 cm/s	$k_\ell a = 0.39\ (1-\frac{\Psi_s}{0.58})\ U_g^{0.67}$ (T-3.1.2a) for 0.05 mm $\le d_p \le$ 1 mm $\Psi_s = \frac{\varepsilon_s}{\varepsilon_\ell+\varepsilon_s}$ $k_\ell a = 13.9\ (d_p)^{0.709} (U_g)^{0.5}$ (T-3.1.2b) for d_p > 1 mm, U_ℓ = 10 cm/s (Wisecarver and Fan, 1988)
Chang et al. (1986)	Chemical absorption-Physical desorption; CO_2-N_2/Na_2CO_3-$NaHCO_3$ buffer/glass beads	d_p = 1.7, 4.0, 6.0 mm U_g = 0.5 ~ 7.0 cm/s U_ℓ = 5.4 ~ 13 cm/s	$k_\ell a = 1597\ U_g^{0.68} U_\ell^{0.63} d_p^{1.21}$ (T-3.1.3) $a = 2.08 \times 10^6\ U_g^{0.35} U_\ell^{0.85} d_p^{0.88}$ (T-3.1.4) (based on data of Ostergaard and Fosbol, 1972; Dhanuka and Stepanek, 1980b; and their own data)

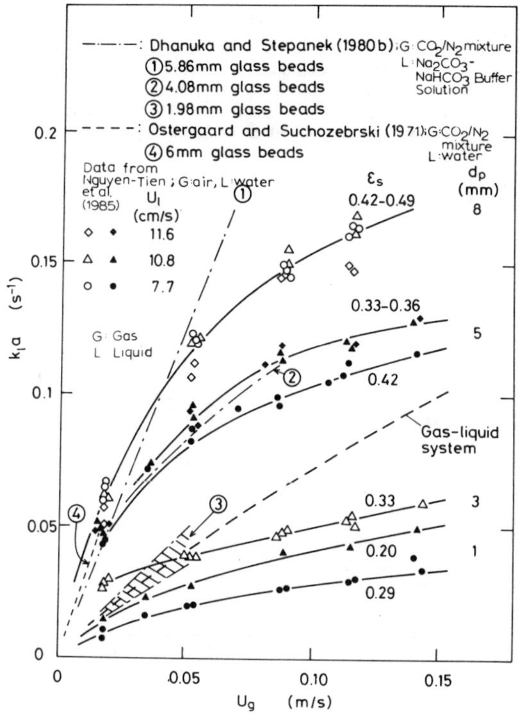

Figure 3.3 Volumetric mass transfer coefficient as a function of gas velocity in fluidized beds of glass beads.

reaches a minimum at a d_p of about 1 ~ 2 mm, and then increases with increasing particle diameter. It is noted that, in contrast to the significant U_g effect, relatively little effect of U_ℓ on $k_\ell a$ is evident based on the data of Dhanuka and Stepanek (1980b), Nguyen-Tien et al. (1985) and Chang et al. (1986) with d_p greater than 1 mm .

The important effect of bubble behavior on mass transfer is also evident in terms of the axial variations in local mass transfer rate. In practice, the distributor and bulk fluidized bed regions of the system have different mass transfer behavior and the relative importance of each region depends on operating parameters such as d_p, U_g, and D_c. In a two-dimensional bed using four 1 cm ID injector nozzles (22 cm apart) as a gas distributor, Alvarez-Cuenca and Nerenberg (1981) and Alvarez-Cuenca et al. (1984) obtained the average local volumetric mass transfer coefficient over the axial interval [0,z], $(k_\ell a)_z$, based on diffused oxygen concentration measurement and a liquid phase plug flow analysis. The distributor region at high gas and/or liquid velocities occupies as much as 25% of

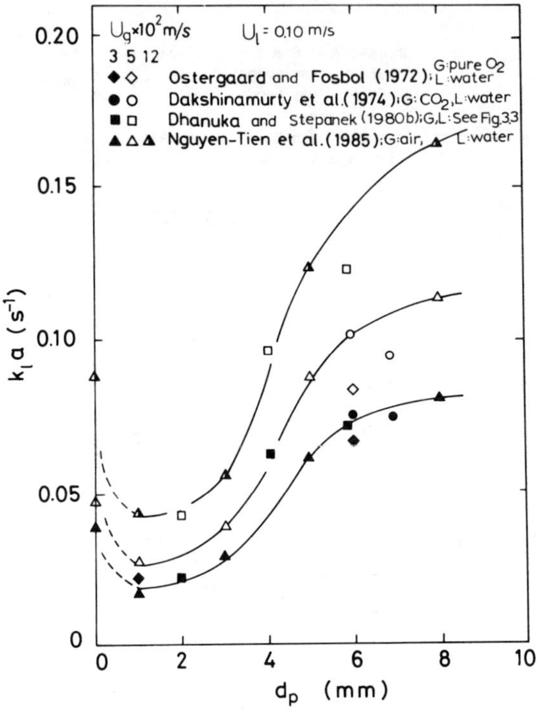

Figure 3.4 Volumetric mass transfer coefficient as a function of particle diameter in fluidized beds of glass beads.

the bed. They reported the axial variation of $(k_\ell a)_z$ at various gas and liquid velocities for 1 mm and 5 mm particles as shown in Figs. 3.5(a) and 3.5(b), respectively. It is seen that at low gas and liquid velocities $(k_\ell a)_z$ for both particles is small. Notably, $(k_\ell a)_z$ increases, reaches a maximum, and then decreases with axial distance for 5 mm particles. This behavior is similar to that observed by Ostergaard and Fosbol (1972) for the same size particles. At high gas and/or liquid velocities, however, nonuniformity in $(k_\ell a)_z$ distribution in the distributor region occurs for both particles. It is noted that for 1 mm particles, $(k_\ell a)_z$ in the distributor region is higher than that in the bulk region. Thus, a shallow bed containing 1 mm particles would exhibit particularly high mass transfer rates. $(k_\ell a)_z$ in the distributor region for 5 mm particles, however, can be lower than that in the bulk region. Thus, the concentration distribution in the liquid phase at high gas and/or liquid velocities can best be accounted for by considering separately the concentration in each of the two regions as described, for example, by the two zone model of Alvarez-Cuenca et al. (1984). When a

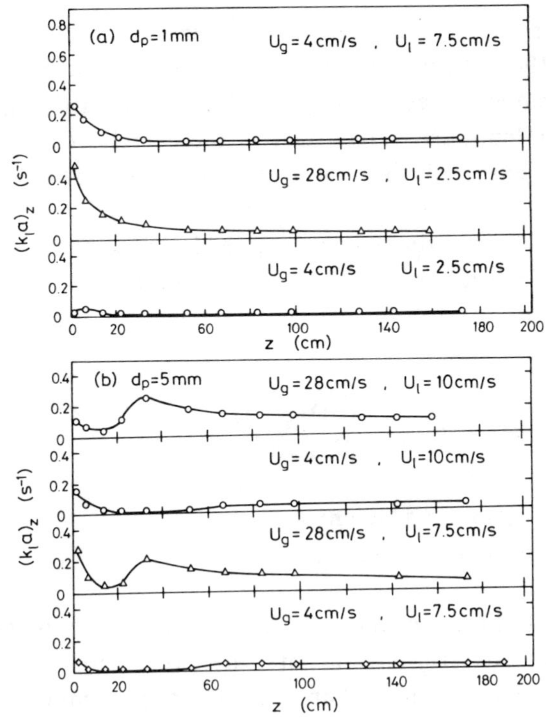

Figure 3.5 Axial variation of $(k_{\ell}a)_z$ in air-water-glass bead fluidized beds for 1 mm and 5 mm particles at various gas and liquid velocities (from Alvarez-Cuenca and Nerenberg, 1981).

perforated plate with orifices considerably smaller than 1 cm is used as a gas distributor, as is commonly encountered, a "one" zone model based on the axial dispersion concept could be used to approximate the overall bed mass transfer behavior at moderate gas and liquid velocities.

Table 3.1 summarizes the empirical correlations for the overall volumetric gas-liquid mass transfer coefficients for various operating conditions. As shown in the table, some correlations are specific in particle size; e.g., those of Ostergaard (1978) and Dhanuka and Stepanek (1980a,b). The correlations of Dakshinamurty et al. (1974a,b) and Dakshinamurty and Rao (1976) considered the effects of particle size, particle terminal velocity, and gas velocity (Eqn. (T-3.1.1)). Nguyen-Tien et al.'s (1985) correlations (Eqns. (T-3.1.2a) and (T-3.1.2b)) apply for particles smaller than 1 mm and greater than 1 mm, respectively; Eqn. (T-3.1.2a) extends from the expanded bed regime to the transport regime of three-phase fluidization. The correlation of Chang et al. (1986) (Eqn. (T-3.1.3)) for $k_\ell a$ has the widest range of applicability among all correlations and is useful although the extent of the U_ℓ dependency is overstressed in the correlations. Care must be exercised when extrapolating these correlations beyond the range of data in which the correlations were obtained.

Note that, for a given system, to estimate $k_\ell a$ for a species with diffusivity D_M from a known $(k_\ell a)_0$ for a species with diffusivity $D_{M,O}$, the relationship according to the penetration model is usually followed, i.e.,

$$k_\ell a = (k_\ell a)_0 \left(\frac{D_M}{D_{M,O}}\right)^{1/2} \tag{3.17}$$

3.1.1.3 Mass Transfer Coefficient k_ℓ and the Interfacial Area "a"

In multi-bubble systems, the typical approach to treating mass transfer is the determination of the volumetric mass transfer coefficient; however, a more fundamental approach offering greater insight is to resolve the volumetric coefficient into k_ℓ and "a". Studies involving the separation of "a" and k_ℓ require a separate measurement of the interfacial area. Lee and Worthington (1974) measured the interfacial area just above the fluidized bed surface by a light transmission method and found k_ℓ to increase with an increase in bubble size, over bubble diameters ranging from 0.4 to 2.7 mm; this behavior is consistent with the trends predicted and observed for a single bubble. Dhanuka and Stepanek (1980a,b) found that k_ℓ for 4.08 mm particle systems was greater than that in 5.86 mm or 1.98 mm particle systems, as shown in Fig. 3.6. Furthermore, for d_p's of 5.86 mm, 4.08 mm and 1.98 mm, the values for k_ℓ were found to be constant, to decrease and to increase, respectively, with increasing U_g. This behavior can be explained, in part, based on the k_ℓ versus bubble size relationship for the small bubble, transition and large bubble regimes defined by Calderbank (1967a). For a d_p of 5.86 mm, the bubble sizes are uniformly distributed and are close to the small bubble regime where k_ℓ is independent of bubble size variation and hence the gas velocity. For a d_p of 4.08 mm, the bubble sizes are in the transition regime where k_ℓ decreases (or increases) with decreasing (or increasing) bubble size; a slight decrease in bubble size with increasing gas velocity which was verified indirectly for this particle

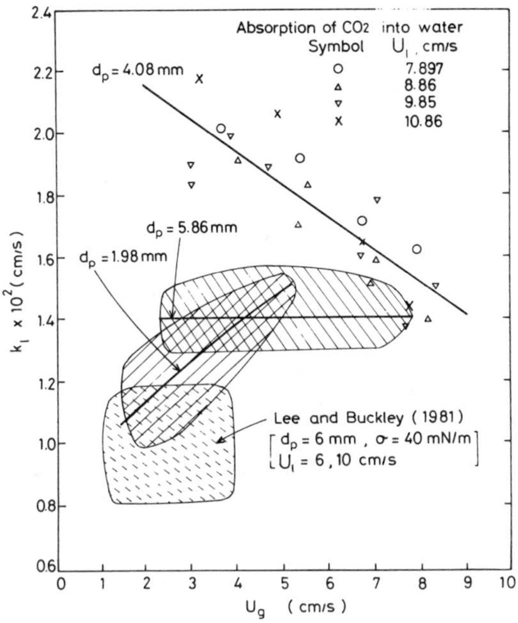

Figure 3.6 Dependence of the mass transfer coefficient k_ℓ on the gas and liquid velocities (from Dhanuka and Stepanek, 1980a,b).

probably contributes to k_ℓ decreasing with gas velocity (Dhanuka and Stepanek, 1980b). For a d_p of 1.98 mm, the bubble size can be very large due to bubble coalescence; an increase in the gas velocity increases the average bubble size and hence k_ℓ. Lee and Buckley (1981) reported that in the fluidization of 6 mm glass beads using an aqueous octanol solution, the bubble size was smaller than that in a pure water system and its distribution was quite uniform. The values of $k_\ell a$ were, however, nearly the same as those for an octanol-free system suggesting that k_ℓ in the octanol solution is lower than that in pure water; this conclusion is consistent with the observations that an increase in bubble size increases the single bubble mass transfer coefficient. Lee and Buckley (1981) also indicated that k_ℓ in a three-phase fluidized bed with 6 mm glass beads using an aqueous octanol solution could be estimated reasonably well by the correlation of Calderbank (1967a) proposed for the small bubble regime.

The significant variation in bubble behavior with particle size suggests that the interfacial area will also show significant variation with particle size. Essentially, any changes in particle size, gas

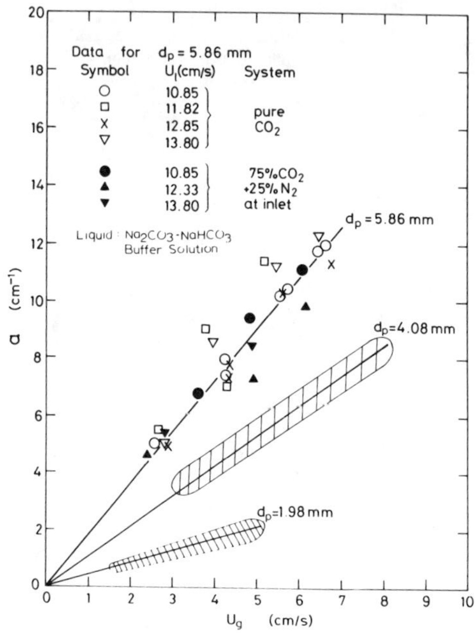

Figure 3.7 Dependence of the interfacial area on the gas and liquid velocities (from Dhanuka and Stepanek, 1980a,b).

velocity, or liquid velocity which alter the prevalent flow regime will also significantly alter the interfacial area. Dhanuka and Stepanek (1980a,b) found that the particle size had a profound effect on "a"; the interfacial area increased markedly with an increase in particle size, as shown in Fig. 3.7. The combined result of a moderate change in k_ℓ and the large increase in "a" results in an increase in $k_\ell a$ with increasing particle diameter. In addition, "a" was found to increase significantly with an increase in gas velocity. There is, however, relatively little effect of the liquid velocity on "a" based on the data of Dhanuka and Stepanek (1980b), Nguyen-Tien et al. (1985) and Chang et al. (1986) with d_p greater than 1 mm. The correlation of Chang et al. (1986) (Eqn. (T-3.1.4)) for "a" shown in Table 3.1 has the widest range of applicability among all correlations and is useful although the extent of the U_ℓ dependency is overstressed as in that for $k_\ell a$.

3.1.1.4 Summary

The complex behavior of volumetric mass transfer coefficients indicates that gas-liquid mass transfer is a strong function of the bubble flow patterns prevalent in the system. The separate evaluation of k_ℓ and "a" indicates that particle size has a profound effect on "a" but only a moderate effect on k_ℓ. In addition, k_ℓ was found to be invariable or to increase with the bubble diameter, but "a" showed the opposite behavior. Correlations for $k_\ell a$ proposed by various investigators can be employed only for the limited conditions under which they were developed. Separate consideration of mass transfer effects in the distributor region from the bulk fluidized bed region is necessary, particularly, at high gas and liquid velocities. For systems containing small particles (less than 1 mm), the correlation equation of Nguyen-Tien et al. (1985) can be used. For large particles (larger than 1 mm) the correlation of Chang et al. (1986) can be used. Further efforts are required to obtain a generalized expression for the mass transfer coefficient.

3.1.2 Liquid-Solid Mass Transfer

Like gas-liquid mass transfer, the rate of liquid-solid mass transfer can be described by the product of three terms: the overall mass transfer coefficient, the liquid-solid interfacial area, and the concentration difference. However, unlike gas-liquid mass transfer, the external surface of the solids is undeformable and the liquid-solid interfacial area remains constant regardless of the hydrodynamic conditions so long as the external particle surface remains completely wetted. Thus, only the mass transfer coefficient depends upon the system hydrodynamics and consequently, is a strong function of the operating variables and physical properties of the individual phases.

Studies of liquid-solid mass transfer in three-phase fluidized beds are limited to date. Mass transfer in a liquid-solid fluidized bed, an important limiting condition of a three-phase fluidized bed, provides some fundamental information on mass transfer behavior.

3.1.2.1 Brief Review on Liquid-Solid Fluidized Beds

Most experimental data on liquid-solid mass transfer have been obtained from dissolution of solid particles into a liquid (McCune and Wilhelm, 1949; Evans and Gerald, 1953; L.T. Fan et al., 1960; Rowe and Claxton, 1965; Couderc et al., 1972; Upadhyay and Tripathi, 1975; Damronglerd et al., 1975; Laguerie and Angelino, 1975; Vanadurongwan et al., 1976; Tournie et al., 1977; Kumar and Upadhyay, 1981; and Ballesteros et al., 1982). Several investigators have used the dissolution of benzoic acid in water since the acid powder can be pelleted relatively easily, the solubility of the acid in water is low, and the physical properties of the system are well known. Other dissolution systems used include 2-naphthol-water, cinnamic acid-water, salicylic acid-water, and steric acid-water. Studies of more complex phenomena such as ion exchange (Koloini et al., 1977; Rahman and Streat, 1981), adsorption of a solute from a liquid solution (Ganho et al., 1975), and crystal growth (Garside and Mullin, 1968; Laguerie and Angelino, 1975) have also been reported. Finally, data have been obtained from experiments involving transfer between a liquid and solid particles arranged in typical arrays in a channel (Rowe and Claxton, 1965), between a liquid and a fixed sphere inserted into a fluidized bed of inert particles of equal diameters, and from studies of semifluidized beds (L.T. Fan et al., 1960).

Researchers have approached the problem of correlating mass transfer data in several ways. A common approach is to use dimensionless expressions for the Sherwood number, Sh, or the Chilton and Colburn factor, J_d. Table 3.2 summarizes typical correlations for liquid-solid mass transfer coefficients presented in the literature. Experimental calculations of the mass transfer in the literature for liquid-solid fluidized beds have traditionally assumed plug flow and neglected time variations of the particle external surface area. Correlations were most often expressed in terms of dimensionless numbers, often in the form of a power law, and may be divided into two groups. In the first group, the void fraction is explicitly involved in the correlation (e.g., Chu et al., 1953). In the second group, the void fraction is not used or has been replaced by the Archimedes number or sometimes a density number based on Richardson and Zaki's law of expansion (1954) (e.g., Calderbank, 1967b; Ballesteros et al., 1982). Kikuchi et al. (1983), in a different approach, employed the concept of energy dissipation to mass transfer between suspended particles and a liquid in liquid-solid fluidized beds. The correlation for the mass transfer coefficient, expressed in terms of the rate of energy dissipation, is advantageous in that the correlation is of a semi-mechanistic nature and is generally accurate in accounting for mass transfer behavior in various flow fields (Ohashi et al., 1979, 1981a, 1981b, 1982).

Table 3.2 Correlations for Liquid-Solid Mass Transfer in a Two-Phase (Liquid-Solid) Fluidized Bed

Investigators	Correlations
Chu et al. (1953)	$J_d = 5.7\ [Re_\ell/(1-\varepsilon)]^{-0.78}$ $(1<Re_\ell/(1-\varepsilon)<30)$ $J_d = 1.77\ [Re_\ell/(1-\varepsilon)]^{-0.44}$ $(30<Re_\ell<(1-\varepsilon)<10000)$
Calderbank (1967a)	$Sh = 0.31\ Sc^{1/3}Ar^{1/3}$ for Stoke's and Newton's regimes
Ballesteros et al. (1982)	$Sh = 0.245\ \phi_s^{1.35}Ar^{0.323}Mv^{0.30}Sc^{0.40}$ for $5 \times 10^5 < Ar < 3 \times 10^6$; $0.7 < \phi_s < 1$ ϕ_s = shape factor Equivalent diameter should be used in calculating Sh and Ar
Kikuchi et al. (1983)	$Sh = 2 + 0.51\ (e^{1/3}d_p^{4/3}/\nu_\ell)^{0.60}Sc^{1/3}$ for $10 < (e^{1/3}d_p^{4/3}/\nu_\ell) < 10^3$ where e is the energy dissipation rate per unit mass of liquid defined for gas-liquid-solid fluidization systems by $\frac{[(U_g+U_\ell)(\varepsilon_s\rho_s+\varepsilon_\ell\rho_\ell+\varepsilon_g\rho_g) - U_\ell\rho_\ell - U_g\rho_g]\ g}{\varepsilon_\ell\rho_\ell}$ and correlated for liquid-solid fluidized systems by Kikuchi et al. (1983) as $(e^{1/3}d_p^{4/3}/\nu_\ell) = (C_{dt}/2)^{1/3}Re_t^{2/3}Re_\ell^{1/3}$ where Re_t is based on particle terminal velocity and C_{dt} is drag coefficient of a single particle moving under the terminal velocity condition.

3.1.2.2 Liquid-Solid Mass Transfer Coefficients in Three-Phase Fluidized Beds

Liquid-solid mass transfer in a three-phase fluidized bed has been studied experimentally by Arters (1984), Arters and Fan (1984, 1988), Hassanien et al. (1984), Prakash et al. (1984, 1987), Nikov and Delmas (1987), Fukuma et al. (1988) and Arters et al. (1989). Arters (1984) and Arters and Fan (1984) examined the effects of gas flow rate, liquid flow rate, and particle size on the liquid-solid mass transfer in three-phase fluidized beds using air, water, and benzoic acid pellets. Hassanien et al. (1984), Nikov and Delmas (1987), and Fukuma et al. (1988) used the diffusion-controlled cathodic reduction of ferricyanide ions on a fixed sphere electrode in a fluidized bed of inert particles in an electrolytic support solution of NaOH. In the diffusion-controlled electrochemical method, the mass transfer coefficient can be readily determined from the limiting current measured (also see Section 3.1.3). Hassanien et al. (1984) varied liquid properties by the addition of polyethylene glycol to the solution and by varying the temperature of the liquid. The effects of gas and liquid velocity and particle diameter and density on the mass transfer coefficient were also studied. Nikov and Delmas (1987) extended the range of particle density studied. Fukuma et al. (1988) used radially-fixed multi-sphere electrodes to measure the liquid-solid mass transfer coefficient, k_s, in various two- and three-phase flow systems involving fixed and fluidized beds. They studied the variation of k_s with radial position and compared the behavior of k_s for various flow systems. Prakash et al. (1984; 1987) used dissolution of both tethered and free floating benzoic acid coated glass beads in a air-water-glass bead fluidized bed to determine the effect of gas and liquid velocity on mass transfer coefficients. Prakash et al. (1987) further studied the effects of restricting the dissolving particle's motion, surfactants, and liquid void fraction on the liquid-solid mass transfer coefficient in the same fluidized bed. Arters and Fan (1988) reproduced and critically examined differences in experimental technique in the prior literature in two-phase and three-phase fluidized beds in order to propose a unified mass transfer correlation.

The conventional plug flow model used in the analysis of liquid-solid fluidized beds cannot be adequately applied to the solute concentration gradient found in dissolution experiments in mass transfer in a three-phase fluidized bed. Arters and Fan (1984) employed an axial dispersion model (see Section 3.2.1) to analyze axial solute concentration profiles in three-phase fluidized beds. The mass transfer coefficient, k_s, was obtained by fitting the analytical solution for the solute concentration to the experimental concentration at the exit of the fluidized bed. In the calculation, the axial liquid dispersion coefficient was estimated based on the correlation of Kim and Kim (1983). Note that this analysis can be simplified in experiments in which the bulk solute concentration is assumed constant, such as those of Prakash et al. (1984, 1987) and Hassanien et al. (1984).

Results from the three different types of mass transfer experiments in three-phase fluidized beds, that is, dissolution into a liquid of non-negligible solute concentration (e.g., Arters and Fan, 1984), dissolution into a liquid of negligible solute concentration (e.g.,

Prakash et al., 1984) and electrochemical reduction (e.g., Hassanien et al., 1984) are consistent with each other and with two-phase fluidized bed results. Each study revealed that k_s is practically independent of liquid velocity and particle size, consistent with results obtained in two-phase fluidized beds (Rowe and Claxton, 1965; Tournie et al., 1979; Ballesteros et al., 1982). Further, the presence of the gas phase has a positive, progressive effect on the mass transfer coefficient. Hassanien et al. (1984) indicated the effect of liquid properties to be similar in two- and three-phase fluidized beds (i.e., k_s proportional to $Sc^{-2/3}$). Hassanien et al. (1984) and Nikov and Delmas (1987) reported the effect of the gas phase decreased with increasing particle density. According to Fukuma et al. (1988), the values of k_s are almost constant radially in the core region of the bed, but in the near-wall region, they are lower by about 20% than those in the core region. Arters and Fan (1988) indicated this effect diminishes with increasing column diameter and disappears for columns of greater than about 12 cm in diameter, due to the greater radial mixing in the larger diameter bed.

Differences in the results of the literature due to experimental conditions also exist. Arters and Fan (1988) delineated limitations in the use of fixed probe systems such as these used by Hassanien et al. (1984), Nikov and Delmas (1987) and Fukuma et al. (1988); fixed probes immersed in a fluidized system yield accurate results for that particular location in the bed, but due to the dynamic, mobile nature of the fluidized particle may not accurately reflect the overall mass transfer coefficient. Mobile (tethered) particles and free floating particles in inert beds, on the other hand, were found to yield results nearly identical to those in beds of 100% active particles. Note that the mass transfer coefficient of a given single particle is insensitive to the properties of the particles around it, as long as good solids mixing is achieved. Differences were also obtained between results for negligible and non-negligible concentrations of surfactants. Arters and Fan (1984, 1988) found little difference in the mass transfer between beds of light particles (ρ_s < 1.5 g/cm^3) in negligible and non-negligible concentrations of benzoic acid, a weak surfactant, while Prakash et al. (1987) found the presence of surfactants benzoic acid and methylisobutyl ketone in concentrations as low as 150 ppm had a pronounced positive effect on the mass transfer coefficient in beds of dense particles (ρ_s = 2.5 g/cm^3). However, Prakash et al. (1987) reported a cocurrent doubling of the gas holdup in their fluidized system in the presence of these particular surfactants not seen in other studies (Gorowara and Fan, 1988), indicating the possibility of foaming or unstable operation. Other differences in experimental results may also be flow regime related. Arters and Fan (1984), operating in the slugging regime, found continuous increases in k_s through gas velocities as high as 25 cm/s, while Prakash et al. (1987), working in the dispersed regime, noted that a constant k_s was obtained at gas velocities above 20 cm/s. A limiting effect with gas velocity has been seen in liquid-solid mass transfer in aerated suspensions as well (Saito and Kobayashi, 1965; Sanger and Deckwer, 1981). Arters and Fan (1988) have also noted that effects on the mass transfer of porous and perforated plate distributors are limited to a relatively small region of the bed while Prakash et al. (1987) found a decrease in the mass transfer in beds utilizing more inefficient spouted gas distributors.

Arters et al. (1989) combined the instantaneous capabilities of the electrochemical reduction method with two-dimensional visualization techniques to identify regions of increased mass transfer of a single mobile particle in the wake region of a single bubble rising in a liquid-solid fluidized bed. They found increased mass transfer in regions of high shear, such as between two shed vortices, along the free shear layer at the edge of the bubble, or in the primary wake directly below the bubble.

3.1.2.3 Correlation of Liquid-Solid Mass Transfer in Three-Phase Fluidized Beds

Fukuma et al. (1988) and Arters and Fan (1988) applied the concept of energy dissipation to three-phase fluidized bed systems and found the correlation proposed by Ohashi et al. (1981b) for particles in various flow fields provides a good prediction of the liquid-solid mass transfer:

$$Sh = 2 + 0.52(e^{1/3}d_p^{4/3}/\nu_\ell)^{0.59}\ Sc^{1/3} \tag{3.18}$$

where e is the energy dissipated per unit mass of liquid (see Table 3.2 for definition). The correlation is considered valid over the range: $1 < e^{1/3}\ d_p^{\ 4/3}/\nu_\ell < 10^4$, where $6.0 < Re_\ell < 2500$, $0 < Re_g < 120$, $8500 < Ar < 2.3 \times 10^7$ and $860 < Sc < 19{,}900$. Note that, as calculated directly from experimental pressure drop and phase holdups by Arters and Fan (1988), the expression for k_s is relatively insensitive to d_p and U_ℓ, in agreement with the mass transfer literature. However, correlations proposed to calculate e based on a separation of phase effects, such as by Kikuchi et al. (1983) or Fukuma et al. (1988), lead to over-emphasized dependence of k_s on d_p and U_ℓ. Note also that the equation does not appear to hold for foaming systems.

3.1.3 Wall-to-Bed Mass Transfer

Wall-to-bed mass transfer occurs in three-phase fluidized bed electrode vessels (Oloman and Watkinson, 1975; Kusakabe et al., 1981). Morooka et al. (1979) measured the wall-to-bed mass transfer coefficient in liquid-solid and gas-liquid-solid fluidized beds using the diffusion controlled reduction of copper ions on a flat copper cathode. Yasunishi et al. (1988) also measured the wall-to-bed mass transfer coefficient using the diffusion-controlled cathodic reduction of ferricyanide ions. In such electrochemical methods, the mass transfer coefficient (k_w) is related to the limiting current (I) and the concentration of copper ions in the bulk liquid (C) by I/n_eFA_wC, where F is Faraday's constant, n_e is the number of electrons per ion reacted, and A_w is the surface area of the electrode. The mass transfer coefficients in both two- and three-phase systems are shown in Fig. 3.8. It is seen that k_w in three-phase fluidized beds is generally larger than or comparable to that in liquid-solid fluidized beds. The value of k_w for either liquid-solid or gas-liquid-solid systems increases with increasing U_ℓ in the lower range of U_ℓ, then passes through a maximum and decreases with further increasing U_ℓ. In the liquid-solid system it finally approaches that in the open pipe liquid flow at a bed porosity of unity. The maximum k_w occurs at a bed porosity of about 0.6 ~ 0.7. Morooka et al. (1979) correlated k_w for both liquid-solid and gas-liquid-solid fluidized beds with the expressions:

$$\epsilon(St_{\ell,g}Sc^{2/3}) = \epsilon\left(\frac{k_w}{U_\ell + U_g}\right)\left(\frac{\mu_\ell}{\rho_\ell D_M}\right)^{2/3} = 0.30\left[\frac{d_p(U_\ell + U_g)\rho_\ell}{\mu_\ell(1-\epsilon)}\right]^{-0.33} \tag{3.19a}$$

for $\dfrac{d_p(U_\ell + U_g)\rho_\ell}{\mu_\ell(1-\epsilon)} > 60$ and

$$\epsilon(St_{\ell,g}Sc^{2/3}) = \epsilon\left(\frac{k_w}{U_\ell + U_g}\right)\left(\frac{\mu_\ell}{\rho_\ell D_M}\right)^{2/3} = 0.60\left[\frac{d_p(U_\ell + U_g)\rho_\ell}{\mu_\ell(1-\epsilon)}\right]^{-0.50} \tag{3.19b}$$

for $\dfrac{d_p(U_\ell + U_g)\rho_\ell}{\mu_\ell(1-\epsilon)} < 60$

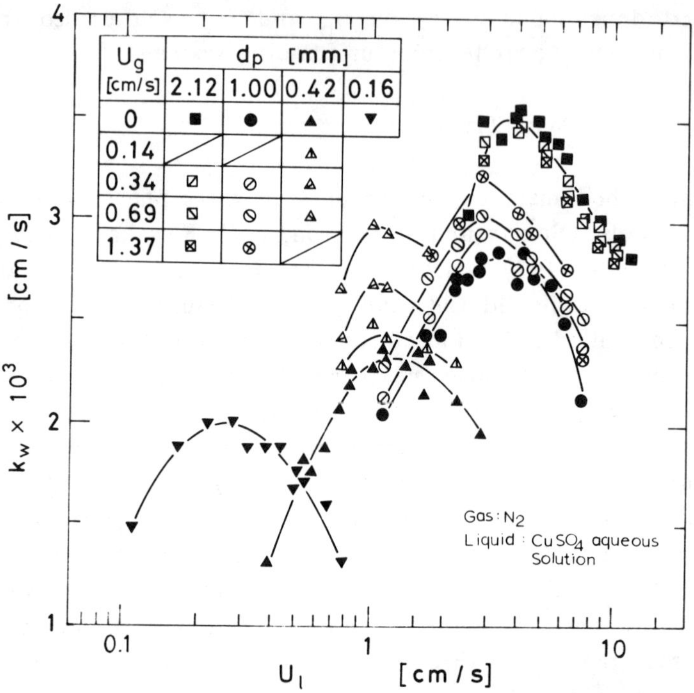

Figure 3.8 Dependence of the wall-to-bed mass transfer coefficient on the liquid velocity (from Morooka et al., 1979).

k_w in a liquid-solid fluidized bed can be estimated by setting U_g and ϵ_g equal to zero. Note that an analogy exists between the wall mass and heat transfer coefficients in both liquid-solid (Muroyama et al., 1986a, Yasunishi et al., 1988) and gas-liquid-solid fluidized beds (Muroyama et al., 1986b). Furthermore, similar to the variation of k_w with respect to U_ℓ, a maximum value of the wall-to-bed heat transfer coefficient, h, was observed in both liquid-solid and gas-liquid-solid fluidized beds (Magiliotou et al., 1987). Clearly, the behavior for mass transfer in such systems can be interpreted by the same mechanisms as that for the heat transfer (see Section 3.3.2.3 for interpretation).

3.2 PHASE MIXING

The extent of backmixing of both the liquid and solid phases depends heavily on the rising motion of the gas bubbles. Generally, in three-phase fluidized beds appreciable backmixing may occur in the liquid phase. However, relatively little backmixing exists in the gas phase, especially for beds containing small particles. Segregation or stratification phenomena of solid particles may occur in a fluidized bed containing particles with a wide size and/or density distribution.

3.2.1 Liquid Mixing

Liquid phase mixing in three-phase fluidized beds can be described using the dispersion model. A two-dimensional model considers both radial and axial dispersion while a one-dimensional model considers only axial dispersion. Both axial and radial dispersion coefficients are strong functions of operating conditions such as liquid and gas velocities and particle properties. Liquid phase axial mixing can approach two ideal situations: complete backmixing, where the liquid is considered perfectly mixed, and plug flow, where no mixing occurs. Evaluations of liquid phase dispersion coefficients are based on a tracer injection method and subsequent analysis of the mean and the variance of the system response curves (e.g., Wen and Fan, 1975).

Based on the assumptions that the velocities and holdups of individual phases are uniform in the radial and axial directions and the axial and radial dispersion coefficients, $E_{z\ell}$ and $E_{r\ell}$, are constant throughout the fluidized bed, the two-dimensional unsteady-state dispersion model is

$$\frac{\partial C}{\partial t} = E_{z\ell}\frac{\partial^2 C}{\partial z^2} + \frac{E_{r\ell}}{r}\frac{\partial}{\partial r}\left(r\frac{\partial C}{\partial r}\right) - \frac{U_\ell}{\epsilon_\ell}\frac{\partial C}{\partial z} \tag{3.20}$$

Equation (3.20) can be reduced to the axial dispersion (one-dimensional) model when the radial dispersion can be neglected in comparison with axial dispersion; i.e.,

$$\frac{\partial C}{\partial t} = E_{z\ell}\frac{\partial^2 C}{\partial z^2} - \frac{U_\ell}{\epsilon_\ell}\frac{\partial C}{\partial z} \tag{3.21}$$

Note that $E_{z\ell}$ in Eqns. (3.20) and (3.21) is defined based on the linear liquid velocity.

To evaluate the axial dispersion coefficient, an imperfect tracer pulse method may be conveniently used in place of the perfect tracer pulse method. Moreover, a transfer function analysis can be readily made to evaluate the mixing parameters based on the axial dispersion model over the conventional moment analysis. The disadvantage in the moment analysis stems from the sensitivity of the calculated first and second moments to small measurement errors in the tails of the response curves (Ostergaard and Michelsen, 1969; Michelsen and Ostergaard, 1970). For an imperfect tracer response system considering only axial dispersion, Eqn. (3.21) is subject to the initial and boundary conditions:

$$t \le 0, \qquad 0 < z < \infty \qquad C = 0 \tag{3.22a}$$

$$t > 0, \qquad z = 0 \qquad U_\ell C_{in} = -\epsilon_\ell E_{z\ell} \frac{\partial C}{\partial z} + U_\ell C \tag{3.22b}$$

$$z = \infty \qquad C = 0 \tag{3.22c}$$

The transfer function for a linear system enclosed between two measuring points, 1 and 2, can be defined as

$$F(s) = \frac{\bar{C}_2(s)}{\bar{C}_1(s)} = \frac{\int_0^\infty C_2(t)\exp(-st)dt / \int_0^\infty C_2(t)dt}{\int_0^\infty C_1(t)\exp(-st)dt / \int_0^\infty C_1(t)dt} \tag{3.23}$$

By Laplace transformation of the model equation, F(s) can be related to the parameters of the axial dispersion model with infinite boundaries by

$$F(s) = \exp\left\{\frac{Pe_z}{2}\left[1 - \left(1 + \frac{4s\tau}{Pe_z}\right)^{1/2}\right]\right\} \tag{3.24}$$

where $Pe_z = U_\ell H/\epsilon_\ell E_{z\ell}$, $\tau = H\epsilon_\ell/U_\ell$. Rearranging Eqn. (3.24) gives

$$\frac{1}{\ln\{1/F(s)\}} = \tau \frac{s}{[\ln\{1/F(s)\}]^2} - \frac{1}{Pe_z} \tag{3.25}$$

If the axial dispersion model is applicable, a plot of

$$\frac{1}{\ln\{1/F(s)\}} \quad \text{versus} \quad \frac{s}{[\ln\{1/F(s)\}]^2}$$

gives a straight line for arbitrary values of s. Optimum parameter values of Pe_z and τ can be obtained in the range $2 < s\tau < 5$ (Hopkins et al., 1969). Thus, the mixing parameters τ and Pe_z can be readily obtained from the slope of the straight line and the intercept on the ordinate, respectively.

To evaluate the radial dispersion coefficient, the two-dimensional dispersion model, Eqn. (3.20) at a steady-state becomes

$$E_{z\ell}\frac{\partial^2 C}{\partial z^2} + \frac{E_{r\ell}}{r}\frac{\partial}{\partial r}\left(r\frac{\partial C}{\partial r}\right) - \frac{U_\ell}{\epsilon_\ell}\frac{\partial C}{\partial z} = 0 \qquad (3.26)$$

For a point-source tracer injection, Eqn. (3.26) is subject to the following boundary conditions and a mass balance equation for the tracer:

$$r = 0, \qquad \frac{\partial C}{\partial r} = 0 \qquad (3.27a)$$

$$r = R, \qquad \frac{\partial C}{\partial r} = 0 \qquad (3.27b)$$

$$z = -\infty, \qquad \frac{\partial C}{\partial z} = 0 \quad \text{and } C = 0 \qquad (3.27c)$$

$$Q = \pi R^2 U_\ell C_0 \qquad (3.27d)$$

where Q is the quantity of tracer injected per unit time and C_0 is the tracer concentration at $z = \infty$. The analytical solution at a downstream position is (Klinkenberg et al., 1953)

$$\frac{C}{C_0} = 1 + \frac{1}{2}\sum_{i=1}^{\infty}\frac{J_0(a_i r/R)}{J_0^2(a_i)}\frac{1}{q_i}\exp\{(\tfrac{1}{2} - q_i)\,Pe'_z\,(\tfrac{z}{R})\} \qquad (3.28)$$

where a_i is the i-th positive root for which $J_1(a_i)$ equals 0, $Pe_r = U_\ell R/\epsilon_\ell E_{r\ell}$, $Pe'_z = U_\ell R/\epsilon_\ell E_{z\ell}$, and q_i is given by

$$q_i = \left(\frac{1}{4} + \frac{a_i^2}{Pe_r Pe'_z}\right)^{1/2} \qquad (3.29)$$

Multiplying Eqn. (3.28) by the ratio of the cross-sectional average concentration to the volumetric flow average concentration (C_0/C_{mix}) yields

$$\frac{C}{C_{mix}} = \frac{C_0}{C_{mix}}\left[1 + \frac{1}{2}\sum_{i=1}^{\infty}\frac{J_0(a_i r/R)}{J_0^2(a_i)}\frac{1}{q_i}\exp\{(\tfrac{1}{2} - q_i)\,Pe'_z\,(\tfrac{z}{R})\}\right] \qquad (3.30)$$

Five terms of the summation are sufficient to correctly express the radial tracer concentration profile at a given axial position. The

Table 3.3 Studies on Axial or Radial Liquid Mixing in Three-Phase Fluidized beds

Investigators	Systems	Column Diameters	Particle Sizes and Densities	Ranges of Gas Velocity	Correlations (in SI Units) (Axial Mixing or Radial Mixing)
Vail et al. (1968)	air/water/molten sludge beads	146 mm ID	$d_p = 0.4 \sim 1.0$ mm (average diam. = 0.87 mm) $\rho_s = 2.7$ g/cm^3	$U_g = 0 \sim 10$ cm/s	None (axial mixing)
Michelsen and Ostergaard (1970)	air/water/glass beads	152.4 mm ID	$d_p = 1$ mm*, $\rho_s = 2.67$ g/cm^3 $d_p = 3$ mm*, $\rho_s = 2.45$ g/cm^3 $d_p = 6$ mm*, $\rho_s = 2.63$ g/cm^3 * nominal size	$U_g = 0 \sim 27$ cm/s	graphical presentation (axial mixing)
Kim et al. (1972)	air/water/glass beads; air/water/irregular gravel; air/water	660.4 mm wide x 25.4 mm thick rectangular column	$d_p = 6.0$ mm, $\rho_s = 2.3$ g/cm^3 $d_p = 2.6$ mm, $\rho_s = 2.52$ g/cm^3	$U_g = 0 \sim 26$ cm/s	$\frac{HMU_\ell}{H_o} - \left(\frac{HMU_\ell}{H_o}\right)_{U_g=0} = 0.068\left(Fr_\ell \frac{\rho_s}{\rho_\ell}\right)^{0.128} \cdot \left(Fr_g \frac{\rho_s}{\rho_g}\right)^{0.168} (Re_\ell\, Re_g)^{0.12}$ with $\left(\frac{HMU_\ell}{H_o}\right)_{U_g=0} = 5.05\left(Fr_\ell \frac{\rho_s}{\rho_\ell}\right)^{0.842} (Re_\ell)^{-0.450}$ and correlations for H/H_o $Fr_g = \frac{U_g^2}{gd_p}$, $Fr_\ell = \frac{U_\ell^2}{gd_p}$, $Re_\ell = \frac{d_p U_\ell \rho_\ell}{\mu_\ell}$, $Re_g = \frac{d_p U_g \rho_g}{\mu_g}$ (axial mixing)
Ohshima et al. (1977)	air/water/glass beads	50 mm ID	$d_p = 1.5 \sim 4.3$ mm	$U_g = 0 \sim 6.5$ cm/s	$\frac{d_p U_\ell}{\varepsilon_\ell E_{z\ell}} = 5.3\,(0.014)^{U_\ell/U_{\ell mf0}} \left(\frac{U_g}{\sqrt{2gd_p}}\right)^{0.67 (U_\ell/U_{\ell mf0})^{-0.23}}$ $0.35 < U_\ell/U_{\ell mf0} < 1.17$ (operation near minimum fluidization) (axial mixing)

Muroyama et al. (1978)	air/water/glass beads; air/water/alumina beads	60, 100 mm ID	$d_p = 0.215 \sim 4.8$ mm $\rho_s = 2.5$ g/cm^3 $d_p = 2.0$ mm $\rho_s = 3.6$ g/cm^3	$U_g = 0 \sim 30$ cm/s	$\frac{D_c U_\ell}{\varepsilon_\ell E_{z\ell}} = 1.01 (U_\ell)^{0.738}(U_g)^{-0.167}(D_c)^{-0.583}$ for coalesced bubble and slug flows $\frac{D_c U_\ell}{\varepsilon_\ell E_{z\ell}} = 26(d_p/D_c)^{1/2}$ $(0.02 < d_p/D_c < 0.12)$ for dispersed bubble flow (axial mixing)
Ostergaard (1978)	air/water/glass beads	216 mm ID	$d_p = 1.1, 3.0, 6.0$ mm	$U_g = 0 \sim 20$ cm/s	graphical presentation
El-Temtamy et al. (1979a,b)	air/water/glass beads	50 mm ID	$d_p = 0.45 \sim 3.0$ mm $\rho_s = 2.6 \sim 3.0$ g/cm^3	$U_g = 0 \sim 7$ cm/s	$\frac{U_\ell d_p}{\varepsilon_\ell E_{z\ell}} = 5.85 \times 10^{-6} (Re_\ell)^{1.156} D_c^{-1.156}$ (axial mixing) graphical presentation (radial mixing)
Joshi (1980)	no experiments				$E_{z\ell} = 0.29 (U_c + U_\ell) D_c$ $U_c = 1.31\{gD_c[U_g + U_\ell - \frac{\rho_\ell U_\ell}{\varepsilon_s\rho_s + \varepsilon_\ell\rho_\ell} - \varepsilon_s (\frac{\rho_s}{\varepsilon_s\rho_s + \varepsilon_\ell\rho_\ell} - 1) U_t - \varepsilon_g V_g]\}^{1/3}$ (axial mixing)
Kim and Kim (1983)	air/water/glass beads; air/water-CMC/glass beads; air/water-glycerol/glass beads; air/water-methanol/glass beads; air/water-Triton X-100/glass beads $\sigma = 38.5 \sim 76$ dyn/cm $\mu_\ell = 1.0 \sim 3.042$ mNs/m^2 (Newtonian fluid) and $7.85 \sim 27.12$ mNs^n/m^2 (non-Newtonian fluid)	145 mm ID	$d_p = 1.7, 3.0, 6.0$ mm $\rho_s = 2.5$ g/cm^3	$U_g = 0 \sim 12$ cm/s	$\frac{d_p U_\ell}{E_{z\ell}} = 20.19(\frac{d_p}{D_c})^{1.66}(\frac{U_\ell}{U_\ell + U_g})^{1.03}$ (T-3.3.1) $0.143 < U_\ell/(U_\ell + U_g) < 0.962$ $0.0022 < d_p/D_c < 0.08$ (axial mixing)

(Table 3.3 continued)

Investigators	Systems	Column Diameters	Particle Sizes and Densities	Ranges of Gas Velocity	Correlations (in SI Units) (Axial Mixing or Radial Mixing)
Nguyen-Tien et al. (1984)	air/water/glass beads (from gas-liquid mass transfer experiments)	140 mm ID	$d_p = 0.05 \sim 8$ mm	$U_g = 0 \sim 16$ cm/s	$\frac{d_p U_\ell}{E_{z\ell}} = 61 \left(\frac{d_p}{D_c}\right)^{1.66} \left(\frac{U_\ell}{U_\ell + U_g}\right)^{1.03}$ (T-3.3.2) $0.25 < \frac{U_\ell}{U_\ell + U_g} < 0.85$ $0.002 < d_p/D_c < 0.007$ for $d_p = 0.3 \sim 1.0$ mm (axial mixing)
Kato et al. (1985)	air/water/glass beads; air/water/porous alumina	120, 190 mm ID	$d_p = 0.52 \sim 5.2$ mm $\rho_s = 2.5$ g/cm^3 $d_p = 1.5 \sim 3.2$ mm $\rho_s = 1.7 \sim 1.8$ g/cm^3	$U_g = 0 \sim 16$ cm/s	$U_g < 9$ cm/s $Pe''_{z\ell} = \frac{U_g D_c}{E_{z\ell}} = Pe''_{z\ell}(BC)\left(1 + \frac{U_t}{U_\ell}\right)^{0.40} (g d_p^2 \rho_\ell/\sigma)^{0.225}$ $U_g > 25$ cm/s $\frac{U_g D_c}{E_{z\ell}} = Pe''_{z\ell}(BC)$ $9 < U_g < 25$ cm/s $Pe''_{z\ell} = Pe''_{z\ell 1} (U_g/9)^{2.31 \log_{10}(Pe''_{z\ell 2}/Pe''_{z\ell 1})}$ with $Pe''_{z\ell}(BC) = \left(\frac{\mu_\ell}{\mu_w}\right)^{0.07} \left(\frac{13(U_g/\sqrt{g D_c})}{1 + 6.5(U_g/\sqrt{g D_c})^{0.8}}\right)$ $Pe''_{z\ell 1}$ corresponds to the value of $Pe''_{z\ell}$ at $U_g = 9$ cm $Pe''_{z\ell 2}$ corresponds to the value of $Pe''_{z\ell}$ at $U_g = 25$ cm (axial mixing) BC : bubble column
Yang (1986)		Source data from Ostergaard (1971), El-Temtamy et al. (1979a), Kim and Kim (1983) and Kato et al. (1972,1985)			$\frac{U_\ell D_c}{E_{z\ell}} = 3.31 (Fr_m)^{0.35}$ where $Fr_m = 1 + \left(\frac{U_t}{U_\ell + U_g}\right)^2 \left(\frac{U_\ell}{U_\ell + U_g}\right) \left(\frac{U_\ell^2}{g D_c}\right)$ $3 \times 10^{-7} < Fr_m < 10^{-1}$ (axial mixing)

Kang and Kim (1986)	air/water/glass beads	102 mm ID	$d_p = 1.7, 4.0, 6.0$ mm $\rho_s = 2.5$ g/cm^3	$U_g = 0 \sim 10$ cm/s	$\frac{d_p U_\ell}{E_{r\ell}} = 28.3 \left(\frac{d_p}{D_c}\right)^{1.0} \left(\frac{U_\ell}{U_\ell + U_g}\right)^{1.16}$ (T-3.3.3) $0.019 < d_p/D_c < 0.06$ $0.33 < U_\ell/(U_\ell + U_g) < 0.86$ (radial mixing)
Yasunishi et al. (1987)	air/water/glass beads; air/water/alumina beads	74, 150 mm ID	$d_p = 1.1 \sim 4.8$ mm $\rho_s = 2.5$ g/cm^3; (glass beads) $d_p = 2.0$ mm $\rho_s = 3.55$ g/cm^3 (alumina beads)	$U_g = 0 \sim 25$ cm/s	$Pe_{r\ell,m} = 5.0(Re_m/Re_{m,mf0})^{-2.4} + 7.2 \times 10^{-4} \{Re_m(\mu_\ell/\mu_w)\}^{1.3}$ (T-3.3.4) $Pe_{r\ell,m} = d_{er}(U_\ell + U_g)/(\varepsilon_\ell E_{r\ell})$ $Re_m = d_{er} U_\ell \rho_\ell / (\varepsilon_\ell \mu_\ell)$ $Re_{m,mfo} = \frac{d_{er} U_{\ell mf0} \rho_\ell}{\varepsilon_{\ell mf0} \mu_\ell}$ $d_{er} = (\varepsilon_g + \varepsilon_\ell) D_c / \{1 + 1.5(D_c/d_p)(1 - \varepsilon_g - \varepsilon_\ell)\}$ (radial mixing)

mixing parameters involved in Eqn. (3.30) are C_0/C_{mix}, Pe_r, and Pe'_z. The values of C_0/C_{mix} and Pe_r may be determined using a least squares fitting method with Pe'_z estimated by the one-dimensional (axial) dispersion model.

The results of mixing studies have been reported in terms of the height of mixing unit or the Peclet number. The height of axial mixing unit is defined as (Michelsen and Ostergaard, 1970; Kim et al., 1972)

$$HMU_\ell = \frac{2H}{Pe_z} = \frac{2E_{z\ell}}{V_\ell} \tag{3.31}$$

Two types of Peclet numbers have been used to correlate the liquid axial dispersion coefficient, $E_{z\ell}$: a Peclet number based on the linear liquid velocity:

$$Pe_{z\ell} = \frac{V_\ell D_c}{E_{z\ell}} = \frac{U_\ell D_c}{\epsilon_\ell E_{z\ell}} \tag{3.32}$$

and a Peclet number based on the superficial liquid or gas velocity:

$$Pe'_{z\ell} = \frac{U_\ell d_p}{E_{z\ell}} \tag{3.33}$$

$$Pe''_{z\ell} = \frac{U_g D_c}{E_{z\ell}} \tag{3.34}$$

$Pe_{z\ell}$ can be readily converted to the column Peclet number (Pe_z) which is usually the primary parameter in design calculations. However, to convert $Pe'_{z\ell}$ or $Pe''_{z\ell}$ into Pe_z, information on ϵ_ℓ is needed.

Most mixing studies have considered axial mixing while only a limited number of studies have considered both axial and radial mixing. Table 3.3 summarizes the various studies and correlations on axial and radial mixing of the liquid phase in the bed. Detailed comments on them are given below.

3.2.1.1 Axial Mixing

Generally, axial liquid mixing in the nearly uniform dispersed bubble regime is limited and the axial dispersion coefficient is small and the Peclet number is large. However, in the coalesced bubble and slugging regimes, the nonuniform flow behavior creates significant axial liquid mixing. Ostergaard and Michelsen (1968) and Ostergaard (1969) reported that the degree of axial liquid mixing in beds of 0.25 and 1 mm particles at high liquid rates was comparable to that in a corresponding gas-liquid flow system both in intensity and trends. Mixing intensity increases with an increase in gas velocity and with a decrease in liquid velocity. However, the axial mixing intensity in a bed of 6 mm particles is low in the dispersed bubble flow regime where the superficial gas velocity is lower than 15 cm/s. A bed of 3 mm particles exhibits rather complex behavior. In the dispersed bubble regime the degree of liquid mixing in the bed was very low and decreases with increasing gas velocity; however, when bubble coalescence becomes significant upon increasing gas velocity a sharp increase in the degree of liquid mixing was observed.

Muroyama et al. (1978) examined the degree of liquid mixing in the coalesced bubble, dispersed bubble, and slugging regimes. The effect of column size on liquid mixing was assessed by comparing the literature data for column sizes ranging from 60 to 152.4 mm. As shown in Fig. 3.9 the axial dispersion coefficient, $E_{z\ell}$, strongly depends on the bubble flow pattern as well as column size. The axial dispersion coefficient may increase with particle size for a given column size in the dispersed bubble regime (Fig. 3.9(b)). This behavior may be attributed to increased liquid phase turbulence caused by the increased liquid velocity required to fluidize a bed of larger particles. The behavior of $Pe_{z\ell}$ in the slugging regime was found to be similar to that in the coalesced bubble flow regime given in Fig. 3.9(a). Separate correlation equations were proposed for different flow regimes by Muroyama et al. (1978) as given in Table 3.3.

Ostergaard (1978) reported that the axial dispersion coefficient for 1 mm particles in a 228 mm ID column was 50% higher than that in a 152.4 mm ID column reported by Michelsen and Ostergaard (1970). Michelsen and Ostergaard (1970) indicated that the axial dispersion coefficient in the bed varies with the column diameter to the power of 1.1 to 1.7. This behavior is in reasonably good agreement with the results obtained by Muroyama et al. (1978) and is comparable to the 1.5 power obtained for gas-liquid bubble columns (Towell and Ackerman, 1972).

Joshi (1980), using an energy balance model originally proposed by Whalley and Davidson (1974), presented

$$E_{z\ell} = 0.29\,(U_\ell + U_c)D_c \tag{3.35}$$

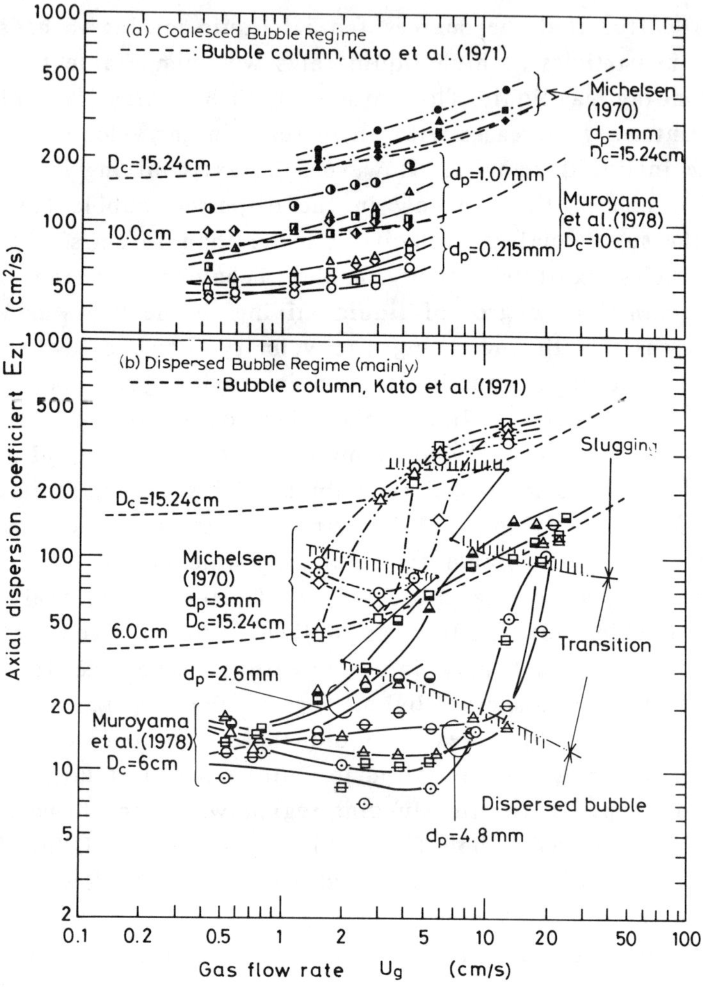

Figure 3.9 Effects of the gas and liquid flow rates on the axial dispersion coefficient for air-water bubble columns and air-water-particle fluidized beds: (a) coalesced bubble regime; and (b) dispersed bubble regime (mainly).

as a unified correlation of axial liquid mixing for gas-liquid columns and three-phase fluidized bed systems. In Eqn. (3.35), U_c is the average liquid circulation velocity (see Section 4.6.1).

Kim and Kim (1981, 1983), investigating the effects of viscosity and surface tension of the liquid on the axial dispersion coefficient, found that the axial dispersion coefficient slightly increases with an increase in liquid viscosity for a bed of large particles (6 mm glass beads), but decreases with an increase in liquid viscosity at high gas velocities in a bed of small particles (d_p < 3 mm). The axial dispersion coefficient increases somewhat with an increase in liquid surface tension. Kim and Kim (1983) correlated a wide range of liquid-phase axial mixing data with an equation given in Table 3.3 (Eqn. (T-3.3.1)); however, this correlation did not include any viscosity and surface tension effects. Nguyen-Tien et al. (1984) found that the predictions of Kim and Kim's (1983) equation agree reasonably well with their data for particle diameters greater than 3 mm in the expanded bed regime, but for particle diameters between 0.3 and 1 mm, they showed that the coefficient 20.19 in Eqn. (T-3.3.1) needs to be replaced by 61 (Eqn. (T-3.3.2)).

Kato et al. (1985) related three-phase axial dispersion coefficients to those in a gas-liquid bubble column over different ranges of gas velocity. Ohshima et al. (1977) found that, in a fluidized bed at liquid velocities close to minimum fluidization conditions, the Peclet number ($\frac{d_p U_\ell}{\epsilon_\ell E_{z\ell}}$) is nearly equal to that in a packed bed (≈ 1.2). At high liquid velocities, ($\frac{d_p U_\ell}{\epsilon_\ell E_{z\ell}}$) decreases markedly with an increase in the gas Froude number (U_g^2/gd_p).

3.2.1.2 Radial Mixing

El-Temtamy et al. (1979b) reported that the radial dispersion coefficient is usually much less than the axial dispersion coefficient in three-phase fluidized beds, as well as in other types of contactors. For example, for beds of 3 mm glass beads, the radial dispersion coefficient is roughly 1/10 to 1/2 of the axial dispersion coefficient, while for beds of 0.45 and 0.96 mm glass beads, the radial dispersion coefficient is roughly 1/30 to 1/20 of the axial dispersion coefficient.

Figure 3.10 shows that the radial liquid dispersion coefficient increases with increasing gas velocity. In the dispersed bubble regime, the radial liquid dispersion coefficient is only slightly affected by gas velocity; however, in the coalesced bubble and slugging regimes, it increases significantly with increasing gas velocity (Yasunishi et al., 1987). In addition, the radial dispersion coefficient increases with liquid velocity for low bed porosities, eventually reaches a maximum in the bed porosity range of 0.6 ~ 0.7, and then decreases with further increase in liquid velocity, as shown in Fig. 3.11. The maximum value of $E_{r\ell}$ at a given gas velocity increases significantly with

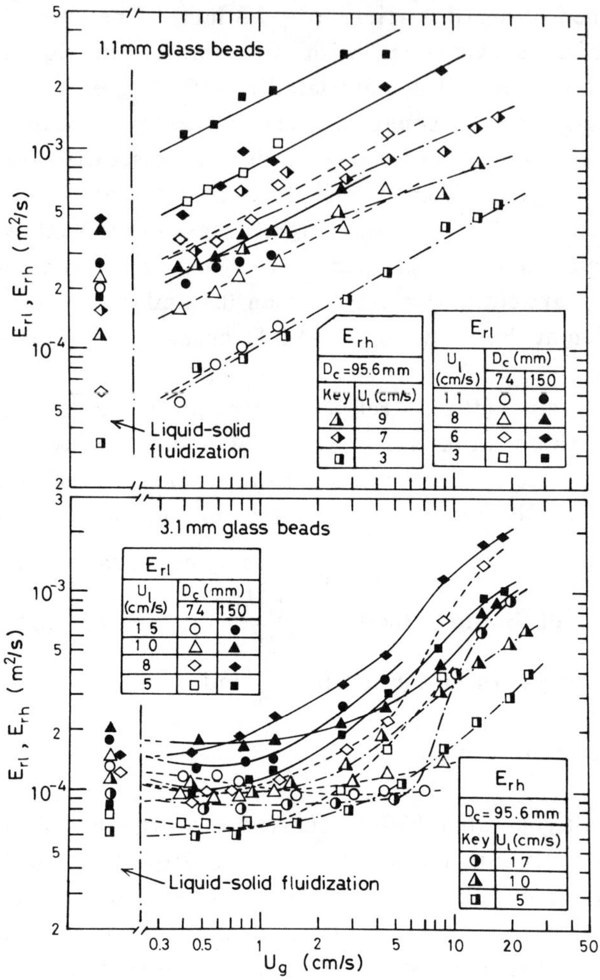

Figure 3.10 Liquid radial dispersion coefficient in air-water-particle fluidized beds containing 1.1 mm and 3.1 mm glass beads (from Yasunishi et al., 1987).

significantly with an increase in gas velocity, and $E_{r\ell}$ increases significantly with increasing column diameter (Yasunishi et al., 1987). The behavior of the radial liquid dispersion coefficient can be associated with variations in the effective radial thermal conductivity (Muroyama et al., 1986a,b). The radial liquid dispersion coefficient (or the radial effective thermal conductivity) for particles with a diameter less than 2 mm is much greater than that for larger particles at low gas velocities and high liquid velocities (Muroyama et al.,

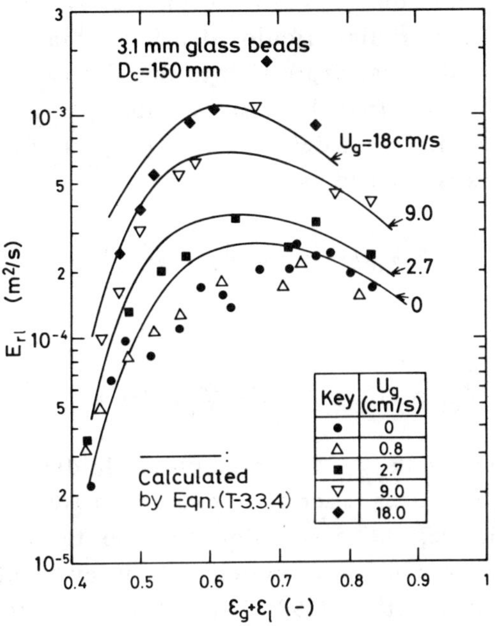

Figure 3.11 Variation in the radial dispersion coefficient with bed voidage in an air-water-glass bead fluidized bed (from Yasunishi et al., 1987).

Yasunishi et al, 1987). This behavior can be interpreted as resulting from differences in prevailing flow regimes. In a bed with small particles, where coalesced bubble flow prevails, a significant stirring is induced in the bed by the rising large bubbles, but in a bed with large particles, where dispersed bubble flow prevails, the passage of small bubbles does not induce much liquid circulation in the bed. At very high gas velocities, where gas passes through the bed as slugs, significant mixing occurs in the bed.

Kang and Kim (1986) indicated that the energy consumption rate, described in terms of an overall energy balance (Kang et al., 1985), increases with both particle size and gas velocity. They suggested that the energy consumption rate may be related to the radial liquid dispersion coefficient. They proposed a simple empirical correlation (Eqn. (T-3.3.3)) based on an isotropic turbulence assumption (Kim and Kim, 1983). This correlation, however, should not be extended to liquid-solid fluidized beds where the Peclet number exhibits a minimum with respect to the liquid velocity (or bed porosity).

The correlation (Eqn. (T-3.3.4)) developed by Yasunishi et al. (1987) to account for radial liquid dispersion in a three-phase fluidized bed has an identical expression to that for two-phase (liquid-solid) fluidized beds except that the Peclet number ($Pe_{r\ell,m}$) for the two different systems has different definitions. $(Pe_{r\ell,m})_2$ for liquid-solid fluidized bed systems is defined as

$$(Pe_{r\ell,m})_2 = d_{er}U_\ell/(\epsilon_\ell E_{r\ell}) \tag{3.36a}$$

where

$$d_{er} = \epsilon D_c/\{1+1.5(D_c/d_p)\,(1-\epsilon)\} \tag{3.36b}$$

whereas the definition of $Pe_{r\ell,m}$ for three-phase fluidized beds is given in Table 3.3. As U_g approaches zero, the expression for $Pe_{r\ell,m}$ reduces to $(Pe_{r\ell,m})_2$ and hence, the correlation for the three-phase system reduces to that of the two-phase system. It is also noted that the correlation of Yasunishi et al. (1987) can also satisfactorily account for the data of El-Temtamy et al. (1979b).

3.2.1.3 Summary

The axial and radial liquid dispersion coefficients are complex functions of gas and liquid velocities, column size and particle properties. The fundamental behavior of the axial liquid dispersion coefficient in a three-phase fluidized bed is similar to that in a gas-liquid bubble column; that is, it increases with increasing gas velocity and is a strong function of the column diameter. The axial dispersion coefficient in a three-phase fluidized bed can be larger or smaller than that in a gas-liquid bubble column depending on particle size and gas and liquid velocities. The axial dispersion coefficient in a non-foaming system can be well predicted by the correlations proposed by Kim and Kim (1983), Kato et al. (1985) or Yang (1986) over a wide range of operating conditions.

Relatively little was reported on the radial dispersion. The correlation of Yasunishi et al. (1987) has a general applicability and can be used for radial dispersion coefficient prediction for both two-phase and three-phase fluidized beds.

3.2.2 Gas Phase Mixing

A radioactive gas tracer with very low solubility can be used to measure the gas phase residence time distribution, thereby obtaining the mixing parameter of the gas. Ostergaard and Michelsen (1969) and Michelsen and Ostergaard (1970) used argon 41 as a gaseous tracer and indirectly measured the system response to a gas phase tracer pulse from outside the column. However, mixing in the gas phase is rather difficult to analyze quantitatively due to the complex effects involved in the absorption and desorption of the tracer material into the liquid phase. Ostergaard and Michelsen (1969) showed a simplified method to analyze an imperfect pulse response of the radioactive gas tracer which diffuses from the gas bubbles into the liquid in a finite amount. They suggested that the initial portion of the response curve, or the peak, contains mainly tracer carried by the gas bubbles, while the latter portion, or the tail, corresponds largely to the tracer absorbed in the liquid. After correcting the measured tracer concentration for the amount of tracer absorbed into the liquid, the axial dispersion coefficient for the gas was calculated in the same manner as that used for the liquid. The height of mixing unit for the gas (HMU_g) was determined by Ostergaard (1969) for a bed of 216 mm ID containing 0.25, 1, and 6 mm glass beads. HMU_g in the beds was found to be comparable to that in gas-liquid bubble columns. Michelsen and Ostergaard (1970) reported that the axial dispersion model could not account for gas phase mixing in systems with rapid coalescence. In such systems, negative dispersion coefficients were observed. Furthermore, accurate determination of the axial gas dispersion coefficient was difficult for beds of large particles since the large gas-liquid interfacial area established in the resulting dispersed bubble regime causes large amounts of tracer to be transferred to the liquid phase.

While the quantitative effect of solid particles on the gas phase axial dispersion coefficient is not fully known, as a rough approximation the gas phase dispersion coefficient for solid-free bubble columns (see Section 4.5.2) can be used.

3.2.3 Solids Mixing and Bed Segregation

Solid catalysts for catalytic operations or solid reactants in noncatalytic reactions may be uniform or nonuniform in size. Also, in bioreactors, both particle size and density may or may not vary due to the growth behavior of biofilm on the particle surfaces (Hirata et al., 1982 and Fan et al., 1984b). In these reactors or bioreactors, the uniformity of the particle properties and hence, the overall rate of reactions are strongly dependent upon the degree of solids mixing. Little work has been done on solids intermixing or segregation among dissimilar particles and mixing among the similar particles. For large bubble systems, the mechanism for solids mixing is mostly dominated by the interaction through the bubble wake as noted in Chapter 2. For small bubble systems or any bubble systems with light particles, solids mixing can be significantly induced by drift effects or random motion of the particles resulting from, respectively, velocity gradients caused by the passing gas bubble and from local nonuniformities in the flow field. The focus of the following discussion is on the phenomenological interpretation of solids intermixing among dissimilar particles. A brief description is given to mixing behavior among similar particles.

A measure of the solids macro-mixing in a three-phase fluidized bed is the solids residence time distribution. The catalyst residence time distribution in the H-Coal PDU unit measured by Bickel and Thomas (1982) could be successfully described by an axial dispersion model. The column Peclet number for the solid catalyst particles was found to be 0.451; this low value indicates the high extent of backmixing (nearly complete mixing) in the industrial three-phase fluidized bed. Euzen and Fortin (1987) studied solids mixing using a layer of magnetic tracer particles placed on top of the bed, and detecting the tracer concentrations at three bed locations. They found that the solids mixing time, i.e., the time required to reach complete solids mixing in the bed, decreases with an increase in gas velocity; the rate of decrease is significantly higher for a low liquid velocity than for a high liquid velocity. Furthermore, at a given gas velocity, the solids mixing time for low liquid velocity is higher than that for high liquid velocity.

Fan et al. (1984a) measured the solids intermixing behavior in a bed containing a binary mixture of glass beads. They used two different mixtures of particles: 3 mm and 4 mm particles and 3 mm and 6 mm particles. They defined three solids intermixing states: complete segregation, partial intermixing, and complete intermixing. The bed remains in a state of complete segregation at high liquid velocities and low gas velocities. At a moderate gas velocity, a partial intermixing state where both large and small particles coexist occurs between two segregated ends. At high gas velocities, complete

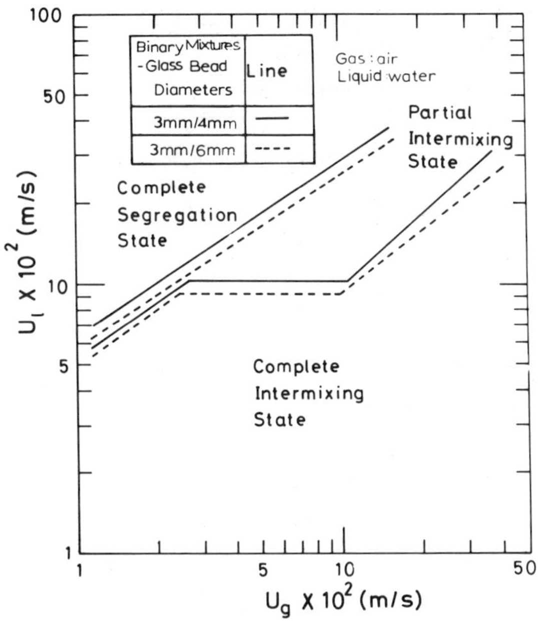

Figure 3.12 Solids intermixing map for binary mixtures of particles in a fluidized bed (from Fan et al., 1984a).

intermixing occurs with large and small particles evenly distributed throughout the entire bed. The higher the liquid velocity, the more complete the bed segregation becomes. Fan et al. (1984a) constructed a solids intermixing map which relates the gas and liquid velocities to the complete intermixing, partial intermixing, and complete segregation states, as shown in Fig. 3.12. A strong correspondence exists between the solids intermixing map and the flow regime map. A complete inter-mixing state occurs largely in the coalesced bubble and slugging regimes and slightly in the transition flow regime. The partial intermixing state occurs largely in the dispersed bubble and transition flow regimes and to a small extent in the coalesced bubble flow regime. The complete segregation state occurs solely in the dispersed bubble regime. Furthermore, particle density variations in the mixture have a considerably stronger effect on the extent of solids intermixing or segregation than do particle size variations.

For a bed with particles of both different sizes and different densities, the solids intermixing behavior is more complicated. Fan et al. (1985) employed a binary mixture of small (2.2 mm), heavy alumina

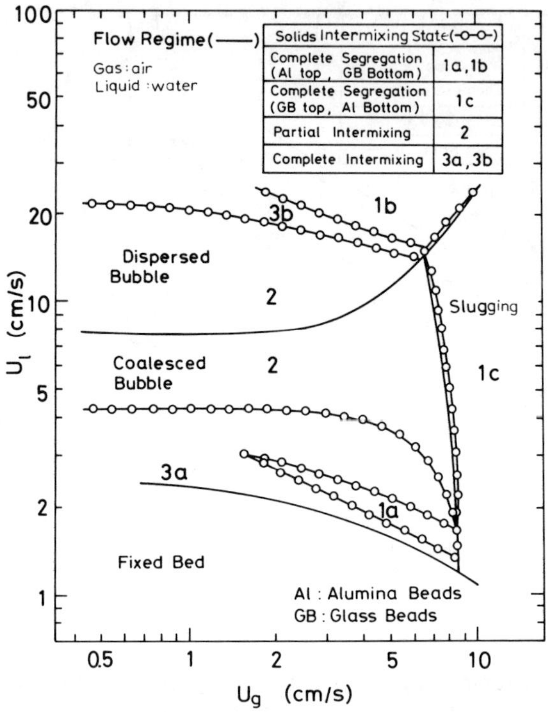

Figure 3.13 Flow regime and solids intermixing state maps for a binary mixture of small, heavy and large, light particles in a fluidized bed (from Fan et al. 1985).

beads and large (6.0 mm), light glass beads and presented the solids intermixing map shown in Fig. 3.13. At low liquid velocities, a complete intermixing state exists (region 3a). At intermediate liquid velocities, an alumina bead-rich layer exists near the bottom of the bed and a glass bead-rich layer was observed to form on top of a distinct glass bead layer (region 2). Note that, for gas velocities between about 2 and 9 cm/s, an additional complete segregation state exists for liquid velocities slightly higher than the minimum fluidization velocity. It is interesting to note that these solids intermixing states were observed only in the coalesced bubble and dispersed bubble regime. In the slugging regime, only a complete segregation state (region 1c) was observed which, however, is opposite to that found in region 1b: in this case the alumina beads were at the bottom and glass beads at the top. This reverse distribution phenomenon is known as solids layer inversion. Clearly, these phenomena observed in the three-phase system are far more complex than those found in a two-phase (liquid-solid) system (e.g., Moritomi et al., 1982; Jean and Fan, 1986a).

In a further investigation, Fan et al. (1987) suggested a phenomenological approach to analyze the axial solids concentration distributions for a binary mixture of particles. They found that a bed containing a binary mixture of particles with similar density, at superficial gas velocities beyond about 0.9 cm/s, forms three distinct regions:

Region A: Constant large and small particle concentration region located at the top of the bed;

Region B: Constant large and small particle concentration region located at the bottom of the bed; and

Region C: Variable concentration region of large and small particles, located between Regions A and B.

The constant particle concentration in Region A is possibly caused by the violent shedding vortices of the wakes when the bubble leaves the bed surface, whereas that in Region B is possibly caused by turbulent motion induced by the gas-liquid jetting around the distributor area.

The simultaneous effects of solids intermixing and segregation in Region C of the three-phase fluidized bed, under the assumptions of continuum in solid phase, and uniform radial concentration yield (Kennedy and Bretton, 1966)

$$E_{zsi}\frac{dC_i}{dz} - \bar{u}_{zi}C_i = 0 \tag{3.37}$$

where subscript i denotes the type of the particle. C_i is defined as $w_i/\rho_{si}\Delta hA$, E_{zsi} is the solids intermixing coefficient. The segregation velocity ($\bar{u}_{zi}$) is defined by

$$\bar{u}_{zi} = \frac{U'_\ell}{1-\epsilon_s} - \frac{U'_{\ell i}}{1-\epsilon_s} \tag{3.38}$$

In Eqn. (3.38) U'_ℓ is the "equivalent" two-phase (liquid-solid) superficial liquid velocity for the binary mixture in a three-phase fluidized bed, and $U'_{\ell i}$ is the "equivalent" two-phase superficial liquid velocity required to fluidize a bed of type i particles at the porosity of the mixture in a three-phase system. U'_ℓ and $U'_{\ell i}$ can be calculated by the methods of Jean and Fan (1986b) (also see Eqns. (T-2.1.13) and (T-2.1.14) for $U'_{\ell i}$ for type i particles).

$\bar{u}_{zi}$ varies linearly with the dimensionless axial distance (z_r) in Region C which is similar to that for a liquid-solid system reported by Juma and Richardson (1983). Thus, $\bar{u}_{zi}$ can be expressed as

$$\bar{u}_{zi} = \bar{a}_i + \bar{b}_i z_r \tag{3.39}$$

where $\bar{a}_i$ and $\bar{b}_i$ are linearity constants for particle i.

Combining Eqns. (3.37) and (3.39) and integrating lead to

$$C_i = k_i \exp[(\tfrac{1}{2}\bar{b}_i z_r^2 + \bar{a}_i z_r)H/E_{zsi}] \tag{3.40}$$

subject to the following boundary conditions:

$$C_1 = C_{10} \quad \text{at } z_r = L_0, \quad \text{for large particles} \tag{3.41}$$

and

$$C_2 = C_{21} \quad \text{at } z_r = L_1, \quad \text{for small particles} \tag{3.42}$$

k_i in Eqn. (3.40) is the integration constant.

Expressions for the individual particle intermixing coefficients can be derived from an overall mass balance on both large and small particles. The mass balance equations are

$$\frac{W_1}{AH\rho_{s1}} = \int_{L_0}^{L_1} C_1 dz_r + C_{10}L_0 + (1-L_1)C_{11} \tag{3.43}$$

$$\frac{W_2}{AH\rho_{s2}} = \int_{L_0}^{L_1} C_2 dz_r + C_{20}L_0 + (1-L_1)C_{21} \tag{3.44}$$

Differentiating Eqn. (3.43) with respect to L_1 and Eqn. (3.44) with respect to L_0 gives

$$C_1 = C_{11} \text{ at } z_r = L_1 \tag{3.45}$$

$$C_2 = C_{20} \text{ at } z_r = L_0 \tag{3.46}$$

And finally, combining Eqns. (3.40), (3.41), (3.42), (3.45) and (3.46) yields expressions for the dispersion coefficients:

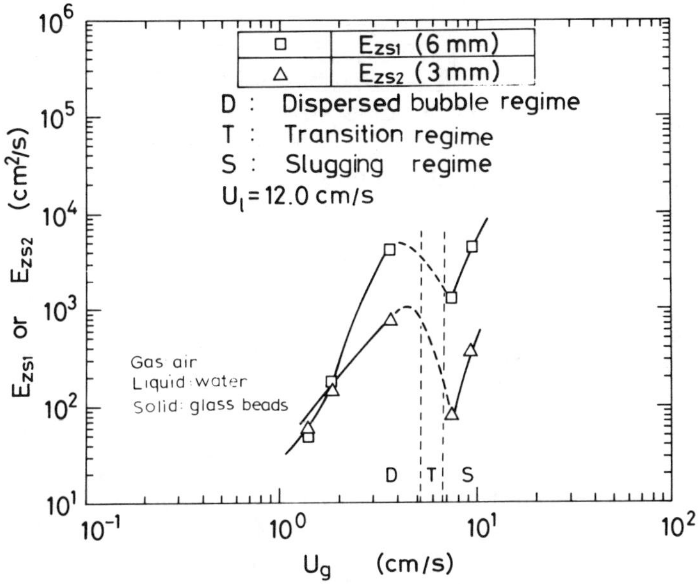

Figure 3.14 Effects of gas velocity and flow regime on the intermixing coefficient for a binary mixture of 3 and 6 mm glass beads in a fluidized bed (from Fan et al., 1987).

$$E_{zs1} = \{1/2\ \bar{b}_1(L_1{}^2 - L_0{}^2) + \bar{a}_1\ (L_1 - L_0)\}H/\ln(C_{11}/C_{10}) \tag{3.47}$$

$$E_{zs2} = \{1/2\ \bar{b}_2(L_0{}^2 - L_1{}^2) + \bar{a}_2\ (L_0 - L_1)\}H/\ln(C_{20}/C_{21}) \tag{3.48}$$

Figure 3.14 shows that, for a binary mixture of 3 mm and 6 mm glass beads, the intermixing coefficient increases with the gas velocity in the dispersed bubble regime, then decreases sharply in the transition regime and finally increases sharply in the slugging regime. Figure 3.15 shows that, for a binary mixture of 3 mm and 4 mm glass beads, the intermixing coefficient decreases with increasing liquid velocity in both the coalesced bubble and the dispersed bubble regimes for a given gas velocity. Note that the intermixing coefficient in the coalesced bubble regime is significantly greater than that in the dispersed bubble regime.

The global mixing behavior can be simply described by the mixing index M defined in the following (Rowe et al., 1972):

$$M = x_w/\bar{x}_w \tag{3.49}$$

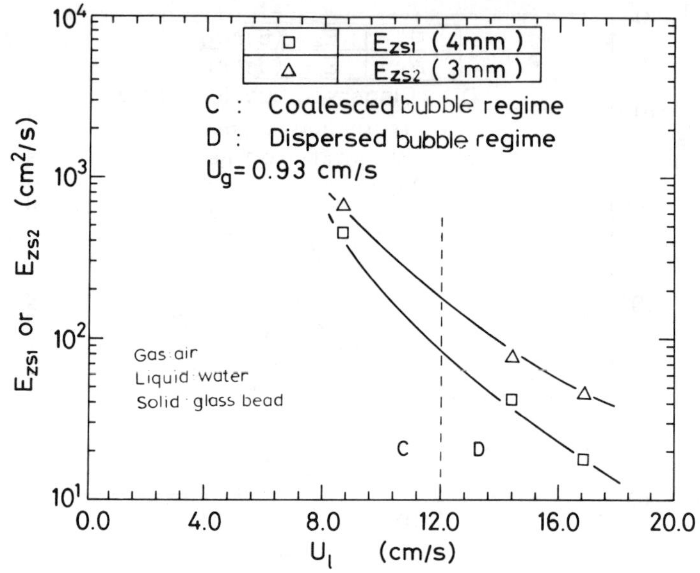

Figure 3.15 Effects of liquid velocity and flow regime on the intermixing coefficient for a binary mixture of 3 and 4 mm glass beads in a fluidized bed (from Fan et al., 1987).

where x_w is the weight fraction of large particles in the upper section of the bed and $\bar{x}_w$ is the weight fraction of large particles in the entire bed. In gas-liquid-solid systems, M varies from 0 for the completely segregated state to 1 for the complete intermixing state. Figure 3.16 shows the effect of gas velocity on M as a function of liquid velocity for a binary mixture. It is seen that M increases with the gas velocity. The increase in liquid velocity generally decreases M.

3.3 HEAT TRANSFER

In three-phase fluidized beds, heat transfer may occur between a particle and the fluid (particularly a particle and the liquid), the liquid and the gas, the wall and the bed, and the bed and internals. While there are no reported particle-liquid or liquid-gas heat transfer studies under three-phase fluidization conditions, the corresponding two-phase information (i.e., for liquid-solid fluidization and gas-liquid bubble columns, respectively) could serve as a good approximation for

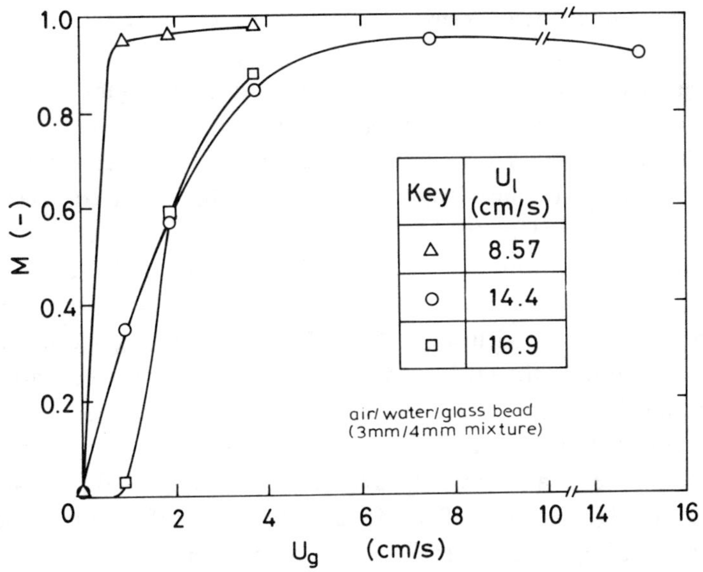

Figure 3.16 Effect of the gas velocity on mixing index for a binary mixture of 3 and 4 mm glass beads in a fluidized bed (from Fan et al., 1987).

the three-phase situation. Fundamental studies have been mainly concerned with heat transfer from the column wall to the bed or from the surface of an immersed object (rod or cylinder) to the bed. In these studies, a uniform radial temperature distribution is usually imposed to permit an expedient analysis of heat transfer; for example for heat transfer between the bed and the column wall (e.g., Viswanathan et al., 1964; Kato et al., 1980) or between the bed and an immersed heater (e.g., Baker et al., 1978). Van Driesen and Stewart (1964) reported that only a small temperature gradient exists throughout the H-Oil ebullated bed reactor; this behavior occurs even when the heat of reaction is sufficient to raise the temperature of the feed by 55.6 K (see Chapter 10). Furthermore, the desired temperature uniformity can be achieved by limiting the heat transfer rate and/or reducing the heat exchanger surface area.

The analysis of surface-to-bed heat transfer can be complicated when the thermal resistance in the bed must be considered. Such a situation exists when a parabolic radial temperature distribution induced by high heat transfer rates and/or large heat exchanger external surface area occurs in the bed (e.g., Muroyama et al., 1981,

1984). In considering the radial heat transfer mechanism in the overall wall-to-bed heat transfer process, the thermal resistance of the fluidized bed must be considered. Typically, the bed thermal resistance is divided into the near-wall and core regions. The near-wall region usually offers the greater resistance.

Wall-to-bed heat transfer in liquid-solid fluidized beds provides suitable background knowledge about wall-to-bed heat transfer in gas-liquid-solid fluidized beds since the entire heat transfer mechanism is similar when the liquid is the continuous phase in both systems. However, gas injection may significantly enhance heat transfer due to the increased mixing within the bed. In the following section the heat transfer in liquid-solid fluidized beds is briefly discussed followed by a comprehensive description of heat transfer in three-phase fluidized beds. Note that heat transfer behavior is strongly dependent on the geometry of the heat transfer surface (e.g., Kato et al., 1984a,b) and the physical significance of the heat transfer coefficient depends on the corresponding heat transfer models employed.

3.3.1 Brief Review on Heat Transfer in Liquid-Solid Fluidized Beds

In liquid-solid fluidized beds, the presence of solids increases the turbulence in the system and provides additional surface renewal through the thermal boundary layer at the wall. Early studies indicated that particles do not appear to contribute to heat transfer through the particle carrier mechanism that plays a prominent role in heat transfer in gas-solid fluidized beds. Recently, however, it has become apparent that, in conjunction with isotropic fluid micro-eddies, particles contribute to heat transfer in both liquid-solid and gas-liquid-solid fluidized beds (Magiliotou et al., 1987). Most heat transfer studies have neglected the effect of interior thermal resistance in the evaluation of the overall wall-to-bed heat transfer process. The overall heat transfer coefficient can be evaluated based on the log mean temperature difference or effective average of the difference between the wall temperature and the bulk bed temperature (Lemlich and Caldas, 1958; Richardson and Mitson, 1958; Richardson and Smith, 1962; Tripathi and Pandey, 1970; Kato et al., 1981). For a nonuniform temperature across the bed, a radially averaged temperature must be used to represent the bulk bed temperature in the evaluation of the overall heat transfer coefficient. The wall-to-bed heat transfer coefficient increases with increasing liquid flow rate at low liquid flow rates, passes through a maximum at a bed porosity near 0.7, and then decreases gradually with further increases in the liquid flow rate. Furthermore, the heat transfer coefficient, in most cases, increases with increasing diameter of the fluidized particles. Various correlation equations derived from these studies for the wall-to-bed heat transfer are summarized in Table 3.4.

Recent studies have considered the effects of the interior thermal (or the bed) resistance to the overall wall-to-bed heat transfer process. A parabolic radial temperature distribution in a liquid-solid fluidized bed indicates considerable thermal resistance in the core portion of the bed (Wasmund and Smith, 1967; Patel and Simpson, 1977). Wasmund and Smith (1967) were the first to provide a theoretical base for the wall-to-bed heat transfer mechanism and to propose a series resistance model which takes into account the core (interior) thermal resistance in series with the wall film resistance. It was shown that the interior resistance is a function of the effective thermal diffusivity or radial Peclet number for heat transfer. Muroyama et al. (1986a) also analyzed the wall-to-bed heat transfer in a liquid-solid fluidized bed based on the series resistance model. They showed that the contribution of the core resistance relative to the total resistance decreases with increasing bed porosity because of increased bed mobility and lateral liquid mixing and decreases with increasing bed diameter, especially for small particles. The modified radial Peclet number obtained from the effective thermal conductivity takes on a minimum at bed porosities ranging from 0.6 to 0.7 and strongly depends on the particle properties and column diameter.

The apparent wall heat transfer coefficient can be satisfactorily correlated using the Chilton and Colburn j-factor versus the modified Reynolds number relationship proposed by Beek (1971) for wall mass transfer (see Table 3.4). It is also interesting to note that a close analogy exists between the j_H'-factor for wall heat transfer based on the correlations of Muroyama et al. (1986a) and the j_d-factor for wall mass transfer based on the correlations of Ito et al. (1973), Smith and King (1975), and Morooka et al. (1979) in liquid-solid fluidized beds. The wall-to-bed heat transfer coefficient was also measured by Kato et al. (1981) using a short wall heater in which the temperature rise in the liquid phase was so small that the temperature profile was flat within the fluidized bed. Thus, the wall-to-bed coefficient obtained by Kato et al. (1981) can be considered to be virtually the same as the wall film coefficient; in fact, the heat transfer coefficients obtained from Kato et al.'s correlation agree well

Table 3.4 Studies on Heat Transfer in a Two-Phase (Liquid-Solid) Fluidized Bed

Investigators	Systems	Size of Heat Transfer Section	Solid Particles	Correlations and Remarks
Richardson and Mitson (1958)	water/miscellaneous particles	50.8 mm ID x 812.8 mm long	d_p = 0.46 mm ~ 4.26 mm ρ_s = 2.67 ~ 11.3 g/cm^3	$\frac{hD_c}{\lambda_\ell} = 55\, Pr_\ell^{0.4} (\frac{C_{ps}}{C_{p\ell}})^{0.28} (\frac{d_p U_i \rho_\ell}{\mu_\ell})^{0.020} (\rho_s/\rho_\ell + 3.45)$ for $0.072 < 1-\varepsilon_\ell < 0.373$ (T-3.4.1) U_i = extrapolated superficial liquid velocity at $\varepsilon_\ell=1$ C_{ps} = specific heat of solid $h = h_e + (h_2 - h_e)(1-\varepsilon_\ell)/0.072$ for $0<1-\varepsilon<0.072$ h_2 = calculated at $1-\varepsilon_\ell=0.072$ by Eqn. (T-3.4.1) h_e = heat transfer coefficient for open pipe flow with liquid velocity U_i (Wall-to-bed heat transfer coefficient)
Tripathi and Pandey (1970)	water/miscellaneous particles	50.8 mm ID x 1365.2 mm long	d_p = 0.499 ~ 6.583 mm ρ_s = 2.276 ~ 11.055 g/cm^3	$\frac{hd_p}{\lambda_\ell} = 0.0173 (\frac{d_p U_\ell \rho_\ell}{\mu_\ell})^{0.73} Pr_\ell^{0.31} (\frac{d_p}{D_c} \frac{\rho_s - \rho_\ell}{\rho_\ell})^{-0.19} \varepsilon_\ell^{-1.6}$ (Wall-to-bed heat transfer coefficient)
Muroyama et al. (1984)	water/glass beads water/activated carbon water/non-porous alumina	95.6 mm ID x 800 mm long	d_p = 0.61 mm ~ 6.9 mm ρ_s = 1.3 ~ 3.55 g/cm^3	$j_H = StPr_\ell^{2/3} = 0.04$ for $(\frac{d_p}{D_c})^{-0.935}[Re_\ell/(1-\varepsilon_\ell)] \leq 1.81 \times 10^4$ $j_H = 3.6 (\frac{d_p}{D_c})^{0.43} [Re_\ell/(1-\varepsilon_\ell)]^{-0.46}$ for $(\frac{d_p}{D_c})^{-0.935} [Re_\ell/(1-\varepsilon_\ell)] > 1.81 \times 10^4$ (T-3.4.2) $30 < Re_\ell/(1-\varepsilon_\ell) < 20000$; $0.0064 < d_p/D_c < 0.072$ (Wall-to-bed heat transfer coefficient)
Muroyama et al. (1986a)	See Muroyama et al. (1984) given above		d_p = 0.61 mm ~ 6.9 mm ρ_s = 1.3 ~ 3.55 g/cm^3	For wall film coefficient $j_H' = \frac{h_w}{U_\ell \rho_\ell C_{p\ell}} \cdot \varepsilon_\ell Pr_\ell^{2/3} = 0.137[Re_\ell/(1-\varepsilon_\ell)]^{-0.271}$ (Series thermal resistance model)

with the wall film heat transfer coefficients obtained by Muroyama et al. (1986a).

The heat transfer from an inserted surface to the liquid-solid fluidized bed was investigated by Richardson et al. (1976) and Khan et al. (1983). These studies provide insight into immersed object surface-to-bed heat transfer in gas-liquid-solid fluidized beds.

3.3.2 Column Wall-to-Bed Heat Transfer in Three-Phase Fluidized Beds

Muroyama et al. (1981, 1986b), using a very long wall heater, found that the radial temperature distribution was a weak parabolic function in the core portion of the bed and very steep near the wall surface. This behavior to a great extent resembles that in the liquid-solid fluidized bed and indicates the existence of considerable thermal resistance near the wall in series with low resistance in the core. Therefore, an accurate evaluation of the heat transfer requires consideration of both the interior thermal resistance and the wall film resistance.

Wall-to-bed heat transfer in a three-phase fluidized bed with a constant wall temperature can be analyzed on the basis of a series resistance model used for two-phase fluidized bed heat transfer as discussed in the previous section. By assuming that any temperature differences between the phases can be ignored, the mass velocity of each phase is uniformly distributed over the cross section, the physical properties of individual phases are invariant within the bed, and the system is at steady state, a differential heat balance yields

$$(G_\ell C_{p\ell} + G_g C_{pg}^*)\frac{\partial T}{\partial z} = k_{er}\left(\frac{\partial^2 T}{\partial r^2} + \frac{1}{r}\frac{\partial T}{\partial r}\right) + k_{ez}\frac{\partial^2 T}{\partial z^2} \tag{3.50}$$

In this formulation, both the axial and radial effective thermal conductivities reflect the macroscopic motion, or mixing, of the liquid and solid phases within the bed. Equation (3.50) is subject to the boundary conditions:

$$z = 0, \quad -k_{ez}\left(\frac{\partial T}{\partial z}\right) = (G_\ell C_{p\ell} + G_g C_{pg}^*)(T_{in} - T) \tag{3.51a}$$

$$z = L, \quad \frac{\partial T}{\partial z} = 0 \tag{3.51b}$$

$$r = 0, \quad \frac{\partial T}{\partial r} = 0 \tag{3.51c}$$

$$r = R, \quad k_{er}\left(\frac{\partial T}{\partial r}\right) = h_w(T_w - T) \tag{3.51d}$$

where the wall temperature is considered constant and the wall heat transfer coefficient (h_w) is a measure of the thermal resistance near the column wall. Note that Eqn. (3.51d) is for the heat balance over the surface of the thermal boundary layer whose thickness is assumed to be negligibly small.

The analytical solution of Eqn. (3.50), given by Wakao and Yagi (1959), is

$$\frac{T_w - T}{T_w - T_{in}} = \frac{4}{R} \sum_{i=1}^{\infty} \frac{\zeta J_0(r\xi_i)}{(\zeta^2 + \xi_i^2) J_0(R\xi_i)} F \tag{3.52}$$

with

$$F = \frac{(1+A_i)\exp\left\{\frac{(1-A_i)z}{2\kappa\delta'}\right\} - (1-A_i)\exp\left\{\frac{(1-A_i)z - 2A_iL}{2\kappa\delta'}\right\}}{(1+A_i)^2 - (1-A_i)^2 \exp\{-A_i L/(\kappa\delta')\}} \tag{3.53}$$

where

$$A_i = (1 + 4\kappa^2\xi_i^2\delta')^{1/2}, \qquad \zeta = h_w/k_{er}$$

$$\kappa = k_{er}/(G_\ell C_{p\ell} + G_g C_{pg}^*), \qquad \delta' = k_{ez}/k_{er}$$

Here, ξ_i is the i-th positive root of the transcendental equation

$$\zeta J_0(R\xi_i) = \xi_i J_1(R\xi_i) \tag{3.54}$$

In order to evaluate k_{er} and h_w based on the above model from a measured temperature profile, k_{ez} must be known.

3.3.2.1 Effective Axial Thermal Conductivity

The effective axial thermal conductivity (k_{ez}) is predominantly determined by the axial mixing of the fluids and solid particles; the molecular conduction through each phase is insignificant. Assuming the gas phase has negligible impact on the axial thermal conductivity, k_{ez} can be expressed as

$$k_{ez} = \rho_\ell C_{p\ell} \epsilon_\ell E_{z\ell} + \rho_s C_{ps} \epsilon_s E'_{zs} \tag{3.55}$$

where $E_{z\ell}$ and E'_{zs} are the liquid and solids axial dispersion coefficients, respectively. Correlation equations for estimating $E_{z\ell}$ were presented in Section 3.2.1; however, no general correlations for estimating E'_{zs} are available. k_{ez} is therefore approximated by assuming equal degrees of solid and liquid mixing and consequently, Eqn. (3.55) becomes

$$k_{ez} = (\rho_\ell C_{p\ell} \epsilon_\ell + \rho_s C_{ps} \epsilon_s) E_{z\ell} \tag{3.56}$$

This assumption leads to a relatively small error in the wall-to-bed heat transfer coefficient. Note that the importance of the axial conductivity term depends on the prevailing flow regime: the axial dispersion coefficient is much larger in the coalesced bubble and slugging regimes than in the dispersed bubble regime.

3.3.2.2 Effective Radial Thermal Conductivity

The effective radial thermal conductivity (k_{er}) qualitatively represents the intensity of radial mixing of the liquid and solid phases due to the stirring action created by passing bubbles. The evaluation of k_{er} from the measured radial temperature profile involves graphic and iterative procedures as described by Muroyama et al. (1986b). As shown in Fig. 3.17 in a bed of small particles where coalesced bubble flow prevails, k_{er} increases with increasing gas velocity and varies over a wide range. The increase in k_{er} with increasing gas velocity is sharp at low liquid velocities, but less significant at high liquid velocities. k_{er} increases with increasing liquid velocity, reaches a maximum, and then decreases with any further increase in the liquid rate in a manner similar to that observed in liquid-solid fluidized beds (Muroyama et al., 1986a). The maximum k_{er} occurs at a porosity of about 0.7, similar to the maximum radial mixing coefficient commonly observed in a liquid-solid fluidized bed (Hanratty et al., 1956). On the other hand, a rather complicated behavior of k_{er} accompanied by flow pattern changes can be observed in the fluidization of a 3.1 mm glass beads bed, as shown in the same figure. In the coalesced bubble regime, at low and moderate liquid and gas velocities ($U_\ell < 8$ cm/s, $U_g < 10$ cm/s), the variation in k_{er} with gas velocity is very small. In the slugging regime, at high gas velocities ($U_g > 10$ cm/s), k_{er} increases sharply with gas velocity.

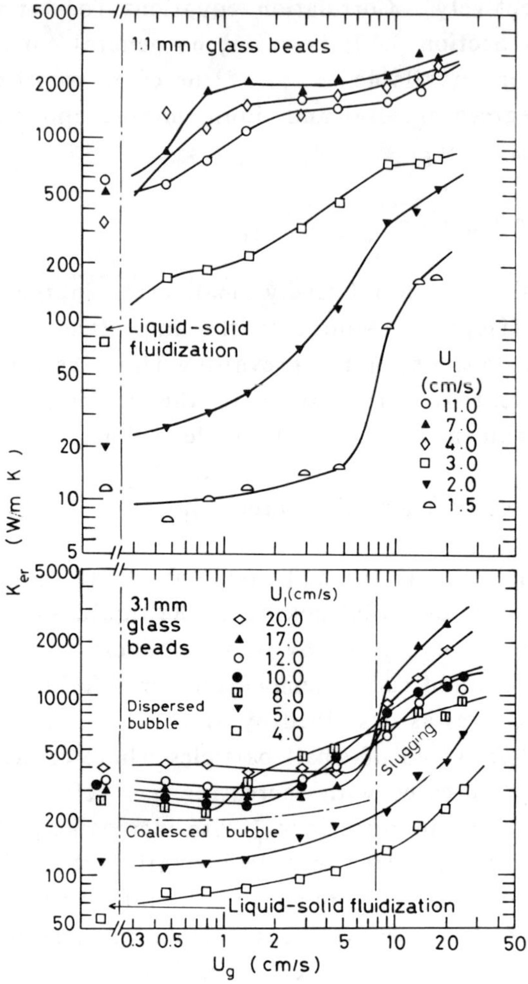

Figure 3.17 Effective radial thermal conductivity as a function of the gas velocity in air-water-particle fluidized beds (from Muroyama et al., 1986b).

Typically, in a three-phase fluidized bed where the contribution of the radial solids mixing to the radial heat transfer is negligible, the radial Peclet number (Pe_{rh}) can be defined as

$$Pe_{rh} = \rho_\ell C_{p\ell} U_\ell d_p / k_{er} \tag{3.57}$$

where k_{er} may be approximated in terms of the radial liquid mixing coefficient ($E_{r\ell}$) as

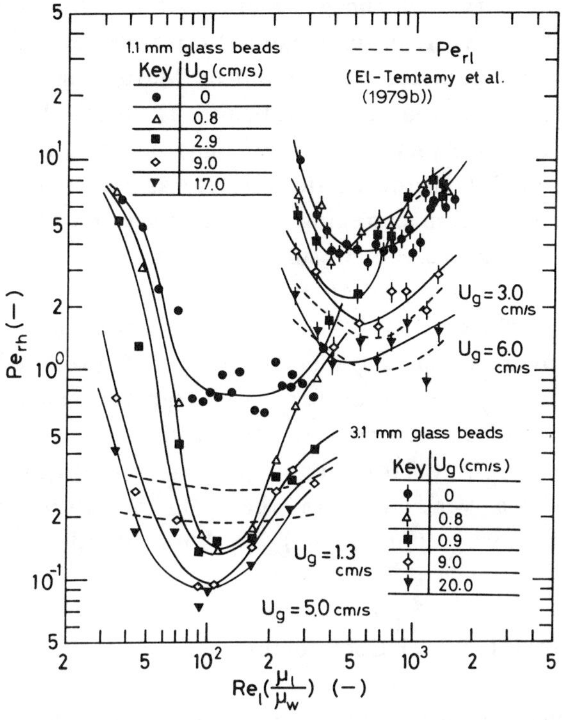

Figure 3.18 Variation in the modified Peclet number with the modified Reynolds number in air-water-particle fluidized beds (from Muroyama et al., 1986b).

$$k_{er} = E_{rh}\epsilon_{\ell}\rho_{\ell}C_{p\ell} \approx E_{r\ell}\epsilon_{\ell}\rho_{\ell}C_{p\ell} \tag{3.58}$$

where E_{rh} and $E_{r\ell}$ are defined based on a unit cross-sectional area for liquid flow.

The relationships between Pe_{rh} and $(Re_{\ell}\mu_{\ell}/\mu_{w})$ are shown in Fig. 3.18 for two sizes of particles and several gas velocities (Muroyama et al., 1986b). Two trends are evident in this figure: Pe_{rh} shows a concave dependency on $(Re_{\ell}\mu_{\ell}/\mu_{w})$ and decreases with increasing gas velocity. In beds of 1.1 mm glass beads, vigorous bed mixing occurs in the coalesced bubble regime and a considerable reduction in Pe_{rh} occurs with an increase in gas velocity. In contrast, in beds of large/heavy particles, such as 3.1 mm glass beads operated at high liquid velocities, the introduction of small amounts of gas results in only a small change in Pe_{rh} (dispersed bubble flow). With increasing gas velocity, however, the flow regime shifts from the dispersed bubble to the slugging and a significant decrease in Pe_{rh} occurs due to the violent stirring action of the gas slugs. At low liquid velocities, the decrease in Pe_{rh} with an increasing liquid velocity follows the trends in liquid-solid fluidized beds found by Muroyama et al. (1986a). However, in contrast to that in liquid-solid fluidized beds, at high liquid velocities in three-phase fluidized beds, a definite increase in Pe_{rh} occurs with an increasing liquid velocity as shown in the figure. This behavior results from changing mixing behavior due to a flow regime transition and a decrease in solids holdup.

The correlation for radial Peclet number for heat transfer proposed by Yasunishi et al. (1987) has the same form as that for the radial Peclet number for mass transfer which they proposed as given in Table 3.3. Their correlation for mass transfer can be used to estimate the radial dispersion coefficient for heat transfer simply by replacing $E_{r\ell}$ with E_{rh} in defining $Pe_{r\ell,m}$. Their radial heat dispersion correlation has been verified with the data of Muroyama et al. (1986a,b) and Yasunishi et al. (1987) for both liquid-solid and gas-liquid-solid fluidized beds.

3.3.2.3 Overall Heat Transfer Coefficient

The analysis of heat transfer from a wall surface to a fluid may also follow a macroscopic approach like that used in mass transfer where the rate is expressed in terms of an overall transfer coefficient, the temperature difference, and a transfer interfacial area. For wall-to-bed heat transfer, the local transfer rate can be defined as

$$dq = 2\pi Rh(T_w - T_B)dz = 2\pi Rh_w(T_w - T_b)dz \qquad (3.59)$$

where T_b is the temperature at infinitesimally small distance from the wall or $T|_{r\approx R}$. In Eqn. (3.59), the overall heat transfer coefficient, h, is less than or equal to h_w. Figure 3.19 defines the relationship among various temperatures used in this analysis. The effective temperature difference (T_w- T_B) in Eqn. (3.59) is defined by

$$T_w - T_B = \frac{\int_0^R U_\ell(r)\,[\,T_w - T(r)]r\,dr}{\int_0^R U_\ell(r)r\,dr} \qquad (3.60)$$

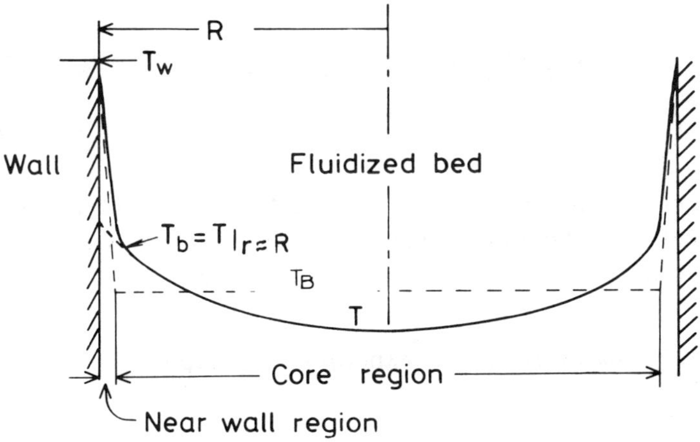

Figure 3.19 Temperature profile within a three-phase fluidized bed.

$U_\ell(r)$ is the superficial liquid velocity at a radial location r. For a uniform velocity distribution and constant wall temperature, T_B equals the cross-sectional average bed temperature, T_m (Chiu and Ziegler, 1983), given by

$$T_m = (2/R^2)\int_0^R T(r)rdr \tag{3.61}$$

Note that other definitions of the temperature difference are possible and consequently, h will be different for different definitions. By assuming plug flow (Viswanathan et al., 1964), the axial temperature gradient can be calculated by integrating Eqn. (3.59) over the reactor length, assuming a constant h. By assuming a completely mixed bed (Baker et al., 1978; Kato et al., 1980, 1981), the heat transfer coefficient can be defined based on an overall heat balance (dq is replaced by q and dz is replaced by L). Actually, either of these assumptions may not be appropriate in three-phase fluidized beds heated through a relatively long column section, and the effect of liquid and solids mixing on the temperature distribution should be considered. For situations involving an axial temperature variation, a one-dimensional heat balance at steady state yields

$$k_{ez}(d^2T_m/dz^2) - (G_\ell C_{p\ell} + G_g C^*_{pg})(dT_m/dz) + (4h/D_c)(T_w - T_m) = 0 \tag{3.62}$$

subject to the boundary conditions:

$$z = 0, \qquad -k_{ez}(dT_m/dz) = (G_\ell C_{p\ell} + G_g C^*_{pg})(T_{in} - T_m) \tag{3.63a}$$

$$z = L, \qquad dT_m/dz = 0 \tag{3.63b}$$

The solution of Eqn. (3.62) is the same as that for isothermal reactors accompanied by a first-order reaction. The solution yields the exit temperature in the heat transfer section ($T_{m,L}$) as

$$\frac{T_w - T_{m,L}}{T_w - T_{in}} = 4a\ \exp(b)/[(1+a)^2\exp(ab) - (1-a)^2\exp(-ab)] \tag{3.64}$$

where

$$a = \{1 + 8k_{ez}h/[R(G_\ell C_{p\ell} + G_g C^*_{pg})^2]\}^{1/2}$$

$$b = (G_\ell C_{p\ell} + G_g C^*_{pg})L/2k_{ez}$$

Equation (3.64) expresses the relationship between h and $T_{m,L}$ where $T_{m,L}$ can be obtained from the measured radial temperature profile at the exit and Eqn. (3.61). Note that the measured axial bed temperature distribution, which shows an abrupt jump at the bed inlet, could be well represented by the model equations, i.e., Eqns. (3.62) and (3.63a,b).

Experimental values of h reported in the literature fall between 1 and 8 $kW/m^2 \cdot K$ depending on particle properties, liquid velocities, etc. Ostergaard (1964) found that the wall-to-bed heat transfer coefficient increases markedly upon introducing a gas flow into a liquid-solid fluidized bed, but increases weakly upon any further increase in gas velocity. Chiu and Ziegler (1983) found the product of h and the liquid holdup for a liquid-solid fluidized bed equals that for a three-phase fluidized bed (see Eqn. (T-3.5.2) in Table 3.5). In addition, at a given gas velocity, h increases as bed porosity (or liquid velocity) increases, passed through a maximum near a porosity of 0.7, and decreases with a further increase in porosity; the maximum occurs at lower porosities with increasing particle size at a given gas velocity and at a lower gas velocity for a given particle size (Chiu and Ziegler, 1983; Muroyama et al. 1984). Also, the heat transfer coefficient increases appreciably as either the particle size or density increases. The variation in h with gas or liquid velocity, however, becomes

Table 3.5 Correlations for Overall Wall-to-Bed Heat Transfer Coefficient in Gas-Liquid-Solid Fluidized Beds

Investigators	Correlations	Dimensionless Variables	Driving Force	Location of Temperature Measurement /Comments
Kato et al. (1981)[1]	$\frac{d_p \varepsilon_\ell h}{\lambda_\ell (1-\varepsilon_\ell)} = 0.044\ [Re'_\ell Pr_\ell]^{0.78} + 2\ Fr_g^{0.17}$ (T-3.5.1)	$Re'_\ell = \frac{U_\ell \rho_\ell d_p}{\mu_\ell (1-\varepsilon_\ell)}$ $Fr_g = U_g^2/(g d_p)$	$(T_w - \bar{T}_0)$	Inlet, Outlet/ Mean value
Chiu and Ziegler (1983)	$St_h Pr_\ell^{2/3} = C\ Re'^{-0.305}_{m2}$ (T-3.5.2) or $h/(h)_{U_g=0} = (\varepsilon_\ell)_{U_g=0}/\varepsilon_\ell$ C = 0.1238 for cylindrical γ alumina particles C = 0.1038 for 1 mm glass beads C = 0.088 for 3 mm glass beads	$St_h = h\varepsilon_\ell/[\rho_\ell C_{p\ell} U_\ell]$ $Re'_{m2} = U_\ell \rho_\ell/[S_p(1-(\varepsilon_\ell)_{U_g=0})\mu_\ell]$	$(T_w - T_m)$	Level of main heater
Muroyama et al. (1984)[3]	$h/(h)_{U_g=0} = 1.0 + 0.0413(\frac{U_g}{U_\ell})^{0.30}[(\frac{d_p}{D_c})(\frac{\rho_s - \rho_\ell}{\rho_\ell})]^{-0.60}$ (T-3.5.3)	————	$(T_w - T_m)$	Six axial locations
Chiu and Zeigler (1985)[2]	$St_h Pr_\ell^{2/3} = 0.127\ Re'^{-0.354}_{m2} \phi_s^{-1} (\frac{U_{\ell mf0}}{U_t})^{0.266}\ Pr_\ell^{0.304}$	$Re'_{m2} = U_\ell \rho_\ell/[S_p(1-(\varepsilon_\ell)_{U_g=0})\mu_\ell]$ $\phi_s = (1.5 d_p^2 L)^{2/3}/[d_p(L + d_p/2)]$	$(T_w - T_m)$	Level of main heater
Saberian-Broudjenni et al. (1985)	$Nu''' = 2.40 \times 10^{-3} Re_{LG}^{0.338} Re_L^{0.540} Pr_\ell^{1.04} \cdot (\frac{d_p}{D_c})^{-0.930} (\frac{\rho_s - \rho_\ell}{\rho_\ell})^{-0.357}$	$Nu''' = hD_c/\lambda_\ell$ $Re_{LG} = \rho_\ell D_c U_g/\mu_\ell$ $Re_L = \rho_\ell D_c U_\ell/\mu_\ell$	$(T_w - T_0)$	Center heater section

1: In this study, h is approximately equal to h_w since the interior resistance is negligible.

2: The correlation was obtained based on data from both wall-to-bed and immersed object-to-bed heat transfer systems.

3: Correction for $(h)_{U_g=0}$ is given in Table 3.4.

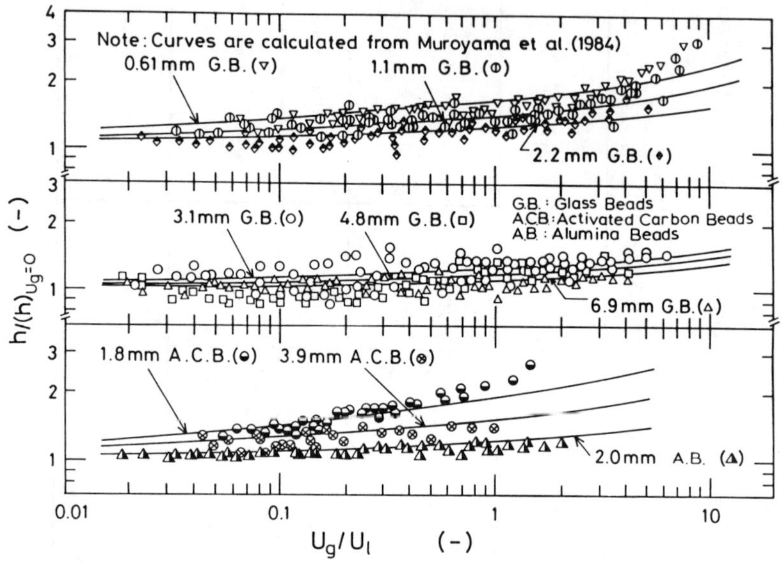

Figure 3.20 Correlation of $h/(h)_{U_g=0}$ as a function of U_g/U_ℓ in air-water-particle fluidized beds (from Muroyama et al., 1984).

smaller as either the size or density of the particles increases. The ratio of the heat transfer coefficient in a gas-liquid-solid fluidized bed to that in a liquid-solid fluidized bed with an identical liquid velocity ($h/(h)_{U_g=0}$) could be expressed as a unique function of U_g/U_ℓ for the data shown in Fig. 3.20. $h/(h)_{U_g=0}$ is close to unity for a low U_g/U_ℓ, but increases progressively with an increase in U_g/U_ℓ. The increase in $h/(h)_{U_g=0}$ is pronounced in a bed of small/light particles. Muroyama et al. (1984) proposed an empirical correlation for $h/(h)_{U_g=0}$ (see Eqn. (T-3.5.2)) which is represented by solid lines in Fig. 3.20; the correlation used in estimating $(h)_{U_g=0}$ is given in Table 3.4.

Specific effects of particle properties are elusive. Nevertheless, for a given type of particle (same ρ_s and C_{ps}), beds of larger particles have higher overall heat transfer coefficients at low gas velocities. Particle size becomes unimportant to heat transfer for particles larger than 3 mm in diameter (Chiu and Ziegler, 1983 and Muroyama et al., 1984).

Most correlations for the heat transfer coefficient in three-phase fluidized beds presented in the literature are strictly empirical.

However, Magiliotou et al. (1987) presented a semi-theoretical approach based on a combination of the surface renewal theory and the isotropic turbulence model; they suggested the heat transfer processes involving liquid micro-eddies and particle movement with Kolmogoroff's theory of isotropic turbulence (Hinze, 1959) to provide the renewal frequencies. Surface renewal caused by the particles is achieved through the moving particle and its associated liquid film displacing the liquid originally adjacent to the wall surface. The solid particles are not directly involved in the heat transfer process through conduction.

The liquid micro-eddy renewal frequency is

$$S_1 \propto u/\eta \tag{3.65}$$

where u, an eddy characteristic velocity, is $(\nu_\ell e'')^{1/4}$ and η, an eddy characteristic length, is $(\nu_\ell^3/e'')^{1/4}$. The energy dissipation rate per unit mass of the system (e"), evaluated through an energy balance on the system, is

$$e'' = [(U_g + U_\ell)(\epsilon_g\rho_g + \epsilon_\ell\rho_\ell + \epsilon_s\rho_s) - U_\ell\rho_\ell]g/(\epsilon_\ell\rho_\ell + \epsilon_s\rho_s) \tag{3.66}$$

The energy dissipation by particles must also be accounted for since the solid mass appears in the denominator of Eqn. (3.66); the energy may be consumed through particle collisions and also through the particles being supported in the bed without involving the micro-eddies.

The particle renewal frequency is proposed to be

$$S_2 \propto u_s/\ell_s \tag{3.67}$$

where u_s is the average particle speed in the bed and ℓ_s is the mean free path of the particle. u_s and ℓ_s were assumed to be proportional to $(U_\ell - U_{\ell mf})$ and $\phi_s d_e/\epsilon_s^{1/3}$, respectively. Thus, S_2 becomes

$$S_2 \propto (U_\ell - U_{\ell mf})\epsilon_s^{1/3}/(\phi_s d_e) \tag{3.68}$$

This approach leads to the following general expression for h:

$$h = c_1[\lambda_\ell\rho_\ell C_{p\ell}(e''/\nu_\ell)^{1/2}]^{1/2} + c_2[\lambda_\ell\rho_\ell C_{p\ell}\{\epsilon_s{}^{1/3}(U_\ell - U_{\ell mf})/(\phi_s d_e)\}]^{1/2} \tag{3.69}$$

Table 3.6 Studies on Immersed Object-to-Bed Heat Transfer in Gas-

Investigators	Systems (gas/liquid/solid)	Solid Properties	Column Diameter
Armstrong et al. (1976) Baker et al. (1978)	air/water/glass beads	d_p = 0.5, 1, 3, 5 mm ρ_s = 2.475-2.955 g/cm^3	D_c = 240 mm
Kato et al. (1984a)	air/water/glass beads air/water-CMC/porous alumina	d_p = 0.52, 1.2, 2.2 mm ρ_s = 2.5g/cm^3 d_p = 3.2 mm ρ_s = 1.62 g/cm^3	D_c = 120, 190 mm
Kato et al. (1984b)	air/water/glass beads air/water-CMC/ porous alumina (μ_ℓ = 1 ~ 15.6 m Pa•s)	d_p = 0.52, 1.2, 2.2 mm ρ_s = 2.5g/cm^3 d_p = 3.2 mm ρ_s = 1.62 g/cm^3	D_c = 120, 190 mm
Kang et al. (1985)	air/water/glass beads air/water-CMC/glass beads (μ_ℓ = 1 ~ 39 m Pa•s) also data taken from Kato et al. (1981) Chiu and Ziegler (1983) Baker et al. (1978)	d_p = 1.7, 4.0, 6.0, 8.0 mm ρ_s = 2.5g/cm^3	D_c = 152 mm
Suh et al. (1985)	Source data from Kato et al. (1981) Chiu and Ziegler (1983) Baker et al. (1978) Kang et al. (1985)	See information given above	
Magiliotou et al. (1987)	air/0.5% tert-pentyl alcohol aqueous soln./ resid hydrotreating catalysts	d_e = 1.5, 1.6, 1.7, 1.9 mm ρ_s = 1.89, 2.00 g/cm^3	D_c = 76.2 mm 152 mm

where c_1 and c_2 are empirical constants (e.g., see Table 3.6 for immersed vertical cylinder to bed heat transfer). In Eqn. (3.69), both terms are of the same order of magnitude in the dispersed bubble condition, thus reflecting the importance of heat transfer effects due to both the fluid eddies and particles. For zero solids concentrations, Eqn (3.69) reduces to Deckwer's equation (Deckwer, 1980) for bubble

Liquid-Solid Fluidized Beds

Heater Dimensions	Gas and Liquid Velocities	Correlations (in SI Units)/(Remarks)/(Vertical or Horizontal Heater)
$d_h \times L =$ 63 x 250 mm	$U_g = 0 \sim 23.8$ cm/s $U_\ell = 0.8 \sim 12.6$ cm/s	$h = 10^4 U_\ell^{0.070} U_g^{0.059} d_p^{0.106}$ (T-3.6.1) (vertical heater)
$d_h \times L =$ 12.7 x 120, 12.7 x 190, 38 x 190 mm	$U_g = 0 \sim 16$ cm/s $U_\ell = 0.1 \sim 8$ cm/s	$Nu' = 0.12\{(\frac{d_{ho}+d_h}{d_h})^{0.5} (Re'\ Pr_\ell)^{0.65+0.12Fr_h^{0.5}}\} + 1.3Fr_g^{0.33}$ $Nu' = hd_p\varepsilon_\ell/(\lambda_\ell(1-\varepsilon_\ell))$; $Fr_h = U_g^2/gd_h$; $Fr_g = U_g^2/gd_p$; $d_{ho} = 0.02$ m (horizontal heater)
$d_h \times L =$ 22 x 196 mm	$U_g = 0 \sim 16$ cm/s $U_\ell = 0.05 \sim 8$ cm/s	$Nu' = 0.058(Re'Pr_\ell)^{0.78} + 2.3$ (for their own data) (T-3.6.2) $Nu' = hd_p\varepsilon_\ell/\lambda_\ell(1-\varepsilon_\ell)$ $Re'Pr_\ell = d_pU_\ell\rho_\ell C_{p\ell}/\lambda_\ell(1-\varepsilon_\ell)$ $Nu' = \{0.058(Re'Pr_\ell)^{0.78} + 2.3\}\ (d_h/\ d_{ho})^{-0.5}$ (corrected for heater diameter) $d_{ho} = 0.022$ m, $0.022<d_h<0.063$ m (veritcal heater)
$d_h \times L =$ 12 x 330 mm	$U_g = 0 \sim 12$ cm/s $U_\ell = 0.6 \sim 16$ cm/s	$Nu'' = 0.036\ Pr_\ell^{0.65} Re''^{0.81}$ $Nu'' = \frac{hd_p(1-\varepsilon_s)}{\lambda_\ell\varepsilon_s}$, $Re'' = \frac{d_p\rho_\ell(U_g+U_\ell)}{\mu_\ell\varepsilon_s}$ (T-3.6.3) (No distinction between wall-to-bed heat transfer and immersed heater-to-bed heat transfer) (vertical heater)
and in Table 3.5 for source data		$h = 0.0647[\lambda_\ell\rho_\ell C_{p\ell}\{[(U_\ell+U_g)(\varepsilon_s\rho_s+\varepsilon_\ell\rho_\ell+\varepsilon_g\rho_g) - U_\ell\rho_\ell]g/(\varepsilon_\ell\mu_\ell)\}^{1/2}]^{1/2}$ (T-3.6.4) (No distinction between wall-to-bed heat transfer and immersed heater-to-bed heat transfer) (vertical heater)
$d_h \times L =$ 18.9 x 66.0 mm	$U_g = 0 \sim 6$ cm/s $U_\ell = 0.8 \sim 6$ cm/s	$h = C_1[\lambda_\ell\rho_\ell C_{p\ell}\{[(U_\ell+U_g)(\varepsilon_s\rho_s+\varepsilon_\ell\rho_\ell+\varepsilon_g\rho_g) - U_\ell\rho_\ell]\cdot g\rho_\ell/[(\varepsilon_\ell\rho_\ell+\varepsilon_s\rho_s)\mu_\ell]\}^{1/2}]^{1/2} + C_2[\lambda_\ell\rho_\ell C_{p\ell}\{\frac{\varepsilon_s^{1/3}(U_\ell-U_{\ell mf})}{\phi_s d_e}\}]^{1/2}$ $C_1 = 0.1$, $C_2 = 0.285$ for dispersed bubble regime $C_1 = 0.071$, $C_2 = 0.01$ for coalesced bubble regime (vertical heater)

columns. For three-phase fluidized beds in the coalesced bubble regime, c_1 and c_2 are lower than those for the dispersed bubble regime. This result indicates that the fluid micro-eddy formation is less efficient and the particle renewal is less important in the coalesced bubble conditions.

Note that a simplified form of Eqn. (3.69) was proposed by Suh et al. (1985) in which the contribution of particle movement was neglected and the energy dissipation rate was evaluated based on a unit mass of the liquid phase alone. The effects of column size and particle properties on h are not readily quantified in Eqn. (3.69), but both affect the energy dissipation rate through the phase holdups. The correlations proposed by Saberian-Broudjenni et al. (1985) and Muroyama et al. (1984) include the column diameter (see Table 3.5); the correlation of Muroyama et al. (1984) predicts $h \propto D_c^{0.60}$, while that of Saberian-Broudjenni et al. (1985) predicts $h \propto D_c^{0.93}$. The correlations of Chiu and Ziegler have gas velocity effects indirectly incorporated through changes in the liquid holdup; the other correlations in Table 3.5 show that h increases with gas velocity to about 0.30 ~ 0.35 power.

3.3.2.4 Apparent Wall Heat Transfer Coefficient

The apparent wall heat transfer coefficient (h_w) represents the resistance to heat transfer in the near wall region (see Fig. 3.21 and Eqn. (3.59)). As shown in Fig. 3.21, h_w increases gradually with increasing gas velocity from the value for a liquid-solid fluidized bed and ultimately approaches that for a bubble column system at very high gas velocities.

At a moderate gas velocity, h_w passes through a maximum as the liquid velocity increases in a way similar to that observed in a liquid-solid fluidized bed. This behavior for the apparent wall heat transfer coefficient is similar to that for the overall wall-to-bed heat transfer coefficient given in Section 3.3.2.3. The noted similarity in behavior of h_w with respect to the liquid velocity in both two-phase and three-phase fluidized beds leads to the approach that h_w in a three-phase fluidized bed may be correlated by a Chilton and Colburn type equation similar to that for a two-phase fluidized bed (Muroyama et al. 1986a):

$$j'_H = 0.137\, Re'^{\,-0.271}_{\ell,g} \tag{3.70}$$

where

$$j'_H = \{h_w/(\rho_s C_{p\ell} U_\ell)\}\epsilon_\ell\, Pr_\ell^{2/3} \tag{3.71a}$$

$$Re'_{\ell,g} = Re_\ell \left\{\frac{\epsilon}{\epsilon_\ell(1-\epsilon)}\right\} \tag{3.71b}$$

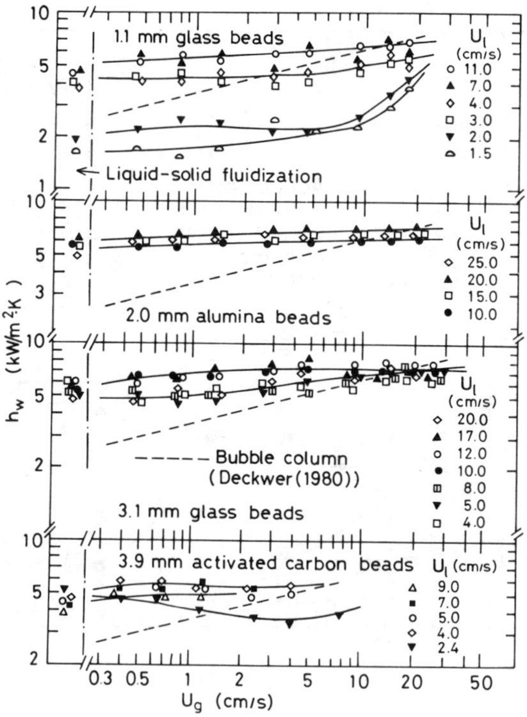

Figure 3.21 Apparent wall heat transfer coefficient versus the gas velocity for several types of particles in an air-water-particle fluidized bed (from Muroyama et al., 1986b).

Note a characteristic diameter for the Reynolds number may be defined as $d_p(\frac{\epsilon}{1-\epsilon})$ based on the total gas-liquid void fraction. Also, Eqn. (3.70) reduces to the correlation for a liquid-solid fluidized bed system (Muroyama et al., 1986a) given in Table 3.4 as $Re'_{\ell,g}$ reduces to $Re_{\ell}/(1-\epsilon_{\ell})$ at a zero gas velocity. This result implies that the increase in the wall film heat transfer induced by the gas phase can be accounted for by correcting the effective diameter and interstitial liquid velocity for a three-phase system.

3.3.2.5 Relative Importance of Bed Resistance to Overall Resistance

The overall resistance to wall-to-bed heat transfer may be represented by the reciprocal of the overall heat transfer coefficient, $1/h$. The thermal resistance near the wall may be given by the reciprocal of the apparent wall heat transfer coefficient, $1/h_w$.

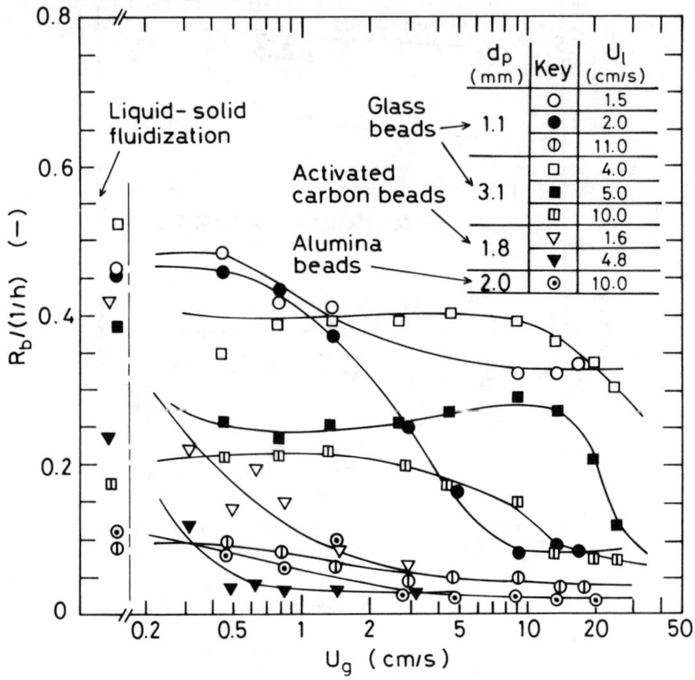

Figure 3.22 Ratio of the bed resistance to the total wall-to-bed heat transfer resistance versus the gas velocity for several types of particles in an air-water-particle fluidized bed (from Muroyama et al., 1986b).

Therefore, considering the additive nature of both the wall and the bed resistances, the ratio of the bed resistance R_b to the overall resistance (1/h) is readily given as

$$R_b/(1/h) = 1 - h/h_w \tag{3.72}$$

In a three-phase fluidized bed, $R_b/(1/h)$ generally decreases with increasing gas velocity as shown in Fig. 3.22 (Muroyama et al., 1986b). For fully expanded beds containing particles with diameters less than 2 mm, $R_b/(1/h)$ reduces considerably upon introducing gas into a liquid-solid fluidized bed; the bed resistance becomes negligibly small at gas velocities greater than about 2 cm/s. In contrast, for a bed containing particles with diameters greater than 3 mm, $R_b/(1/h)$ becomes constant, ranging from 0.2 to 0.4, for a given liquid velocity at low to moderate gas velocities; the value increases with decreasing liquid velocity. Thus, the relative importance of the bed resistance to the overall resistance depends on the degree of bed expansion; i.e., $R_b/(1/h)$ decreases as bed expansion, and hence bed mobility, increases.

3.3.2.6 Summary

To evaluate the overall wall-to-bed heat transfer coefficient, it is essential to take into account the interior thermal resistance, which varies with the scale and extent of fluid mixing in the system. Because of the similarity in the behavior of wall-to-bed heat transfer coefficient in a three-phase fluidized bed and that in a liquid-solid fluidized bed, several correlation equations for the wall-to-bed heat transfer coefficient in three-phase fluidized beds were developed by modifying existing correlations for two-phase fluidized beds to include enhancement effects due to the gas phase. The correlations given by Eqns. (T-3.5.1) and (T-3.5.2) for the wall-to-bed coefficient and the correlation given by Eqn. (T-3.5.3) for the apparent wall film coefficient can be recommended when the interior thermal resistance is negligibly small. The correlation given by Eqn. (T-3.4.2) can be used when the interior thermal resistance is significant.

3.3.3 Immersed Vertical Cylinder-to-Bed Heat Transfer

Heat transfer from an immersed vertical cylinder (or tube) surface to the three-phase fluidized bed has been investigated by several authors. The heat transfer coefficient can be related to the total heat input (q) and the heat transfer surface area (A_h) by

$$h = q/A_h(\bar{T}_w - \bar{T}_B) \tag{3.73}$$

In Eqn. (3.73), $\bar{T}_w$ is the average temperature of the cylinder surface, and $\bar{T}_B$ is the temperature of the bed; both are taken as the average of the temperatures upstream and downstream of the heater.

Armstrong et al. (1976), Baker et al. (1978) and Kang et al. (1983) measured the effect of particle size, gas velocity, and liquid velocity on the cylinder surface-to-bed heat transfer coefficient. The variation in the heat transfer coefficient with respect to gas flow rate is similar to that observed for the wall-to-bed heat transfer coefficient. Plots of h against particle size d_p are shown in Fig. 3.23 (Baker et al., 1978). A similar trend was noted for the wall-to-bed heat transfer coefficient versus particle diameter (Chiu and Ziegler, 1983; Muroyama et al., 1986b). An increase in particle size (or

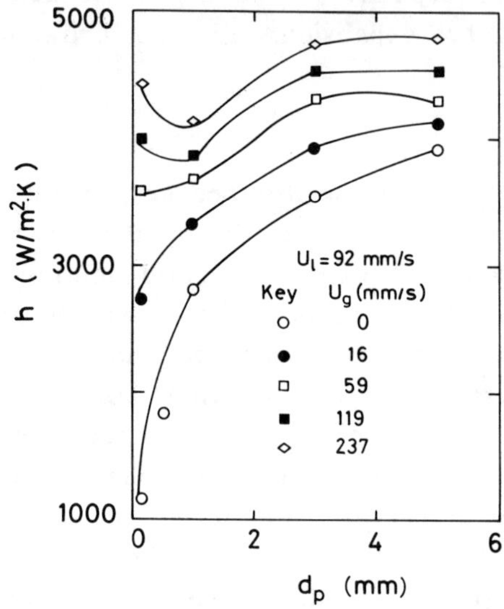

Figure 3.23 Relationship between the heat transfer coefficient and particle size in an immersed vertical cylinder to bed heat transfer in an air-water-glass bead fluidized bed (from Baker et al., 1978).

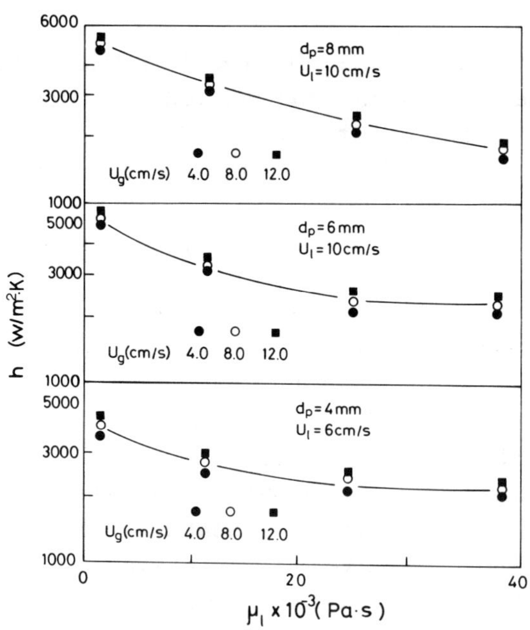

Figure 3.24 Effect of liquid viscosity on h in air-liquid-glass bead fluidized beds (from Kang et al., 1983).

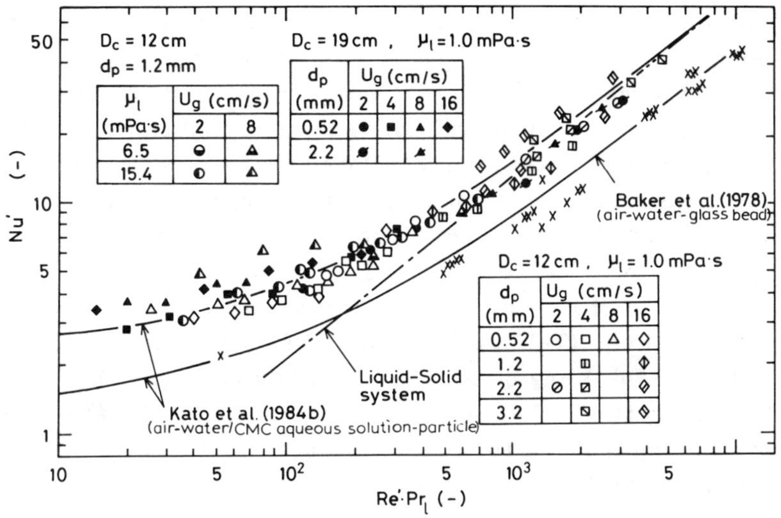

Figure 3.25 Dependence of the Nu′ on the $Re'Pr_\ell$ in gas-liquid-solid fluidized beds with vertical immersion cylinder-to-bed heat transfer (from Kato et al., 1984b).

introducing solid particles in a gas-liquid column) generally increases the heat transfer coefficient, except for small particles with diameters of about 1 mm at high gas velocities. Kang et al. (1983) found that the heat transfer coefficient decreases gradually with increasing liquid viscosity, as shown in Fig. 3.24. The decrease in the heat transfer coefficient with liquid viscosity was commonly observed in the liquid-solid fluidized bed. This decrease can be further enhanced by a resulting decrease in gas holdup (Kang et al., 1985), and suppression of particle movement (Kato et al., 1984b). Heat transfer coefficients in surfactant systems with high gas holdup and small bubble size conditions were found to be higher than those in air-water systems by Magiliotou et al. (1987). The high gas holdup increases the energy dissipation rate due to the liquid micro-eddies (the first term on the right hand side of Eqn. (3.69)) and thus, increases the heat transfer coefficient.

Kato et al. (1984b) found the column size has no effect on the heat transfer coefficient, yet the heat transfer coefficient decreases with increasing diameter of the heater in a manner similar to that in a solid-free bubble column, as shown in Fig. 3.25. Note that heat transfer coefficients for a vertical immersed cylinder (heater) are considerably greater than those for wall-to-bed heat transfer. This may be due to the fact that gas bubbles tend to rise in the core region and the turbulent movement of the liquid and solid phases is more intensive in the center of the column where the heater is located.

Correlations for the heat transfer coefficient from the cylindrical vertical heater, or vertical cylinder, to the three-phase fluidized bed have been developed by several investigators. The equations and the corresponding range of applicabilities are summarized in Table 3.6. The correlation given by Enq. (T-3.6.1) proposed by Baker et al. (1978) is purely empirical and cannot be extended to the conditions of no gas flow. Generalized correlations given by Enqs. (T-3.6.2), (T-3.6.3) and (T-3.6.4) were presented by Kato et al. (1984b), Kang et al. (1985), and Suh et al. (1985), which are valid for no gas flow. Note that Eqns. (T-3.6.3) and (T-3.6.4) were developed by lumping the data for wall-to-bed heat transfer and immersed rod-to-bed heat transfer.

3.3.4 Immersed Horizontal Cylinder-to-Bed Heat Transfer

Kato et al. (1984a) measured the effect of column diameter, heater diameter, particle diameter, liquid viscosity, and both gas and liquid velocities on the heat transfer coefficient for heat transfer from an immersed horizontal cylinder surface to a three-phase fluidized bed. The effect of the liquid velocity on the heat transfer coefficient for various conditions is shown in Fig. 3.26. The heat transfer coefficient increases with increasing liquid and gas velocities. The heat transfer coefficient increases with decreasing liquid viscosity, and with decreasing heater diameter, but is unaffected by column diameter. The heat transfer coefficient for the horizontal immersed cylinder with a small diameter is significantly greater than the wall-to-bed heat transfer coefficient under similar hydrodynamic conditions as compared in the figure. Note that the behavior of the heat transfer coefficient with respect to the heater diameter for horizontal and vertical heaters is the same (Kato et al., 1984a). The correlation equation taking into account the effect of heater diameter is given in Table 3.6.

3.4 CONCLUDING REMARKS

The importance of the bubble characteristics to the bed hydrodynamics evidenced in Chapter 2 extends to mass transfer,

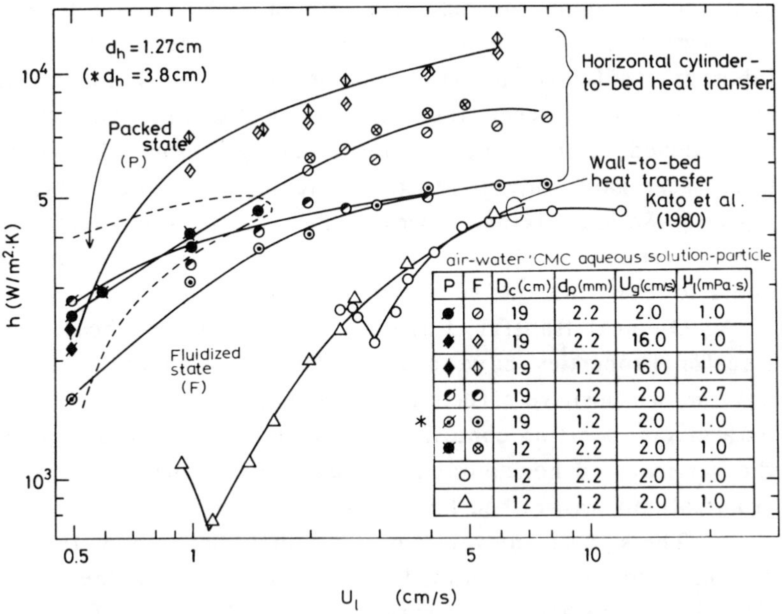

Figure 3.26 Effect of the liquid velocity on the heat transfer coefficient for various operating conditions (from Kato et al., 1984a).

mixing behavior and heat transfer. So are the general concluding remarks made on hydrodynamics and in research needs in Chapter 2.

For the gas-liquid mass transfer, the gas-side mass transfer resistance can usually be neglected, thus leading to the simplification of the overall volumetric mass transfer coefficient to the volumetric liquid-side mass transfer coefficient. Extensive information is available which indicates that the particle size has profound effects on gas-liquid interfacial area but has only small effects on the gas-liquid mass transfer coefficient. Furthermore, the gas-liquid mass transfer coefficient increases with increasing bubble size while the gas-liquid interfacial area exhibits opposite behavior. For liquid-solid mass transfer, the liquid-solid mass transfer coefficient is independent of superficial liquid velocity in a manner similar to that for a liquid-solid fluidized bed. The liquid-solid mass transfer coefficient increases with increasing gas velocity and appears to be independent of particle size over a wide range of particle size and low gas flow rates.

Extensive information is available on axial liquid mixing analyzed based on the dispersion model. The axial liquid dispersion increases with increasing gas velocities. The liquid dispersion is inherently affected by flow regime, as is the solids mixing. The complete solids intermixing state among dissimilar particles tends to be achieved in the coalesced and slugging regime while the complete solids segregate state tends to be achieved in the dispersed bubble regime. Solids mixing, however, is a complex phenomenon. Solids layer inversion involved in a system with particles of varying density and size represents one such complex phenomenon.

On heat transfer, wall-to-bed and immersed heater-to-bed heat transfer properties have been moderately explored. Gas velocities have shown positive effects in enhancing the heat transfer coefficient. At a given operating condition, the heat transfer coefficient from both the vertical and horizontal immersed object-to-bed is higher than that from column wall-to-bed heat transfer.

The significant variations in bubble behavior with particle size, column size, and gas and liquid velocities, and the corresponding significant variations in mass transfer, heat transfer, and mixing behavior generally indicate the pitfalls involved in extrapolating correlations beyond their range of applicability (specifically compatible flow regimes). The far reaching implications of the bubble behavior on all important aspects of the fluidized bed behavior, such as phase holdups and mixing characteristics, indicate the importance and the need of predicting accurately the operating flow regime for a given bed size and operating parameters. Generalization of correlations for mass and heat transfer is preferably made based on the well-established correlations for a liquid-solid fluidized bed system, which is a limiting system of a three-phase fluidized bed system.

3.5 NOMENCLATURE

a	: gas-liquid interfacial area per unit volume of reactor
a_b	: semi-major axis of a bubble
A	: cross-sectional area of the column
A_f	: surface area of bubble base; πa_b^2 for a spherical-cap bubble; $2a_b w$ for a circular-cap bubble
A_h	: effective heat transfer surface area

A_r : surface area of bubble roof; $4\pi(E^2+4)b^2/4$ for a spherical-cap bubble: $[wb(E^2+4)/2]\sin^{-1}[4E/(E^2+4)]$ for a circular-cap bubble

A_t : total surface area of a bubble

A_w : wall surface area

Ar : Archimedes number, $d_p{}^3\rho_\ell(\rho_s-\rho_\ell)g/\mu_\ell{}^2$

b : semi-minor axis of a bubble

C : measured concentration of dissolved species (tracer) of solids or gas in liquid; C_1 at point 1; C_2 at point 2

$\bar{C}(s)$: Laplace transform of $\bar{C}(t)$; $\bar{C}_1(s)$ at point 1; $\bar{C}_2(s)$ at point 2

$\bar{C}(t)$: normalized tracer concentration, $C(t)/\int_0^\infty C(t)dt$

C^* : saturation concentration of solute in solvent

C_0 : radially integrated average outlet concentration of tracer

C_b : bulk concentration of solute in solvent

C_{dt} : drag coefficient for a single particle falling with terminal velocity

C_i : local solids concentration for particle i; C_{i0} at $z_r = L_0$; C_{i1} at $z_r = L_1$

C_{in} : inlet concentration

C_{mix} : volumetric flow average outlet concentration of tracer

C^*_{pg} : enthalpy increase of gas saturated with liquid per unit increase in temperature

$C_{p\ell}$: specific heat of liquid

C_{ps} : specific heat of solid

d_e : equivalent particle diameter

d_{er} : effective hydraulic diameter for liquid channel, $\epsilon D_c/[1+1.5(D_c/d_p)(1-\epsilon)]$

d_h : diameter of cylindrical heater

d_{ho} : reference heater diameter

d_p : particle diameter

D_c : column diameter

D_M : molecular diffusivity of dissolved species in liquid; $D_{M,O}$ for a reference species

e : energy dissipation rate per unit mass of liquid

e"	: energy dissipation rate per unit mass of the system
E	: eccentricity or width-to-height ratio of a bubble
E_{rh}	: radial dispersion coefficient for heat transfer defined based on V_ℓ
$E_{r\ell}$	: liquid radial dispersion coefficient defined based on V_ℓ
E_{zg}	: gas axial dispersion coefficient defined based on V_g
$E_{z\ell}$	: liquid axial dispersion coefficient defined based on V_ℓ
E'_{zs}	: solids axial dispersion coefficient defined based on the linear solids velocity
E_{zsi}	: solids intermixing coefficient for particle i
F	: Faraday's constant
F(s)	: general transfer function
Fr_g	: gas Froude number, U_g^2/gd_p
Fr_h	: Froude number, U_g^2/gd_h
Fr_ℓ	: liquid Froude number, U_ℓ^2/gd_p
g	: gravitational acceleration
G_g	: gas mass velocity, $\rho_g U_g$
G_ℓ	: liquid mass velocity, $\rho_\ell U_\ell$
h	: wall-to-bed or immersed object-to-bed heat transfer coefficient
h_e	: heat transfer coefficient for open pipe flow with liquid velocity U_i
h_ℓ	: thickness of a stable liquid wake
h_w	: apparent wall heat transfer coefficient
Δh	: distance between two shutter plates
H	: total bed height
H_0	: static bed height
HMU_g	: height of mixing unit, $2E_{zg}\epsilon_g/U_g$
HMU_ℓ	: height of mixing unit, $2H/Pe_z$
I	: limiting electrical current
j_d	: j factor for wall mass transfer
j_H	: j factor for wall heat transfer, St $Pr_\ell^{2/3}$

j'_H : modified j factor for wall heat transfer, $(h_w/\rho_\ell C_{p\ell} U_\ell)\epsilon_\ell Pr_\ell^{2/3}$

J_d : j factor for particle mass transfer, $(k_s/U_\ell)(\mu_\ell/\rho_\ell D_M)^{2/3}$

J_0, J_1 : Zero and first order Bessel functions of first kind

k_i : integration constant in Eqn. (3.40)

k_{er} : radial effective thermal conductivity

k_{ez} : axial effective thermal conductivity

k_ℓ : liquid-side mass transfer coefficient for gas-liquid mass transfer

k_{Lf} : liquid phase mass transfer coefficient through bubble floor

k_{Lr} : liquid phase mass transfer coefficient through bubble roof

k_{Lt} : overall mass transfer coefficient of a single bubble

k_s : solid-liquid mass transfer coefficient

k_w : wall-to-bed mass transfer coefficient

$k_\ell a$: volumetric gas-liquid mass transfer coefficient

$(k_\ell a)_z$: average volumetric mass transfer coefficient over the axial interval [0,z]

$(k_\ell a)_0$: volumetric gas-liquid mass transfer coefficient for a reference system

ℓ_s : particle mean free path

L : length of heated wall or length of cylindrical heater

L_0 : lower bound of Region C in a bed of a binary mixture of particles

L_1 : upper bound of Region C in a bed of a binary mixture of particles

M : mixing index

Mv : density number, $(\rho_s - \rho_\ell)/\rho_\ell$

n : rheological index

n_e : number of electrons per ion reacted

N_f : mass transfer rate over bubble floor

N_r : mass transfer rate over bubble roof

Nu′ : Nusselt number, $hd_p\epsilon_\ell/\lambda_\ell(1-\epsilon_\ell)$

Nu" : Nusselt number, $hd_p(1-\epsilon_s)/\epsilon_s\lambda_\ell$

Nu"′	: Nusselt number, hD_c/λ_ℓ
Pe	: Peclet number, $2r_eU_B/D_M$
Pe_r	: radial Peclet number, $U_\ell R/\epsilon_\ell E_{r\ell}$
Pe_{rh}	: radial Peclet number for heat transfer, $U_\ell \rho_\ell C_{p\ell} d_p/k_{er}$
$Pe_{r\ell,m}$	: modified radial Peclet number for a three-phase system, $d_{er}(U_g + U_\ell)/\epsilon_\ell E_{r\ell}$
$(Pe_{r\ell,m})_2$	: modified radial Peclet number for two-phase systems
Pe_z	: axial Peclet number, $U_\ell H/\epsilon_\ell E_{z\ell}$
Pe'_z	: axial Peclet number, $U_\ell R/\epsilon_\ell E_{z\ell}$
$Pe_{z\ell}$	: axial Peclet number, $U_\ell D_c/\epsilon_\ell E_{z\ell}$
$Pe'_{z\ell}$	: axial Peclet number, $U_\ell d_p/E_{z\ell}$
$Pe''_{z\ell}$	: axial Peclet number, $U_g D_c/E_{z\ell}$
Pr_ℓ	: Prandtl number, $\mu_\ell C_{p\ell}/\lambda_\ell$
q	: heat released from the column wall or immersed heater per unit time
Q	: quantity of point tracer injection per unit time
r	: radial distance
r_e	: equivalent circular radius of a circular-cap bubble or equivalent spherical radius of a spherical-cap bubble
R	: column radius
R_b	: bed resistance for wall-to-bed heat transfer
Re′	: Reynolds number, $U_\ell d_p \rho_\ell/\mu_\ell(1-\epsilon_s)$
Re"	: Reynolds number, $[(U_\ell + U_g)d_p\rho_\ell]/\mu_\ell\epsilon_s$
Re_b	: Reynolds number, $2r_eU_B\rho_\ell/\mu_\ell$
Re_g	: gas Reynolds number, $U_g d_p \rho_g/\mu_g$
Re_ℓ	: liquid Reynolds number, $U_\ell d_p \rho_\ell/\mu_\ell$
$Re'_{\ell,g}$	: modified liquid Reynolds number, $Re_\ell[(\epsilon_g+\epsilon_\ell)/\epsilon_\ell(1-\epsilon_g-\epsilon_\ell)]$
Re_m	: modified Reynolds number, $d_{er}U_\ell\rho_\ell/\epsilon_\ell\mu_\ell$
Re'_{m2}	: modified Reynolds number (two-phase), $U_\ell\rho_\ell/S_p(1-\epsilon_\ell)\mu_\ell$
$Re_{m,mf0}$	: modified Reynolds number at minimum fluidization (liquid-solid), $d_{er}U_{\ell mf0}\rho_\ell/\epsilon_{\ell mf0}\mu_\ell$
Re_t	: particle Reynolds number, $d_pU_t\rho_\ell/\mu_\ell$
s	: parameter for Laplace transformation

S_1 : liquid eddies renewal frequency

S_2 : particle packets renewal frequency

S_p : surface area per unit particle volume

Sc : liquid Schmidt number, $\mu_\ell/\rho_\ell D_M$

Sh : Sherwood number, $k_s d_p/D_M$

Sh_b : Sherwood number, $k_{Lt} 2r_e/D_M$

Sh_r : Sherwood number, $k_{Lr} 2r_e/D_M$

St : Stanton number, $h/\rho_\ell C_{p\ell} U_\ell$

$St_{\ell,g}$: $k_w/(U_\ell + U_g)$

St_h : Stanton number, $h\epsilon_\ell/\rho_\ell C_{p\ell} U_\ell$

t : time

T : local temperature

T_b : temperature at the boundary between the core region and near-wall region in the bed

T_B : effective bed temperature at a given axial location

$\bar{T}_B$: average bed temperature

T_{in} : inlet temperature

T_m : integrated average bed temperature; $T_{m,L}$ at the exit

T_w : column wall temperature

$\bar{T}_w$: average column wall or heater surface temperature

T_0 : centerline temperature

$\bar{T}_0$: average centerline temperature

u : eddy characteristic velocity, $(\nu_\ell e'')^{1/4}$

u_s : average particle speed

$\bar{u}_{zi}$: segregation velocity of particle i

U_B : bubble rise velocity relative to the linear liquid velocity

U_b : bubble rise velocity relative to the wall

U_c : average liquid circulation velocity

U_g : superficial gas velocity

U_i : extrapolated superficial liquid velocity as ϵ approaches 1

U_ℓ : superficial liquid velocity

U'_ℓ : equivalent two-phase (liquid-solid) superficial liquid velocity for a binary mixture in a three-phase fluidized bed

$U'_{\ell i}$: equivalent two-phase superficial liquid velocity required to fluidize type i particles at the porosity of the mixture in a three-phase system

$U_{\ell mf}$: minimum fluidization velocity for a three-phase fluidized bed

$U_{\ell mf0}$: minimum fluidization velocity for a two-phase (liquid-solid) system

U_r : radial velocity component

U_t : terminal velocity of solid particle

U_θ : tangential velocity component

$U_{\theta 1}$: $U_{\theta s}/U_B$

$U_{\theta s}$: interfacial tangential velocity on bubble roof

V_g : linear gas velocity

V_ℓ : linear liquid velocity

w : width of a circular-cap bubble

w_i : mass of solids i in the subsection of the column

W_i : total mass of solids i in the column

x : distance along the bubble surface from the front stagnation point

x_w : weight fraction of large particles in the upper section of the bed

$\bar{x}_w$: overall weight fraction of large particles in the bed

x_1 : dimensionless distance, x/r_e

X : semi-circumference of a bubble from the front stagnation point to bubble rim

X_1 : dimensionless semi-circumference, X/r_e

y : distance from, but normal to, the bubble surface to the bulk phase

z : axial distance in the bed

z′ : vertical downward distance from the bubble base

z_r : dimensionless axial distance, z/H

Greek Letters

δ : penetration depth of the solute in solvent

ϵ	: bed porosity, $\epsilon_g + \epsilon_\ell$
ϵ_f	: bed voidage immediately behind bubble floor
ϵ_g	: gas holdup
ϵ_ℓ	: liquid holdup
$\epsilon_{\ell mf0}$	: liquid holdup at minimum fluidization in liquid-solid systems
ϵ_s	: solids holdup
λ_ℓ	: liquid thermal conductivity
μ_g	: gas viscosity
μ_ℓ	: liquid viscosity
μ_w	: water viscosity at room temperature
ν_ℓ	: kinematic liquid viscosity
ρ_g	: gas density
ρ_ℓ	: liquid density
ρ_s	: solid density
ρ_{si}	: solid density of particle i
σ	: surface tension
τ	: liquid mean residence time, $H\epsilon_\ell/U_\ell$
τ_b	: bubble residence time
ϕ_s	: shape factor
θ	: coordinate for a circular-cap bubble as shown in Fig. 3.1
ψ_s	: $\epsilon_s/(\epsilon_\ell+\epsilon_s)$
η	: eddy characteristic length, $(\nu_\ell{}^3/e'')^{1/4}$

3.6 REFERENCES

Alvarez-Cuenca, M., and M.A. Nerenberg, Can. J. Chem. Eng., 59, 739 (1981).

Alvarez-Cuenca, M., M.A. Nerenberg, and A.-F.A. Asfour, I&EC Fundam., 23, 381 (1984).

Armstrong, E.R., C.G.J. Baker, and M.A. Bergougnou, in "Fluidization Technology," ed. by D.L. Keairns, I, p. 453, Hemisphere (1976).

Arters, D.C., M.S. Thesis, The Ohio State University (1984).

Arters, D.C., and L.-S. Fan, paper presented at the AIChE Annual Meeting, San Fransisco, Nov. 25-30 (1984); Chem. Eng. Sci., 41, 107 (1986).

Arters, D.C., and L.-S. Fan, AIChE J. (in review, 1988).

Arters, D.C., K. Tsuchiya, and L.-S. Fan, paper presented at the 6th Engineering Foundation Conference on Fluidization, Banff, Canada, May 7-12 (1989).

Azbel, D., "Two-Phase Flows in Chemical Engineering," Cambridge University Press (1981).

Baker, C.G.J., E.R. Armstrong, and M.A. Bergougnou, Powder Technol., 21, 195 (1978).

Ballesteros, R.L., J.P. Riba, and J.P. Couderc, Chem. Eng. Sci., 37, 1639 (1982).

Beek, W.J., in "Fluidization," ed. by J.F. Davidson and D. Harrison, p. 341, Academic Press (1971).

Bickel, T.C., and M.G. Thomas, I&EC Process Des. & Dev., 21, 377 (1982).

Calderbank, P.H., Chem. Eng., p. CE209 (1967a).

Calderbank, P.H., in "Mixing: Theory and Practice," II, ed. by V.W. Uhl and J.B. Gray, p. 66, Academic Press (1967b).

Calderbank, P.H., D.S.L. Johnson, and J. London, Chem. Eng. Sci., 25, 235 (1970).

Chang, S.K., Y. Kang, and S.D. Kim, J. Chem. Eng. Japan, 19, 524 (1986).

Chiu, T.-M., and E.N. Ziegler, AIChE J., 29, 677 (1983).

Chiu, T.-M., and E.M. Ziegler, AIChE J., 31, 1504 (1985).

Chu, J.C., J. Kalil, and W.A. Wetteroth, Chem. Eng. Prog., 49, 141 (1953).

Couderc, J.P., H. Gibert, and H. Angelino, Chem. Eng. Sci., 27, 11 (1972).

Cussler, E.L., "Diffusion - Mass Transfer in Fluid Systems," Cambridge International Press (1984).

Dakshinamurty, P., C. Chiranveji, V. Sabrahmanyam, and P. Kameswara Rao, in "Fluidization and Its Applications," ed. by H. Angelino, J.P. Couderc, H. Gibert, and C. Laguerie, p.429, Cepadues-Editions, Toulouse (1974a).

Dakshinamurty, P., P. Kameswara Rao, and K. Veerabhadra Rao, Indian J. Technol., 12, 276 (1974b).

Dakshinamurty, P., and K. Veerabhadra Rao, Indian J. Technol., 14, 9 (1976).

Damronglerd, S., J.P. Couderc, and H. Angelino, Trans. Inst. Chem. Engrs., 53, 175 (1975).

Deckwer, W.-D., Chem. Eng. Sci., 35, 1341 (1980).

Dhanuka, V.R., and J.B. Stepanek, in "Fluidization," ed. by J.R. Grace and G.M. Matsen, p.261, Plenum Press (1980a).

Dhanuka, V.R., and J.B. Stepanek, AIChE J., 26, 1029 (1980b).

El-Temtamy, S.A., Y.O. El-Sharnoubi, and M.M. El-Halwagi, Chem. Eng. J., 18, 151 (1979a).

El-Temtamy, S.A., Y.O. El-Sharnoubi, and M.M. El-Halwagi, Chem. Eng. J., 18, 161 (1979b).

Euzen, J.P., and Y. Fortin, Chem.-Ing.-Tech., 59, 416 (1987).

Evans, G.C., and C.F. Gerald, Chem. Eng. Prog., 49, 135 (1953).

Fan, L.-S., S.H. Chern, and K. Muroyama, AIChE J., 30, 858 (1984a).

Fan, L.-S., K. Fujie, and T.R. Long, AIChE Symposium Series, 241, No. 80, 102 (1984b).

Fan, L.-S., A. Matsuura, and S.-H. Chern, AIChE J., 31, 1801 (1985).

Fan, L.-S., T. Yamashita, and R.H. Jean, Chem. Eng. Sci., 42, 17 (1987).

Fan, L.T., Y.C. Yang, and C.Y. Wen, AIChE J., 6, 482 (1960).

Friedlander, S.K., AIChE J., 3, 43 (1957).

Fukuma, M., M. Sato, K. Muroyama, and A. Yasunishi, J. Chem. Eng. Japan, 21 (in press, 1988).

Fukushima, S., J. Chem. Eng. Japan, 12, 489 (1979).

Ganho, R., H. Gibert, and H. Angelino, Chem. Eng. Sci., 30, 1231 (1975).

Garside, J., and J.W. Mullin, Trans. Inst. Chem. Engrs., 46, T11 (1968).

Gorowara, R., and L.-S. Fan, paper presented at the AIChE Annual Meeting, Washington, DC, Nov. 27 - Dec. 2 (1988).

Hanratty, T.J., G. Latinen, and R.H. Wilhelm, AIChE J., 2, 372 (1956).

Hassanien, S., H. Delmas, and J.P. Riba, Entropie, No. 119, 17 (1984).

Hinze, J.O., "Turbulence," McGraw-Hill, New York (1959).

Hirata, A., Y. Hosaka, H. Mukai, and M. Kaiho, Water Purification and Liquid Wastes Treatment, Japan, 23, 15 (1982).

Hopkins, M.J., A.J. Sheppard, and P. Eisenklam, Chem. Eng. Sci., 24, 1131 (1969).

Ito, R., Y. Komagawa, K. Omodaka, and M. Yamamoto, Kagaku Kogaku, 37, 497 (1973).

Jean, R.H., and L.-S. Fan, Chem. Eng. Sci., 41, 2811 (1986a).

Jean, R.H., and L.-S. Fan, Chem. Eng. Sci., 41, 2823 (1986b).

Jean, R.H., and L.-S. Fan, paper presented at the AIChE Annual Meeting, Washington, DC, Nov. 27 - Dec. 2 (1988).

Joshi, J.B., Trans. Inst. Chem. Engrs., 58, 155 (1980).

Juma, A.K.A., and J.F. Richardson, Chem. Eng. Sci., 38, 955 (1983).

Kang, Y., I.-S. Seo, and S.D. Kim, in Proceedings of the 3rd Pacific Chemical Engineering Conference, II, p. 1, Seoul (1983).

Kang, Y., I.S. Suh, and S.D. Kim, Chem. Eng. Commun., 34, 1 (1985).

Kang, Y., and S.D. Kim, I&EC Process Des. & Dev., 25, 717 (1986).

Kato, Y., T. Kago, K. Uchida, and S. Morooka, Kagaku Kogaku Ronbunshu, 6, 5790 (1980).

Kato, Y., K. Uchida, T. Kago, and S. Morooka, Powder Technol., 28, 173 (1981).

Kato, Y., M. Koyama, T. Kago, and S. Morooka, Kagaku Kogaku Ronbunshu, 10, 294 (1984a).

Kato, Y., Y. Tamura, T. Kago, and S. Morooka, Kagaku Kogaku Ronbunshu, 10, 427 (1984b).

Kato, Y., S. Morooka, M. Koyama, T. Kago, and S.-Z. Yang, J. Chem. Eng. Japan, 18, 313 (1985).

Kennedy, S.C., and R.H. Bretton, AIChE J., 12, 24 (1966).

Khan, A.R., A.K.A. Juma, and J.F. Richardson, Chem. Eng. Sci., 38, 2053 (1983).

Kikuchi, K., T. Sugawara, and H. Ohashi, J. Chem. Eng. Japan, 16, 426 (1983).

Kim, S.D., C.G.J. Baker, and M.A. Bergougnou, Can. J. Chem. Eng., 50, 695 (1972).

Kim, S.D., and C.H. Kim, in Proceedings of the 2nd World Congress of Chemical Engineering, III, p. 1, Montreal (1981).

Kim, S.D., and C.H. Kim, J. Chem. Eng. Japan, 16, 172 (1983).

Kitano, K., and L.-S. Fan, Chem. Eng. Sci., 43, 1355 (1988).

Klinkenberg, A., H.J. Krajenbrink, and H.A. Lauwerier, I&EC, 45, 1202 (1953).

Koloini, T., M. Sopcic, and M. Zumer, Chem. Eng. Sci., 32, 637 (1977).

Kumar, S., and S.N. Upadhyay, I&EC Fundam., 20, 186 (1981).

Kusakabe, K., S. Morooka, and Y. Kato, J. Chem. Eng. Japan, 14, 208 (1981).

Laguerie, C., and H. Angelino, Chem. Eng. J., 10, 41 (1975).

Lamb, H., "Hydrodynamics," Dover (1945).

Lee, J.C., and H. Worthington, Inst. Chem. Engrs. Symposium Series, No. 38, "Multiphase Flow Systems," I, Paper B2, Inst. Chem. Engrs., London (1974).

Lee, J.C., and P.S. Buckley, in "Biological Fluidized Bed Treatment of Water and Wastewater," ed. by P.F. Cooper and B. Atkinson, Chap. 4, p. 62, Ellis Horwood, England (1981).

Lemlich, R., and I. Caldas, Jr., AIChE J., 4, 376 (1958).

Lochiel, A.C., and P. H. Calderbank, Chem. Eng. Sci., 19, 471 (1964).

Magiliotou, M., Y.M. Chen, and L.-S. Fan, paper presented at the AIChE Annual Meeting, New York, Nov. 15-20 (1987); AIChE J., 34, 1043 (1988).

Massimilla, L., N. Majuri, and P. Signorini, La Ricerca Scientifica, 29, 1934 (1959).

McCune, L.K., and R.H. Wilhelm, I&EC, 41, 1124 (1949).

Michelsen, M.L., and K. Ostergaard, Chem. Eng. J., 1, 37 (1970).

Moritomi, H., T. Iwase, and T. Chiba, Chem. Eng. Sci., 37, 175 (1982).

Morooka, S., K. Kusakabe, and Y. Kato, Kagaku Kogaku Ronbunshu, 5, 162 (1979); Int. Chem. Eng., 20, 433 (1980).

Muroyama, K., K. Hashimoto, T. Kawabata, and M. Shiota, Kagaku Kogaku Ronbunshu, 4, 622 (1978).

Muroyama, K., M. Fukuma, and A. Yasunishi, paper presented at the 1981 ASME-AIChE National Heat Transfer Conference, Milwaukee, WI, Aug. 2-5 (1981); Abstracted in AIChE Symposium Series, 77, No. 208, 385 (1981).

Muroyama, K., M. Fukuma, and A. Yasunishi, Can. J. Chem. Eng., 62, 199 (1984).

Muroyama, K., M. Fukuma, and A. Yasunishi, Can. J. Chem. Eng., 64, 399 (1986a).

Muroyama, K., M. Fukuma, and A. Yasunishi, Can. J. Chem. Eng., 64, 409 (1986b).

Nguyen-Tien, K., A.N. Patwari, A. Schumpe, and W.-D. Deckwer, J. Chem. Eng. Japan, 17, 652 (1984).

Nguyen-Tien, K., A.N. Patwari, A. Schumpe, and W.-D. Deckwer, AIChE J., 31, 194 (1985).

Nikov, I., and H. Delmas, Chem. Eng. Sci., 42, 1089 (1987).

Nishikawa, M., K. Kosaka, and K. Hashimoto, in Proceedings of the 2nd Pacific Chemical Engineering Congress, I, p. 389, Denver (1977).

Ohashi, H., T. Sugawara, K. Kikuchi, and T. Henmi, J. Chem. Eng. Japan, 12, 190 (1979).

Ohashi, H., T. Sugawara, and K. Kikuchi, J. Chem. Eng. Japan, 14, 489 (1981a).

Ohashi, H., T. Sugawara, K. Kikuchi, and H. Konno, J. Chem. Eng. Japan, 14, 433 (1981b).

Ohashi, H., T. Sugawara, K. Kikuchi, and M. Takeda, J. Chem. Eng. Japan, 15, 311 (1982).

Ohshima, S., M. Suzuki, K. Shimada, T. Takematsu Y. Kuriki, and J. Kato, Kagaku Kogaku Ronbunshu, 3, 406 (1977).

Oloman, C., and A.P. Watkinson, Can. J. Chem. Eng., 53, 268 (1975).

Ostergaard, K., Discussion in the Paper of R. Turner in "Fluidization," p. 58, Soc. Chem. Ind., London (1964).

Ostergaard, K., and M.L. Michelsen, Joint AIChE-IIQPR Meeting, Tampa, FL (1968).

Ostergaard, K., "Studies of Gas-Liquid Fluidization," Danish Technical Press, Copenhagen (1969).

Ostergaard, K., and M.L. Michelsen, Can. J. Chem. Eng., 47, 107 (1969).

Ostergaard, K., and W. Suchozebrski, in Proceedings of the 4th European Symposium on Chemical Reaction Engineering, p. 21, Pergamon Press, Oxford (1971).

Ostergaard, K., and P. Fosbol, Chem. Eng. J., 3, 105 (1972).

Ostergaard, K., AIChE Symposium Series, 74, No. 176, 82 (1978).

Patel, R.D., and J.M. Simpson, Chem. Eng. Sci., 32, 67 (1977).

Prakash, A., C.L. Briens, and M.A. Bergougnou, in Proceedings of the 34th Canadian Chemical Engineering Conference, p. 478, Sep. 30 - Oct. 3 (1984).

Prakash, A., C.L. Briens, and M.A. Bergougnou, Can. J. Chem. Eng., 65, 228 (1987).

Rahman, K., and M. Streat, Chem. Eng. Sci., 36, 293 (1981).

Richardson, J.F., and W.N., Zaki, Trans. Inst. Chem. Engrs., 32, 35 (1954).

Richardson, J.F., and A.E. Mitson, Trans. Inst. Chem. Engrs., 36, 270 (1958).

Richardson, J.F., and J.W. Smith, Trans. Inst. Chem. Engrs., 40, 13 (1962).

Richardson, J.F., M.N. Romani, and K.J. Shakiri, Chem. Eng. Sci., 31, 619 (1976).

Rigby, G.R., and C.E. Capes, Can. J. Chem. Eng., 48, 343 (1970).

Rowe, P.N., and K.T. Claxton, Trans. Inst. Chem. Engrs., 43, T321 (1965).

Rowe, P.N., A.W. Nienow, and A.J. Agbim, Trans. Inst. Chem. Engrs., 50, 310 (1972).

Saberian-Broudjenni, M., G. Wild, N. Midoux, and J.C. Charpentier, Can. J. Chem. Eng., 63, 553 (1985).

Saito, H., and T. Kobayashi, Kagaku Kogaku, 29, 237 (1965).

Sanger, P., and W.-D. Deckwer, Chem. Eng. J., 22, 179 (1981).

Smith, J.W., and D.H. King, Can. J. Chem. Eng., 53, 41 (1975).

Suh, I.S., G.T. Jin, and S.D. Kim, Int. J. Multiphase Flow, 11, 255 (1985).

Tournie, P., C. Laquerie, and J.P. Couderc, Chem. Eng. Sci., 32, 1259 (1977).

Tournie, P., C. Laquerie, and J.P. Couderc, Chem. Eng. Sci., 34, 1247 (1979).

Towell, G.D., and G.H. Ackerman, in Proceedings of International Symposium on Chemical Reaction Engineering, B, p. 3, Elsevier, Amsterdam (1972).

Tripathi, G., and G.N. Pandey, Indian J. Technol., 8, 285 (1970).

Upadhyay, S.N., and G. Tripathi, J. Chem. Eng. Data, 20, 20 (1975).

Vail, Yu K., N. Kh. Manakov, and V.V. Manshilin, Int. Chem. Eng., 8, 293 (1968).

Vanderschuren, J., J.P. Schrayen, and S. Roland, in "Fluidization and Its Applications," ed. by H. Angelino, J.P. Couderc, H. Gibert, and C. Laguerie, p. 351, Cepadues-Editions, Toulouse (1974).

Vanadurongwan, V., C. Laguerie, and J.P. Couderc, Chem. Eng. J., 12, 29 (1976).

Van Driesen, R.P., and N.C. Stewart, Oil and Gas J., 62, 100, May 18 (1964).

Viswanathan, S., A.S. Kakar, and P.S. Murti, Chem. Eng. Sci., 20, 903 (1964).

Wakao, V., and S. Yagi, Kagaku Kogaku, 23, 161 (1959).

Wasmund, B., and J.W. Smith, Can. J. Chem. Eng., 45, 156 (1967).

Wen, C.Y., and L.T. Fan, "Models for Flow Systems and Chemical Reactors," Marcel Dekker, New York (1975).

Whalley, P.B., and J.F. Davidson, Chem. Eng. Symposium Series, No. 38, J5 (1974).

Wisecarver, K., and L.-S. Fan, Chem. Eng. Sci., 43, 2741 (1988).

Yang, S.-Z., in Proceedings of the 3rd World Congress of Chemical Engineering, III, p. 496, Tokyo (1986).

Yasunishi, A., M. Fukuma, and K. Muroyama, Kagaku Kogaku Ronbunshu, 13, 208 (1987).

Yasunishi, A., M. Fukuma, and K. Muroyama, J. Chem. Eng. Japan, 21, 522 (1988).

Chapter 4*

Slurry Bubble Column Systems (Modes E-I-a-2, E-III-a, and T-I-a-2; Figure 1.3)

This chapter describes slurry bubble column systems with liquid as the continuous phase; in these systems particle terminal velocities in the liquid media are relatively small (due to small size or low density) compared with those encountered in fluidized beds. Operations are classified as Mode E-I-a-2 or T-I-a-2 when particles are supported by an upward flow of gas and liquid or as Mode E-III-a when particles are supported by an upward flow of gas alone. For Mode E-I-a-2, solid particles can be in a batch (i.e., solids-batch) condition whereas for Mode E-III-a, both solid particles and liquid are in a batch (i.e., solids-liquid-batch or slurry-batch) condition. Since the analyses of transport behavior for Modes E-I-a-2 and T-I-a-2 bear a close resemblance, they are considered jointly throughout this chapter unless otherwise noted.

More specifically, the transport behavior in a slurry bubble column covered in this chapter includes flow regimes, gas holdup, the critical gas velocity for complete solids suspension, liquid flow patterns, phase mixing, gas-liquid and liquid-solid mass transfer, and heat transfer. Furthermore, the behavior of a gas-liquid bubble column, or a bubble column in brief, is extensively illustrated as the limiting case (solid-free) of the slurry bubble column. Interrelation of the behavior in the slurry bubble column with that in a fluidized bed system given in Chapters 2 and 3 is discussed.

4.1 FLOW REGIMES

As in Mode E-I-a-1 fluidized beds, three different flow regimes are observed in slurry bubble or bubble columns: the dispersed bubble, coalesced bubble, and slugging regimes (see Section 2.1.2 for a general

*: by L.-S. Fan and K. Muroyama

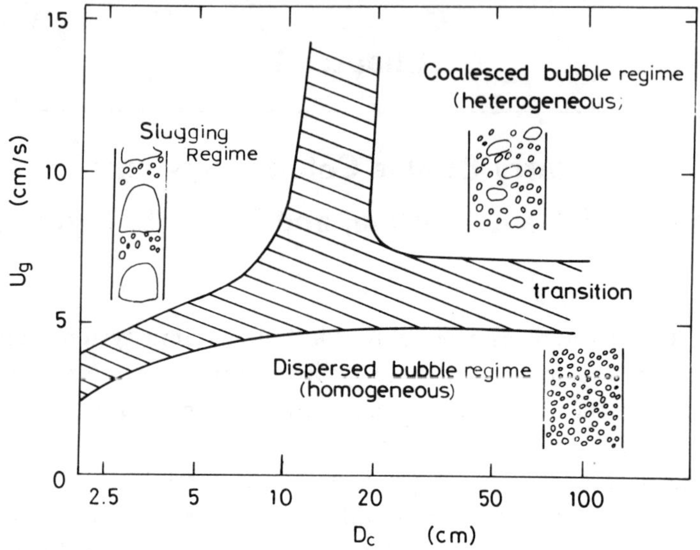

Figure 4.1 Flow regime map for a liquid-batch bubble column or a slurry-batch slurry bubble column containing a low viscosity liquid phase (from Deckwer et al. 1980).

description of each regime). The terminology used for these regimes differs in bubble and slurry bubble column literature. The dispersed bubble regime is often referred to as homogeneous or homogeneous bubble flow and the coalesced bubble regime as heterogeneous or churn-turbulent flow.

In bubble or slurry bubble columns, the dispersed bubble regime occurs at low gas velocities, while at high gas velocities either the coalesced bubble or slugging regime occurs depending on the column diameter. For highly viscous fluids slug flow is observed even at relatively low gas velocities (Schumpe, 1981). Deckwer et al. (1980) proposed the flow regime map given in Fig. 4.1 for both bubble and slurry bubble columns with a batch liquid phase. Figure 4.1 applies, in particular, for systems having efficient gas distributors and containing low viscosity liquids. For slurry systems where liquid is not in a batch condition, the flow regime behavior is expected to be similar to that of Mode E-I-a-1 operation (e.g., see Fig. 2.3).

The specific operating conditions for the transitions between the three regimes depend on the gas distributor design, liquid properties, particle properties, and column size. Bubble coalescence and breakup tendencies are of obvious importance; for related discussions see Sections 2.1.2, 2.2.1.1, and 2.2.1.2. In systems where either bubble coalescence or breakage phenomena dominate, system properties associated with bubble formation, such as gas distributor design, become less important to flow regime behavior. However, in liquid-surfactant systems where bubble coalescence is inhibited partially due to formation of an electrically charged layer at the gas-liquid interface (see Section 2.1.7), the gas distributor design becomes more important since the initial bubble size distribution controls the bubble flow pattern.

4.2 GAS HOLDUP AND BUBBLE SIZE DISTRIBUTION

The average gas holdup can be measured using methods similar to those described in Chapter 2 for Mode E-I-a-1. For example, gas holdup can be measured directly from the difference between the aerated liquid level and the settled liquid level or, in the case of bubble columns, by the pressure gradient method. Similarly, the local gas holdup or bubble size can be measured using electrical resistivity probe techniques (e.g., Hills (1974) for a bubble column, and Smith et al. (1984) and Yasunishi et al. (1986b) for a slurry bubble column).

4.2.1 Gas Holdup

Since gas holdup reflects the prevailing bubble size and rise velocity behavior, any system property which affects the bubble size or rise velocity affects the gas holdup. As in Mode E-I-a-1 fluidized bed systems, the radial holdup profile is nearly flat in the dispersed bubble regime and in the coalesced bubble regime shows a maximum at the axis, decreases radially, and approaches zero at the wall. The radial gas holdup distribution can be empirically expressed by Eqn. (2.104) for both a bubble column (Ueyama and Miyauchi, 1979) and a slurry bubble column with $\bar{\epsilon}_S < 0.2$ (Yasunishi et al., 1986b). However, for high solids content ($\bar{\epsilon}_S > 0.2$) the radial gas holdup distribution deviates significantly from Eqn. (2.104) (Yasunishi et al., 1986b). Smith et al. (1984) used a cubic polynomial equation to approximate the radial gas holdup in their slurry bubble column system.

A large number of empirical correlations are available to predict the average gas holdup in bubble columns and slurry bubble columns (e.g., Hughmark, 1967; Akita and Yoshida, 1973; Gestrich and Rahse, 1975; Kumar et al., 1976; Mersmann, 1978; Hikita et al., 1980;

Table 4.1 Correlations for Gas Holdup in Bubble Columns and Slurry

Investigators	Correlation Equations (in SI Units)	
Hughmark (1967)	$\varepsilon_g = \{2+(0.35/U_g)[(\rho_\ell/1000)(\sigma/0.072)]^{1/3}\}^{-1}$	(T-4.1.1)
Akita and Yoshida (1973)	$\frac{\varepsilon_g}{(1-\varepsilon_g)^4} = C\left(\frac{D_c^2\rho_\ell g}{\sigma}\right)^{1/8}\left(\frac{D_c^3\rho_\ell^2 g}{\mu_\ell^2}\right)^{1/12}\left(\frac{U_g}{\sqrt{gD_c}}\right)$ C = 0.2 for pure liquids and non-electrolyte aqueous solns. C = 0.25 for electrolyte aqueous solns.	(T-4.1.2)
Kumar et al. (1976)	$\varepsilon_g = 0.728U - 0.485U^2 + 0.0975U^3$ where $U = U_g[\rho_\ell^2/\sigma(\rho_\ell-\rho_g)g]^{1/4}$	
Mersmann (1978)	$\frac{\varepsilon_g}{(1-\varepsilon_g)^4} = 0.14\, U_g\left(\frac{\rho_\ell^2}{\sigma(\rho_\ell-\rho_g)g}\right)^{1/4}\left(\frac{\rho_\ell^2\sigma^3}{\mu_\ell^4(\rho_\ell-\rho_g)g}\right)^{1/24}\left(\frac{\rho_\ell}{\rho_g}\right)^{1/36}\left(\frac{\rho_\ell}{\rho_\ell-\rho_g}\right)^{1/3}$	
Kara et al. (1982)	$\varepsilon_g = \frac{Re_g^*}{C_1 + C_2 Re_g^* + C_3 Re_{s\ell} + C_4 \psi_s}$ where $Re_{s\ell} = D_c U_{s\ell}^* \rho_{s\ell}/\mu_{eff}$, $Re_g^* = D_c U_g \rho_g/\mu_g$ C_1, C_2, C_3, C_4 = adjustable parameters for specified particles	(T-4.1.3)
Kelkar et al. (1983)	$\varepsilon_g = \frac{0.96\, U_g^{0.58} C_N^{0.26}}{1+2.6\, U_\ell}$ for continuous liquid flow systems $\varepsilon_g = 0.75\, U_g^{0.50} C_N^{0.20}$ for batch liquid systems C_N = carbon number	
Koide et al. (1984)	For coalesced bubble regime: $\frac{\varepsilon_g}{(1-\varepsilon_g)^4} = \frac{k_1(U_g\mu_\ell/\sigma)^{0.918}(g\mu_\ell^4/\rho_\ell\sigma^3)^{-0.252}}{1+4.35(\bar{C}_s/\rho_s)^{0.748}[(\rho_s-\rho_\ell)/\rho_\ell]^{0.88}(D_c U_g/\rho_\ell)^{-0.168}}$	(T-4.1.4)
	k_1 = 0.227 for water and aqueous solutions of glycerol & glycol k_1 = 0.364 for aqueous soltuions of inorganic electrolytes	
	For transition regime: $\frac{\varepsilon_g}{\varepsilon_{go}} = \frac{1 + M\exp\{-[\ln(U_g/U_{gM})]^2\}}{1+49.1(\bar{C}_s/\rho_s)^{0.619}[(\rho_s-\rho_\ell)/\rho_\ell]^{1.14}(D_c U_g \rho_\ell/\mu_\ell)^{-0.496}}$	(T-4.1.5)
	$M = 2.52\times10^{-2}\left[\left(\frac{d_o}{L_p}\right)\left(\frac{D_c^2 g\rho_\ell}{N\sigma}\right)\right]^{-0.398}(g\mu_\ell^4/\rho_\ell\sigma^3)^{-0.139}$	(T-4.1.6)
	$\frac{U_{gM}\mu_\ell}{\sigma} = 9.59\left[\left(\frac{d_o}{L_p}\right)\left(\frac{D_c^2 g\rho_\ell}{N\sigma}\right)\right]^{-0.521}\left(\frac{g\mu_\ell^4}{\rho_\ell\sigma^3}\right)^{0.384}$	(T-4.1.7)
	$\frac{\varepsilon_{go}}{(1-\varepsilon_{go})^4} = 0.277\left(\frac{U_g\mu_\ell}{\sigma}\right)^{0.918}\left(\frac{g\mu_\ell^4}{\rho_\ell\sigma^3}\right)^{-0.252}$	(T-4.1.8)

Bubble Columns

Size of Column and Gas Distributor	Range of Variables	Systems
$D_c \le 0.1$ m (multi-orifice)	$0.004 \le U_g \le 0.45$ m/s $780 \le \rho_\ell \le 1700$ kg/m^3 $0.9 \le \mu_\ell \le 152$ mPa·s $25 \le \sigma \le 76$ mN/m	gas: air liquid: water, kerosene, light oil, glycerol aqueous soln., Na_2SO_3 aqueous soln., $ZnCl_2$ aqueous soln.
$0.152 \le D_c \le 0.60$ m $2.0 \le H \le 3.0$ m (single nozzle) d_o = 5 mm	$0.005 \le U_g \le 0.419$ m/s $790 \le \rho_\ell \le 1590$ kg/m^3 $0.58 \le \mu_\ell \le 21.1$ mPa·s $22.3 \le \sigma \le 74.2$ mN/m	gas: air, O_2, He, CO_2 liquid: water, glycol, methanol, glycol aqueous soln., methanol aqueous soln.
$0.05 \le D_c \le 0.1$ m (single nozzle and multi-orifice) $0.87 \le d_o \le 2.65$ mm	$0.0014 \le U_g \le 0.14$ m/s $790 \le \rho_\ell \le 1100$ kg/m^3 $31.2 \le \sigma \le 74.5$ mN/m	gas: air liquid: water, kerosene, 40% glycerol aqueous soln.

semi-theoretical equation for bubble column

D_c = 0.152 m H = 3.35 m	$0.03 \le U_g \le 0.30$ m/s $0 \le U^*_{s\ell} \le 0.10$ m/s $0 \le \bar{C}_s/\rho_{s\ell} \le 0.40$ d_p = 10,30,70 μm	gas: air liquid: water solid: coal, dried mineral
D_c = 0.154, 0.30 m H = 2.44, 3.35 m (perforated plate)	$0.01 \le U_g \le 0.30$ m/s $0 \le U_\ell \le 0.15$ m/s	gas: air liquid: alcohol aqueous soln.
D_c = 0.14, 0.218, 0.30 m (perforated plate) $0.5 \le d_o \le 2.5$ mm	$0.01 \le U_g \le 0.18$ m/s $U^*_{s\ell}$ = 0 m/s $997 \le \rho_\ell \le 1178$ kg/m^3 $0.894 \le \mu_\ell \le 17.6$ mPa·s $5.5 \le \sigma \le 73$ mN/m $47.5 \le d_p \le 192$ μm ρ_s = 2500, 8770 kg/m^3	gas: air liquid: water, glycerol aqueous soln., ethylene glycol aqueous soln. solid: glass sphere, bronze sphere

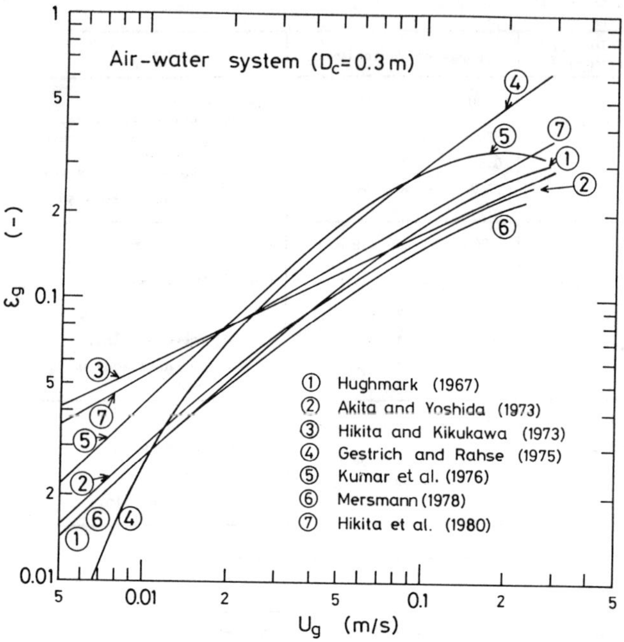

Figure 4.2 Comparison of correlations for the gas holdup in bubble columns.

Kara et al., 1982; Kelkar et al., 1983; Koide et al., 1984; Chang and Morsi, 1988). Table 4.1 lists several important ones. For a bubble column containing non-Newtonian liquid, correlation equations by Hughmark (1967), Akita and Yoshida (1973), and Mersmann (1978) are generally applicable for average gas holdup prediction; the Akita and Yoshida equation, among all, gives the best prediction. For the case of low solids concentrations and within the same flow regime, correlations for ϵ_g in bubble columns can be used to predict gas holdup in slurry bubble columns by replacing the liquid properties with the slurry properties. For example, in a system with a slurry density less than 1150 kg/m^3, Smith and Ruether (1985) confirmed that the Hughmark (1967) correlation (Eqn. (T-4.1.1)) developed for a bubble column could be applied to the slurry system when the liquid density, ρ_ℓ, in Eqn. (T-4.1.1) is replaced with the slurry density, $\rho_{s\ell}$. Figure 4.2 shows a comparison of gas holdups predicted from seven correlations (with four of them given in Table 4.1) for an air-water system in a 0.3 m diameter column. Generally, as shown in Fig 4.2, gas holdup in a bubble or slurry bubble column increases with increasing superficial gas velocity; the actual relationship depends on the flow regime and distributor design. In coalesced bubble systems, $\epsilon_g/(1-\epsilon_g)^4$ was found to be proportional to the superficial gas velocity by Akita and Yoshida (1973) in a bubble column with a single nozzle gas distributor (Eqn. (T-4.1.2)) and by Koide et al. (1984) in a slurry bubble column (Eqn. (T-4.1.4)). Note that in these two cases, and in most correlations presented in Table 4.1, gas holdup is almost independent of column diameter and height.

At low superficial liquid velocities, gas holdup is independent of the liquid velocity and coincides with that in a batch-liquid system (Mode E-III-a). At moderately high superficial liquid velocities, the liquid velocity does affect gas holdup; an increase in liquid velocity may increase the absolute bubble rise velocity, decrease the initial bubble size formed at the distributor, or change the coalescence/ breakage tendencies. Under moderately high liquid velocity conditions, the gas holdup behavior can be satisfactorily described by the drift flux model given by Zuber and Findley (1965) (also see Section 2.3.6).

Liquid properties affect the gas holdup behavior, in part, through affecting both bubble formation and bubble coalescing tendencies. An increase in liquid viscosity will decrease gas holdup; the increase in viscosity leads to larger stable bubbles and thus higher bubble rise velocities. Gas holdup data obtained for slurry bubble columns containing various organic solvents are available from Hammer (1979). Liquid interfacial properties play a very significant role in gas holdup. Adding a small amount of surface-active material to water, such as a short chain alcohol, produced significantly higher gas holdup (Kelkar et al., 1983). The presence of electrolytes or impurities in water can also lead to higher gas holdup (Hikita et al., 1980; Kelkar et al., 1984; Morooka et al., 1986; Anderson and Quinn, 1970) (also see Sections 2.1.7 and 6.1.2.1). In both cases, the higher gas holdup was attributed, at least in part, to a decrease in or suppression of coalescing tendencies. Foaming properties of the liquid are also important and can lead to a hysteresis phenomenon. Experimental studies designed to simulate conditions prevailing in the Fischer-Tropsch slurry reactor by using paraffin waxes as the liquid phase (Calderbank et al., 1963; Deckwer et al., 1980; Quicker and Deckwer, 1981; Smith et al., 1984; Kuo et al., 1985; Bukur and Daly, 1987; Bukur et al., 1987) were particularly susceptible to foaming. Bukur and Daly (1987) observed hysteresis in the gas holdup behavior for a molten wax system as shown in Fig. 4.3. In liquid-surfactant systems, a transition to a foaming system is observed under some conditions and can be

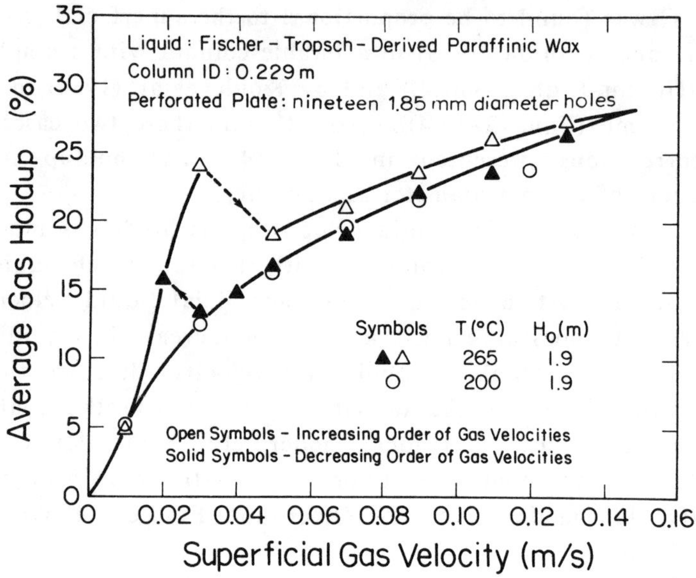

Figure 4.3 Hysteresis phenomenon for a bubble column (from Bukur and Daly, 1987).

detrimental to system operability; however, these systems provide important data for industrial bubble columns which experience small bubble sizes and large gas holdup.

In a slurry bubble column using fine particles (< 100 μm), in general, the gas holdup decreases with an increase in solids concentration (Koide et al., 1984), but the magnitude of this decrease reflects the complex nature of the prevailing flow regime: (1) Deckwer et al. (1980) found that, in a homogeneous dispersion of fine bubbles (Al_2O_3/paraffin slurry at a relatively high temperature), a decrease in gas holdup occurred for up to 5 wt% solids, but further increases up to about 16 wt% produced little effect; (2) Quicker et al. (1984) found that, for a solid suspension in a sodium sulfite solution, the decrease was significant; (3) Kara et al. (1982) reported that, in a coal-water slurry, the magnitude of the decrease increased with particle size; and (4) Ying et al. (1980b), using a silica-water slurry, observed only a slight effect for solids concentrations up to 45 wt% (note that the data for this system followed Akita and Yoshida's (1973) correlation (Eqn. (T-4.1.2)) indicating an established coalesced bubble regime). The most plausible explanation for this seemingly varied behavior is that increasing solids concentration increases the "pseudo-viscosity" of the

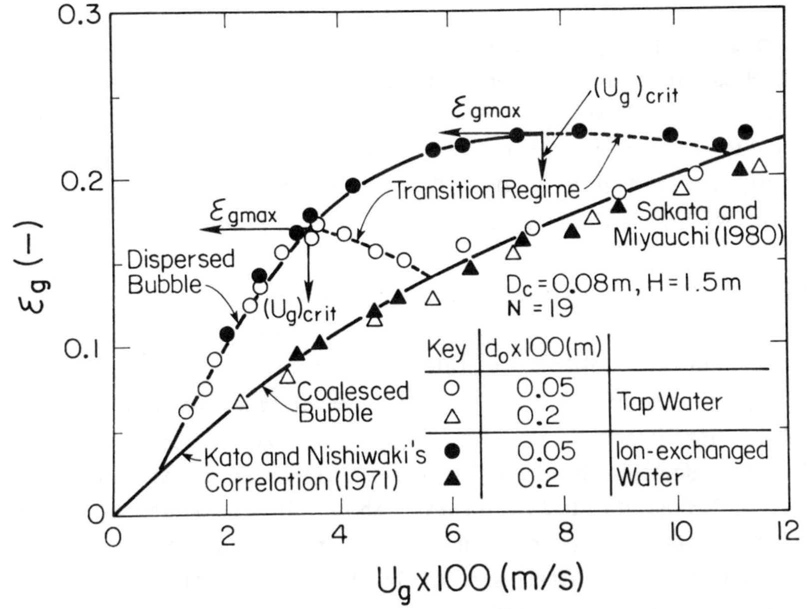

Figure 4.4 Effects of water properties and distributors on the gas (air) holdup of a bubble column (from Koide, 1982).

suspension which in turn promotes the bubble coalescence, resulting in an increase in bubble size and a decrease in gas holdup. Furthermore, systems which undergo flow regime transition from dispersed bubble to coalesced bubble regime as a result of changing solids concentration show significant changes in gas holdup behavior. In the established coalesced bubble regime, on the other hand, the gas holdup changes only slightly with increasing solids concentration.

Koide et al. (1984) indicated that in a slurry bubble column the distributor affected gas holdup in the transition regime as well as the onset conditions for the transition regime (see Eqns. (T-4.1.5) through (T-4.1.8)). The effect of the distributor on the gas holdup in a bubble column is manifested in Fig. 4.4. For columns with a perforated plate distributor having an orifice diameter (d_0) smaller than 0.2 cm or having a pitch greater than 0.25 cm, gas holdup increases approximately linearly with gas velocity up to a maximum value in the dispersed bubble regime, then decreases in the transition regime, and finally increases again in the coalesced bubble regime. Yamashita and Inoue (1975) correlated the maximum gas holdup by

$$\epsilon_{gmax} = 0.0083 \{NL_p/(d_o D_c^2)\}^{0.29} \tag{4.1}$$

for an air-water system using a perforated plate distributor. In Eqn. (4.1), N is the number of orifices on the distributor, L_p is the pitch of orifices on the distributor, and D_c is the column diameter. Maruyama et al. (1981) found that the gas velocity at which the maximum gas holdup occurred depended only on the unaerated liquid height for systems with distributors generating uniform stable gas distribution.

4.2.2 Bubble Size and Size Distribution

In bubble column or slurry bubble column systems, bubbles are formed at the sparger (or distributor) orifices or nozzles. As the bubbles rise, however, they may grow in size through coalescence or split into smaller bubbles by turbulent shear in the liquid phase. Akita and Yoshida (1974), observed that bubbles formed at the orifices or nozzles are almost uniform in size for given conditions. They also indicated that the initial size of bubbles from single orifice spargers and perforated plate spargers is affected only by the orifice diameter and the gas velocity through the orifice, and is insensitive to system properties such as surface tension, liquid viscosity, and liquid and gas densities. Ueyama et al. (1980) reported that the bubble size in a bubble column was largest near the column center and decreased radially; the minimum size occurred near the column wall. This profile reflects the radial gas holdup profile. Note, however, that in their system the gas stream is directed through nozzles toward the column center.

Bubble size and size distribution measurement has been extensively discussed in Section 2.2.1.1; the reader should refer to this section for details. The arithmetic mean of the maximum and minimum chord length of projected bubbles determined by photographic methods follows a log-normal distribution (Akita and Yoshida, 1974). In addition, the chord length data from the conductivity probe of Yasunishi et al. (1986b) and Fukuma et al. (1987a) also followed a log-normal distribution. Yamashita et al. (1979) compared the photographic and conductivity methods for determining bubble size in the dispersed bubble regime for a bubble column; they indicated that, in this regime where ellipsoidal bubbles predominated, the mean chord length could be approximately converted to the mean bubble size by multiplying by 1.5. This result was also suggested by Ueyama and Miyauchi (1976).

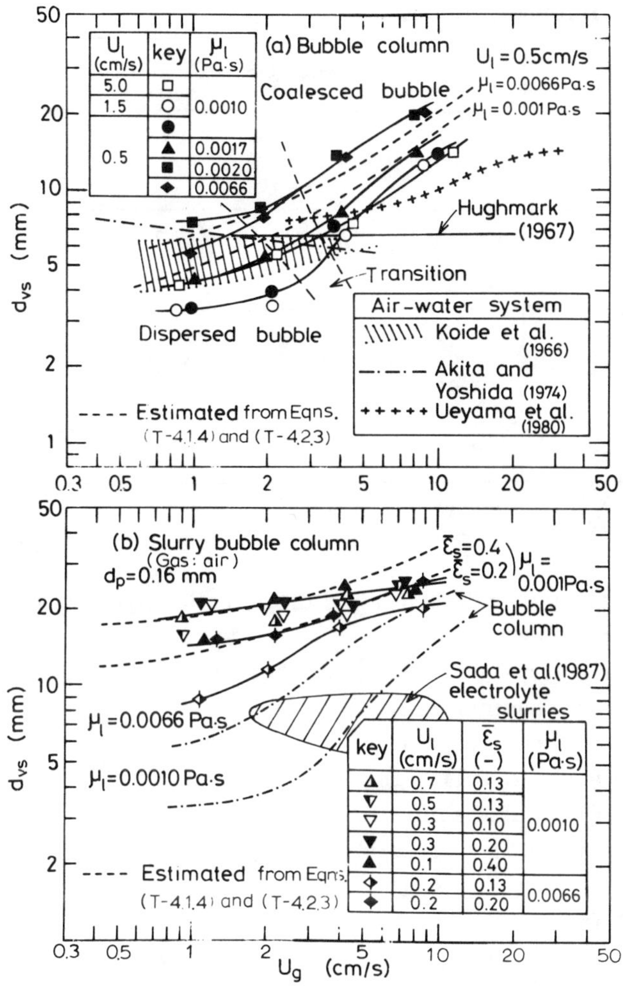

Figure 4.5 Variation of d_{vs} with U_g: (a) in a bubble column and (b) in a slurry bubble column (from Fukuma et al., 1987a).

The effect of gas velocity on bubble size in a bubble column, as shown in Fig. 4.5(a), depends on flow regime. In the dispersed bubble regime, bubble size is small (d_{vs}< 6 mm) and increases only slightly with gas velocity. In the coalesced bubble regime, bubble size is large (d_{vs}> 1 cm) and increases more rapidly with gas velocity. Akita and Yoshida (1974) found that bubble size measured by photographs at the column wall decreased with gas velocity. However, as confirmed by the data of Ueyama et al. (1980), the bubble size measured at the

Table 4.2 Correlations for Sauter Mean Bubble Diameter, d_{vs}, in Bubble Columns

Investigators	Correlation Equations (in SI Units)	
Hughmark (1967)	$d_{vs} = 0.00635 \left(\frac{\sigma}{0.072}\right)^{0.6} \left(\frac{1000}{\rho_\ell}\right)^{0.2}$	(T-4.2.1)
Akita and Yoshita (1973, 1974)	$\frac{d_{vs}}{D_c} = 26 \left(\frac{D_c^2 \rho_\ell g}{\sigma}\right)^{-0.5} \left(\frac{D_c^3 \rho_\ell^2 g}{\mu_\ell^2}\right)^{-0.12} \left(\frac{U_g}{\sqrt{gD_c}}\right)^{-0.12}$	(T-4.2.2)
Fukuma et al. (1987a)	$d_{vs} = 0.59\,(j_{g\ell}/\varepsilon_g)^2/g$	(T-4.2.3)
	$j_{g\ell} = U_g(1-\varepsilon_g) - U_\ell\varepsilon_g(1-\varepsilon_g)/\varepsilon_\ell$	(T 4.2.4)
	ε_g can be obtained from Eqn. (T-4.1.4).	

column wall does not reflect the cross-sectionally averaged bubble size in the system.

For slurry bubble columns, both the solids holdup and particle size may affect bubble size. For 160 μm glass beads at a given liquid velocity, Fukuma et al. (1987a) indicated that increasing the solids holdup, $\bar{\epsilon}_s$, up to 0.05 significantly increased the mean bubble size. However, beyond a solids holdup of 0.10 for this particle, there is essentially no effect of solids holdup on the mean bubble diameter, d_{vs}, as shown in Fig. 4.5(b). It is noted in the figure that the mean bubble diameter in this slurry bubble column system increases only slightly with U_g. For 56 μm glass beads, however, a slightly different behavior was observed; in this case a significant change in bubble size occurred when $\bar{\epsilon}_s$ was increased from about 0.05 to 0.18; at $\bar{\epsilon}_s$ below 0.05 or above 0.18 the mean chord length was insensitive to solids holdup.

Several correlations for the bubble diameter are provided in Table 4.2. The correlations of Hughmark (1967) and Akita and Yoshida (1974) were developed for bubble columns, while the correlation of Fukuma et al. (1987a) was developed for a slurry bubble column. Note that the bubble size can be reported as many different types of equivalent diameters and careful consideration should be used when comparing different values.

4.3 CRITICAL GAS VELOCITY FOR COMPLETE SOLIDS SUSPENSION

For a given solids concentration in a slurry bubble column where particles are not solely supported by the liquid flow, a critical gas velocity exists beyond which solid particles in the column are in complete suspension. Similarly, for a given gas velocity, a critical solids concentration exists below which solid particles in the column are in complete suspension. Different criteria have been used to determine the critical gas velocity or the critical solids concentration. Kojima and Asano (1982) considered that all particles are in a suspended state so long as the solids concentration at the bottom of the column is lower than that in a packed condition. Koide et al. (1983) used a pressure drop method to determine the critical condition for solids suspension.

Many workers investigated the critical gas velocity (Roy et al., 1964; Narayanan et al., 1969; Ying et al., 1980a; Kojima and Asano, 1982; Koide et al., 1983; Furchner and Mersmann, 1983). Table 4.3 presents the experimental conditions and the proposed correlations reported by two of them. It is noted that most of these studies concern Mode E-III-a operation.

The critical gas velocity depends on the particle properties and solids concentration. In general, the critical gas velocity increases with increasing solids concentration as indicated by correlations of Koide et al. (1983). In addition, Koide et al. (1983) observed that the critical gas velocity increased with increasing particle terminal velocity. Note that the ratio of the critical gas velocity to the particle terminal velocity usually ranges from 2 to 4 as observed experimentally by Kato (1963), Koide et al. (1983), and Ying et al. (1980a).

Other important factors influencing the critical gas velocity are the distributor design and column size. Koide et al. (1983) and Smith et al. (1986) reported that the critical gas velocity increased with increasing column diameter; Furchner and Mersmann (1983), however, found the opposite trend. This disparity may be attributed to the effects of column and distributor design. Furchner and Mersmann (1983) obtained a graphical correlation, as shown in Fig. 4.6, between the Froude number defined based on the liquid circulation velocity and the ratio of particle to column diameters, d_p/D_c, at the critical condition for solids suspension. It is seen that the Froude number

Table 4.3 Correlations for Critical Solids Suspension in Slurry Bubble Column

Investigators	Correlation Equations (in SI Units)	Size of Column and Gas Distributor	Systems	Range of Variables
Roy et al. (1964)	$\bar{C}_{sc}/\rho_{s\ell} = 6.84 \times 10^{-4} (Re_g)_c^{1.0} \left(\frac{\sigma}{U_b \mu_\ell}\right)^{-0.23} \left(\frac{U_t}{U_b}\right)^{-0.18} (X)^{-3.0} C_\mu$ for $(Re_g)_c < 500$ $\bar{C}_{sc}/\rho_{s\ell} = 1.072 \times 10^{-1} (Re_g)_c^{0.2} \left(\frac{\sigma}{U_b \mu_\ell}\right)^{-0.23} \left(\frac{U_t}{U_b}\right)^{-0.18} (X)^{-3.0} C_\mu$ for $(Re_g)_c > 600$ $C_\mu = 0.1558 + 0.0264 (\log_{10}\mu_\ell) + 0.1026 (\log_{10}\mu_\ell)^2$ X = ratio of wettability of a particle to that of quartz.	D_c = 0.051m H = 1.27 mm (perforated plate)	gas: air liquid: alcohol aqueous soln. diesel oil, til oil, compressor oil solid: coal (d_p = 511 μm, ρ_s = 1440 kg/m^3); quartz (138 ≤ d_p ≤ 511 μm, ρ_s = 2630 kg/m^3); Ni-Al alloy (d_p = 138 μm, ρ_s = 3466 kg/m^3); Fischer-Tropsch catalysts (d_p = 282 μm, ρ_s = 2601 kg/m^3)	$0.06 \le U_g \le 0.6$ m/s $U^*_{s\ell} = 0$ m/s $0 \le \bar{C}_s/\rho_{s\ell} \le 0.30$ kg/kg-slurry
Koide et al. (1983)	$\frac{U_{gc}}{U_t} = 0.801 \left(\frac{\rho_s-\rho_\ell}{\rho_\ell}\right)^{0.600} \left(\frac{\bar{C}_s}{\rho_s}\right)^{0.146} \left(\frac{\sqrt{gD_c}}{U_t}\right)^{0.240}$ $\cdot \left[1+807 \left(\frac{g\mu_\ell^4}{\rho_\ell \sigma^3}\right)^{0.578}\right] \left\{1-1.20\left(1-\frac{D_d}{D_c}\right)^{0.0301 (D_c^2 g\rho_\ell/\sigma)^{0.559}}\right.$ $1.12 \le [(\rho_s-\rho_\ell)/\rho_\ell] \le 7.79$, $4.76\times10^{-3} \le \bar{C}_s/\rho_s \le 3.39\times10^{-1}$, $15.6 \le (\sqrt{gD_c}/U_t) \le 278$, $2.57\times10^{-11} \le (g\mu_\ell^4/\rho_\ell\sigma^3) \le 5.65\times10^{-6}$, $0.18 \le D_d/D_c \le 1$, $1.35\times10^3 \le (D_c^2 g\rho_\ell/\sigma) \le 1.21\times10^4$ and $3.33 \le H/D_c \le 10.7$	D_c = 0.10, 0.14, 0.30 m H = 1.9, 2 m (perforated plate) $0.18 \le D_d/D_c \le 1.0$	gas: air liquid: water, glycerol aqueous soln., ethylene glycol aqueous soln. solid: glass (77 ≤ d_p ≤ 846 μm, ρ_s = 2500 kg/m^3) bronze (d_p = 167 μm, ρ_s = 8770 kg/m^3)	$0.015 \le U_g \le 0.30$ m/s $U^*_{s\ell} = 0$ m/s $1.002 \le \mu_\ell \le 21.7$ mPa·s $998 \le \rho_\ell \le 1181$ kg/m^3 $25 \le \bar{C}_s \le 400$ kg/m^3-slurry

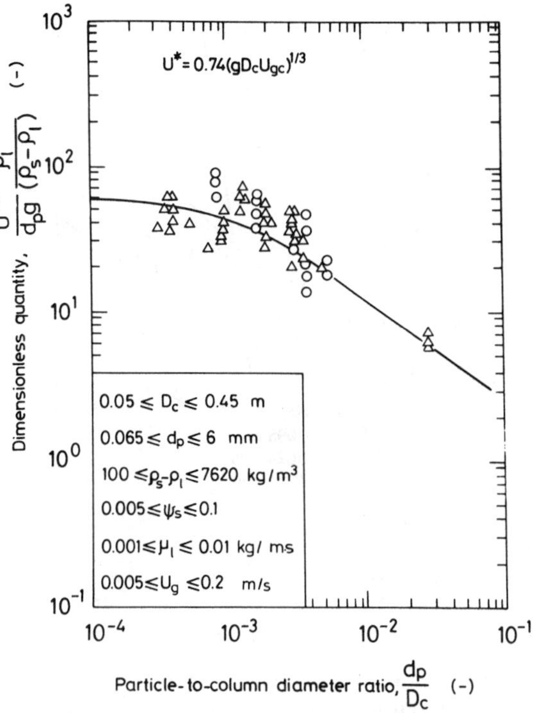

Figure 4.6 Correlation of critical gas velocity for complete solids suspension in slurry-batch slurry bubble columns (from Furchner and Mersmann, 1983). (Note: Conditions for symbols were not specified in the original paper).

decreases with an increase in d_p/D_c. Koide et al. (1983) found that the critical gas velocity was independent of the column height.

Figure 4.7 shows a comparison of various correlations for the critical gas velocity. Significant disagreements exist among these correlations, which may be due to the apparatus geometry and dimensions or possibly due to methods used or degree of subjectivity involved in determining the critical velocity. Koide et al.'s (1983) correlation is recommended for design calculations since this correlation was based on data obtained by the pressure drop method (involving little subjectivity) and collected over the widest range of experimental conditions.

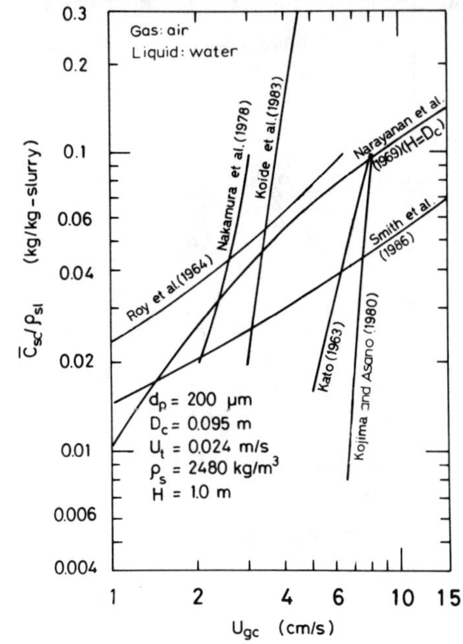

Figure 4.7 Comparison of predicted values of critical solids holdup from various correlations for a slurry bubble column (from Kojima, 1981).

4.4 LIQUID FLOW PATTERNS

Coalesced bubbles ascending in the column lead to a gross circulation flow pattern for the liquid (or slurry) with an upward flow in the core region and a downward flow near the wall. Bubble columns with internal multistaging also may experience liquid circulation causing an apparent liquid backflow through the weirs. The information in the literature regarding liquid velocity measurements and distributions has been mainly concerned with bubble column systems; however, this information should qualitatively reflect that in slurry bubble systems also. In the following, liquid velocity measurements and distributions in bubble column systems are described.

Typical liquid velocity profiles have been measured using a Pavlov (or modified Pitot) tube (Hills, 1974), a tracer injection technique (Ueyama and Miyauchi, 1976; Morooka et al., 1986), a hot-film anemometer (Franz et al., 1984), and a laser-doppler anemometer (Franz et al., 1984). A typical liquid velocity profile is depicted in

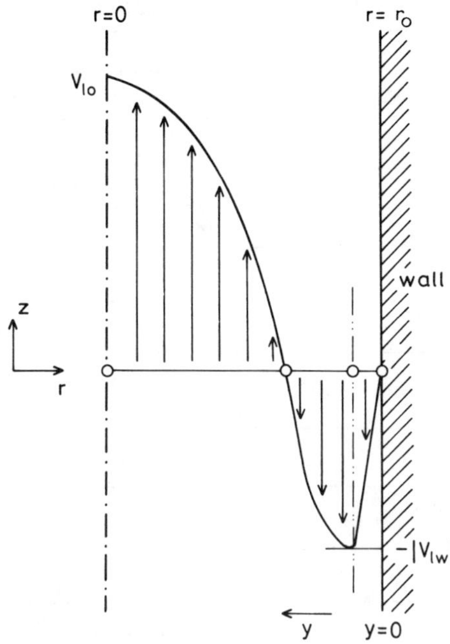

Figure 4.8 Recirculating flow pattern of liquid in a bubble column (from Ueyama and Miyauchi, 1979).

Fig. 4.8. Ueyama and Miyauchi (1979) reported a correlation for the liquid velocity profile in a bubble column as

$$\frac{V_{\ell r}+|V_{\ell w}|}{V_{\ell o}+|V_{\ell w}|} = [1-(\frac{r}{r_o})^2]^2 \tag{4.2}$$

where $V_{\ell o}$ and $V_{\ell w}$ are defined in Fig. 4.8.

Franz et al. (1984) reported that the radial distribution of the axial mean liquid velocity is not necessarily axisymmetric. The skewness of the radial distribution of the axial liquid velocity varies progressively with the axial position indicating a steep helical upward motion of bubbles in the central zone. A non-axisymmetric distribution may also be contributed to by imperfect vertical alignment of the column or imperfect horizontal positioning of the gas distributor (Bergougnou, 1988).

Flow models, namely the energy balance and momentum balance models, are available to describe the liquid circulation velocity and liquid velocity profiles in a bubble column. These

Table 4.4 Correlations for Liquid Phase Axial Dispersion Coefficient in Bubble Columns and Slurry Bubble Columns

Investigators	Correlation Equations (in SI Units)	Range of Variables	Systems
Kato and Nishiwaki (1971)	$Pe_{z\ell} = 13\, Fr_g^{0.5}/(1+6.5Fr_g^{0.4})$ (T-4.4.1)	$1.6\times10^{-5} \le Fr_g \le 0.16$, $D_c \ge 0.122$ m	gas-liquid
Kato et al. (1972a)	$Pe_{z\ell} = 13\, Fr_g^{0.5}/(1+8Fr_g^{0.425})$ (T-4.4.2)	$0.02 \le U_g \le 0.30$ m/s $0 \le U_\ell \le 0.022$ m/s $48 \le \bar{C}_s \le 202$ kg/m^3-slurry $63 \le d_p \le 177$ μm $0.066 \le D_c \le 0.214$ m	gas-liquid-solid
Towell and Ackerman (1972)	$E_{z\ell} = 1.225\, D_c^{1.5}\, U_g^{0.5}$	$0.009 \le U_g \le 0.09$ m/s $4 \le U_\ell \le 14$ mm/s	gas-liquid
Deckwer et al. (1974)	$E_{z\ell} = 0.678\, D_c^{1.4}\, U_g^{0.3}$	————	gas-liquid
Hikita and Kikukawa (1975)	$\left(\frac{E_{z\ell}}{\sqrt{D_c^3 g}}\right)\left(\frac{\mu_\ell^4 g}{\rho_\ell \sigma^3}\right) = 0.037 + 0.188\, Fr_g^{0.36}$	$2.63\times10^{-11} \le \mu_\ell^4 g/\rho_\ell\sigma^3 \le 2.51\times10^{-6}$ $1.6\times10^{-3} \le Fr_g \le 0.16$	gas-liquid
Baird and Rice (1975)	$E_{z\ell} = 0.35\, D_c^{4/3}\, (g\, U_g)^{1/3}$	$0.003 \le U_g \le 0.45$ m/s $0.082 \le D_c \le 1.53$ m	gas-liquid
Joshi (1980)	$E_{z\ell} = 0.29\, D_c\, (U_\ell + U_c)$ $U_c = 1.31\, \{gD_c[U_g + U_\ell - \frac{U_\ell \rho_\ell}{\varepsilon_s\rho_s + \varepsilon_\ell\rho_\ell} - \varepsilon_s(\frac{\rho_s}{\varepsilon_s\rho_s + \varepsilon_\ell\rho_\ell} - 1)\, U_p - \varepsilon_g U_{b\infty}]\}^{1/3}$	$0.01 \le U_g \le 0.39$ m/s $0 \le U_\ell \le 0.12$ m/s $0.10 \le D_c \le 1.067$ m	gas-liquid and gas-liquid-solid
Riquarts (1981)	$Pe_{z\ell} = 14.7\, (Fr_g^3/Re_g)^{1/8}$	————	gas-liquid in heterogeneous bubble flow regime
Kara et al. (1982)	$E_{z\ell} = A\, U_g^B\, U_{s\ell}^{*\,C}$ d_p (μm) / A / B / C air-water / 240.1 / 0.205 / -0.235 10 / 179.6 / 0.336 / -0.126 30 / 285.3 / 0.262 / -0.450 70 / 165.9 / 0.249 / -0.465	$D_c = 0.152$ m $0.03 \le U_g \le 0.30$ m/s $0 \le U_{s\ell}^* \le 0.10$ m/s	gas-liquid and gas-liquid-solid, continuous slurry flow
Kelkar et al. (1983)	$E_{z\ell}(1-\varepsilon_g) = 1.42\, D_c^{1.33}[U_g - \frac{\varepsilon_g U_\ell}{1-\varepsilon_g}]^{0.73}$	$D_c = 0.154$, 0.30 m $0.01 \le U_g \le 0.30$ m/s $0 \le U_\ell \le 0.15$ m/s	gas-liquid with low surface tension liquid, continuous liquid flow
Kawase and Moo-Young (1986)	$Pe_{z\ell} = 2.92\, n^{8/3} Fr_g^{1/3}$	$0.625 \le n \le 1.0$ $2\times10^{-5} \le Fr_g \le 5\times10^{-1}$	gas-liquid with Newtonian and non-Newtonian liquids

Dimensionless parameters:

$$Pe_{z\ell} = \frac{U_g D_c}{E_{z\ell}},\quad Fr_g = \frac{U_g^2}{gD_c},\quad Re_g = \frac{U_g D_c \rho_\ell}{\mu_\ell}$$

n = power-law index for liquid viscosity

models, which can be extended to slurry bubble column systems, are given in Sections 4.6.1 and 4.6.2.

4.5 PHASE MIXING

The axial mixing behavior of the liquid phase, gas phase, and solids phase is given below.

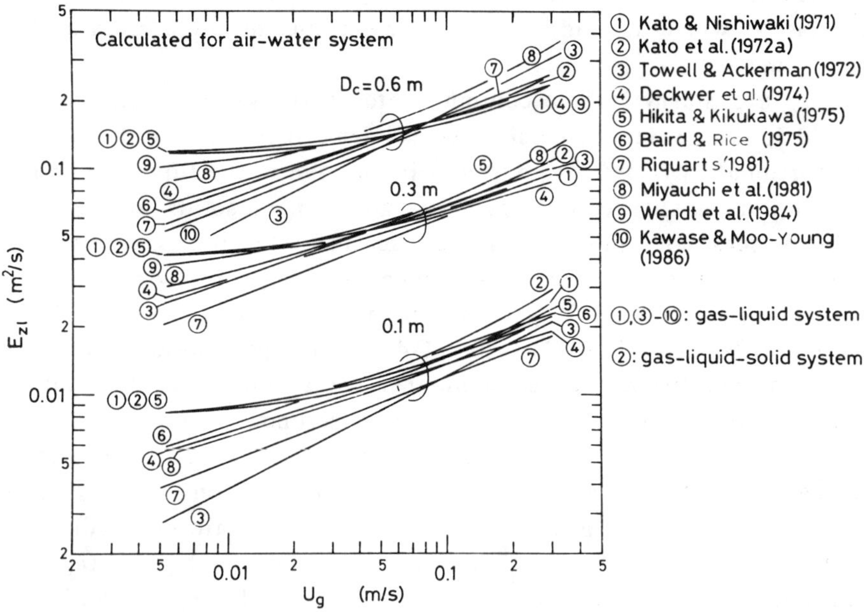

Figure 4.9 Comparison of $E_{z\ell}$ values estimated from various correlations for bubble columns.

4.5.1 Liquid Phase Axial Mixing

A large number of correlations are available for the liquid phase axial dispersion coefficient for gas-liquid systems. Correlations for the liquid phase axial dispersion coefficient (based on the linear liquid velocity) that have wide applicability are listed in Table 4.4 together with those specifically for slurry bubble columns. Figure 4.9 compares axial dispersion coefficients in bubble columns calculated from ten correlations (with eight of them given in Table 4.4) for 0.1 m, 0.3 m, and 0.6 m column diameters. In the coalesced bubble regime (i.e., at high gas velocities), $E_{z\ell}$ predicted by the various correlations agree fairly well, especially for a column diameter of 0.3 m.

In general, the liquid axial dispersion coefficient in a bubble column is proportional to the 0.3 ~ 0.5 power of superficial gas velocity, proportional to the 1.25 ~ 1.5 power of column diameter, decreases slightly with an increase in liquid viscosity, and is essentially unaffected by the liquid surface tension in the coalesced bubble regime. Note that most experimental work has been done mainly for batch-liquid systems and thus, only a few correlations consider the effect of the superficial liquid velocity. Caution must be exercised when extending any correlation to high liquid flow conditions.

In a slurry bubble column, Kato et al. (1972a) found that the liquid phase axial dispersion coefficient increased with U_g, was proportional to $D_c^{1.0\sim1.5}$, and was insensitive to the liquid velocity at low liquid velocity conditions. The addition of solids to the bubble column does not appear to affect significantly the axial dispersion coefficient as can be seen by comparing this correlation (Eqn. (T-4.4.2)) (for slurry bubble systems) with the same type of correlation by Kato and Nishiwaki (1971) (Eqn. (T-4.4.1)) (for gas-liquid systems). At large gas velocities, the Peclet number for a slurry bubble column is slightly greater than that for a bubble column. Kara et al. (1982) obtained correlations with adjustable parameters which show a strong dependency of the axial dispersion coefficient on the superficial slurry velocity. For a slurry bubble column containing a surfactant and particles with $d_p \leq 113$ μm, a well dispersed bubble regime is encountered at low gas velocities. The axial dispersion coefficient under these conditions is significantly lower than that in an air-water system, but the difference becomes small at high gas velocities ($U_g > 8$ cm/s).

Several investigators have studied liquid mixing and liquid phase residence time distribution (RTD) in multistage bubble columns. In general, the axial dispersion coefficients in multistage bubble columns were smaller than those in single stage bubble columns (Sekizawa and Kubota, 1974).

4.5.2 Gas Phase Axial Mixing

Information available in the literature on gas phase mixing is mostly limited to studies of bubble columns (without suspended solids). Gas phase axial dispersion coefficients have been measured based on pulse, step, or frequency response techniques using a low solubility gas as the tracer injected into the main operating gas stream. Gases used as the tracer are H_2, freon, O_2 and argon 41. Gas phase axial dispersion coefficients increase with increasing superficial gas velocity to the power of 1 ~ 3.56 and column diameter to the power of 1.33 ~ 2 and are roughly 2 ~ 10 times greater than liquid phase dispersion coefficients. Correlation equations for the gas phase axial dispersion coefficient (defined based on linear gas velocity) are summarized in Table 4.5. Figure 4.10 compares the values of E_{zg} calculated from the

Table 4.5 Correlations for Gas Phase Axial Dispersion Coefficient in a Bubble Column

Investigators	Correlation Equations (in SI Units)	Range of Variables	Data Source	Systems
Towell and Ackerman (1972)	$E_{zg} = 19.7\, D_c^2\, U_g$ (T-4.5.1)	$D_c = 0.406,\ 1.067$ m $0.0085 \leq U_g \leq 0.13$ m/s $0.0072 \leq U_l \leq 0.0135$ m/s	Their own data	air-water
Field and Davidson (1980)	$E_{zg} = 56.4\, D_c^{1.33} \left(\frac{U_g}{\varepsilon_g}\right)^{3.56}$ (T-4.5.2)	$0.076 \leq D_c \leq 3.2$ m $0.0085 \leq U_g \leq 0.13$ m/s $0 \leq U_l \leq 0.0135$ m/s	Their own data Literature data	air-water
Mangartz and Pilhofer (1980)	$E_{zg} = 50\, D_c^{1.5} \left(\frac{U_g}{\varepsilon_g}\right)^{3.0}$ (T-4.5.3)	$0.092 \leq D_c \leq 1.067$ m $0.015 \leq U_g \leq 0.13$ m/s $0.0072 \leq U_l \leq 0.0135$ m/s	Their own data Literature data	gas-liquid air/water N_2/n-propanol air/glycol
Joshi (1982)	$E_{zg} = 110\, D_c^2 \left(\frac{U_g^2}{\varepsilon_g}\right)$ (T-4.5.4)	$0.092 \leq D_c \leq 1.067$ m $0.0076 \leq U_g \leq 0.13$ m/s $0 \leq U_l \leq 0.06$ m/s	Literature data	———

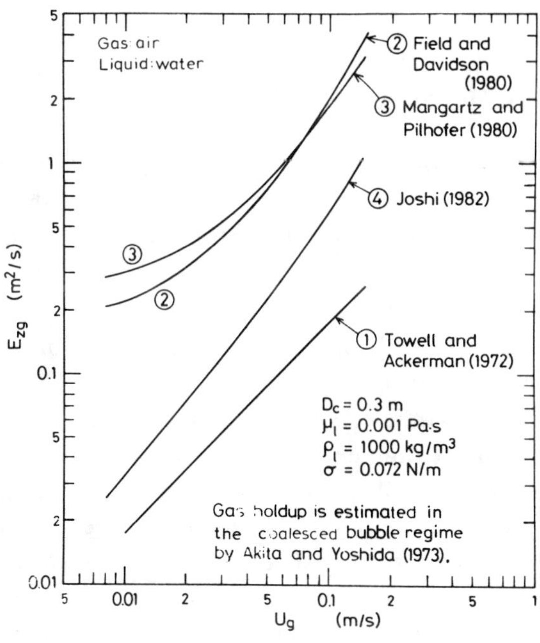

Figure 4.10 Comparison of E_{zg} values estimated from various correlations for a bubble column.

correlation equations listed in Table 4.5 for a 0.3 m column diameter and shows significant deviations in the estimated values. The correlation of Towell and Ackerman (1972) (Eqn. (T-4.5.1)) predicts the lowest values of E_{zg}, those of Field and Davidson (1980) (Eqn. (T-4.5.2)) and Mangartz and Pilhofer (1980) (Eqn. (T-4.5.3)) yield similar, high predictions, and that of Joshi (1982) (Eqn. (T-4.5.4)) predicts intermediate values. For a 1 m column diameter, the deviations among the predicted values of E_{zg}, except that from the correlation of Towell and Ackerman (1972), become much smaller. The correlation proposed by Joshi (1982) covers a wide range of operating conditions and is the most widely applicable of all the correlations.

Magiera (1984) found that the axial gas phase dispersion coefficients in a multistage slurry bubble column with continuous slurry flow depended strongly on the superficial slurry velocity and solids concentration. The dispersion coefficient increased significantly with increasing solids concentration and superficial gas velocity. The increase in the gas phase axial dispersion coefficient with an increase in solids concentration may be due to increased bubble coalescence.

4.5.3 Solid Phase Mixing

The solids mixing behavior in a slurry bubble column under Modes E-I-a-2 and E-III-a, to a great extent, resembles that in the freeboard region of a gas-liquid-solid fluidized bed containing large/heavy particles. The solids concentration in these systems shows a general decreasing trend with axial distance. A slurry bubble column under Mode T-I-a-2, on the other hand, exhibits uniformity of solids concentration with axial distance. For Mode E-I-a-2, when the density of the solid particle is close to that of the liquid, a constant solids concentration region exists immediately above the distributor; beyond this region, the solids concentration decreases gradually with axial distance. The flow regime has a strong effect on the profile of the axial solids concentration distribution. Within the same flow regime, either dispersed or coalesced bubble regime, the effect of the gas velocity on the axial solids concentration distribution is not very significant. The profile of the axial solids concentration distribution, however, varies considerably between these two regimes. Liquid properties such as density, viscosity, and surface tension affect the axial solids dispersion in slurry bubble column systems. The effect is especially obvious in the presence of surfactants due to their strong

Table 4.6 Correlations for Solid Phase Axial Dispersion Coefficient in Slurry Bubble Columns

Investigators	Correlation Equations for E_{zs} and U_p (in SI Units)	Size of Column and Gas Distributor	Systems	Range of Variables
Kato et al. (1972a)	$\frac{U_g D_c}{E^*_{zs}} = 13Fr_g^{0.5}\{1+0.009Re_t Fr_g^{-0.4}\}/(1+8Fr_g^{0.425})$ $U_p = 1.33\ U_t(U_g/U_t)^{0.25}\Psi_\ell^{1.5}$	$0.0066 \le D_c \le 0.214$ m $2.01 \le H \le 4.05$ m (perforated plate)	gas: air liquid: water solid: glass beads ($75.5 \le d_p \le 163$ μm, $\rho_s = 2520$ kg/m^3)	$0.02 \le U_g \le 0.30$ m/s $0.020 \le U^*_{s\ell} \le 0.022$ m/s $48 \le \bar{C}_s \le 202$ kg/m^3-slurry
Smith and Ruether (1985) O'Dowd et al. (1987)	$Pe_{zs} = 9.6(Fr_g^3/Re_g)^{0.1114} + 0.019\ Re_t^{1.1}$ (T-4.6.1) $U_p = 1.91\ U_g^{0.26}U_t^{0.80}\Psi_\ell^{3.5}$ (T-4.6.2) $C_s^1 = 1.27\ C_s^f$	$D_c = 0.108$ m $H = 1.94$ m (single bubble cap)	gas: N_2 liquid: water, ethanol solid: glass beads ($48.5 \le d_p \le 164$ μm, $\rho_s = 2420,\ 3990$ kg/m^3)	$0.03 \le U_g \le 0.2$ m/s $0.0071 \le U^*_{s\ell} \le 0.020$ m/s $0 \le \bar{C}_s \le 150$ kg/m^3-slurry
Morooka et al. (1986)	$E^*_{zs} = E_{z\ell}$ (a graphical correlation for $E_{z\ell}$) $\Psi_\ell U_p/U_t = 1 + 3.5\ \varepsilon_g\ Re_t^{-0.55}$ (for homogeneous bubble flow) $\Psi_\ell U_p/U_t = 1 + 10\ \varepsilon_g\ Re_t^{-0.25}$ (for heterogeneous bubble flow)	$D_c = 0.12$ m $H = 2.05,\ 2.265$ m (perforated plate, single nozzle)	gas: air liquid: aqueous soln. of surfactants solid: glass beads ($44 \le d_p \le 230$ μm, $\rho_s = 2500$ kg/m^3)	$0.02 \le U_g \le 0.20$ m/s $0.002 \le U^*_{s\ell} \le 0.040$ m/s $50 \le \bar{C}_s \le 100$ kg/m^3-slurry

Parameters, E^*_{zs}, U_p, C_s^1 are defined based on the following equations:

$$-E^*_{zs}\frac{dC_s}{dz} + V_pC_s = -E^*_{zs}\frac{dC_s}{dz} + (V^*_{s\ell}-\Psi_\ell U_p)C_s = V^*_{s\ell}C_s^f$$

$$C_s = \{C_s^1 + \frac{V^*_{s\ell}}{\Psi_\ell U_p - V^*_{s\ell}}C_s^f\}\exp\left[\frac{\Psi_\ell U_p - V^*_{s\ell}}{E^*_{zs}}(H-z)\right] - \frac{V^*_{s\ell}}{\Psi_\ell U_p - V^*_{s\ell}}C_s^f$$

Dimensionless variables:

$$Pe_{zs} = \frac{U_g D_c}{E^*_{zs}},\quad Fr_g = \frac{U_g^2}{gD_c},\quad Re_t = \frac{d_p U_t \rho_\ell}{\mu_\ell},\quad \Psi_\ell = \frac{\varepsilon_\ell}{1-\varepsilon_g}$$

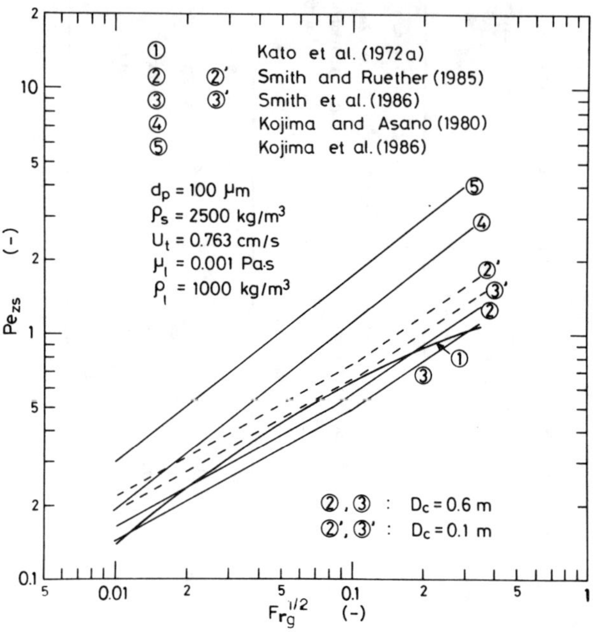

Figure 4.11 Comparison of Pe_{zs} values predicted from various correlations for a slurry bubble column.

effects on the bubble properties. Some experimental results are available regarding the effects of liquid properties (Badgujar et al., 1986) and surfactants (Morooka et al., 1986; Kelkar et al., 1984) on solids dispersion.

Quantitative analysis of the axial solids concentration profile in continuous solids flow or solids-batch slurry bubble column systems has been extensively based on the sedimentation-dispersion model (see Section 4.6.3). The model in its most generally used form is characterized by two parameters, namely the axial solids dispersion coefficient, E_{zs} (or E^*_{zs}), and the solids settling velocity. There are a large number of empirical correlations proposed in the literature to account for these two parameters. However, due to the inconsistent physical interpretation of these parameters special attention must be paid to the specific definitions used in accounting for the parameters when applying any correlation. Table 4.6 lists three correlations based on unified definitions given in Eqns. (4.39) and (4.40) of Section 4.6.3. The correlations cover a wide range of operating conditions. Figure 4.11 shows a comparison of five correlations for E^*_{zs} (with two given in Table 4.6). The Pe_{zs} increases with increasing Fr_g (Kato et al.,

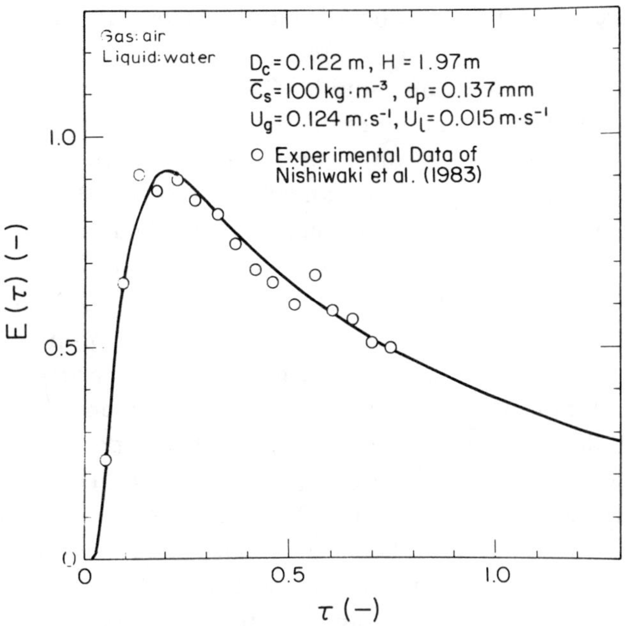

Figure 4.12 Residence time distribution of solid particles in a slurry bubble column.

1972a; Smith and Ruether, 1985; Smith et al., 1986; O'Dowd et al., 1987); the Pe_{zs} increases with decreasing column diameter (Smith and Ruether, 1985; Smith et al., 1986; O'Dowd et al., 1987). The correlations given by Kojima and Asano (1980) and Kojima et al. (1986) predict Pe_{zs} independent of column diameter and show higher values relative to other correlations.

The axial mixing of solids in a slurry bubble column was also studied from the aspect of solids residence time distribution by Nishiwaki et al. (1983). Figure 4.12 shows a typical residence time distribution of solid particles which was obtained based on an impulse input of solid particle tracers. Nishiwaki et al. (1983) demonstrated that the solids residence time distribution can be well fitted by the sedimentation-dispersion model.

4.6 FLOW MODELS

Two types of models, namely, the pressure balance and energy balance models, have been proposed to account for the liquid circulation velocity and liquid velocity profile in a bubble column (also see Section 4.4). Freedman and Davidson (1969), based on the pressure balance concept, predicted the liquid velocity profile and bubble path in a quiescent liquid. Though the model satisfactorily predicted the experimental results, its evaluation procedures were tedious, and the model was only applicable to a shallow pool of liquid involving $H_0/D_c \leq 1.0$. Whalley and Davidson (1974) used the energy balance concept to provide an improved method over the pressure balance method. To account for the liquid velocity profile through a deep pool of batch liquid, Joshi and Sharma (1979) modified the Whalley and Davidson (1974) approach and proposed multiple circulation cells in the axial direction. In this model, the average liquid circulation velocity was shown to be independent of liquid height, which is consistent with considerable experimental data for $H_0/D_c > 1.0$. To account for the liquid velocity profile in a deep pool of liquid, Ueyama and Miyauchi (1979) proposed a different approach, based on the momentum balance on a gas-liquid mixture. Note that the energy balance model and the momentum balance model developed originally for the bubble column systems are extendible to dilute slurry bubble column systems. The axial solids concentration distribution can be successfully described by the sedimentation-dispersion model, which is phenomenological in nature, or the axial solids holdup distribution (batch) model (developed by Tang and Fan, 1987), which is mechanistic in nature. There are two other mechanistic models (Dayan and Zalmanovich, 1982; Turi and Ng, 1986) proposed for description of the axial solids concentration distribution in a solids-batch slurry bubble column. The model proposed by Dayan and Zalmanovich (1982) is applicable only under the situation when solid dispersion is due to entrainment in the closed wakes associated with bubbles. This situation is not commonly encountered in slurry bubble column operations. Turi and Ng (1986) proposed a model accounting for solid dispersion due to drift effects resulting from bubble motion. Their model involves complicated fluid mechanic parameters which are difficult to evaluate. In the following, the energy balance models proposed by Whalley and Davidson (1974) and modified by Joshi and Sharma (1979), and the momentum balance model of Ueyama and Miyauchi (1979) are discussed. Furthermore, a general derivation of the sedimentation-dispersion model, and the axial solids holdup distribution (batch) model of Tang and Fan (1987) are given.

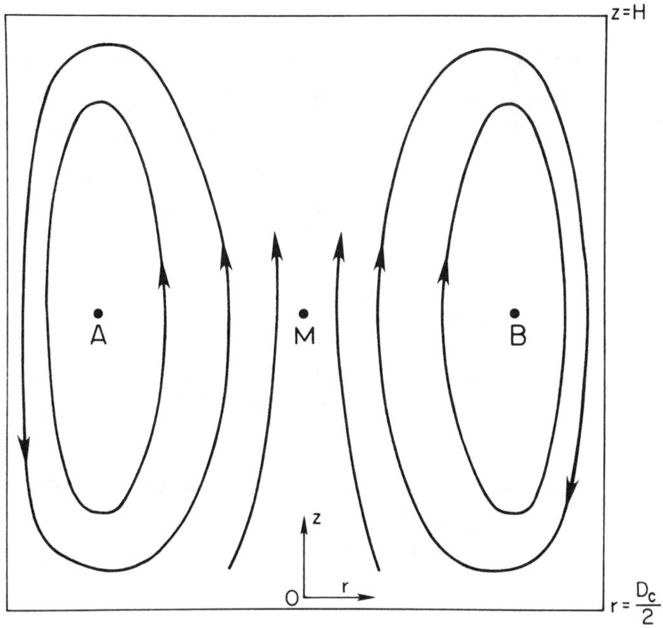

Figure 4.13 An axisymmetric cell for the liquid circulation in a bubble column.

4.6.1 Energy Balance Model

The energy balance model is developed under conditions for the liquid involving both the single circulation cell and multiple circulation cells as described below.

4.6.1.1 Single Circulation Cell

For the circulation regime in a bubble column, Whalley and Davidson (1974) adopted the vorticity and stream function relationship in inviscid fluids for steady vortex motion developed by Lamb (1945) as shown in Eqn. (4.3) (referring to the coordinates in Fig. 4.13).

$$\zeta = k_2 r\Psi = -\frac{1}{r}\left(\frac{\partial^2\Psi}{\partial z^2} + \frac{\partial^2\Psi}{\partial r^2} - \frac{1}{r}\frac{\partial\Psi}{\partial r}\right) \tag{4.3}$$

with boundary conditions

$$r = 0 \text{ and } r = \frac{D_c}{2} = r_o \; ; \; \Psi = 0 \tag{4.4a}$$

$$z = 0 \text{ and } z = H_c = H_A \quad ; \quad \Psi = 0 \tag{4.4b}$$

Here the whole column is treated as consisting of a single circulation cell and thus, H_c is equal to H_A. Equation (4.3) can be solved by the method of separation of variables. Express Ψ, in dimensionless form, as

$$\frac{\Psi}{\Psi_o} = R^*(r^*)\, Z^*(z^*) \tag{4.5}$$

where $R^*(r^*)$ satisfies

$$\frac{d^2R^*}{dr^{*2}} - \frac{1}{r^*}\frac{dR^*}{dr^*} + R^*(\alpha^2 r^{*2} - m_s^2\pi^2 r_o^2/H_c^2) = 0 \tag{4.6}$$

Using the boundary condition Eqn. (4.4b), $Z^*(z^*)$ becomes

$$Z^* = \sin(m_s \pi z^*) \qquad (m_s = 1,2,3...) \tag{4.7}$$

where

$$r^* = \frac{r}{r_o}, \; z^* = \frac{z}{H_c}, \; \alpha^2 = k_2 r_o^4$$

For a single cell m_s is equal to 1. Whalley and Davidson (1974) originally solved Eqn. (4.6) for k_2 numerically, using Eqn. (4.4a) as the boundary condition. However, Eqn. (4.6) can be transformed into Coulomb's Wave equation of order zero, and the complete solution can be obtained analytically as (Field and Davidson, 1980)

$$R^* = B(\alpha r^{*2}/2) \sum_{k=1}^{\infty} A_k(\beta) \left(\frac{\alpha r^{*2}}{2}\right)^{k-1} \tag{4.8}$$

$$\frac{\Psi}{\Psi_o} = B \sin(\pi z^*)(\alpha r^{*2}/2) \sum_{k=1}^{\infty} A_k(\beta)(\alpha r^{*2}/2)^{k-1} \tag{4.9}$$

where

$$\beta = \left(\frac{\pi D_c}{H_c}\right)^2 /(16\alpha), \; A_1 = 1 \; , \; A_2 = \beta$$

and

$$k(k-1)A_k = 2\beta A_{k-1} - A_{k-2} \qquad \text{for } k > 2$$

Without losing generality, B can be set to 1.0. Using Eqns. (4.4a), (4.5), and (4.8), α (or k_2) can be evaluated. It should be noted that there are various values of α (or k_2) satisfying Eqn. (4.4a), each corresponding to a certain number of circulation cells in the radial direction. Here, only one cell is chosen.

The velocity profile can be expressed as

$$V_{\ell r} = \frac{\Psi_o}{r_o}\{[\frac{\pi R^*}{H_c r^*}\cos(\pi z^*)]^2 + [\frac{1}{r_o r^*}\frac{dR^*}{dr^*}\sin(\pi z^*)]^2\}^{1/2} \tag{4.10}$$

As will be shown later, the only unknown in Eqn. (4.10), i.e., Ψ_o, can be obtained using the energy balance method. Note that the distance between the vortex center and the axis of symmetry (points A and M in Fig. 4.13), $\bar{A}$ or A^* ($A^* = \bar{A}/r_o$), can be obtained from the following equation:

$$\sum_{k=1}^{\infty} A_k(\beta)k(\alpha A^{*2}/2)^{k-1} = 0 \tag{4.11}$$

as $dR^*/dr^* = 0$ at $r^* = A^* = \bar{A}/r_o$

As pointed out by Whalley and Davidson (1974), the energy dissipation in the system is due to (i) the wakes behind the bubbles, E_1, (ii) the net bubble breakup, E_2, (iii) hydraulic jump at the liquid surface, E_3, (iv) the turbulence near the nozzle, E_4, (v) the viscous drag at the walls, E_5, and (vi) kinetic energy loss during the downward flow of liquid, E_6. They have shown that only E_1 and E_3 are important, and can be expressed, respectively, as

$$E_1 = \frac{\pi}{4} D_c^2 H_c \epsilon_g(1-\epsilon_g)U_{g\ell} g \rho_\ell \tag{4.12}$$

$$E_3 = \frac{\pi\rho_\ell(1-\epsilon_g)\Psi_o^3}{r_o^4}\int_0^{A^*}\frac{1}{r^{*2}}\left(\frac{dR^*}{dr^*}\right)^3 dr^* \tag{4.13}$$

In Eqn. (4.13), it is assumed that the hydraulic jump is equal to the kinetic energy associated with the upward-moving liquid. Neglecting kinetic energy input by the gas, the energy input rate can be given by

$$E_i = \frac{\pi}{4} D_c^2 U_g \rho_\ell g H_c \tag{4.14}$$

Equating the rate of energy input to the column by the gas to the energy dissipation in the bubble column we have

$$E_i = E_1 + E_3 \tag{4.15}$$

Hence, Ψ_0 can be given as

$$\Psi_0 = \left[\frac{D_c^4 (E_i - E_1)}{16\pi \rho_\ell (1-\epsilon_g) \int_0^{A^*} \frac{1}{r^{*2}} \left(\frac{dR^*}{dr^*}\right)^3 dr^*}\right]^{1/3} \tag{4.16}$$

Therefore, Eqn. (4.16) can be used to evaluate Ψ_0 if $U_{g\ell}$ is known. The average circulation velocity, U_c, is obtained as follows:

$$U_c = \int_0^{\bar{A}} U' 2\pi r/(\pi \bar{A}^2) dr \tag{4.17}$$

where U' is the liquid velocity at $H_c/2$ and can be expressed as

$$U' = \frac{\Psi_0}{r_0^2} \frac{1}{r^*} \frac{dR^*}{dr^*} \tag{4.18}$$

Therefore, the liquid velocity profile and average circulation upward velocity can be obtained by Eqns. (4.10) and (4.17), respectively.

4.6.1.2 Multiple Circulation Cells

According to Joshi and Sharma (1979), direct application of the energy balance method of a single circulation cell to deep beds involving $H_0/D_c > 4$ encountered the following setbacks: (a) the values of liquid velocity at any point predicted by the energy balance method are much higher than the experimental values, and (b) the method predicts that the liquid circulation is proportional to the cubic root of liquid height (see Eqns. (4.12), (4.14), (4.16), (4.17) and (4.18)). Consequently, other relevant parameters, such as the axial dispersion coefficient and mass transfer coefficient, should also vary with the liquid height. However, sufficient data from the literature show that the above parameters are essentially independent of the liquid height in the range $3 < H_0/D_c < 12$. Thus, they conceived that multiple circulation cells should exist for $H_0/D_c > 1.0$. Their modification of the energy balance method adjusts the number of circulation cells in the axial direction to minimize the vorticity strength ζ_o, (or $k_2\bar{A}\Psi_o$). They further showed that ζ_o is a minimum at an H_c/D_c close to unity and that the H_c/D_c at which ζ_o is minimum was independent of column diameter and superficial gas velocity. Thus, the bubble column can be regarded as having multiple circulation cells in the axial direction in which each circulation cell has the same height as the column diameter. For multiple circulation cells, the hydraulic jump is possible only for the top cell; for any intermediate cell it is assumed that no energy is dissipated in the hydraulic jump (E_3 equal to zero). The kinetic energy during the downward flow is assumed not recoverable as the pressure energy and is lost through turbulent dissipation. Therefore, E_6 should be considered. The energy equations for each cell in the multiple circulation cells case corresponding to Eqns. (4.12) through (4.16) are

$$E_1 = \frac{\pi}{4} D_c^3 \epsilon_g (1-\epsilon_g) g \rho_\ell U_{g\ell} \tag{4.19}$$

$$E_6 = \frac{\pi \rho_\ell \Psi_o^3}{r_o^4} \int_{A^*}^{1} \frac{1}{r^{*2}} \left(\frac{dR^*}{dr^*}\right)^3 (1-\epsilon_{gr}) dr^* \tag{4.20}$$

or

$$E_6 = \frac{\pi \rho_\ell \Psi_o^3}{r_o^4} (1-\epsilon_g) \int_{A^*}^{1} \frac{1}{r^{*2}} \left(\frac{dR^*}{dr^*}\right)^3 dr^* \tag{4.21}$$

$$E_i = \frac{\pi}{4} D_c^3 U_g g \rho_\ell (1-\epsilon_g) \tag{4.22}$$

and

$$\Psi_o = \left[\frac{D_c^4 (E_i - E_1)}{16\pi(1-\epsilon_g)\, \rho_\ell \int_{A^*}^{1} \frac{1}{r^{*2}} \left(\frac{dR^*}{dr^*}\right)^3 dr^*}\right]^{1/3} \tag{4.23}$$

ϵ_{gr} in Eqn. (4.20) is the radial gas holdup and can be obtained from Eqn. (2.104). Equation (4.21) uses an average gas holdup ϵ_g.

Note that the maximum difference in the values of Ψ_o calculated based on Eqns. (4.20) and (4.21) was found to be 5%. Therefore, Eqn. (4.21) was used in the derivation of Eqn. (4.23) because of its simplicity. From numerical integration, the integral term in Eqn. (4.21) was found to be 24.22 for an H_c/D_c of 1.

Combining Eqns. (4.19), (4.22) and (4.23) and substituting the numerical value for the integral term in Eqn. (4.23) give

$$\Psi_0 = \frac{D_c^{\,2}}{5.4} [D_c(U_g - \epsilon_g U_{g\ell})]^{1/3} \tag{4.24}$$

The average liquid circulation upward velocity can be obtained, similar to Eqn. (4.17), as

$$U_c = \int_A^{r_0} U' 2\pi r/[\pi(r_0^2 - \bar{A}^2)]dr = 3.0\,[D_c(U_g - \epsilon_g U_{g\ell})]^{1/3} \tag{4.25}$$

From Eqn. (4.25), it is seen that U_c is independent of liquid height, H_0. The above approach may be extended to a dilute slurry bubble column to predict the liquid circulation velocity since the assumption of inviscid vortex flow used in Eqn. (4.3) may be satisfied.

4.6.2 Momentum Balance Model

The basic concepts for a momentum balance model to describe the liquid velocity distribution, which consider the shear stress in the core region of the bubble column to result from both the liquid kinematic viscosity (ν_ℓ) and a turbulent kinematic viscosity (ν_T), have been described in Section (2.3.4). Equation (2.102) can be used for slurry bubble column systems; with ρ_e replaced with ρ_ℓ, it becomes the equation derived by Hills (1974) for bubble column systems.

The shear stress is evaluated solely through the turbulent kinematic viscosity (ν_T) since the liquid kinematic viscosity (ν_ℓ) can be neglected. The turbulent kinematic viscosity is found to be almost independent of gas velocity and is affected mainly by the column diameter (Miyauchi and Shyu, 1970) as given by

$$\nu_T = 0.0322\, D_c^{\,1.7} \tag{4.26}$$

Figure 4.14 shows the variation in ν_T with D_c based on data obtained by various investigators. As shown, both Eqn. (4.26) and the correlation of Sekizawa et al. (1983) satisfactorily correlate the data.

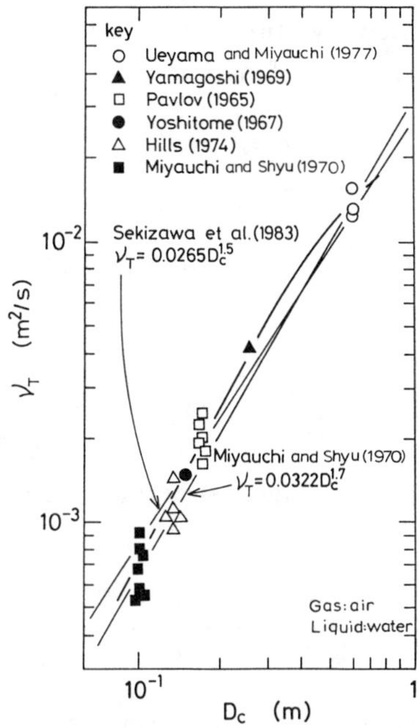

Figure 4.14 Variation of turbulent kinematic viscosity ν_T with D_c (from Ueyama and Miyauchi, 1979).

4.6.3 Sedimentation-Dispersion Model

Quantitative description of the axial solids concentration distribution has been mainly based on the so-called sedimentation-dispersion model, originally proposed by Cova (1966) and Suganuma and Yamanishi (1966). In this model, the concept of the widely used axial dispersion model was applied to the slurry phase, which was treated as a pseudo-homogeneous phase. The model considers a solids axial "dispersion" flux and a solids "sedimentation" flux superimposed on the average slurry convection flux, and thus, is termed the "sedimentation-dispersion" model. In the following, its original form is first briefly presented. Comments on the sedimentation flux are then given; it is followed by detailed formulation of the model in its revised form.

For a continuous solids flow, the net mass flux of solids is given by Cova (1966) and Parulekar and Shah (1980) among others as

$$n_s = \frac{C_s(n_s+n_\ell)}{\rho_{s\ell}} - C_s U_{st} - E_{zs}\frac{dC_s}{dz} \tag{4.27}$$

where n_s and n_ℓ are the mass fluxes of the solids and liquid based on the cross section occupied by slurry, U_{st} is the solids settling velocity, and $\rho_{s\ell}$ is the slurry density defined as

$$\rho_{s\ell} = \frac{\rho_\ell \epsilon_\ell + \rho_s \epsilon_s}{1 - \epsilon_g} \tag{4.28}$$

The sedimentation-dispersion model in its original form is thus given as (e.g., Parulekar and Shah, 1980)

$$\frac{\partial C_s}{\partial t} = \frac{\partial}{\partial z}\left[E_{zs}\frac{\partial C_s}{\partial z} - \left(\frac{U_{s\ell}}{1-\epsilon_g} - U_{st}\right)C_s\right] \tag{4.29}$$

where $U_{s\ell}$ is the weight-averaged superficial slurry velocity, defined as

$$\frac{U_{s\ell}}{1-\epsilon_g} = \frac{n_\ell + n_s}{\rho_{s\ell}} = \frac{V_p \epsilon_s \rho_s + V_\ell \epsilon_\ell \rho_\ell}{\epsilon_s \rho_s + \epsilon_\ell \rho_\ell} \tag{4.30}$$

The model is also expressed in terms of the volume-averaged superficial slurry velocity, $U^*_{s\ell}$, as given by (e.g., Kato et al., 1972a; Smith and Ruether, 1985)

$$\frac{\partial C_s}{\partial t} = \frac{\partial}{\partial z}\left[E_{zs}\frac{\partial C_s}{\partial z} - \left(\frac{U^*_{s\ell}}{1-\epsilon_g} - U_{st}\right)C_s)\right] \tag{4.31}$$

where $U^*_{s\ell}$ is defined by

$$U^*_{s\ell} = V_p \epsilon_s + V_\ell \epsilon_\ell \tag{4.32}$$

It is noted that Eqns. (4.29) and (4.31) can also be expressed in terms of ϵ_s simply by replacing C_s with ϵ_s. This is because C_s can be related to ϵ_s by

$$C_s = \left(\frac{\rho_s}{1-\epsilon_g}\right)\epsilon_s \tag{4.33}$$

It is evident from experimental data of various studies, such as those of Smith and Ruether (1985), that ϵ_g is essentially invariant with z for a given operating condition. Thus, C_s is essentially linearly proportional to ϵ_s. It is important to indicate that the numerical values for E_{zs} are the same for the model expressed in terms of C_s, where E_{zs} is interpreted as being based on the cross section occupied by the slurry phase (Smith and Ruether, 1985), and those expressed in terms of ϵ_s, where E_{zs} can be based on either the cross section occupied by the slurry phase or the cross section of the column. This is due to the fact that the model parameters, i.e., E_{zs} and U_{st}, in either model expression are, in general, obtained indirectly from empirical fitting of the model with data relating C_s or ϵ_s to z at steady state conditions.

Superimposing a solids settling flux onto a solids convective flux as employed in the sedimentation-dispersion model in its original form is an phenomenological approach in model formulation and has resulted in various discrepant physical interpretations of the U_{st} term in Eqns. (4.29) and (4.31) in the literature. For example, U_{st} has been interpreted either as the particle terminal velocity (e.g., Cova, 1966; Nakamura et al., 1978; Kojima and Asano, 1980; Kojima et al., 1984) or as the hindered settling velocity of a particle swarm relative to liquid (e.g., Imafuku et al., 1968; Brian and Dyer, 1984; Ueyama et al., 1985). Parulekar and Shah (1980) defined U_{st} as the hindered solids settling velocity relative to the column. Smith and Ruether (1985) defined U_{st} in Eqn. (4.31) as $\psi_\ell U_p$ where U_p was interpreted inconsistently as either the solids velocity relative to slurry or that relative to liquid velocity. The different interpretations of U_{st} perceived by different investigators are indeed confusing. In the following, a formulation of the sedimentation-dispersion model in its revised form based on Jean et al. (1988) is presented to qualify the physical implication of the model parameters. Comparisons of this model with its original form are also made.

The axial mixing of solids in a multiphase system can be expressed by directly extending the approach proposed by Cussler (1984) for describing the axial dispersion of solutes in turbulent flow systems. By treating the solid phase as a continuum, the continuity equation for solids flow can be written as

$$\frac{\partial \tilde{\epsilon}_s}{\partial t} = - \vec{\nabla} \cdot (\underline{\tilde{V}}_p \tilde{\epsilon}_s) \tag{4.34}$$

where $\underline{\tilde{V}}_p$ is the instantaneous velocity vector of solids and $\tilde{\epsilon}_s$ is the instantaneous solids holdup. Both $\underline{\tilde{V}}_p$ and $\tilde{\epsilon}_s$ fluctuate with time and can be represented as the sum of their average values and fluctuating components

$$\tilde{\underline{V}}_p = \underline{V}_p^* + \underline{v}_p' \quad \text{and} \quad \tilde{\epsilon}_s = \epsilon_s + \epsilon_s' \tag{4.35}$$

Substituting Eqn. (4.35) into Eqn. (4.34), then applying the averaging rules results in

$$\frac{\partial \epsilon_s}{\partial t} = - \vec{\nabla} \cdot (\underline{V}_p^* \epsilon_s) - \vec{\nabla} \cdot (\overline{\underline{v}_p' \epsilon_s'}) \tag{4.36}$$

where $(\overline{\underline{v}_p' \epsilon_s'})$ is the time average of the product of the fluctuations. The flux $(\overline{\underline{v}_p' \epsilon_s'})$ bears similar meaning to the diffusional flux vector, and can thus be defined as (in cylindrical coordinates)

$$\overline{\underline{v}_p' \epsilon_s'} = - E_{rs}^* \frac{\partial \epsilon_s}{\partial r} \vec{r} - \frac{E_{\theta s}^*}{r} \frac{\partial \epsilon_s}{\partial \theta} \vec{\theta} - E_{zs}^* \frac{\partial \epsilon_s}{\partial z} \vec{z} \tag{4.37}$$

Combining Eqns. (4.36) and (4.37), and taking surface integration over the cross-sectional area of the column, along with the assumption of constant ϵ_s with respect to r and θ, yield

$$\frac{\partial \epsilon_s}{\partial t} = \frac{\partial}{\partial z}[E_{zs}^* \frac{\partial \epsilon_s}{\partial z} - V_p \epsilon_s] \tag{4.38}$$

Note that E_{zs}^* is defined based on the cross-sectional area of the column. At steady state and under the condition of constant V_p, Eqn. (4.38) reduces to

$$E_{zs}^* \frac{d^2 \epsilon_s}{dz^2} - V_p \frac{d \epsilon_s}{dz} = 0 \tag{4.39}$$

V_p can be related to the slip velocity between the liquid and solids, U_p, by

$$U_p = V_\ell - V_p \tag{4.40}$$

From Eqns. (4.30), (4.32) and (4.40), the average solids convective velocity can be related to the average slurry velocity and the solid-liquid relative velocity by

$$V_p = \frac{U_{s\ell}}{(1 - \epsilon_g)} - \omega_\ell U_p \quad \text{or} \quad V_p = \frac{U_{s\ell}^*}{(1 - \epsilon_g)} - \psi_\ell U_p \tag{4.41}$$

Combining Eqns. (4.33), (4.38), and (4.41) and considering ϵ_g as invariant with z yields

$$\frac{\partial C_s}{\partial t} = \frac{\partial}{\partial z}\{E_{zs}^* \frac{\partial C_s}{\partial z} - [\frac{U_{s\ell}}{1-\epsilon_g} - \omega_\ell U_p] C_s\} \tag{4.42a}$$

and

$$\frac{\partial C_s}{\partial t} = \frac{\partial}{\partial z}\{E_{zs}^* \frac{\partial C_s}{\partial z} - [\frac{U_{s\ell}^*}{1-\epsilon_g} - \psi_\ell U_p] C_s\} \tag{4.42b}$$

For low solids concentration in the slurry, the slurry velocity, $U_{s\ell}^*/(1-\epsilon_g)$, can be approximated by the linear liquid velocity, V_ℓ. Comparing Eqns. (4.42a) and (4.42b) with Eqns. (4.29) and (4.31), respectively, the physical meaning of the U_{st} term and definition for E_{zs} in the sedimentation-dispersion model become apparent.

At steady state, using Eqn. (4.33), Eqn. (4.38) can be integrated to give the net flux of solid particles

$$E_{zs}^* \frac{dC_s}{dz} - V_p C_s = - \frac{U_{s\ell}^*}{1-\epsilon_g} C_s^f = - V_{s\ell}^* C_s^f \tag{4.43a}$$

for constant continuous slurry flow condition, and

$$E_{zs}^* \frac{dC_s}{dz} - V_p C_s = 0 \tag{4.43b}$$

for slurry-batch and solids-batch conditions.

The boundary condition for Eqns. (4.43a) and (4.43b) is the same and can be expressed as

$$z = 0 ; \quad C_s = C_s^0 \tag{4.44a}$$

or

$$z = H ; \quad C_s = C_s^1 \tag{4.44b}$$

At the column top, a slight accumulation of solid particles may occur relative to the solids concentration in the effluent slurry due to flow restrictions caused by the weir used for slurry discharge. Consequently, the solids concentration at the column top is frequently higher than the feed concentration. The relationship between C_s^1 and

C_s^f depends on the column design and operating conditions. V_p does not vary significantly in the axial direction in the slurry bubble column system (e.g., Smith and Ruether, 1985). However, to be more precise, an axially averaged value should be used for V_p in Eqn. (4.43a). Using Eqn. (4.44b) as the boundary condition, Eqns. (4.43a) and (4.43b) can be readily integrated, respectively, to

$$C_s = \{C_s^1 + \frac{V_{s\ell}^*}{(-V_p)} C_s^f\} \exp\{\frac{(-V_p)}{E_{zs}^*} (H-z)\} - \frac{V_{s\ell}^*}{(-V_p)} C_s^f \tag{4.45a}$$

for constant continuous slurry flow conditions, and

$$C_s = C_s^1 \exp\{\frac{(-V_p)}{E_{zs}^*} (H-z)\} \tag{4.45b}$$

for slurry-batch and solids-batch conditions. In Eqn. (4.45b), $(-V_p)$ is equal to $\psi_\ell U_p$. In addition, H is referred to as the height of the batch system, and C_s^1 is the solid concentration at H. Following the same procedure as in the original form of the model, E_{zs}^*, U_p (or V_p) and C_s^1 are to be obtained by fitting the model to the experimental axial solids concentration profile. Jean et al. (1988) noted that, for operation involving slurry-batch or solids-batch under the coalesced bubble flow regime and slow slurry continuous flow conditions where the sedimentation-dispersion model in its original form applies, the numerical values for the model parameters obtained in this manner are not physically compatible with those defined in the model formulation. Consequently, the parameters of the model when applied to such operating conditions must be regarded as being solely empirical.

A critical review of the sedimentation-dispersion model in its original forms (Eqns. (4.29) and (4.31)) and in its revised forms (Eqns. (4.42a) and (4.42b)) for both the slurry continuous flow and slurry-batch conditions is given in Jean et al. (1988).

4.6.4 Axial Solids Holdup Distribution (Batch) Model

Based on the axial solids distribution mechanisms conceptually similar to that underlying the Stagewise Partition Process Model for three-phase fluidized bed systems (see Section 2.3.3), Tang and Fan (1987) presented a mechanistic model in a single differential form which satisfactorily accounts for the axial solids holdup distribution in a slurry bubble column containing low density particles in a batch condition. The model considers three distinct phases in the column, namely the gas bubble phase, the wake phase, and the liquid-solid emulsion phase. Assuming that the interaction among the gas bubbles and hence their associated wakes is negligible, a mass balance around the wake phase associated with each bubble gives rise to

$$V_R \frac{\epsilon_w}{N_b} \frac{d\epsilon_{sw}}{dt} = V_R \bar{U}_b \frac{\epsilon_w}{N_b} \frac{d\epsilon_{sw}}{dz} = V_e A_e \epsilon_{sf} - V_d A_d \epsilon_{sw} \tag{4.46}$$

where V_R is the total volume of the column, N_b is the total number of bubbles, and ϵ_w is the overall volume fraction of the wake phase. The first and second terms on the right-hand side of Eqn. (4.46) account for the rate of solids entrainment into and de-entrainment out of the wake phase, respectively. Both rates are approximated as linear functions of the solids concentrations in the liquid-solid emulsion and wake phases (ϵ_{sf} and ϵ_{sw}), respectively. V_e and V_d are the average velocity of solids relative to the bubble rise velocity and the average solids discharge velocity across the effective areas, A_e and A_d, respectively.

At steady state, the upward flow of solids in the wake phase must be balanced by a net downward flow of solids in the liquid-solid emulsion phase. Integrating the equation from a mass balance around a differential section of the column results in

$$\bar{U}_b \epsilon_w \epsilon_{sw} = |V_{sf}| \epsilon_{\ell\text{-}s} \epsilon_{sf} \qquad \text{at any } z \tag{4.47}$$

where $|V_{sf}|$ is the absolute magnitude of the linear net downward velocity of solids in the liquid-solid emulsion phase. Combining Eqns. (4.46) and (4.47) gives

$$V_R \bar{U}_b \frac{\epsilon_w}{N_b} \frac{d\epsilon_{sw}}{dz} = \left(\frac{V_e A_e}{|V_{sf}|} - \frac{V_d A_d \epsilon_{\ell\text{-}s}}{\bar{U}_b \epsilon_w} \right) |V_{sf}| \epsilon_{sf} \tag{4.48}$$

It can be considered that x, defined by Eqn. (2.60), is only a weak function of the axial position, and may be assumed constant with respect to the axial position (El-Temtamy and Epstein, 1980). Substituting Eqn. (2.60) into Eqn. (4.48) and rearranging gives

$$\frac{V_R \bar{U}_b \epsilon_w x}{N_b[V_e A_e/|V_{sf}| - V_d A_d \epsilon_{\ell\text{-}s}/(\bar{U}_b \epsilon_w)]} \frac{d\epsilon_{sf}}{dz} = |V_{sf}| \epsilon_{sf} \tag{4.49}$$

Define E'_{zs} as

$$E'_{zs} = - \frac{V_R \bar{U}_b x \epsilon_w}{N_b[V_e A_e / |V_{sf}| - V_d A_d \epsilon_{\ell\text{-}s} / (\bar{U}_b \epsilon_w)]} \tag{4.50}$$

E'_{zs} may be a weak function of z, but is assumed to be constant. The solids concentrations, ϵ_{sf} and ϵ_s, are interrelated by the mass balance as given by

$$\epsilon_{sf} = \frac{\epsilon_s}{x\epsilon_w + \epsilon_{\ell\text{-}s}} = \frac{\epsilon_s}{xk\epsilon_g + (1 - \epsilon_g - k\epsilon_g)} \tag{4.51}$$

where k is defined as ϵ_w/ϵ_g. Using Eqn. (4.51) and with x and k assumed constant with respect to axial position, Eqn. (4.49) can be readily converted to

$$E'_{zs} \frac{d\epsilon_s}{dz} + |V_{sf}|\epsilon_s = E'_{zs} \frac{d\epsilon_s}{dz} + (|V_{\ell s}| - V_{\ell f})\epsilon_s = 0 \tag{4.52}$$

where $V_{\ell f}$ is the linear liquid velocity in the liquid-solid emulsion phase, $|V_{\ell s}|$ is the relative velocity between the solids and the liquid in this phase, and the velocities are expressed in terms of absolute scalar magnitudes. $|V_{\ell s}|$ can be related to the liquid holdup, $\epsilon_{\ell f}$, by any appropriate equation for particulate fluidization, e.g., the Richardson-Zaki expression. The equation of the model (Eqn. (4.52)) has a form identical to Eqn. (4.43b). It is important to note, however, that for solids-batch conditions, the interpretation of the parameters appearing as coefficients of $d\epsilon_s/dz$ or dC_s/dz and ϵ_s or C_s between these two model equations is distinctly different. It is also noted that this model can be directly extended to account for the axial solids holdup distribution in a slurry bubble column with continuous slurry flow conditions. Furthermore, when the particle size is large, the effect due to drift may significantly contribute to the axial solids concentration distribution. Little, however, is known regarding this effect.

4.7 GAS-LIQUID MASS TRANSFER

The behavior of mass transfer from gas bubbles to the surrounding liquid medium in bubble or slurry bubble column systems parallels that in fluidized bed systems with a continuous liquid phase. When mass transfer occurs from gas bubbles to the surrounding liquid in a bubble or slurry bubble column, the gas-side resistance can often be neglected in determining the overall gas-liquid mass transfer resistance due to the low solubility of gas in the liquid. Thus, the gas-liquid mass transfer rate per unit volume of aerated liquid (or slurry) can often be defined in terms of the liquid-side volumetric mass transfer coefficient, $k_\ell a$. In this section, the overall gas-liquid mass transfer behavior is discussed in terms of $k_\ell a$, k_ℓ and "a". In addition, the gas-side mass transfer coefficient is briefly discussed.

4.7.1 Overall or Liquid-Side Volumetric Mass Transfer Coefficient

The volumetric mass transfer coefficient in bubble columns or slurry bubble columns can be measured through physical absorption of carbon dioxide into water or various aqueous solutions or through chemical absorption carried out such that a mass transfer enhancement factor of unity is obtained. In a liquid-batch bubble column system (Mode E-III-a) or in large diameter columns ($D_c \geq 1.0$ m), a complete mixing of the liquid phase may be assumed when evaluating the volumetric mass transfer coefficient from transient physical absorption experiments. However, for conditions with liquid (or slurry) through flow, particularly for those with a large H/D_c, liquid phase axial mixing must be considered in evaluating the volumetric coefficient.

For bubble or slurry bubble column systems in general, the liquid-side volumetric coefficient depends on the superficial gas velocity, the liquid physical properties, the solute diffusivity in the liquid, the gas distributor design, and the solids concentration and physical properties; in particular, those properties which affect the bubble behavior, and hence the interfacial area, are the most important. Table 4.7 summarizes several correlations for predicting $k_\ell a$. In most of these correlations, the dependence on molecular diffusivity, D_M, approaches the square root relationship predicted by the penetration model (Danckwerts, 1970). For low liquid velocities (e.g., $U_\ell < 5$ cm/s), the effect of liquid velocity on $k_\ell a$ is usually negligible.

The gas velocity affects $k_\ell a$ through both the interfacial area and the liquid-side mass transfer coefficient, since both are inherently related to bubble size and gas holdup. At low or intermediate gas velocities, $k_\ell a$ increases somewhat less than linearly with gas velocity ($k_\ell a \propto U_g^{0.6 \sim 1}$), while at high gas velocities, this dependency on U_g becomes less. The dependency of $k_\ell a$ on gas velocity qualitatively reflects the gas holdup behavior described previously. In fact, both

Table 4.7 Correlations for Gas-Liquid Mass Transfer Variables ($k_\ell a$, k_ℓ, and "a") in Bubble Columns (See Table 2.1 for ϵ_g).

Investigators	Correlation Equations (in SI Units)	Systems
Calderbank and Moo-Young (1961)	For large bubbles (d_{vs} > 3mm) $\frac{k_\ell d_{vs}}{D_M} = 0.42\left(\frac{\mu_\ell}{\rho_\ell D_M}\right)^{1/2}\left(\frac{g d_{vs}^3 \rho_\ell^2}{\mu_\ell^2}\right)^{1/3}$ (T-4.7.1) For small bubbles (d_{vs} < 0.7 mm) $\frac{k_\ell d_{vs}}{D_M} = 0.31\left(\frac{\mu_\ell}{\rho_\ell D_M}\right)^{1/3}\left(\frac{g d_{vs}^3 \rho_\ell^2}{\mu_\ell^2}\right)^{1/3}$ (T-4.7.2) For intermediate bubbles (0.7 < d_{vs} < 3mm), k_ℓ is proportional to d_{vs}.	gas-liquid
Hughmark (1967)	$\frac{k_\ell d_{vs}}{D_M} = 2+0.0187\left[\left(\frac{d_{vs} U_g \rho_\ell}{\varepsilon_g \mu_\ell}\right)^{0.484}\left(\frac{d_{vs} g^{1/3}}{D_M^{2/3}}\right)^{0.072}\left(\frac{\mu_\ell}{\rho_\ell D_M}\right)^{0.339}\right]^{1.61}$ (T-4.7.3) d_{vs} can be obtained from Eqn. (T-4.2.1).	gas-liquid
Akita and Yoshida (1973, 1974)	$\frac{k_\ell a D_c^2}{D_M} = 0.6\left(\frac{D_c^2 \rho_\ell g}{\sigma}\right)^{0.62}\left(\frac{D_c^3 \rho_\ell^2 g}{\mu_\ell^2}\right)^{0.31}\left(\frac{\mu_\ell}{\rho_\ell D_M}\right)^{0.5} \varepsilon_g^{1.1}$ $a D_c = \frac{1}{3}\left(\frac{D_c^2 \rho_\ell g}{\sigma}\right)^{0.5}\left(\frac{D_c^3 \rho_\ell^2 g}{\mu_\ell^2}\right)^{0.1} \varepsilon_g^{1.13}$ $\frac{k_\ell d_{vs}}{D_M} = 0.5\left(\frac{d_{vs}^2 \rho_\ell g}{\sigma}\right)^{3/8}\left(\frac{d_{vs}^3 \rho_\ell^2 g}{\mu_\ell^2}\right)^{1/4}\left(\frac{\mu_\ell}{\rho_\ell D_M}\right)^{1/2}$ d_{vs} can be obtained from Eqn. (T-4.2.2)	gas-liquid
Nakanoh and Yoshida (1980)	$\frac{k_\ell a' D_c^2}{D_M} = 0.09\left(\frac{\mu_\ell}{\rho_\ell D_M}\right)^{0.5}\left(\frac{g D_c^2 \rho_\ell}{\sigma}\right)^{0.75}\left(\frac{g D_c^3 \rho_\ell^2}{\mu_\ell^2}\right)^{0.39}\left(\frac{U_g}{\sqrt{g D_c}}\right) \cdot \left[1+0.13\left(\frac{\bar{U}\theta}{d_{vs}}\right)^{0.55}\right]^{-1}$ where a' = interfacial area per unit volume of liquid $\bar{U} = U_g(1+\varepsilon_g)/\varepsilon_g$ θ = characteristic material time for viscoelastic liquid	gas-liquid
Hikita et al. (1981b)	$\frac{k_\ell a U_g}{g} = 14.9\left(\frac{U_g \mu_\ell}{\sigma}\right)^{1.76}\left(\frac{\mu_\ell^4 g}{\rho_\ell \sigma^3}\right)^{-0.248}\left(\frac{\mu_g}{\mu_\ell}\right)^{0.243}\left(\frac{\mu_\ell}{\rho_\ell D_M}\right)^{-0.604}$ Correction factor, $f = (k_\ell a)_{elec.}/(k_\ell a)_{non\text{-}elec.}$ is given as $f = 10^{0.068I}$ for $0 < I < 1.0$ kg-ion/m^3 $f = 1.114\times10^{0.021I}$ for $I > 1.0$ kg-ion/m^3	gas-liquid
Quicker et al. (1984)	$a = 583.6\, U_g^{0.81} \mu_{eff}^{-0.22}$ for activated carbon and Kieselghur slurries $a = 2.06\times10^{-6}\, U_g^{0.65} \mu_{eff}^{-3.09}$ for Al_2O_3 slurries	gas-liquid-solid
Koide et al. (1984)	$\frac{k_\ell a \sigma}{\rho_\ell D_M g} = \frac{2.11 (\mu_\ell/\rho_\ell D_M)^{0.5} (g \mu_\ell^4/\rho_\ell \sigma^3)^{-0.159} \varepsilon_g^{1.18}}{1+1.47\times10^{4} (\bar{C}_s/\rho_s)^{0.612} (U_t/\sqrt{g D_c})^{0.486} (D_c^2 g \rho_\ell/\sigma)^{-0.477} (D_c U_g \rho_\ell/\mu_\ell)^{-0.345}}$ (T-4.7.4) $3.71\times10^2 \le \frac{\mu_\ell}{\rho_\ell D_M} \le 1.06\times10^5$, $0 \le (\bar{C}_s/\rho_s) \le 0.08$ $1.69\times10^{-11} \le \left(\frac{g\mu_\ell^4}{\rho_\ell \sigma^3}\right) \le 2.84\times10^{-6}$ $2.18\times10^{-4} \le \left(\frac{U_t}{\sqrt{g D_c}}\right) \le 2.48\times10^{-2}$ $0.0291 \le \varepsilon_g \le 0.253$, $1.36\times10^3 \le \left(\frac{D_c^2 g \rho_\ell}{\sigma}\right) \le 1.22\times10^4$ $3.16\times10^2 \le \left(\frac{D_c U_g \rho_\ell}{\mu_\ell}\right) \le 4.18\times10^4$	gas-liquid-solid

(Table 4.7 continued)

Investigators	Correlation Equations (in SI Units)	Systems
Yasunishi et al. (1986a) Fukuma et al. (1987b)	$k_\ell a = 3.09\times10^{-7}[(1-\Psi_s/0.60)/\mu_\ell]^{1.50}\,U_g$ (T-4.7.5) $0.16\times10^{-3} < d_p < 0.23\times10^{-3}$ m; $0 < \Psi_s < 0.4$; $0.8\times10^{-2} < U_g < 8\times10^{-2}$ m/s $0.001 < \mu_\ell < 0.002$ Pa·s $\frac{k_\ell d_{vs}}{D_M} = 4.5\times10^{-4}\left(\frac{\mu_\ell}{\rho_\ell D_M}\right)^{0.5}\left(\frac{g d_{vs}^3\rho_\ell^2}{\mu_\ell^2}\right)^{0.8}\left(\frac{d_{vs}^2\rho_\ell g}{\sigma}\right)^{-0.2}$ (T-4.7.6) $480 \le \frac{\mu_\ell}{\rho_\ell D_M} \le 1600$, $3.2\times10^5 \le \frac{g d_{vs}^3\rho_\ell^2}{\mu_\ell^2} \le 2.6\times10^8$ $1.4 \le \frac{d_{vs}^2\rho_\ell g}{\sigma} \le 120$, $0 \le \varepsilon_s \le 0.4$ $a = 6\varepsilon_g/d_{vs}$	gas-liquid-solid
Ozturk and Schumpe (1987)	$k_\ell a = k\,U_g^{0.75}\,\mu_{eff}^{-0.42}$ $k = 0.042$ for tetralin $k = 0.031$ for ligroin	gas-liquid-solid

Akita and Yoshida (1973) and Sada et al. (1986b, 1987) found the volumetric coefficient to vary nearly linearly with gas holdup. In a slurry system, the effect of the gas velocity on the liquid-side volumetric mass transfer coefficient reflects the complex influence of particle size, density, and concentration. Figure 4.15 shows the relationship between $k_\ell a$ and U_g reported by various investigators and, in part, depicts the complexity involved.

Liquid properties also have an important impact on $k_\ell a$, especially when comparing values from a water system with those from an electrolyte solution system. $k_\ell a$ decreases with increasing μ_ℓ (Akita and Yoshida, 1973; Koide et al., 1984). For low viscosity liquids, the viscosity effect was only slight (Akita and Yoshida, 1973; Hikita et al., 1981b), but for high viscosity non-Newtonian liquids the effect is significant (Schumpe and Deckwer, 1982; Godbole et al., 1984). An increase in surface tension also decreases $k_\ell a$ as indicated by Koide et al. (1984). The presence of electrolytes or surfactants within the liquid phase greatly affects mass transfer behavior.

Electrolytes or surfactants affect the behavior at the bubble-liquid interface, as discussed in Section 2.1.7. In terms of mass transfer behavior, the effects are evident particularly through changes in the interfacial area. As shown in Fig. 4.15, the values of $k_\ell a$ obtained by Kato (1963) and Kato et al. (1972b) by sulfite oxidation in the electrolyte solutions, as indicated by the shaded area, are greater than those obtained for aqueous non-electrolyte slurries by Koide et al. (1984) and Yasunishi et al. (1986a). For electrolyte slurry systems

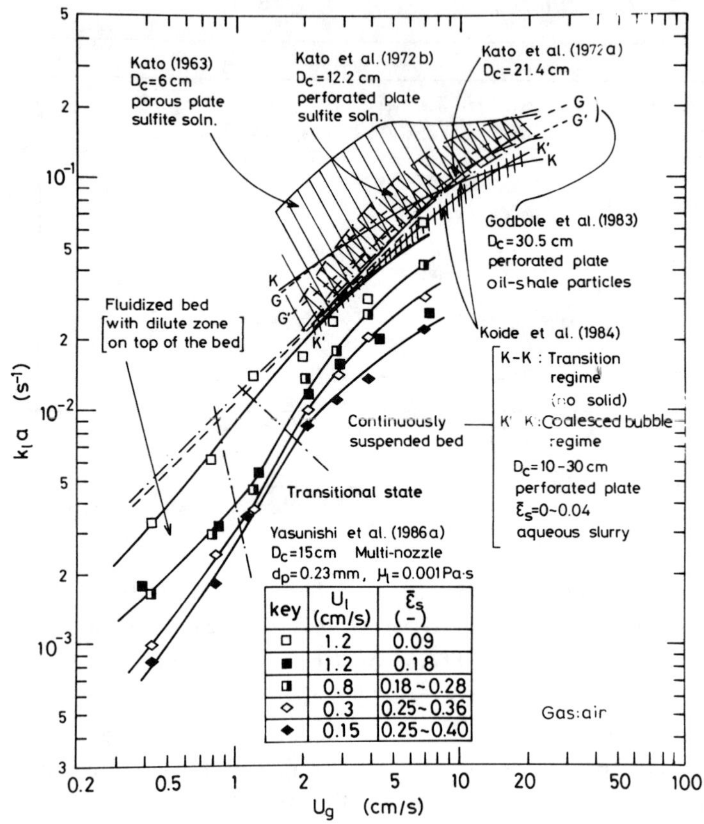

key	U_l (cm/s)	$\bar{\varepsilon}_s$ (-)
□	1.2	0.09
■	1.2	0.18
◨	0.8	0.18 ~ 0.28
◇	0.3	0.25 ~ 0.36
◆	0.15	0.25 ~ 0.40

Figure 4.15 Variation of $k_\ell a$ with U_g in various three-phase fluidization systems.

where bubble coalescence is hindered, the relationship between $k_\ell a$ and ϵ_g is almost independent of electrolyte concentration (Sada et al., 1986b, 1987) and follows the previously mentioned near linear dependence. The electrolyte effects on $k_\ell a$ are in agreement with those in draft-tube three-phase fluidized bed systems given in Section 6.1.2.1.

The concentration of particles and their physical properties also have a significant impact on the volumetric mass transfer coefficient; the effects, however, are complex. Koide et al. (1984) found that, at low gas velocities, $k_\ell a$ decreases with particle terminal velocity. Furthermore, they indicated that the presence of solids generally enhances bubble coalescence and decreases $k_\ell a$ and "a". At low solids concentrations, the effect of particles larger than about 40 μm can be neglected and the design parameters estimated for the gas-liquid system can be used for the slurry bubble column. Sada et al. (1985) and Schumpe et al. (1987) reported that $k_\ell a$ was independent of slurry concentration up to a solids concentration in the slurry of 3 ~ 5 wt% and decreases with slurry concentration at higher concentrations. This effect of solids concentration on $k_\ell a$ becomes small in a larger diameter column (Kato et al., 1972b; Koide et al., 1984). In contrast, Sada et al. (1986a,b, 1987) found that the presence of a very low concentration (< 1.0 wt%) of fine particles (< 12 μm) actually enhances gas-liquid mass transfer; at these low solids concentrations the volumetric mass transfer coefficient, gas holdup, and interfacial area were higher than those in corresponding solid-free systems. For electrolyte slurry systems in particular, suspended particles have less influence on gas holdup and hence, less influence on the volumetric mass transfer coefficient than for non-electrolyte slurries (Sada et al., 1986b).

Since the gas distributor design and column size affect bubble size, they also affect the volumetric mass transfer coefficient. The general effects of distributor design and column size on bubble behavior are discussed in Sections 2.2.1.2 and 4.2.1.

Summary

Among correlation equations presented in Table 4.7, those of Koide et al. (1984) (Eqn. (T-4.7.4)), Nguyen-Tien et al. (1985) (Eqn. (T-3.1.2a)) and Yasunishi et al. (1986a) (Eqn. (T-4.7.5)) have the widest applicability. Figure 4.16 shows the comparisons between these three correlation equations and the experimental data of Yasunishi et al. (1986a) for non-electrolyte slurries. Koide et al.'s (1984) correlation satisfactorily predicts $k_\ell a$ at zero or low solid concentrations, but predictions deviate significantly at high solids concentrations. Nguyen-Tien et al.'s and Yasunishi et al.'s correlations predict similar trends in $k_\ell a$ with respect to ψ_S $(=\epsilon_S/(\epsilon_S + \epsilon_\ell))$. At low gas velocities, Nguyen-Tien et al.'s correlation overpredicts $k_\ell a$. The overprediction may be due to the fact that the experiments of Nguyen-Tien et al. were conducted at high slurry velocities, where bubble coalescence was restrained, in comparison to the case of Yasunishi et al. (1986a), where the system was operated at solids-batch conditions with significant bubble coalescence. Note, however, that Eqns. (T-3.1.2a) and (T-4.7.5) were obtained from the absorption of oxygen into water and thus, correction for the liquid-phase solute diffusivity is required when these equations are applied to other gas absorption systems. More

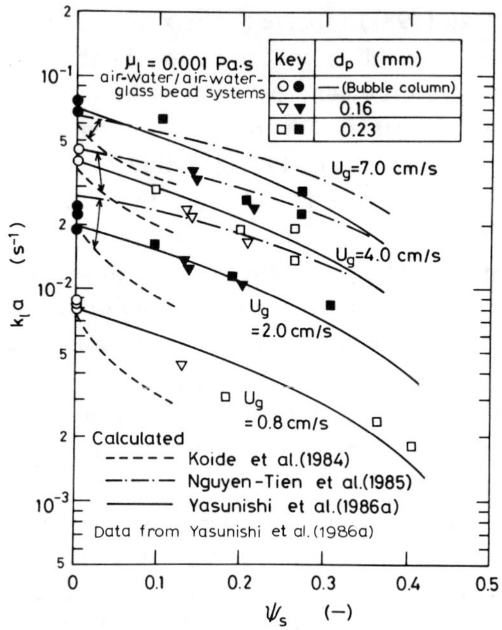

Figure 4.16 Effect of solids concentration on $k_\ell a$ in a bubble column and a slurry bubble column (from Yasunishi et al., 1986a).

research on $k_\ell a$ is needed for systems with physical properties of gas and liquid phases significantly deviating from air-water systems.

4.7.2 Gas-Side Volumetric Mass Transfer Coefficient

For a highly soluble gas such as sulfur dioxide in flue gas, the gas-side resistance to mass transfer cannot be neglected. Limited information is available in the literature on the gas-side mass transfer coefficient in bubble and slurry bubble columns (Botton et al., 1980; Sharma and Mashelkar, 1968; Sada et al., 1985). $k_g a'$ was measured in a slurry bubble column for an electrolytic slurry by Sada et al. (1985) where a' is defined as $a/(1-\epsilon_g)$. $k_g a'$ for the slurry bubble column agreed well with those for a solid-free bubble column reported previously by Botton et al. (1980) and Sharma and Mashelkar (1968). The following empirical equation was proposed by Sada et al. (1985) for $k_g a'$ in a slurry bubble column:

$$k_g a' = 4.9\, U_g^{0.73} \tag{4.53}$$

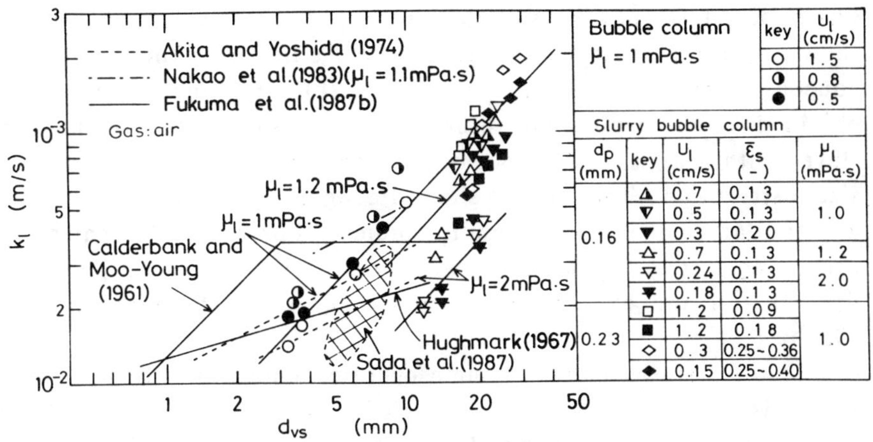

Figure 4.17 Variation of k_ℓ with d_{vs} in a bubble column and a slurry bubble column (from Fukuma et al., 1987b).

4.7.3 Liquid-Side Mass Transfer Coefficient

Akita and Yoshida's (1974) correlation indicates that k_ℓ in a bubble column is proportional to the square root of both the solute diffusion coefficient and the volume-surface average bubble diameter, but inversely proportional to the 1/4 power of the liquid surface tension. Hughmark's (1967) correlation agrees fairly well with that of Akita and Yoshida's for a water-O_2 system as shown in Fig. 4.17. According to Calderbank and Moo-Young (1961), for $d_{vs} > 3$ mm k_ℓ is large and independent of d_{vs}, for $0.7 < d_{vs} < 3$ mm k_ℓ is proportional to d_{vs}, and for $d_{vs} < 0.7$ mm k_ℓ is relatively small and independent of d_{vs}; this behavior is also illustrated in Fig. 4.17.

The mass transfer coefficient in slurry bubble columns has been measured by several investigators. Zaidi et al. (1979) and Deckwer et al. (1980) determined k_ℓ via absorption of CO_2 from 0.7 mm diameter bubbles into a paraffin-Al_2O_3 slurry; their k_ℓ values are satisfactorily predicted by the correlations of Hughmark (1967) and Calderbank and Moo-Young (1961) for bubble columns. Sada et al. (1985) obtained k_ℓ from chemical absorption of CO_2 into an aqueous $Ca(OH)_2$ slurry. They found k_ℓ to be independent of the gas velocity in the dispersed bubble regime ($U_g < 5$ cm/s) and to increase significantly in the transition from the dispersed bubble to the coalesced bubble regime. In addition, they reported that, at low solids concentrations ($C_s/\rho_{s\ell} < 2$ wt%), k_ℓ decreased significantly with the solids concentration. Sada et al. (1987) also determined k_ℓ in the slurry of Al_2O_3 powder-sulfite solution. The range of their data is shown by the shaded area in Fig. 4.17 together with the data for the aqueous slurries of Fukuma et al. (1987b). Fukuma et al. (1987b) obtained k_ℓ from physical absorption of O_2 into a water-glass beads slurry with a high solids concentration and measurement of the bubble size distribution. Their result showed a significant increase in k_ℓ with d_{vs} similar to that for a bubble column. For a bubble coalescence-promoting system, k_ℓ decreases with liquid viscosity, is independent of d_p and ϵ_s , and can be correlated by Eqn. (T-4.7.6) (Fukuma et al., 1987b).

For a slurry bubble column operated under the Fischer-Tropsch synthesis conditions, the correlations of Hughmark (Eqn. (T-4.7.3)) and Calderbank and Moo-Young (Eqns. (T-4.7.1) and (T-4.7.2)) have been found to be useful in the calculation of the gas-liquid mass transfer coefficient (Zaidi et al., 1979; Kolbel and Ralek, 1980).

4.7.4 Interfacial Area

The gas-liquid interfacial area is strongly affected by the column geometry, the gas distributor, the hydrodynamic conditions, and the liquid or slurry physical properties. In a bubble column, the variation in $k_\ell a$ primarily results from variations in the interfacial area. Physical and chemical methods have been employed to measure the interfacial area (see Section 3.1.1.3). For example, in the physical method the interfacial area is determined from the gas holdup and bubble size distribution. The gas-liquid interfacial area can be related to the gas holdup and the volume-surface mean bubble diameter, d_{vs}, as given by

$$a = 6\epsilon_g/d_{vs} \tag{4.54}$$

with d_{vs} calculated by Eqn. (2.39). In the chemical method, a gas-liquid reaction, such as sulfite oxidation, of known kinetics can be employed (Linek and Vacek, 1981).

For a given bubble column system, since d_{vs} varies as a weak function of U_g (Fig. 4.5a) over a moderate range of U_g, the interfacial area becomes almost proportional to ϵ_g; this is exemplified by the data of Akita and Yoshida (1974) which showed a dependency on ϵ_g to the 1.13 power. Note that d_{vs} can also be calculated by the correlation of Akita and Yoshida (1974) as given in Table 4.7. Though this

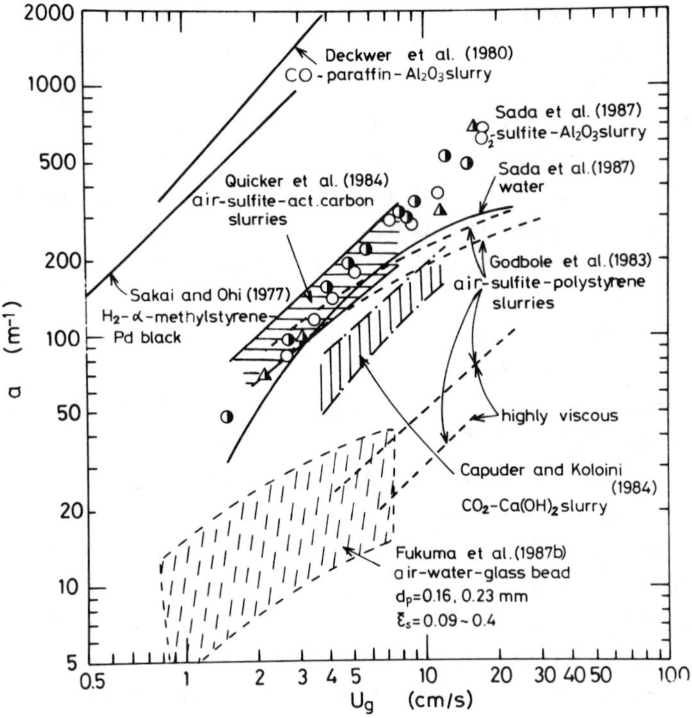

Figure 4.18 Interfacial area "a" versus U_g for various slurry bubble columns.

correlation shows a dependence of d_{vs} on the column diameter, this dependence vanishes for column diameters larger than 0.3 m.

The variation in "a" with U_g obtained by various investigators is shown in Fig. 4.18. As shown, "a" varies significantly with operating conditions: e.g., "a" varies from about 10 to 1500 m^{-1} at a gas velocity of about 3 cm/s. More specifically, "a" reported by Deckwer et al. (1980) for a paraffin-Al_2O_3 slurry in the Fischer-Tropsch synthesis is one order of magnitude higher than that for a system with fine solids suspended in an aqueous sulfite solution. The large "a" found by Deckwer et al. (1980) is due to small bubble (average diameter of 0.7 mm) formation in their foaming slurry system (Zaidi et al., 1979). Similar behavior is also found from the data of Sakai and Ohi (1977) involving a suspension of palladium black catalyst in α-methylstyrene. Contrary to the behavior in aqueous slurries in which "a" decreases with increasing solids concentration, Sada et al. (1985) reported that in an aqueous slurry containing fine $Ca(OH)_2$ particles "a" increases with particle concentration for solids concentrations up to

2 wt% and beyond 2 wt% "a" levels off. Sada et al. (1987) also reported that, in an aqueous Na_2SO_3 and fine Al_2O_3 particles (up to 0.5 wt%) slurry, "a" was greater than that in the solid-free system. This behavior may be attributed to the fine particles dispersed in the liquid film inhibiting bubble coalescence (Sada et al., 1986b).

Quicker et al. (1984) measured "a" in slurry bubble columns containing fine particles using the sulfite oxidation method. They suggested that the decrease in "a" with solids concentration was due to the increase in apparent viscosity of the suspension and proposed correlations for activated carbon and Kieselguhr slurries and Al_2O_3 slurries as given in Table 4.7. Capuder and Koloini (1984) measured "a" by chemical absorption of CO_2 into a $Ca(OH)_2$ solution in the presence of fine $Ca(OH)_2$ (or $CaCO_3$) particles. They obtained a correlation for "a" similar to that proposed by Akita and Yoshida (1974) for bubble columns. Fukuma et al. (1987b) showed that "a" in a slurry bubble column with high solids concentration could be satisfactorily predicted by Eqn. (4.54) using values of d_{vs} predicted by Eqns. (T-4.2.3) and (T-4.2.4), and ϵ_g predicted by the correlation of Koide et al. (1984) (Eqn. (T-4.1.4)).

Figure 4.19 compares the relationship between "a" and ϵ_g reported by various investigators. As shown, "a" in aqueous electrolyte slurries with fine particles obtained by Godbole et al. (1983), Capuder and Koloini (1984), Quicker et al. (1984) and Sada et al. (1987) has similar values at a given ϵ_g and is satisfactorily expressed by the correlation of Akita and Yoshida (1974) developed for bubble columns. Data of "a" obtained for non-electrolyte and high viscosity slurries containing large amounts of solids (Fukuma et al., 1987b) or for CMC solutions (Schumpe and Deckwer, 1982; Godbole et al., 1983; Godbole et al., 1984) are significantly lower than the values predicted by Akita and Yoshida's (1974) correlation; this deviation may result from increased bubble coalescence and decreased gas holdup in such viscous liquid or slurry systems.

4.8 LIQUID-SOLID MASS TRANSFER

The rate of mass transfer through the interface between the bulk liquid and solid particles suspended in a bubble column can be measured by means of physical dissolution of a granular solid matter into a solvent or by ion exchange of an acid or base solution using a resin (also see Section 3.1.2.1). The treatment in this section is limited

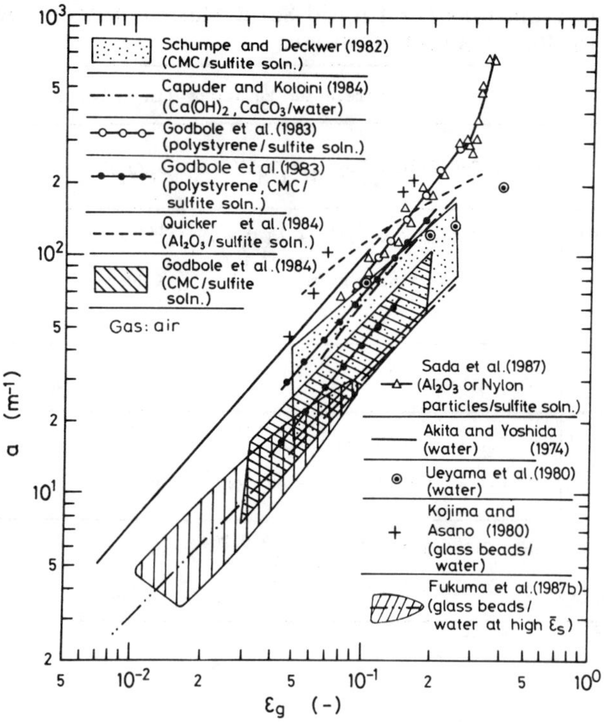

Figure 4.19 Variation of "a" with ϵ_g for a bubble column and a slurry bubble column (from Fukuma et al., 1987b).

to liquid-solid mass transfer in slurry bubble columns in which the solids are in a complete suspension state.

Various correlations have been proposed to describe the liquid-solid mass transfer coefficient, k_s, in slurry bubble columns as given in Table 4.8. Basically, two models were used in correlating the liquid-solid mass transfer coefficient in slurry bubble columns. The first is the slip velocity model, which was discussed by Harriot (1962) for mechanically agitated vessel and developed further by Satterfield (1970) for slurry systems. The slip velocity model assumes that the completely suspended particles move at their terminal settling velocity relative to the surrounding liquid. Satterfield (1970) proposed a correlation equation, Eqn. (T-4.8.3) in Table 4.8, for estimating the liquid-solid mass transfer coefficient in slurry reactors. If the particle terminal velocity follows Stokes' law, Pe^* in Eqn. (T-4.8.3) can be replaced by (Ar Sc/18). Figure 4.20 illustrates the relationship between the liquid-solid mass transfer coefficient and the particle diameter calculated based on Eqn. (T-4.8.3). As shown, the second

Table 4.8 Correlations for Liquid-Solid Mass Transfer Coefficient (k_S) in Slurry Bubble Columns

Investigators	Correlation Equations (in SI Units)	
Kobayashi and Saito (1965)	$Sh = 2.0+0.212(gd_p^3\Delta\rho_{s-\ell}/\mu_\ell D_M)^{1/3}(d_pU_g\rho_\ell/\mu_\ell)^{0.112}$	(T-4.8.1)
Kawamura et al. (1965)	$Sh = 0.72[(ed_p^4/\nu_\ell^3)^{1/3} + (d_pU_t\rho_\ell/\mu_\ell)]^{1/2}Sc^{1/3}$ $1270 \le \rho_s \le 6160\ kg/m^3$, $5 \le d_p \le 60\ \mu m$, $0.004 \le U_g \le 0.2\ m/s$	(T-4.8.2)
Satterfield (1970)	$Sh^2 = 4.0+(Pe^*)^{2/3}$	(T-4.8.3)
Sano et al. (1974)	$Sh = \phi_s[2+0.4\ Sc^{1/3}(ed_p^4/\nu_\ell^3)^{1/4}]$ for $217 \le Sc \le 1410$ $1 \le ed_p^4/\nu_\ell^3 \le 10^7$	(T-4.8.4)
Sanger and Deckwer (1981)	$Sh = 2+0.545\ Sc^{1/3}(ed_p^4/\nu_\ell^3)^{0.264}$ for $137 \le Sc \le 50000$ $10^{-3} \le ed_p^4/\nu_\ell^3 \le 10^7$	(T-4.8.5)

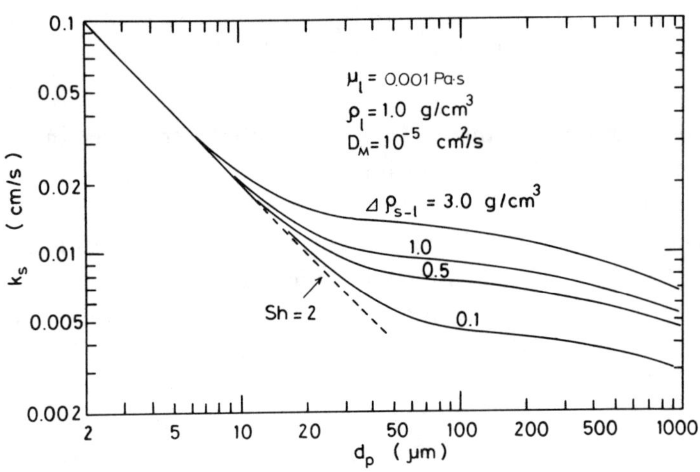

Figure 4.20 Effect of d_p on k_s based on the slip velocity model (from Muroyama, 1985, 1986).

term on the right hand side of the equation becomes insignificant for small particles (d_p< ~10 μm) with differences between particle and liquid densities near 1 g/cm^3. This is consistent with Deckwer et al.'s

(1980) limiting case assumption of a Sh equal to 2 for their paraffin/Al_2O_3 slurry in which the size of the Al_2O_3 particles suspended was less than 5 μm. A similar slip velocity approach was used by Kobayashi and Saito (1965) as given in Eqn. (T-4.8.1) in Table 4.8. The gas Reynolds number in Eqn. (T-4.8.1), however, considers the enhancement effect of the gas velocity on the mass transfer rate.

The second model is based on Kolmogoroff's theory of local isotropic turbulence as described previously in reference to heat transfer (see Section 3.3.2.4). The local energy dissipation rate per unit mass of liquid, e, at low solids holdups can be approximated by

$$e = P_w/W_\ell = AU_g(1-\epsilon_g)\rho_\ell H_A g/A(1-\epsilon_g)\rho_\ell H_A = U_g g \tag{4.55}$$

Note that for relatively larger solids holdups, e should be multiplied by a factor of $[1 + \epsilon_s\rho_s/\epsilon_\ell\rho_\ell]$, though researchers using the energy dissipation approach have traditionally neglected this term. The correlation for mass transfer coefficient may have the general form

$$Sh = 2 + K_1\, Sc^p(e\, d_p{}^4/\nu_\ell{}^3)^q \tag{4.56}$$

Based on the boundary layer theory, p is usually assumed to be 1/3.

Some pertinent correlations and ranges of application based on the isotropic turbulence model were given by Kawamura et al. (1965) (Eqn. (T-4.8.2)), Sano et al. (1974) (Eqn. (T-4.8.4)) and Sanger and Deckwer (1981) (Eqn. (T-4.8.5)). Note that Eqn. (T-4.8.2) neglects the term for molecular diffusion in the infinite medium, but includes an additional term relating to the enhancement due to particle settling relative to the medium. The Sherwood numbers predicted from the above correlations are compared in Fig. 4.21, together with some correlations for stirred tank slurry systems. It is interesting to note that the values of Sh predicted from the correlations for slurry bubble columns agree fairly well with those for the stirred tank slurry systems despite the differences in hydrodynamic properties and solid suspension mechanisms between the two systems.

Generally, the mass transfer coefficient increases with U_g to about the 1/4 power and in some cases, becomes independent of gas velocity at higher gas velocities (Kobayashi and Saito, 1965; Sanger and Deckwer, 1981). It decreases with increasing liquid viscosity (Kawamura et al., 1965), and decreases with increasing particle diameter for particles less than 0.1 cm in diameter (Kobayashi and Saito, 1965; Kawamura et al., 1965). Figure 4.22 compares the values

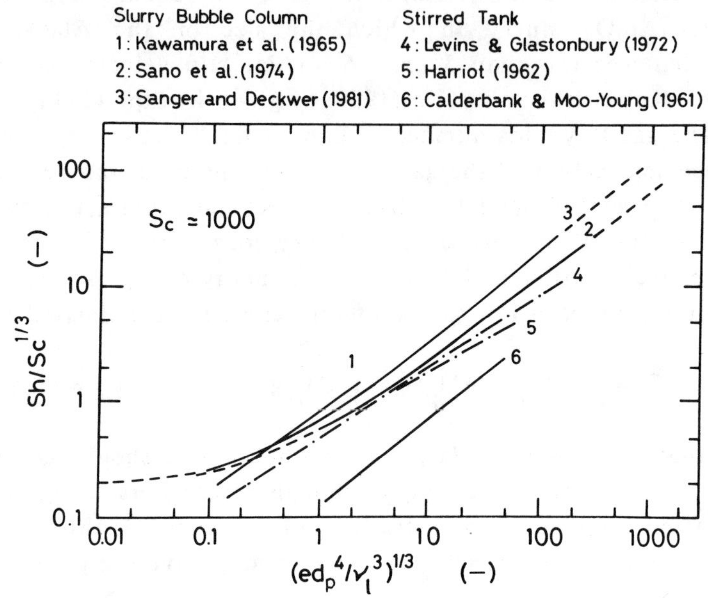

Figure 4.21 Comparison of correlations for liquid-solid mass transfer in terms of energy dissipation rate based on Kolmogoroff's Theory of local isotropic turbulence (from Muroyama, 1985, 1986).

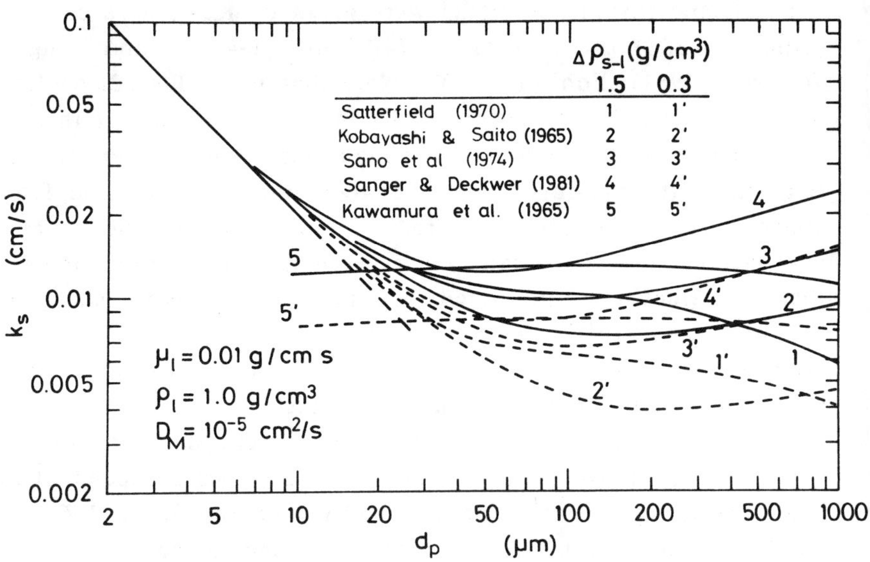

Figure 4.22 Comparison of k_S values estimated from various correlations under a complete solids suspension state in a slurry bubble column (from Muroyama, 1985, 1986).

of k_s estimated based on various correlations for $\Delta\rho_{s-\ell}$ of 1.5 and 0.3 g/cm^3. Sanger and Deckwer's correlation predicts the highest k_s among the correlations in Table 4.8 and Kawamura et al.'s correlation does not converge to the limiting condition that Sh equals 2 at small d_p. It is clear, however, that in all correlations, k_s is more sensitive to d_p at small d_p than at large d_p; the behavior for large d_p is consistent with that observed in the three-phase fluidized bed system (see Section 3.1.2.2).

Summary

Both the slip velocity and energy dissipation models exhibit shortcomings in view of the general observations given above. According to the slip velocity model, Sh depends on d_p and $\Delta\rho_{s-\ell}$, but is independent of U_g; the energy dissipation model as it has been traditionally applied, on the other hand, indicates a dependency on U_g, but not on $\Delta\rho_{s-\ell}$ when the solids holdup is low. Despite these apparent differences both models yield similar values for the Sh in so far as a complete solids suspension state is satisfied. Application of these correlations should be restricted to d_p less than 1 mm.

4.9 HEAT TRANSFER

Various important industrial processes involving large heat effects are carried out in bubble column or slurry bubble column reactors. The necessary heat transfer surface may be provided by various means: internal coils, vertically or horizontally arranged tubes, or a jacketed column wall. Internal heat transfer within the column occurs mainly due to liquid or slurry mixing.

4.9.1 Submerged Surface to Liquid (or Slurry) Heat Transfer

Investigations into heat transfer behavior from immersed cylinder or column wall heaters in bubble (or slurry bubble) columns have interpreted the behavior in terms of heat transfer coefficients similar to those in fluidized bed systems. In this section, the general behavior of the heat transfer coefficient is described and the correlations proposed are reviewed and analyzed.

In a bubble column where heat transfer occurs from a heated surface to an aerated liquid phase, the liquid temperature varies sharply only near the heater surface. Kolbel et al. (1958a) found the temperature profile was flat beyond about 2 mm from the heater surface because of extensive mixing; consequently, a simplified approach based on Eqn. (3.72) can be used to evaluate the heat transfer coefficient, h, in a bubble column by assuming a completely mixed liquid phase. Table 4.9 summarizes the major experimental conditions and correlations for heat transfer coefficients in bubble or slurry bubble columns. The results of these studies lead to the following observations:

(1) The effect of heater geometry on the heat transfer coefficient is generally small. Yoshitome et al. (1965) measured the heat transfer coefficients for spherical, cylindrical, and ring shaped heating elements and found only slight differences in the heat transfer coefficients, although the largest differences occurred at low gas velocities. According to Nishikawa et al. (1977), the heat transfer coefficients agreed with each other for coil-to-bed heat transfer and wall-to-bed heat transfer. Lewis et al. (1982) showed that the heat transfer rate decreases with increasing vertical dimension of a cylindrical heater whether it is inserted vertically or horizontally, but becomes independent of heater geometry and direction of insertion when the axial length of the vertical heater or the diameter of the horizontal heater is greater than about 40 mm.

(2) The heat transfer coefficient is independent of both column diameter and height.

(3) The heat transfer coefficients vary with the gas distributor design. Yoshitome et al. (1965) showed that columns equipped with perforated plates have a higher heat transfer coefficient at high gas velocities (> 5 cm/s) than those equipped with porous plate distributors. The heat transfer coefficient is independent of the type of porous plate used but depends strongly on the perforated plate design.

(4) Increases in gas velocity intensify liquid phase turbulence and increase the heat transfer coefficient. Several investigators, however, observed that the increase tapered off at high gas velocities (Yoshitome et al., 1965; Nishikawa et al., 1977).

(5) The superficial liquid velocity also significantly affects the heat transfer coefficient at high liquid velocities. Nishikawa et al. (1977) showed that for a bubble column containing either Newtonian or non-Newtonian liquid and aerated at U_g of 27.8 cm/s, the heat transfer coefficient was independent of liquid velocity for $U_\ell < 0.015$ m/s, but, for $U_\ell > 0.015$ m/s, increased with increasing liquid velocity.

Table 4.9 Correlations for Heat Transfer Coefficient in Bubble Columns and Slurry Bubble Columns

Investigators	Correlation Equations (in SI Units)	Systems/ Range of variables
Kolbel et al. (1960)	$\frac{hD}{\lambda_\ell} = 222.8\left(\frac{DU_g\rho_\ell}{\varepsilon_g\mu_\ell}\right)^{0.16}\left(\frac{d_p}{4\times10^{-5}}\right)^{0.05}$ $\quad \left(\frac{DU_g\rho_\ell}{\varepsilon_g\mu_\ell} < 6300\right)$ $\frac{hD}{\lambda_\ell} = 350.8\left(\frac{DU_g\rho_\ell}{\varepsilon_g\mu_\ell}\right)^{0.108}\left(\frac{d_p}{4\times10^{-5}}\right)^{0.05}$ $\quad \left(\frac{DU_g\rho_\ell}{\varepsilon_g\mu_\ell} > 6300\right)$ where $D = \{(D_c^2-D_h^2)(1-\varepsilon_g)\}^{1/2}$	gas-liquid-solid quarts sand/water $40 \leq d_p \leq 200$ μm $0 \leq \bar{C}_s/\rho_{s\ell} \leq 0.4$ $0 \leq U_g \leq 0.12$ m/s
Nishikawa et al. (1977)	$St=0.418\,Pr_\ell^{-2/3}\left(\frac{U_g^3\rho_\ell}{\mu_\ell g}\right)^{-1/3}\left(\frac{\Delta\rho_{\ell-g}}{\rho_\ell}\right)^{1/3}\left(\frac{\mu_{\ell w}}{\mu_\ell}\right)^{1/4}$ ($U_\ell \leq 0.015$ m/s, $U_g \leq 0.278$ m/s) $St=0.3\,Pr_\ell^{-2/3}\left(\frac{U_g^3\rho_\ell}{\mu_\ell g}\right)^{-1/3}\left(\frac{\Delta\rho_{\ell-g}}{\rho_\ell}\right)^{1/3}\left(\frac{\mu_{\ell w}}{\mu_\ell}\right)^{0.05}$ ($U_\ell \leq 0.01$ m/s, $U_g > 0.278$ m/s) $St=0.418\,Pr_\ell^{-2/3}\left(\frac{U_g^3\rho_\ell}{\mu_\ell g}\right)^{1/3}\left(\frac{\Delta\rho_{\ell-g}}{\rho_\ell}\right)^{1/3}\left(\frac{\mu_{\ell w}}{\mu_\ell}\right)^{1/4}$ ($0.015 < U_\ell < 0.138$ m/s)	gas-liquid water, aqueous glycerin Millet jelly soln., CMC soln. $0.47 \leq \mu_\ell \leq 700$ mPa·s $0.05 \leq U_g \leq 0.35$ m/s $D_c = 0.051, 0.15$ m
Deckwer (1980)	$St = 0.1\,Pr_\ell^{-0.5}\left(\frac{U_g^3\rho_\ell}{\mu_\ell g}\right)^{-0.25}$ (T-4.9.1)	gas-liquid
Deckwer et al. (1980)	$St_{s\ell} = 0.1\,Pr_{s\ell}^{-0.5}\left(\frac{U_g^3\rho_{s\ell}}{\bar{\mu}\,g}\right)^{-0.25}$ (T-4.9.2)	gas-liquid-solid $U_g \leq 0.1$ m/s
Hikita et al. (1981a)	$St=0.411\,Pr_\ell^{-2/3}\left(\frac{U_g^3\rho_\ell}{\mu_\ell g}\right)^{-0.2837}\left(\frac{\mu_\ell^4 g}{\rho_\ell\sigma^3}\right)^{0.02433}$	gas-liquid water, aqueous sucrose soln., n-butanol, aqueous methanol solns. $0.05 \leq U_g \leq 0.35$ m/s D_c=0.1, 0.19 m
Michael and Reichert (1981)	$St_{s\ell} = 0.11\,Pr_{s\ell}^{-0.5}\left(\frac{U_g^3\rho_{s\ell}}{\mu\,g}\right)^{-0.25}$ (T-4.9.3)	gas-liquid-solid Exsol D_{80} $201 \leq$ B.P. ≤ 233 °C suspended with high density polyethylene $47 \leq d_p \leq 241$ μm $0 \leq \bar{C}_s/\rho_{s\ell} \leq 0.34$ $0.01 \leq U_g \leq 0.12$ m/s D_c=0.1 m

$St = \frac{h}{C_{p\ell}\rho_\ell}$, $St_{s\ell} = \frac{h}{C_p\,\rho_{s\ell}}$, $Pr_\ell = \frac{C_{p\ell}\mu_\ell}{\lambda_\ell}$, $Pr_{s\ell} = \frac{C_p\bar{\mu}}{\lambda}$

$\Delta\rho_{\ell-g} = \rho_\ell-\rho_g$, $\mu_{\ell w}$: viscosity of liquid at wall temperature

4.9.2 Heat Transfer Coefficient Correlations

In developing a correlation for the heat transfer coefficient in bubble columns where the effect of liquid rate is negligible, a reasonable hypothesis is that the hydrodynamic conditions in bubble agitated systems may be satisfactorily represented by the superficial gas velocity and the liquid viscosity, density, specific heat, and thermal conductivity. The heat transfer coefficient may be expressed in terms of these variables as

$$h = f(U_g, \mu_\ell, \rho_\ell, C_{p\ell}, \lambda_\ell) \tag{4.57}$$

or simply by an empirical power function as

$$h = K_2 \, U_g^{a_1} \, \lambda_\ell^{a_2} \, \rho_\ell^{a_3} \, C_{p\ell}^{a_4} \, \mu_\ell^{a_5} \tag{4.58}$$

According to Deckwer (1980), the exponents in Eqn. (4.58) proposed in various correlations for heat transfer in a bubble column are: a_1 = 0.19 ~ 0.34, a_2 = 0.44 ~ 0.60, a_3 = 0.67 ~ 0.78, a_4 = 0.40 ~ 0.50, and a_5 = (-0.17) ~ (-0.35). Note the significance of the liquid thermal conductivity and viscosity to the heat transfer coefficient.

By means of a dimensional analysis the following formula may be derived from Eqn. (4.57) (Akita, 1985):

$$h/(\rho_\ell C_{p\ell} U_g) = C(C_{p\ell}\mu_\ell/\lambda_\ell)^p[\rho_\ell U_g{}^3/(g\mu_\ell)]^q \tag{4.59}$$

or

$$St = C(Pr_\ell)^p(Re' Fr'_g)^q \tag{4.60}$$

Many correlations listed in Table 4.9 are similar to Eqn. (4.60). Note that p in Eqn. (4.60) varies from -1/2 to -2/3. The assumption of p equal to -2/3, as used by Hart (1976), Nishikawa et al. (1977) and Hikita et al. (1981a), is based on Pohlhausen's boundary layer theory (see Bennett and Myers, 1974), while setting p equal to -1/2, as done by Deckwer (1980), is, on the other hand, based on the surface renewal theory of Higbie (1935). Additionally, note that the quantity $Re' Fr'_g$ is independent of the characteristic length of the heater (D_h). Kolmogoroff's theory of isotropic turbulence (Hinze, 1975) in combination with the surface renewal theory of Higbie (1935), as previously described in Section 3.3.2.3, has also been applied

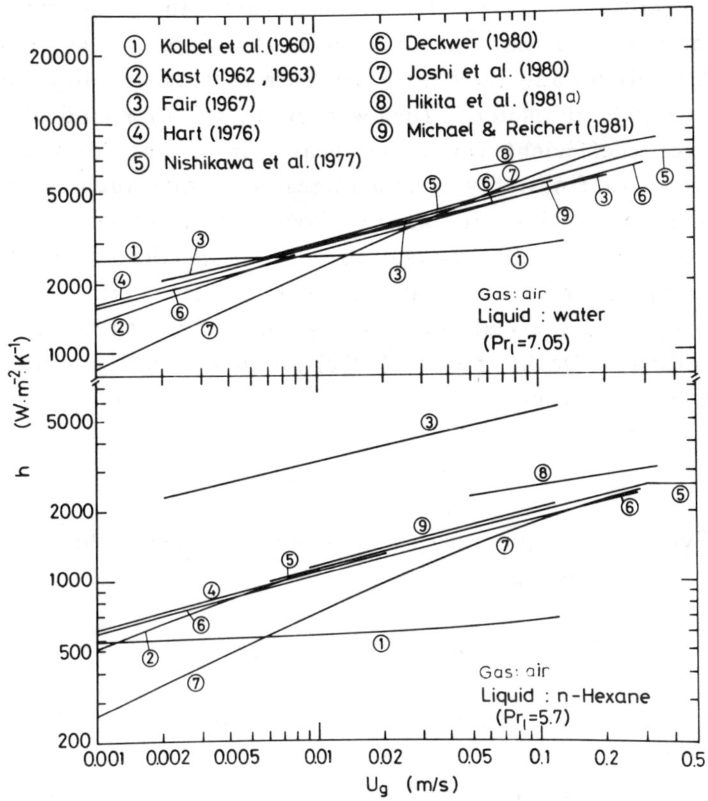

Figure 4.23 Comparison of heat transfer coefficient in bubble columns, predicted from various correlations.

to heat transfer in bubble columns and slurry bubble column systems (Deckwer, 1980) as given in Eqns. (T-4.9.1) and (T-4.9.2) in Table 4.9. Figure 4.23 compares the values of h calculated from various correlations (five of them are listed in Table 4.9) for particular systems using water and n-hexane as the liquid phase. For water systems, the predicted values of h agree fairly well, but for a system using n-hexane, the predicted values showed significant deviations. Note that the correlations by Nishikawa et al. (1977) and Deckwer (1980) yield nearly equal values of h over a wide range of liquid properties.

Heat transfer in slurry bubble columns has been investigated by several workers (Kolbel et al., 1958b, 1960; Deckwer et al., 1980; Michael and Reichert, 1981). Kolbel et al. (1960) measured the effect of solids concentration and particle size on the heat transfer coefficient for a rod immersion heater in a quartz sand-water suspension. The heat transfer coefficients were greater than those in a bubble column and they increased with increasing solids concentration and particle diameter. Deckwer et al. (1980) found that the heat transfer coefficient for a rod heater in a fine ($d_p \leq 5\ \mu m$) Al_2O_3 powder-paraffin slurry system increased significantly with increasing solids concentration. It was conceived that the variation in h might be the result of changing physicochemical properties of the slurry phase; an increase in solids concentration increased the slurry density, viscosity, and thermal conductivity, but decreased the heat capacity. According to Deckwer et al. (1980) the mean density of the slurry can be calculated from

$$\rho_{s\ell} = \psi_s \rho_s + \psi_\ell \rho_\ell \tag{4.61}$$

The viscosity of the slurry can be obtained using the Einstein (1906) type of relationship

$$\bar{\mu} = \mu_\ell (1 + K_3 \psi_s) \tag{4.62}$$

where K_3 is a constant depending on the solid and liquid phases involved.

The heat capacity of the suspension is given by

$$C_p = \omega_s C_{ps} + \omega_\ell C_{p\ell} \tag{4.63}$$

Tareef's equation (Tareef, 1940) was recommended by Deckwer et al. to estimate the thermal conductivity of the suspension:

$$\bar{\lambda} = \lambda_\ell \frac{2\lambda_\ell + \lambda_s - 2\psi_s(\lambda_\ell - \lambda_s)}{2\lambda_\ell + \lambda_s + \psi_s(\lambda_\ell - \lambda_s)} \tag{4.64}$$

Deckwer et al. (1980) showed that Eqn. (T-4.9.2) using the effective thermophysical properties of the slurry predicted fairly well the available heat transfer coefficient data for slurry bubble columns in the Fischer-Tropsch synthesis application.

Michael and Reichert (1981) found that the heat transfer coefficient for a rod immersion heater in various high density polyethylene powder-organic solvent slurries strongly decreased with increasing solids concentration, especially beyond 22 wt% for the dispersion of two smaller sizes of high density polyethylene powders. It was found that this behavior strongly related to the average viscosity of the suspension which increased sharply with solids concentration. Michael and Reichert (1981) confirmed that Eqn. (T-4.9.2) could adequately correlate their own data using average properties of the suspension; they used a coefficient of 0.11 instead of 0.10 originally given by Deckwer et al. (1980).

Several correlations, including those by Hart (1976), Nishikawa et al. (1977), and Deckwer (1980), originally developed for gas-liquid systems may also be applicable to gas-liquid-solid systems by using effective thermophysical properties of the slurry evaluated with Eqns. (4.61) through (4.64).

4.9.3 Internal Heat Transfer

Though radially uniform, the temperature may not be axially uniform in a relatively long cylindrical bubble column heated through its wall (Hart, 1976). Radial and axial heat transport within a bubble column occurs mainly due to turbulent mixing in the liquid or slurry induced by ascending gas bubbles, since the contribution of conduction through the phases involved are negligibly small. For a bubble column the radial heat transfer can usually be neglected compared to the axial heat transfer for design purposes.

Axial thermal dispersion in a bubble column has been studied by several authors (Chen and McMillan, 1982; Kara et al., 1982; Kelkar et al., 1984). The axial thermal dispersion rate, q, can be expressed by

$$q = -k_{ez} A \frac{dT}{dz} \tag{4.65}$$

where k_{ez} is the axial effective thermal conductivity, A is the column cross-sectional area, and dT/dz is the axial temperature gradient. k_{ez} can be related to the axial dispersion coefficient of liquid when the contributions of gas phase and solid phase are neglected, i.e.,

$$k_{ez} = E_{z\ell} \rho_\ell C_{p\ell} \epsilon_\ell \tag{4.66}$$

The relationship of Eqn. (4.66) permits evaluation of the liquid phase axial dispersion coefficient, $E_{z\ell}$, from experimental results of axial dispersion of heat or vice versa. The former approach has been employed by Kara et al. (1982) and Kelkar et al. (1984) for single stage bubble columns and Sekizawa and Kubota (1974) for a multistage bubble column.

4.10 CONCLUDING REMARKS

The flow regimes reflecting the behavior of gas holdup for bubble column or slurry bubble column systems (Modes E-I-a-2, E-III-a, and T-I-a-2) are generally classified according to dispersed bubble regime (or homogeneous flow regime), coalesced bubble regime (or heterogeneous flow regime; churn turbulent flow regime) and slugging regime, in a manner similar to those for three-phase fluidized bed systems (Mode E-I-a-1). For slurry bubble column systems, the distinct decreasing axial solids concentration variation for Modes E-I-a-2 and E-III-a and uniform axial solids concentration distribution for Mode T-I-a-2 are established largely due to the effect of bubble wake; such effects are predominant for the solids mixing in the three-phase fluidized bed systems. When large, light particles are encountered in the slurry bubble column system, however, the effect of drift may also becomes important; little is known regarding this effect. Significant differences exist between the slurry bubble column and fluidized bed systems concerning effects of particle properties on flow regime transition.

The circulating liquid flow pattern in bubble columns with an upward flow in the core region and a downward flow near the wall is inherently associated with the bubble flow pattern. The mixing behavior and the heat and mass transfer characteristics are also dependent on the prevailing bubble flow pattern. Liquid phase dispersion in a slurry bubble column in the coalesced bubble regime is only slightly different from that in a bubble column operated in the same flow regime, reflecting a limited effect due to the presence of solid particles.

The general modeling approach to describing the transport behavior of the three-phase fluidized bed systems treats each of the three phases separately with slip relationships accounting for the interaction among the phases. The approach is also generally applicable to slurry bubble column systems. When the slip velocity is sufficiently low, the liquid and solid phases in the slurry bubble column system, however, can be treated as one pseudo-homogeneous phase. Fundamental understanding of the slurry bubble column systems and hence a more fluid mechanics based model development requires comprehensive knowledge of the bubble-wake behavior, turbulent interaction of the liquid-solid emulsion phase with bubble and wake phases, and intrinsic rheological behavior of the slurry. More research to gain such fundamental understanding is deemed necessary.

Transport behavior also varies significantly with the properties of the liquid medium. Addition of surfactants alters the average bubble size, gas holdup, and bubble-particle interaction, possibly resulting in frothing or foaming flow regimes. Flotation effects can occur with the proper combination of surfactants and particles. Knowledge of dynamic interfacial tension values and surface adsorption rates is required to fully predict the effects.

Heat and mass transfer properties in the slurry bubble column systems are reasonably well explored. For practical interest, more research is needed in general areas of transport behavior addressing the effects of fluid or slurry properties (e.g., organic liquids, liquid mixtures, viscous slurries, and non-Newtonian liquids), high temperature and pressure, distributor design and scale up.

4.11 NOMENCLATURE

a	: gas-liquid interfacial area per unit volume of aerated slurry
a'	: gas-liquid interfacial area per unit volume of liquid defined by $a/(1-\epsilon_g)$
A	: cross-sectional area of the column
$\overline{A}$	: distance between the axis of symmetry and the center of vortex
A^*	: dimensionless distance, $\overline{A}/r_o$
A_d	: effective cross-sectional area for solids discharge from the bubble-wake phase in the solids-batch mixing model
A_e	: effective cross-sectional area for solids intake into the bubble-wake phase in the solids-batch mixing model
A_k	: coefficients in Eqns. (4.8) and (4.9)
Ar	: Archimedes number, $d_p{}^3 g(\rho_s-\rho_\ell)\rho_\ell/\mu_\ell{}^2$
B	: constant defined in Eqn. (4.8)
C_N	: carbon number, the number of carbon atoms in the straight chain of an alcohol
C_p	: heat capacity of slurry

$C_{p\ell}$	: heat capacity of liquid
C_{ps}	: heat capacity of solid
C_s	: solids concentration in the slurry, mass per volume
$\bar{C}_s$	: average solids concentration in the slurry
C_s^0	: solids concentration at the bottom of the column
C_s^1	: solids concentration at the top of the column
C_s^f	: solids concentration in the influent slurry
$\bar{C}_{sc}$	: critical solids concentration in the slurry for complete solids suspension
d_o	: diameter of orifices in gas distributor plate
d_p	: particle diameter
d_{vs}	: volume-surface mean bubble diameter
D_c	: column diameter
D_d	: diameter of perforated plate gas distributor
D_h	: diameter of immersion heater
D_M	: molecular diffusivity in the liquid phase
e	: energy dissipation rate per unit mass of liquid
E_1	: energy dissipation due to wakes behind bubbles
E_2	: energy dissipation due to net bubble breakup
E_3	: energy dissipation due to hydraulic jump at the liquid surface
E_4	: energy dissipation due to turbulence near the nozzle
E_5	: energy dissipation due to viscous drag
E_6	: energy dissipation due to kinetic energy loss during the downward flow of liquid
E_i	: energy input rate for gas
E_{zg}	: gas phase axial dispersion coefficient based on the linear gas velocity
$E_{z\ell}$	: liquid phase axial dispersion coefficient based on the linear liquid velocity
E_{zs}	: solid phase axial dispersion coefficient defined in Eqns. (4.29) and (4.31)
E'_{zs}	: parameter defined by Eqn. (4.50)
E^*_{rs}	: solid phase radial dispersion coefficient defined in Eqn. (4.37)

E_{zs}^* : solid phase axial dispersion coefficient defined in Eqn. (4.37)

$E_{\theta s}^*$: solid phase tangential dispersion coefficient defined in Eqn. (4.37)

$E(\tau)$: external residence time distribution function

Fr_g : gas Froude number, U_g^2/gD_c

Fr_g' : gas Froude number, U_g^2/gd_p

g : gravitational acceleration

h : heat transfer coefficient

H : column height

H_0 : static liquid height in the bed

H_A : liquid height under aeration

H_c : height of circulation cell

I : ionic strength

$j_{g\ell}$: drift flux of gas relative to liquid flow

k : ratio of wake volume to bubble volume, ϵ_w/ϵ_g

k_2 : parameter defined in Eqn. (4.3)

k_{ez} : axial effective thermal conductivity

k_g : gas-side mass transfer coefficient for gas-liquid mass transfer

k_ℓ : liquid-side mass transfer coefficient for gas-liquid mass transfer

$k_g a'$: gas phase volumetric mass transfer coefficient

$k_\ell a$: liquid phase volumetric coefficient for gas-liquid mass transfer based on unit volume of liquid or slurry

k_s : liquid-solid mass transfer coefficient

L_P : pitch of orifices on the perforated gas distributor

m_s : parameter defined in Eqn. (4.6)

n : power-law index for liquid viscosity

n_ℓ : total mass flux of liquid based on the cross section of slurry

n_s : total mass flux of solids based on the cross section of slurry

N : number of orifices in the perforated gas distributor

N_b : total number of bubbles in a slurry bubble column

P_w : energy dissipation rate due to isotropic turbulence

Pe^*	: Peclet number, $U_t d_p/D_M$
$Pe_{z\ell}$	: Peclet number for liquid axial dispersion, $U_g D_c/E_{z\ell}$
Pe_{zs}	: Peclet number for solid axial dispersion, $U_g D_c/E^*_{zs}$
Pr_ℓ	: liquid Prandtl number, $C_{p\ell}\mu_\ell/\lambda_\ell$
$Pr_{s\ell}$	: Prandtl number based on slurry properties, $C_p\overline{\mu}/\overline{\lambda}$
q	: rate of axial heat transfer
r	: radial coordinate in cylindrical system
$\vec{r}$	: unit vector in r-direction
r^*	: dimensionless radial distance, r/r_o
r_o	: radius of the column
R^*	: radial component of the dimensionless stream function (Ψ/Ψ_o)
Re'	: Reynolds number, $d_p U_g \rho_\ell/\mu_\ell$
Re_g	: gas Reynolds number, $D_c U_g \rho_\ell/\mu_\ell$
Re^*_g	: gas Reynolds number, $D_c U_g \rho_g/\mu_g$
$(Re_g)_c$	: critical gas Reynolds number, $D_c U_{gc} \rho_\ell/\mu_\ell$
Re_t	: particle terminal Reynolds number, $d_p U_t \rho_\ell/\mu_\ell$
$Re_{s\ell}$	: slurry Reynolds number, $D_c U^*_{s\ell} \rho_{s\ell}/\mu_{eff}$
Sc	: Schmidt number, ν_ℓ/D_M
Sh	: Sherwood number, $k_s d_p/D_M$
St	: Stanton number, $h/\rho_\ell C_{p\ell} U_g$
$St_{s\ell}$	: Stanton number based on properties of slurry, $h/\rho_{s\ell} C_p U_g$
t	: time
$\overline{t}$	: mean residence time of solid particles
T	: temperature
U'	: liquid velocity at $H_c/2$
U_b	: bubble rise velocity relative to column
$\overline{U}_b$	: mean bubble rise velocity relative to column
$U_{b\infty}$	: terminal rise velocity of a single bubble
U_c	: average liquid circulation velocity
U_g	: superficial gas velocity
U_{gc}	: critical gas velocity for the complete solids suspension

$(U_g)_{crit}$: critical gas velocity for transition from dispersed bubble to coalesced bubble regime.

$U_{g\ell}$: slip velocity of a bubble in the liquid recirculation region

U_{gM} : gas velocity where maximum value of $(\epsilon_{goT}/\epsilon_{go} - 1)$ is observed

U_ℓ : superficial liquid velocity

U_p : slip velocity between liquid and solids

$U_{s\ell}$: weight-averaged superficial slurry velocity

$U^*_{s\ell}$: volume-averaged superficial slurry velocity

U_{st} : a general term for solids settling velocity used in the so-called sedimentation-dispersion model by various investigators

U_t : terminal settling velocity of a single particle

$\underline{v}'_p$: fluctuation velocity vector of solid particles relative to $\underline{V}^*_p$

V_d : average solids discharge velocity in the solids-batch mixing model

V_e : average velocity of solids relative to the bubble rise velocity $(\overline{U}_b)$ in the solids-batch mixing model

V_ℓ : cross-sectionally averaged linear liquid velocity

$V_{\ell f}$: linear liquid velocity in the liquid-solid emulsion phase

$V_{\ell o}$: the value of V_ℓ at column axis

$V_{\ell r}$: linear liquid velocity at radius position r in the coalesced bubble flow condition

$V_{\ell s}$: linear relative velocity between the liquid and solid phases in the liquid-solid emulsion phase

$V_{\ell w}$: value of V_ℓ defined in Fig. 4.8

V_p : average convective velocity of solid particles over the cross-sectional area of the column

$\underline{V}^*_p$: time-averaged velocity vector of solid particles

$\underline{\tilde{V}}_p$: instantaneous velocity vector of solid particles

V_R : total volume of a slurry bubble column

V_{sf} : linear solids velocity in the liquid-solid emulsion phase

$V^*_{s\ell}$: linear volume-averaged slurry velocity, given as $U^*_{s\ell}/(1-\epsilon_g)$

W_ℓ : mass of liquid in column

x : ratio of solids concentration in the wake phase to that in the liquid-solid emulsion phase, $\epsilon_{sw}/\epsilon_{sf}$

z : axial coordinate in cylindrical system; axial distance from bottom of the column

$\vec{z}$: unit vector in z-direction

z^* : z/H_c

Z^* : axial component of the dimensionless stream function (Ψ/Ψ_o)

Greek Letters

α : defined as $\sqrt{k_2 r_o{}^4}$

β : constant given in Eqn. (4.9)

ϵ_g : gas holdup

ϵ_{gmax} : maximum gas holdup

ϵ_{go} : gas holdup in the coalesced bubble regime without particles

ϵ_{goT} : gas holdup in the transition regime without particles

ϵ_{gr} : radial gas holdup

ϵ_ℓ : liquid holdup

$\epsilon_{\ell f}$: liquid holdup in the liquid-solid emulsion phase

$\epsilon_{\ell\text{-}s}$: volume fraction of the liquid-solid emulsion phase

ϵ_s : time-averaged solids holdup

ϵ_s' : fluctuation of solids holdup from the time-averaged value

$\tilde{\epsilon}_s$: instantaneous solids holdup

$\bar{\epsilon}_s$: average solids holdup in the column

ϵ_{sf} : solids holdup in the liquid-solid emulsion phase

ϵ_{sw} : average solids holdup in the wake phase

ϵ_w : wake volume fraction

ζ : vorticity; ζ_o, vorticity at the vortex center

$\bar{\lambda}$: thermal conductivity of slurry

λ_ℓ : thermal conductivity of liquid

λ_s : thermal conductivity of solid

$\bar{\mu}$: effective viscosity of slurry defined by Eqn. (4.62)

μ_{eff} : effective viscosity of slurry calculated based on the Ostwald-de Waele relation with respect to shear rate

μ_g : gas viscosity

μ_ℓ	: liquid viscosity
$\mu_{\ell w}$	: liquid viscosity at wall temperature
ν_ℓ	: liquid kinematic viscosity
ν_T	: turbulent kinematic viscosity
ρ_e	: effective density of the liquid-solid mixture in a three-phase fluidized bed
ρ_g	: gas density
ρ_ℓ	: liquid density
ρ_s	: solid density
$\rho_{s\ell}$	: slurry density
$\Delta\rho_{\ell-g}$	: density difference, $\rho_\ell - \rho_g$
$\Delta\rho_{s-\ell}$	: density difference, $\rho_s - \rho_\ell$
σ	: surface tension
Ψ	: stream function; Ψ_o, stream function at the vortex center
ψ_ℓ	: liquid volume fraction in the slurry
ψ_s	: solid volume fraction in the slurry
ω_ℓ	: liquid weight fraction in the slurry
ω_s	: solid weight fraction in the slurry
τ	: normalized time, $t/\bar{t}$
ϕ_s	: sphericity of particles
θ	: θ component in cylindrical coordinates
$\vec{\theta}$	: unit vector in θ-direction

4.12 REFERENCES

Akita, K., and F. Yoshida, I&EC Process Des. & Dev., 12, 76 (1973).

Akita, K., and F. Yoshida, I&EC Process Des. & Dev., 13, 84 (1974).

Akita, K., in Chemical Engineering Technical Report No. 9, "Design of Bubble Column and Slurry Bubble Column Reactors," p. 37, Soc. Chem. Engrs. Japan (1985).

Anderson, J.L., and J.A. Quinn, Chem. Eng. Sci., 25, 373 (1970).

Badgujar, M.N., A. Deimling, B.I. Morsi, Y.T. Shah, and N.L. Carr, Chem. Eng. Commun., 48, 127 (1986).

Baird, M.H.I., and R.G. Rice, Chem. Eng. J., 9, 171 (1975).

Bennett, C.O., and J.E. Myers, "Momentum, Heat, and Mass Transfer," 2nd ed., McGraw-Hill (1974).

Bergougnou, M.A., personal communication (1988).

Botton, R., D. Cosserat, and J.C. Charpentier, Chem. Eng. J., 20, 87 (1980).

Brian, B., and P.N. Dyer, ACS Symposium Series, No. 237, 107 (1984).

Bukur, D.B., and J.G. Daly, Chem. Eng. Sci., 42, 2967 (1987).

Bukur, D.B., D. Petrovic, and J.G. Daly, I&EC Res., 26, 1087 (1987).

Calderbank, P.H., and M.B. Moo-Young, Chem. Eng. Sci., 16, 39 (1961).

Calderbank, P.H., F. Evans, R. Farley, G. Jepson, and A. Poll, in "Catalysis in Practice," p. 66, Inst. Chem. Engrs. (1963).

Capuder, E., and T. Koloini, Chem. Eng. Res. Des., 62, 255 (1984).

Chang, M.Y., and B.I. Morsi, Chem. Eng. J. (in press, 1988).

Chen, B.H., and A.F. McMillan, Can. J. Chem. Eng., 60, 436 (1982).

Cova, D.R., I&EC Process Des. & Dev., 5, 20 (1966).

Cussler, E.L., "Diffusion," pp. 93-95, Cambridge University Press (1984).

Danckwerts, P.V., "Gas-Liquid Reactions," McGraw-Hill (1970).

Dayan, A., and S. Zalmanovich, Chem. Eng. Sci., 37, 1253 (1982).

Deckwer, W.-D., R. Burckhart, and G. Zoll, Chem. Eng. Sci., 29, 2177 (1974).

Deckwer, W.-D., Chem. Eng. Sci., 35, 1341 (1980).

Deckwer, W.-D., Y. Louisi, A. Zaidi, and M. Ralek, I&EC Process Des. & Dev., 19, 699 (1980).

Einstein, A., Annalen der Physik, 19, 289 (1906).

El-Temtamy, S.A., and N. Epstein, in "Fluidization," ed. by J.R. Grace and G.M. Matsen, p. 519, Plenum Press (1980).

Fair, J.R., Chem. Eng., 74(14), 67 (1967); ibid., 74(15), 207 (1967).

Field, R.W., and J.F. Davidson, Trans. Inst. Chem. Engrs., 58, 228 (1980).

Franz, K., T. Borner, H.J. Kantorek, and R. Buchholz, Ger. Chem. Eng., 7, 365 (1984).

Freedman, W., and J.F. Davidson, Trans. Inst. Chem. Engrs., 47, T251 (1969).

Fukuma, M., K. Muroyama, and A. Yasunishi, J. Chem. Eng. Japan, 20, 28 (1987a).

Fukuma, M., K. Muroyama, and A. Yasunishi, J. Chem. Eng. Japan, 20, 321 (1987b).

Furchner, B., and A. Mersmann, Chem.-Ing.-Tech., 55, 972 (1983).

Gestrich, W., and W. Rahse, Chem.-Ing.-Tech., 47, 8 (1975).

Godbole, S.P., A. Schumpe, and Y.T. Shah, Chem. Eng. Commun., 24, 235 (1983).

Godbole, S.P., A. Schumpe, Y.T. Shah, and N.L. Carr, AIChE J., 30, 213 (1984).

Hammer, H., in "Chemical Engineering of Gas-Liquid Solid Catalyst Reactions," ed. by G.A. L'Homme, p. 18, CEBEDOC, Liege, Belgium (1979).

Harriot, P., AIChE J., 8, 93 (1962); ibid., 8, 101 (1962).

Hart, W.F., I&EC Process Des. & Dev., 15, 109 (1976).

Higbie, R., Trans. AIChE, 31, 365 (1935).

Hikita, H., and H. Kikukawa, Bull. University of Osaka Prefecture, Ser. A, 22, 76 (1973).

Hikita, H., and H. Kikukawa, J. Chem. Eng. Japan, 8, 412 (1975).

Hikita, H., S. Asai, K. Tanigawa, K. Segawa, and M. Kitao, Chem. Eng. J., 20, 59 (1980).

Hikita, H., S. Asai, H. Kikukawa, T. Zaike, and M. Ohue, I&EC Process Des. & Dev., 20, 540 (1981a).

Hikita, H., S. Asai, K. Tanigawa, K. Segawa, and M. Kitao, Chem. Eng. J., 22, 61 (1981b).

Hills, J.H., Trans. Inst. Chem. Engrs., 52, 1 (1974).

Hinze, J.O., "Turbulence," 2nd ed., McGraw-Hill, New York (1975).

Hughmark, G.A., I&EC Process Des. & Dev., 6, 218 (1967).

Imafuku, K., T.-Y. Wang, K. Koide, and H. Kubota, J. Chem. Eng. Japan, 1, 153 (1968).

Jean, R.-H., W.-T. Tang, and L.-S. Fan, AIChE J. (in press, 1988).

Joshi, J.B., Trans. Inst. Chem. Engrs., 58, 155 (1980).

Joshi, J.B., Chem. Eng. J., 24, 213 (1982).

Joshi, J.B., and M.M. Sharma, Trans. Inst. Chem. Engrs., 57, 244 (1979).

Kara, S., G. Kelkar, Y.T. Shah, and N.L. Carr, I&EC Process Des. & Dev., 21, 584 (1982).

Kast, W., Int. J. Heat Mass Transfer, 5, 329 (1962); idem, Chem.-Ing.-Tech., 35, 785 (1963).

Kato, Y., Kagaku Kogaku, 27, 7 (1963); ibid., Abr. English eds., 1, 3 (1963).

Kato, Y., and A. Nishiwaki, Kagaku Kogaku, 35, 912 (1971); idem, Int. Chem. Eng., 12, 182 (1972).

Kato, Y., A. Nishiwaki, T. Fukuda, and S. Tanaka, J. Chem. Eng. Japan, 5, 112 (1972a).

Kato, Y., A. Nishiwaki, T. Kago, T. Fukuda, and S. Tanaka, Kagaku Kogaku, 36, 1333 (1972b); idem, Int. Chem. Eng., 13, 562 (1973).

Kawamura, K., T. Sasano, and A. Mifune, Kagaku Kogaku, 29, 693 (1965).

Kawase, Y., and M. Moo-Young, Chem. Eng. Sci., 41, 1969 (1986).

Kelkar, B.G., S.P. Godbole, M.F. Honath, Y.T. Shah, N.L. Carr, and W.-D. Deckwer, AIChE J., 29, 361 (1983).

Kelkar, B.G., Y.T. Shah, and N.L. Carr, I&EC Process Des. & Dev., 23, 308 (1984).

Kobayashi, T., and Saito, H., Kagaku Kogaku, 29, 237 (1965); idem, Abr. English eds., 3, 210 (1965).

Koide, K., T. Hirata, and H. Kubota, Kagaku Kogaku, 30, 712 (1966).

Koide, K., in "Advances in Chemical Engineering 16," Chap. 3, ed. by Soc. Chem. Engrs. Japan (1982).

Koide, K., T. Yasuda, S. Iwamoto, and E. Fukuda, J. Chem. Eng. Japan, 16, 7 (1983).

Koide, K., A. Takazawa, M. Komura, and H. Matsunaga, J. Chem. Eng. Japan, 17, 459 (1984).

Kojima, H., and K. Asano, Kagaku Kogaku Ronbunshu, 6, 46 (1980); idem, Int. Chem. Eng., 21, 473 (1981).

Kojima, H., Petrotech, 4, 219 (1981).

Kojima, H., and K. Asano, J. Chem. Eng. Japan, 15, 321 (1982).

Kojima, H., A. Iguchi, and K. Asano, Can. J. Chem. Eng., 62, 346 (1984).

Kojima, H., H. Anjyo, and Y. Mochizuki, J. Chem. Eng. Japan, 19, 232 (1986).

Kolbel, H., W. Siemes, R. Maas, and K. Muller, Chem.-Ing.-Tech., 30, 400 (1958a).

Kolbel, H., E. Borchers, and K. Muller, Chem.-Ing.-Tech., 30, 729 (1958b).

Kolbel, H., E. Borchers, and J. Martins, Chem.-Ing.-Tech., 32, 84 (1960).

Kolbel, H., and M. Ralek, Rev. Sci. Eng., 21, 225 (1980).

Kumar, A., T.E. Degaleesan, G.S. Laddha, and H.E. Hoelscher, Can. J. Chem. Eng., 54, 503 (1976).

Kuo, J.C.W., F.P. Di Sanzo, W.E. Garwood, K.M. Gupte, T.M. Leib, and J. Smith, Final Report to U.S. Department of Energy, Contract No. DE-AC22-83PC60019 (1985).

Lamb, H., "Hydrodynamics," Dover, New York (1945).

Levins, D.M., and J.R. Glastonbury, Chem. Eng. Sci., 27, 537 (1972).

Lewis, D.A., R.W. Field, A.M. Xavier, and D. Edwards, Trans. Inst. Chem. Engrs. 60, 40 (1982).

Linek, V., and V. Vacek, Chem. Eng. Sci., 36, 1747 (1981).

Magiera, J., Chem. Eng. Commun., 30, 119 (1984).

Mangartz, K.H., and T. Pilhofer, Verfahreustechnik, 14, 40 (1980).

Maruyama, T., S. Yoshida, and T. Mizushina, J. Chem. Eng. Japan, 14, 352 (1981).

Mersmann, A., Ger. Chem. Eng., 1, 1 (1978).

Michael, R., and K.H. Reichert, Can. J. Chem. Eng., 59, 602 (1981).

Miyauchi, T., and C. N. Shyu, Kagaku Kogaku, 34, 958 (1970).

Morooka, S., T. Mizoguchi, T. Kago, Y. Kato, and N. Hidaka, J. Chem. Eng. Japan, 19, 507 (1986).

Muroyama, K., in Chemical Engineering Technical Report No. 9, "Design of Bubble Column and Slurry Bubble Column Reactors," Soc. Chem. Engrs. Japan (1985).

Muroyama, K., in Proceedings of ROC-Japan Joint Symposium on Three-Phase Reactors, p. 39, Taipei, Taiwan, Jan. 8-10 (1986).

Nakamura, M., K. Hioki, A. Takahashi, H. Tanahashi, and S. Watarai, Kagaku Kogaku Ronbunshu, 4, 473 (1978).

Nakanoh, M., and F. Yoshida, I&EC Process Des. & Dev., 19, 190 (1980).

Nakao, K., H. Takeuchi, H. Kataoka, H. Kaji, T. Otake, and T. Miyauchi, I&EC Process Des. & Dev., 22, 577 (1983).

Narayanan, S., V.K. Bhatia, and D.K. Guha, Can. J. Chem. Eng., 47, 360 (1969).

Nguyen-Tien, K., A.N. Patwari, A. Schumpe, and W.-D. Deckwer, AIChE J., 31, 194 (1985).

Nishikawa, M., H. Kato, and K. Hashimoto, I&EC Process Des. & Dev., 16, 133 (1977).

Nishiwaki, A., S. Tanaka, T. Fukuda, and Y. Kato, J. Chem. Eng. Japan, 16, 526 (1983).

O'Dowd, W., D.N. Smith, J.A. Ruether, and S.C. Saxena, AIChE J., 33, 1959 (1987).

Ozturk, S.S., and A. Schumpe, Chem. Eng. Sci., 42, 1781 (1987).

Parulekar, S.J., and Y.T. Shah, Chem. Eng. J., 20, 21 (1980).

Pavlov, V.P., Khim. Prom., No. 9, 698 (1965).

Quicker, G., and W.-D. Deckwer, Chem. Eng. Sci., 36, 1577 (1981).

Quicker, G., A. Schumpe, and W.-D. Deckwer, Chem. Eng. Sci., 39, 179 (1984).

Riquarts, H.P., Ger. Chem. Eng., 4, 18 (1981).

Roy, N.K., D.K. Guha, and M.N. Rao, Chem. Eng. Sci., 19, 215 (1964).

Sada, E., H. Kumazawa, C.-H. Lee, and N. Fujiwara, I&EC Process Des. & Dev., 24, 255 (1985).

Sada, E., H. Kumazawa, C.-H. Lee, and T. Iguchi, I&EC Process Des. & Dev., 25, 472 (1986a).

Sada, E., H. Kumazawa, and C.-H. Lee, AIChE J., 32, 853 (1986b).

Sada, E., H. Kumazawa, C.-H. Lee, and H. Narukawa, I&EC Res., 26, 112 (1987).

Sakai, T., and N. Ohi, Nippon Kagaku Kaishi, 421 (1977).

Sakata, M., and T. Miyauchi, Kagaku Kogaku Ronbunshu, 6, 428 (1980).

Sanger, P., and W.-D. Deckwer, Chem. Eng. J., 22, 179 (1981).

Sano, Y., N. Yamaguchi, and T. Adachi, J. Chem. Eng. Japan, 7, 255 (1974).

Satterfield, C.N., "Mass Transfer in Heterogeneous Catalysis," MIT Press (1970).

Schumpe, A., Ph.D. Dissertation, Universitat Hannover (1981).

Schumpe, A., and W.-D. Deckwer, I&EC Process Des. & Dev., 21, 706 (1982).

Schumpe, A., A.K. Saxena, and L.K. Fang, Chem. Eng. Sci., 42, 1787 (1987).

Sekizawa, T., and H. Kubota, J. Chem. Eng. Japan, 7, 441 (1974).

Sekizawa, T., H. Kubota, and W.C. Chung, J. Chem. Eng. Japan, 16, 327 (1983).

Sharma, M.M., and R.A. Mashelkar, Inst. Chem. Engrs. Symposium Series, No. 28, 10, Inst. Chem. Engrs., London (1968).

Smith, D.N., W. Fuchs, R.J. Lynn, and D.H. Smith, ACS Symposium Series, No. 237, 125 (1984).

Smith, D.N., and J.A. Ruether, Chem. Eng. Sci., 40, 741 (1985).

Smith, D.N., J.A. Ruether, Y.T. Shah, and M.N. Badgujar, AIChE J., 32, 426 (1986).

Suganuma, T., and T. Yamanishi, Kagaku Kogaku, 30, 1136 (1966).

Tang, W.-T., and L.-S. Fan, paper presented at the AIChE Annual Meeting, New York (1987); AIChE J. (in press, 1988).

Tareef, B.M., Colloid J. U.S.S.R., 6, 545 (1940); cited in Chem. Eng., 64(3), 273 (1957).

Towell, G.D., and G.H. Ackerman, in Proceedings of the 5th European (2nd International) Symposium on Chemical Reaction Engineering, B3-1, Elsevier, Amsterdam (1972).

Turi, E., and K.M. Ng, Chem. Eng. Commun., 46, 323 (1986).

Ueyama, K., and T. Miyauchi, Kagaku Kogaku Ronbunshu, 2, 430 (1976).

Ueyama, K., and T. Miyauchi, Kagaku Kogaku Ronbunshu, 3, 19 (1977).

Ueyama, K., and T. Miyauchi, AIChE J., 25, 258 (1979).

Ueyama, K., S. Morooka, K. Koide, H. Kaji, and T. Miyauchi, I&EC Process Des. & Dev., 19, 592 (1980).

Ueyama, K., N. Tamura, and S. Furusaki, in "Fluidization '85 Science and Technology," ed. by M. Kwauk, D. Kunii, J. Zheng, and M. Hasatani, p. 274, Elsevier, Amsterdam (1985).

Wendt, R., A. Steiff, and P.-M. Weinspach, Ger. Chem. Eng., 7, 267 (1984).

Whalley, P.B., and J.F. Davidson, Inst. Chem. Engrs. Symposium Series, No. 38, J5, "Multi-Phase Flow Systems," Inst. Chem. Engrs., London (1974).

Yamagoshi, T., B.S. Thesis, University of Tokyo, Japan (1969).

Yamashita, F., and H. Inoue, J. Chem. Eng. Japan, 8, 334 (1975).

Yamashita, F., Y. Mori, and S. Fujita, J. Chem. Eng. Japan, 12, 5 (1979).

Yasunishi, A., M. Fukuma, and K. Muroyama, Kagaku Kogaku Ronbunshu, 12, 420 (1986a).

Yasunishi, A., M. Fukuma, and K. Muroyama, J. Chem. Eng. Japan, 19, 444 (1986b).

Ying, D.H., R. Sivasubramanian, and E.N. Givens, Report to U.S. Department of Energy, Contract No. DE-AC01-79ET14801, Jan. (1980a).

Ying, D.H., E.N. Givens, and R.F. Weimer, I&EC Process Des. & Dev., 19, 635 (1980b).

Yoshitome, H., Y. Mannami, K. Mukai, N. Yashikoshi, and T. Kanazawa, Kagaku Kogaku, 29, 19 (1965).

Yoshitome, H., Dr. of Eng. Dissertation, Tokyo Institute of Technology, Japan (1967).

Zaidi, A., Y. Louisi, M. Ralek, and W.-D. Deckwer, Ger. Chem. Eng., 2, 94 (1979).

Zuber, N., and J.A. Findley, Trans. ASME J. Heat Transfer, 87, 453 (1965).

Chapter 5*

Countercurrent (and Liquid-Batch) Fluidized Bed Systems (Modes E-II-a-1, E-II-a-2, E-II-b, and E-III-b; Figure 1.3)

This chapter deals with countercurrent gas-liquid-solid fluidized bed systems in which low density particles are fluidized by a downward liquid flow and an upward gas flow. These systems were identified in Fig. 1.3 as Modes E-II-a-1 and E-II-a-2 when liquid is the continuous phase and as Mode E-II-b when gas is the continuous phase. Also considered in this chapter is a liquid-batch fluidized bed system with gas as the continuous phase (Mode E-III-b).

In Mode E-II-a-1, known as inverse gas-liquid-solid fluidization (Fan et al., 1982a), the liquid density exceeds the particle density and the bed expands downward due to drag induced by the downward liquid flow. For this mode, only limited information is available concerning bed hydrodynamics. In inverse gas-liquid-solid fluidization with the bed bounded by upper and lower retaining grids, semifluidization may result depending on the gas and liquid flow rates (Fan et al., 1982a,b; Chern et al., 1983).

Mode E-II-b has been denoted by a variety of names. Kielback (1959) was the first to demonstrate this mode for industrial purposes and gave the contactor the name "floating bed scrubber" (also see Chem. Eng., 1959). In Western literature this mode has also been called "turbulent bed contactor" (TBC) (Douglas, 1964), "mobile bed contactor" (Tichy et al., 1972; Tichy and Douglas, 1972), "fluidized packing contactor" (O'Neill et al., 1972), and "three-phase fluidization" (Strumillo and Kudra, 1977). In Russian literature, the name "tower with fluidized packing" is used. Countercurrent gas-continuous three-phase fluidization systems have been widely used in industry to absorb gases into a liquid phase, and the term "turbulent contact absorber" (TCA) has become the prevalent English designation for these types of

*: by L.-S. Fan and K. Muroyama

operational systems. Considering other important applications of these systems (see Chapter 1), a generic term, i.e., turbulent bed contactor (TBC) is adopted throughout the text.

In Mode E-II-b TBC operation the liquid density usually significantly exceeds the particle density and the continuous gas phase supports the expanded bed. It is, however, important to note that, while maintaining constant gas and liquid velocities in a TBC, by simply increasing the particle density it is possible to change the operation mode from Mode E-II-b to Mode E-II-a-2 (see Section 5.1.1). Thus, TBC operation can be conducted under both Modes E-II-b and E-II-a-2, although this operation is commonly associated only with Mode E-II-b. When the TBC is bounded by an upper retaining grid, under sufficient gas flow, particles may form a packed bed at the upper retaining grid, thus forming a semifluidized bed.

In the following, hydrodynamics, mixing, and mass and heat transfer in a TBC are given. Furthermore, hydrodynamics and mass transfer in an inverse fluidized bed are described. Note that the behavior of Mode E-III-b operation is illustrated in the context of TBC operation.

5.1 TURBULENT BED CONTACTOR (TBC)

Published works on the fundamental behavior of TBCs have examined bed hydrodynamics, and numerous correlations are now available to predict pressure drop, minimum fluidization conditions, liquid holdup, gas holdup, and bed expansion. Little, however, has been reported concerning back mixing of the liquid and solid phases. Since the TBC can be utilized as an efficient gas absorber, mass and heat transfer phenomena have also been investigated.

5.1.1 Hydrodynamics and Mixing in a TBC

Three hydrodynamic states are observed in a TBC for a constant irrigation (liquid feed) rate: fixed bed, fluidized bed, and "true" flooding bed. At low gas velocities the bed remains in a stationary or fixed bed state and the pressure drop increases steeply with increasing gas velocity. As the gas flow rate exceeds the minimum fluidization velocity, the particles become fluidized; however, the transition from the fixed bed to the fluidized bed proceeds gradually as observed by Tichy et al. (1972) and Kito et al.

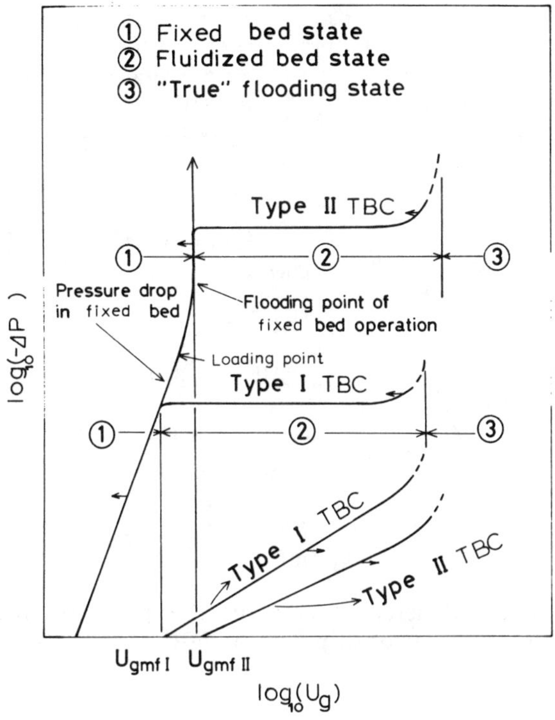

Figure 5.1 Pressure drop and bed height as functions of superficial gas velocity at a given liquid velocity for a TBC (after O'Neill et al., 1972).

(1976d). When the entire bed becomes fluidized with very vigorous particle movement, the pressure drop across the bed reaches a nearly constant value with increasing gas velocity (Tichy et al., 1972, Rama et al., 1983). The bed height progressively increases with gas velocity. At high gas velocities, a "true" flooding bed state can occur when liquid flow in the bed is impeded by the upward flow of gas. In such a state the pressure drop sharply increases with increasing gas velocity, and a significant amount of liquid is entrained in the exhaust gas stream as observed by Douglas et al. (1963) and Barile and Meyer (1971). The typical variation in pressure drop of TBC columns is shown qualitatively in Fig. 5.1 (O'Neill et al., 1972; Vunjak-Novakovic et al., 1980).

O'Neill et al. (1972) observed two different types of minimum fluidization behavior in a TBC depending on the density of the particle. For beds with very low density particles, fluidization occurs at a gas velocity lower than the flooding velocity for an equivalent

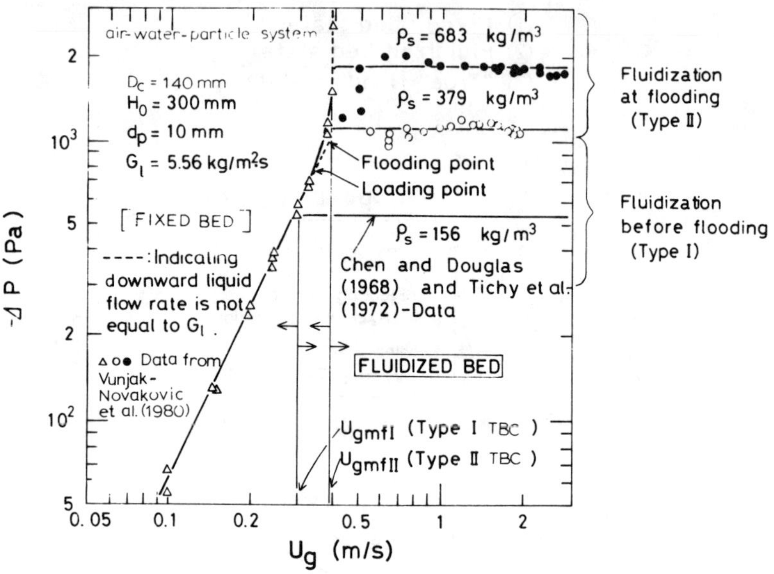

Figure 5.2 Particle density effects on the pressure drop versus gas velocity relationship for a TBC (from Vunjak-Novakovic et al., 1987a).

countercurrent fixed bed. For beds with relatively high density particles, fluidization always occurs at the flooding point (the point with a sharp increase in pressure drop between states 1 and 2 in Fig 5.1). In any case, once fluidization occurs, the liquid can flow downward with the aid of vigorous particle movement. The first class of TBC is called Type I TBC - fluidization without incipient flooding - and the latter is called Type II TBC - fluidization due to incipient flooding[1]. The behavior of Types I and II TBCs strongly depends on the particle size and density and gas flow rate.

Quantitative pressure drop behavior with varying flow rate, based on the data of Chen and Douglas (1968), Tichy et al. (1972) and Vunjak-Novakovic et al. (1980), is shown in Fig 5.2. These data were obtained using particles with densities of 156, 379, and 683 kg/m^3. As shown, fluidization of the lowest density particles takes place without

1: In this context, the term flooding refers to points between states 1 and 2 in Fig. 5.1 for Type II TBCs and the term "true" flooding refers to points between states 2 and 3. In addition, depending on the particle size and density, liquid flow rate, and design of the distributor and retaining grid, the flooding point may, in some conditions, represent a transition from a continuous gas phase to a continuous liquid phase (i.e., a transition from Mode E-II-b to Mode E-II-a-2). Little has been reported regarding such a transition.

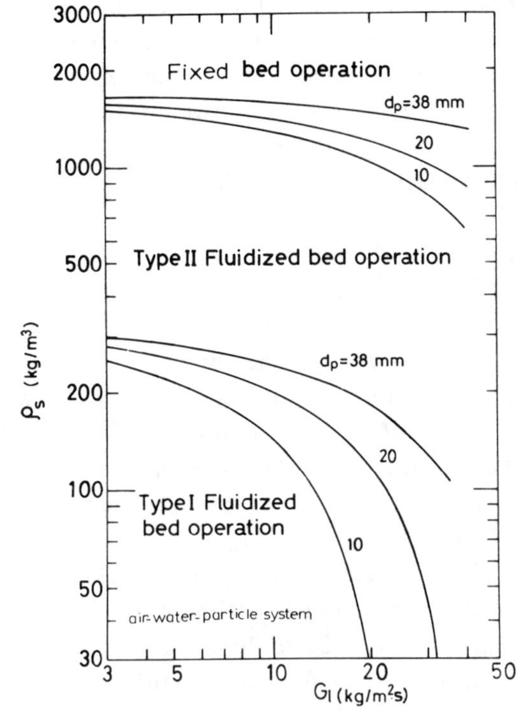

Figure 5.3 Operating regimes for Types I and II TBCs (from Vunjak-Novakovic et al., 1987a).

incipient flooding and hence, is a Type I TBC operation; fluidization of medium density particles takes place with incipient flooding and hence, is a Type II TBC operation. The dashed line in Fig 5.2 indicates a region of decreasing outlet liquid flow while accommodating the flooding condition (O'Neill et al., 1972). The loading point represents a break point in the slope on a pressure drop versus gas velocity plot prior to the flooding point. A more precise account of Type I and II TBC operations, relative to operating variables including particle size and density and liquid flow rate and operation of a packed bed, is given in Fig. 5.3 for air-water systems (Vunjak-Novakovic et al., 1987a). Typical particle sizes are considered to be 10, 20, and 38 mm in diameter. As shown, the particle density has a significant effect on the demarcation between these operations. For a given particle size, the required particle density for demarcation of operations decreases with increasing liquid flow rate; furthermore, it decreases with decreasing particle size for a given liquid flow rate.

Uchida et al. (1977) inferred two possible reasons to account for the true flooding phenomena in TBCs. The first possibility is that, for a TBC with an upper retaining grid, the expanded bed height is restricted to the distance between the two grids so that the particles reaching the upper retaining grid form a packed bed. This condition is, essentially, the formation of a semifluidized bed in the TBC. Once the packed bed is formed at the upper retaining grid virtually no liquid is allowed to flow down through the packed bed. The second possibility is that flooding takes place when the gas velocity exceeds the terminal velocity of the wetted particle. Hence, the wetted particles are transported up by the upward flow of the gas and accumulate under the upper retaining grid. Uchida et al. (1980) showed that the onset of true flooding in Type I TBCs can be estimated from the flooding chart for a countercurrent fixed bed. On the other hand, Gel'perin et al. (1968) claimed that the flooding condition in a TBC is also dependent upon the open area and geometry of the lower supporting grid that restricts through flow. They proposed an empirical correlation to estimate the flooding condition of a TBC with grid open areas of 25 to 70%.

The extent of bed homogeneity of fluidization is considerably different between shallow beds and deep beds. Uniform fluidization occurs in beds with static bed heights lower than the column diameter, while slugging commonly occurs in deep beds (Barile and Meyer, 1971). A conventional flow regime map for TBCs using 13 mm and 19 mm packing presented by Barile et al. (1974) showed that the operable range narrows with increasing static bed height.

The hydrodynamic characteristics of a TBC system strongly depend on the open area of the lower supporting grid (Gel'perin et al., 1968; Kito et al., 1976a-d, 1978). Achieving a smooth drainage of liquid from the fluidized bed depends on the percentage of open area in the lower retaining grid. The downflow of liquid by gravity must overcome the upward frictional drag of the gas flow. The design of the gas distributor (bottom grid) also affects the behavior of fluidization. If the open area of the gas distributor is not sufficiently large, flooding initiates at the distributor, thus significantly reducing the flexibility in gas and liquid flow rates. Many studies carried out by Russian investigators refer to flooding as the one that initiates at the distributor (open areas of 30 to 60%) and caution should be exercised when interpreting such data (Gel'perin et al., 1966; Blyakher et al., 1967; Aksel'rod and Yakovenko, 1969; Balabekov et al., 1969a,b). For industrial purposes the open area in the lower retaining grid is usually greater than 70%. When a grid with relatively small open area is used, a crosscurrent scheme for liquid flow is also possible in TBC operation. Mrowiec and Laszuk (1974) sprayed liquid through side holes in the column wall, located just above the retaining grid, and withdrew it from the bottom of the column.

Wall effects may also be significant for units having small column diameter to particle diameter ratios. Tichy et al. (1972) observed that when the column diameter to particle diameter ratio was as low as 11, a relatively static monolayer of spheres formed around the wall. This phenomenon reduces particle mobility, resulting in gas channeling, and occurs more significantly in deep beds. Tichy and Douglas (1973) showed that the pseudo-static monolayer of particles formed more easily for a particle diameter of 13 mm than for 19 mm. At high liquid flow rates more particles are held to the wall via action of liquid surface tension and maldistribution of gas flow. Recently, Rama et al. (1983) also observed the formation of a static monolayer of particles around the wall in TBC columns containing spherical and irregular shaped particles with equivalent diameters from 23.7 to 38 mm and with densities from 53 to 183 kg/m^3.

5.1.1.1 Pressure Drop

The pressure drop through the bed, the most important parameter in TBC operation, determines the energy consumption required during operation and uniquely characterizes the hydrodynamic behavior of the system. A steady-state macroscopic force balance on a TBC column yields an expression for the pressure drop in a TBC (Wozniak, 1977)

$$(-\Delta P) = (\rho_s \epsilon_s + \rho_\ell \epsilon_\ell + \rho_g \epsilon_g) g H + \Sigma(-\Delta P_f) \qquad (5.1)$$

where $(-\Delta P)$ denotes the total pressure drop through the column, H is the fluidized bed height, and $\Sigma(-\Delta P_f)$ is the summation of all pressure drops excluding the hydrostatic head terms of each individual phase. $\Sigma(-\Delta P_f)$ for a typical TBC can be represented by

$$\Sigma(-\Delta P_f) = (-\Delta P_g) + (-\Delta P_w) \qquad (5.2)$$

where $(-\Delta P_g)$ and $(-\Delta P_w)$ are the pressure drop due to flow through the grids and that due to the drag force exerted on the wall, respectively.

The impact of these parameters becomes negligible for large open grid areas and large column diameters.

$(-\Delta P_g)$ and $(-\Delta P_w)$ are not easily determined from basic principles. Several empirical correlations for $(-\Delta P_g)$ and $(-\Delta P_w)$ have been developed. A number of authors (e.g., Blyakher et al. (1967) and Gel'perin et al. (1968)) have observed that $(-\Delta P_g)$ is strongly affected by grid geometry. Correlations for $(-\Delta P_g)$ were developed for lower retaining grids with small open areas by Blyakher et al. (1967), Gel'perin et al. (1968), Levsh et al. (1968b), and Mayak and Matrozov (1969). $(-\Delta P_g)$ was accounted for in pressure drop correlations by several authors (Gel'perin et al., 1968; Kito et al., 1976e; Levsh et al., 1968b,c). Tichy et al. (1972), however, found that $(-\Delta P_g)$ was negligible for a grid with an opening of 83% and Kito et al. (1976e) stated that $(-\Delta P_g)$ can be neglected when a stagnant liquid layer is not present on the lower retaining grid. Wozniak (1977) and Uysal (1978) concluded that for large diameter columns the contribution of $(-\Delta P_w)$ is very small and can be neglected. By neglecting $(-\Delta P_g)$, $(-\Delta P_w)$, and the hydrostatic head of the gas, Eqn. (5.1) reduces to

$$(-\Delta P) = (\rho_s \epsilon_s + \rho_\ell \epsilon_\ell) gH \tag{5.3}$$

or in terms of static bed properties

$$(-\Delta P) = (\rho_s \epsilon_{s,st} + \rho_\ell \epsilon_{\ell,st})\, gH_0 \tag{5.4}$$

where

$$\epsilon_{\ell,st} = (H/H_0)\epsilon_\ell \tag{5.5a}$$

$$\epsilon_{s,st} = (H/H_0)\epsilon_s \tag{5.5b}$$

The pressure drop due to the weight of dry particles is always constant, while that due to liquid holdup depends strongly on the operating conditions and system geometry. Consequently, the pressure drop in a TBC mainly depends on the liquid holdup and can be predicted based on correlations of liquid holdup. The pressure drop through a bed with a grid open area greater than 70% is almost independent of gas velocity as shown in Fig. 5.4 (Rama et al., 1983). Tichy et al. (1972) found the pressure drop in this state to be nearly independent of particle size. Note, as indicated by Eqn. (5.4), the pressure drop increases with increasing liquid holdup. Static bed

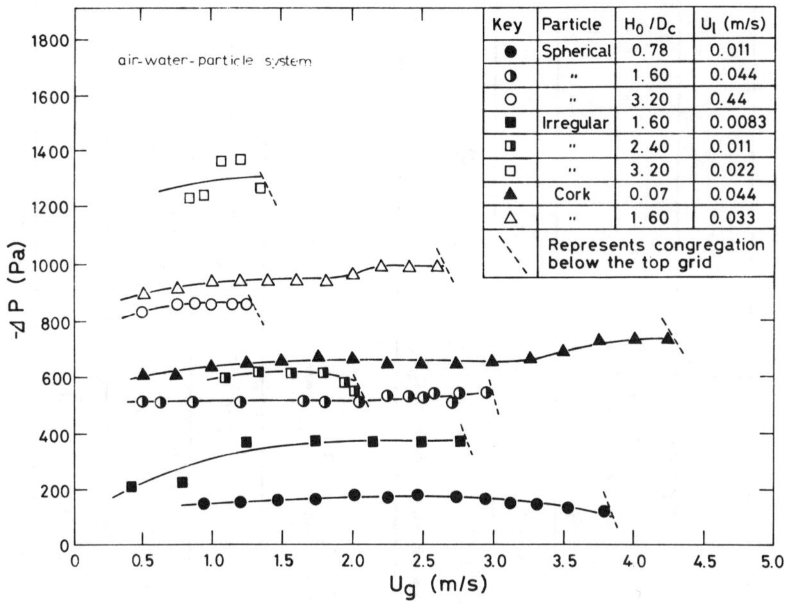

Figure 5.4 Typical plots of ΔP versus U_g for a TBC with a large grid open area (from Rama et al., 1983).

height also affects the pressure drop (Wozniak, 1977; Kito et al., 1976e) implying that liquid holdup depends on the static bed height, as will be shown later. For small grid open areas the pressure drop sharply increases with gas velocity (e.g., Blyakher et al., 1967; and Balabekov et al., 1969b) reflecting the behavior of the liquid holdup in the column.

Experimental data on pressure drop in TBCs have been reported by several investigators (Douglas et al., 1963; Gel'perin et al., 1966; Balabekov et al., 1969b; Miconnet et al., 1982). Some investigators have attempted to correlate the pressure drop through a TBC using macroscopic models other than the simple hydrostatic head model described above (Blyakher et al., 1967; Gel'perin et al., 1968; Levsh et al., 1968a; Tichy et al., 1972; Wozniak, 1977). Recent studies support the hydrostatic head model in terms of the pressure drop in the TBC (Barile and Meyer, 1971; Kito et al., 1976e; Vunjak-Novakovic et al., 1980; Rama et al., 1983). Table 5.1 summarizes the correlations for pressure drop in TBC columns. Obviously liquid holdup correlations are required to estimate the pressure drop from Eqn. (5.3) or (5.4) and the accuracy of the pressure drop estimation mainly depends on the precision in estimating the liquid holdup.

Table 5.1 Correlation Equations for Pressure Drop in a Single Stage TBC

Investigators	f [-]	ρ_s [kg/m^3]	D_c/d_p [-]	Type of** Fluidization	Correlation Equations (in SI Units)
Levsh et al. (1968a)*	———	1083	* ———	Type II	$(-\Delta P) = (1-\varepsilon_0)\rho_s g H_0 + 14.0 g U_\ell^{0.55} U_g (14.29 H_0)^m + (-\Delta P_g)$ for $U_\ell < 0.008$ m/s , $m = 0.75 \sim 0.8$ for $0.008 < U_\ell < 0.017$ m/s , $m = 1$
Barile and Meyer (1971)	0.82	109, 160	7.5, 15	Type I	$\frac{(-\Delta P)}{g} = (1-\varepsilon_0)\rho_s H_0 + \varepsilon_{\ell mf}\rho_\ell H_0$
Tichy et al. (1972)	0.78	155 ± 39%	11.4, 7.5	Types I and II	$\log_{10}\left(\frac{2\rho\frac{(-\Delta P)}{g}}{G_g^2}\frac{D_c}{H_0}\frac{1}{1+CG_\ell}\right) = 4.003 - 2.240\,G_g + 0.840\,G_g^2 - 0.127\,G_g^3$ C is a constant which depends on H_0/D_c.
Kito et al. (1976e)	0.705, 0.712, 0.84	170 ~ 1250	3.6 ~ 10.3	Type II	$\frac{(-\Delta P)}{g} = (1-\varepsilon_0)\rho_s H_0 + \varepsilon_{\ell,st}\rho_\ell H_0$ Correlation equations for $\varepsilon_{\ell,st}$ are given in Table 5.3.
Vunjak-Novakovic et al. (1980)	0.36, 0.52, 0.78	200, 400, 700	3.68 ~ 29	Types I and II	
Uchida et al. (1977)	———	———	———	Type I	$(-\Delta P) = (1-\varepsilon_0)\rho_s g H_0 + 9.38\times10^8 \mu_\ell^{2.3} f^{-0.42} \left(\frac{d}{D_c}\right)^{-0.84} d_p^{-0.84} \rho_s^{0.18} H_0 U_\ell + 788\, U_g^{2.91} U_\ell^{1.27}$
Wozniak (1977)	0.6	266	10.2	Types I and II	$\frac{(-\Delta P)}{\rho_g U_g^2} = 4.672\times10^5 \left(\frac{H_0}{D_c}\right)^{0.4515} \left(\frac{d_p U_g \rho_g}{\mu_g}\right)^{-1.798} \left(\frac{d_p U_\ell \rho_\ell}{\mu_\ell}\right)^{0.8261} + \frac{(1-\varepsilon_0)\rho_s g H_0}{\rho_g U_g^2}$
Rama et al. (1983)	0.7	53, 112, 183	3.95, 4.80, 6.33	Type I	$\frac{(-\Delta P)}{g} = (1-\varepsilon_0)(\rho_s-\rho_g) H_0 + \varepsilon_{\ell,st}\rho_\ell H_0$
Vunjak-Novakovic et al. (1987a,b)	0.36, 0.52, 0.78	182 ~ 980	3.68 ~ 29	Types I and II	$\frac{(-\Delta P)}{g} = (1-\varepsilon_0)\rho_s H_0 + (\varepsilon_{\ell,st}-0.02)\rho_\ell H_0$ Correlation equations for $\varepsilon_{\ell,st}$ are given in Table 5.3

* A special type of packing (plastic ring) was used.
** From Figure 5.3.

5.1.1.2 Minimum Fluidization Velocity

The transition between the fixed and fluidized bed states is rather gradual and somewhat ill-defined. Consequently, several definitions can be found in the literature for the minimum fluidization velocity in a TBC (U_{gmf}). Chen and Douglas (1968) and Tichy and Douglas (1972, 1973) defined the minimum fluidization velocity as the gas velocity at which bed expansion begins. This point is determined by a break point on a plot of the dimensionless bed height (H/H_0: the ratio of the expanded bed height to the static bed height) versus gas velocity. Kito et al. (1976d,e) expressed the minimum fluidization velocity based on the gas velocity at which the rapid increase in pressure drop through the fixed bed region levels off and approaches a constant value regardless of gas flow rates. Kito et al. (1976d,e) observed that the bed progressively expanded prior to reaching a fully fluidized state; therefore, the minimum fluidization velocity based on the pressure drop variation is always greater than that based on the initiation of bed expansion. Note that the minimum fluidization velocity denoted in Fig. 5.2 is obtained based on the initial bed expansion.

The minimum fluidization velocity generally decreases with increasing liquid flow rate and liquid viscosity and with decreasing particle diameter (Chen and Douglas, 1968; Kito et al., 1976d; Vunjak-Novakovic et al., 1980; Kuroda and Tabei, 1979), but is independent of static bed height (Balabekov et al., 1969a; Kito et al., 1976d). In addition, the minimum fluidization velocity increases with increasing particle density for both TBC types when calculated based on the pressure drop variation method (Kito et al., 1976d); whereas, when calculated based on initial bed expansion, it is independent of particle density for Type II TBCs. Gel'perin et al. (1968), Balabekov et al. (1969a), and Kito et al. (1976d) showed that the minimum fluidization velocity increases with the open area of the supporting grid, but, as shown by Kito et al. (1976d), the minimum fluidization velocity becomes independent of grid open area if the parameter fd/D is greater than 0.05.

A number of correlations for predicting U_{gmf} are available in the literature. The geometry of the experimental system, especially the open area of the supporting grid, as well as the operating conditions, such as liquid velocity and physical properties of liquid and gas, must be taken into account to develop a reliable correlation. Correlations for estimating U_{gmf} are summarized in Table 5.2. Most of the

Table 5.2 Correlation Equations for Minimum Fluidization Velocity in TBC Columns

Investigators	f [-]	ρ_s [kg/m^3]	D_c/d_p [-]	Correlation Equations (in SI Units)
Chen and Douglas (1968)	See Table 5.3			$G_{gmf} = 120.4\, d_p^{1.15}\, 10^{-0.03812\, G_\ell}$
Gel'perin et al. (1968)	See Table 5.1			$\frac{U_{gmf}^2}{g d_p} = C\, f^{1.54} \exp[-12.6\, (\frac{U_\ell}{U_{gmf}})^{0.25}]$ for $d = 3, 4, 6$ mm, $C = 4.6 \times 10^3$; for $d = 2$ mm, $C = 2.8 \times 10^3$
Balabekov et al. (1971)	0.3 ~ 0.6	356 ~ 1036	29.2 ~ 7.95	$Re_{gmf} = (\frac{d_p U_{gmf} \rho_g}{\mu_g}) = 142\, (f \frac{\rho_s}{\rho_\ell})^{0.33} (\frac{0.005}{d_p})^{-2.2}$ $\cdot \exp[-7.3 \times 10^{-6} (\frac{\rho_s}{\rho_\ell})^{-0.33} (\frac{0.005}{d_p})^{1.14} (\frac{2}{3} \cdot \frac{d_p G_\ell}{\mu_\ell (1-\varepsilon_0)})]$
Khanna (1971) (quoted from Uysal, 1978)	0.7	155 ~ 458	3.7 ~ 11	$G_{gmf} = 526.47\, d_p^{1.5}\, 10^{-0.0317\, G_\ell}$
Tichy and Douglas (1972)	0.7	155, 234, 458	3.7 ~ 11	$G_{gmf} = 0.36355 + 57.90\, d_p - 1.848\, G_\ell^{0.6}\, d_p^{0.5}$
Leung et al. (1972)	——	——	——	$G_{gmf} = [0.5002 (d_p \rho_\ell \rho_g)^{1/4} - G_\ell^{1/2} (\rho_g/\rho_\ell)^{1/4}]^2$ (T-5.2.1)
Strumillo et al. (1974)	0.65	470 ~ 1050	8.5, 11.33, 17.0	$U_{gmf} = (13.3\, D_c + 2.24) \exp(-\frac{3.6\, U_\ell}{D_c^{0.014}})$

Kito et al. (1976d)	0.04 ~ 0.84	170 ~ 1160	3.5 ~ 8.62	$\frac{U_{gmf}}{U_{gmf0}} = 0.5983\ (f\frac{d}{D})^{0.46}\ U_{\ell}^{-0.36}$ for $f\frac{d}{D} \leq 0.05$ $\frac{U_{gmf}}{U_{gmf0}} = 0.1486 U_{\ell}^{-0.36}$ for $f\frac{d}{D} > 0.05$ U_{gmf0} can be estimated by using the Ergun equation (Kunii and Levenspiel, 1969).
Uysal (1978)	0.87	157	7.6 ~ 15.3	$G_{gmf} = 10.86\ d_p^{0.488}\ 10^{-0.01985\ G_{\ell}}$
Kuroda and Tabei (1979)	0.04 ~ 0.84	170 ~ 760	3.51 ~ 8.62	$C_{fi}(\frac{\rho_g U_{gmf}^2}{d_p})(\frac{1}{\varepsilon_{g\,mf}^3}) = (\rho_{ws} - \rho_g)\ g$ $\rho_{ws} = (\frac{\varepsilon_{\ell,st}}{1 - \varepsilon_0})\ \rho_{\ell} + \rho_s$ $C_{fi} = C_{fs}(1 + 800\ \varepsilon_{\ell\,mf}^3)$ $C_{fs} = 1.75 + 150\ (1 - \varepsilon_g)/Re_g$ where C_{fi} = apparent friction coefficient C_{fs} = friction coefficient in a gas-liquid packed bed
Vunjak-Novakovic et al. (1987a,b)	See Table 5.1			for Type I TBC: $U_{gmf} = 11.3\ d_p^{1.2}\{(1-\varepsilon_0)(\rho_s-\rho_g) + 4.903\times10^{-5}\rho_{\ell}\ d_p^{-0.568} G_{\ell}^{0.719} 10^{-0.04788\ G_{\ell}}\}^{0.5}$ (T-5.2.2a) for Type II TBC: $\log_{10}(\frac{U_{gmf}^2}{g}\ \frac{a_p}{\varepsilon^3}\ \frac{\rho_g}{\rho_{\ell}}) = 0.247 - 1.615\ (\frac{G_{\ell}}{G_{gmf}})^{0.25}\ (\frac{\rho_g}{\rho_{\ell}})^{0.125}$ (T-5.2.2b)

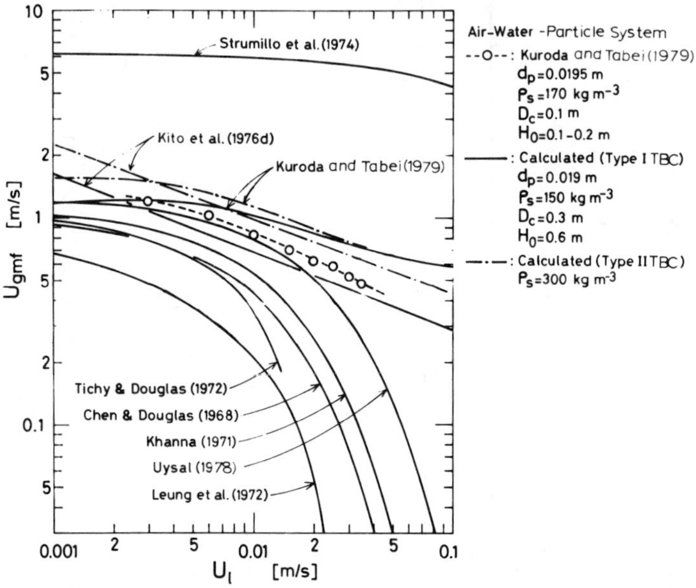

Figure 5.5 Comparisons of U_{gmf} predictions based on various correlations given in Table 5.2 for a TBC.

equations listed in Table 5.2 are purely empirical and considerable disagreements exist in the predictions of these equations. Clearly, these disagreements are predominantly attributed to differences in the method of determining the minimum fluidization velocity.

Figure 5.5 compares U_{gmf} estimated from the correlations given in Table 5.2 as a function of liquid velocity for specific particle properties and gas velocity. The correlations proposed by Chen and Douglas (1968), Tichy and Douglas (1972), Khanna (1971), Uysal (1978), and Leung et al. (1972) underestimate U_{gmf} for liquid velocities greater than about 0.02 m/s. Note that the correlations of Kito et al. (1976d) and Kuroda and Tabei (1979) yield different estimations for Type I and Type II TBC systems. According to these correlations, U_{gmf} for a Type II TBC utilizing heavier particles is generally greater than that for a Type I TBC at the same gas and liquid velocities.

Models and Correlations

Models proposed to date overwhelmingly resemble those for two-phase fluidization which are established on the equivalent relationship between the dynamic pressure drop across the fixed bed and the effective weight of the solid particles in the fluidized condition.

Thus, in essence, these models are consistent with the method of measurement of the minimum fluidization velocity based on pressure drop variations. A more precise distinction between model and method of measurement is presented in the recent work of Vunjak-Novakovic et al. (1987b). They verified experimentally that when the minimum fluidization velocity is defined based on initial bed expansion, the following holdup and pressure drop relationship must be satisfied under minimum fluidization conditions:

$$\frac{(-\Delta P)}{g} = (1 - \epsilon_0)\rho_s H_0 + \epsilon_{\ell o,st}\rho_\ell H_0 \tag{5.6}$$

Note that the liquid holdup in Eqn. (5.6) is based on the operational liquid holdup (see Section 5.1.1.3) which is empirically related to the liquid holdup ($\epsilon_{\ell,st}$) by

$$\epsilon_{\ell o,st} = \epsilon_{\ell,st} - 0.02 \tag{5.7}$$

This finding suggests that the minimum fluidization velocity based on pressure drop variations exceeds that based on initial bed expansion by a magnitude corresponding to the (-ΔP)/g value of $0.02(\rho_\ell H_0)$. Some pertinent work on modeling is given below.

Chen and Douglas (1968) suggested that at the point of incipient fluidization the pressure drop in a non-flooded fixed bed, which could usually be expressed by a Leva type equation (Leva, 1959), equals the equivalent weight of irrigated particles. Thus, for a Type I TBC, the incipient fluidization gas velocity can be obtained by equating the dynamic pressure drop through the fixed bed, $(-dP_d/dz)$ (also see Eqn. (2.5) for fluidization with liquid as the continuous phase), with the effective weight of wetted particles in fluidized conditions as follows:

$$\left(-\frac{dP_d}{dz}\right) = \left(-\frac{dP}{dz}\right) - \rho_g g = (1-\epsilon_{gmf})(\rho_{ws} - \rho_g)g \tag{5.8}$$

where ϵ_{gmf} is the gas holdup at minimum fluidization conditions, and ρ_{ws} is the density of the wetted particles.

Kuroda and Tabei (1979) applied the one-dimensional, two-phase flow model proposed by Wallis (1969) to obtain the dynamic pressure drop in terms of an apparent friction coefficient (C_{fi}) for both a countercurrent fixed bed and a TBC close to minimum fluidization conditions.

$$(-\frac{dP_d}{dz}) = C_{fi} \left(\frac{\rho_g U_g^2}{d_p}\right) \left(\frac{1-\epsilon_g}{\epsilon_g^3}\right) \tag{5.9}$$

where C_{fi} can be empirically correlated by

$$C_{fi} = C_{fs}(1 + 800\, \epsilon_{\ell mf}^3) \tag{5.10}$$

with $\epsilon_{\ell mf}$ expressed by

$$\epsilon_{\ell mf} = \epsilon_{\ell,st} \frac{H_0}{H_{mf}} = \epsilon_{\ell,st} \frac{1-\epsilon_{gmf}}{1+\epsilon_{\ell,st} - \epsilon_0} \tag{5.11}$$

In Eqn. (5.11), $\epsilon_{\ell,st}$ is the static liquid holdup and ϵ_0 is the porosity of the bed without gas-liquid flow. The friction coefficient in the gas-solid packed bed, C_{fs} can be estimated by the Ergun equation, given by

$$C_{fs} = 1.75 + 150 \frac{(1-\epsilon_g)}{\left(\frac{d_p U_g \rho_g}{\mu_g}\right)} \tag{5.12}$$

which reduces to 1.75 at large $(\frac{d_p U_g \rho_g}{\mu_g})$.

Combining Eqns. (5.8) and (5.9) at incipient fluidization conditions gives

$$U_{gmf} = \{ \frac{d_p}{C_{fi}\, \rho_g} \epsilon_{gmf}^3 (\rho_{ws} - \rho_g) g \}^{1/2} \tag{5.13}$$

Assuming that the liquid covers the entire surface of the particles with a uniform thickness, ρ_{ws} can be evaluated by

$$\rho_{ws} = \frac{\epsilon_{\ell,st}\, \rho_\ell + \rho_s (1-\epsilon_0)}{(1-\epsilon_0) + \epsilon_{\ell,st}} \tag{5.14}$$

Kuroda and Tabei (1979) recommended Eqn. (T-5.3.4a) (see Section 5.1.1.3) to estimate $\epsilon_{\ell,st}$ while ϵ_0 and ϵ_{gmf} are experimentally given as 0.48 ~ 0.50 (depending on the size of the particle) and 0.43, respectively.

For a Type II TBC the incipient fluidization gas velocity can be determined from the flooding chart for countercurrent fixed bed operation as proposed by Lobo et al. (1945). Leung et al. (1972) and

O'Neill et al. (1974) showed that any suitable correlation for predicting flooding flow rates can be employed to predict the minimum fluidization velocity, for example, the correlations of Lobo et al. (1945) and Wallis (1969). Leung et al. (1972) obtained Eqn. (T-5.2.1) as a correlation for the minimum fluidization velocity based on the Wallis (1969) correlation which predicts the flooding condition in the fixed bed. Vunjak-Novakovic et al. (1987a,b) proposed the Eqns. (T-5.2.2a) and (T-5.2.2b) as correlations for the minimum fluidization velocity of Type I and Type II TBCs, respectively.

5.1.1.3 Liquid Holdup

The liquid holdup consists of an operational liquid holdup and a static liquid holdup. The operational liquid holdup can be measured by collecting the accumulated liquid upheld at the bottom of the bed after simultaneously shutting off the gas and liquid flows. The static holdup represents the liquid upheld by adhesive forces on the particles and can be measured by weighing the wetted and dry particles.

Chen and Douglas (1968), measuring the liquid holdup using a transient tracer response technique (Schiesser and Lapidus, 1961), found that the liquid holdup was independent of gas flow rate, increased with increasing liquid flow rate, and increased with decreasing packing size. Gel'perin et al. (1968), using grids with low open areas, distinguished two separate regions of liquid holdup: one near the grid and the other in the fluidized bed proper. They found that the liquid holdup in the fluidized bed region increased with both increasing liquid velocity and static bed height, but was independent of gas flow rate.

Figure 5.6 shows the variation in operational liquid holdup with the gas velocity for both TBC types (Vunjak-Novakovic et al., 1987a). As shown, for a Type I TBC (ρ_s = 156 kg/m^3), $\epsilon_{\ell o,st}$ remains almost constant for gas velocities comprising both fixed and fluidized beds; for a Type II TBC, $\epsilon_{\ell o,st}$ increases significantly at U_{gmf} with increasing gas velocities and the extent of the increase depends on particle density. Note the gas velocity in region A-B on the curve for a Type II TBC represents a partial fluidization region where particles are fluidized layer by layer as the gas velocity increases; this behavior reflects a varying $\epsilon_{\ell o,st}$ with increasing gas velocity. At point B the bed is fully fluidized, beyond which $\epsilon_{\ell o,st}$ becomes independent of gas velocity.

Table 5.3 Correlation Equations for Liquid and Gas Holdups in a TBC

Investigators	f [-]	ρ_s [kg/m^3]	D_c/d_p [-]	Correlation Equations (in SI Units)	
Chen and Douglas (1968)	0.46	156, 158, 170	8, 12, 24	$\varepsilon_\ell = 0.02 + 2.371 \times 10^{-3}\, G_\ell^{0.6}\, d_p^{-0.5}$	
Gel'perin et al. (1968)	0.35 ~ 0.70	160 ~ 900	1.7 ~ 61.7	$\varepsilon_{\ell,st} = 0.138\, U_\ell^{0.5}\, H_0^{-0.25}\, d_p^{-0.6}$ $\varepsilon_g = 0.93 \left(\frac{d_p U'_g \rho_g}{\mu_g}\right)^{0.4} \left(\frac{d_p^3(\rho_s-\rho_g)\rho_g g}{\mu_g^2}\right)^{-0.2}$ $U'_g = \frac{U_g}{1-\varepsilon_\ell} = \frac{U_g}{1-\varepsilon_{\ell,st}\frac{H_0}{H}} = \frac{U_g}{1-\varepsilon_{\ell,st}\frac{1-\varepsilon_g}{\varepsilon_{\ell,st}+(1-\varepsilon_0)}}$	(T-5.3.1)
Balabekov et al. (1971)	0.3 ~ 0.6	356 ~ 1036	8 ~ 29.2	$\varepsilon_{\ell,st} = c\left[\left(\frac{G_\ell}{G_g}\right)^2 \frac{\rho_g}{\rho_\ell}\right]^{k+1} (1-\varepsilon_o)\, \frac{\rho_s}{\rho_\ell}$ for partial fluidization: $c = 0.65$ $k = -0.80$ for developed fluidization: $c = \frac{0.695}{f^3} \left(\frac{0.005}{d_p}\right) \left(\frac{\rho_s}{\rho_\ell}\right)^{-0.74} \left(\frac{G_\ell}{10^2}\right)^2$ $k = -\frac{1.58}{f^{-0.16}} \left(\frac{0.005}{d_p}\right)^{-0.1}$	
Barile and Meyer (1971)	0.3 ~ 0.6	356 ~ 1036	7.95 ~ 29.1	$\varepsilon_{\ell,st} = 1160 \left(\frac{G_\ell^2}{d_p g \rho_\ell^2}\right)^{0.78} \left(\frac{d_p G_\ell}{\mu_\ell}\right)^{-0.51} \left(\frac{H_0}{d_p}\right)^{-0.36}$	
Kito et al. (1976a,b)*	0.0127 ~ 0.315	300 ~ 1100	1.89 ~ 9.09	$\frac{\varepsilon_g}{\left(\varepsilon_g(1-\varepsilon_g)^2\right)^{0.44}} = 0.5 \left(\frac{D_c U_g^2 \rho_\ell}{\sigma}\right)^{0.11} \left(\frac{U_g}{\sqrt{g D_c}}\right)^{0.22}$	(T-5.3.2)

Kito et al. (1976e)	0.71 ~ 0.84	170 ~ 1250	3.6 ~ 10.3	$\varepsilon_{\ell,st} = 0.06 + 3.018\times10^{-3}\left(f\frac{d}{D}\right)^{-0.84} d_p^{-0.84} \rho_s^{0.18} H_0^{-0.4} U_\ell$ $\varepsilon_g = 0.417\, U_g^{0.44}$ (T-5.3.3)
Gel'perin and Kruglyakov (1977)	0.37 ~ 0.6	430	11.1	$\varepsilon_{\ell,st} = \frac{9.73}{\rho_\ell} G_\ell^{0.31} H_0^{-0.78} d_p^{-0.07} (c+4.54\, H_0^{0.78})$ for partial fluidization: $c = 0$ for developed fluidization: $c = \frac{U_g - U_{gmf}}{f^{0.86}}$
Kito et al. (1978)	Same as those of Kito et al. (1976e)			$\varepsilon_{\ell,st} = 12.8 \left(\frac{H_0}{d_p}\right)^{-0.4} \left(f\frac{d}{D}\right)^{-0.58} \left(\frac{g d_p^3 \rho_\ell \rho_s}{\mu_\ell^2}\right)^{0.09} \left(\frac{U_\ell}{\sqrt{g d_p}}\right)^{1.66} \left(\frac{d_p U_\ell \rho_\ell}{\mu_\ell}\right)^{-0.34} \left(\frac{d_p U_\ell^2 \rho_\ell}{\sigma}\right)^{-0.34}$ (T-5.3.4a) $\varepsilon_g = 0.19 \left(\frac{d_p U_g^2 \rho_\ell}{\sigma}\right)^{0.11} \left(\frac{U_g}{\sqrt{g d_p}}\right)^{0.22}$ (T-5.3.4b)
Uysal (1978)	See Table 5.2			$\varepsilon_{\ell,st} = 1.15 \times 10^{-4} G_\ell^{0.826} d_p^{-1.289}$
Rama et al. (1983)	See Table 5.1			$\varepsilon_{\ell,st} = 0.086 + 11\left(\frac{H_0}{d_p}\right)^{-0.4} \left(f\frac{d}{D_c}\right)^{-0.58} \left(\frac{g d_p^3 \phi_s^3 \rho_s^2}{\mu_\ell^2}\right)^{0.09} \left(\frac{U_\ell}{\sqrt{g d_p \phi_s}}\right)^{1.66} \cdot \left(\frac{d_p \phi_s U_\ell \rho_\ell}{\mu_\ell}\right)^{-0.34} \left(\frac{d_p \phi_s U_\ell^2 \rho_\ell}{\sigma}\right)^{-0.34}$ (T-5.3.5)
Vunjak-Novakovic et al. (1987a,b)	0.36, 0.52 0.78	182, 195, 203, 320, 374, 379, 407, 413, 683, 980	3.6 ~ 29	$\varepsilon_{\ell,st} = 2.48\times10^{-3} \left(\frac{H_0}{D_c}\right)^{-0.567} d_p^{-0.568} G_\ell^{0.719} + 0.02$ (Type I TBC) (T-5.3.6a) $\varepsilon_{\ell,st} = 7.33\, Re_\ell^{-0.059} Fr_\ell^{0.435} \left(\frac{H_0}{D_c}\right)^{-0.433} (\rho_s/\rho_\ell)^{0.090} + 0.02$ (Type II TBC) (T-5.3.6b) $\varepsilon_g = 0.628\, U_g^{0.237}$ (Type I and Type II TBCs)

* Batch liquid system.

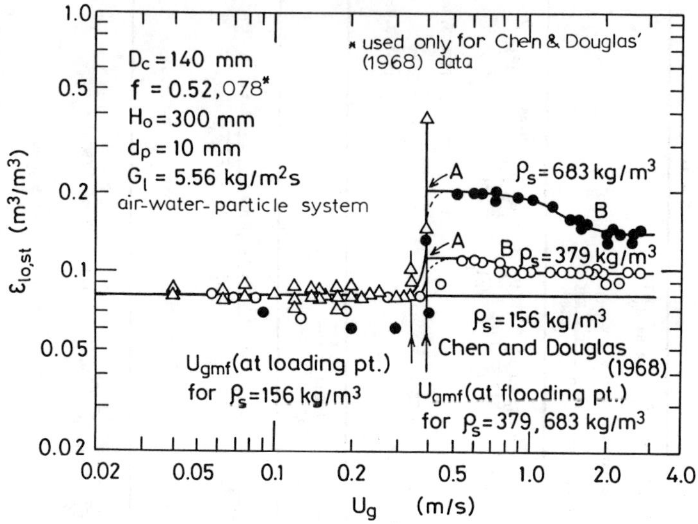

Figure 5.6 Particle density effects on the operational liquid holdup versus gas velocity relationship for a TBC (from Vunjak-Novakovic et al., 1987a).

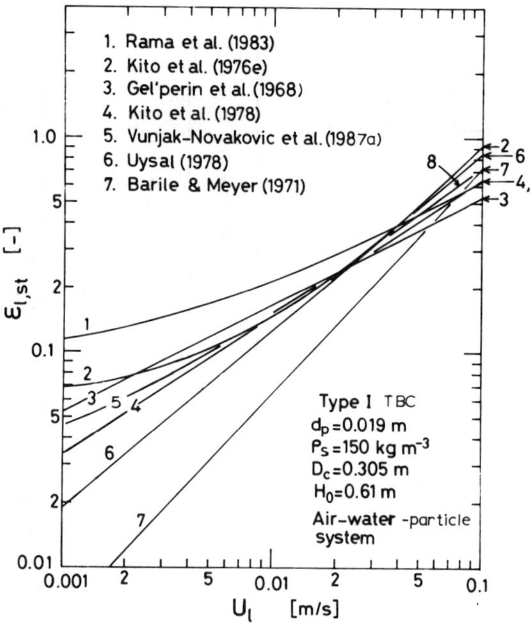

Figure 5.7(a) Comparisons of $\epsilon_{\ell,st}$ variation with U_ℓ based on various correlations given in Table 5.3 for a Type I TBC.

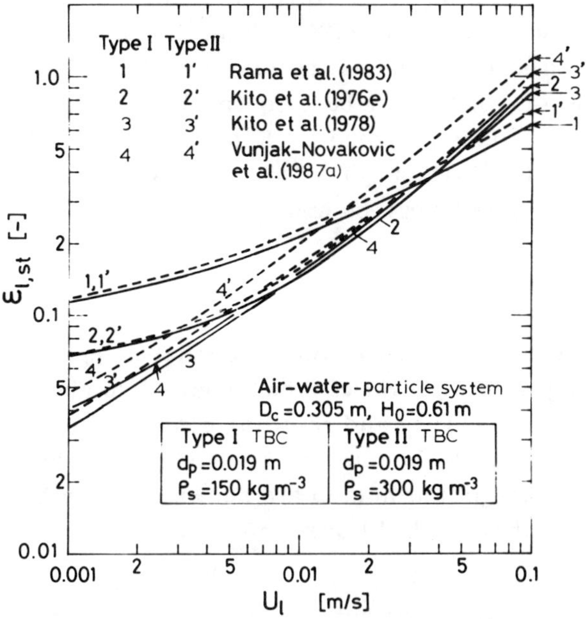

Figure 5.7(b) Comparisons of $\epsilon_{\ell,st}$ variations with U_ℓ based on various correlations given in Table 5.3 for Types I and II TBCs.

Table 5.3 summarizes the correlations of liquid holdup: Kito et al.'s (1978) correlation (Eqn. T-5.3.4a) considered the effect of the grid open area; Rama et al. (1983) modified Kito et al.'s correlation to account for effects of particle shape (Eqn. T-5.3.5); and Vunjak-Novakovic et al. (1987a,b) correlated liquid holdup based on the TBC type and differentiated between contributions of the static and operational liquid holdups (Eqns. (T-5.3.6a) and (T-5.3.6b)).

Figures 5.7(a) and 5.7(b) compare the liquid holdups estimated from the correlations in Table 5.3 for operating conditions representing both Type I and Type II fluidization. As seen in Fig. 5.7a, for Type I TBCs the relative deviation between predictions for these correlations is significantly higher at a low U_ℓ than at a high U_ℓ. The correlations of Barile and Meyer (1971) and Rama et al. (1983) in general give respectively lower and higher values than the other correlations. The correlation of Uysal (1978) yields low liquid holdup for low liquid velocities. In Fig. 5.7(b), it is seen that the differences between $\epsilon_{\ell,st}$ for Type I and for Type II TBCs predicted by these correlations increase as U_ℓ increases.

Based on validity and range of applicability, the correlations proposed by Gel'perin et al. (1968) (Eqn. (T-5.3.1)), Kito et al. (1978) (Eqn. (T-5.3.4a)), Rama et al. (1983) and Vunjak-Novakovic et al. (1987a) (Eqn. (T-5.3.6a)) can be used for Type I TBCs. For Type II TBCs, the correlations proposed by Kito et al. (1978) (Eqn. (T-5.3.4a)), Rama et al. (1983) (Eqn. (T-5.3.5)), and Vunjak-Novakovic et al. (1987a) (Eqn. (T-5.3.6b)) can be used.

5.1.1.4 Gas Holdup

Relatively little information is available regarding gas holdup. The gas holdup in a TBC varies from about 0.3 to about 0.8 depending mainly on the gas velocity and resulting continuity or discontinuity of the gas phase. The gas holdup can be determined when the expanded bed height and the liquid holdup are known as given by the following equation (Balabekov et al., 1969a):

$$\epsilon_g = 1 - \frac{h_\ell}{H} - \frac{(1-\epsilon_0)H_0}{H} = 1 - [\epsilon_{\ell,st} + (1-\epsilon_0)]\frac{H_0}{H} \tag{5.15}$$

where h_ℓ is the liquid height in the empty column corresponding to the liquid holdup in the bed.

Early studies of gas holdup were conducted by Gel'perin et al. (1968), Krainev et al. (1968), and Balabekov et al. (1969a,b) using rather heavy particles in columns with grids having small open areas. Balabekov et al. (1969a) showed that gas holdup increased with increasing gas velocity, but was almost independent of liquid velocity at a constant gas velocity. Kito et al. (1978) measured gas holdup in TBCs over a wide range of grid opening areas, particle sizes, particle densities and liquid properties, and observed effects of gas and liquid velocities on the gas holdup similar to those reported by Balabekov et al. (1969a). They, however, concluded that gas holdup is nearly independent of particle density, liquid viscosity, static bed height, and the open area and the orifice diameter of the supporting grid.

A list of correlations for gas holdup in a TBC is also given in Table 5.3. Kito and his co-workers have developed several correlations for gas holdup in a TBC. In a system with batch liquid (Kito et al., 1976b) the gas holdup is not affected as the open area of the supporting grid varies between 1.27 and 31.5 percent in the fluidized region (U_g > 0.2 ~ 0.3 m/s). They proposed Eqn. (T-5.3.2) to correlate gas holdup. Kito et al. (1976e), in developing Eqn. (T-5.3.3), a correlation for gas holdup in the TBC with countercurrent gas-liquid

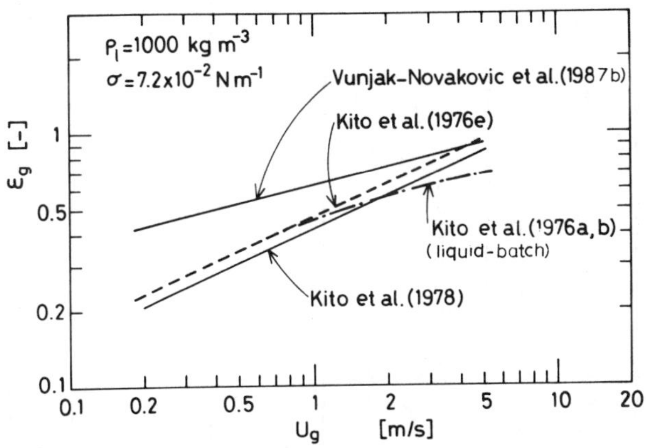

Figure 5.8 Comparisons of ϵ_g predictions based on various correlations given in Table 5.3 for a TBC.

flow, found that gas holdup was independent of liquid flow rate. In addition, Kito et al. (1978) proposed a correlation for gas holdup in a TBC in terms of dimensionless variables (Eqn. (T-5.3.4b)).

Figure 5.8 compares the gas holdup predicted from various correlations previously proposed under specific conditions by Kito et al. (1976a,b,e, 1978) and Vunjak-Novakovic et al. (1987b). The equations presented by Kito et al. yield similar gas holdups, while the correlation proposed by Vunjak-Novakovic et al. (1987b) overestimates the gas holdup at low gas velocities. The correlation by Kito et al. (1978) provides the best estimate for design purposes because it is based on a wide range of data and expressed in a dimensionless form.

5.1.1.5 Bed Expansion

The expanded bed height, or bed expansion, can be determined visually by averaging the maximum and minimum heights between which the bed surface fluctuates or from the axial pressure distribution as discussed in Section 2.1.5.1 (also see Tichy et al., 1972). Bed expansion data were first reported for a grid of small open area by Gel'perin et al. (1966). Chen and Douglas (1968), for low density particles, showed that the bed height linearly increased with increasing gas velocity and increased with increasing liquid velocity. Tichy and Douglas (1972) reported that H/H_0 is independent of both the static bed height and the particle density for low density particles. Rama et al. (1983) observed that a bed of cork particles with tapered

Table 5.4 Correlation Equation for Bed Expansion in a TBC

Investigators	f [-]	ρ_s [kg/m^3]	D_c/d_p [-]	Correlation Equations (in SI Units)/Comments
Levsh et al. (1968a)*	See Table 5.1			$\frac{H}{H_0} = 4.40\, U_\ell^{0.43}\, U_g^2$ ($U_\ell < 0.0078$ m/s, $U_g < 2.5$ m/s) $\frac{H}{H_0} = 2.28\, U_\ell^{0.35}\, U_g^2$ ($U_\ell > 0.0078$ m/s, $U_g > 2.5$ m/s)
Aksel'rod and Yakovenko (1969)	0.3 ~ 0.65	380	9.4	$H = 0.933\, U_\ell^{0.3}\, H_0^{0.6}\, (\frac{U_g}{f})^{0.93}$ for $\frac{U_g}{f} \geqslant 6$ m/s , $H = 2.16\, U_\ell^{0.3}\, H_0^{0.6}\, (\frac{U_g}{f})^{0.43}$ for $\frac{U_g}{f} < 6$ m/s
Khanna (1971) (quoted from Uysal, 1978)	See Table 5.2			$\frac{H}{H_0} = 1 + 0.414\, (G_g - G_{gmf})\, G_{gmf}^{0.2}$
Balabekov et al. (1971)	See Table 5.3			
Gel'perin et al. (1968)	See Table 5.3			
Kito et al. (1976d)	Same as those of Kito et al. (1976e) given in Table 5.3			$\frac{H}{H_0} = \frac{(1-\varepsilon_0) + \varepsilon_{\ell,st}}{1-\varepsilon_g} = \frac{\varepsilon_{s,st} + \varepsilon_{\ell,st}}{1-\varepsilon_g}$ Correlation equations for $\varepsilon_{\ell,st}$ and ε_g are given in Table 5.3
Uchida et al. (1977)	See Table 5.1			
Vunjak-Novakovic et al. (1980; 1987a,b)	See Table 5.1			

Tichy and Douglas (1972)	See Table 5.2			$\frac{H}{H_0} = 0.8849 + 0.3166\, G_g - 18.33\, d_p + 0.5852\, G_\ell^{0.6} d_p^{0.5}$
O'Neill et al. (1974)	———			$\frac{H}{H_0} = \frac{1 - \varepsilon_0}{1 - \varepsilon} = \frac{\varepsilon_{s,st}}{\varepsilon_s}$ for Type II TBC Correlation for ε was proposed as $\varepsilon = \{K[(27 + 4K)/108]^{1/2} + K/2\}^{1/3} - \{K[(27 + 4K)/108] - K/2\}^{1/3}$ $K = \frac{U_g^{1/2} + U_\ell^{1/2}(\rho_\ell/\rho_g)^{1/4}}{0.775\,[g d_p \rho_\ell / 6\rho_g]^{1/4}}$
Uysal (1978)	See Table 5.2			$\frac{H}{H_0} = 1 + 0.147\, \frac{(G_g - G_{gmf})}{H_0}$
Rama et al. (1983)	0.7	53, 112, 183	3.95, 4.81, 6.33	$\frac{H}{H_0} = 2.132\, d_p^{0.12} U_\ell^{0.31} + 1.02\, d_p^{-1.7} \rho_s^{-1.2} U_\ell^{0.2} U_g$ (Spherical and irregular particles) $\frac{H}{H_0} = -1.07 + 43\, U_\ell + 1.545\, U_g$ (Cork particles)

* Special type of packing(plastic ring) was used

cylindrical shape, expanded more than a bed of spherical or other irregular shaped particles with similar densities. Experimental data on bed expansion for relatively high density particles ($470 < \rho_s < 2600$ kg/m^3) were obtained by Balabekov et al. (1969a) and Strumillo et al. (1974). A rapid increase in bed height has been observed for gas velocities approaching the true flooding point (Tichy and Douglas, 1973).

A simultaneous volume balance on the liquid and solid phases yields an expression for H/H_0:

$$\frac{H}{H_0} = \frac{\epsilon_{\ell,st} + \epsilon_{s,st}}{1 - \epsilon_g} = \frac{\epsilon_{\ell,st} + 1 - \epsilon_0}{1 - \epsilon_g} \tag{5.16}$$

where $\epsilon_{s,st}$ is the solids holdup based on static bed volume. Application of Eqn. (5.16) requires knowledge of both ϵ_g and $\epsilon_{\ell,st}$. The original form of Eqn. (5.16) was introduced by Gel'perin et al. (1968). Since then, several investigators have used Eqn. (5.16) to correlate H/H_0.

An alternative expression for H/H_0, based on a volume balance of the solid phase, is (O'Neill et al., 1974)

$$\frac{H}{H_0} = \frac{\epsilon_{s,st}}{\epsilon_s} = \frac{1-\epsilon_0}{1-\epsilon} \tag{5.17}$$

where the bed porosity ϵ can be obtained from correlations like those developed by O'Neill et al. (1974) as given in Table 5.4.

Empirical equations for correlating H/H_0, proposed by numerous investigators, are summarized in Table 5.4. Note that the correlation by Aksel'rod and Yakovenko (1969) was obtained for distributors with small open areas and that by Levsh et al. (1968a) was obtained for plastic ring particles. Since these correlations apply only for limited conditions, caution must be exercised in using them.

The gas distributing grid strongly affects the bed expansion behavior; in a grid of small open area a liquid layer builds up immediately above the grid, creating vertical variations in the liquid holdup. The behavior of such beds, to some degree, resembles that of a bubble column with high gas holdup or with a sieve tray. An attempt to correlate the height of the liquid layer immediately above the grid was made by Levsh et al. (1968a). Kito et al. (1976a,b,c) showed that TBCs with grids of very small open areas can be operated under liquid-batch conditions.

Several empirical equations have also been developed for systems having distributors with large open areas and for low density

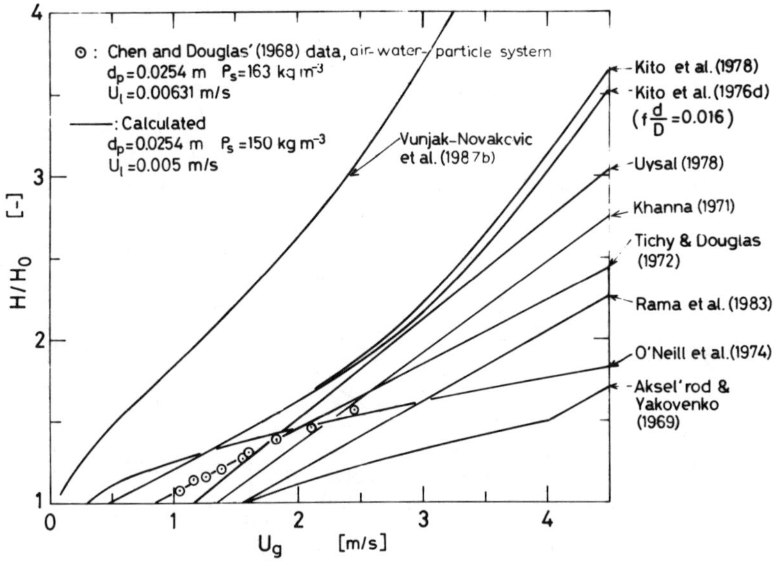

Figure 5.9 Comparisons of H/H_0 prediction based on data of Chen and Douglas (1968) using various correlations given in Table 5.4 for a TBC.

particles similar to those employed industrially (Khanna, 1971; Tichy and Douglas, 1972; Uysal, 1978; Rama et al., 1983). Because of the purely empirical nature of these equations, application of these equations should be limited only to the conditions from which each equation was developed. Kito et al. (1976e, 1978) and Vunjak-Novakovik et al. (1987b) used Eqn. (5.16) for correlating the relative bed expansion.

Figure 5.9 compares the relative bed expansion calculated from various correlation equations with the bed expansion data of Chen and Douglas (1968) for typical operating conditions using a low liquid velocity and low density particles. The figure shows that the empirical correlations of Tichy and Douglas (1972) and Uysal (1978) and the indirect expression of Kito et al. (1976e, 1978) predict H/H_0 fairly well. The correlation by O'Neill et al. (1974) agrees fairly well with the data points except at low gas velocities. The correlations by Khanna (1971) and Rama et al. (1983) moderately underestimate the experimental H/H_0. The correlation by Vunjak-Novakovic et al. (1987b) significantly overpredicts H/H_0. This overprediction possibly results from the overestimation of the gas holdup by their correlation.

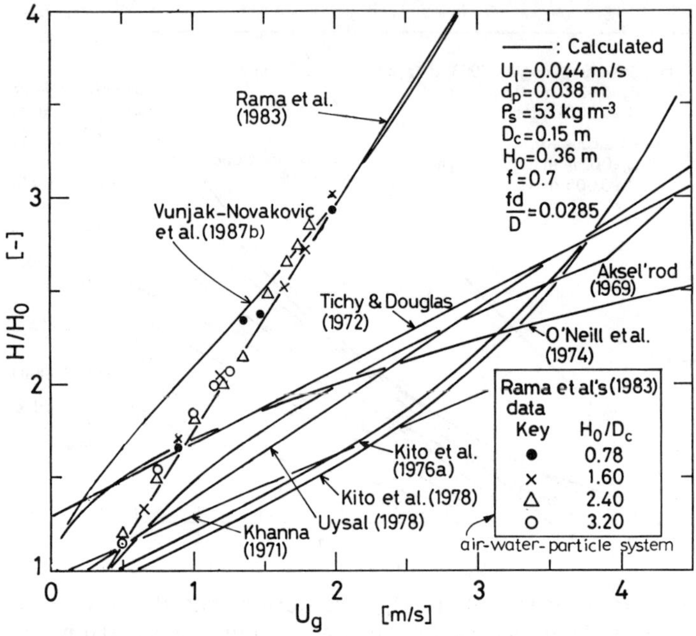

Figure 5.10 Comparisons of H/H_0 prediction based on data of Rama et al. (1983) using various correlations given in Table 5.4 for a TBC.

Correlations proposed by Blyakher et al. (1967) and Aksel'rod and Yakovenko (1969) also give rather poor predictions.

Figure 5.10 shows a comparison of the relative bed expansion calculated from various correlations with the data of Rama et al. (1983). The calculations are performed for typical conditions at a high liquid velocity (U_ℓ = 0.044 m/s) and low density particles. In this case all the best correlations under low liquid velocity conditions (the correlations of Khanna (1971), Tichy and Douglas (1972), Uysal (1978), Kito et al. (1976e), and Kito et al. (1978)) yield values markedly lower than the data. Note that the correlation of Vunjak-Novakovic et al. (1987b) satisfactorily predicts the relative bed expansion at high liquid velocities. Rama et al.'s (1983) correlation agrees well with their own experimental data.

5.1.1.6 Fluid Mixing

In the proper design and evaluation of a TBC, fluid mixing effects must be considered. Phase mixing can be adequately characterized by an axial dispersion coefficient for each phase. Several studies have been reported on axial mixing of the liquid phase (Chen and Douglas, 1969; Koval' et al., 1975a,b; Rama et al., 1985) while little has been done on gas phase mixing.

Liquid phase axial mixing in a TBC was first investigated by Chen and Douglas (1969) who measured the axial dispersion coefficient of liquid by introducing a step change of an inert liquid to the flowing tracer liquid. The basic equation describing tracer concentration in the bed is given by Eqn. (3.20) which is subject to boundary and initial conditions of

$t > 0$

$$z = 0; \qquad V_\ell C - E_{z\ell}\frac{\partial C}{\partial z} = 0 \tag{5.18a}$$

$$z = H; \qquad \frac{\partial C}{\partial z} = 0 \tag{5.18b}$$

$t = 0$

$$z < 0; \qquad C = 0 \tag{5.18c}$$

$$0 < z < H; \qquad C = C_0 \tag{5.18d}$$

Here C is the tracer concentration, t is the time, $E_{z\ell}$ is the liquid axial dispersion coefficient based on the linear liquid velocity, z is the axial distance, and V_ℓ is the linear velocity of liquid. The analytical solution is

$$\left(\frac{C}{C_0}\right)\bigg|_{z=H} = 2\sum_{i=1}^{\infty}\frac{\delta_i \sin\delta_i}{\delta_i^2+M^2+M}\exp\left[M-\left(\frac{\delta_i^2+M^2}{2M}\right)\Theta\right] \tag{5.19}$$

where $M = V_\ell H/2E_{z\ell}$, $\Theta = t/\bar{\Theta}$, $\bar{\Theta} = H/V_\ell$, and the values of δ_i are the roots of the two transcendental equations:

$$\frac{\delta_i}{2}\tan\left(\frac{\delta_i}{2}\right) = \frac{M}{2} \qquad i = 1, 3, 5, ... \tag{5.20a}$$

$$\frac{\delta_i}{2}\cot\left(\frac{\delta_i}{2}\right) = -\frac{M}{2} \qquad i = 2, 4, 6, ... \tag{5.20b}$$

Chen and Douglas (1969) obtained an approximated solution by considering only the first term of the infinite series in Eqn. (5.19) as given below:

Table 5.5 Experimental Details of Liquid-Phase Axial Mixing Studies in a TBC (from Rama et al., 1985)

References	Details of contactor				Characteristics of particles					Ranges of variables				
	D_c [m]	H_c [m]	f [%]	number of stages	Shape	d_p [mm]	ρ_s [kg/m^3]	ϕ_s	ε_0	H_0 [mm]	H_0/D_c	H/H_0	$U_\ell \times 10^2$ [m/s]	U_g [m/s]
Chen and Douglas (1969)	0.305	0.61	70	1	Spherical	12.7 25.4 38.1	156 174 158	1 1 1	—	305	1	1.5	0.20 ~ 1.5	0.56 ~ 3.04
Koval' et al. (1975a)	0.10	1.02	46	1	Spherical	9.0 16.0	830 283	1 1	—	80 90 100 160 300	0.8 3.0	—	0.39 ~ 1.54	1 ~ 5.22
Koval' et al. (1975b)	0.10	1.02	46	1	Spherical	9.0	830 490	1 1	—	80 300	0.8 3.0	a	0.39 ~ 1.54	0 ~ 5.22
	0.20	1.02	46	1	Spherical	16.0 16.0	283 283	1 1		80 130	0.40 0.65	a	0.39 ~ 0.93	2.9 ~ 3.4
Rama et al. (1985)	0.15	1.1	70	1	Spherical	38	53	1	0.475	120 240 360 480	0.78 1.60 2.40 3.20	a	1.11 ~ 4.4	0 ~ 5
					Irregular	31.2	112	0.675	0.470	120 240 360 480	0.78 1.60 2.40 3.20	a	1.11 ~ 4.4	0 ~ 5
					Cork	23.7	183	0.822	0.480	120 240	0.78 1.60	a	1.11 ~ 4.4	0 ~ 5

Note: a - The study was carried out up to the point of congregation of the particles below the top grid or inception of semifluidization.

$$\left(\frac{C}{C_0}\right)\Bigg|_{z=H} = \frac{2\delta_i \sin \delta_i}{\delta_i^2 + M^2 + M} \exp\left[M - \left(\frac{\delta_1^2 + M^2}{2M}\right)\Theta\right] \tag{5.21}$$

The logarithm of Eqn. (5.21) is linear in Θ and thus the slope of the straight line on the semi-logarithmic plot of $(C/C_0)|_{z=H}$ versus Θ yields M. The average residence time of liquid $\bar{\Theta}$, can be obtained from a plot of $(C/C_0)|_{z=H}$ versus t. $E_{z\ell}$ can then be evaluated from M.

Table 5.5 summarizes the experimental conditions used in axial mixing studies of TBCs. The excess gas flow rate over that required for incipient fluidization, $(G_g - G_{gmf})$, provides some measure of the intensity of bed motion. Chen and Douglas (1969) found that the liquid axial dispersion coefficient increased significantly with both an increase in $(G_g - G_{gmf})$ and an increase in liquid velocity for a given particle. The ratio of the Peclet number in a TBC to that in a fixed bed, Pe''/Pe''_0, is virtually independent of liquid velocity. The Peclet number ratio is a function of Δ defined as $\frac{(G_g - G_{gmf})}{G_{gmf}}$ and varies slightly with particle diameter. They suggested the following correlation for the liquid Peclet number:

$$\frac{Pe''}{Pe''_0} = \left(\frac{d_p}{D_c}\right)^{-0.304} F(\Delta) \tag{5.22}$$

where $F(\Delta)$ is given graphically in Fig. 5.11.

Koval' et al. (1975a,b) measured liquid axial dispersion coefficients in TBCs of 100 and 200 mm ID with a grid having an open area of 46%. The particles used had diameters of 9 and 16 mm and densities varying from 283 to 830 kg/m^3. Using the pulse response technique, Koval' et al. evaluated the liquid axial dispersion coefficient from the first and second moments of the response curve and found that an increase in liquid flow rate resulted in a significant increase in $E_{z\ell}$. In addition, they indicated that the axial dispersion coefficient, or the column Peclet number ($Pe_c = V_\ell H/E_{z\ell}$), was virtually independent of particle density despite a nearly twofold variation in density. Particle diameter also appeared not to affect either Pe_c or $E_{z\ell}$.

Rama et al. (1985) also measured $E_{z\ell}$ in a TBC using the step response method. They studied the behavior of Pe_c and $E_{z\ell}$ for low density particles and a grid with an opening of 70%. At low static bed heights ($H_0/D_c \leq 1.6$), Pe_c decreases with an increase in liquid velocity but remains independent of gas velocity, as was also observed by Chen

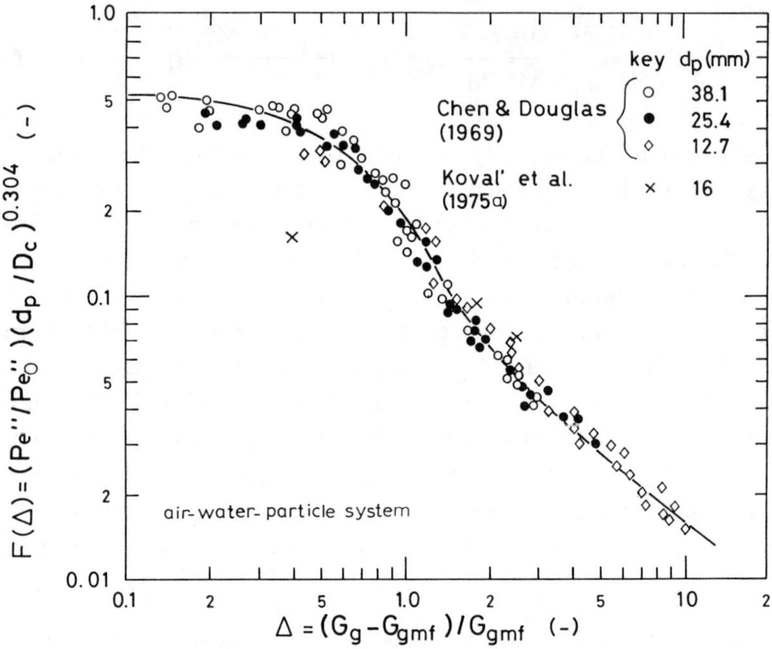

Figure 5.11 Variation of TBC Peclet number with reduced gas velocity (from Chen and Douglas, 1969).

and Douglas (1969) and Koval' et al. (1975b). At higher static bed heights, ($H_0/D_c \geq 2.4$), Pe_c decreases with an increase in both gas and liquid velocities. Note that high values of Pe_c were obtained where aggregation of particles occurred at the wall for low density, nonspherical particles, such as cork particles, in beds of high static bed height. In this case, most of the liquid rains down through the particle-free core, resulting in little axial liquid phase dispersion.

Rama et al. (1985) also found that particle shape affects the dependency of $E_{z\ell}$ on U_g, i.e., $E_{z\ell} \propto U_g^{2.26}$ for spherical particles, $E_{z\ell} \propto U_g^{2.41}$ for irregular particles and $E_{z\ell} \propto U_g^{2.56}$ for tapered cylindrical particles. However, particle shape does not affect the dependency of $E_{z\ell}$ on V_ℓ ($= \frac{U_\ell}{\epsilon_\ell}$), i.e., $E_{z\ell} \propto V_\ell^{2.92}$ for spherical and irregular shaped particles. They suggested the following empirical correlation for Pe_c:

$$Pe_c = 10^{4.49} Re_\ell^{-0.47} Re_g^{-0.308} (Ar^*)^{-0.165} (H_0/d_p)^{0.824} (d_p/D_c)^{1.464} \tag{5.23}$$

Note that Eqn. (5.23) is obtained based only on air-water-particle systems. The equation predicts well the data of Chen and Douglas (1969) at gas velocities far above the minimum fluidization velocity, but appreciably underpredicts their data at gas velocities close to the minimum fluidization velocity. This discrepancy results because the correlation of Rama et al. was obtained from data of high bed expansion (for H/H_0 up to 4). Rama et al. (1985) pointed out that the values of Pe_c obtained by Koval' et al. (1975a,b) are significantly lower than those estimated from Eqn. (5.23). This behavior may be attributed to the small open area of the grid used by Koval' et al.; the small open area of the grid supports large amounts of liquid and the increased liquid holdup intensifies the mixing in the bed, hence reducing the value of Pe_c.

5.1.1.7 Solids Mixing

Very little information is available on the axial movement of particles in a TBC. Tabei et al. (1988) studied the particle trajectories using cine photography and reported quantitatively an increase of the root mean square of the absolute axial particle velocities with increasing U_g. The behavior of particle mixing was characterized by the solids axial dispersion coefficient E_{zs}, defined similarly to the fluid axial dispersion coefficient based on the dispersion model (Eqn. (3.20)) except that there was no convective solids transfer in the model when defining E_{zs}.

Tabei et al. (1988) obtained E_{zs} by analyzing the transient distribution of tracer particle concentrations in the bed where a layer of tracer particles was initially placed at the bottom of the bed. E_{zs} was found to increase with increasing U_g, U_ℓ and D_c and can be correlated by

$$Pe_s \left(= \frac{d_p U_g}{\epsilon_g E_{zs}}\right) = 315\, \epsilon_s^{1.13}\, \epsilon_{\ell,st}^{-0.46} \left(\frac{d_p}{D_c}\right)^{0.7} \tag{5.24}$$

Equation (5.24) is valid for $U_g > 1.3\, U_{gmf}$, $1 \le U_g \le 2.5$ m/s, and $0.01 \le U_\ell \le 0.02$ m/s.

5.1.2 Mass Transfer

TBC operation usually involves inert plastic particles. Consequently, mass transfer through liquid-solid interfaces are of little or no concern and studies of mass transfer in the literature are limited to gas-liquid interfacial phenomena.

In a TBC, fluidized low density particles create a high specific interfacial area and a rapid renewal of the gas-liquid interface through their violent, turbulent motion. Both of these phenomena greatly increase the contacting intimacy between the gas and liquid phases resulting in a high volumetric gas-liquid mass transfer coefficient, a high heat transfer rate accompanied by vaporization or condensation of water, and a high particulate collection efficiency. These characteristics are desirable in applying TBCs for absorbing or scrubbing for removing gaseous pollutant, cooling and dehumidification towers for saturated gases, and scrubbers to recover small particles, respectively (see Chapter 9).

The overall mass transfer coefficient and height of a transfer unit are the primary parameters used to account for mass transfer performance. Early studies of gas-liquid mass transfer in TBCs usually centered around these two parameters; however, knowledge of both the gas-side and liquid-side mass transfer resistances and the specific interfacial area are essential to the characterization of intrinsic mass transfer mechanisms.

5.1.2.1 Overall Volumetric Gas-Liquid Mass Transfer Coefficient

Douglas (1964) defined an overall volumetric gas-side absorption coefficient, $K_g a_{st}$, as

$$K_g a_{st} = G_{Mg}/H_{og}P \tag{5.25}$$

This definition is based on the static bed volume rather than the real working bed volume due to the uncertainty in the actual bed height. H_{og}, the height of a mass transfer unit for static bed height (HTU), is

$$H_{og} = H_0/N_{og} \tag{5.26}$$

and N_{og}, the number of transfer units, is

$$N_{og} = (y_1-y_2)/(y-y^*)_{lm} \tag{5.27}$$

where $(y-y^*)_{lm}$ is the logarithmic mean driving force; and y_1 and y_2 are the mole fraction in the gas phase at the bottom and the top of the bed, respectively. Equations (5.25) through (5.27) are derived based on ideal contacting schemes which assume plug flow for both phases and that both the equilibrium and operating lines are linear. Douglas et al. (1963) found that for the absorption of CO_2 the volumetric mass transfer coefficients in a TBC were about 70 times that obtained in a coke packed mill tower. For the absorption of SO_2 into NaOH solution, using a single absorption zone with 3.6×10^4 cm^3 of packing particles, both $K_g a_{st}$ and the absorption efficiency increased appreciably with increasing liquid flow rate, and an efficiency of essentially 100% was obtained after lengthening the column with two additional absorption zones. In addition, the concentration of hydrogen sulfide (150 to 250 ppm) in the digester effluent gas from the alkaline pulp mill was reduced to essentially zero in a single stage TBC when cold effluent water from the pulp mill containing residual chlorine was brought into contact with the gas in the column.

Subsequently, Douglas (1964) found, for NH_3 absorption, H_{og} decreased with increasing gas mass velocity, increased with increasing liquid mass velocity, and increased with increasing static bed height. Douglas (1964) compared the data for a TBC with a fixed bed in which Fiberglas packings were vertically stacked in a special arrangement to permit the high liquid and gas mass velocities comparable to those in a TBC (Williams et al., 1947). The H_{og} found for the TBC was approximately 1/3 to 1/2 of that in the fixed bed.

Barile et al. (1974) evaluated the height of a transfer unit, based on Eqn. (5.26), in terms of the static bed height and found $K_g a_{st}$ increased with increasing G_g and G_ℓ, decreased with increasing H_0, and slightly decreased with increasing d_p. They proposed a correlation for $K_g a_{st}$ based on the cooling performance using particles of 38 mm diameter and 160 kg/m^3 density:

$$\frac{M_B P K_g a_{st} H_0}{G_\ell} = 0.0819 \left(\frac{d_p G_\ell}{\mu_\ell}\right)^{0.025} \left(\frac{G_\ell^2}{d_p g \rho_\ell^2}\right)^{-0.169} \cdot \left(\frac{G_\ell}{G_g}\right)^{-0.248} \left(\frac{H_0}{d_p}\right)^{0.309} \tag{5.28}$$

where M_B, the molecular weight of air, is 0.029 kg/mol. Equation (5.28) is applicable for air-water system only. Assuming the functional dependence of $K_g a_{st}$ on the dimensionless variables given in Eqn. (5.28) is valid, Visvanathan and Leung (1985) proposed the following extrapolation procedure in applying Eqn. (5.28) to different TBCs: (1) Select a $K_g a_{st}$ value for a system (system 1) which is identical, or similar, to the one under study. $K_g a_{st}$ in a system with different mass transfer species (system 2) can be obtained through the relationship of the penetration model (also see Eqn. (3.16)):

$$\frac{(K_g a_{st})_{system\ 1}}{(K_g a_{st})_{system\ 2}} = \left[\frac{(D_{Mg})_{system\ 1}}{(D_{Mg})_{system\ 2}}\right]^{0.5} \tag{5.29}$$

(2) Re-evaluate the constant in Eqn. (5.28) for system 2 and apply the resulting equation to calculate $K_g a_{st}$ under various design process conditions for system 2.

5.1.2.2 Liquid-Side and Gas-Side Volumetric Mass Transfer Coefficients

The liquid-side volumetric mass transfer coefficient, $k_\ell A$, based on a unit cross section of the column was measured by Elenkov and Kossev (1970). Note that $k_\ell A$ is equivalent to $k_\ell a_{st} H_0$ in which a_{st} is defined as the interfacial area per unit volume of a static bed. They employed desorption of oxygen from a supersaturated water stream to air to measure the liquid-side volumetric coefficient in a TBC with a grid opening of 41.7, 60, or 79%. The experiments used hollow polystyrene spheres 18 mm in diameter and 167 kg/m^3 in density, solid polystyrene spheres 17 mm in diameter and 930 kg/m^3 in density and solid spheres of a rosin-paraffin mixture 17 mm in diameter and 1090 kg/m^3 in density. $k_\ell A$ increases with both increasing gas flow rate and increasing static bed height, but decreases as the open area of the supporting grid increases. $k_\ell A$ exhibits complicated behavior with respect to the liquid velocity. At higher gas velocities, increasing the liquid rate generally causes an increase in $k_\ell A$, but at lower gas velocities $k_\ell A$ passes through a maximum in some cases. These authors suggested an empirical correlation for $k_\ell A$ of

$$k_\ell A/U_\ell = 0.024(D_M/h_0 U_\ell)(h_0 U_\ell \rho_\ell/\mu_\ell)^{0.1}(h_0 U_g \rho_g/\mu_g)^{1.5}(\mu_\ell/\rho_\ell D_M)^{0.5} \tag{5.30}$$

where h_0 is the head of liquid accounting for pressure loss through the TBC excluding the pressure loss through the grid.

Kossev and Elenkov (1973) used the evaporation of water to measure the gas-side volumetric mass transfer coefficient based on a unit column cross section, $k_{cg}A$. $k_{cg}A$ is obtained using a gas phase concentration difference defined based on the relative mass fraction for absorbing species as the driving force instead of the more commonly used partial pressure difference. The experiments were conducted under conditions similar to those in their previous study of the liquid-side mass transfer coefficient. Note that $k_{cg}A$ is equivalent to $(k_g a_{st})(H_0)(RT)$. $k_{cg}A$ increases significantly with increasing gas velocity, and increases moderately with increasing liquid velocity and static bed height and with decreasing grid opening. Kossev and Elenkov proposed a correlation for $k_{cg}A$ of

$$k_{cg}Ah_0/D_{Mg} = k_g a_{st} H_0 RT h_0/D_{Mg}$$

$$= 0.11(U_g h_0 \rho_g/\mu_g)(\mu_g/\rho_g D_{Mg})^{0.5}(g h_0^3 \rho_g^2/\mu_g^2)^{0.1} \qquad (5.31)$$

Note that Eqns. (5.30) and (5.31) were developed based on limited operating conditions. Future study needs to examine more widely the effects of static bed height and particle properties on both the gas- and liquid-side mass transfer coefficients.

5.1.2.3 Gas-Liquid Interfacial Area

One of the primary means of increasing the heat transfer and gas-liquid mass transfer rates is to increase the gas-liquid interfacial area. Only a very few investigations have been reported which examined the variation in gas-liquid interfacial area with respect to gas flow rate, liquid flow rate, particle properties, and supporting grid geometry.

Chemical absorption methods have been used to measure the gas-liquid interfacial area in a TBC (see Section 3.1.1.3). The technique for measuring the interfacial contact area is based on several assumptions. A chemical reaction with pseudo-first order irreversible kinetics takes place in the liquid phase at a rate fast enough so the concentration of dissolved gaseous reactant in the bulk liquid is zero while the liquid phase reactant is not depleted near the interface. Under such conditions, the mass transfer rate through the liquid film is virtually independent of liquid phase hydrodynamics. Kossev et al. (1971), Wozniak and Ostergaard (1973), and Wozniak (1977) expressed the overall mass transfer resistance, which resides entirely in the gas phase, as

$$\frac{1}{K_g a} = \frac{1}{k_g a} + \frac{He}{a(D_M k_r C_B)^{1/2}} \qquad (5.32)$$

where $K_g a$ is the overall volumetric mass transfer coefficient, $k_g a$ is the gas-side volumetric mass transfer coefficient, He is the Henry's law constant, "a" is the effective interfacial area per unit volume of the bed, k_r is a pseudo-first order reaction rate constant, and C_B is the concentration of the liquid phase reactant in the bulk liquid. $K_g a$ can be obtained experimentally from an overall mass balance for either the dissolved gaseous or the liquid phase reactant. Equation (5.32) holds when (Kossev et al., 1971)

$$5 < (D_M k_r C_B)^{1/2}/k_\ell < 0.2\ C_B/z_r C_{Ai} \tag{5.33}$$

where z_r is the stoichiometric coefficient for the liquid phase reactant species and C_{Ai} is the concentration of the gaseous reactant at the liquid-side interface. From Eqn. (5.32) it is evident that the interfacial area can be determined from the slope of the line on a plot of $1/K_g a$ versus $He/(D_M k_r C_B)^{1/2}$. Kossev et al. (1971), Wozniak and Ostergaard (1973), Kito et al. (1976a,c), and Wozniak (1977) used absorption of CO_2 into a NaOH aqueous solution and applied the same technique as described here to obtain the gas-liquid interfacial area.

When the gas-side mass transfer resistance can be neglected, the procedure for determining "a" becomes easier. Gel'perin et al. (1972) used directly the simplified equation, i.e., Eqn. (5.32) without the gas-side mass transfer resistance term and Strumillo and Kudra (1977) used the simplified equation coupled with Danckwerts' plot technique (Danckwerts, 1970). In Danckwerts' plot technique, the quantity $N_A^2/C_{Ai}^2 D_M$ is plotted against k_r to obtain a straight line whose slope yields the square of "a". Here, N_A is the absorption rate of the gaseous reactant.

Three kinds of interfacial areas have been defined so far: the effective interfacial area per unit cross section of the column, "A" (Kossev et al., 1971; Gel'perin et al., 1972; Strumillo and Kudra, 1977); the effective interfacial area per unit volume of static bed, "a_{st}" (Wozniak and Ostergaard, 1973; Wozniak, 1977); and the effective interfacial area per unit volume of expanded bed, "a" (Kossev et al., 1971; Kito et al., 1976a,c). They can be related to the effective interfacial area per unit volume of liquid, a^* (Kito et al., 1976a,c), by

$$A = a_{st}\ H_0 = a\ H = (a^* \epsilon_\ell)\ H \tag{5.34}$$

Moreover, the overall volumetric mass transfer coefficient, the gas-side mass transfer coefficient, and the liquid-side volumetric mass transfer

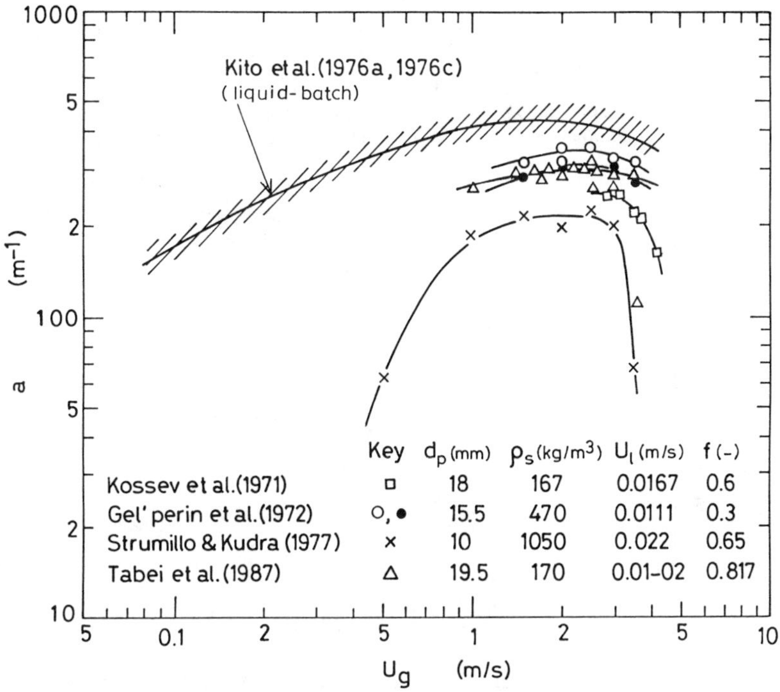

Figure 5.12 Gas-liquid interfacial area versus superficial gas velocity for a TBC (from Muroyama and Fan, 1985).

coefficient are uniquely defined for each interfacial area. Thus, in order to avoid confusion, caution must be exercised in interpreting the data for interfacial areas and mass transfer coefficients. "A" increases with increasing gas flow rate, passes through a maximum, and finally decreases with further increases in gas flow rate (Kossev et al., 1971; Strumillo and Kudra, 1977). In addition, "A" increases with an increase in static bed height and liquid velocity, and decreases with increasing particle size (Strumillo and Kudra, 1977). Figure 5.12 shows the variation in "a" with gas velocity. In this figure, the "a" of Gel'perin et al. (1972) and Strumillo and Kudra (1977) are obtained from their original data of "A" by estimating the expanded bed height from the correlation of Kito et al. (1978).

Kito et al. (1976a,c) measured "a" in a TBC with batch liquid using three kinds of plastic spheres. The "a" obtained by Kito et al. (1976a,c) are greater than those obtained by other investigators, especially at low gas flow rates, as shown in Fig. 5.12. These high values may result from the small open area of the distributing grid preventing seepage of the liquid through it. In reality, the fractional

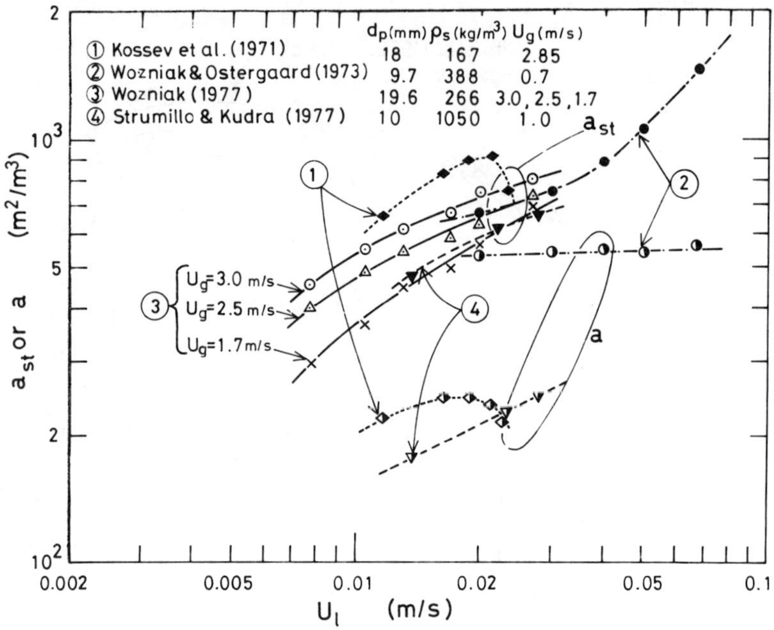

Figure 5.13 Gas-liquid interfacial area versus superficial liquid velocity for a TBC (from Muroyama and Fan, 1985).

liquid holdup in their system was much greater than that in a typical TBC countercurrent system. Moreover, the hydrodynamic behavior was similar to that of a bubble column with a perforated plate.

Figure 5.13 shows the variation of "a" (or a_{st}) with respect to liquid velocity. The values of "a" plotted in Fig. 5.13 were calculated from the experimental values of a_{st} using Eqn. (5.34). It is apparent that the data of a_{st} from various authors lie in a relatively narrow region in the figure although the experimental conditions for each of the data groups are rather different in gas velocity and particle density. a_{st} generally increases significantly with increasing liquid velocity since the increased liquid velocity increases the wetted area on the particle surface and increases the liquid holdup, resulting in greater bed expansion. Also shown in Fig. 5.13 are the relative magnitudes of "a" and a_{st}; note that "a" increases more gradually with an increase of liquid velocity than a_{st}. In some cases, "a" is virtually independent of liquid velocity (Wozniak and Ostergaard, 1973).

Empirical correlations for the interfacial area per unit cross section of the column have been developed by Gel'perin et al. (1972) and Strumillo and Kudra (1977). Gel'perin et al. correlated "A" by

$$A = 1.6\, U_\ell^{0.44-0.68H_0} U_g^{0.92-1.72H_0} f^{-0.93+2.54H_0} \exp(11.30H_0) \tag{5.35}$$

It is evident from Eqn. (5.35) that "A" is a strong function of the supporting grid open area. Strumillo and Kudra correlated A by

$$A = 16.28\, U_g^{0.92}\, U_\ell^{0.34}\, H_0^{0.83}\, d_p^{-0.94} \tag{5.36}$$

Equation (5.36) was developed from data taken for gas velocities up to 3.0 m/s, liquid velocities from 0.009 to 0.0306 m/s, bed heights up to 0.16 m, particle diameters from 5 to 10 mm, and particle densities of 1050 kg/m^3. Equation (5.36) shows that "A" increases almost linearly with static bed height but a_{st} ($=A/H_0$) is a weak function of the static bed height.

Recently, Tabei et al. (1987) reported a very simple correlation to predict a^* in terms of the gas holdup and liquid holdup, i.e.,

$$\frac{a^*}{\epsilon_\ell} = 2100 \left(\frac{\epsilon_g}{\epsilon_\ell}\right)^{1.25} \tag{5.37}$$

Equation (5.37) is applicable to $0.3 < U_g < 4.0$ m/s, $0.008 < U_\ell < 0.0675$ m/s, $0.012 < f < 0.6$, $0.0050 < d_p < 0.0265$ m, and $167 < \rho_s < 1050$ kg/m^3.

The interfacial area can be estimated from Eqns. (5.35) through (5.37) or from Figs. 5.12 and 5.13, but is important to note that each of these equations was developed from data obtained under a very limited range of experimental conditions. Consequently, they should not be applied beyond that range. The estimation of a_{st} may be more reliable than that of "a" since when using a_{st}, estimation of the expanded bed height can be avoided.

5.1.3 Heat Transfer

Little has been reported on heat transfer behavior of countercurrent fluidization. As noted earlier, the TBC was demonstrated as an effective device in cooling tower applications by Barile and Meyer (1971) and Barile et al. (1974). Simultaneous heat and mass transfer takes place between liquid droplets and air which can be analyzed by considering the dominant transport mechanism based on the convective mass transfer.

5.2 INVERSE FLUIDIZED BED

Hydrodynamics and mass transfer behavior of an inverse fluidized bed are given below.

5.2.1 Hydrodynamics

Only a limited number of studies are available on inverse gas-liquid-solid fluidized beds. Inverse fluidization in a gas-liquid-solid system was first reported by Page (1970); Werner and Schugerl (1972) examined mass transfer performance in this system. A series of studies have been systematically conducted on the hydrodynamic characteristics of inverse fluidization in liquid-solid and gas-liquid-solid systems by Fan et al. (1982a,b) and Chern et al. (1983).

The static pressure gradient (-dP/dz), and consequently individual phase holdups and the expanded bed height, can be measured in the same way as that reported for cocurrent upward three-phase fluidization in Chapter 2. In the following, flow regime, bed expansion, and gas holdup of an inverse fluidized bed are given.

5.2.1.1 Flow Regimes

Four distinct flow regimes can be classified: fixed bed (with and without partial fluidization) with dispersed bubble regime, inverse fluidized bed regime, transition regime, and inverse slugging fluidized bed regime (Fan et al., 1982b). Flow regime maps for inverse fluidization for two types of particles (i.e., Particle I: d_p = 10 mm and ρ_s = 388 kg/m^3 and Particle II: d_p = 9.53 mm and ρ_s = 822 kg/m^3) are given in Fig. 5.14.

Taking the bed of Particle I as an example, when both liquid and gas velocities are low (point A in Fig. 5.14), the drag force exerted on the particles by the liquid flow and the turbulence generated by the bubbles do not exceed the net buoyancy force, thus the bed remains in a fixed bed state. When the gas velocity increases to near incipient fluidization conditions, the lower portion of the bed begins to fluidize. As the gas velocity further increases, the remaining fixed portion progressively expands until the entire bed is fluidized. The conditions under which the entire bed initially becomes fluidized, the incipient fluidization condition, is indicated by point B in the figure; this point approximately corresponds to the break point in the curve

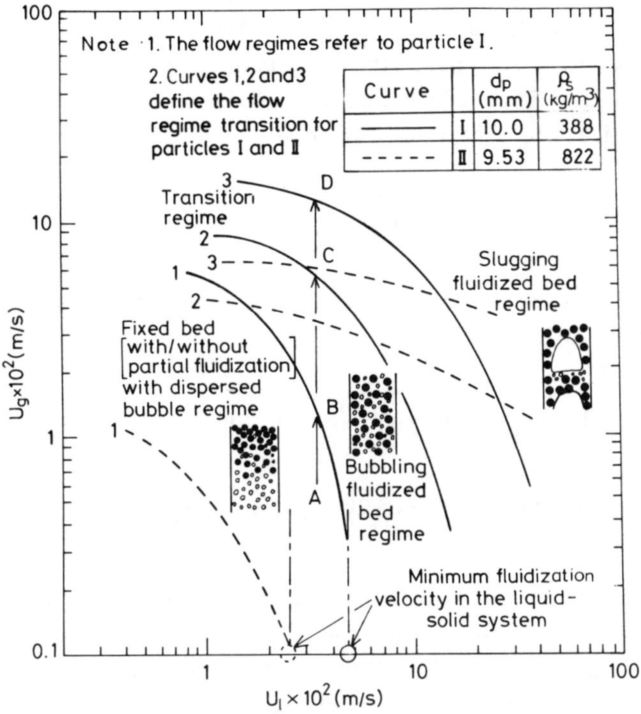

Figure 5.14 Flow regime map of the inverse air-water-particle fluidization (from Fan et al., 1982b).

on a plot of the dynamic pressure gradient versus the liquid velocity at a given gas velocity.

In the bubbling fluidized bed regime (between points B and C), the bubble size is uniform and the bed expands downward uniformly by an increase in either the gas or liquid velocity. As the gas velocity increases beyond point C, a transition regime is reached where significant bubble coalescence occurs. Point D marks the beginning of the slugging regime where the bed of particles moves up aggregatively with the gas slug and then falls as the slug breaks at the bed surface.

5.2.1.2 Bed Expansion

The hydrodynamic behavior of inverse fluidization in liquid-solid systems represents a limiting case of three-phase inverse fluidization operation. Bed expansion data for inverse liquid-solid fluidization can be adequately described by modified models for bed expansion in an upward liquid-solid system. Fan et al. (1982a) applied two approaches to the correlation of bed expansion data. First, bed expansion was correlated in terms of a relationship between a reduced velocity (U_R) and bed porosity in a manner similar to Lewis and Bowerman (1952), Richardson and Zaki (1954), and Letan (1974) for upward fluidization. The relationship is

$$U_R = \frac{U_\ell}{U_0} = \epsilon^n \tag{5.38}$$

Fan et al. (1982a) suggested n can be evaluated empirically from the following correlation:

$$n = 15\ Re_t^{-0.35} \exp\left(3.9 \frac{d_p}{D_c}\right) \qquad \text{for } 350 < Re_t < 1250 \tag{5.39a}$$

$$n = 8.6\ Re_t^{-0.2} \exp\left(-0.75 \frac{d_p}{D_c}\right) \qquad \text{for } Re_t > 1250 \tag{5.39b}$$

A second approach involved modifying the drag coefficient for a single particle to evaluate that for a multi-particle system. Based on the approach of Wen and Yu (1966), $\bar{f}$, the ratio of the multi-particle drag coefficient to a single particle drag coefficient in an inverse fluidized bed can be expressed in terms of a modified Archimedes number ($Ar' = d_p^3(\rho_\ell - \rho_s)\rho_\ell g/\mu_\ell^2$) and Re_ℓ as

$$\bar{f} = \frac{Ar'}{13.9\ Re_\ell^{1.4}} \qquad (2 < Re_\ell < 500) \tag{5.40a}$$

$$\bar{f} = 3\frac{Ar'}{Re_\ell^2} \qquad (Re_\ell > 500) \tag{5.40b}$$

For upward liquid-solid fluidization, Eqns. (5.40a) and (5.40b) can be used with Ar, defined as ($d_p^3(\rho_s - \rho_\ell)\rho_\ell g/\mu_\ell^2$), in place of Ar′. Ramamurthy and Subbaraju (1973) and Riba and Couderc (1977) expressed $\bar{f}$ in terms of $\phi(\epsilon)$ ($=[1-1.21(1-\epsilon)^{2/3}]^{-1}$). Wen and Fan (1974) showed that in addition to $\phi(\epsilon)$, $\bar{f}$ is also a function of Ar. A plot of $\bar{f}$ versus $\log_{10} \epsilon$ for an inverse liquid-solid fluidization system is given in Fig. 5.15. As noted by Wen and Fan (1974) for upward liquid-solid fluidization, a significant effect of Ar is evident. Fan et al. (1982a) suggested the following empirical correlation for $\bar{f}$, developed for an inverse liquid-solid fluidized bed in terms of ϵ, Ar′, and d_p/D_c:

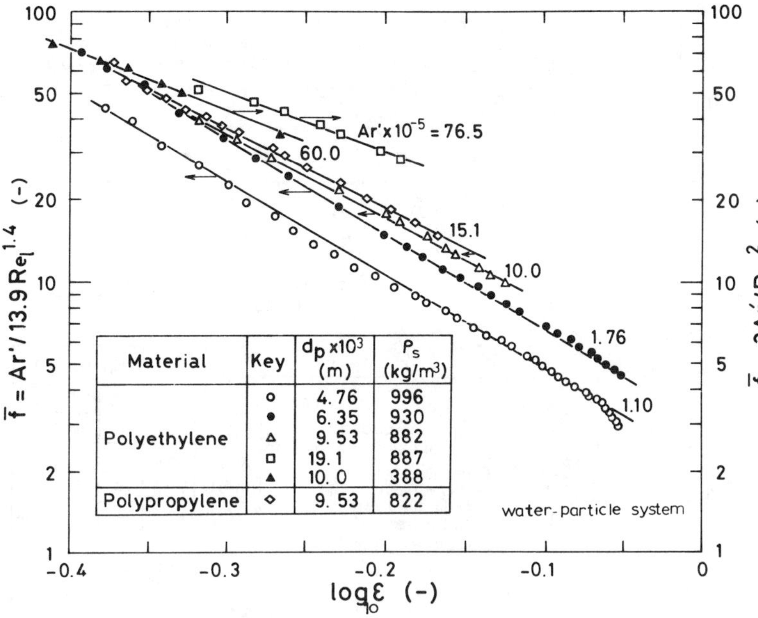

Figure 5.15 Relationship between $\bar{f}$ and ϵ for inverse liquid-solid fluidization (from Fan et al., 1982a).

$$\bar{f} = 3.21\,\epsilon^{-4.05} Ar'^{-0.07} \exp\left(3.5\,\frac{d_p}{D_c}\right) \tag{5.41}$$

for $0.40 < \epsilon < 0.88$; $1.1\times10^5 < Ar' < 7.65\times10^6$; and $0.062 < d_p/D_c < 0.250$. The hydrodynamic effect of introducing gas into the liquid-solid fluidized bed has not been analyzed theoretically and for gas-liquid-solid inverse fluidization only strictly empirical correlations have been presented to date.

In gas-liquid-solid inverse fluidization, Fan et al. (1982a) found that, as shown in Fig. 5.16, the bed voidage increased with gas velocity in the bubbling regime and reached a constant value when the downward bed expansion is restrained by a lower retaining grid[2]. Depending on the liquid velocity, further increases in the gas velocity causes a reduction in bed expansion during a flow transition to the slugging regime. Once the slugging regime is obtained, bed expansion again increases with increasing gas velocity.

2: The fluidized bed operated under this condition is denoted as a constrained fluidized bed.

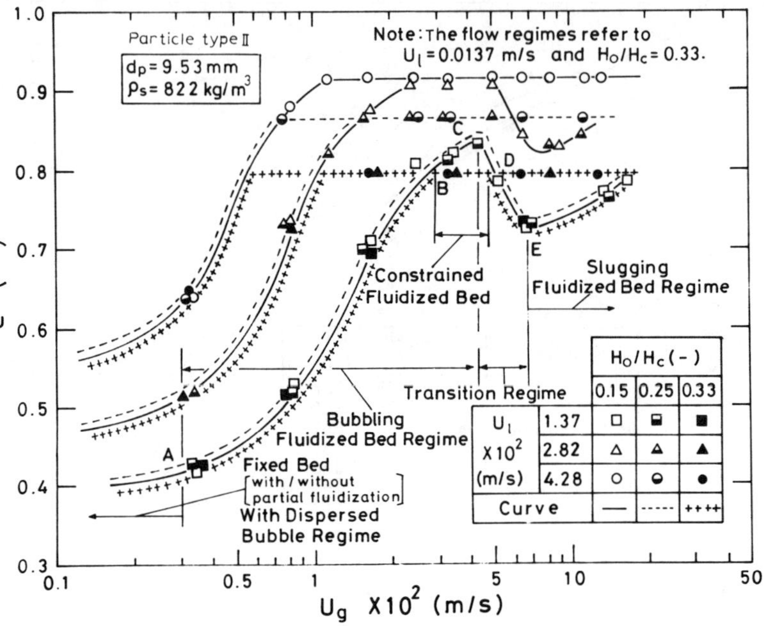

Figure 5.16 Effects of U_ℓ and H_0/H_c on the ϵ versus U_g relationship for inverse air-water-particle fluidization (from Fan et al., 1982b).

5.2.1.3 Gas Holdup

The behavior of the gas holdup in an inverse gas-liquid-solid fluidized bed resembles that in a cocurrent upward gas-liquid-solid fluidized bed (Mode E-I-a-1) in the manner in which it varies with the behavior of the bubble flow and properties of the particles. Figure 5.17 shows the variation in gas holdup, ϵ_g, with the superficial gas velocity, U_g, for Particle I (d_p = 10 mm, ρ_s = 388 kg/m^3). The individual flow regimes are identified in the figure. The effect of U_ℓ depends appreciably upon which flow regime prevails and whether or not the bed is in a constrained condition. When free bed expansion downwards is allowed, as is the case for a bed of Particle I, U_ℓ has a negligible effect on ϵ_g in the bubbling fluidized bed regime, while ϵ_g increases as U_ℓ increases in the slugging fluidized bed regime. Empirical equations for gas holdup in beds of Particle I vary with flow regime. For Particle I, for example, gas holdup could be correlated by (Fan et al., 1982b)

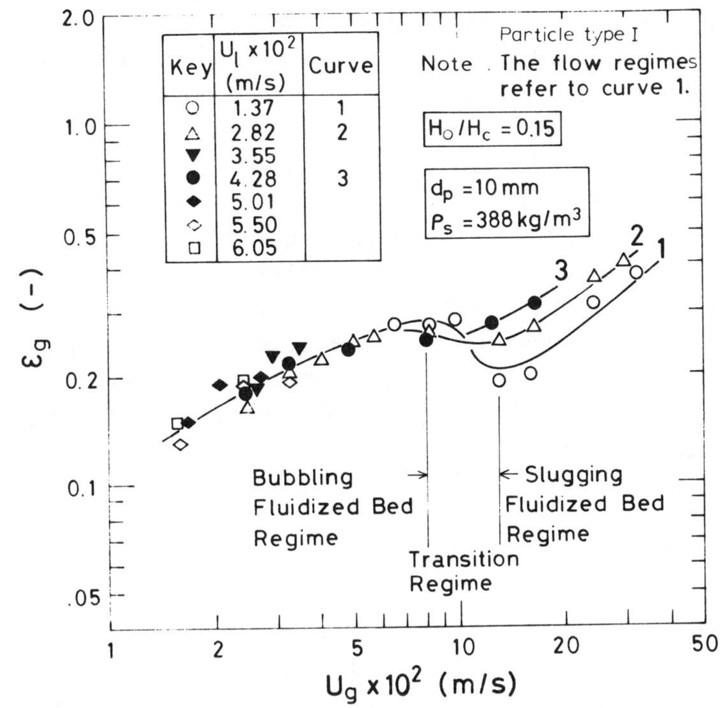

Figure 5.17 Effects of U_ℓ on the ϵ_g versus U_g relationship for an inverse air-water-particle fluidized bed (from Fan et al., 1982a).

$$\epsilon_g = 0.322\ \epsilon^{1.35}(U_g/U_\ell)^{0.18} \tag{5.42a}$$

for the bubbling fluidized bed regime, and

$$\epsilon_g = 2.43\ U_g^{0.704} U_\ell^{0.25} \tag{5.42b}$$

for the slugging fluidized bed regime. In Eqn. (5.42a), the effect of U_ℓ on ϵ_g cancels through ϵ.

5.2.2 Mass Transfer

Little work has been reported on mass transfer behavior in an inverse gas-liquid-solid fluidized bed. With an interest in aerobic wastewater treatment applications, Shimodaira and Yushina (1983) and Yushina and Sato (1983) reported the gas-liquid mass transfer coefficient, $k_\ell a$, in an inverse fluidized bed using polypropylene tubular particles. The sodium sulfite oxidation method was used to determine $k_\ell a$. $k_\ell a$ obtained for the inverse fluidized bed was correlated empirically with the volumetric air feed rate, G_V, and the fluidized bed volume, V_R, as

$$k_\ell a = 0.432\ (G_V/V_R)^{0.658} \tag{5.43}$$

Note that the reported $k_\ell a$, other than those used to establish Eqn. (5.43), by Shimodaira and Yushina (1983) and Yushina and Sato (1983) involve considerable error. The error was caused by an excessive sodium sulfite concentration present in the liquid phase, which resulted in bubble size reduction and chemical absorption effects in the determination of the pure physical gas-liquid mass transfer coefficient (also see Section 6.1.2.1).

5.3 CONCLUDING REMARKS

Understanding the transport behavior of a TBC requires a clear distinction between the characteristics of fluidization without incipient flooding (Type I TBC) and fluidization with incipient flooding (Type II TBC). Particle density plays a dominant role in determining which type of TBC operation occurs. The TBC is usually operated with gas as the continuous phase (Mode E-II-b); however, when flooding occurs under high density particle conditions, the flow transition to a continuous liquid phase operation (Mode E-II-a-2) may take place. For both TBC types, extensive investigation has been performed on the hydrodynamic properties including pressure drop, minimum fluidization velocity, bed expansion, liquid and gas holdups, and the overall volumetric gas-liquid mass transfer. Numerous correlations were proposed to account for these properties. However, they are design specific and thus suffer from a lack of generality in applications. Caution should be exercised in applying them. Information on liquid-side mass transfer at the gas-liquid interface for physical operations is needed to describe overall mass transfer behavior involving chemical reactions. A TBC operating under Type II conditions exhibits intense turbulence in the liquid-phase yielding excellent mass transfer behavior for process applications involving or without involving chemical reactions.

Continuous gas phase operation with a batch liquid phase (Mode E-III-b), in general, includes conditions where particles are relatively small (100 μm to 2 mm) and dense (1.5 to 3 g/cm^3) as well as those used in a TBC. When the amount of liquid upheld in the bed is very small, the bed behavior approaches that of a gas-solid fluidized bed. Extensive work is available under these operating conditions (e.g., Kunii and Levenspiel, 1969 and Davidson et al., 1985). Extensive hydrodynamic studies have also been performed for Mode E-III-b under Type II TBC conditions.

Inverse fluidization (Mode E-II-a-1) represents one relatively new mode of operation. The gas holdup behavior in inverse fluidization resembles that in cocurrent upward fluidization (Mode E-I-a-1) in that it varies with the behavior of bubble flow and properties of the particles. Because of its important applications in biochemical processing more work is required to understand the mass and heat transfer behavior.

5.4 NOMENCLATURE

a	: effective interfacial area per unit volume of expanded bed
a^*	: effective interfacial area per unit volume of liquid
a_{st}	: effective interfacial area per unit volume of static bed
A	: effective interfacial area per unit cross section of the column
Ar	: Archimedes number, $d_p^3(\rho_s-\rho_\ell)\rho_\ell g/\mu_\ell^2$
Ar'	: modified Archimedes number for inverse liquid-solid fluidized bed, $d_p^3(\rho_\ell-\rho_s)\rho_\ell g/\mu_\ell^2$
Ar^*	: modified Archimedes number, $g(d_p\phi_s)^3\rho_s^2/\mu_\ell^2$
C	: tracer concentration in liquid phase
C_0	: initial tracer concentration
C_{Ai}	: concentration of gaseous reactant at the liquid side interface
C_B	: concentration of liquid phase reactant in the bulk of liquid
C_{fi}	: apparent coefficient of friction for the gas phase in the countercurrent flow through irrigated fixed and fluidized bed
C_{fs}	: coefficient of friction for single phase flow through fixed bed
d	: equivalent diameter of slots or orifices in supporting grid

d_p : particle diameter

D : equivalent diameter for free open area of supporting grid

D_c : column diameter

D_M : molecular diffusivity in liquid phase

D_{Mg} : molecular diffusivity in gas phase

$E_{z\ell}$: liquid axial dispersion coefficient based on the linear liquid velocity

E_{zs} : solids axial dispersion coefficient

f : free opening of supporting grid (ratio of the open area to the total cross section of the column)

$\bar{f}$: ratio of the drag force on particles in a multi-particle system to that in a single particle system

g : gravitational acceleration

G_g : mass flow rate of gas per unit cross section

G_{gmf} : mass flow rate of gas per unit cross section at incipient fluidization

G_ℓ : mass flow rate of liquid per unit cross section

G_{Mg} : gas molar velocity per unit cross section of the column

G_V : volumetric gas flow rate

h_0 : pressure loss through the TBC expressed by water manometer height excluding the pressure loss of the grid

h_ℓ : liquid height in the empty column corresponding to the liquid holdup

H : expanded bed height

H_0 : static bed height

H_c : column height

H_{mf} : bed height at minimum fluidization

H_{og} : height of gas-side overall mass transfer unit

He : Henry's law constant

k_{cg} : gas-side mass transfer coefficient defined by Eqn. (5.31)

k_g : gas-side mass transfer coefficient at the gas-liquid interface

k_ℓ : liquid-side mass transfer coefficient (without chemical reaction) or physical mass transfer coefficient in the liquid side

$k_\ell a$: liquid-side volumetric coefficient for gas-liquid mass transfer

$k_\ell A$	: liquid-side volumetric mass transfer coefficient based on the unit cross section of the column
k_r	: pseudo-first order reaction rate constant
K_g	: overall gas-liquid mass transfer coefficient
$K_g a_{st}$	: overall volumetric gas-phase mass transfer coefficient based on the static bed height
M	: column Peclet number, $V_\ell H/2E_{z\ell}$
M_B	: molecular weight of air
n	: index for Richardson-Zaki relationship for bed porosity
N_A	: absorption rate of the gaseous reactant
N_{og}	: number of overall gas-phase transfer unit
P	: total pressure
Pe''	: axial Peclet number, $V_\ell d_p/E_{z\ell}$
Pe''_0	: axial Peclet number in the fixed bed, $V_\ell d_p/E_{z\ell}$
Pe_c	: column Peclet number, $V_\ell H/E_{z\ell}$
Pe_s	: Peclet number, $U_g d_p/\epsilon_g E_{zs}$
$-\frac{dP}{dz}$	: total pressure gradient
$-\frac{dP_d}{dz}$	: dynamic pressure gradient
$-\Delta P$	: total pressure drop through the column
$-\Delta P_g$	: pressure drop through the grid
$-\Delta P_w$	: pressure drop due to drag force exerted on the wall
R	: ideal gas constant
Re_g	: gas Reynolds number, $(d_p\phi_s)U_g\rho_g/\mu_g$
Re_{gmf}	: gas Reynolds number at minimum fluidization, $d_pU_{gmf}\rho_g/\mu_g$
Re_ℓ	: liquid Reynolds number, $(d_p\phi_s)U_\ell\rho_\ell/\mu_\ell$
Re_t	: particle Reynolds number, $d_pU_t\rho_\ell/\mu_\ell$
t	: time
T	: temperature
U_0	: superficial liquid velocity at ϵ=1 in an inverse liquid-solid fluidized bed
U_g	: superficial gas velocity
U_{gmf}	: minimum fluidization gas velocity

U_{gmf0} : minimum fluidization gas velocity for dry particles

U_{ℓ} : superficial liquid velocity

U_R : ratio of superficial liquid velocity to the asymptotic liquid velocity at ϵ=1 for inverse liquid-solid fluidization, U_{ℓ}/U_0

U_t : terminal velocity of a single particle

V_{ℓ} : linear liquid velocity, U_{ℓ}/ϵ_{ℓ}

V_R : fluidized bed volume

y : mole fraction of absorption species in gas phase; y_1 at the bottom of the bed; y_2 at the top of the bed

y^* : mole fraction of absorption species in equilibrium with the bulk liquid phase

z : axial distance

z_r : stoichiometric coefficient for liquid phase reactant species

Greek Letters

δ_i : i-th root of the transcendental equations, Eqns. (5.20a,b)

ϵ : bed porosity for the expanded bed

ϵ_0 : porosity of the static bed without gas-liquid flow

ϵ_g : gas holdup

ϵ_{gmf} : gas holdup at minimum fluidization

ϵ_{ℓ} : liquid holdup

$\epsilon_{\ell mf}$: liquid holdup at minimum fluidization

$\epsilon_{\ell,st}$: liquid holdup based on the static bed volume

$\epsilon_{\ell o,st}$: operational liquid holdup based on the static bed volume

ϵ_s : solids holdup

$\epsilon_{s,st}$: solids holdup based on static bed volume

Θ : dimensionless time, $t/\overline{\Theta}$

$\overline{\Theta}$: average residence time of liquid in the expanded bed

μ_g : gas viscosity

μ_{ℓ} : liquid viscosity

ρ_g : gas density

ρ_{ℓ} : liquid density

ρ_s : particle density

ρ_{ws} : density of wetted particles

σ : surface tension

$\phi(\epsilon)$: voidage function, $[1 - 1.21(1-\epsilon)^{2/3}]^{-1}$

ϕ_s : sphericity of particle

5.5 REFERENCES

Aksel'rod, L.S., and M.M. Yakovenko, Theor. Found. Chem. Eng., 3, 124 (1969).

Balabekov, O.S., P.G. Romankov, E. Ya. Tarat, and M.F. Mikhalev, J. Appl. Chem. U.S.S.R., 42, 1454 (1969a).

Balabekov, O.S., E. Ya. Tarat, P.G. Romankov, and M.F. Mikhalev, J. Appl. Chem. U.S.S.R., 42, 2128 (1969b).

Balabekov, O.S., E. Ya. Tarat, P.G. Romankov, and M.F. Mikhalev, J. Appl. Chem. U.S.S.R., 44, 1068 (1971).

Barile, R.G., and D.W. Meyer, Chem. Eng. Prog. Symposium Series, 67, No. 119, 134 (1971).

Barile, R.G., J.L. Dengler, and T.A. Hertwig, AIChE Symposium Series, 70, No. 138, 154 (1974).

Blyakher, I.G., L. Ya. Zhivaikin, and N.A. Yurovskaya, Int. Chem. Eng., 7(3), 485 (1967).

Chen, B.H., and W.J.M. Douglas, Can. J. Chem. Eng., 46, 245 (1968).

Chen, B.H., and W.J.M. Douglas, Can. J. Chem. Eng., 47, 113 (1969).

Chern, S.-H., K. Muroyama, and L.-S. Fan, Chem. Eng. Sci., 38, 1167 (1983).

Danckwerts, P.V., "Gas-Liquid Reactions," McGraw-Hill (1970).

Davidson, J.F., R. Clift, and D. Harrison, "Fluidization," Academic Press (1985).

Douglas, H.R., I.W.A. Snider, and G.H. Tomlinson, Chem. Eng. Prog., 59(12), 85 (1963).

Douglas, W.J.M., Chem. Eng. Prog., 60(7), 66 (1964).

Elenkov, D., and A. Kossev, Theor. Found. Chem. Eng., 4, 100 (1970).

Fan, L.-S., K. Muroyama, and S.-H. Chern, Chem. Eng. J., 24, 143 (1982a).

Fan, L.-S., K. Muroyama, and S.-H. Chern, Chem. Eng. Sci., 37(10), 1570 (1982b).

Gel'perin, N.I., V.Z. Grishko, V.I. Savchenko, and V.M. Shchedro, Khimicheskoe i Neftyanoe Mashinostroenie, No. 1, Jan. 22 (1966).

Gel'perin, N.I., V.I. Savchenko, and V.Z. Grishko, Theor. Found. Chem. Eng., 2, 65 (1968).

Gel'perin, N.I., V.Z. Grishko, and V.A. Mikhailov, Theor. Found. Chem. Eng., 6, 479 (1972).

Gel'perin, N.I., and B.S. Kruglyakov, Sov. Chem. Ind., 53, 66 (1977).

Kielback, A.W., Chem. Eng. Prog. Symposium Series, 57, No. 35, 51 (1959).

Khanna, R.T., Ph.D. Dissertation, McGill University (1971).

Kito, M., M. Shimada, T. Sakai, S. Sugiyama, and C.Y. Wen, in "Fluidization Technology," ed. by D.L. Keairns, I, p. 411, Hemisphere (1976a).

Kito, M., M. Sawada, M. Shimada, M. Takata, T. Sakai, and S. Sugiyama, Kagaku Kogaku Ronbunshu, 2(1), 12 (1976b).

Kito, M., M. Shimada, R. Iijima, T. Sakai, M. Takata, and S. Sugiyama, Kagaku Kogaku Ronbunshu, 2(1), 18 (1976c).

Kito, M., Y. Kayama, T. Sakai, and S. Sugiyama, Kagaku Kogaku Ronbunshu, 2(1), 21 (1976d).

Kito, M., T. Monma, Y. Kayama, T. Sakai, and S. Sugiyama, Kagaku Kogaku Ronbunshu, 2(5), 476 (1976e).

Kito, M., K. Tabei, and K. Murata, I&EC Process Des. & Dev., 17(4), 568 (1978).

Kossev, A., G. Peev, and D. Elenkov, Verfahrenstechnik, 5(8), 340 (1971).

Kossev, A., and D. Elenkov, Theor. Found. Chem. Eng., 7, 795 (1973).

Koval', Zh.A., A.V. Bespalov, and O.G. Kuleshov, Theor. Found. Chem. Eng., 9, 289 (1975a).

Koval', Zh.A., A.V. Bespalov, O.G. Kuleshov, and A.P. Zhukov, Theor. Found. Chem. Eng., 9, 837 (1975b).

Krainev, N.I., M.I. Niyazov, I.P. Levsh, and S.U. Umarov, J. Appl. Chem. U.S.S.R., 41, 1961 (1968).

Kunii, D., and O. Levenspiel, "Fluidization Engineering," John Wiley, New York (1969).

Kuroda, M., and K. Tabei, Kagaku Kogaku Ronbunshu, 5, 449 (1979); Int. Chem. Eng., 21(2), 219 (1981).

Letan, R., Chem. Eng. Sci., 29, 621 (1974).

Leung, L.S., B.K. O'Neill, and D.J. Nicklin, in Proceedings of the 1st Pacific Area Chemical Engineering Congress (PACHEC), III, p. 82, Tokyo (1972).

Leva, M., "Fluidization," McGraw-Hill (1959).

Levsh, I.P., N.I. Krainev, and M.I. Niyazov, Int. Chem. Eng., 8(2), 311 (1968a).

Levsh, I.P., M.I. Niyazov, N.I. Krainev, and F.F. Ganikhanova, Int. Chem. Eng., 8, 379 (1968b).

Levsh, I.P., N.I. Krainev, and M.I. Niyazov, Int. Chem. Eng., 8(4), 619 (1968c).

Lewis, E.W., and E.W. Bowerman, Chem. Eng. Prog., 48, 603 (1952).

Lobo, W.E., L. Friend, F. Hashmall, and F. Zenz, Trans. AIChE, 41, 693 (1945).

Mayak, V.I., and V.I. Matrozov, Theor. Found. Chem. Eng., 3, 64 (1969).

Miconnet, M., P. Guigon, and J.-F. Large, Int. Chem. Eng., 22, 133 (1982).

Mrowiec, M., and A. Laszuk, "Fluidization and Its Applications," ed. by H. Angelino, J.P. Couderc, H. Gibert, and C. Laguerie, p. 417, Cepadues-Editions, Toulouse (1974).

Muroyama, K., and L.-S. Fan, AIChE J., 31, 1 (1985).

O'Neill, B.K., D.J. Nicklin, N.J. Morgan, and L.S. Leung, Can. J. Chem. Eng., 50, 595 (1972).

O'Neill, B.K., D.J. Nicklin, and L.S. Leung, "Fluidization and Its Applications," ed. by H. Angelino, J.P. Couderc, H. Gibert, and C. Laguerie, p. 365, Cepadues-Editions, Toulouse (1974).

Page, R.E., Ph.D. Thesis, Appendix D, p. 137, University of Cambridge (1970).

Rama, O.P., D.P. Rao, and V. Subba Rao, Can. J. Chem. Eng., 61, 863 (1983).

Rama, O.P., D.P. Rao, and V. Subba Rao, Can. J. Chem. Eng., 63, 443 (1985).

Ramamurthy, K., and K. Subbaraju, I&EC Process Des. & Dev., 12, 184 (1973).

Riba, J.P., and J.P. Couderc, Can. J. Chem. Eng., 55(2), 118 (1977).

Richardson, J.F., and W.N. Zaki, Trans. Inst. Chem. Engrs., 32, 35 (1954).

Schiesser, W.E., and L. Lapidus, AIChE J., 7(1), 163 (1961).

Shimodaira C., and Y. Yushina, in Proceedings of the 3rd Pacific Chemical Engineering Congress, IV, p. 237, Seoul (1983).

Strumillo, C., J. Adamiec, and T. Kudra, Int. Chem. Eng., 14(4), 652 (1974).

Strumillo, C., and T. Kudra, Chem. Eng. Sci., 32, 229 (1977).

Tabei, K., M. Hasatani, and M. Kuroda, Kagaku Kogaku Ronbunshu, 13, 442 (1987).

Tabei, K., M. Hasatani, and M. Kuroda, Kagaku Kogaku Ronbunshu, 14, 26 (1988).

Tichy, J., A. Wong, and W.J.M. Douglas, Can. J. Chem. Eng., 50, 215 (1972).

Tichy, J., and W.J.M. Douglas, Can. J. Chem. Eng., 50, 702 (1972).

Tichy, J., and W.J.M. Douglas, Can. J. Chem. Eng., 51, 618 (1973).

Uchida, S., C.S. Chang, and C.Y. Wen, Can. J. Chem. Eng., 55, 392 (1977).

Uchida, S., T. Suzuki, and H. Maejima, Can. J. Chem. Eng., 58, 406 (1980).

Uysal, B.Z., Ph.D. Thesis, McGill University (1978).

Visvanathan, C., and L.S. Leung, I&EC Process Des. & Dev., 24, 677 (1985).

Vunjak-Novakovic, G.V., D.V. Vukovic, A. Obermayer, and A. Vogelpohl, in "Fluidization," ed. by G.R. Grace and J.M. Matsen, p. 253, Plenum Press, New York (1980).

Vunjak-Novakovic, G.V., D.V. Vukovic, and H. Littman, I&EC Product Res. & Dev., 26(5), 958 (1987a).

Vunjak-Novakovic, G.V., D.V. Vukovic, and H. Littman, I&EC Product Res. & Dev., 26(5), 967 (1987b).

Wallis, G.B., "One-Dimensional Two-Phase Flow," McGraw-Hill, New York (1969).

Wen, C.Y., and Y.H. Yu, Chem. Eng. Prog. Symposium Series, 62, No. 62, 100 (1966).

Wen, C.Y., and L.-S. Fan, I&EC Process Des. & Dev., 13, 194 (1974).

Werner, G.C., and K. Schugerl, in Proceedings of the 5th European (2nd International) Symposium on Chemical Reaction Engineering, B9-1, Elsevier, Amsterdam, May 2-4 (1972).

Williams, G.C., R.B. Akell, and C.P. Talbott, Chem. Eng. Prog., 43, 585 (1947).

Wozniak, M., and K. Ostergaard, Chem. Eng. Sci., 28, 167 (1973).

Wozniak, M., Int. Chem. Eng., 17(3), 553 (1977).

Yushina, Y., and H. Sato, in Proceedings of the Muroran Meeting, p. 131, Soc. Chem. Engrs. Japan (1983).

——, Chem. Eng., 66, 106, Dec. 14 (1959).

Chapter 6[*]

Miscellaneous Systems

This chapter describes gas-liquid-solid fluidization systems with internals and those involving transport operations. More specifically, Mode E-I-a-1[1] and T-I-a-1[1] draft tube systems, Mode E-I-a-1 annular systems, Mode E-I-a-1 and E-II-a-1 semifluidized bed systems, and Mode T-I-a-1, T-I-b, T-III-a and T-III-b vertical and cocurrent horizontal three-phase flow systems are considered (see Fig. 1.3 for mode designations). Fundamental information on each of the above necessary to permit optimum design or operation is still lacking.

6.1 DRAFT TUBE SYSTEMS

The draft tube gas-liquid-solid fluidization system (DTFB)[2] has recently emerged as a promising technology for three-phase operations. It is already of considerable industrial importance as evidenced by its applications in biological wastewater treatment (see Chapter 8). In a DTFB, a draft tube is commonly concentrically located inside the column and the liquid is the continuous phase. Gas and liquid can be introduced either into the draft tube region, creating a cocurrent upflow pattern in the draft tube, or into the annular region, creating a cocurrent downward flow pattern in the draft tube. The gas and liquid can be discharged from the side of the annular region or top of the column. In small columns and in systems without solids, annular region aeration may result in the formation of gas slugs in the draft tube (due to bubble coalescence), impeding the downward liquid flow in the draft tube and ultimately causing an upward flow in the draft tube (de Nevers, 1968). In this situation, annular region aeration

---

*: by L.-S. Fan and K. Muroyama

1: The mode designation given here is based on the draft tube region of the system.

2: It is also known as a draft tube gas-liquid-solid fluidized bed (DTFB) or spouted bed, a draft tube gas-liquid-solid slurry bubble column, an internal loop gas-liquid-solid reactor, or an internal air-lift gas-liquid-solid reactor.

Table 6.1 Various Designs and Operational Variables on a DTFB

Investigators	Draft Tube		Column Diameter	Static Liquid Height Above Distributor	Liquid Height Above Draft Tube (H_u)	Liquid Height Below Draft Tube (H_l)	Distributor		Particle Properties		Gas-Liquid Properties	Base (Cone) Angle
	Height	Diameter					Gas	Liquid	Density	Diameter		
Sawai et al. (1982)	300 cm	50 cm	70 cm	300 cm	—	—	—	—	1.02~1.2 g/cm^3 (bioparticle: sand covered with biofilms)	0.4~0.9 mm (sand 0.1-0.3 mm and covered with biofilm 0.15-0.3 mm)	air/wastewater	—
Matsuura et al. (1982) Matsuura and Akehata (1985)	67 ~ 74 cm	5.9~9.2 cm	15 cm	83.5 cm	5.3~11.5 cm	4~7.2 cm	porous plate (outer tube of a concentric tube distributor)	tubular injector (inner tube of a concentric tube distributor)	2.4 g/cm^3	1.2 mm	air/water or aqueous glycerol solution	60°
Fan et al. (1984a)	96 cm	10.2 cm	15.2 cm	120.6 cm	19 cm	2.5 cm	perforated plate	single orifice nozzle (1.27 mm ID)	2.47~3.73 g/cm^3	2.2~4.0 mm	air/water	90°
Fan et al. (1984b) Fujie and Fan (1986)	120 cm	5.0 cm	7.6 cm	132 cm	5.5 cm	2.5 cm	perforated and porous plates	single orifice nozzle	1.45~2.5 g/cm^3	307~1500 μm	air/aqueous sodium sulfite solution	90°

Hwang and Fan (1985)	96 cm	10.2 cm	15.2 cm	120.6 cm	19 cm	2.5 cm	perforated plate	single orifice nozzle (1.27 mm ID)	1.26 g/cm^3	1.8~4 mm	air/aqueous benzoic acid solution	90°
Muroyama et al. (1985)	50~125 cm	6.4~9.9 cm	10~14.9 cm	67~138 cm	7.9~13.9 cm	3.1 cm	single orifice nozzle injector (4 and 6 mm ID) and packed bed (30 mm ID) distributor	—	1.7~1.9 g/cm^3	250 μm ~ 2.2 mm	air/water	60°
Koide et al. (1984a,1985)	70~210 cm	6.0~19 cm	10~30 cm	84~224 cm	7.9~21.8 cm	0.34~6.2 cm	perforated plate	—	2.5~8.7 g/cm^3	79~498 μm	air/water solution of glycerol, ethylene glycol, barium chloride or sodium sulfate	60°
Wiley (1986)	120 cm	5.0 cm	7.6 cm	132 cm	5.5 cm	2.5	porous plate	single orifice nozzle	1.5~2.5 g/cm^3	300~1500 μm	air/aqueous sodium sulfite solution	90°
Fan et al. (1987b)	37.5 cm for both stages	5.0 cm for both stages	7.6 cm for both stages	50 cm for 1st stage; 54 cm for 2nd stage	8 cm for 1st stage; 12 cm for 2nd stage	5.5 cm for both stages	porous plate for 1st stage	side tube injector for 1st stage	1.5~2.5 g/cm^3	300~714 μm	air/water	90° for both stages
							single tube orifice for 2nd stage					

is less stable than draft tube aeration, but provides better column-to-wall heat transfer (Cooper et al., 1975). Unless otherwise noted, discussion in this section will focus on systems in which gas and liquid are introduced in the draft tube region and exit at the column top; these systems have a conical base to assist in liquid and solid circulation and prevent the formation of dead space. Note that the bulk of the work on mass transfer involves liquid-batch conditions.

6.1.1 Hydrodynamics

A number of factors, such as the draft tube height and diameter, column diameter, spacings above and below the draft tube, and gas and liquid distributors, combine to produce the observed hydrodynamic behavior in a draft tube system. Fundamental studies on a draft tube (solid-free) bubble column (DTBC) have been extensively reported in the literature regarding hydrodynamic characteristics (e.g., Kubota et al., 1978; Keitel and Onken, 1981; Fields and Slater, 1983; Koide et al., 1983, 1984a,b; Weiland, 1984; Jones, 1985) and gas-liquid mass transfer behavior (e.g., Hatch, 1973; Malfait et al., 1981; Weiland and Onken, 1981a,b, Kawase and Moo-Young, 1986, Koide et al., 1983, 1984b). However, relatively little has been reported regarding the DTFB. The information reported is sketchy and, in most cases, applying these findings must be limited only to a specific DTFB design. That is, the correlations reported are not adequately universal to allow broad application. In the following, the hydrodynamic behavior of a DTFB is described. Wherever appropriate, the DTBC is described as a limiting case of the DTFB. Comparisons of the behavior of a DTFB or DTBC with a BC (a bubble column without a draft tube) are given. Specifications of DTFBs employed by various investigators discussed here are summarized in Table 6.1.

6.1.1.1 Flow Regimes

The flow regime of the draft tube system can be classified as packed bed, fluidized bed, or circulated bed regime. As an example, a flow regime map for various solids loadings of 3 mm glass beads is shown in Fig. 6.1. The figure reflects that at low gas and liquid velocities, the bed is in the packed bed (PB) regime in which all the particles are packed at the bottom of the draft tube and annular regions of the column. As the gas or liquid velocity increases, the

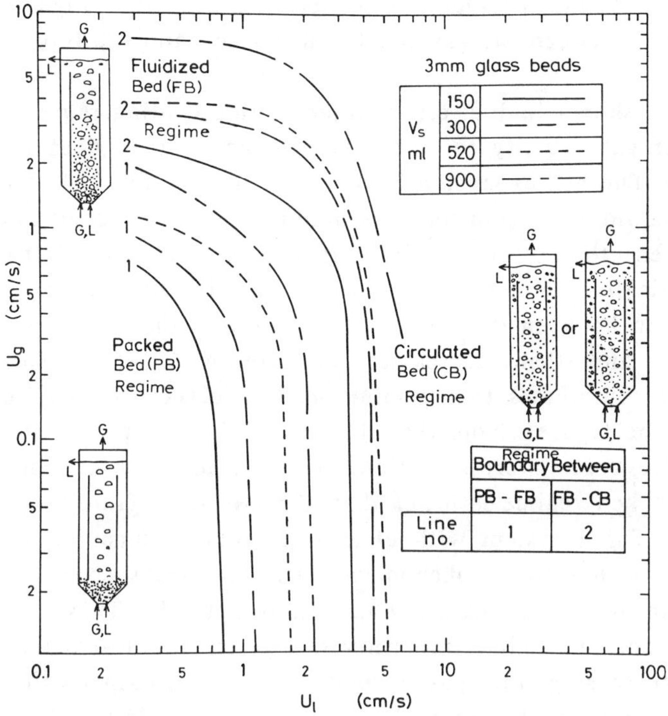

Figure 6.1 Flow regime map for a DTFB as a function of solids loadings (from Fan et al., 1984a). See Table 6.1 for unit dimensions.

solid particles under the draft tube become fluidized. The onset of the fluidized bed regime is defined as the point where a global downward movement of the solid particles occurs around the wall area of the conical section of the bed. By increasing the gas or liquid velocity beyond the onset point, the solid particles under the annular region begin to massively migrate to the draft tube region to form a fluidized bed. In the fluidized bed (FB) regime, considerable liquid circulation occurs and primarily small bubbles are circulated into the annular region. As the gas or liquid velocity is further increased, the height of the fluidized bed increases. When the fluidized bed height exceeds the draft tube height or when the liquid velocity exceeds the particle terminal velocity in the gas-liquid medium, the solid particles are carried over into the annular region and then circulated back to the draft tube: this flow regime is known as the circulated bed (CB) regime. In this regime, a large number of bubbles are observed in the annular region and an intense solid particle circulation is also evident.

The CB regime provides high gas holdup and high contacting efficiency between the gas and liquid phases, and the liquid and solid phases.

As shown in Fig. 6.1, a higher solids loading requires a higher liquid or gas velocity to reach the CB regime than does a low solids loading. The bed expansion in the draft tube region is directly related to the liquid flow rate and gas holdup. For flow conditions in Fig. 6.1, when the solids loading increases, the overall gas holdup monotonically decreases (see Section 6.1.1.4) and the rate of liquid circulation between the draft tube region and the annular region first increases and then decreases (see Section 6.1.1.5). This coupled behavior contributes to the solids loading effects on the flow regime map in the figure. Note that the superficial fluid velocity (based on the entire column cross section) required to reach the onset of the fluidized bed regime is lower than the minimum fluidization velocity required for a system without a draft tube. Matsuura et al. (1982) reported a hysteresis phenomenon for the transition from the FB regime to the CB regime using 1.2 mm glass beads. They found the gas velocity for the regime transition at a given liquid velocity was lower when decreasing the gas velocity than when increasing the gas velocity. No such hysteresis behavior was observed for the 3 mm glass bead system on which Fig. 6.1 is based. Note that the flow in the draft tube region of a DTFB for most operating conditions given in Table 6.1 is Mode T-I-a-1. Unless otherwise indicated, the DTFB in the following text is considered as in the CB regime.

The minimum gas velocity for complete solids suspension in a bubble column slurry system with or without a draft tube is commonly determined based on the pressure drop measurement (also see Section 4.3). In a DTFB, hysteresis behavior was also observed for the column pressure drop variation with increasing or decreasing gas velocity in determining the minimum gas velocity for complete solids suspension (Heck and Onken, 1987). This results in the minimum gas velocity for complete solids suspension determined based on increasing gas velocity being higher than that based on the decreasing gas velocity. Based on the decreasing gas velocity approach, Koide et al. (1984a) reported that the minimum gas velocity required for complete suspension of small particles increased with increasing solids loading, liquid surface tension, and difference between solid and liquid density, but decreased with increasing column diameter and liquid viscosity. Solids suspension behavior can be characterized by three states of solids circulation based on the total pressure drop versus gas velocity

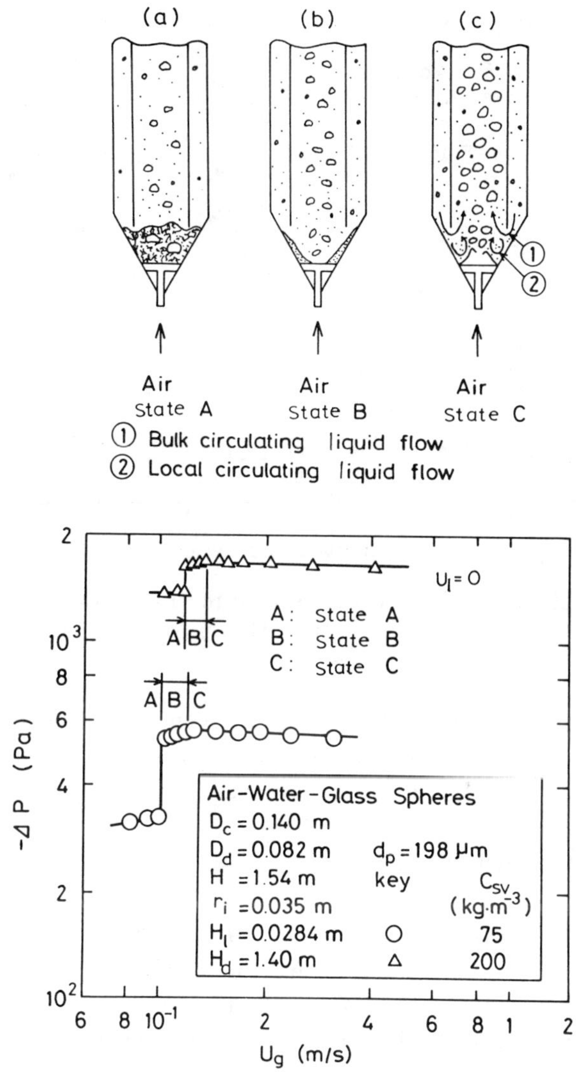

Figure 6.2 Variation of the pressure drop with the gas velocity for a DTFB (from Koide et al., 1984a).

relationship, as shown in Fig. 6.2. State A corresponds to both the PB and FB regimes given in Fig. 6.1. States B and C are in the CB regime where State B involves solids forming a dead space around the conical base and State C involves solids in complete suspension. The effects of solids loading on the transition velocity from State B to State C, corresponding to the minimum gas velocity for complete solids suspension, for liquid-batch systems with and without a draft tube are

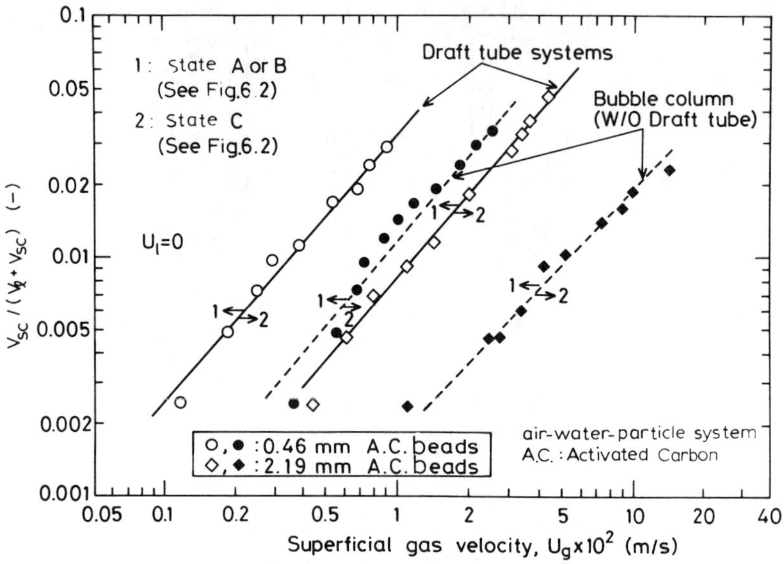

Figure 6.3 Solids loading effects on state transition for systems with and without a draft tube (from Muroyama et al., 1985).

shown in Fig. 6.3. It is seen that for a given solids loading, the system without a draft tube requires a considerably larger gas velocity to reach State C, or complete solids suspension.

6.1.1.2 Pressure Distribution

Typical pressure distributions in the draft tube and annular regions are shown in Fig. 6.4. At the draft tube top, the pressure within the draft tube is higher than that in the annular region; however, at the draft tube base, the pressure within the draft tube is lower than that in the annular region. This pressure difference provides an adequate driving force for liquid circulation from the draft tube region to the annular region (at the draft tube top). The pressure variation in the annular region is approximately linear with axial distance along the length of the draft tube. This is not the case, however, for the pressure distribution in the draft tube region due to the drastic variation in the pressure profile near the draft tube base, as shown in Fig. 6.4. The variation can be attributed to liquid and gas jetting effects which are strongly associated with the physical properties of the particles involved.

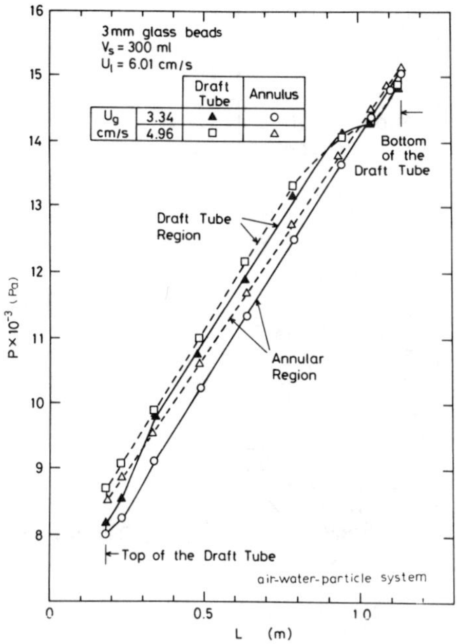

Figure 6.4 Pressure distribution in the draft tube region and the annular region of a DTFB (from Fan et al., 1984a). L is the downward distance from the retaining grid.

Consider a DTFB which consists of four regions, i.e., a region below the draft tube, the draft tube itself, a region above the draft tube, and an annulus as shown in Fig. 6.5. The pressure drop in the bed can be described based on momentum balances for each of the four regions as given below (Fan et al., 1984a):

(i) Below draft tube region

$$(\rho_g V_{g1}^2 A_{g1} + \rho_\ell V_{\ell 1}^2 A_{\ell 1}) - (\rho_g V_{g2}^2 A_{g2} + \rho_\ell V_{\ell 2}^2 A_{\ell 2} + \rho_s V_{s2}^2 A_{s2})$$
$$- (\rho_g V_{g3}^2 A_{g3} + \rho_\ell V_{\ell 3}^2 A_{\ell 3} + \rho_s V_{s3}^2 A_{s3}) + (P_1 A_1 - P_2 A_2 - P_3 A_3)$$
$$- (\rho_g \epsilon_{gB} + \rho_\ell \epsilon_{\ell B} + \rho_s \epsilon_{sB}) V_B g - F_B + F_N = 0 \tag{6.1}$$

(ii) Draft tube region

$$(\rho_g V_{g3}^2 A_{g3} + \rho_\ell V_{\ell 3}^2 A_{\ell 3} + \rho_s V_{s3}^2 A_{s3}) - (\rho_g V_{g4}^2 A_{g4} + \rho_\ell V_{\ell 4}^2 A_{\ell 4} + \rho_s V_{s4}^2 A_{s4})$$
$$+ (P_3 A_3 - P_4 A_4) - (\rho_g \epsilon_{gD} + \rho_\ell \epsilon_{\ell D} + \rho_s \epsilon_{sD}) V_D g - F_D = 0 \tag{6.2}$$

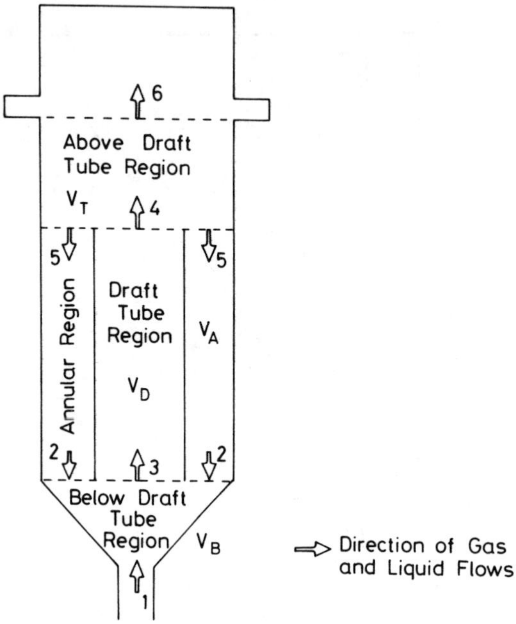

Figure 6.5 Schematic representation of notations used in the analysis of a DTFB.

(iii) Above draft tube region

$$(\rho_g V_{g4}^2 A_{g4} + \rho_\ell V_{\ell 4}^2 A_{\ell 4} + \rho_s V_{s4}^2 A_{s4}) + (\rho_g V_{g5}^2 A_{g5} + \rho_\ell V_{\ell 5}^2 A_{\ell 5} + \rho_s V_{s5}^2 A_{s5})$$
$$- (\rho_g V_{g6}^2 A_{g6} + \rho_\ell V_{\ell 6}^2 A_{\ell 6} + \rho_s V_{s6}^2 A_{s6}) + (P_4 A_4 + P_5 A_5 - P_6 A_6)$$
$$- (\rho_g \epsilon_{gT} + \rho_\ell \epsilon_{\ell T} + \rho_s \epsilon_{sT}) V_T g - F_T = 0 \tag{6.3}$$

(iv) Annular region

$$-(\rho_g V_{g5}^2 A_{g5} + \rho_\ell V_{\ell 5}^2 A_{\ell 5} + \rho_s V_{s5}^2 A_{s5}) + (\rho_g V_{g2}^2 A_{g2} + \rho_\ell V_{\ell 2}^2 A_{\ell 2} + \rho_s V_{s2}^2 A_{s2})$$
$$+ (P_2 A_2 - P_5 A_5) - (\rho_g \epsilon_{gA} + \rho_\ell \epsilon_{\ell A} + \rho_s \epsilon_{sA}) V_A g + F_A = 0 \tag{6.4}$$

Combining Eqns. (6.1) through (6.4) yields

$$(\rho_g V_{g1}^2 A_{g1} + \rho_\ell V_{\ell 1}^2 A_{\ell 1}) - (\rho_g V_{g6}^2 A_{g6} + \rho_\ell V_{\ell 6}^2 A_{\ell 6}) + (P_1 A_1 - P_6 A_6)$$
$$- (\rho_g \epsilon_g + \rho_\ell \epsilon_\ell + \rho_s \epsilon_s) V g - F = 0 \tag{6.5}$$

where

$$\rho_g\epsilon_g+\rho_\ell\epsilon_\ell+\rho_s\epsilon_s=(\rho_g\epsilon_{gA}+\rho_\ell\epsilon_{\ell A}+\rho_s\epsilon_{sA})\frac{V_A}{V}+(\rho_g\epsilon_{gB}+\rho_\ell\epsilon_{\ell B}+\rho_s\epsilon_{sB})\frac{V_B}{V}$$

$$+(\rho_g\epsilon_{gD}+\rho_\ell\epsilon_{\ell D}+\rho_s\epsilon_{sD})\frac{V_D}{V}+(\rho_g\epsilon_{gT}+\rho_\ell\epsilon_{\ell T}+\rho_s\epsilon_{sT})\frac{V_T}{V} \tag{6.6}$$

and

$$F = F_B + F_D + F_T - F_A - F_N \tag{6.7}$$

F can also be expressed as

$$F = (-\Delta P_{fr})(\frac{\pi}{4} D_c^2) \tag{6.8}$$

where $(-\Delta P_{fr})$ is the pressure drop in the bed due to force interaction with the wall. The wall interaction factor (f) can be defined based on the modified Fanning equation as

$$(-\Delta P_{fr}) = 4f(\frac{H}{D_c})(\frac{1}{2}\rho_m U_c^2) \tag{6.9}$$

where H is the distance from the top of the distributor to the top liquid level, D_c is the column diameter, ρ_m is the mixture density defined as $(\rho_g\epsilon_g + \rho_\ell\epsilon_\ell + \rho_s\epsilon_s)$, and U_c is the apparent liquid circulation velocity based on the cross-sectional area of the column. Fan et al. (1984a) found that f was a unique function of the apparent liquid circulation Reynolds number $(Re)_c$ for a given DTFB operated with or without solid particles and can be expressed by

$$f = k_1(Re)_c^{k_2} \tag{6.10}$$

where $(Re)_c$ is defined as $\frac{D_c U_c \rho_\ell}{\mu_\ell}$, and k_1 and k_2 are empirical constants. Fan et al. (1984a) reported the values for k_1 and k_2 as 1.28×10^9 and -1.74, respectively.

6.1.1.3 Bubble Penetration Depth in the Annular Region

Fields and Slater (1983) observed that in a DTBC a front formed in the annular region which separated bubbly liquid from

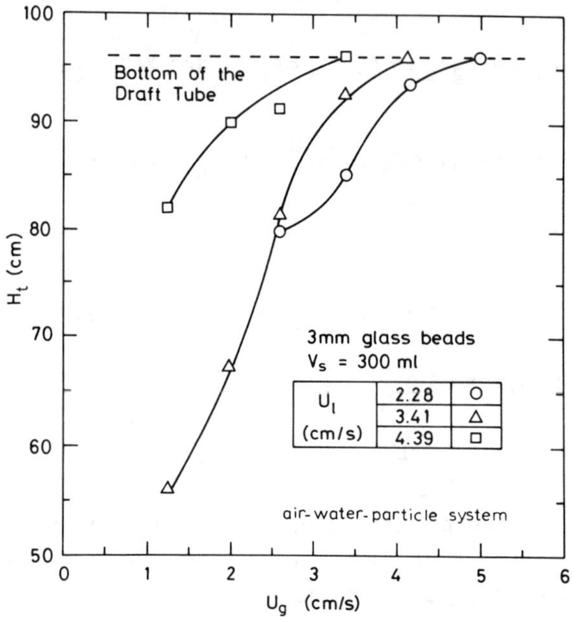

Figure 6.6 Effects of gas velocity on the bubble penetration depth in the annular region, H_t, as a function of liquid velocity (from Fan et al., 1984a). H_t is evaluated based on the downward distance from the top of the draft tube.

bubble-free liquid. A similar, although apparently unstable, bubbly liquid front was also observed in a DTFB by Fan et al. (1984a). Figure 6.6 shows the gas and liquid velocity effects on the bubble penetration depth using large particles, i.e., 3 mm glass beads. In this figure, the bubble penetration depth was defined as the distance from the draft tube top to the bubbly liquid front; the front location is taken as an arithmetic mean of the maximum and minimum penetration depths for given operating conditions. As shown in Fig. 6.6, the bubble penetration depth increases as the liquid velocity increases. In addition, it increases as the gas velocity increases, similar to the behavior of gas-liquid systems reported by Fields and Slater (1983). As the gas velocity increases, a critical gas velocity will be reached at which the bubbly liquid front reaches the draft tube base. With an increase in gas velocity beyond the critical velocity, complete bubble circulation is achieved. Note, as shown in Fig. 6.6, the critical gas velocity decreases with an increase in liquid velocity.

Counter to the large particle behavior observed by Fan et al. (1984a), Wiley (1986) reported that increasing the liquid velocity at a constant gas velocity decreases the bubble penetration depth in the annulus for small and/or light particles (300 μm glass beads, 300 and 1500 μm activated carbon). Wiley (1986) found the linear liquid velocity in the annulus decreased with an increase in the superficial liquid velocity over the operating conditions considered; the decrease in linear liquid velocity is responsible for the decrease in the bubble penetration depth.

Note that the behavior of the bubble penetration in the annulus resembles bed expansion in inverse gas-liquid-solid fluidization (see Chapter 5). For a bubble to be entrained to an infinite depth, the liquid circulation velocity must exceed the bubble terminal velocity. Thus, at low liquid circulation velocities, only a small number of bubbles are entrained into the annular region; as the liquid circulation velocity increases, more bubbles are entrained. It is further noted that the bubble size distributions in both the annular and draft tube regions of a DTFB (Fan et al., 1984a) as well as those in a DTBC (Miyahara et al., 1986), follow a log-normal distribution function resembling that in a three-phase fluidized bed (without a draft tube) (see Chapter 2). The average bubble size in the annular region is comparable to that in the dispersed bubble regime of a three-phase fluidized bed independent of whether the draft tube region is in the dispersed or coalesced bubble regime.

6.1.1.4 Overall Gas Holdup

Gas holdup behavior depends significantly on the gas velocity; an increase in the gas velocity increases gas holdup. Similar behavior was also observed in a DTBC (e.g., Hatch, 1973; Fields and Slater, 1983) or an external circulating bubble column (e.g., Weiland and Onken, 1981a). Figure 6.7 shows various studies on gas holdup behavior in a DTFB and contrasts the behavior with that found in a DTBC and a BC. A list of the design and operating variables for these studies is given in Table 6.2. The overall gas holdup behavior is also significantly affected by the physical properties of the particle. For low density particles, e.g., activated carbon particles, gas holdup is rather unaffected by particle size, solids loading (curves 3 and 7), and the presence of the draft tube. This is not the case, however, for relatively heavy particles, e.g., glass beads and bronze particles, where gas holdup decreases with an increase in solids loading (curves 1a, 1b

Table 6.2 Legend for Figure 6.7

Curve No.	Type of Contactor	Air Diffuser	Solid Particles	References/ Comments
1	DTFB	Concentric injector with perforated plate as gas diffuser	Glass beads d_p = 3 mm 1a*: L_s=0.9 vol% 1b*: L_s=1.8 vol% 1c*: L_s=3.6 vol% * Curve No.	Fan et al. (1984a)/ U_ℓ=3.41 cm/s
2	DTFB	Perforated plate	Glass beads d_p=198 μm L_s=4 vol%	Koide et al. (1985)/ batch liquid
3	DTFB	Single orifice nozzle	Activated carbon beads d_p=0.25 - 2.2 mm L_s= Variable(crital holdup for solid suspension)	Muroyama et al. (1985)/ batch liquid
4	DTFB	Porous plate	Glass beads d_p=1.2 mm L_s=2.12 vol%	Matsuura et al. (1982)/ U_ℓ=0.3 cm/s
5	DTBC	Concentric injector with perforated plate as gas diffuser	—	Fan et al. (1984a)/ U_ℓ=3.41 cm/s
6	DTBC	Multi-nozzles	—	Kawase and Moo-Young(1986)/ batch liquid
7	BC	Single nozzle	—	Muroyama et al. (1985)/ batch liquid
8	BC	Single nozzle	—	Akita and Yoshida (1973)/ batch and continuous liquid flow

and 1c). For a given DTFB or DTBC, an increase in liquid velocity increases the gas holdup (curve 1a, 1b or 1c and curve 2 or 4; curves 5 and 6) It is noted that compared to a fluidization system without a draft tube, the gas holdup in the DTFB is significantly higher.

The overall gas holdup is also affected by system design variables such as draft tube size, top spacing, bottom spacing, and type of distributor (or injector). Hwang and Fan (1986) examined three ratios of draft tube to column diameter for a given DTFB (0.5, 0.67, and 0.83) and found a ratio of 0.5 yielded the highest overall gas

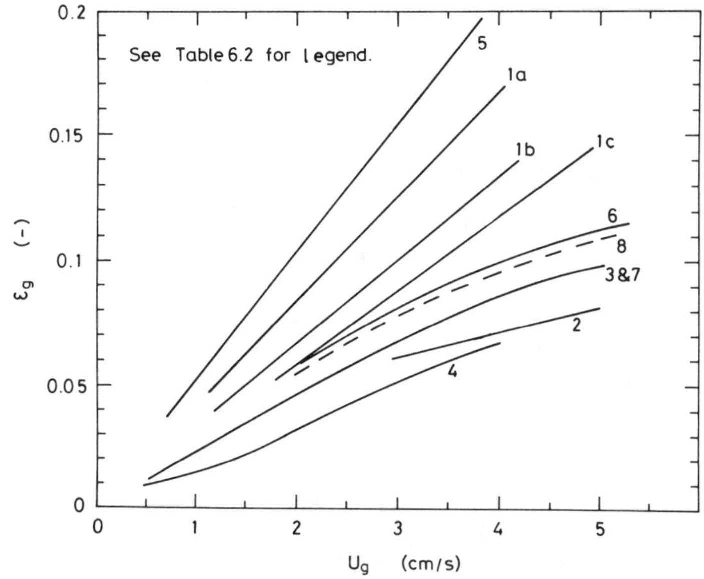

Figure 6.7 Relationship between ϵ_g and U_g for various draft tube systems and bubble columns.

holdup. In addition, they studied four types of concentric tube injectors shown in Fig. 6.8: injectors 1 through 3 have liquid flowing through the inner tube and gas flowing through the outer tube while injector 4 has liquid flowing through the outer tube and gas flowing through the inner tube. They found that injector 2 gave rise to the highest overall gas holdup. In addition, an optimum bottom spacing exists for maximum overall gas holdup.

Fan et al. (1984a) found the overall gas holdup in a DTFB can be expressed as a unique function of its liquid circulation Reynolds number $(Re)_c$ as

$$\epsilon_g = k_3 + k_4\,(Re)_c \tag{6.11}$$

where k_3 and k_4 are empirical constants. Fan et al. (1984a) reported the values for k_3 and k_4 as -0.115 and 1.045×10^{-5}, respectively. A correlation which relates ϵ_g to the operating variables for relatively small particles (see Table 6.1) were given by Koide et al. (1985) as

$$\frac{\epsilon_g}{(1-\epsilon_g)^4} = 0.127\left(\frac{U_g\mu_\ell}{\sigma}\right)^{0.934}\left(\frac{\rho_\ell\sigma^3}{g\ \mu_\ell^4}\right)^{0.292}\left(\frac{D_d}{D_c}\right)^{0.300}$$

$$\cdot\left\{1+0.17\left(\frac{C_{sv}}{\rho_s}\right)^{0.091}\left(\frac{\rho_\ell\sigma^3}{g\mu_\ell^4}\right)^{0.043}\left(\frac{U_t}{U_g}\right)^{0.067}\right\}^{-1} \tag{6.12}$$

where $C_{sv}=\rho_s\epsilon_s/(\epsilon_\ell+\epsilon_s)$. The equation is applicable to a wide range of operating conditions including: $3.59\times10^{-4}\le(U_g\mu_\ell/\sigma)\le3.70\times10^{-2}$, $1.69\times10^{-11}\le(g\mu_\ell^4/\rho_\ell\sigma^3)\le2.55\times10^{-6}$, $0.471\le(D_d/D_c)\le0.743$, $0\le(C_{sv}/\rho_s)\le0.08$, and $0.0117\le(U_t/U_g)\le0.844$.

Equation (6.12) is applicable only to non-frothing liquid in a batch system. For a frothing liquid, Eqn. (6.12) can be modified by including the Marrucci parameter (Marrucci, 1969) to account for the effect of liquid frothing ability on ϵ_g (Koide et al., 1985).

6.1.1.5 Apparent Volumetric Liquid Circulation Rate

The true liquid circulation rate between the draft tube region and the annular region of a DTFB is represented by the liquid flow rate in the annular region. However, it is convenient to characterize the true liquid circulation rate based on an "apparent" rate, Q_ℓ, defined by Fan et al. (1984a):

$$Q_\ell = V_\ell/\bar{t} \tag{6.13}$$

where $\bar{t}$ is the average time for liquid elements to travel a single loop from the annulus to the draft tube and back to the annulus. When the liquid is in a batch condition, the apparent liquid circulation rate is equivalent to the true liquid circulation rate.

The effects of the gas and liquid velocities on the apparent liquid circulation rate are shown in Fig. 6.9 for 1.2 mm and 3 mm glass beads. The solids loading effects on Q_ℓ depend significantly on the particle properties. For 1.2 mm glass beads, Q_ℓ decreases with increasing solids loading. A sharp increase in Q_ℓ is seen in the transition from State A to B (defined in Fig. 6.2) while a smooth transition is seen from State B to C. For 3 mm glass beads at gas velocities beyond 2 cm/s, as the solids loading increases, Q_ℓ first increases then decreases; apparently, under this gas velocity range an optimal solids loading exists which gives rise to a maximum apparent liquid circulation rate. Also, as shown, when the solids loading is less than 600 mℓ and at gas velocities beyond 3.6 cm/s, Q_ℓ for a solid-free system is lower than that in a DTFB.

Fan et al. (1984a) proposed a mathematical model to account for the apparent liquid circulation rate in a DTFB; the model has been

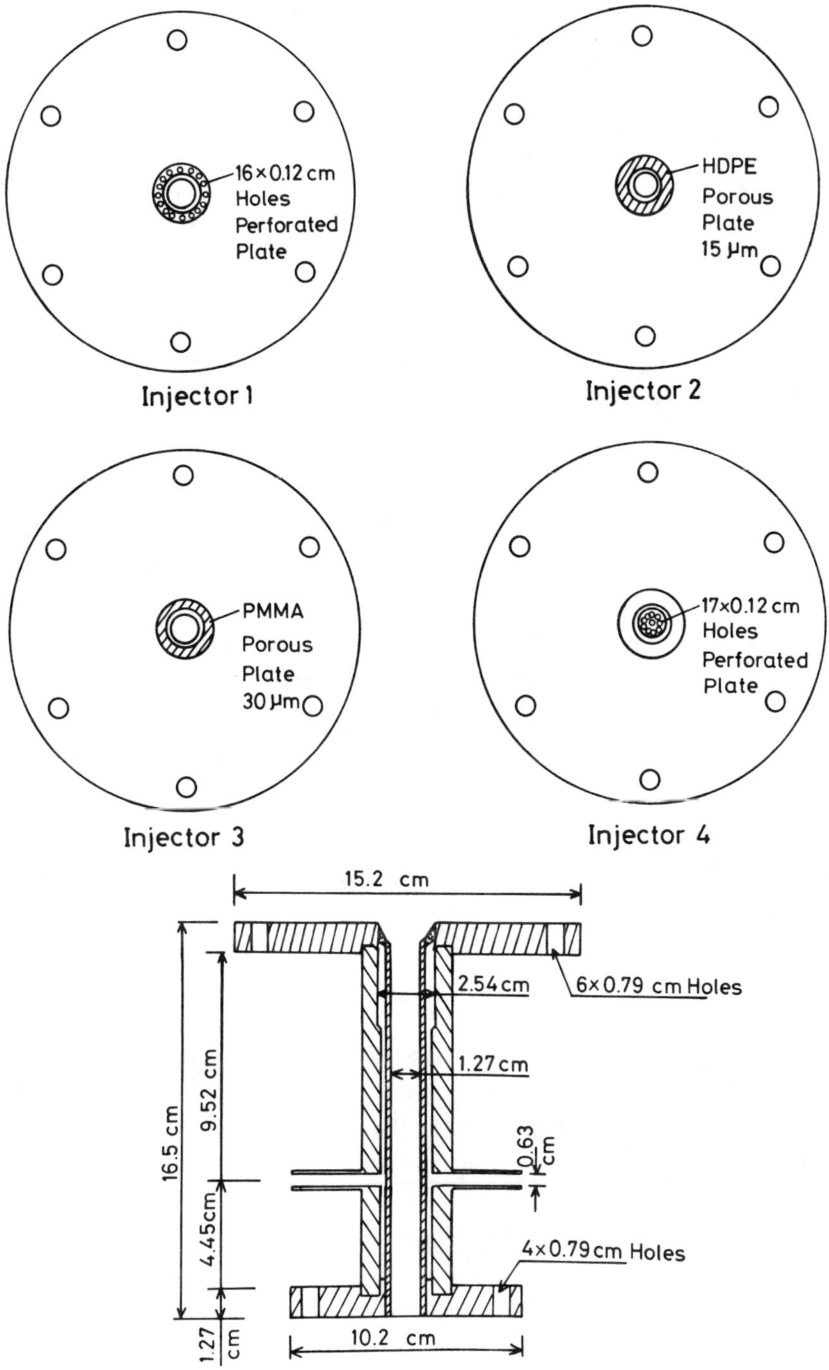

Figure 6.8 Schematic diagram of various gas-liquid injectors for a DTFB (from Hwang and Fan, 1986).

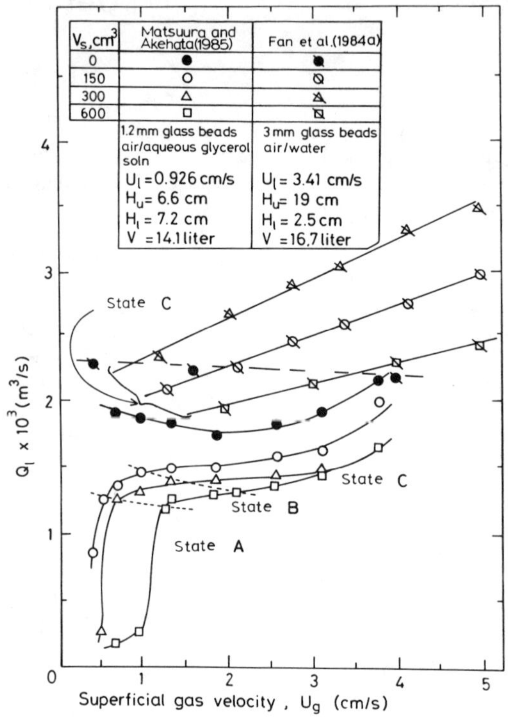

Figure 6.9 Effects of gas velocity and solids loading on the liquid circulation rate for a DTFB.

verified with experimental data. Referring to Fig. 6.5, the pressure drop in the draft tube region (P_3 - P_4) can be expressed as

$$P_3 - P_4 = -\Delta P_{Df} - \Delta P_{Da} - \Delta P_{Dh} \tag{6.14}$$

where $-\Delta P_{Df}$, $-\Delta P_{Da}$, and $-\Delta P_{Dh}$ are the pressure drops in the draft tube due to friction, fluid acceleration, and to the hydrostatic head, respectively. By neglecting the frictional and acceleration contributions, Eqn. (6.14) becomes

$$P_3 - P_4 = -\Delta P_{Dh} \tag{6.15}$$

Similarly, the pressure drop in the annulus is

$$P_2 - P_5 = -\Delta P_{Ah} \tag{6.16}$$

where $-\Delta P_{Ah}$ is the pressure drop in the annular region due to hydrostatic head. Subtracting Eqn. (6.15) from Eqn. (6.16) yields

$$(P_4 - P_5) + (P_2 - P_3) = \Delta P_{Dh} - \Delta P_{Ah} \tag{6.17}$$

The right-hand side of Eqn. (6.17) can be expressed in terms of the bulk density difference between the annular and draft tube regions as

$$\Delta P_{Dh} - \Delta P_{Ah} = [(\rho_g \epsilon_{gA} + \rho_\ell \epsilon_{\ell A} + \rho_s \epsilon_{sA}) - (\rho_g \epsilon_{gD} + \rho_\ell \epsilon_{\ell D} + \rho_s \epsilon_{sD})]gH_d \tag{6.18}$$

where H_d is the draft tube height. The pressure loss due to fluid circulation from points 4 to 5 defined in Fig. 6.5 can be divided into two components: pressure loss due to fluid expansion from the draft tube region to the above draft tube region and that due to fluid contraction from the above draft tube region to the annular region. Accordingly, $P_4 - P_5$ can be expressed as

$$(P_4 - P_5) = \frac{1}{2}\rho_\ell U_D^2 (1 - A_D/A_T)^2 + \frac{1}{2} K_c \rho_\ell U_A^2 \tag{6.19}$$

where U_D is the liquid velocity in the draft tube region based on the draft tube area, U_A is the liquid velocity in the annular region based on the annulus area, A_T is the cross sectional area of the column, and K_c is the sudden contraction loss coefficient. Note that A_D is equal to A_3 or A_4 for a straight draft tube. K_c in Eqn. (6.19) is a function of A_T/A_D and can be approximated by that for a single-phase liquid flow; the correlation for K_c can be found in Perry and Chilton (1973).

The pressure loss due to fluid circulation from points 2 to 3 defined in Fig. 6.5 can also be divided into two components: the pressure loss due to gradual contraction from the annular region to the injector and that due to liquid jetting from the injector to the draft tube region. Accordingly, $P_2 - P_3$ can be expressed as

$$(P_2 - P_3) = \frac{1}{2} \kappa \rho_\ell U_B^2 + \frac{1}{2} \rho_\ell (U_I^2 - U_D^2) \tag{6.20}$$

where U_B is the liquid velocity in the annular region based on the cross-sectional area of the mid-outer tube of the injector; i.e., the area with radius of $(r_i + r_o)/2$ where r_i and r_o are the radii of the inner and outer tubes of the injector, respectively, and κ is the gradual contraction loss coefficient. U_I is the liquid velocity from the injector. Note that for the second term on the right-hand side of Eqn. (6.20), the kinetic energy loss is considered to dominate the pressure loss due to liquid jetting from the injector to the draft tube region. κ in Eqn. (6.20) can be related to the angle of the conical section (θ) (Zhou, 1979) as

$$\kappa = \frac{\lambda}{8 \sin(\theta/2)} (1 - 1/n^2) \tag{6.21}$$

Here n is the area ratio of the annular region to the mid-outer tube of the injector; note that the cross-sectional area of the injector is very small compared to that of the annular region and, therefore, the term $(1-1/n^2)$ in Eqn. (6.21) is practically equal to 1. The Darcy friction factor (λ) characterizes the flow condition prior to entering the conical contraction section and can be evaluated by the Blasius equation (Blasius, 1913):

$$\lambda = 0.3164(D_e U_A \rho_\ell / \mu_\ell)^{-0.25} \tag{6.22}$$

In this equation, D_e, the equivalent diameter of the annular region, is defined as four times the hydraulic radius. The use of an equivalent diameter expressed in terms of the hydraulic radius is a reasonable way to represent the present draft tube system.

Material balances on the liquid phase at the below draft tube and above draft tube regions give

$$U_D A_3 = U_A A_2 + U_I A_1 \tag{6.23}$$

$$U_D A_4 = U_A A_5 + U_E A_6 \tag{6.24}$$

The liquid velocities in the draft tube region (U_D) and in the annular region (U_A) can be obtained by solving Eqns. (6.17) through (6.24). The liquid circulation time in each region can be calculated by

$$t_D = H_d \epsilon_{\ell D} / U_D \tag{6.25}$$

$$t_A = H_d \epsilon_{\ell A} / U_A \tag{6.26}$$

$$t_B = V_B \epsilon_{\ell B} / U_D A_3 \tag{6.27}$$

$$t_T = V_T \epsilon_{\ell T} / U_D A_4 \tag{6.28}$$

The total liquid circulation time is the sum of the liquid circulation times in each region:

$$t_{Total} = t_A + t_B + t_D + t_T \tag{6.29}$$

Note that in calculating t_A, t_B, t_D and t_T from Eqns. (6.25) through (6.28), experimental values for the liquid holdup in each region ($\epsilon_{\ell A}$, $\epsilon_{\ell B}$, $\epsilon_{\ell D}$, and $\epsilon_{\ell T}$) are required. The apparent liquid circulation rate can then be calculated by

$$Q_\ell = V_\ell / t_{Total} \tag{6.30}$$

Note that for a DTBC, a maximum volumetric liquid circulation rate was found to occur for a draft tube to column diameter ratio of 0.6 by Koide et al. (1984b) and 0.5 by Jones (1985).

6.1.2 Overall Gas-Liquid Mass Transfer

Like for Mode E-I-a-1 (see Chapter 3), the overall gas-liquid mass transfer coefficient for a DTFB reported in the literature reflects the liquid-side mass transfer coefficient. Typically, gas-liquid mass transfer experiments involve oxygen transfer from the gas phase to a liquid and possibly subsequent liquid phase oxidation reactions with sodium sulfite (Na_2SO_3). Fan et al. (1984b) and Fujie and Fan (1986) incorporated these phenomena into a model to account for gas-liquid mass transfer and considered systems in which the liquid phase is completely mixed, but the gas phase can either be in plug flow or be completely mixed.

For plug flow, a mass balance on oxygen in terms of the fraction of oxygen utilization (η) gives

$$G_0 \frac{d\eta}{dz} = k_\ell a(C_0 Z - C) A_T \tag{6.31}$$

where

$$G_0 = \frac{U_g A_T Z_i}{v_m}$$

$$C_0 = C_0^* \{1 + \alpha(H - z)\}$$

where α is the pressure conversion factor which is equal to 9.8 ρ_m/P^*. Note that ρ_m, P^*, H and z are in SI units. Equation (6.31) is subject to the boundary condition:

$$\text{at } z = 0; \quad \eta = 0 \tag{6.32}$$

Z can be related to η by

$$Z = Z_i\,(1 - \eta)/(1 - Z_i\eta) \tag{6.33}$$

From Eqn. (6.33), η can be given as

$$\eta = (Z_i - Z)/[Z_i(1-Z)] \tag{6.34}$$

To simplify the first order nonlinear differential equation, Eqn. (6.31), C_0 can be approximated by a mean value evaluated at the column midpoint (z = H/2):

$$\bar{C}_0 = C_0^*\,(1 + \alpha H/2) \tag{6.35}$$

With this approximation, Eqn. (6.31) can be integrated to give

$$k_\ell a = \frac{U_g[Z_i\eta_e(\bar{C}_0\text{-}C)\text{-}\bar{C}_0(1\text{-}Z_i)\ \ln\{1 - \eta_e[\bar{C}_0\text{-}C]/[\bar{C}_0\text{-}C/Z_i]\}]}{(\bar{C}_0\text{-}C)^2 v_m H} \tag{6.36}$$

For complete mixing, a mass balance on oxygen in the gas phase gives

$$U_g\eta_e = k_\ell a^\$(\bar{C}_0 Z_e - C)Hv_m/Z_i \tag{6.37}$$

With $\bar{C}_0$ given by Eqn. (6.35), $k_\ell a^\$$ can be expressed as

$$k_\ell a^\$ = \frac{U_g\eta_e Z_i}{v_m H[C_0^*\,(1+\alpha H/2)Z_e\text{-}C]} \tag{6.38}$$

Both models provide good estimates of the overall gas-liquid mass transfer. More specifically, based on the sodium sulfite oxidation method (see Section 6.1.2.1) coupled with an analysis of the off-gas concentration, $k_\ell a^\$$ calculated using Eqn. (6.38) under the completely mixed assumption is nearly equal to or slightly larger than $k_\ell a$ using Eqn. (6.36) under the plug flow assumption.

6.1.2.1 Sodium Sulfite Concentrations Effects

In mass transfer experiments, the sodium sulfite concentration in the liquid phase needs to be restricted to allow determination of a purely physical gas-liquid mass transfer coefficient. An excessive amount of Na_2SO_3 in the liquid phase increases the gas-liquid oxygen transfer rate per unit gas-liquid interfacial area due to an increased Na_2SO_3 oxidation rate or oxygen consumption rate in the liquid film surrounding the gas bubbles. The ubiquitous nature of this method in the mass transfer literature and the representative interrelated effects on the hydrodynamics and mass transfer warrant an in depth look at this phenomenon.

Fuller and Crist (1941) reported that in the absence of catalysts the sodium sulfite oxidation rate in an aqueous solution follows first-order reaction kinetics with respect to the sodium sulfite concentration over a wide range of reaction conditions, i.e.,

$$-r_B = -2r_{O_2} = KC_B \tag{6.39}$$

where $-r_B$ and $-r_{O_2}$ are the Na_2SO_3 oxidation rate and the oxygen consumption rate per unit volume of the liquid phase, respectively, and C_B is the Na_2SO_3 concentration. The reaction rate constant (K) was evaluated as 0.013 s^{-1} at 25°C (Fuller and Crist, 1941). Based on Eqn. (6.39), the Hatta number (γ), which is defined as the ratio of the oxygen consumption rate due to sodium sulfite oxidation to the maximum available rate for physical absorption, can be expressed by

$$\gamma = \frac{1}{k_\ell} \sqrt{D_{M,O_2} KC_B/C_{sa}} \tag{6.40}$$

For γ less than 0.3, an enhancement factor, defined as the ratio of the amount of gas absorbed into a reacting liquid in a given time to the amount which would be absorbed if there were no reaction, of about 1 is expected (Astarita, 1967) and the oxygen transfer through the gas-liquid interface can be regarded as pure physical absorption. The upper limit of Na_2SO_3 concentration in the liquid allowable to maintain an enhancement factor of about 1, corresponding to the value for γ of 0.3, is estimated as 8550 mg/ℓ or 0.067 mol/ℓ; the estimation is based on a value for k_ℓ in a bubble column of 0.03 cm/s (Fujie et al., 1980a, 1980b) and is conservative since k_ℓ for a DTFB is expected to be higher than that for a bubble column.

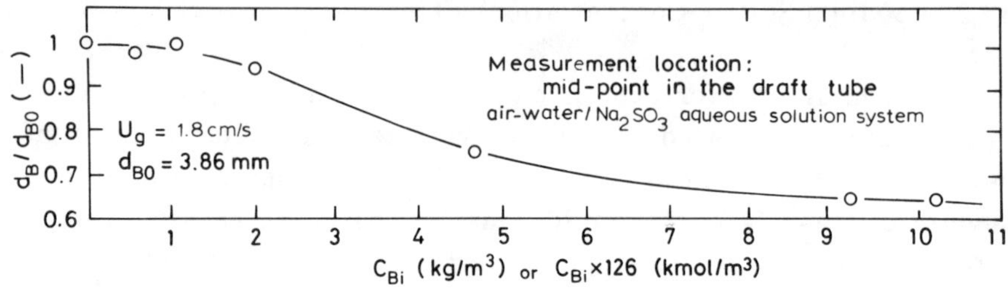

Figure 6.10 Effects of influent Na_2SO_3 concentration on bubble size reduction in a DTBC (from Fujie and Fan, 1985).

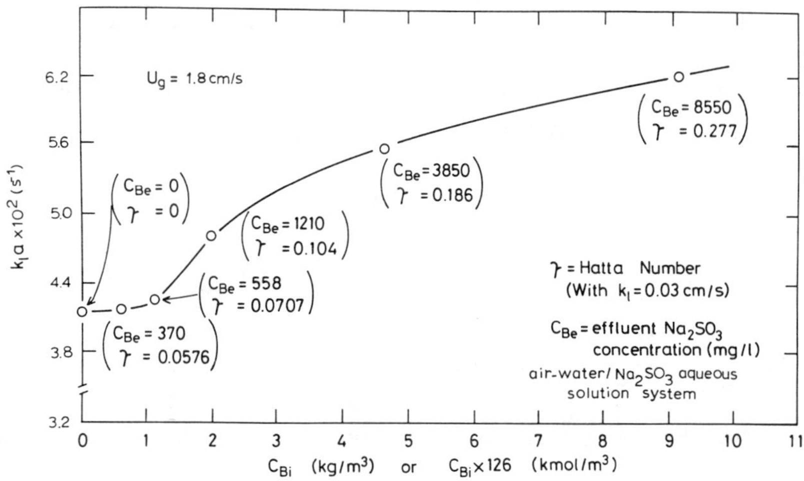

Figure 6.11 Effects of influent Na_2SO_3 concentration on volumetric gas-liquid mass transfer coefficient in a DTBC (from Fujie and Fan, 1985).

It is further known that the presence of electrolytes, such as Na_2SO_3, appreciably reduces the size of bubbles generated by an air distributor (Koide et al., 1972; Fujie et al., 1980a). This phenomenon is more pronounced when using porous plates (Miyahara and Takahashi, 1986) or jet nozzles which generate fine bubbles. For a given air flow rate, a reduction in the bubble size yields an increase in $k_\ell a$ due to an increase in gas-liquid interfacial area.

Figures 6.10 and 6.11 show the effects of the influent sodium sulfite concentration on the bubble size and $k_\ell a$ in a DTBC, respectively, for a U_g of 1.8 cm/s. It is seen that over a wide range of Na_2SO_3 concentrations for which the enhancement factor is near 1, a substantial reduction in bubble size, and hence an increase in the volumetric mass transfer coefficient, occurs at the higher Na_2SO_3 concentrations. The influent Na_2SO_3 concentration (in mol/ℓ) indicated in these figures also represents the total electrolyte concentration in the DTBC. Also indicated in Fig. 6.11 are values of γ as well as the effluent Na_2SO_3 concentrations (in mg/ℓ) which are calculated based on the mass balance and stoichiometric relationship for Na_2SO_3 oxidation. The electrolyte effect is thus apparent. It is noted in the figure that when the concentration of Na_2SO_3 in the system is higher than 0.012 mol/ℓ or the influent concentration of Na_2SO_3 is higher than 1500 mg/ℓ, significant electrolyte effects which amount to 7% experimental error on $k_\ell a$ measurement would occur. Influent Na_2SO_3 concentrations of less than 1500 mg/ℓ, which yields a smaller experimental error, can reasonably be used in $k_\ell a$ measurement. Bubble size reduction similar to that shown in Fig. 6.10 was also observed in a bubble column using a jet nozzle with NaCl as an electrolyte (Fujie et al., 1980a). Similar electrolyte effects on $k_\ell a$ are expected in a DTFB.

6.1.2.2 Operating and Design Parameter Effects

Draft tube systems include operating and design effects on $k_\ell a$ both common to fluidized bed and bubble column systems (e.g., U_ℓ, U_g, and d_p) and unique (e.g., D_d/D_c). Figure 6.12 shows various studies on $k_\ell a$ in a DTFB, and the design and operating parameters for these studies are given in Table 6.3. As seen in the figure, $k_\ell a$ is significantly affected by gas velocity. Increasing gas velocity monotonically increases $k_\ell a$. Effects of solids properties and loadings on $k_\ell a$ are more complicated. The general behavior indicates that for small particles (smaller than 300~400 μm), $k_\ell a$ decreases as the solids loading increases, as evidenced in curves 8a and 8b and curves 5 and 7 of Fig. 6.12. Koide et al. (1985) showed that particle diameter also affects $k_\ell a$ (curves 9a and 9b of Fig. 6.12): a DTFB using 79 μm glass beads has a higher $k_\ell a$ than that using 198 μm glass beads.

Fujie and Fan (1986), working with larger particles (700 μm activated carbon and 300 and 600 μm glass beads), reported that $k_\ell a$ increases, reaches a maximum, and then decreases with increasing solids loading (L_s) at high gas velocities (2.8 ~ 7.5 cm/s). The maximum occurs at an L_s of 2 vol% to 6 vol% for maximum liquid circulation rate between the draft tube and the annulus (Fan et al., 1984a). At low gas velocities (e.g., 1.3 cm/s), $k_\ell a$ decreases

Table 6.3 Legend for Figure 6.12

Curve No.	Type of Contactor	Air Diffuser	Influent Na_2SO_3 Concentration (mol/ℓ)	Solid Particles	References/ Comments
1	BC	Plastic porous plate	0.01	———	Fujie and Fan (1986)/ batch liquid
2	BC	Nozzles	0	———	Urza and Jackson (1975)/ batch liquid
3	BC	Single nozzle	0.15	———	Akita and Yoshida (1973)/ batch and continuous liquid Chatib et al. (1981)/
4	SBC	Domed distri- butor	0	Crushed brick granules d_p=0.83mm, ρ_s=1.7 g/cm^3	Chatib et al. (1981) 4a*: $U_ℓ$=0.35 cm/s 4b*: $U_ℓ$=1.06 cm/s * Curve No.
5	DTBC	Plastic porous plate	0.01	———	Fujie and Fan (1986)/ batch liquid
6	DTBC	Perforated plate	6a*: 0.4 6b*: 0 * Curve No.	———	Koide et al. (1983)/ batch liquid
7	DTFB	Plastic porous plate	0.01	Spherical activated carbon(Kureha beads) d_p=307 μm, L_s=10 vol%	Fujie and Fan (1986)/ batch liquid
8	DTFB	Single orifice nozzle	0.0	Activated carbon beads d_p=0.46 mm 8a*: L_s=1.48 vol% 8b*: L_s=5.94 vol% * Curve No.	Muroyama et al. (1985)/ batch liquid
9	DTFB	Perforated plate	0.0	Glass beads 9a*: d_p=79 μm 9b*: d_p=198μm L_s=4 vol% * Curve No.	Koide et al. (1985)/ batch liquid
10	DTFB	Perforated plate	0.02	Spherical activated carbon(Kureha beads) d_p=714 μm L_s=2.0 vol%	Fan et al. (1984b)/ batch liquid
11	ETFB	———	0	Spherical activated carbon(Kureha beads)	Suzuki et al. (1983)/ batch liquid

Notations
BC: Bubble Column
SBC: Slurry Bubble Column
DTBC: Draft Tube Bubble Column
DTFB: Draft Tube Three-Phase Fluidized Bed
ETFB: External Circulating Three-Phase Fluidized Bed

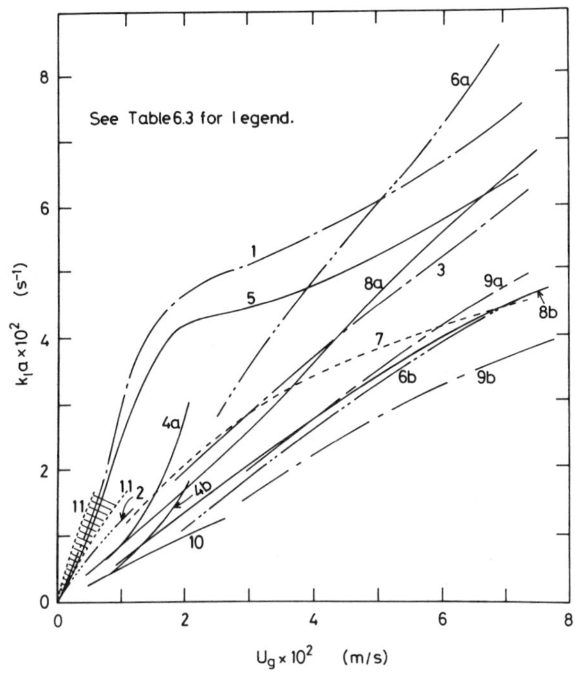

Figure 6.12 Relationship between $k_\ell a$ and U_g for various draft tube systems and bubble columns.

monotonically with increasing L_S. For 640 μm activated carbon particles, as reported by Muroyama et al. (1985), $k_\ell a$ reaches a maximum at an L_S of 1.5 vol% at high gas velocities, and little effect of L_S on $k_\ell a$ was observed at low gas velocities. In general, for larger particle (d_p > 300 μm) systems, $k_\ell a$ exhibits a maximum at solids loading between 1 vol% and 4 vol%, depending on particle properties.

Wiley (1986) found that the liquid velocity has negligible effects on $k_\ell a$ in a DTBC, but does affect $k_\ell a$ in a DTFB depending on the particle properties and the gas velocity. The gas velocity effect on $k_\ell a$ is considerably greater. For example, for glass beads of 300 μm (GB-300), 600 μm (GB-600) and 1000 μm (GB-1000) with 1 vol% loading, $k_\ell a$ increases with an increase in U_ℓ; however, it does not vary significantly among the glass beads at a U_g of 2.74 cm/s, as shown in Fig. 6.13. For activated carbon particles of 300 μm, (AC-300), 700μm (AC-700) and 1500 μm (AC-1500), on the other hand, the liquid velocity has little effect on $k_\ell a$ at low liquid velocities whereas it yields an increase in $k_\ell a$ at high liquid velocities at a U_g of 2.74 cm/s, as shown in Fig. 6.14. It is seen in the figure that $k_\ell a$ for small

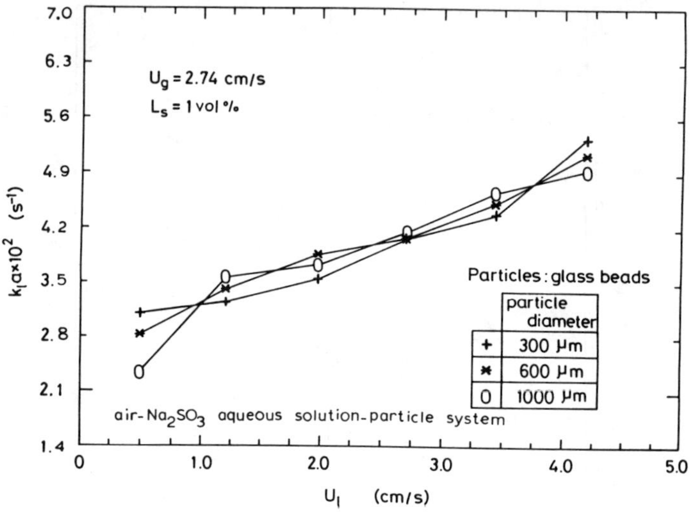

Figure 6.13 Effects of liquid velocity on $k_\ell a$ for various glass beads in a DTFB (from Wiley, 1986).

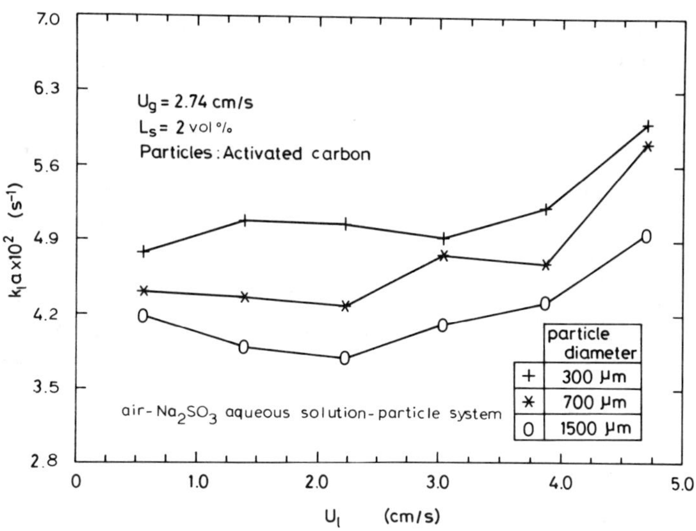

Figure 6.14 Effects of liquid velocity on $k_\ell a$ for various activated carbon particles in a DTFB (from Wiley, 1986).

activated carbon particles is higher than that for large activated carbon particles.

Effects of D_d/D_c, one of the most important design variables, on $k_\ell a$ are complex and cannot easily be generalized. For a DTBC using air-water, Koide et al. (1983) reported that $k_\ell a$ decreases slightly with an increase in D_d/D_c varying from 0.47 to 0.74, whereas Weiland (1984) concluded that $k_\ell a$ increases then decreases with an increase in D_d/D_c varying from 0.59 to 1.0, with an optimum value of about 0.85. For a DTFB using 170 μm sand particles, Sawai et al. (1982) reported an optimum D_d/D_c of 0.4 for maximum $k_\ell a$. The variation in the effects of D_d/D_c on $k_\ell a$ are a result of the relative differences in flow states of draft tube systems (see Fig. 6.2) caused by variations in system design, including distributors. The effects can be further complicated due to differences in liquid properties. More studies are indeed needed to identify the critical role of D_d/D_c on gas-liquid mass transfer.

The empirical correlations proposed by Fujie and Fan (1986) can be used to estimate $k_\ell a$ for systems with particles larger than 300 μm and a porous plate gas distributor operated under liquid-batch conditions:

$$k_\ell a/(k_\ell a)_0 = 1 + F(Re_p, L_s) \tag{6.41a}$$

where

$$F(Re_p, L_s) = (0.0029Re_p - 16.64L_s)Re_p^{-0.2521}(1-L_s)^{7.216} \tag{6.41b}$$

In Eqn. (6.41a), $(k_\ell a)_0$, $k_\ell a$ in a solid-free column, can be estimated by

$$(k_\ell a)_0 = A\,U_g^{\,B} \tag{6.42}$$

where for $U_g < 1.8$ cm/s

$$A = 1.49 \;,\; B = 1.12;$$

for $1.8 < U_g < 3.6$ cm/s

$$A = 81.5 \;,\; B = 0.145;$$

and for $U_g > 3.6$ cm/s

$$A = 12.6 \;,\; B = 0.523$$

It is noted that the lowest velocity range given above represents the dispersed bubble regime in the draft tube region, the intermediate velocity range represents the transition from dispersed to coalesced bubble regime, and the highest velocity range represents the coalesced bubble regime. Equation (6.41a) is applicable over the following operating conditions: $1.3 < U_g < 7.5$ cm/s, $0 < Re_p < 163$, and $0 < L_s < 0.2$. Equation (6.41a) reduces to that for a DTBC when L_s and Re_p reduce to zero. $k_\ell a$ obtained in a bioactive DTFB (treating phenolic wastewater containing bioflocs of 407 μm) was reported to be slightly lower than that predicted by Eqn. (6.41a), possibly due to the effect of the substrate and microbial metabolites on the bubble properties. Despite this discrepancy, Eqn. (6.41a) remains a reasonably good estimate for $k_\ell a$ in a bioactive DTFB.

Incorporating column size effects, Muroyama et al. (1985) proposed a simplified correlation equation (in SI units) which neglects the L_s effects and can be applicable to systems with a nozzle type gas injector operated under liquid-batch conditions for small or low density particles (~1.3 g/cm^3):

$$\frac{k_\ell a D_c^2 \rho_\ell}{\mu_\ell} = 3.87 \left(\frac{D_c U_g \rho_\ell}{\mu_\ell}\right)^{1.0} D_c^{1.8} \tag{6.43}$$

The equation is valid for $300 < D_c U_g \rho_\ell / \mu_\ell < 20{,}000$. Note that $k_\ell a$ predicted by Eqn. (6.41a) is higher than that predicted by Eqn. (6.43) due to gas distributors effects.

Comparisons of the behavior of a DTFB or DTBC with a BC can be made by re-examining Fig. 6.12. The figure indicates that introducing a draft tube into a BC reduces $k_\ell a$ in air-water or Newtonian liquid systems (curves 1 and 5 in the figure), in general, by about 10-30%. Insertion of a draft tube into the BC reduces the overall bubble retention time and hence, reduces $k_\ell a$ compared to that in a BC. For non-Newtonian liquids, on the other hand, the reverse is observed: the presence of a draft tube enhances $k_\ell a$ (Kawase and Moo-Young, 1986). Note that for relatively large particle systems, $k_\ell a$ in a DTFB under optimum solids loading conditions is significantly higher than that in a three-phase fluidized bed (without a draft tube) (Fan et al., 1984b).

6.1.2.3 Staging Effects

Staging of a DTFB has merit in biological applications (Orazem and Erickson, 1979 or see Chapter 7). Fan et al. (1987b) investigated a two-stage DTFB (see Table 6.1) and developed the following model equations for evaluating the overall mass transfer coefficient, based on the assumption of complete liquid mixing in each of the two stages of the DTFB:
for a gas phase in plug flow

$$k_\ell a^* = \frac{U_g[Z_i\eta_2\{\bar{C}_0 - (C_1 + C_2)/2\} - \bar{C}_0(1 - Z_i) \cdot \ln(1-\eta_2\{\bar{C}_0 - (C_1+C_2)/2\}/\{\bar{C}_0 - (C_1+C_2)/2Z_i\})]}{[\bar{C}_0 - (\frac{C_1 + C_2}{2})]^2 v_m H} \tag{6.44a}$$

and for a gas phase in complete mixing

$$k_\ell a^\# = \frac{U_g Z_i \eta_2}{[(\frac{(C_0)_1 Z_1 + (C_0)_2 Z_2}{2}) - (\frac{C_1+C_2}{2})] v_m H} \tag{6.44b}$$

Fan et al. (1987b) noted that, as in the single stage DTFB given in Section 6.1.2, the different model equations of the mass transfer coefficients ($k_\ell a^*$ and $k_\ell a^\#$) yield similar results. In the following, the behavior of the volumetric mass transfer coefficient expressed by $k_\ell a^*$ is presented.

Figures 6.15 and 6.16 show the effect of solids loading on $k_\ell a^*$ and $k_\ell a$ for several gas velocities where $k_\ell a$ is evaluated by Eqn. (6.41a). The mass transfer is expressed in terms of the volumetric mass transfer coefficient in the absence of solids, i.e., $k_\ell a^*/(k_\ell a^*)_0$ or $k_\ell a/(k_\ell a)_0$. For AC-300 particles, a maximum $k_\ell a^*/(k_\ell a^*)_0$ occurs at an L_s of 2 vol% for gas velocities above 5.2 cm/s; below 5.2 cm/s the maximum occurs at an L_s of 1 vol%. Then in general, $k_\ell a^*/(k_\ell a^*)_0$ decreases monotonically with increasing L_s beyond the maximum point. In addition, $k_\ell a^*/(k_\ell a^*)_0$ is greater than $k_\ell a/(k_\ell a)_0$ for all values of U_g over almost the entire range of L_s shown in Fig. 6.15. For GB-300 particles, a maximum $k_\ell a^*/(k_\ell a^*)_0$ occurs at a solids loading of about 1 vol% for most of the range of U_g studied. However, $k_\ell a^*/(k_\ell a^*)_0$ is significantly lower than $k_\ell a/(k_\ell a)_0$ over the entire range of solids loading for the lowest U_g employed (1.7 cm/s). They concluded that $k_\ell a^*/(k_\ell a^*)_0$ in the two-stage DTFB is higher than

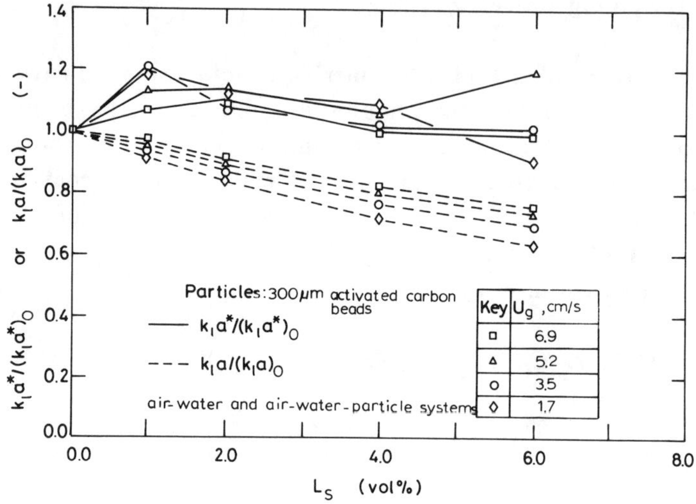

Figure 6.15 Variations of volumetric mass transfer coefficient with solids loadings for one-stage and two-stage DTFBs with activated carbon particles (from Fan et al., 1987b).

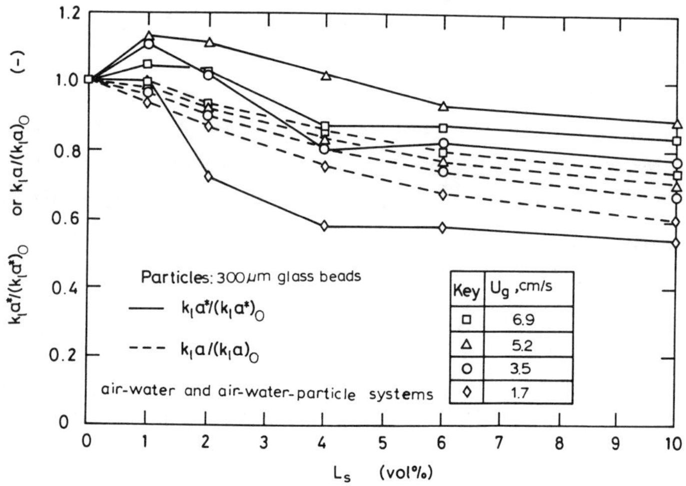

Figure 6.16 Variations of volumetric mass transfer coefficient with solids loadings for one-stage and two-stage DTFBs with glass beads (from Fan et al., 1987b).

$k_\ell a/(k_\ell a)_0$ in the single stage DTFB for low density and small diameter particles. As the particle density and diameter increase, the improvement of mass transfer performance of the two-stage DTFB over the single stage DTFB progressively decreases.

6.1.3 Liquid-Solid Mass Transfer

Utilizing the benzoic acid dissolution method, Hwang and Fan (1985) measured the overall liquid-solid mass transfer coefficient for a DTFB using solid particles ranging from 2 to 4.5 mm in diameter. They found that the liquid-solid mass transfer coefficient increases with an increase in gas velocity or solids loading, but decreases with an increase in liquid velocity. The mass transfer coefficient was found to be independent of particle diameter over the moderately broad size range studied, similar to that in a liquid-solid fluidized bed. The mass transfer coefficients in the DTFB are higher than those in the draft tube liquid-solid fluidized bed at a given liquid velocity. Furthermore, the mass transfer coefficients in the draft tube liquid-solid fluidized bed are lower than those in the liquid-solid fluidized bed (without a draft tube) at a given liquid velocity. At low gas and liquid velocities, the mass transfer coefficient in the DTFB is higher than that in the three-phase fluidized bed. However, the opposite is true at high liquid and low gas velocities. At high gas velocities, the mass transfer coefficient in the DTFB is higher than that in the three-phase fluidized bed at any liquid velocity.

6.2 SEMIFLUIDIZED BED SYSTEMS

The phenomenon of semifluidization was first reported by L.T. Fan and Wen (1959) in the context of investigating the mass transfer in a two-phase (liquid-solid) semifluidized bed. A semifluidized bed is characterized by a fluidized bed and a packed bed in series within a single containing vessel. It is formed when a mass of fluidized particles is compressed by the fluidizing media against a porous retaining grid which constrains the bed. Three-phase semifluidization can be established in cocurrent systems with gas or liquid as the continuous phase where the packed bed is formed under the top retaining grid (for Modes E-I-a-1, E-I-a-2, and E-I-b) or in countercurrent systems with gas or liquid as the continuous phase where the packed bed is formed under the top retaining grid (for Mode E-II-b) or above the bottom retaining grid (for Mode E-II-a-1). Studies of semifluidization have been mainly limited to two-phase (gas-solid or liquid-solid) systems (e.g., Wen et al., 1963). Little, however, is known regarding the gas-liquid-solid systems (Chern et al., 1983). In the following, fundamental hydrodynamic properties for Modes E-I-a-1 and E-II-a-1 semifluidization are described.

6.2.1 Pressure Drop

The pressure drop through the semifluidized bed is the sum of the individual pressure drops in the packed and fluidized bed sections:

$$-\Delta P = (-\Delta P)^p + (-\Delta P)^f \tag{6.45}$$

Consider the solids in the packed bed section, with one-dimensional gas and liquid flow, to be completely wetted by the liquid; thus, no direct contact between the gas and solid phases occurs. Momentum balances on the liquid and gas phases, at steady state and neglecting any acceleration effects, yield

$$\left(-\frac{dP}{dz}\right)^p = \rho_\ell g + \left(-\frac{dP_f}{dz}\right)^\ell \tag{6.46}$$

$$\left(-\frac{dP}{dz}\right)^p = \rho_g g + \left(-\frac{dP_f}{dz}\right)^g \tag{6.47}$$

In each case, the pressure drop results from a static head and a frictional loss term.

The frictional pressure drop in the liquid phase results from interactions with the solid and gas phases and can be expressed as

$$\left(-\frac{dP_f}{dz}\right)^\ell = \left(-\frac{dP_f}{dz}\right)^{\ell\text{-}s} + \left(-\frac{dP_f}{dz}\right)^{\ell\text{-}g} \tag{6.48}$$

The frictional pressure drop in the gas phase results solely from interaction with the liquid phase and can be rewritten as

$$\left(-\frac{dP_f}{dz}\right)^g = \left(-\frac{dP_f}{dz}\right)^{g\text{-}\ell} \tag{6.49}$$

Since the mutual forces at the gas-liquid interface cancel, the following equation holds:

$$\delta_\ell\left(-\frac{dP_f}{dz}\right)^{\ell\text{-}g} + \delta_g\left(-\frac{dP_f}{dz}\right)^{g\text{-}\ell} = 0 \tag{6.50}$$

where δ_g and δ_ℓ are the volumetric gas and liquid fractions on the solid-free basis defined by

$$\delta_\ell = \frac{\epsilon_\ell^p}{1-\epsilon_s^p} \tag{6.51}$$

and

$$\delta_g = \frac{\epsilon_g^p}{1-\epsilon_s^p} \tag{6.52}$$

Combining Eqns. (6.46) through (6.50) yields an expression for the pressure drop through the packed bed section:

$$\left(-\frac{dP_f}{dz}\right)^{\ell\text{-}s} = \frac{1}{\delta_\ell}\left[\left(-\frac{dP}{dz}\right)^p - \delta_\ell \rho_\ell g - \delta_g \rho_g g\right] \tag{6.53}$$

Equation (6.53) relates the pressure drop to the frictional pressure loss between the liquid and solid and the individual phase holdups.

The frictional pressure loss between the liquid and solid can be expressed in terms of a friction factor as

$$\left(-\frac{dP_f}{dz}\right)^{\ell\text{-}s} = 4C_f' \frac{1}{D_e}\left[\frac{1}{2}\rho_\ell\left\{\frac{U_\ell}{(1-\epsilon_s^p)\delta_\ell}\right\}^2\right] \tag{6.54}$$

where C_f' is the friction factor and D_e, the equivalent diameter for the assumed liquid channel, is (Chern et al., 1983)

$$D_e = \frac{2}{3}\left(\frac{1-\epsilon_s^p}{\epsilon_s^p}\right) d_p \left[1 - \sqrt{\delta_g}\right] \tag{6.55}$$

Two correlations for the friction factor exist in the literature for three-phase packed bed sections in semifluidized bed systems:
for Mode E-I-a-1 operation (Chern et al., 1984)

$$C_f' = 0.1075 + 6.359\overline{Re}_\ell^{\,-0.368} \tag{6.56}$$

and for Mode E-II-a-1 operation (Chern et al., 1983)

$$C_f' = 36.94\overline{Re}_\ell^{\,-0.681} \tag{6.57}$$

where

$$\overline{Re}_{\ell} = \frac{D_e \rho_\ell U_\ell}{(1 - \epsilon_s^p) \delta_\ell \mu_\ell}$$

Note that Ergun's equation satisfactorily predicted Chern et al.'s (1983) data for both one- and two-phase flow through packed beds at high $\overline{Re}_\ell$ (~ 500), but was unsatisfactory at relatively low $\overline{Re}_\ell$ (< 100).

The pressure drop in the fluidized bed section can be obtained from the fundamental relationship between pressure drop, phase holdups, and phase densities. The total pressure drop through a semifluidized bed can be estimated by combining the pressure drops in the fixed and fluidized bed sections (Eqn. (6.45)). The pressure drop thus estimated agrees well with experimental data (Chern et al., 1983, 1984).

6.2.2 Onset Liquid Velocity

The onset liquid velocity for semifluidization, $(U_\ell)_{osf}$, can be evaluated visually as the liquid velocity at which the retaining grid is first packed with particles. It can also be determined with comparable results through a plot of the dynamic pressure drop versus liquid velocity curve at a constant gas velocity; the break point on the curve defines $(U_\ell)_{osf}$. Based on the former approach, $(U_\ell)_{osf}$ can be calculated from a relationship in which the solids holdup in a fully fluidized bed is equal to $(\epsilon_s)_{osf}$ in a semifluidized bed, defined as

$$(\epsilon_s)_{osf} = \frac{W_s}{\rho_s A_T H_c} \tag{6.58}$$

The solids holdup in Mode E-I-a-1 fluidization can be evaluated by various models given in Chapter 2. An empirical correlation (in SI units) was also proposed for $(U_\ell)_{osf}$, obtained based on the latter approach, for Mode E-II-a-1 semifluidization as (Chern et al., 1983)

$$(U_\ell)_{osf} = 2.4 \times 10^{-4} \, Ar^{0.138} \, Fr_g^{-0.0820} \, L_s^{-1.50} \tag{6.59}$$

The equation is applicable over $7.4 \times 10^4 < Ar < 1.2 \times 10^7$ and $6.2 \times 10^{-5} < Fr_g < 0.15$.

6.2.3 Packed Bed Section Height

From a material balance on the solid phase, the height of the packed bed section in a semifluidized bed is

$$H_{pa} = \frac{W_s/A_T - H_c(1-\epsilon^f)\,\rho_s}{[\epsilon_s^p - (1-\epsilon^f)]\rho_s} \qquad (6.60)$$

where ϵ^f, the bed voidage in the fluidized bed section, can be approximated by the bed voidage in a fully fluidized bed under the same operating conditions and ϵ_s^p, the solids holdup in the packed bed section, can be approximated as 0.6 when spherical particles are used. With Eqn. (6.60), determining H_{pa} has been reduced to determining ϵ^f, which can be readily accomplished based on information given in Chapter 2 for Mode E-I-a-1 fluidization.

For Mode E-II-a-1 semifluidization, H_{pa} was correlated by Chern et al. (1983) as

$$\frac{H_{pa}}{H_c} = 66.2\ Ar^{0.012} Fr_\ell^{\,0.590} Fr_g^{\,0.0980} L_s^{\,3.32} \qquad (6.61)$$

The ranges of applicability of Eqn. (6.61) are $9 \times 10^{-4} < Fr_\ell < 4 \times 10^{-2}$ and those given for Eqn. (6.59).

6.2.4 Phase Holdups

The hydrodynamic behavior for a fluidized bed described previously can be used to account for the behavior of the fluidized bed section in the Mode E-I-a-1 semifluidized bed. For example, the correlation for k_0 by Chern et al. (1984) given in Table 2.4 and correlations of ϵ_g given in Table 2.1 for different flow regimes can be used.

The holdup of the gas or liquid phase in a semifluidized bed can be given by

$$\epsilon_g = \left(\frac{H_{pa}}{H_c}\right)\epsilon_g^p + \left(1 - \frac{H_{pa}}{H_c}\right)\epsilon_g^f \qquad (6.62a)$$

$$\epsilon_\ell = \left(\frac{H_{pa}}{H_c}\right)\epsilon_\ell^p + \left(1 - \frac{H_{pa}}{H_c}\right)\epsilon_\ell^f \qquad (6.62b)$$

As noted, ϵ_g^f and ϵ_ℓ^f can be estimated based on the information provided in Chapter 2 for Mode E-I-a-1 fluidization and Chapter 5 for Mode E-II-a-1 fluidization. Very limited information is available on ϵ_g^p and ϵ_ℓ^p under high gas and/or liquid velocity conditions underlying semifluidization. Chern et al. (1984) found ϵ_g^p for cocurrent gas-liquid upflow can be determined via the correlation as (in SI units)

$$\delta_g = 0.0296 U_\ell^{-0.350} U_g^{0.977} \tag{6.63}$$

and for countercurrent flow as (Chern et al., 1983)

$$\delta_g = 0.084 U_\ell^{0.253} U_g^{0.878} \tag{6.64}$$

Equation (6.63) is applicable to $1.69 \le U_\ell \le 2.2$ cm/s and $0 \le U_g \le 26$ cm/s; Eqn. (6.64) is applicable to $1.2 \le U_\ell \le 2.5$ cm/s and $0.3 \le U_g \le 8.5$ cm/s.

6.3 ANNULAR SYSTEMS

A combination of factors in the three-phase fluidized bed (Mode E-I-a-1) hydrotreating of petroleum resids has led to the development of internal fluid recycle; the resulting reactor design confines the fluidized bed within an annular region, allowing fluid recycle through the reactor center (see Chapter 10). Since these annular type systems are among the most important commercial applications of three-phase fluidized beds, a discussion of the differences in bed behavior between conventional (non-annular) and annular systems is appropriate.

Confining the fluidized bed within an annulus imposes physical constraints not occurring in conventional systems. The wetted perimeter increases, thus increasing the importance of wall effect; the thickness of the annular space restricts the maximum bubble size; and the more complicated system configuration restricts, somewhat, the freedom of bubble motion. Consequently, these additional constraints are expected to produce changes in bubble motion and size and bed expansion behavior.

Ramamurthy and Subbarju (1973) reported that in liquid-solid fluidized beds the addition of an inner column had little effect if $(D_h/d_p) > 8$, where D_h is the hydraulic diameter and d_p is the particle diameter. Dakshinamurty et al. (1972) found that the presence of an inside column slightly altered bed expansion characteristics.

Confining a fluidized bed in an annular region should, in some way, be related to a reduction in column diameter. The fact that the column diameter affects the transition between flow regimes (Matsuura and Fan, 1984; Fan et al., 1986; Fan et al., 1987a) and gas holdup behavior (e.g., Michelsen and Ostergaard, 1970) suggests that gas holdup will also depend on the annulus dimensions. Fan et al. (1987a) reported that in an air-water-glass bead system, an increase in the annulus size at a given column (outer) size decreased the gas holdup for the coalesced bubble flow regime, but had little effect on the gas holdup in the dispersed bubble flow regime. Furthermore, the presence of the annular region promotes bubble coalescence, and thus a higher liquid velocity is required to reach the dispersed bubble flow regime in an annular system than in the same column without an annular region. When simulating hydrotreating conditions involving high gas holdup and small bubbles, however, the effects of the annular region on bed behavior is negligible (Fan et al., 1986).

6.4 TRANSPORT SYSTEMS

Gas-liquid-solid flow has been associated with air-lift pump operation. In this operation, drag reduction phenomena were reported to occur when gas is introduced into the flow of slurries such as clay suspensions and aqueous polymer solutions (Giot, 1982). Chemical reactions or heat and mass transfer operations in gas-liquid-solid systems often involve the transport or flow of gas-liquid-solid mixtures through vertically or horizontally arranged tubes. Three-phase flows are often complicated by heat transfer and chemical reactions, such as those experienced in a preheater for coal liquefaction or in the inner column of ebullated bed reactors used in hydrotreating of petroleum resids (see Chapter 10). Fundamental understanding of a simple three-phase flow phenomena is, however, essential to the basic design and control of three-phase flow systems. This section describes fundamental transport behavior of vertical upflow, vertical downflow, and cocurrent horizontal flow.

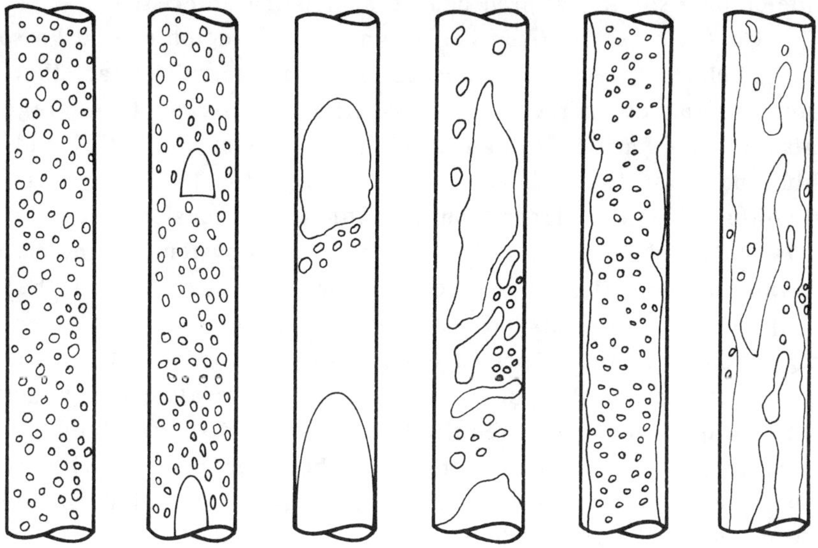

Figure 6.17 Gas-liquid flow regimes in vertical upflow (after Hewitt, 1978).

6.4.1 Vertical Upflow

In three-phase vertical upflow (Modes T-I-a-1 and T-I-b) the observed flow regimes resemble the bubble, bubble-slug, slug (or plug), churn, annular and wispy-annular flows in gas-liquid flow systems (Toda et al., 1982; Miller and Cain, 1986). The prevailing regime depends on the relative magnitude of the gas and liquid/slurry velocities. Figure 6.17 shows schematically the various flow regimes for gas-liquid vertical upflow (Hewitt, 1978). Miller and Cain (1986) found that the bubble to bubble-slug and bubble-slug to slug flow transitions occurred at lower gas velocities in three-phase systems than in two-phase systems because of increased bubble coalescing tendencies, due to the presence of solid particles.

Flow regime transition from bubble flow to slug flow occurs at low gas velocities as indicated by Toda et al. (1982) in the study of an air-water-glass bead system. They found that the transition from bubble flow to slug flow with solids concentration in the slurry of up

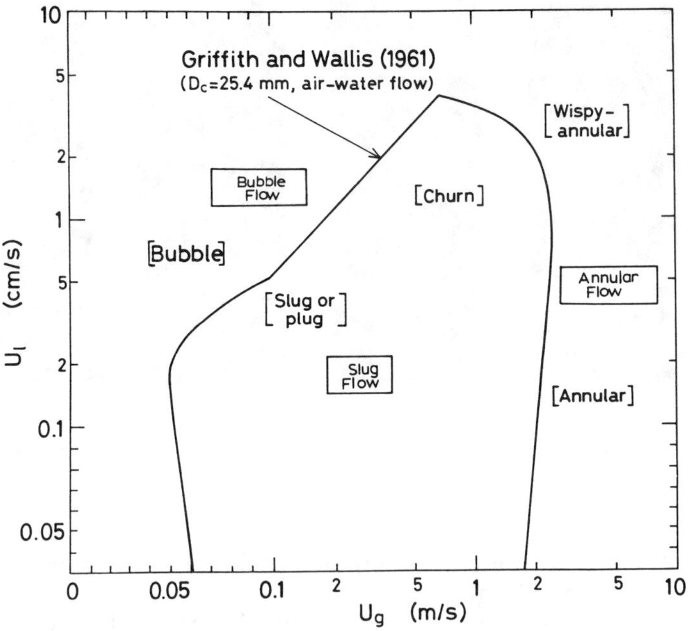

Figure 6.18 Flow regime map for a gas-liquid upflow (from Griffith and Wallis, 1961).

to 10 vol%, can be satisfactorily predicted by the two-phase (liquid-solid) flow chart of Griffith and Wallis (1961) using the superficial slurry velocity instead of the superficial liquid velocity. For reference, operating regimes for air-water cocurrent vertical upflow through a 2.5 cm ID pipe based on the correlation by Griffith and Wallis (1961) are shown in Fig. 6.18. In this figure, churn flow regime is lumped together with the slug flow regime. Figure 6.19 presents a photograph showing the effects of the velocity of water-glass beads slurry, $U_{s\ell}$, on the flow regime in a vertical three-phase upflow system. $U_{s\ell}$ varies from 0.2 to 1.5 m/s at a constant U_g of 0.1 m/s. The volume fraction of solids in the slurry, ψ_s, is 0.01. It is seen that at $U_{s\ell}$ of 0.2 m/s, the flow is in the slug flow regime whereas at other $U_{s\ell}$, the flow is in the bubble flow regime. The flow regime and transition are consistent qualitatively with those for a two-phase upflow system given in Fig. 6.18. It is noted that the solids concentration distribution in the three-phase upflow system appears to be uniform due to significant liquid phase turbulence induced by gas bubbles. In contrast, for a liquid-solid upflow system, solids concentration is low in the core region and high in the near-wall

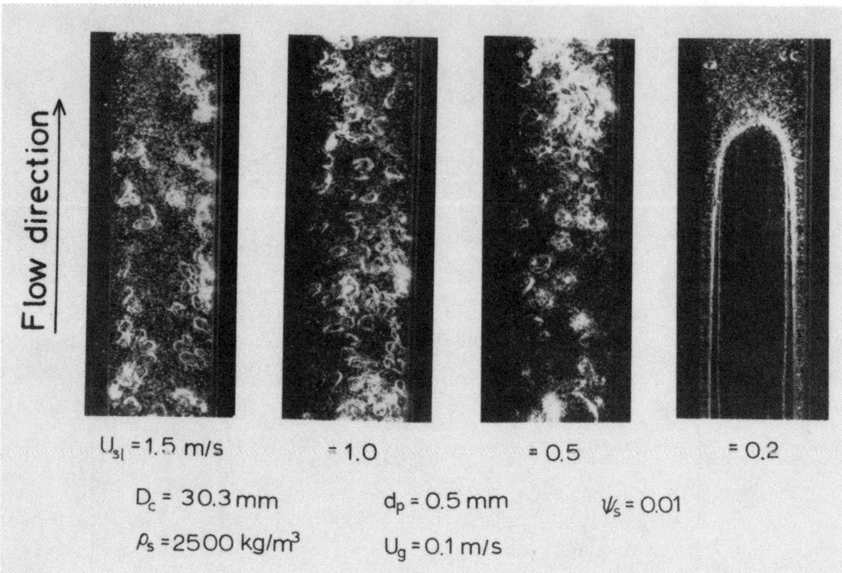

Figure 6.19 Photographic representation of an air-water-glass bead upflow (from Toda et al., 1982; with permission from Kagaku Kogaku Ronbunshu; the original kindly supplied by M. Toda of Yamagata University, Japan).

region due to nonuniformity of radial liquid velocity distribution. Readers are referred to Hewitt (1982), Weisman and Kang (1981) and Crawford et al. (1985) for more details on flow regimes in gas-liquid two-phase upflow.

6.4.1.1 Phase Holdups

Gas and slurry holdup behavior has been studied by several investigators: gas holdup by Kato et al. (1975), Weber and Dedegil (1976), Ying et al. (1980), Giot (1982), Toda et al. (1982), Hatate et al. (1986a) and Kato et al. (1986); and slurry holdup by Kim and Choi (1984) and Kato et al. (1986). When the slurry (or liquid) velocity is low ($U_{s\ell}$ < 0.2 cm/s), Ying et al. (1980) reported that the effect of slurry (or liquid) velocity on the gas holdup is virtually negligible and the gas holdup can be satisfactorily represented by Akita and Yoshida's (1973) correlation for a gas-liquid bubble column (see Eqn. (T-4.1.2)). When the slurry velocity exceeds the free bubble rise velocity due to buoyancy, both slurry and gas velocities affect the phase holdups; under such conditions the holdups can be correlated in a manner similar to that for a two-phase (gas-liquid) flow system as given below. It is noted that the presence of suspended solid particles reduces gas holdup, especially in a large column.

Toda et al. (1982) showed that the gas holdup in an air-water-glass bead system can be represented well by the two-phase correlation equation of Hughmark (1962) in terms of average superficial velocity and physical properties of the slurry. Kim and Choi (1984) conducted experiments using pulverized coal particles with an average diameter of 0.112 mm suspended in oils at concentrations up to 30 wt%. Their slurry holdup data showed poor agreement with Hughmark and Pressburg's correlation (1961) for liquid holdup in gas-liquid flow. Kim and Choi (1984) presented a dimensionless equation for slurry holdup over a wide range of liquid viscosities:

$$\epsilon_{s\ell} = 1.06(U_g/U_\ell)^{-0.21}Re_{s\ell}^{-0.09} \tag{6.65}$$

where $Re_{s\ell}$ is the slurry (or liquid) Reynolds number, $D_cU_{s\ell}\rho_{s\ell}/\mu_{s\ell}$. Equation (6.65) covers $0.1 < U_g/U_{s\ell} < 1571$ and $1.0 < Re_{s\ell} < 5160$. Note that Equation (6.65) may not be applicable at very low gas velocities.

Hatate et al. (1986a) measured the gas holdup in an air-water-glass bead system for a wide range of gas velocities. They found that the gas holdup data could be correlated well by the equation of Nicklin et al. (1962) for slug flow in gas-liquid systems. Moreover, the gas holdup in vertical upflow is virtually independent of tube diameter (15 ~ 26 mm), average particle size (28 ~ 94 μm), and particle concentration (C_s = 0 ~ 0.65). More recently, Kato et al. (1986) measured slurry holdup over a much wider range of gas and slurry velocities. They found the ratio of slurry holdup to gas holdup, $\epsilon_{s\ell}/(1-\epsilon_{s\ell})$, increases with increasing apparent slurry viscosity as shown in Fig. 6.20. A correlation equation for the slurry holdup was developed by them by modifying the correlation of Nicklin et al. (1962) to take into consideration the effect of increased viscosity of the slurry relative to a pure liquid.

$$\frac{\epsilon_{s\ell}}{1-\epsilon_{s\ell}} = \frac{1-\epsilon_g}{\epsilon_g} = \left\{1.2(1+\frac{U_{s\ell}}{U_g}) + 0.35\frac{\sqrt{g\,D_c}}{U_g}\right\}\left(\frac{\mu_{s\ell}}{\mu_w}\right)^{0.04} - 1 \tag{6.66}$$

Equation (6.66) covers $0.2 \le \epsilon_{s\ell}/(1-\epsilon_{s\ell}) \le 40$, $18 \le D_c \le 40$ mm, $1 \le \mu_{s\ell} \le 133$ mPa•s, $0 \le C_s \le 0.45$, $0 \le \psi_s \le 0.25$, $0.01 \le U_g \le 18$ m/s, and $0 \le U_{s\ell} \le 2.6$ m/s. The data of Kim and Choi (1984) for coal-oil slurries can be satisfactorily predicted by Eqn. (6.66). Equation (6.66) also yields a sufficiently accurate prediction for gas holdup because of the

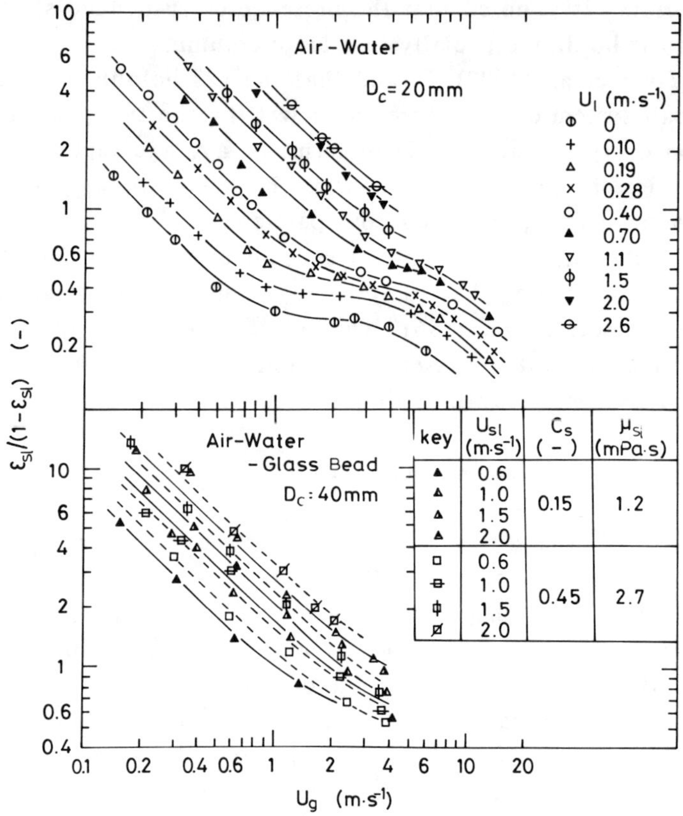

Figure 6.20 Relationship between $\epsilon_{s\ell}/(1-\epsilon_{s\ell})$ and U_g for an air-water or air-water-solid vertical upflow (from Kato et al., 1986).

well defined variable $(1-\epsilon_g)/\epsilon_g$. Note that for $(1-\epsilon_g)/\epsilon_g < 0.2$, where annular flow prevails, Eqn. (6.66) over-predicts the slurry holdup and thus, underestimates the gas holdup.

When the superficial slurry velocity is low, the solids concentration in the slurry flowing in a vertical pipe becomes greater than the concentration of the influent or effluent slurry (Toda et al., 1982). This is due to the sedimentation effect which becomes negligible when the parameter $U_{s\ell}/(gD_c)^{0.5}$ is greater than about two (Toda et al., 1982).

6.4.1.2 Pressure Drop

The pressure drop due to wall friction in gas-liquid-solid upflow, $(-\Delta P/\Delta z)_{GLS}$, can be defined in terms of the total pressure drop corrected for the static head (Toda et al., 1982):

$$(-\frac{\Delta P}{\Delta z})_{GLS} = (-\frac{\Delta P}{\Delta z}) - (\epsilon_g \rho_g + \epsilon_\ell \rho_\ell + \epsilon_s \rho_s)g \quad (6.67)$$

When the superficial liquid or slurry velocity is low (U_ℓ or $U_{s\ell}$ < about 0.2 m/s), the values for $(-\Delta P/\Delta z)_{GLS}$ tend to be negative (Toda et al., 1982). Toda et al. (1982) found that a simple modification of the Lockhart-Martinelli relationship (Lockhart and Martinelli, 1949) developed for gas-liquid flow yielded a good correlation for the pressure drop due to wall friction in a three-phase upflow. More specifically, considering the liquid and solid phase as a pseudo-homogeneous (slurry) phase, Toda et al. (1982) adopted the identical form of the original Lockhart-Martinelli correlation with proper inclusion of the pressure drop terms for the three-phase flow $(-\Delta P/\Delta z)_{GLS}$, for the system if the gas were flowing alone $(-\Delta P/\Delta z)_G$, and for the system if the slurry were flowing alone $(-\Delta P/\Delta z)_{SL}$. The modified Lockhart-Martinelli parameters Φ_G, Φ_{SL} and X are defined by

$$\Phi_G = \sqrt{(-\Delta P/\Delta z)_{GLS}/(-\Delta P/\Delta z)_G} \quad (6.68)$$

$$\Phi_{SL} = \sqrt{(-\Delta P/\Delta z)_{GLS}/(-\Delta P/\Delta z)_{SL}} \quad (6.69)$$

$$X = \Phi_G/\Phi_{SL} = \sqrt{(-\Delta P/\Delta z)_{SL}/(-\Delta P/\Delta z)_G} \quad (6.70)$$

Figure 6.21 shows the relationship between Φ_G and X and comparisons with the calculation based on the original Lockhart-Martinelli equation and the data of Toda et al. (1982) for three-phase upflow. It is seen in the figure that the comparison is satisfactory. In the calculation of parameters, Φ_G or Φ_{SL} and X in a gas-slurry system, Thomas' equation (Thomas, 1963) is recommended to estimate the slurry viscosity:

$$\mu_{s\ell} = \mu_\ell\{1 + 2.5\psi_s^2 + 0.062\exp[\frac{1.875\psi_s}{1-1.595\psi_s}]\} \quad (6.71)$$

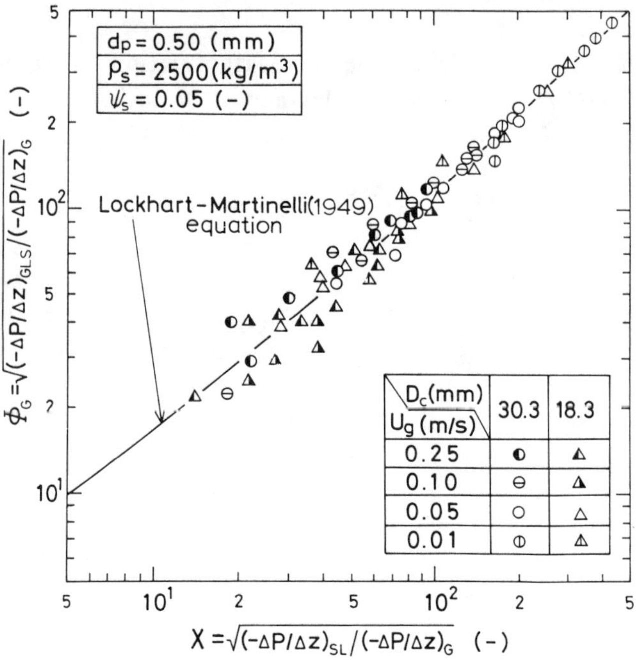

Figure 6.21 Correlation of friction loss in terms of Lockhart-Martinelli parameters, Φ_G and X, for an air-water-particle vertical upflow (from Toda et al., 1982).

where ψ_s is the solid volume fraction in the slurry.

From the experimental results for a coal-oil slurry system, Kim and Choi (1984) developed an empirical correlation for the friction factor for the total pressure drop as follows:

$$f_t = \frac{(-\frac{\Delta P}{\Delta z})D_c}{2U_{s\ell}^2 \rho_{s\ell}} = 21.55(U_g/U_{s\ell})^{0.42}Re_{s\ell}^{0.33}We_{s\ell}^{-0.55} \tag{6.72}$$

where $We_{s\ell}$ is the Weber number defined by $\rho_{s\ell}U_{s\ell}^2D_c/\sigma$. Equation (6.72) covers $0.1 \le U_g/U_\ell < 1571$, $1.0 \le Re_{s\ell} \le 5160$, and $0.01 \le We_{s\ell} \le 220$.

6.4.1.3 Heat Transfer

Hatate et al. (1987) found that for heat transfer from an immersed heater, the heat transfer coefficient increases slightly with increasing liquid and gas velocities, passes through a maximum with increasing slurry concentration at given gas and liquid velocities, and is independent of particle size. The heat transfer coefficient in the gas-liquid-solid flow agrees with that in a corresponding gas-liquid flow and both are much higher than the values of single liquid flow and liquid-solid flow, due to the turbulence created by the gas flow. Hatate et al. (1987) developed the following empirical correlation for both the gas-liquid and gas-liquid-solid systems:

$$\mathrm{Nu}\,\mathrm{Pr}_{s\ell}^{-0.33}\left(\frac{\mu_{s\ell}}{\mu_w}\right)^{-0.14}\mathrm{Re}_{s\ell}^{-0.2}\mathrm{Re}_g^{-0.055} = 10.5\exp\{-3.18\times10^{-4}(100C_s - 22)^2\} \tag{6.73}$$

where Nu is the Nusselt number ($hD_c/\lambda_{s\ell}$), $Re_{s\ell}$ is the slurry (or liquid) Reynolds number ($D_cU_{s\ell}\rho_{s\ell}/\mu_{s\ell}$), and Re_g is the gas Reynolds number ($D_cU_g\rho_g/\mu_g$). The ranges of applicability are $2500 \le Re_{s\ell} \le 25{,}000$, $500 \le Re_g \le 25{,}000$ and $0 \le C_s \le 0.56$.

6.4.2 Vertical Downflow

Little information is available in the literature for vertical downflow in gas-liquid and gas-liquid-solid systems (Perlmutter et al., 1984). Oshinowo and Charles (1974a) presented a flow regime map for vertical gas-liquid downflow. Six flow regimes were described by them, namely corning bubble, bubbly slugs, falling film, falling bubbly film, froth and annular regimes. The flow regime or gas holdup for upflow and downflow systems can be significantly different. For example, for coal liquid slurry and hydrogen flows in a 2.8 cm ID tube at 252 g/s and 19 g/s, respectively, with temperatures 93 ~ 427°C, and pressure 137 atm, Perlmutter et al. (1984) predicted an upward flow mainly in the frothy slug regime whereas the downflow mainly in the falling bubbly film regime. At low liquid velocities and intermediate gas velocities, falling film and falling bubbly film regimes occur, both of which yield lower gas holdup than would be expected in vertical upflow systems under similar flow conditions. In the following, the gas holdup and pressure drop behavior for three-phase downflow is described, which covers both the continuous liquid phase (Mode T-III-a) and the continuous gas phase (Mode T-III-b) operation conditions.

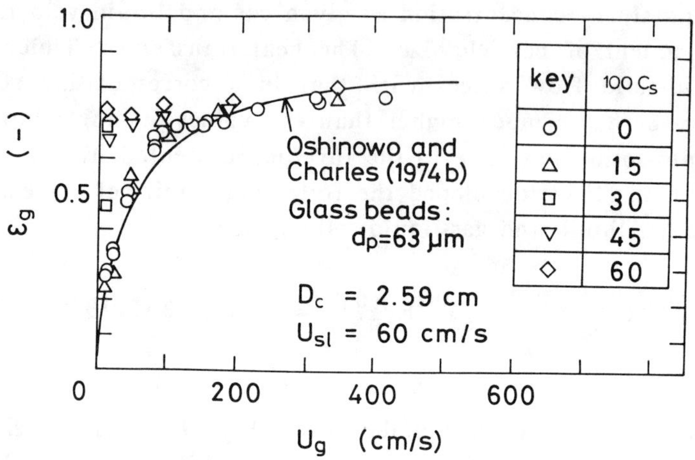

Figure 6.22 Relationship between gas holdup and gas velocity in an air- water-particle vertical downflow (From Hatate et al., 1986a).

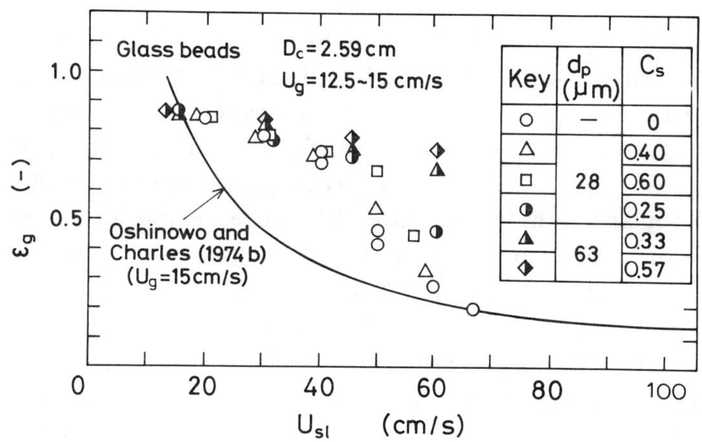

Figure 6.23 Effect of slurry velocity on gas holdup in an air-water-particle vertical downflow (from Hatate et al., 1986a).

6.4.2.1 Gas Holdup

Gas holdup at low liquid and gas velocities is significantly higher in the downflow than that in a corresponding upflow and the effect of slurry solid concentration on the gas holdup is complex (Hatate et al., 1986a). Figures 6.22 and 6.23 show the variation in gas holdup with gas and slurry velocity, respectively. At high gas velocities (U_g > 200 cm/s), gas holdup in gas-liquid-solid systems are virtually the same in magnitude as those in gas-liquid systems when the liquid velocity (for gas-liquid systems) is the same as the slurry velocity (for gas-liquid-solid systems). For flow through a 2.59 cm ID pipe, higher gas holdups are observed at higher solids concentrations and low gas and slurry velocities (< 60 cm/s). Gas holdup is, however, shown to be satisfactorily correlated by an equation derived from Oshinowo and Charles' correlation (Oshinowo and Charles, 1974b) originally proposed for gas-liquid vertical downflow in slug flow as given by

$$\epsilon_g = \frac{U_g/(U_g + U_{s\ell})}{1.1 - 0.33\,[\sqrt{gD_c}\,/(U_g + U_{s\ell})]} \tag{6.74}$$

Shah et al. (1983) examined the effects of surfactants and electrolytes on the gas holdup in a downflow air-water column. Both additives lower the gas holdup because of their coalescence-suppressing effect, which reduces the mean residence time of bubbles. For air-water-coal and air-water-shale systems at C_s up to 0.15, the solids concentration has little effect on the gas holdup. At low gas velocities, the gas holdup in a three-phase system is slightly lower than that in a gas-liquid system. The gas holdup sharply decreases with increasing liquid velocity.

6.4.2.2 Pressure Drop

The dependence of frictional pressure drop in a downflow system on superficial gas and slurry velocities is similar to that in the upflow system (Hatate et al., 1986a). The frictional pressure drop is independent of particle size but increases slightly with increasing slurry solid concentration. Hatate et al. (1986a) examined the applicability of the simple single-flow model by Hughmark (1965) to the pressure drop data of downflow and found good agreement.

6.4.3 Cocurrent Horizontal Flow

Flow regimes of a cocurrent horizontal three-phase flow through tubes were classified similarly to the two-phase flow as bubble, plug, slug flows, and transition states between them. Figure 6.24 shows the schematics of flow regimes for gas-liquid horizontal flow (Hewitt, 1978) where six flow regimes are classified: bubble, plug,

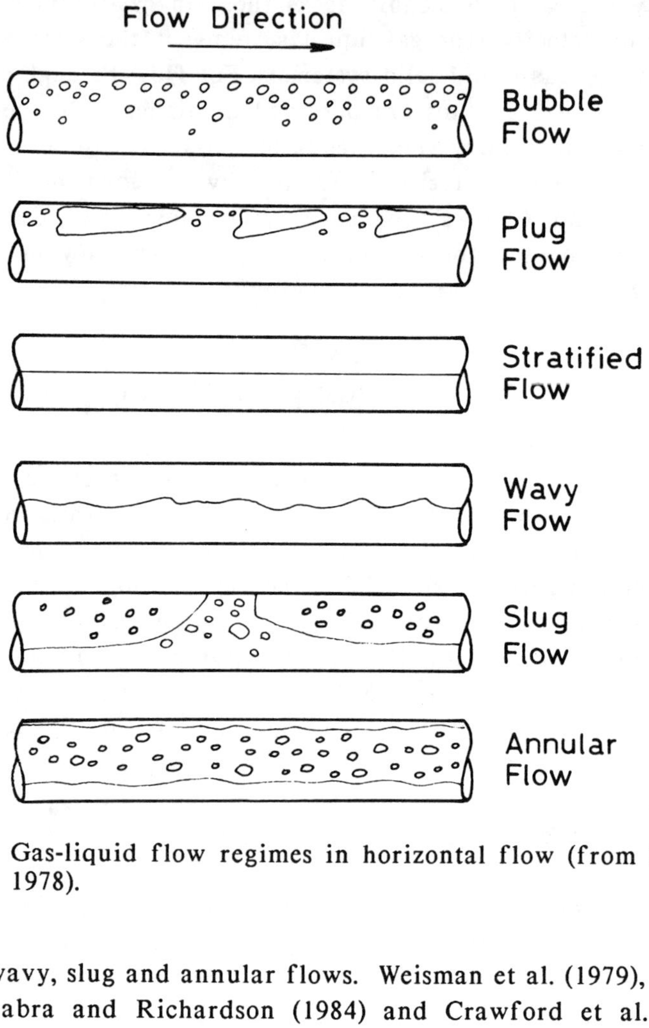

Figure 6.24 Gas-liquid flow regimes in horizontal flow (from Hewitt, 1978).

stratified, wavy, slug and annular flows. Weisman et al. (1979), Hewitt (1982), Chhabra and Richardson (1984) and Crawford et al. (1985) presented detailed explanations of flow regime transition in gas-liquid horizontal flow systems. Chhabra and Richardson's (1984) flow regime map for horizontal gas-liquid flow is presented in Fig. 6.25. Bubble, plug, and slug flows are lumped as intermittent flow. Based on this flow regime map, available information on three-phase horizontal flow in the literature is found to be mainly concerning the intermittent flow.

Different flow behavior for the solid phase are reported for the three-phase system (Scott and Rao, 1971; Toda et al., 1978). Figure 6.26 presents a photograph showing the effects of U_g on the flow regime in a horizontal three-phase flow. The water-glass bead slurry

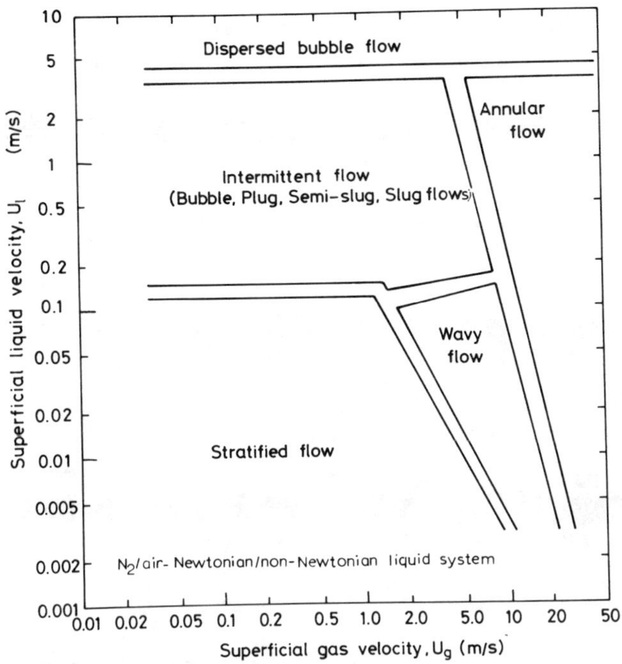

Figure 6.25 Flow regime map for a gas-liquid horizontal flow (from Chhabra and Richardson, 1984).

velocity, $U_{s\ell}$, and ψ_s are set at 0.51 m/s and 0.075, respectively. Five values of U_g are considered including 0, 0.2, 0.83. 1.3 and 4.4 m/s corresponding to photos 1 through 5, respectively. At zero gas velocity, the solid particles tend to form a wavy layer sliding forward on the bottom wall of the tube (photo 1). At low, but non-zero, gas velocities where bubble and plug flows prevail, the bubbling does not significantly improve the uniformity of solids concentration across the tube cross section; the solid particles still flow as a wavy sliding layer (photos 2 and 3). Nevertheless, a critical gas velocity exists beyond which the solid particles are carried up near the upper wall of the tube by the eddies induced at the wake of a bubble (photo 4). At a high gas velocity, an annular flow occurs (photo 5).

Liquid velocity also affects the flow behavior of solids. Scott and Rao (1971) defined a saltation velocity as the slurry velocity at which particles begin to settle on the pipe wall and remain stationary for a moment during a progressive reduction of the slurry velocity. Although the saltation velocity decreases slightly with increasing gas velocity in gas-liquid-solid systems, Scott and Rao (1971) found that

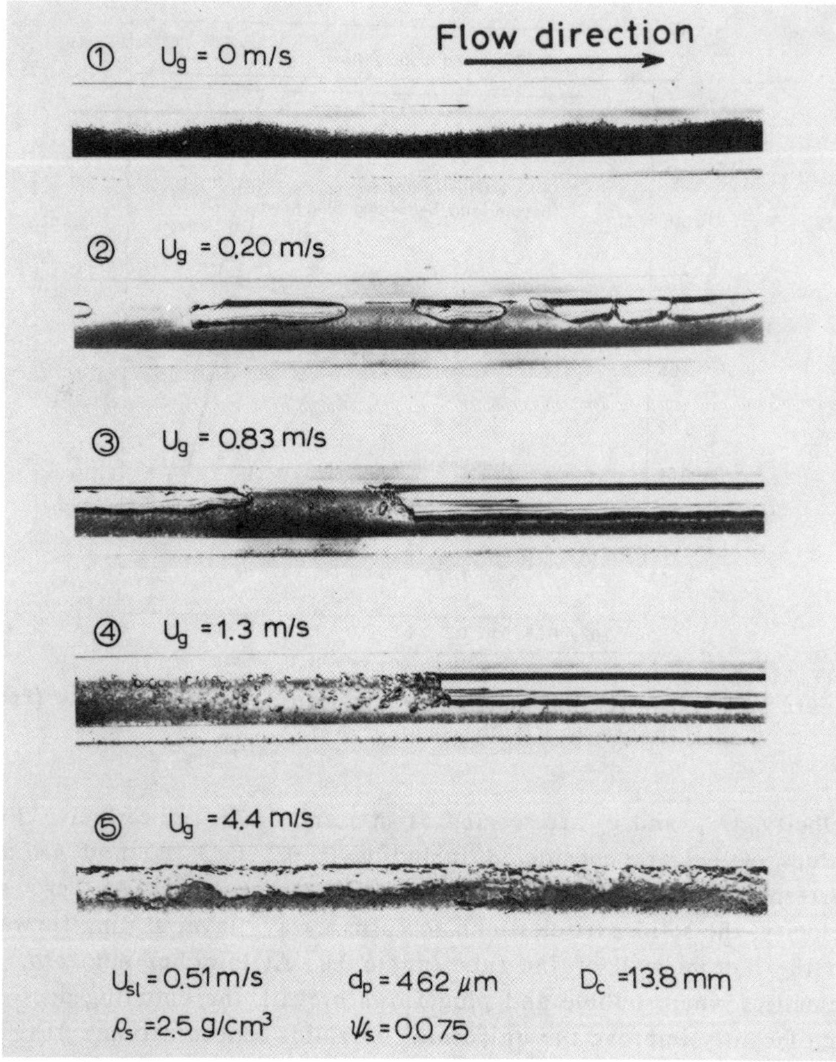

Figure 6.26 Photographic representation of a horizontal air-water-glass bead flow (from Toda et al., 1978; with permission from Kagaku Kogaku Ronbunshu; the original kindly supplied by M. Toda of Yamagata University, Japan).

the saltation velocity can be satisfactorily determined by the correlations proposed for liquid-solid systems.

6.4.3.1 Gas Holdup

Hatate et al. (1986b) measured the effects of various parameters on the gas holdup for horizontal air-water-glass bead flow systems. The gas holdup in these systems is generally smaller than that in the corresponding gas-liquid flow. In a small diameter tube (1.55 cm ID), the solids tend to distribute homogeneously and the effects of particle size and solid concentration on the gas holdup are virtually negligible. In this case, the gas holdup in the three-phase flow is distinctly smaller than that in the two-phase flow. The large diameter tube (2.59 cm ID) yields small gas holdups due to the inhomogeneity of the solids flow. Hatate et al. (1986b) defined the criterion of a homogeneous flow as being that if

$$\frac{U_{s\ell}^2\sqrt{C_d}}{\psi_s D_c g\,[(\rho_s/\rho_\ell)-1]} > 70 \tag{6.75}$$

is satisfied. For a homogeneous flow, the gas holdup in a three-phase flow was found equivalent to that in a gas-liquid flow. The following correlation for ϵ_g was developed by them:

for $V_{s\ell}^2\sqrt{C_d}/\{\psi_s D_c g[(\rho_s/\rho_\ell)-1]\} < 40$

$$\epsilon_g/\epsilon_{g0} = 0.601[V_{s\ell}^2\sqrt{C_d}/\{\psi_s D_c g[(\rho_s/\rho_\ell)-1]\}]^{0.138} \tag{6.76a}$$

and for $V_{s\ell}^2\sqrt{C_d}/\{\psi_s D_c g[(\rho_s/\rho_\ell)-1]\} \geq 40$

$$\epsilon_g/\epsilon_{g0} = 1.0 \tag{6.76b}$$

In these equations, ϵ_{g0} is the gas holdup in a gas-liquid horizontal flow, $V_{s\ell}$ is the slurry linear velocity, and C_d is the particle drag coefficient at terminal velocity. Note that the gas holdup in the gas-liquid horizontal flow can be predicted by the correlation (in SI units) of Yokota (1965) given below:

$$1 - \epsilon_{g0} = \exp\{-\sqrt{0.0466 + (1.9\times10^{-4}\beta + 0.1050)(\ln\beta)^2}\} \tag{6.77a}$$

where

$$\beta = \left(\frac{\rho_g U_g^2}{\rho_\ell U_\ell^2}\right)^{0.5} \frac{G_\ell^{0.4}}{\mu_\ell^{0.3}} \tag{6.77b}$$

Kago et al. (1986) also measured the gas holdup in the three-phase flow by varying the liquid viscosity and surface tension. They reported that the gas holdup increases with increasing gas velocity and decreasing liquid velocity. When the liquid viscosity increases up to 55 times that of water, the gas holdup decreases to as low as 50% of that in water. No effect of liquid surface tension on the gas holdup was reported. Additionally, the gas holdup in the three-phase flow is actually the same as that obtained in gas-liquid flow with the same liquid viscosity. Kago et al. (1986) proposed the following empirical correlation for gas holdup:

$$\frac{1}{\epsilon_g} = 1 + 1.6\{(\frac{U_{s\ell}}{U_g})(\frac{\mu_{s\ell}}{\mu_{\ell 0}})^{0.1}\}^{1.1} + 0.55\{(\frac{U_{s\ell}}{U_g})(\frac{\mu_{s\ell}}{\mu_{\ell 0}})^{0.1}\}^{0.25} \tag{6.78}$$

where $\mu_{\ell 0}$ is set as 1.0 mPa•s. Equation (6.78) is also applicable to gas-liquid flow and agrees closely with the Lockhart-Martinelli correlation for $(U_{s\ell}/U_g)(\mu_{s\ell}/\mu_{\ell 0})^{0.1} < 0.7$.

6.4.3.2 Pressure Drop

Pressure drop in horizontal three-phase flow has been studied by several investigators (Scott and Rao, 1971; Toda et al., 1978; Hatate et al., 1986b; Kago et al., 1986). When the diameter of the particles is smaller than 100 μm, the pressure drop due to friction can be correlated fairly well by the Lockhart-Martinelli correlation with Φ_{SL} and X as parameters except for low X values (Scott and Rao, 1971; Hatate et al., 1986b). Kago et al. (1986) proposed the following equation to correlate the pressure drop data in terms of the Lockhart-Martinelli correlation:

$$\Phi_{SL} = \sqrt{1 + 12/X + 1/X^2} \tag{6.79}$$

where X can be correlated by:

for $Re_{s\ell} > 2000$ and $Re_g > 2000$

$$X = \left(\frac{\rho_{s\ell}}{\rho_g}\right)^{0.5}\left(\frac{U_{s\ell}}{U_g}\right)^{0.9}\left(\frac{\mu_{s\ell}}{\mu_g}\right)^{0.1} \qquad (6.80a)$$

and for $Re_{s\ell} < 1000$ and $Re_g > 2000$

$$X = 18.7\left(\frac{D_c U_g \rho_g}{\mu_g}\right)^{-0.4}\left(\frac{U_{s\ell}\mu_{s\ell}}{U_g\,\mu_g}\right)^{0.5} \qquad (6.80b)$$

Hatate et al. (1986b) showed that the measured pressure drop was slightly higher than that calculated by the Lockhart-Martinelli correlation for a gas-liquid system (Chisholm and Laird, 1958) and obtained an empirical correlation. Toda et al. (1978) measured the frictional pressure drop in three-phase horizontal flows containing large particles (d_p of 462 and 985 μm) and showed that the obtained pressure drop was much lower than that predicted by the Lockhart-Martinelli correlation for a gas-liquid system (Chisholm and Laird, 1958). Furthermore, the correlation showed a dependency on particle concentration. Toda et al. (1978) attempted to analyze the pressure drop in terms of a simple separated flow model, and derived the following semi-theoretical equation:

$$\Phi_{SL}^{\ 2} = \left(-\frac{\Delta P}{\Delta z}\right)_{GL}\Big/\left(-\frac{\Delta P}{\Delta z}\right)_{SL} + (1-\epsilon_g)^{-(3\bar{n}+2)}\left\{1-\left(-\frac{\Delta P}{\Delta z}\right)_L\Big/\left(-\frac{\Delta P}{\Delta z}\right)_{SL}\right\} \qquad (6.81)$$

where $\bar{n} = -1.1$. Toda et al. also developed an empirical correlation to predict $(-\Delta P/\Delta z)_{SL}$.

Figure 6.27 shows a comparison of correlations for Φ_{SL} versus X based on available data. Equation (6.79) satisfactorily represents the pressure drop data of Toda et al. (1978) as well as those of Kago et al. (1986); however, Hatate et al.'s (1986b) data deviates significantly from the correlation when X is small.

6.4.3.3 Heat Transfer

Heat transfer characteristics in a horizontal three-phase flow were measured by Kago et al. (1986). The experiments were conducted using air/water or aqueous solution of carboxymethyl cellulose/porous silica alumina catalyst in a 51.5 mm ID pipe heater with 49 cm long. The slurry viscosity varied from 0.8 ~ 55 mPa•s. The results show that the heat transfer coefficient increases with increasing liquid velocity and decreasing liquid viscosity. In the range of superficial gas velocity less than 1 m/s, the heat transfer coefficient is nearly

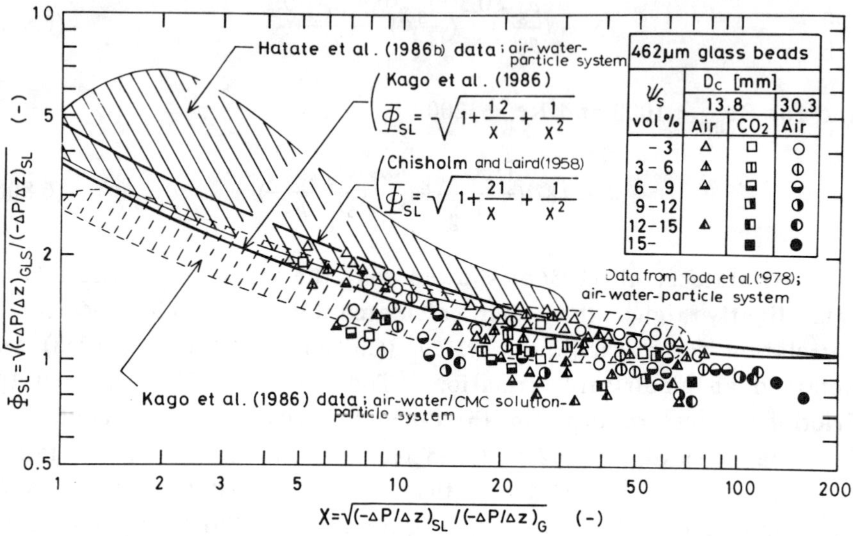

Figure 6.27 Correlation of friction loss in terms of Lockhart-Martinelli parameters, Φ_{SL} and X, for gas-liquid-solid horizontal flow.

independent of the gas velocity, while in the range of gas velocity greater than 1 m/s, it increases gradually with increasing gas velocity. A correlation for the heat transfer coefficient in the horizontal three-phase flow developed by Kago et al. (1986), given below, is also applicable to horizontal two-phase flow:

$$Nu = (0.021Re'^{0.8}_{s\ell} + 4.5)Pr_{s\ell}^{1/3}(\mu_{s\ell}/\mu_w)^{0.14}$$

$$\times \{1+0.3 \exp[-0.5(Fr_{s\ell}-1.0)^2]\} \tag{6.82}$$

where the slurry Reynolds number $Re'_{s\ell}$ is defined by $(D_c U_{s\ell} \rho_{s\ell}/\mu_{s\ell})\epsilon_{s\ell}^{-0.5}$. Equation (6.82) is valid for plug and slug flow regimes in the range of $3\times10^2 \le Re'_{s\ell} \le 3\times10^5$.

6.5 CONCLUDING REMARKS

The hydrodynamic and mass transfer behavior for a draft tube three-phase fluidization system (DTFB) (Modes E-I-a-1 and T-I-a-1) is complex and is strongly dependent on underlying system design configuration such as D_d/D_c, spacing between the bottom of the draft tube and gas-liquid distributor, physical properties of the liquid and solid particles and superficial gas and liquid velocities. The internal liquid circulation rate has dominant effects on flow regime (packed bed regime, fluidized bed regime, and circulated bed regime), bubble penetration depth in the annulus and hence the mass transfer. In general, based on gas-liquid mass transfer performance, DTFBs have merits over three-phase fluidization systems without a draft tube under large particle conditions commonly encountered in fluidized bed systems; the optimum solids loading in a DTFB varies from 1 vol% to 4 vol%. The reverse is true, however, under small particle conditions commonly encountered in slurry bubble column systems.

Three-phase semifluidized beds behave as a composite of a three-phase fluidized bed and a three-phase packed bed. The behavior is generally predictable when the behavior of a single fluidized bed and packed bed under a given flow condition is known. However, more work is needed for high gas and liquid velocity conditions. Similarly, more work is needed to account for various effects of internals on three-phase fluidization as demonstrated by annular three-phase fluidization.

Vertical three-phase flow systems involving Modes T-I-a-1 and T-III-a as well as cocurrent horizontal three-phase flow systems usually utilize particles of less than 0.5 mm in diameter supported by a continuous liquid phase. The common approach in treating three-phase flow systems under low solids loading conditions regards the liquid and solid phases as a pseudo-homogeneous phase and contrasts it with corresponding two-phase flow systems. Such approaches include the Lockhart-Martinelli correlation for phase holdup and pressure drop, and Hughmark's single flow model for pressure drop. These approaches have shown some success. More work is needed on transport properties, however, for the three-phase flow under high solids loading conditions where the deviation from the two-phase flow become significant.

6.6 NOMENCLATURE

a : gas-liquid interfacial area based on system volume

A_D : cross-sectional area of the draft tube

A_{gi} : area for gas flow across plane i

A_i : total area across plane i

$A_{\ell i}$: area for liquid flow across plane i

Ar : Archimedes number, $d_p^3(\rho_\ell-\rho_s)\rho_\ell g/\mu_\ell^2$

A_{si} : area for solid flow across plane i

A_T : cross-sectional area of the column

C : dissolved oxygen concentration in the liquid; C_1 and C_2 for the first stage and the second stage of the 2-stage DTFB, respectively

C_0 : saturated value of C in the liquid using pure oxygen C_0^* at the column outlet; $(C_0)_1$ for the first stage of the 2-stage DTFB; $(C_0)_2$ for the second stage of the 2-stage DTFB

$\bar{C}_0$: average value of C_0 defined by Eqn. (6.35)

C_B : Sodium sulfite concentration in the liquid; C_{Bi} at the inlet; C_{Be} at the outlet

C_d : drag coefficient of a particle falling at its terminal velocity

C_f' : friction factor in a packed bed section of a semifluidized bed

$C_{ps\ell}$: heat capacity of slurry

C_s : solids concentration (solids weight per unit weight of slurry)

C_{sa} : saturated concentration of C at gas-liquid interface

C_{sv} : solids concentration (solids weight per unit volume of slurry)

d_B : bubble diameter in an electrolyte solution; d_{B0} in an electrolyte-free solution

d_p : particle diameter

D_c : column or tube diameter

D_d : draft tube diameter

D_e : equivalent diameter of the annular region or equivalent diameter for a liquid channel in a packed bed

D_h : hydraulic diameter for a draft tube system

D_M : molecular diffusivity in a liquid medium; D_{M,O_2} for oxygen

f : wall interaction factor defined in Eqn. (6.9)

f_t : friction factor defined by Eqn. (6.72)

F : total force interacting with the wall

F_A : friction force in the annular region

F_B : friction force in the below draft tube region

F_D : friction force in the draft tube region

F_N : vertical component of the normal force acting on the mixture from the conical wall in the below draft tube region

F_T : friction force in the above draft tube

Fr_g : gas Froude number, $U_g{}^2/gd_p$

Fr_ℓ : liquid Froude number, $U_\ell{}^2/gd_p$

$Fr_{s\ell}$: slurry Froude number, $U_{s\ell}^2/gD_c$

g : gravitational acceleration

G_0 : mass flow rate of oxygen at the column inlet

G_ℓ : mass flow rate of liquid per unit cross-sectional area of the tube

h : heat transfer coefficient

H : liquid depth above distributor

H_c : height of column between two retaining grids of a semifluidized bed

H_d : draft tube height

H_ℓ : Vertical clearance between bottom of the draft tube and wall of conical bottom

H_{pa} : height of the packed bed section of a semifluidized bed

H_t : bubble penetration depth in the annular region

H_u : liquid height above the draft tube

k_ℓ : gas-liquid mass transfer coefficient

$k_\ell a$: volumetric mass transfer coefficient for a one-stage DTFB based on gas plug flow; $k_\ell a^\$$ based on gas complete mixing

$(k_\ell a)_0$: $k_\ell a$ in a solid-free column

$k_\ell a^*$: volumetric mass transfer coefficient for a 2-stage DTFB based on plug flow of gas phase

$(k_\ell a^*)_0$: $k_\ell a^*$ in a solid-free column

$k_\ell a^\#$: volumetric mass transfer coefficient for a 2-stage DTFB based on complete mixing of gas phase

K : first-order reaction rate constant

K_c : sudden contraction-loss coefficient

L : downward distance from retaining grid in a DTFB

L_s : solids loading expressed in terms of the volume fraction of particles in a DTFB or a semifluidized bed

n : the area ratio of the annular region to the mid-outer tube of the injector

Nu : Nusselt number, $hD_c/\lambda_{s\ell}$

P : pressure

P^* : atmospheric pressure

P_i : pressure on plane i

$-\Delta P$: total pressure drop in the bed

$(-\Delta P)^f$: total pressure drop in the fluidized bed section of a semifluidized bed

$(-\Delta P)^p$: total pressure drop in the packed bed section of a semifluidized bed

$-\Delta P_{Ah}$: total pressure drop in the annular region due to hydrostatic head

$-\Delta P_{Da}$: total pressure drop in the draft region due to acceleration of fluid

$-\Delta P_{Df}$: total pressure drop in the draft tube region due to friction

$-\Delta P_{Dh}$: total pressure drop in the draft tube region due to hydrostatic head

$-\Delta P_{fr}$: total pressure drop due to force interaction with the wall

$(-\frac{dP_f}{dz})^g$: frictional axial pressure gradient of the gas phase in the packed bed section of a semifluidized bed

$(-\frac{dP_f}{dz})^{g-\ell}$: gas-liquid interaction contribution to the gas phase frictional axial pressure gradient in the packed bed section of a semifluidized bed

$(-\frac{dP_f}{dz})^{\ell}$: frictional axial pressure gradient of the liquid phase in the packed bed section of a semifluidized bed

$(-\frac{dP_f}{dz})^{\ell-g}$: gas-liquid interaction contribution to the liquid phase frictional axial pressure gradient in the packed bed section of a semifluidized bed

$(-\frac{dP_f}{dz})^{\ell-s}$: liquid-solid interaction contribution to the liquid phase frictional axial pressure gradient in the packed bed section of a semifluidized bed

$(-\frac{dP}{dz})^p$: total axial pressure gradient in the packed bed section of a semifluidized bed

$-\frac{\Delta P}{\Delta z}$: total pressure gradient
$(-\frac{\Delta P}{\Delta z})_G$: pressure gradient due to friction for gas flow
$(-\frac{\Delta P}{\Delta z})_{GL}$: pressure gradient due to friction for gas-liquid flow
$(-\frac{\Delta P}{\Delta z})_{GLS}$: pressure gradient due to friction for gas-liquid-solid flow
$(-\frac{\Delta P}{\Delta z})_L$: pressure gradient due to friction for liquid flow
$(-\frac{\Delta P}{\Delta z})_{SL}$: pressure gradient due to friction for solid-liquid flow
$Pr_{s\ell}$: slurry Prandtl number, $C_{ps\ell}\rho_{s\ell}/\lambda_{s\ell}$
Q_ℓ : apparent liquid circulation rate
$-r_B$: Na_2SO_3 oxidation rate
r_i : radius of the inner tube of the injector
$-r_{O_2}$: oxygen consumption rate
r_o : radius of the outer tube of the injector
Re_g : Reynolds number, $D_c U_g \rho_g/\mu_g$
$\overline{Re}_\ell$: Reynolds number, $U_\ell D_e \rho_\ell/(1-\epsilon_s^p)\delta_\ell \mu_\ell$
Re_p : Reynolds number, $d_p \rho_s U_g/\mu_\ell$
$Re_{s\ell}$: slurry Reynolds number, $D_c U_{s\ell}\rho_{s\ell}/\mu_{s\ell}$
$Re'_{s\ell}$: modified slurry Reynolds number, $(D_c U_{s\ell}\rho_{s\ell}/\mu_{s\ell})\epsilon_{s\ell}^{0.5}$
$(Re)_c$: apparent liquid circulation Reynolds number, $D_c U_c \rho_\ell/\mu_\ell$
$\bar{t}$: average time for a tracer particle to make a single loop
t_A : liquid circulation time in the annular region
t_B : liquid circulation time in the below draft tube region
t_D : liquid circulation time in the draft tube region
t_T : liquid circulation time in the above draft tube region
t_{Total} : total liquid circulation time
U_A : liquid velocity in the annular region based on the cross-sectional area of the annular region
U_B : liquid velocity in the annular region based on the cross-sectional area of mid-outer tube of the injector
U_c : apparent liquid circulation velocity
U_D : liquid velocity in the draft tube region based on the cross-sectional area of the draft tube
U_E : column outlet liquid velocity based on the cross-sectional area of the column

U_g : superficial gas velocity

U_I : column inlet liquid velocity in the column based on the cross-sectional area of the injector

U_ℓ : superficial liquid velocity based on the cross-sectional area of the column

$(U_\ell)_{osf}$: onset velocity for semifluidization

$U_{s\ell}$: superficial slurry velocity

U_t : terminal velocity of a particle

v_m : specific volume of oxygen

V : total volume of the bed

V_A : total volume of the annular region

V_B : total volume of the below draft tube region

V_D : total volume of the draft tube region

V_{gi} : linear gas velocity across plane i

V_ℓ : volume of liquid in the bed

$V_{\ell i}$: linear liquid velocity across plane i

V_s : volume of solids in the bed

V_{sc} : volume of solids at the critical condition for solids suspension

V_{si} : linear solid velocity across plane i

$V_{s\ell}$: linear velocity of slurry, $U_{s\ell}/\epsilon_{s\ell}$

V_T : total volume of the above draft tube region

W_s : weight of solids in a fluidized bed

$We_{s\ell}$: Weber number of slurry, $U_{s\ell}^2\rho_{s\ell}D_c/\sigma$

X : Lockhart-Martinelli parameter defined by Eqn. (6.70)

z : axial distance

Z : mole fraction of oxygen in the gas phase; Z_i at the column inlet; Z_e at the column exit for one-stage DTFB; Z_1 and Z_2 at the exit of the first stage and the second stage for two-stage DTFB, respectively

Greek letters

α : pressure conversion factor

β : parameter defined by Eqn. (6.77b)

γ	: Hatta number defined by Eqn. (6.40)
δ_g	: solid-free gas volume fraction
δ_ℓ	: solid-free liquid volume fraction
ϵ^f	: bed porosity in the fluidized bed section of a semifluidized bed
ϵ_g	: gas holdup
ϵ_g^f	: gas holdup in the fluidized bed section of a semifluidized bed
ϵ_g^p	: gas holdup in the packed bed section of a semifluidized bed
ϵ_{g0}	: gas holdup for solid-free gas-liquid system
ϵ_{gA}	: gas holdup in the annular region
ϵ_{gB}	: gas holdup in the below draft tube region
ϵ_{gD}	: gas holdup in the draft tube region
ϵ_{gT}	: gas holdup in the above draft tube region
ϵ_ℓ	: liquid holdup in the bed
ϵ_ℓ^f	: liquid holdup in the fluidized bed section of a semifluidized bed
ϵ_ℓ^p	: liquid holdup in the packed bed section of a semifluidized bed
$\epsilon_{\ell A}$	: liquid holdup in the annular region
$\epsilon_{\ell B}$	: liquid holdup in the below draft tube region
$\epsilon_{\ell D}$	: liquid holdup in the draft tube region
$\epsilon_{\ell T}$	: liquid holdup in the above draft tube region
ϵ_s	: solids holdup
$(\epsilon_s)_{osf}$	: solids holdup at onset liquid velocity for semifluidization
ϵ_s^p	: solids holdup in the packed bed section of a semifluidized bed
ϵ_{sA}	: solids holdup in the annular region
ϵ_{sB}	: solids holdup in the below draft tube region
ϵ_{sD}	: solids holdup in the draft tube region
$\epsilon_{s\ell}$	: slurry holdup
ϵ_{sT}	: solids holdup in the above draft tube region
Φ_{SL}	: Lockhart-Martinelli parameter defined by Eqn. (6.69)
Φ_G	: Lockhart-Martinelli parameter defined by Eqn. (6.68)

θ : angle of the conical section

κ : gradual contraction-loss coefficient

λ : Darcy's friction factor

$\lambda_{s\ell}$: thermal conductivity of slurry

η : fraction of oxygen utilization; η_e at the column exit for one-stage DTFB; η_1 and η_2 for the first stage and the second stage of the 2-stage DTFB, respectively

μ_g : gas viscosity

μ_ℓ : liquid viscosity

$\mu_{\ell 0}$: standard liquid viscosity, 1.0 mPa•s

$\mu_{s\ell}$: slurry viscosity

μ_w : water viscosity at room temperature

ρ_g : gas density

ρ_ℓ : liquid density

ρ_m : mixture density

ρ_s : solid density

$\rho_{s\ell}$: slurry density

σ : surface tension of liquid

ψ_s : solid volume fraction in the slurry

6.7 REFERENCES

Akita, K., and F. Yoshida, I&EC Process Des. & Dev., 12, 76 (1973).

Astarita, G., "Mass Transfer With Chemical Reaction," Elsevier, Amsterdam (1967).

Blasius, H., Forschg. Arb. Ing.-Wes., No. 131 (1913).

Chatib, B., A. Grasmick, S. Elmaleh, and R. Ben Aim, in "Biological Fluidized Bed Treatment of Water and Wastewater," ed. by P.F. Cooper and B. Atkinson, p. 192, Ellis Horwood, Chichester, England (1981).

Chern, S.-H., K. Muroyama, and L.-S. Fan, Chem. Eng. Sci., 38, 1167 (1983).

Chern, S.-H., L.-S. Fan, and K. Muroyama, AIChE J., 30, 288 (1984).

Chhabra, R.P., and J.F. Richardson, Can. J. Chem. Eng., 62, 449 (1984).

Chisholm, D., and A.D.K. Laird, Trans. ASME J. Heat Transfer, 80, Ser. C, 276 (1958).

Cooper, P.G., R.S. Silver, and J.P. Boyle, in "Single-Cell Protein II," ed. by D.I.C. Wang and S.R. Tannenbaum, p. 454, MIT Press, Cambridge, MA (1975).

Crawford, T.J., C.B. Weinberger, and J. Weisman, Int. J. Multiphase Flow, 11, 761 (1985).

Dakshinamurty, P., K. Veerabhadra Rao, R.V. Subbaraju, and V. Subrahmanyam, I&EC Process Des. & Dev., 11, 318 (1972).

de Nevers, N., AIChE J., 14, 222 (1968).

Fan, L.-S., S.-J. Hwang, and A. Matsuura, Chem. Eng. Sci., 39, 1677 (1984a).

Fan, L.-S., K. Fujie, and T.-R. Long, AIChE Symposium Series, 80, No. 241, 102 (1984b).

Fan, L.-S., F. Bavarian, R.L. Gorowara, B.E. Kreischer, R.D. Buttke, and L.E. Peck, paper presented at the AIChE Annual Meeting, Miami Beach, Nov. 2-7 (1986); Powder Technol., 53, 285 (1987).

Fan, L.-S., K. Kitano, and B.E. Kreischer, AIChE J., 33, 225 (1987a).

Fan, L.-S., T.S. Ramesh, W.-T. Tang, and T.-R. Long, Chem. Eng. Sci., 42, 543 (1987b).

Fan, L.T., and C.Y. Wen, AIChE J., 5, 407 (1959).

Fields, P.R., and N.K.H. Slater, Chem. Eng. Sci., 38, 647 (1983).

Fujie, K., M. Takaine, H. Kubota, and Y. Miyaji, J. Chem. Eng. Japan, 13, 188 (1980a).

Fujie, K., M. Takaine, H. Kubota, and Y. Miyaji, Hakko Kogaku, Japan, 58, 179 (1980b).

Fujie, K., and L.-S. Fan, unpublished results (1985).

Fujie, K., and L.-S. Fan, in "Fluidization," ed. by K. Ostergaard and A. Sorensen, p. 369, Engineering Foundation Press (1986).

Fuller, E.C., and R.H. Crist, J. Am. Chem. Soc., 63, 1644 (1941).

Giot, M., in "Handbook of Multiphase Systems," ed. by G. Hetsroni, p. 7-29, McGraw-Hill (1982).

Griffith, P., and G.B. Wallis, Trans. ASME J. Heat Transfer, 83, Ser. C, 307 (1961).

Hatate, Y., H. Nomura, T. Fujita, S. Tajiri, N. Hidaka, and A. Ikari, J. Chem. Eng. Japan, 19, 56 (1986a).

Hatate, Y., H. Nomura, T. Fujita, S. Tajiri, and A. Ikari, J. Chem. Eng. Japan, 19, 330 (1986b).

Hatate, Y., S. Tajiri, T. Fujita, T. Fukumoto, A. Ikari, and T. Hano, J. Chem. Eng. Japan, 20, 568 (1987).

Hatch, R.T., Ph.D. Thesis, Massachusetts Institute of Technology (1973).

Heck, J., and U. Onken, Chem. Eng. Sci., 42, 1211 (1987).

Hewitt, G.F., "Measurement of Two-Phase Flow Parameters," Academic Press (1978).

Hewitt, G.F., in "Handbook of Multiphase Systems," ed. by G. Hetsroni, p. 2-3, McGraw-Hill (1982).

Hughmark, G.A., and B.S. Pressburg, AIChE J., 7, 677 (1961).

Hughmark, G.A., Chem. Eng. Prog., 58, 62 (1962).

Hughmark, G.A., Chem. Eng. Sci., 20, 1007 (1965).

Hwang, S.-J., and L.-S. Fan, paper presented at the 16th Annual Meeting of the Fine Particle Society, Miami, April 22-26 (1985).

Hwang, S.-J., and L.-S. Fan, Chem. Eng. J., 33, 49 (1986).

Jones, A.G., Chem. Eng. Sci., 40, 449 (1985).

Kago, T., T. Saruwatari, M. Kashima, S. Morooka, and Y. Kato, J. Chem. Eng. Japan, 19, 125 (1986).

Kato, H., S. Tamiya, and T. Miyazawa, in Proceedings of the 2nd Symposium on Jet Pumps, Ejectors, and Air-Lift Techniques, p. 37 (1975).

Kato, Y., N. Hidaka, and H. Kamimura, Kagaku Kogaku Ronbunshu, 12, 727 (1986).

Kawase, Y., and M. Moo-Young, Chem. Eng. Commun., 40, 67 (1986).

Keitel, G., and U. Onken, Ger. Chem. Eng., 4, 250 (1981).

Kim, S.D., and J.H. Choi, Can. J. Chem. Eng., 62, 85 (1984).

Koide, K., T. Hayashi, M. Noro, Y. Takemura, N. Kawamata, and H. Kubota, J. Chem. Eng. Japan, 5, 236 (1972).

Koide, K., K. Kurematsu, S. Iwamoto, Y. Iwata, and K. Horibe, J. Chem. Eng. Japan, 16, 413 (1983).

Koide, K., K. Horibe, H. Kawabata, and S. Ito, J. Chem. Eng. Japan, 17, 368 (1984a).

Koide, K., S. Iwamoto, Y. Takasaka, S. Matsuura, E. Takahashi, M. Kimura, and H. Kubota, J. Chem. Eng. Japan, 17, 611 (1984b).

Koide, K, K. Horibe, H. Kawabata, and S. Ito, J. Chem. Eng. Japan, 18, 248 (1985).

Kubota, H., Y. Hosono, and K. Fujie, J. Chem. Eng. Japan, 11, 319 (1978).

Lockhart, R.W., and R.C. Martinelli, Chem. Eng. Prog., 45(1), 39 (1949).

Malfait, J.L., D.L. Wilcox, D.G. Mercer, and L.D. Baker, Biotech. Bioeng., 23, 863 (1981).

Marrucci, G., Chem. Eng. Sci., 24, 975 (1969).

Matsuura, A., J. Ogura, and T. Akehata, paper (preprint SA209) presented at the Niigata Meeting of Soc. Chem. Engrs. Japan, Niigata Prefecture, July 30 (1982).

Matsuura, A., and L.-S. Fan, AIChE J., 30, 894 (1984).

Matsuura, A., and T. Akehata, paper (preprint SK307) presented at the 19th Fall Meeting of Soc. Chem. Engrs. Japan (1985).

Michelsen, M.L., and K. Ostergaard, Chem. Eng. J., 1, 37 (1970).

Miller, R.L., and M.B. Cain, Chem. Eng. Commun., 43, 147 (1986).

Miyahara, T., and T. Takahashi, Memoirs of the School of Eng., Okayama University, 20(2), 1 (1986).

Miyahara, T., M. Hamaguchi, Y. Sukeda, and T. Takahashi, Can. J. Chem. Eng., 64, 718 (1986).

Muroyama, K., Y. Mitani, and A. Yasunishi, Chem. Eng. Commun., 34, 87 (1985).

Nicklin, D.J., J.O. Wilkes, and J.F. Davidson, Trans. Inst. Chem. Engrs., 40, 61 (1962).

Orazem, M.E., and L.E. Erickson, Biotech. Bioeng., 21, 69 (1979).

Oshinowo, T., and M.E. Charles, Can. J. Chem. Eng., 52, 25 (1974a).

Oshinowo, T., and M.E. Charles, Can. J. Chem. Eng., 52, 438 (1974b).

Perlmutter, M., R.M. Kornosky, and J.A. Ruether, in "Multi-Phase Flow and Heat Transfer III. Part B: Applications," ed. by T.N. Veziroglu and A.E. Bergles, Elsevier, Amsterdam (1984).

Ramamurthy, K., and K. Subbarju, I&EC Process Des. & Dev., 12, 184 (1973).

Perry, J.H., and C.H. Chilton, in "Chemical Engineers' Handbook," 5th ed., p. 5, McGraw-Hill (1973).

Sawai, T., K. Sugata, and T. Miki, The Mitsubishi Heavy Industry Technical Report, 19(5), 510 (1982).

Scott, D.S., and P.K. Rao, Can. J. Chem. Eng., 49, 302 (1971).

Shah, Y.T., A.A. Kulkarni, J.H. Wieland, and N.L. Carr., Chem. Eng. J., 26, 95 (1983).

Suzuki, M., S. Momonoi, and H. Harada, in Proceedings of the Symposium on Biological Wastewater Treatment, p. 34, Soc. Chem. Engrs. Japan, Sendai, Japan (1983).

Thomas, D.G., I&EC, 55(11), 18 (1963).

Toda, M., K. Shimazaki, and S. Maeda, Kagaku Kogaku Ronbunshu, 4, 56 (1978).

Toda, M., E. Harada, M. Kuriyama, S. Saruta, and H. Konno, Kagaku Kogaku Ronbunshu, 8, 380 (1982).

Urza, I.J., and M.L. Jackson, I&EC Process Des. & Dev., 14, 106 (1975).

Weber, M., and Y. Dedegil, paper presented at the 4th International Conference on Hydraulic Transport Solids Pipes, BHRA Fluid Engineering, Cranfield, Bedford, U.K., May 1 (1976).

Weiland, P., and U. Onken, Ger. Chem. Eng., 4, 42 (1981a).

Weiland, P., and U. Onken, Ger. Chem. Eng., 4, 174 (1981b).

Weiland, P., Ger. Chem. Eng., 7, 374 (1984).

Weisman, J., D. Duncan, J. Gibson, and T. Crawford, Int. J. Multiphase Flow, 5, 437 (1979).

Weisman, J., and S.Y. Kang, Int. J. Multiphase Flow, 7, 271 (1981).

Wen, C.Y., S.C. Wang, and L.T. Fan, AIChE J., 9, 317 (1963).

Wiley, P., M.S. Thesis, The Ohio State University (1986).

Ying, D.H., E.N. Givens, and R.E. Weimer, I&EC Process Des. & Dev., 19, 635 (1980).

Yokota, T., Kagaku Kogaku, 29, 687 (1965).

Zhou, M.R., "Fluid Mechanics," Chinese Architectural Industry Press, China (1979).

Part III

Applications

Chapter 7

Fermentation

Application of three-phase fluidization systems to fermentation in the food, chemical, and pharmaceutical industries is a relatively new and challenging area. Despite the fact that demonstrated commercial operation using a three-phase fluidized bed bioreactor exists, for example in the production of ethanol (Michaels, 1984), the majority of the work is still at the laboratory level. This is due to the complicated physiological and energetic behavior of microorganisms under process conditions, which has compounded the complexity of bioreactor design, operation, and control.

This chapter describes basic features of three-phase fluidization systems used as bioreactors for fermentation. These fluidized bed bioreactors[1] most commonly involve Modes E-I-a-1, E-I-a-2, E-II-a-1, E-II-b, E-III-a and T-I-a-1 as noted in Chapter 1 (see Fig. 1.3 for mode designations). This chapter also addresses pertinent operational considerations essential to the successful application of fluidized bed bioreactors, including cell immobilization, substrate diffusivities, optimal design consideration, and economics. Note that the focus of the discussion is on immobilized cell systems. Plant cells, mammalian cells (including hybridoma), or flocculated microbial cells of size greater than 10 μm in suspended cell systems are treated fluid mechanically as being heterogeneous to the liquid phase and hence, these systems are also considered to be in the realm of three-phase fluidization. However, little is known regarding the fluidization behavior involving plant cells or mammalian cells (e.g., Adamson and Schmidli, 1986). The treatment of transport processes given here is generally extendible to systems involving flocculated microbial cells.

1: For convenience, a general term "fluidized bed bioreactor" is used throughout this chapter to represent various modes of fluidization system operation in fermentation applications.

7.1 BASIC PROCESSING FEATURES

In a three-phase fluidized bed fermentor or bioreactor, the solid phase comprises immobilized cell (or enzyme) particles in which the desired biochemical reactions occur. The culture medium necessary for cell growth and maintenance constitutes the liquid phase. The constituents of the gas phase depend upon whether the fermentation process is aerobic or anaerobic: for aerobic fermentation, air or oxygen used for microbial respiration and/or CO_2 evolved from metabolism constitute the gas phase; for anaerobic fermentation, an inert gas, such as nitrogen, or gases produced during fermentation, such as CO_2, H_2, etc., induce solids mixing and constitute the gas phase.

The term immobilized cells (enzymes) is defined as "cells (enzymes) that are physically confined or localized in a certain defined region of space with retention of their catalytic activity and/or viability, and which can be used repeatedly and continuously" (Klein and Wagner, 1983). Utilizing immobilized cells presents apparent advantages over using immobilized enzymes in the following aspects: (1) No extraction and purification of the enzymes are necessary; (2) In whole cells, enzymes are maintained in their natural environment; consequently, their stabilities are higher and the loss of catalytic activity is greatly reduced; (3) Cofactors can be regenerated in immobilized living cells; and (4) Multi-step enzyme reactions can be carried out using immobilized cells. This chapter focuses mainly on immobilized cell systems. Note that with some modifications, the analysis presented can readily be applied to immobilized enzyme systems.

Typically, immobilized cell particles are made of solid particles having cells attached to their surface or entrapped within their internal porous structure. Cells which physically attach to the external solid surface will grow into a matrix consisting of multilayers of cells and biopolymers excreted by cells. This multilayer matrix is termed a biofilm. In some cases, the immobilized cell particles may be so-called bioflocs, which are formed by the self-flocculation of cells without the support of solid carriers.

Due to the small liquid throughput ordinarily encountered in fermentation processes, the particles are mainly supported by the gas phase. However, particles can also be supported by the liquid phase when liquid is recycled externally via a pump or internally via a propeller. Three-phase fluidized beds with gas as the continuous phase have been used for production of ethanol (Moebus and Teuber, 1982; Bauer, 1986), glutathione (Bauer, 1986), and enzymes including α-amylase and alkaline proteinase (Akao and Okamoto, 1984). Three-phase fluidized beds with liquid as the continuous phase have been used for the production of amino acids, enzymes, and antibiotics: examples include production of L-isoleucine (Wada et al., 1980), penicillin (Morikawa et al., 1979; Endo et al., 1986; Kalogerakis et al., 1986; and Jones et al., 1986), bacitracin (Morikawa et al., 1980), and patulin (Berk et al., 1984a, 1984b).

Based on operational objectives and requirements, application of a three-phase fluidized bed to fermentation is far more difficult and complex than to wastewater treatment (see Chapter 8). This is partly due to the fact that the primary objective in wastewater treatment is simply biodegradation of waste chemicals while that for fermentation is biosynthesis of useful products. Thus, extra-operational considerations are required for fermentation, such as sterilization, pure culture preparation, separation of intermediate or product inhibitors, bioreactor control, and microbial immobilization. Venkatasubramanian and Karkare (1983) qualitatively compared fluidized bed bioreactors with fixed bed and hollow fiber bioreactors and stirred tank bioreactors with mechanical agitation as shown in Table 7.1. It can be seen from the table that the fluidized bed reactor and the hollow fiber reactor offer some unique advantages for biological processing using immobilized cells. The fluidized bed, however, provides better solids and liquid mixing, higher oxygen transfer, easier CO_2 removal, easier cell regeneration, and a more uniform cell population in the reactor than the hollow fiber reactor. The operation of a fluidized bed, however, is restricted by the inherent shear effects on the immobilized cell particles caused by bubble motion and liquid flow. Thus, the gas and liquid flow rates in the reactor need to be properly controlled to minimize particle shear effects while maintaining adequate bed expansion, solids mixing and gas-liquid mass transfer. Since successful operation of a three-phase fluidized bed bioreactor relies heavily on the stability and viability of cells in an immobilized state, it is of high relevance to examine closely immobilization methods and transport properties within the immobilized cell particles.

Table 7.1 Characteristics of Different Types of Immobilized Cell Bioreactor for Fermentation (from Venkatasubramanian and Karkare, 1983)

	Fixed bed reactors	Mechanical stirred reactors	Fluidized bed reactors	Hollow fiber reactors
Ease of operation	Easy	Easy	Difficult	Easy
Ease of catalyst replacement	Difficult	Easy	Easy	Easy
Pressure drop	High	Low	Low	Low
Ease of cell removal	Difficult	Easy	Easy	Easy
Mixing characteristics	Poor	Good	Good	Fair
Oxygen transfer	Poor	Good	Good	Fair
Distribution of viable cells	Uneven	Uniform	Uniform	Uneven
Catalyst attrition	Low	High	Low	Low
Ease of scale-up	Easy	Easy	Difficult	Difficult
Cost	Low	Low	High	Low
Suitability for				
1. Product inhibited kinetics	Good	Poor	Fair	Good
2. Substrate inhibited kinetics	Poor	Good	Fair	Poor
Ease of reactor control	Difficult	Easy	Easy	Difficult

7.2 IMMOBILIZATION METHODS

Methods for immobilization of microbial cells can be roughly classified as (1) covalent coupling, including cross-linking; (2) adsorption; (3) entrapment in a three-dimensional network; and (4) confinement in a two-phase system.

When selecting an immobilization method, it is necessary to consider one or more of the following factors: characteristics of the fermentation reaction(s) and microbial species; retention of high cell metabolic activity; maintenance of cell stability and, in most cases, viability; the availability and cost of the immobilization carriers; and the mechanical strength of the resulting immobilized cell particles. In addition, the ease of liquid phase product separation or biomass harvesting from immobilized cell particles should be considered in some fermentation processes.

A brief overview of the first three immobilized methods given above is described in the following sections. The immobilization by the confinement of cells in a two-phase systems does not involve a solid phase and is, therefore, not discussed here; an extensive review on this subject can be found in Mattiasson and Hahn-Hagerdal (1983). More detailed reviews on immobilization methods can be found in Jack and Zajic (1977a), Durand and Navarro (1978), Klein and Wagner (1983), Chibata and Tosa (1981), Mattiasson (1983), Kennedy and Cabral (1983), Chibata et al. (1986), and Mosbach (1976, 1987a, 1987b, 1988).

Table 7.2 Covalent Immobilization Methods

Carrier		Activating/ Coupling Agent	Cell Viability	References
Inorganics	Glass & ceramics	γ-Amino propyltri-methoxysilan/ glutaraldehyde	No	Navorro and Durand (1977)
	Glass & ceramics	Polyisocyanate	No	Messing and Oppermann (1979)
	Titanium & zirconium oxides	——— *	Yes	Kennedy (1979); Kennedy et al. (1976)
Organics	Carboxymethyl-cellulose & agarose	Carbodiimide	No	Jack and Zajic (1977b)
	Poly(maleic anhydride/ styrene)-co-polyethylene	Maleic anhydride moiety	———	Beddows et al. (1985)
	Poly(maleic anhydride/vinyl acetate)-co-polyethylene	Maleic anhydride moiety	———	Beddows et al. (1985)
	Hydroxyalkyl methacrylate	1,6-Diaminohexane/glut-araldehyde	———	Jirku et al. (1979)
	Bovine serum albumin	Glutaraldehyde	No	Petre et al. (1978)
	Cellulose	Cyanuric chloride	Yes	Okita et al. (1985)
No carrier (direct cell-cell linkage)		Glutaraldehyde	No	Poulsen and Zittan (1976)
		Diazotized diamine	No	Lartigue and Weetall (1976)
		Toluene diisocyanate	No	Chibata et al. (1974)

* Activating/coupling agent is not required.

7.2.1 Covalent Coupling (Cross-Linking)

Covalent coupling techniques are based on covalent bond formation between cells and insoluble support materials. Both inorganic and organic carriers have been used for this immobilization method. Table 7.2 summarizes covalent immobilization techniques. Frequently, the surface structure of the solid carriers needs to be chemically modified to provide active chemical groups. These active groups can then bond either with chemical groups on the cell surface or with a bifunctional reagent which in turn bonds to the cell surface in a similar manner.

Covalent coupling immobilization has the advantages of being free of diffusional limitations and having great operational stability due to the nature of its monolayer coverage and strong binding, respectively. Several drawbacks, however, reduce the acceptance of this method for cell immobilization. Firstly, bifunctional agents such as glutaraldehyde, polyisocyanate, and carbodiimide are deleterious to cells and enzymes. Frequently, cell viability is lost and only 60 ~ 75% of the enzyme activity is retained after these treatments. Secondly, although viable cells were reported in some covalent coupling techniques, e.g., Kennedy et al. (1976), Kennedy (1979), and Okita et al. (1985), a substantial leakage of cells from the particles was also observed (Kennedy et al., 1976). The cell leakage implies an increase in downstream separation cost. Finally, the biomass loading per unit reactor volume is low when using the covalent coupling method compared with other methods.

7.2.2 Adsorption

Adhesion of microbial cells to the solid surface is a rather complicated process which involves van der Waals forces, ionic and hydrogen bonds, interfacial energy and preconditioning of the inert solid surface by extracellular macromolecules. Comprehensive information on the mechanisms of cell adhesion to solid surface can be found in Bryers (1987).

A great variety of microbial species belonging to Bacillus, Pseudomonas, Saccharomyces, Penicillium, Aspergillus, Streptomyces, and Azotobacter genera have been immobilized on various solid supports by adsorption (Messing and Oppermann, 1979; Atkinson et al., 1979; Messing et al., 1979; Sitton and Gaddy, 1980; Ghose and Bandyopadhyay, 1980; Briffaud and Engasser, 1979; Ryu et al., 1982; Gainer et al., 1981). Inorganic carriers such as brick, porous glass, fritted glass, ceramics, sand, clay, asbestos, stainless steel, etc., and organic carriers such as ionic exchange resin, wood chips, activated carbon, polypropylene, etc., have been used.

The adsorption phenomenon greatly depends upon cellular properties as well as the physical and chemical characteristics of the carriers. Both the charges on the cell wall surface and the cell wall composition provide the necessary electrostatic and ionic sites for interaction with the solid carrier. Cell age also plays an important role in cell adsorption. The charge on the carrier surface, carrier composition, and carrier morphology are three major carrier properties influencing cell attachment. Environmental conditions such as pH, ionic strength, and the presence of metal ions also influence the cell adsorption.

The charge properties of cells and/or the solid support can be modified to facilitate cell adhesion. For example, Van Haecht et al. (1985) neutralized the negative charge on the surface of yeast cells with an $Al(NO_3)_3$ solution, thus allowing adhesion of a dense regular layer of cells on the negatively-charged glass or polycarbonate support materials.

Composition of certain carriers may affect cell adsorption by providing ion-exchanging groups for noncovalent linkage between the surfaces of cells and carriers. Thus, the metal ions contained in glass and ceramics can form the metal hydroxide resulting from an ion exchange between the glass or ceramic and a buffer solution. The hydroxyl group in the metal hydroxide may then be replaced by a suitable ligand on the cell surface, resulting in noncovalent linkage (Kolot, 1980).

Carrier morphology, mainly the pore size and surface area available for cells to adhere to, determines the amount and types of cells that can be adsorbed and accumulated. According to Messing and Oppermann (1979), the optimum pore size for maximum accumulation of stable and viable bacteria (e.g., E. coli, S. marcescens, and B. subtilis) is 1 to 5 times the maximum cell dimension, whereas for yeast (e.g., S. cerevisiae, and S. amurcae) the optimum pore size is 4 times the maximum cell dimensions.

Adsorptive immobilization has two distinct advantages: operational simplicity and retention of cell viability and activity. Cell loading can often be easily carried out by simply pumping a cell suspension through a bed of solid carrier. Furthermore, since no chemicals are used, the immobilization occurs under normal cellular physiological conditions, thus maintaining the viability of the cell and the catalytic ability of its enzymes. The main disadvantage, however, arises from the weak linkage between the adsorbed cells and carriers. Substantial leakage of cells can be expected, especially when cells multiply or when cells are subject to shear due to fluid flow. Such a cell leakage increases the cost of downstream separation.

7.2.3 Entrapment

The entrapment method is based on the confinement of cells in a polymeric gel lattice which prevents the release of cells while allowing substrates and products to diffuse to and from the entrapped cells. Consequently, application of entrapment immobilization is restricted to cellular reactions involving low molecular weight substrates and products.

Cells immobilized by the entrapment method are free, i.e., not chemically bound to the gel matrix, within the gel compartments. As a result, cells as well as other biocatalysts (i.e., enzymes and organelles) can be easily immobilized by this method with little destruction of their biological activity as compared to covalent coupling methods. Another advantage is that cells, when entrapped within the gel lattice, are protected from macromolecular inhibitors, including hydrolytic enzymes and bacteriophages. The major concern in using entrapment-immobilized cells, however, is the mechanical stability of the gel particles in a reactor.

Cells can be entrapped within a gel lattice formed by polymerization of synthetic monomers or oligomers, by ionotropic gelation of polyelectrolytes, by precipitation of collagen, gelatin and agar, or by crosslinking of fiber-forming polymers. Each of the above methods will be discussed in the following sections.

7.2.3.1 Entrapment by Polymerization

Entrapment of microbial cells in a synthetic polymer gel has been the most widely used method for cell immobilization since it was first introduced by Mosbach and Mosbach (1966). Such a polymeric network can be formed either from monomeric precursors or prepolymers.

(A) Polymerization from Monomeric Precursors

Several water-soluble monomers, acrylamide, methacrylamide, hydroxyethylmethacrylate (HEMA), vinyl-pyrrolidon, 2-vinylpyridine, and acrylic acid have been used as precursors for the polymeric network in entrapment immobilization (Klein and Wagner, 1983). Among these monomers, acrylamide is used most widely.

An insoluble polymeric gel network is prepared from the polymerization of monomers in an aqueous solution. Such a polymerization requires initiation by either persulfate (e.g., $(NH_4)S_2O_8$) or UV-light and riboflavin, with tetra-methyl-ethylenediamine (TEMED) as a co-initiator. The linear polymer formed is water-soluble, and thus requires the use of bifunctional crosslinking agents, mainly, N,N′-methylene-bis-acrylamide (BIS), to insolubilize the polymer to produce the desired three-dimensional matrix structure. The polymerization is commonly carried out in the absence of oxygen and at low temperatures (below 10°C) to minimize cell destruction. Typical entrapment procedures can be easily found in the literature, e.g., Mattiasson et al. (1977). When glass-forming monomers such as 2-hydroxyethyl acrylate are used, polymerization is induced by γ-ray irradiation at a very low temperature, e.g., -78°C (Kumakura et al., 1978; Kumakura and Kaetsu, 1983).

The mechanical stability of the gels thus obtained is not very high. A limited improvement can be achieved by incorporating inactive, mechanically stable particles such as ceramics, glass or silicates as fillers into the polymerizing solution (Klein and Schara, 1980). The volume fraction of the filler should not be so large as to prevent the formation of a continuous polymeric network. Addition of the filler limits the maximum cell loading capacity to about 20% on a net-weight basis (Klein and Wagner, 1983).

A major disadvantage of this entrapment method is the toxicity of the monomer, the cross-linking agent (e.g., BIS), and the initiator (e.g., TEMED) and the heat evolution during polymerization. Thus a loss in cell viability and a decrease in enzymatic activities are frequently encountered.

(B) Polymerization from Prepolymer

Entrapment of cells in a gel polymerized from prepolymer has the following features: (1) Entrapment procedures are simple, fast, and under mild conditions; (2) Prepolymers do not contain monomers, and thus the toxicity to enzymatic and cellular activities is low; (3) The network structure and physicochemical properties, e.g., hydrophobicity-hydrophilicity balance and ionic nature, of the gels can be controlled using prepolymers of optional chain-length and different types of prepolymers, respectively. Prepolymer methods for entrapment of cells include the use of photo-cross-linkable resin prepolymers, urethane prepolymers, and prepolymers of which polymerization is initiated by irradiation (Fukui and Tanaka, 1984).

Each of the photo-cross-linkable resin prepolymers consists of a linear skeleton of optional length, with photosensitive functional groups, e.g., acryloyl and methacryloyl, attached to the end of the skeleton. Widely used linear skeletons include poly(ethylene glycol), hydrophilic; poly(propylene glycol), hydrophobic; poly(vinyl alcohol), emulsion-type; and trimellitic poly(propylene glycol), anionic (Fukui and Tanaka, 1984). A typical immobilization procedure using a photo-crosslinkable prepolymer was given in Tanaka et al. (1979). This immobilization method has low cell loading. Enzyme catalytic activity is moderate and viability is observed.

Urethane prepolymers are composed of a polyether-diol moiety such as poly(ethylene glycol) or poly(propylene glycol) and end isocyanate groups. All the urethane prepolymers are water miscible.

The hydrophilic/hydrophobic character of the urethane polymers, however, can be manipulated by changing the ratio of the poly(ethylene glycol) to the poly(propylene glycol) in the polyether-diol moiety. In the presence of water, cross-linking among prepolymers occurs via the urea linkage between the end isocyanate groups together with liberation of CO_2. The CO_2 gas liberated during polycondensation gives the polyurethane gel a porous structure and elastic property. This entrapment method gives low cell loading because only a limited amount of water can be tolerated. The retention of enzymatic activity is moderate, and cell viability is lost (Klein and Wagner, 1983).

Glass-forming prepolymers such as poly(ethylene glycol) diacrylate, poly(ethylene glycol) dimethacrylate and poly(vinyl alcohol) can be polymerized by γ-irradiation. The polymeric gels thus produced are rigid and have good mechanical stability. Entrapment of cells is usually carried out at low temperatures ($<-24\,°C$) with 10% prepolymer and at an irradiation of 1 Mrad. A retention of 74 to 87% enzymatic activity was reported (Tanaka et al., 1982).

(C) Other Methods

Freeman and Aharonowitz (1981) added a cell suspension into a cold, aqueous solution of prepolymerized linear polyacrylamide partially substituted with acylhydrazide groups. The polymerization was then carried out by adding a controlled amount of dialdehydes as a crosslinking agent under cold and neutral physiological conditions. The gel thus obtained resembled the common polyacrylamide gel in its physical properties, but cell damage caused by the acrylamide monomer was avoided.

7.2.3.2 Entrapment by Ionotropic Gelation

Polyanions, mainly alginate, and polycations, mainly chitosan, can form an ionic network when multivalent counterions are present. Alginates, which are polymers consisting of D-mannuronic and L-guluronic acids, can be gelled by multivalent ions such as Ca^{2+}, Co^{2+}, Zn^{2+}, Mn^{2+}, Al^{3+}, Fe^{3+} and Fe^{2+}, but not Mg^{2+}. A typical immobilization procedure can be found in Vorlop and Klein (1983).

Good mechanical properties and a cell loading of up to 0.3 g-wet cells/mℓ-catalyst can be obtained (Vorlop and Klein, 1983). According to Cheetham et al. (1979), the mechanical properties of alginate gels are related to the distribution of the D-mannuronic and

L-guluronic residues, and to the molecular weight and degree of polydispersity within the gels. The highest breaking strengths are obtained when alginates used contain a high concentration of polyguluronate blocks. The mechanical stability of alginate gel beads can be further improved without loss of porosity by introducing an intermediate drying step which results in an irreversible decrease in the bead diameter. This decrease in particle size results in an increase in cell concentration per unit volume, up to 0.8 g-cells/mℓ-catalyst (Klein and Wagner, 1979).

Another advantage of this method is its extremely mild immobilization condition which preserves enzyme activity and cell viability. However, this simple method has some disadvantages. The gel matrix network formation is reversible. Disruption of the ionotropic gel occurs due to the solubilization of the bound Ca^{2+} in the presence of highly concentrated electrolyte solutions (e.g., K^+ and Na^+). The alginate gel is also unstable in phosphate buffer solutions, where the counterions (e.g., Ca^{2+}) will be precipitated, and the gel redissolves. Buffer solutions such as zwitterion, Tris, imidazole, barbital, acetate-barbital, and acetate solutions can be used for the alginate biocatalyst (Vorlop and Klein, 1983). Treating the alginate gel with polyamines and cross-linking can provide a stabilized peripheral layer so that the alginate gel can be stable in phosphate-containing media (Birnbaum et al., 1981). Cell leakage is another problem of the alginate biocatalyst.

Polycations, such as chitosan, can also be used for formation of ionotropic gels. Chitosan, a partially de-acetylated chitin, is a high molecular weight linear polymer consisting of glucosamine in 1,4-linkage. Counterions used for the ionotropic gelation of chitosan include pyrophosphate, tri-, tetra-, octa-polyphosphate, hexametaphosphate, and ferri- and ferro-cyanate, etc. Selection of the counterions mainly depends on the compatibility of cells to these ions (Vorlop and Klein, 1983). A typical immobilization procedure has been provided by Vorlop and Klein (1983).

The chitosan biocatalyst has good mechanical properties and a cell loading up to 0.3 g wet-cells/mℓ-biocatalyst. An intermediate drying step, as used in alginate biocatalyst production, can also be applied to this gel to improve its mechanical stability. In addition, the chitosan biocatalyst, in contrast to the alginate, is stable in the phosphate buffer solution (Vorlop and Klein, 1983).

Dissolved chitosan was found to increase the permeability of membranes of soybean (Young et al., 1982), amaranth, and common milkweed cells (Knorr et al., 1985). Since cell disruption or membrane permeabilization are usually required to release the plant metabolites stored in cellular vacuoles, the chitosan carrier which has the combined immobilization/permeabilization functions is advantageous in plant cell immobilization (Knorr et al., 1985).

7.2.3.3 Entrapment by Precipitation

Gel formation by precipitation of soluble polymers can be accomplished by controlling temperature, pH, or solvent systems.

(A) Precipitation by Temperature Change

Agar, agarose, gelatin and κ-carrageenan are soluble in aqueous solution at high temperature. When the temperature is lowered to their gelation temperature, usually around 40°C, the gels formed may be used to entrap cells. A general procedure can be obtained from Mattiasson (1983).

κ-carrageenan is a polyanion polymer, and thus can also be gelled by cations such as NH_4^+, K^+, and Cs^+, amines, or water-miscible organic solvents such as methanol or acetone (Chibata, 1979). The mechanical properties of the κ-carrageenan gel can be improved by treatment with hardening agents such as glutaraldehyde and 1,6-diaminohexane, or by incorporating locust bean gum (Takata et al., 1978; Nishida et al., 1979). The nontoxic, mild, and simple method of entrapment using κ-carrageenan retains high enzyme activity and cell viability. However, cell leakage was observed (Chibata, 1979).

Entrapment of cells in agar gel has several disadvantages, including gelation temperatures too high for labile biological structures, poor mechanical properties, and high diffusional resistance. Recently, however, agar with a low gelation temperature has become available, with increasing importance as an immobilization carrier. Examples of the application of this method can be found in Toda and Shoda (1975), and Middelhoven and Bakker (1982).

(B) Precipitation by pH Change

Collagen is the major protein constituent of connective tissue of higher vertebrates. It can be solubilized at low pH, and reprecipitated at high pH. Entrapment of cells in a collagen matrix by the so-called complexation method was developed by Vieth and Venkatasubramanian (1976; 1979). The complexation method is based on the formation of noncovalent physicochemical bonds, such as salt linkage, hydrogen bonding and van der Waals force, between the carrier-protein and cell-protein. The procedure for entrapment of cells by the complexation method can be found in Constantinides (1980).

Glutaraldehyde, a crosslinking bifunctional agent, can be used to improve the mechanical strength of the collagen matrix. A retention of 50 ~ 55% activity in the complexation method was reported. A cells-to-collagen ratio of 1:1 (dry weight basis) was found to yield a membrane of good activity and strength (Vieth and Venkatasubramanian, 1976).

(C) Precipitation by Solvent

Physical entrapment of enzymes inside the microcavities of porous fibers has been developed by Dinelli et al. (1976). This method consists of the following steps: (1) dissolving a fiber-forming polymer in an organic solvent not miscible with water, (2) emulsifying this solution with an aqueous solution of enzymes or cells, and (3) extruding the emulsion through fine holes into a liquid coagulant which precipitates the polymer in filamentous form. Cellulose triacetate and cellulose have been used as the fiber-forming polymers, and typical solvents used are dimethyl formamide, acetone, methylene chloride or a mixture of N-ethylpyridinium chloride and dimethylformamide. The coagulant liquid can be toluene or water. The same methodology was applied to entrap cells by Linko et al. (1979), in which a general procedure for cell entrapment in α-cellulose can be found.

While the fibers are resistant to weak acids, alkalis, high ionic strength and some organic solvents, the organic solvents used can damage the cells, reducing their activity and viability.

In summary, the covalent coupling, adsorption, and entrapment immobilization methods are suitable for application in gas-liquid-solid fluidized bed bioreactors. The strengths and weaknesses of each method are briefly indicated in Table 7.3.

7.3 EFFECTS OF IMMOBILIZATION

Immobilization has been reported to affect cellular behavior. The reported results, however, are not consistent. Depending upon the immobilization techniques, the microbial species, and the biochemical reactions of interest, expressed metabolic activities of immobilized cell systems have been reported to decrease, remain the same, or even

Table 7.3 Comparison of the Attributes of Different Immobilization Methods (from Kennedy and Cabral, 1983)

Characteristics	Covalent coupling	Adsorption	Entrapment
Preparation	Difficult	Simple	Intermediate
Binding force	Strong	Weak	Intermediate
Retention of activity	Low	High	Intermediate
Regeneration of carriers	Rare	Possible	Impossible
Cost of immobilization	Intermediate	Low	Low
Stability	High	Low	High
Viability	Rare	Yes	Yes

increase as compared with those of freely suspended cells. These deviations are largely ascribed to the intraparticle diffusional resistance in the immobilized cell particles, the partitioning effects of the carriers, and the changes in cellular physiology and morphology.

7.3.1 Intraparticle Diffusional Resistance

Aside from the toxicity of the carriers and of the chemicals used in immobilization processes which results in cell damage, intraparticle diffusional resistance is the major factor responsible for the decrease in expressed metabolic activities generally encountered upon immobilization. In the production of penicillin G by Penicillium chrysogenum immobilized in polyacrylamide beads, the rate of penicillin production of the immobilized mycelia was found to be lower than that of free mycelia due to the damage caused by the toxicity of the monomers and the cross-linking agent, N,N′-methylene-bis-acrylamide (BIS). However, by increasing the relative concentration of BIS to 15%, which was found to increase the pore size of the particle without affecting the particle rigidity, the rate of penicillin production increased. This indicates the effect of intraparticle diffusional resistance on the apparent activity of the immobilized cell system. In a study on α-amylase production by Bacillus amyloliquefaciens entrapped in κ-carrageenan beads, Shinmyo et al. (1982) found that gel beads containing cells less than 1.4 mg/g-wet gel have the same specific respiratory activity (i.e., respiratory rate per unit cell mass) as that of suspended cells released from the carrageenan beads. Beads containing cells greater than 1.4 mg/g-wet gel, on the other hand, exhibited a repressed specific respiratory activity compared to that of the suspended cells. However, for these two types of beads, the oxygen uptake rate per unit volume of beads remained constant despite the repression of the respiratory activity for beads containing high cell population. This behavior suggests that the oxygen diffusion within the gel is the limiting step in the respiration of entrapped cells. The existence of the intraparticle diffusional resistance in immobilized cell systems can also be realized from the fact that entrapped cells mainly grow near the carrier surface as seen from the electron micrograph of sectioned alginate beads entrapping E. coli shown in Fig. 7.1. Similar results were obtained in carriers entrapping Saccharomyces cerevisiae, Aspergillus niger, and Claviceps purpurea (Burrill et al., 1983; Eikmeier et al., 1984; and Kopp and Rehm, 1984).

7.3.2 Partitioning Effects

Partitioning effects arise from electrostatic or hydrophobic interactions between the carrier matrix and the low molecular weight compounds present in the solution. Such effects are strongly related to the chemical nature of the carrier particles, e.g., carrier particles that are positively or negatively charged. When a negatively charged cell particle is used, for example, positively charged substrates tend to be more concentrated in the local environment (i.e., microenvironment) than in the bulk solution, thus resulting in a higher relative activity than when a neutrally charged matrix or a suspended cell system is used. Shifts in optimum pH values and in optimum ion and substrate concentrations are therefore observed in the immobilized cell system. A quantitative expression of the partitioning of charged compounds, substrates or products, between a charged cell particle and the bulk solution can be given by the following equation (Kennedy and Cabral, 1983):

$$S_l = S_b \exp(-Ze\psi/(k_B T)) \tag{7.1}$$

where S_l and S_b are local and bulk concentrations, respectively, of substrate or product. e and ψ are the electrostatic charge and potential, respectively. T is the temperature and Z is the number of charges. However, due to the difficulty in experimentally determining the concentrations of substrates in the microenvironment of

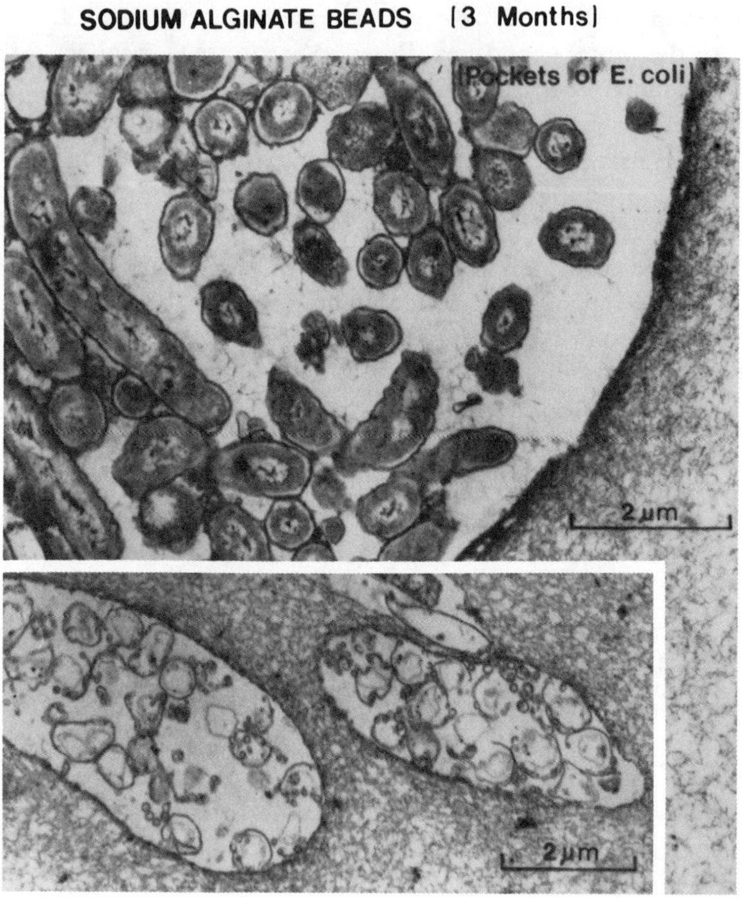

Figure 7.1 An electron micrograph of E. coli immobilized in 2.5 mm Ca^{++} alginate beads after 90 days. The formation of pockets of cells from an initially uniform distribution of cells is apparent. The top picture shows a pocket near the surface while the inset picture is of a pocket further inside the bead. The viability in the inner pocket is believed to be the result of diffusional limitations of nutrients into the bead (kindly supplied by J.J. Chalmers of The Ohio State University and J.N. Telford, M.L. Shuler and D.B. Wilson of Cornell University).

immobilized cell particles, the partitioning effects cannot be verified or quantitatively evaluated.

7.3.3 Cellular Physiological Behavior

While partitioning effects can enhance the overall product synthesis or substrate consumption, increases in specific rates of product synthesis or substrate consumption have also been demonstrated (Navarro and Durand, 1977; Vijayalakshmi et al., 1979). In addition, a change in metabolic pattern was reported for Serratia marcescens upon entrapment in κ-carrageenan gel particles: the production activity of L-isoleucine increased while the oxygen demand decreased by 90% as compared with a suspended cell system for the same amount of living cells (Fig. 7.2). These effects have been claimed to be due to a switch in metabolic routes or caused by some unknown mechanism. Little effort, however, has been made to explain these observations.

Mattiasson and Hahn-Hagerdal (1982) hypothesized that changes in microenvironmental stresses due to immobilization, especially decreased water activity and oxygen deficiency, result in changed yields and/or new metabolic behavior of the immobilized cells. Oxygen deficiency in immobilized cell systems results from the diffusional resistance inside the immobilized cell particles. A decrease in water activity occurs especially in an entrapment immobilization system in which cells are encaged in an environment containing a high macromolecule concentration. The macromolecules are capable of organizing water molecules via electrostatic and/or hydrophobic interactions and thereby decrease the availability of free water to entrapped cells.

When exposed to a low water activity environment, free suspended cells have been reported to change their intracellular pools of certain metabolites to compensate for intracellular osmotic pressure. Examples include an increase of glycerol production in halophilic algae Dunaliella (Ben-Amotz and Avron, 1981) and in Saccharomyces cerevisiae (Edgley and Brown, 1978), the accumulation of amino acids in various non-halophilic bacteria (Measures, 1975), and the increased production of betaine in Ectothiorhodospira halochloris (Galinski and Truper, 1982). The change in the intracellular pools of metabolites due to decreased water activity may alter the already operating metabolic flow and/or trigger the resting metabolism. This, in turn, results in changing yields of metabolites in the maintenance metabolism or changing the transfer of redox potential (e.g. the balance between NADH and NADPH) which leads to the expression of new metabolic behavior. Hahn-Hagerdal et al. (1982) were able to

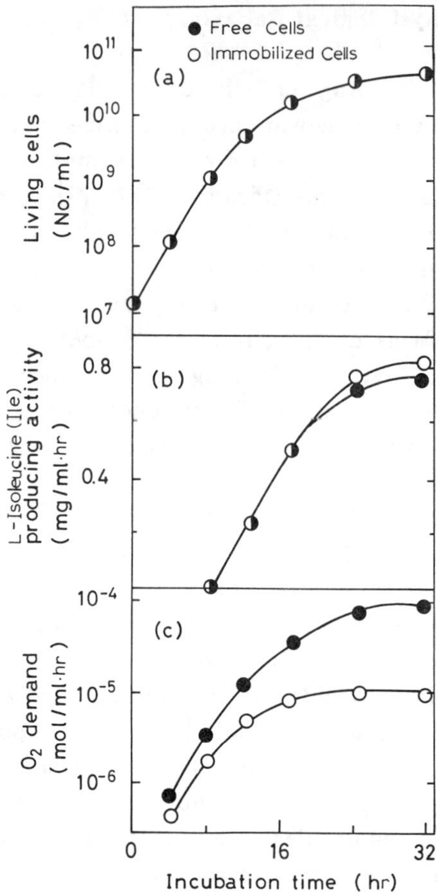

Figure 7.2 (a) Growth; (b) L-isoleucine (Ile) producing activity; and (c) oxygen demand of free and immobilized cells of S. marcescens. The cells are entrapped within κ-carrageenan gel particles. The Ile producing activity increases while the oxygen demand decreases upon immobilization, indicating a change in cellular physiology (from Wada et al., 1980).

demonstrate an increase in the rate of ethanol production by 50% in suspended Saccharomyces cerevisiae by decreasing the water activity in a fermentation broth with the addition of dextran and polyethylene glycol. Oxygen deficiency affects the metabolic behavior in a manner similar to that resulting from changes in intracellular pools of metabolites.

Immobilization can affect the metabolic behavior in an even more drastic manner than just described. Doran and Bailey (1986) immobilized Saccharomyces cerevisiae cells onto gelatin-coated glass beads using the bifunctional agent glutaraldehyde. The specific rate of ethanol production by the immobilized cells was 40 to 50% greater than that by the suspended cells. The glucose consumption rate by the immobilized cells was twice as fast as that by the suspended cells, while the specific growth rate of the immobilized yeast was reduced by 45%. Cellular composition was found to be significantly altered by immobilization. The immobilized yeast cells stored larger quantities of reserve carbohydrates and contained more structural polysaccharide than did the suspended cells. The average DNA content of the immobilized cells and their progeny was 3.8 and 3.1 times, respectively, those of the cells in suspension; the immobilized cells contained only about one-fourth the amount of stable double-stranded RNA of that found in the suspended cells, and the immobilized cell progeny had about 25% higher RNA content than the immobilized cells. The quantities of protein per unit size for both the immobilized and the suspended cells, however, are about the same. Budding emergence seemed to be partially inhibited by immobilization. The fraction of budded cells of immobilized yeast is about half of that of cells in suspension. Doran and Bailey (1986) suggested that the decoupling of budding development and DNA replication in the immobilized yeast results in the polyploidy in the immobilized yeast; the polyploidy is related to the lower cell growth and superior fermentative capabilities.

The physiological properties of Streptomyces tendae entrapped in calcium alginate gel, on the other hand, were found almost the same as those of suspended cells (Veelken and Pape, 1984). The DNA and protein content were the same in both free and immobilized cells. Immobilization did not change the relative amount of nikkomycin compounds in the culture broth. The specific activity of fructosediphosphate aldolase, one of the enzymes in the primary metabolism, was lower for entrapped cells than for suspended cells.

From the diverse results described above, it appears that changes in the metabolic behavior of immobilized cells depend on the magnitude of stresses in the microenvironment created by the immobilization processes and the type and species of cells involved.

7.3.4 Morphological Characteristics

Immobilization was also found to affect the morphology of the immobilized cells. A change in cell size upon covalent immobilization was reported (Doran and Bailey, 1986). Formation of micropellets surrounded by a membrane-like alginate gel coat was observed by Mahmoud and Rehm (1986) with cells of Streptomyces aureofaciens germinating from spores entrapped within calcium alginate gel. The micropellet structure has never been found with free cells. Micropellets were also found with cells of Penicillium chrysogenum germinating from conidia immobilized within calcium alginate gel. The branching frequency of the immobilized mycelia was found to be higher than that of the suspended mycelia (El-Sayed and Rehm, 1986). Immobilization seems also to facilitate morphological development for strains capable of differentiation, probably due to the increase in microenvironmental stresses upon immobilization. Kopp and Rehm (1984) observed a strong tendency for the mycelia of Claviceps purpurea entrapped within calcium alginate gel to break down into arthrosporoid-like structures uniformly composed of vacuolated elements. The development of the arthrosporoid-like cells were probably responsible for the higher and longer lasting productive capacity of immobilized Claviceps purpurea.

7.4 SUBSTRATE DIFFUSIVITIES IN IMMOBILIZED CELL SYSTEMS

Transport of substrates from the bulk liquid phase to each individual cell inside an immobilized cell system (a biofilm, a biofloc, or an entrapping carrier) comprises two processes in series: the external mass transfer, in which substrates in the bulk liquid phase are transported through a thin stagnant liquid film to the surface of an immobilized cell system; and the internal mass transfer, which involves the diffusion of substrates in the voids among cell aggregates (in a biofilm or a biofloc system) or among cell mass and the internal structure of the carrier material (in an entrapped cell system) to each individual cell. In an immobilized cell system, the internal diffusional resistance is usually significantly higher than the external mass transfer resistance. Its importance is apparent as entrapped cells seldom grow in the core of the carrier and an anoxic or anaerobic region forms deep inside a thick biofilm. Consequently, the diffusivity is a crucial parameter which characterizes the performance of an immobilized cell system and must be known if the overall reaction rates of immobilized cell systems are to be accurately predicted. The excreted products generated inside the immobilized cell particles undergo the same mass transfer process as the substrates except the flux is in the opposite direction. The following analysis describes the effective diffusivity and mass balance of substrates and oxygen within an immobilized cell particle, which can be directly extended to that of any solute, including the excreted products.

The diffusivities of substrates or oxygen are generally determined experimentally by either transient or steady-state methods. Transient methods involve initially imposing a known substrate or oxygen concentration in the bulk solution and subsequently measuring the transient concentration decay in the bulk solution. Steady-state methods determine the steady-state diffusional flux of substrates through a simulated immobilized cell system. The diffusivities are then evaluated using a model based on a mass balance and the transient or steady-state data.

For any substrate, a generalized mass balance equation for substrate concentration inside the immobilized cell particle or biofilm, S, which incorporates both diffusion and reaction is

$$\frac{\partial S}{\partial t} + \vec{v} \cdot \vec{\nabla} S - \vec{\nabla} \cdot (D_M \vec{\nabla} S) + (-r_{sv}) = 0 \tag{7.2}$$

Where $-r_{sv}$ is the substrate consumption rate per unit volume of the immobilized cell particle or the biofilm attached to the solid particle. Equation (7.2) assumes that the diffusional process within an immobilized cell system is governed by the Fickian diffusion. The convective flow, $\vec{v}$, is usually negligible, and, with the assumption that the diffusivity is independent of the spatial coordinate, Eqn. (7.2) can be simplified to

$$\frac{\partial S}{\partial t} - D_M \vec{\nabla}^2 S + (-r_{sv}) = 0 \tag{7.3}$$

Evaluation of the diffusivity can be greatly simplified if the unsteady-state term and/or the substrate consumption term $(-r_{sv})$ can be eliminated and if the diffusional process is one-dimensional. The substrate consumption term can be eliminated by deactivating cells or enzymes with chemicals, e.g., $HgCl_2$. When both diffusion and substrate consumption have to be considered, an intrinsic kinetic expression must be known. The additional assumptions made for the kinetic expression of substrate consumption by cells would reduce the accuracy in the diffusivity evaluation.

7.4.1 Effective Substrate Diffusivities in Entrapped Cell Systems

Several investigators have conducted experiments on measurement of effective diffusivities of various nutrients within a polyacrylamide or a calcium alginate gel. The effective diffusivities of substrates within a gel containing entrapped living or dead cells, however, have not been extensively studied.

The diffusion characteristics into spherical calcium alginate gel beads of substrates of various molecular sizes were investigated by Tanaka et al. (1984). Calcium alginate beads free of the substrate of interest were suspended in a solution with a known initial substrate concentration in a well-stirred and thermostatically controlled vessel. The substrate concentration decay with time in the solution was monitored. The effective diffusivity was then determined from the following equation:

$$\frac{S_b(t)}{S_b(0)} = \frac{\alpha}{(1+\alpha)}\left\{1 + \sum_{n=1}^{\infty} \frac{6(1+\alpha)\exp(-D_M q_n^2 t/r_p^2)}{9+9\alpha+q_n^2\alpha^2}\right\} \tag{7.4}$$

where q_n are the positive roots of

$$\tan q_n = \frac{3q_n}{3 + \alpha q_n^2} \tag{7.5}$$

In the above equations, α is defined as $V/(N4\pi r_p^{\ 3}/3)$; V is the total liquid volume; N is the total number of Ca-alginate gel particles; and r_p is the radius of the immobilized cell particles. Their results showed that the effective diffusivities of small molecules (glucose, L-tryptophan, and α-lactoalbumin [molecular weight (MW) less than 2×10^4]) agreed with those in a pure water system. In addition, these diffusivities were not affected by the calcium alginate concentration and the $CaCl_2$ concentration used in gel preparation. For large molecules (albumin with MW of 6.9×10^4, γ-globulin with MW of 1.54×10^5, and fibrinogen with MW of 3.41×10^5) no diffusion into the 2% calcium alginate gel was observed.

Hannoun and Stephanopoulos (1986) investigated the effective diffusivities of glucose and ethanol in both cell-free and cell-entrapped slabs of calcium alginate gels. A slab of calcium alginate gel supported by a wire mesh was placed between two magnetically stirred cylindrical chambers and sealed. Initially, the upper chamber contained a solution of glucose or ethanol of known concentration and the lower chamber contained distilled water. Glucose or ethanol concentration changes in the lower chamber were monitored. By solving Eqn. (7.3) for one-dimensional diffusion and no substrate consumption conditions, the total flux of substrate which has passed through the membrane at any time t is

$$F = AD_M t \frac{S_{b,1}}{\delta} - \frac{AS_{b,1}\delta}{6} - \frac{2AS_{b,1}\delta}{\pi^2} \sum_{n=1}^{\infty} \frac{(-1)^n}{n^2} \exp \frac{-D_M n^2 \pi^2 t}{\delta^2} = VS_{b,2} \quad (7.6)$$

where F is the total mass of glucose or ethanol transferred through the Ca-alginate gel slab at any given time; $S_{b,1}$ and $S_{b,2}$ are the bulk glucose or ethanol concentrations in the upper and lower chambers, respectively; δ is the thickness of the Ca-alginate gel; A is the effective area for diffusion; and V here represents the liquid volume in the lower chamber. For sufficiently large times, the summation term in Eqn. (7.6) becomes negligible. The total amount of glucose or ethanol transferred through the gel slab then increases linearly with time:

$$F = \frac{AS_{b,1}D_M}{\delta}\left(t - \frac{\delta^2}{6D_M}\right) \quad (7.7)$$

A plot of F versus t approaches a straight line; this line intercepts the time axis at $t = \delta^2/6D_M$, which is referred to as the lag time. The diffusivity can then be easily calculated from the slab thickness and the lag time.

The diffusivities of glucose and ethanol in a 2% calcium alginate gel slab are 91% of those in water. In contrast to the results of Tanaka et al. (1984), Hannoun and Stephanopoulos (1986) found that an increase in calcium alginate concentration decreases the effective diffusivities of both glucose and ethanol. Increasing the calcium alginate concentration from 1 to 4% decreases the glucose and oxygen diffusivities by approximately 25 and 50%, respectively. In addition, the presence of 20 wt% dead cells of _Saccharomyces cerevisiae_ in the slab of calcium alginate gels has no effect on the diffusivities.

Sato and Toda (1983) measured the effective oxygen diffusivity in a 2(wt/vol)% agar gel slab placed in a two-compartment vessel. The agar gel slab was supported by a stainless mesh and separated the upper and lower chambers. Air-saturated water was fed to the well-stirred lower chamber at a constant rate and was withdrawn out of the chamber at the same flow rate. Pure nitrogen was introduced into the upper chamber continuously to strip off dissolved oxygen such that a large driving force for oxygen diffusion between the two chambers was maintained. The effective diffusivity of oxygen in the agar gel was calculated from the steady state diffusional flux through the agar gel as

$$Q(C_s - C_2) = A \frac{D_M}{\delta} (C_2 - C_1) \qquad (7.8)$$

where Q is the volumetric liquid flow rate; C_s is the dissolved oxygen (DO) concentration saturated with air; and C_1 and C_2 are the DO concentrations in the upper and lower chambers, respectively. Sato and Toda (1983) obtained the effective diffusivity of oxygen in the 2(wt/vol)% agar gel as 0.107 cm^2/hr at 30°C, corresponding to 70% of that in water. They also reported that the diffusivity was little affected by the presence of inactivated *Candida lipolytica* cells up to a concentration of 8.3×10^{-3} g-dry cells /cm^3.

7.4.2 Effective Substrate Diffusivities in Biofilm, Biofloc, and Cell Aggregate Systems

As mentioned previously, a biofilm in this context refers to multiple layers of cells and biopolymers produced by the cells attached to a solid support. A biofloc, on the other hand, refers to aggregates of cells packed together under natural conditions without a solid support. Both biofilm and biofloc systems have complex irregular geometries which greatly complicate the evaluation of substrate diffusivities. Consequently, aggregate cell systems, which are formed via external physical forces such as gravity, centrifugation, or compression, are often used to simulate biofilm or biofloc systems in the determination of substrate diffusivities. The advantage of aggregate cells is that they can be packed in a container of a well-defined geometry. In the following sections, the methods to evaluate the diffusivity of solutes within biofilm, biofloc, and aggregate cell systems are presented. These same methods can also be utilized to characterize the diffusivity in systems involved in biological wastewater treatment applications (also see Sections 8.3.2 and 8.3.3).

7.4.2.1 Biofloc Systems

Mueller et al. (1968) determined the diffusivity of oxygen in bioflocs of Zoogloea ramigera based on an oxygen uptake curve and an anoxic core model. In the model, an anoxic core, a core with zero dissolved oxygen concentration, is assumed to develop inside the biofloc when the surface oxygen concentration decreases below a limiting value C_ℓ. For the above assumption to be valid, the critical oxygen concentration within the biofloc, defined as the minimum oxygen concentration required at an individual cell surface to give a maximum oxygen uptake rate, must be zero. The critical oxygen concentration of this organism was found to be below 0.1 mg/ℓ and the limiting oxygen concentrations of the biofloc were much greater than this value, thus justifying the model to be a good approximation. By assuming the bioflocs are approximately spherical and applying nominal diameter measurements, an oxygen balance gives the following equation:

$$\frac{D_M}{r^2}\frac{d}{dr}\left(r^2\frac{dC}{dr}\right) - (-r_{cv}) = 0 \tag{7.9}$$

with boundary conditions:

$$\frac{dC}{dr} = 0 \quad \text{at} \quad r = r_c \tag{7.10}$$

$$C = 0 \quad \text{at} \quad r = r_c \tag{7.11}$$

$$C = C^* \quad \text{at} \quad r = r_p \tag{7.12}$$

where $-r_{cv}$ is the oxygen uptake rate per unit biofloc volume. Equations (7.9) through (7.12) yield

$$6D_M C^*/[(-r_{cv,max})r_p^2] = 1 - 3x_c^2 + 2x_c^3\,, \quad 0 \le x_c \le 1 \tag{7.13}$$

where r_c is the radius of the anoxic core; r_p is the radius of the biofloc; and x_c is defined as r_c/r_p, which can be obtained from the oxygen uptake curve and the following equation:

$$(-r_{cw})/(-r_{cw,max}) = 1 - x_c^3\,, \qquad 0 \le x_c \le 1 \tag{7.14}$$

where $(-r_{cw})$ is oxygen uptake rate based on the dry weight of biofloc, and $(-r_{cw,max})$ is the maximum value of $(-r_{cw})$ when C^* is greater than C_ℓ. The oxygen diffusivity can thus be evaluated using Eqns. (7.13) and (7.14).

Mueller et al. (1968) reported that the anoxic model appropriately fitted the experimental oxygen uptake curve and thus was adequate for the determination of oxygen diffusivities in bioflocs. They also found that the oxygen diffusivity varied only slightly when assuming different biofloc shapes (including a sphere, a cylinder, and a triangular prism). The diffusivity did, however, vary greatly when the diffusional path was evaluated by different biofloc sizing techniques (i.e., nominal diameter measurement and specific surface measurement techniques). Typically, the oxygen diffusivity in Z. ramigera bioflocs is 2.1×10^{-6} cm^2/s at 30°C and 1.8×10^{-6} cm^2/s at 20°C.

7.4.2.2 Biofilm Systems

Bungay et al. (1969) used a microprobe electrode to measure the dissolved oxygen concentrations near the surface and within a bacterial slime film about 200 μm thick grown in a continuous flow cell supplied with a continuous flow of nutrient solution. The oxygen diffusivity within the slime film was evaluated from the measured dissolved oxygen profile within the film. The oxygen diffusivity within the slime film obtained was 4.0×10^{-7} cm^2/s, about 1.5% of that of oxygen in pure water.

The use of a microprobe electrode by Bungay et al. (1969) provides an elegant method to measure the oxygen concentration profile in a biofilm. However, their data treatment techniques, which were mainly graphical differentiation to obtain derivatives and the selection of a reasonable respiration rate, are not free of subjectivity, and would influence the accuracy of the values obtained.

7.4.2.3 Microbial Cell Aggregates

Matson and Characklis (1976) devised a Milipore apparatus to measure oxygen diffusivity through aggregates of activated sludges. The Milipore apparatus consisted of two well-mixed chambers separated by a membrane template assembly through which oxygen diffuses. The membrane template assembly (MTA) consisted of a perforated brass template of uniform thickness with a Silastic silicone membrane (0.75 μm thick) attached on each side. Activated sludge was inactivated by $HgCl_2$ or ultraviolet radiation, and then filled into the template. The uniformity of the packed template was checked visually for consistent translucency through the whole plate. The lower chamber of the apparatus was filled with deionized water while the upper chamber was aerated with pure oxygen to maintain a dissolved oxygen concentration (DO) at 30 ~ 40 mg/ℓ. Oxygen diffuses from the upper chamber through the packed template into the lower chamber. The DO concentration in the lower chamber was monitored by a dissolved oxygen probe. Since the DO in the upper chamber, C_1, is very high, it can be considered constant at C_s. The overall diffusivity of oxygen through the packed template, D_t, can be calculated based on a mass balance on oxygen:

$$D_t = \frac{V\ (dC_2/dt)}{A(C_s - C_2)/L} \tag{7.15}$$

where L is the template assembly thickness and C_s and C_2 are the bulk DO concentrations in the upper and lower chambers, respectively.

The overall diffusional resistance through the template assembly consists of the resistance due to both the aggregates of activated sludges and the membranes on both sides of the template:

$$\left(\frac{L}{AD_t}\right)_{\text{overall}} = 2\left(\frac{L}{AD_m}\right)_{\text{membrane}} + \left(\frac{L}{AD_M}\right)_{\text{microbial aggregates}} \tag{7.16}$$

The membrane resistance $(L/(AD_m))$, in particular the oxygen diffusivity through the membranes (D_m), can be obtained from experiments in which the membrane is placed in the apparatus without microbial aggregates. The oxygen diffusivity through the aggregates of activated sludges can then be calculated from Eqns. (7.15) and (7.16).

Diffusivities of oxygen and glucose through aggregates of activated sludges from different sources determined by Matson and Characklis (1976) are shown in Table 7.4. The oxygen and glucose diffusivities were found to decrease with increasing sludge age and decreasing C/N ratio in the growth media.

Smith and Coackley (1984) measured oxygen diffusivity through aggregates of activated sludges of different solid concentrations. The diffusion cell consisted of two well-stirred chambers separated by a central compartment sealed on both sides with oxygen permeable membranes. The central compartment was

Table 7.4 Diffusivity of Oxygen and Glucose in Floc From Various Biological Wastewater Treatment Plants (from Matson and Characklis, 1976)

Source of floc Type of plant	Diffusivity ($cm^2 \cdot s^{-1}$)		Sludge age (day)
	Oxygen $\times 10^{-5}$	Glucose $\times 10^{-5}$	
Municipal (domestic)	1.37 ± 2%	0.25 ± 20%	8 ~ 10
Municipal (combined domestic-industrial)	1.17 ± 21%	0.35 ± 10%	5
Industrial (petrochemicals)	0.85 ± 25%	0.18 ± 14%	20+
Pure water	2.2 ± 25%	0.67 ± 10%	———

filled with activated sludge inactivated by $HgCl_2$ and thickened by gravity or different degrees of centrifugation. The liquid in the left chamber was saturated with pure oxygen, while the right chamber was filled with water and DO concentration was monitored with time. Smith and Coackley (1984) reported that over a dried sludge concentration range of about 3.0 to 9.0%, the measured diffusivities were very similar with an average value, from four activated sludges, of 7.42×10^{-10} m^2/s.

The diffusivity obtained from the thickened activated sludge, however, does not necessarily readily apply to the biofloc situation. Smith and Coackley (1984) therefore further demonstrated that the physical structures including the density, porosity, specific surface area, mean pore radius, and tortuosity of the bulk thickened activated sludges were similar to those of the activated sludge bioflocs. This provides a sound foundation for adopting the values of oxygen diffusivity, i.e., 7.42×10^{-10} m^2/s, determined from thickened activated sludge as the diffusivity in activated sludge bioflocs.

Onuma et al. (1985) filtered bioflocs onto a membrane filter, and placed the membrane filter in a two-compartment vessel to study the diffusivities of oxygen and glucose through these microbial aggregates. The diffusivities of both oxygen and glucose in the microbial aggregates were found to depend upon the C/N ratio in the sewage and the cell concentration. They presented a generalized correlation equation for both the oxygen and glucose diffusivities at a fixed temperature as given by

$$D_M = D_w \exp(-bX) \tag{7.17}$$

where D_w is the diffusivity of oxygen or glucose in water at a fixed temperature, b is a positive constant depending on the C/N ratio, and X is the biomass concentration in the aggregate (kg/m^3). For glucose at 20°C, "b" has values of 0.77×10^{-3}, 3.93×10^{-3}, and 5.47×10^{-3} at C/N ratios of 12, 39 and 83, respectively; and for oxygen, "b" is equal to 0.80×10^{-3}, 4.77×10^{-3}, and 5.88×10^{-3} at a C/N ratio of 12, 39, and 83, respectively. The temperature dependency of the effective diffusivities in their microbial aggregate system was found to follow

$$D_M|_T \;/\; D_M|_{T=293} = 1.008^{T-293} \quad \text{(glucose)} \tag{7.18}$$

$$D_M|_T \;/\; D_M|_{T=293} = 1.023^{T-293} \quad \text{(oxygen)} \tag{7.19}$$

where $D_M|_{T=293}$ and $D_M|_T$ are diffusivities at a temperature of 293 K and T (K), respectively.

7.4.2.4 Summary

Substrate diffusivities in immobilized cell systems, including bioflocs (or microbial aggregates), biofilms, and carrier particles with or without entrapped cells, are summarized in Table 7.5. Significant variations in the diffusivities obtained by different investigators and different methods are observed, especially in biofloc and biofilm systems. It appears that substrate diffusivities within bioflocs or biofilms are influenced by biofloc (biofilm) age, microbial species constituting the bioflocs or biofilms, medium composition, and growth conditions. In addition, the cell density within bioflocs or biofilms has considerable effect on diffusivity (also see Fig. 8.6). Caution, therefore, needs to be exercised when the diffusivity values are to be used to model the simultaneous biochemical reactions and diffusion phenomena in a biofilm or biofloc system.

It has been shown in a biofilm or biofloc system that cells and/or polymers excreted by cells can impose resistance to substrate transport (diffusion). However, the presence of up to 20 wt% dead cells within a calcium alginate gel has been reported to have no effect on the substrate diffusivities within the gel. More studies on the effects of living cells and polymers excreted by the cells on the substrate diffusivity within entrapped cell systems are needed.

Table 7.5 Summary of Solute Diffusivity in Immobilized Cell Systems

Solutes		Diffusion Matrix	Diffusivity $\times 10^5$ (cm^2/s)	$\frac{D_M}{D_W}$	References
Entrapped cell system	Glucose	2% Ca-alginate without cells	0.68	1.00	Tanaka et al. (1984)
	Sucrose	2% Agar (with cells)	0.67	0.72	Sato and Toda (1983)
	Oxygen	2% Agar	1.9	0.70	Toda and Shoda (1975)
	Oxygen	2% Barium chloride	0.70	0.25	Hiemstra et al. (1983)
	Glucose	2% Ca-alginate	0.61	0.90	Hannoun and Stephanopoulos (1986)
	Ethanol	With 20% dead cells	1.0	0.91	Hannoun and Stephanopoulos (1986)
Biofilm, biofloc, or microbial aggregate system	Oxygen	Bacterial slime	1.5	0.70	Tomlinson and Snaddon (1966)
	Oxygen	*Z. ramigera* floc	0.21	0.08	Mueller et al. (1968)
	Glucose	*Z. ramigera* floc	0.048	0.08	Baillod and Boyle (1970)
	Oxygen Ammonium Nitrate	Microbial aggregate of nitrifying culture	2.2 1.3 1.4	0.90 0.80 0.90	Williamson and McCarty (1973)
	Glucose	Microbial aggregate	0.06~0.38	0.1~0.57	Matson and Characklis (1976)
	Oxygen	Activated sludges	0.742	0.34	Smith and Coackley (1984)
	Oxygen	Bacterial slime	0.04	0.015	Bungay et al. (1969)

7.5 BIOREACTOR APPLICATIONS

Three-phase fluidized bed bioreactors have been used in the production of ethanol, antibiotics, and enzymes. Several examples are given in this section.

7.5.1 Ethanol Fermentation

Moebus and Teuber (1982) applied a countercurrent three-phase fluidized bed bioreactor (Mode E-II-b) involving CO_2 as the continuous gas phase for the production of alcohol using immobilized *Saccharomyces cerevisiae*. The particle density in this bioreactor is higher than the liquid. Figure 7.3 illustrates a laboratory scale process. Glucose solution is sprayed down to a fluidized bed of immobilized cell particles and CO_2 gas is externally recycled to support the particles in the bed. The ethanol produced evaporates with water and can be recovered continuously in a condenser. Ethanol accumulated intracellularly is obtained through the drying of yeast

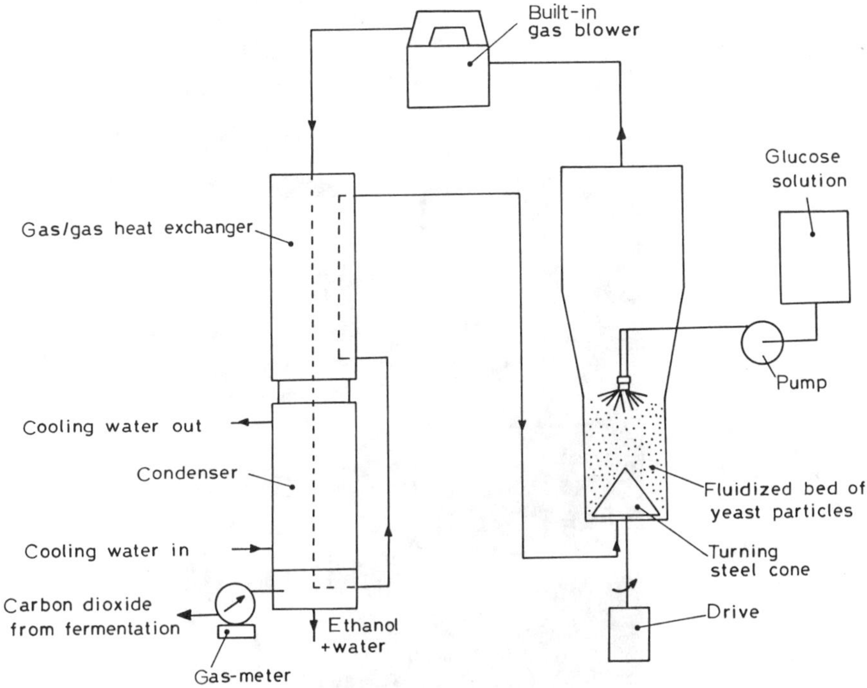

Figure 7.3 Scheme of the developed closed circuit fluidized bed bioreactor. The gas flow was about 0.7 m/s. The fractional pressure drop along the fluidized bed of 0.3 kg of solid yeast particles was between 100 and 110 mm water column; the power input by the blower was around 200 W (from Moebus and Teuber, 1982).

cells and subsequent condensation in the cooling system after stopping the fermentation. The fermentation rate was found to be comparable to that obtained from suspended cell fermentation.

Ethanol production is, however, more commonly conducted in a cocurrent three-phase fluidized bed with liquid as the continuous phase (Mode E-I-a-2) e.g., the laboratory studies by Davidson and Scott (1987) and the pilot to semi-commercial scale production by Nagashima et al. (1983), Samejima et al. (1984), and Hanai (1986). Davidson and Scott (1987) examined the operability and feasibility of ethanol fermentation in a tapered bioreactor using Z. mobilis entrapped within κ-carrageenan gel beads. They were able to obtain a more than ten-fold increase in the ethanol productivity over that from a commercial suspended yeast batch bioreactor.

Figure 7.4 Photograph of one of the two in-series three-phase fluidized bed bioreactors at a semi-commercial scale for continuous ethanol production. The process was developed by the Research Association for Petroleum Alternatives Development and Kyowa Hakko Kogyo Co., Ltd., Japan and is located in Hofu, Japan (kindly supplied by T. Fujio of Kyowa Hakko Kogyo Co., Ltd., Japan).

Figure 7.4 shows a view of a semi-commercial scale ethanol fermentation plant consisting of two three-phase fluidized bed bioreactors in series (Samejima et al., 1984). Ca-alginate gel beads entrapping yeast cells are used as fluidized particles. The immobilized cell particle loading can be up to about 60% of the total reactor volume in each bioreactor. The first bioreactor is aerated with air to maintain a microaerobic environment in the system. The CO_2 gas bubbles evolved from fermentation provide strong mixing, and constitute the bulk part of the gas phase. More than 8 (wt/vol)% of ethanol production over 1000 hour operation was reported.

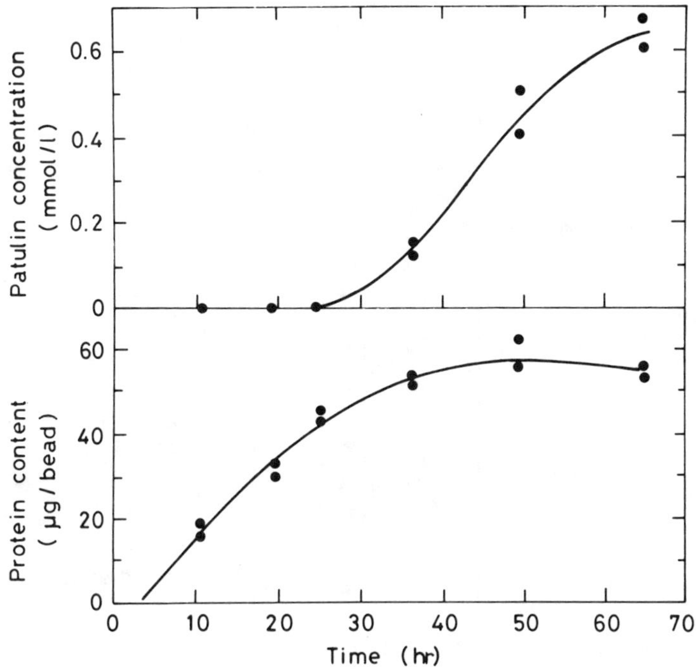

Figure 7.5 Batch growth of immobilized Penicillium urticae cells as shown by the increase in the protein content of the beads. The increase in the concentration of patulin in the medium after the exponential growth phase indicates that patulin is a non-growth associated product (from Berk et al., 1984a).

7.5.2 Antibiotics

Antibiotics account for a major portion of the industrial biotechnological production. Presently, industrial production of antibiotics is mostly limited to the use of free suspended cell systems. However, considerable merit exists for use of the immobilized cell system for antibiotics, as evidenced in the following three examples, two conducted in a bench scale and one in a pilot scale.

7.5.2.1 Patulin Production

In the production of the antibiotic patulin by immobilized Penicillium urticae, the batch growth behavior shown in Fig. 7.5 reflects the increase in patulin concentration in the medium which

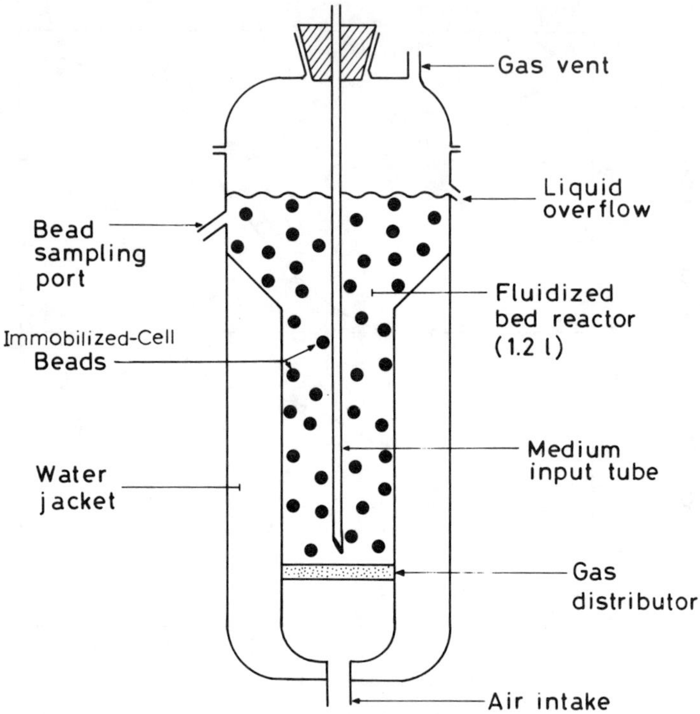

Figure 7.6 Schematic diagram of the three-phase fluidized bed bioreactor for the production of patulin by immobilized Penicillium urticae cells (from Berk et al., 1984a).

takes place in the stationary phase rather than the exponential growth phase (Berk et al., 1984a). Thus, in the operation of a bioreactor, it is preferable that the operation begins with cells which have reached their secondary metabolite production state without undergoing unproductive incubation steps in the reactor. Since there are complex pathways regulated by living microorganism in the biosynthesis of secondary metabolites, proper control of the microorganism's living environment to enhance secondary metabolism for continuous production of secondary metabolites is deemed necessary.

The three-phase fluidized bed (Mode E-III-a) at the laboratory scale used for the production of patulin by Berk et al. (1984a and 1984b) is given in Fig. 7.6. In the figure, glucose solution, which provides the required nutrients to microorganisms, serves as the continuous liquid phase in the reactor. The κ-carrageenan particle immobilized with Penicillium urticae has a diameter of 3.2 mm. The working volume for the reactor is 1.2 liters. Proper limitation of the nitrogen source via intermittent pulse addition of nutrients to minimize cell growth and thus to enhance the longevity of the cell particle is the key factor for ultimate success of the bioreactor operation. A pathway efficiency, η, can be defined which expresses an empirical indicator of the ability of cells to convert the first intermediate of the biosynthesis path in the secondary metabolism, i.e., 6-methyl salicylic acid to patulin. A value of 0.35 for η was found for a continuous three-phase fluidized bed operation, indicating the biosynthesis pathway is operating at a good catalytic efficiency.

7.5.2.2 Penicillin

Production of penicillin using a three-phase fluidized bed bioreactor (Mode E-III-a) was attempted by Endo et al. (1986). In their study, the spores of P. chrysogenum are entrapped within urethane foam and are allowed to germinate under optimum conditions. The hydrophilic property, and rugged surface and internal porous structure of the urethane foam allow the stable attachment and formation of filamentous biofilms on the surface. The urethane particles have a density close to water. Thus, they can be easily suspended and mixed merely by the gas bubble induced turbulence in the bioreactor. The soft and elastic property of urethane foam also reduces the damage and sloughing-off of the biofilm due to particle-particle collision. A penicillin production rate 3 to 4 times higher than that of a suspended system was shown to be achieved in the three-phase fluidized bed bioreactor. Figure 7.7 is a photograph of a pilot scale three-phase fluidized bed bioreactor employed by Endo et al. (1986) and Endo (1986) for penicillin production. The volume of the bioreactor is 200 liters with a working volume of 160 liters. A seed tank situated above the bioreactor provides the fresh medium and urethane foam particles to the fermentor.

7.5.2.3 Bacitracin Production

Morikawa et al. (1980) produced bacitracin using cells of Bacillus sp. entrapped within polyacrylamide gel particles in a three-phase fluidized bed bioreactor (Mode E-III-a). Figure 7.8 shows the time variation in bacitracin productivity, the pH of the fermentation broth, and the effluent optical density of suspended cells at 660 nm (OD_{660}), which is an equivalent measure of the concentration of suspended cells. As shown, the amount of bacitracin produced

Figure 7.7 Photograph of the pilot scale three-phase fluidized bed bioreactor for the production of penicillin. The working volume of this reactor is 160 liters (kindly provided by I. Endo of the Institute of Physical and Chemical Research, Japan).

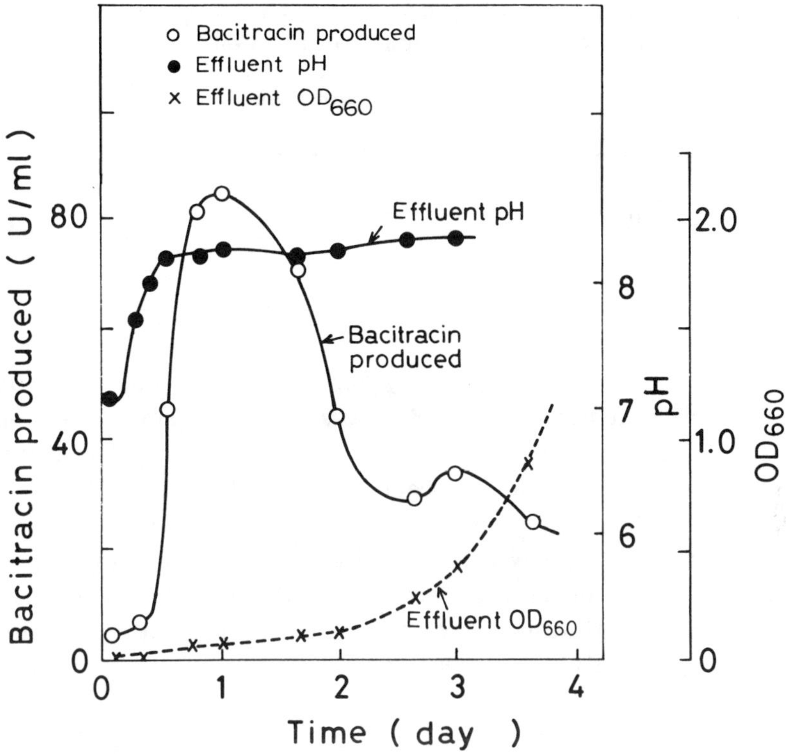

Figure 7.8 Continuous production of bacitracin by cells of Bacillus sp. entrapped within polyacrylamide gel beads. OD_{660} (optical density at 660 nm) reflects the concentration of suspended cells leaching from the gel beads and growing in the medium (from Morikawa et al., 1980).

increases sharply, reaches a peak after 1 day operation, and decreases sharply with further increasing operation time, while the effluent OD_{660} increases gradually with time. The half-life, which is defined as the time elapsed for the biocatalytic activity of the particles to drop to half of its maximum value, of the immobilized cells in such an operation is two days. Electron microscopic examination of the immobilized cell particles reveals that after 50 hours of incubation the surface of the gel particles is completely covered with Bacillus colonies, which augments the diffusional resistance of oxygen and nutrients to the inner part of the gel particles. The oxygen diffusion limitation is evident from the fact that increasing the aeration rate increases the maximum bacitracin concentration (Table 7.6). The short half-life is therefore presumably due to diffusional limitations on the

Table 7.6 Effects of Aeration Rate on Maximum Bacitracin Production (from Morikawa et al., 1980)

Aeration rate (ℓ/min)	Maximum bacitracin concentration (U/mℓ)
1.0	32
1.5	51
2.0	70
3.0	81
3.5	86
3.9	85

oxygen and nutrients availability resulting from the outgrowth of cells. By implementing periodic washing to remove cells on the surface of gel particles, the half-life of the immobilized Bacillus cells for continuous production of bacitracin can be extended to 10 days (Fig. 7.9).

Figure 7.10 shows the relationship between the bacitracin production and dilution rate for both immobilized cells and free cells. At any given dilution rate, the immobilized cells yield higher bacitracin productivity than do the free cells.

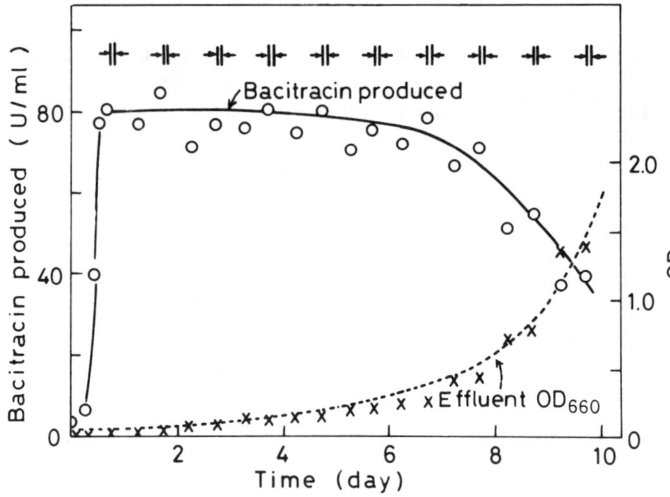

Figure 7.9 Continuous production of bacitracin by cells of Bacillus sp. entrapped within polyacrylamide gel beads can be prolonged (compared to that in Fig. 7.8) by periodic washing of the outgrowing cells on the surface of the beads with saline. Time intervals between two arrows indicate a period when immobilized cell particles were washed (from Morikawa et al., 1980).

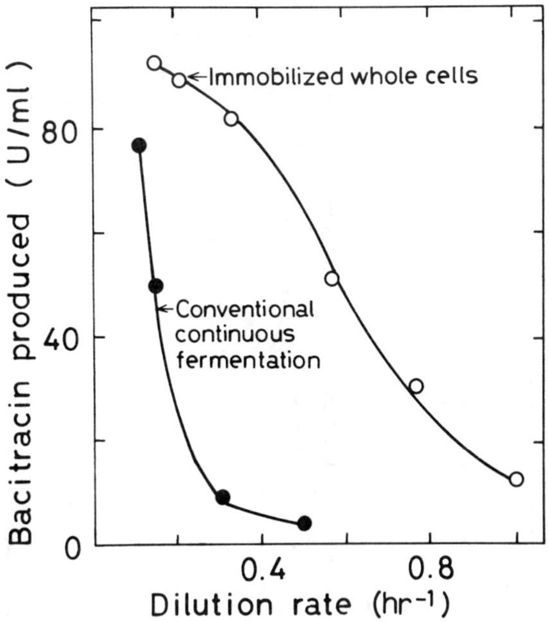

Figure 7.10 Comparison of bacitracin productivity at different dilution rates in the immobilized cell and conventional continuous (freely suspended cell) fermentation systems (from Morikawa et al., 1980).

7.5.3 Enzyme Production

Using a 10 liter gas-liquid-solid fluidized bed fermentor, (Mode E-I-a-1) or more precisely a spouted bed fermentor, Fukuda et al. (1984) studied continuous cellulase production by Trichoderma viride immobilized on 6 mm spherical porous particles made from stainless steel wires (see Section 8.2.2). Glucose is used as the main carbon source for the microorganisms. A liquid spout, or jet, is generated by recycling broth in the fermentor and is introduced through the base of a bed of particles. Due to high shear stresses generated by the high velocity spout, excessive cell mass on the particle is removed and the biofilm thickness on the particle is controlled. Figure 7.11 shows the time variation in cell concentrations in both the immobilized and free suspended states, glucose and dissolved oxygen concentrations and cellulase concentration assayed as filter paper activity. In the experiments, the batch operation was carried out for about 60 hours before continuous operation taking place at a dilution rate of 0.15

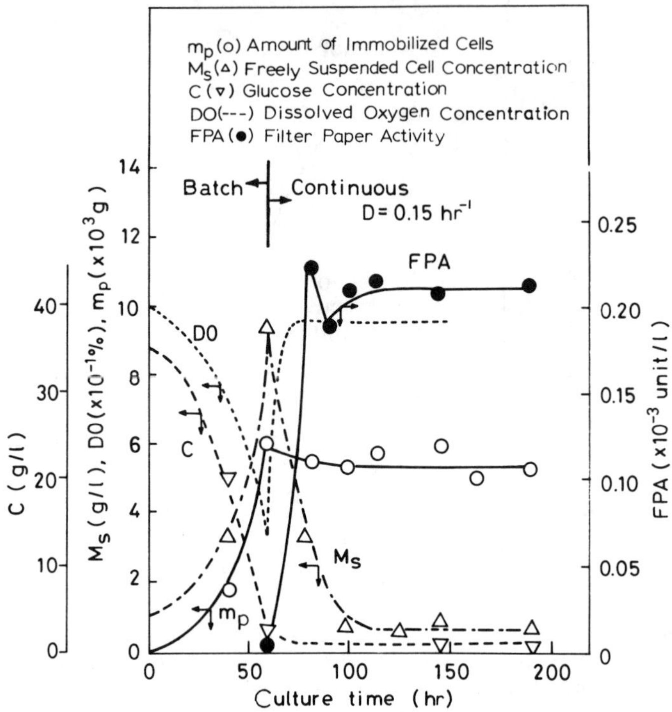

Figure 7.11 Time course of the culture of immobilized Trichoderma viride cells for cellulase production. The concentration of cellulase is assayed as the filter paper activity (FPA). A steady state cellulase production is established after the inception of continuous operation for 40 hours (from Fukuda et al., 1984).

hr^{-1}. In batch operation, as shown in the figure, cell concentrations for both the immobilized and free suspended states increase sharply and the dissolved oxygen concentration decreases rapidly. Following the start of continuous operation, the immobilized cell and dissolved oxygen concentrations rapidly become constant, the free suspended cell concentration sharply decreases to a minimum, and the filter paper (enzyme) activity increases sharply then becomes constant at a high value. The maximum filter paper activity observed was 0.21×10^3 unit/ℓ. Note that the dilution rate used ($0.15\ hr^{-1}$) to obtain this maximum filter paper activity is 12.5 times that of freely suspended cell system (Brown and Zainudeen, 1977) under which maximum filter paper activity and enzyme productivity are obtained.

7.6 OPTIMUM DESIGN AND OPERATION CONSIDERATIONS

In the design and operation of a gas-liquid-solid fluidized bed bioreactor, cellular physiology, control mechanisms in the cellular metabolism, and the growth phase during which the desired products are produced must be considered in addition to transport phenomena, mixing behavior, control of the reactor, and metabolic intermediate or product separation. The products derived from a fermentation process can be the biomass and primary or secondary metabolites. The primary metabolites are the chemicals produced during the exponential growth phase (e.g., amino acids, nucleotides, proteins, nucleic acids, lipids, and carbohydrates) and are essential to cell growth. Thus, both the biomass and primary metabolites are growth associated. Secondary metabolites, however, are synthesized late during the exponential growth phase or in the stationary growth phase and do not have an obvious function in cell metabolism. Some secondary metabolites may be produced in a continuous culture at slow growth rates, and thus are characteristics of both slow-growing and nongrowing cells.

7.6.1 Media Formulation and Feeding Strategy

Media formulation and the feeding strategy are two key factors in the optimal operation of an immobilized cell reactor. If biomass or primary metabolites are the desired products of the fermentation process, the media should be formulated to permit maximum growth potential. In the case where a catabolite repression situation is encountered, intermittent or gradient feeding of glucose or use of slowly fermentable carbohydrates is a vital strategy. When the production of secondary metabolites is desired, initial maximum growth is still required to provide the maximum allowable biomass. After a sufficiently high biomass concentration is obtained, the medium is then switched to a so-called production medium which maximizes the production of the desired metabolites while maintaining the cells in a relatively slow or nongrowing condition. Production media are usually deficient in one or more nutrients crucial for primary metabolisms such as phosphorus, nitrogen, or trace metals. A common problem in immobilized cell systems, especially one which involves the production of secondary metabolites, is that cell productivity declines with operating time due to deactivation of cells by aging; consequently, intermittent feeding of the key nutrients required for primary metabolisms is necessary to rejuvenate the aging cells. Note that differences in media formulation affect the rheological properties and surface tension of the fermentation broth, which in turn influence the reactor design.

One interesting phenomenon in immobilized cell reactor operation is the periodic oscillation (or fluctuation) in cell and product concentrations (Bandyopadhyay and Ghose, 1982). Such behavior probably results from changes in physiological state that occur with a frequency possibly responding to cell regeneration time and an amplitude possibly relating to the dilution rate (Venkatasubramanian and Karkare, 1983). Thus, in a strict sense, operation of such an immobilized cell reactor can never be in a steady state condition.

It is important to note that the media formulation will affect the reactor design. The rheological properties and surface tension of the media affect the bubble properties, such as size and rise velocity, and consequently, influence the phase holdup distribution, heat transfer, solid-liquid and gas-liquid mass transfer, and the mixing states of the phases.

7.6.2 Rheological Properties of Fermentation Broths

The typical viscosity of the fermentation broths containing free suspended microbial species is given in Table 7.7. Note that the presence of immobilized cell particles in a fermentation broth may affect its rheological properties in the following aspects. First, a fermentation broth containing suspended solids (immobilized cell particles) may have different rheological characteristics from the broth without solids. The degree of deviation in the rheological properties depends on the size, density, compressibility, and concentration of the immobilized cell particles. Little information regarding this deviation is available. Second, the immobilizing carrier matrix confines not only the cells, but also the high molecular weight metabolites excreted by the cells and thus, the fermentation broth formed in these systems have a different composition from those formed using free suspended cells under the same conditions. For systems where the rheological behavior of the fermentation broth is predominantly determined by cell concentrations or biopolymers produced by the cells, the rheological behavior of the fermentation broth containing surface-immobilized cell particles involving substantial cell leakage or biopolymer excretion from particles to the

Table 7.7 Viscosity Values of Fermentation Broth Containing Suspended Cells Typically Encountered (from Wang, 1986)

Types of organisms	Apparent viscosity (Pa•s)
Unicellular organisms	0.01 ~ 0.02
Filamentous organisms	5 ~ 20
Biopolymer producing organisms	10 ~ 100

Table 7.8 Critical Stress (Including Shear and Normal Stress) Values of Several Suspended Microorganisms in Free Liquid Jet and Stirring Experiments in a 1.74 Liter Fermentor (from Bronnenmeier and Markl, 1982)

Test organisms	Free-jet experiments	Stirring experiments
Chlamydomonas reinhardi (cell-wall defect mutant strain)	$0 \sim 2 \times 10^5$ N/m^2	300 rpm
Chlamydomonas reinhardi (wild strain)	$15 \sim 20 \times 10^5$ N/m^2	2400 rpm
Spirulina platensis	$2 \sim 4 \times 10^5$ N/m^2	900 rpm

bulk liquid phase resembles that of the fermentation broth containing free suspended cells.

7.6.3 Hydrodynamic Stresses

Another important physiological aspect to be considered for bioreactor design is the capacity of cells to sustain hydrodynamic stresses (shear stress mainly). This capability of cells usually dictates the operating conditions, e.g., flow rates of liquid and/or gas, or stirring speed of turbine, and the types of reactors that can be used for cultivation and production. Hydrodynamic stresses have been demonstrated to cause changes in morphology, a decrease in growth rate, retardation of metabolism, release of intracellular substances, and even destruction of cells (Bronnenmeier and Markl, 1982; Stathopoulos and Hellums, 1985). Table 7.8 lists the critical hydrodynamic stress values under which no essential damage to cell cultures occurs for three freely suspended microorganisms in both free liquid jet and

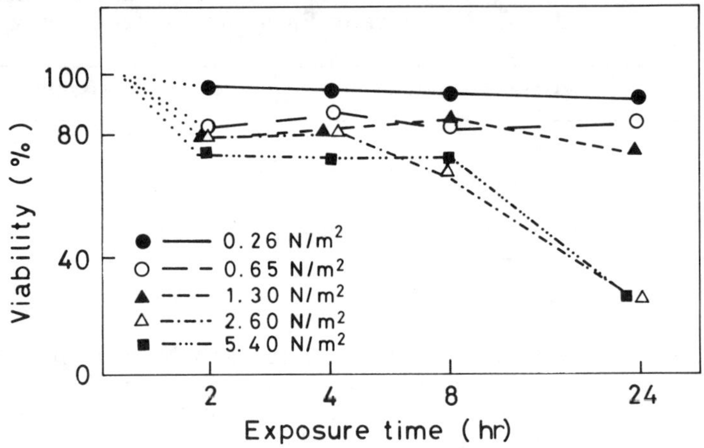

Figure 7.12 Effects of duration of exposure and shear stress level on the viability of human embryonic kidney cells (from Stathopoulos and Hellums, 1985).

stirring turbine experiments (Bronnenmeier and Markl, 1982). Note that as cells are immobilized onto a carrier surface via covalent binding or physical adhesion, they gain the terminal velocity of the carrier particles which is significantly larger than that of free suspended cells. When exposed to the same flow or stirring conditions, the relative velocity between the liquid and immobilized cells is far greater than that between the liquid and free suspended cells, thus surface-immobilized cells actually experience far greater shear stress than do free cells. The shear effects on cells are much more severe for fragile cells such as animal and plant cells. Stathopoulos and Hellums (1985) tested the shear effects on the confluent culture of human embryonic kidney cells attached to a glass substrate. The variation in cell viability with shear stress is shown in Fig. 7.12. Shear stresses of 0.26 N/m^2 and less yield no apparent effect on cell viability and on post-shear urokinase release, but create cell morphology changes. An increase in shear stress higher than 0.26 N/m^2 increases the loss of viability, morphological changes, and post-shear urokinase release. A loss of 75 ~ 85% viability, cell disruption, and a reduced level of urokinase release occur for the exposure of the cells under a shear stress higher than 2.6 N/m^2 for 24 hours. The extremely low critical stress value of the animal cells restricts the use of high liquid and/or gas flow rates or stirring speed, which in turn

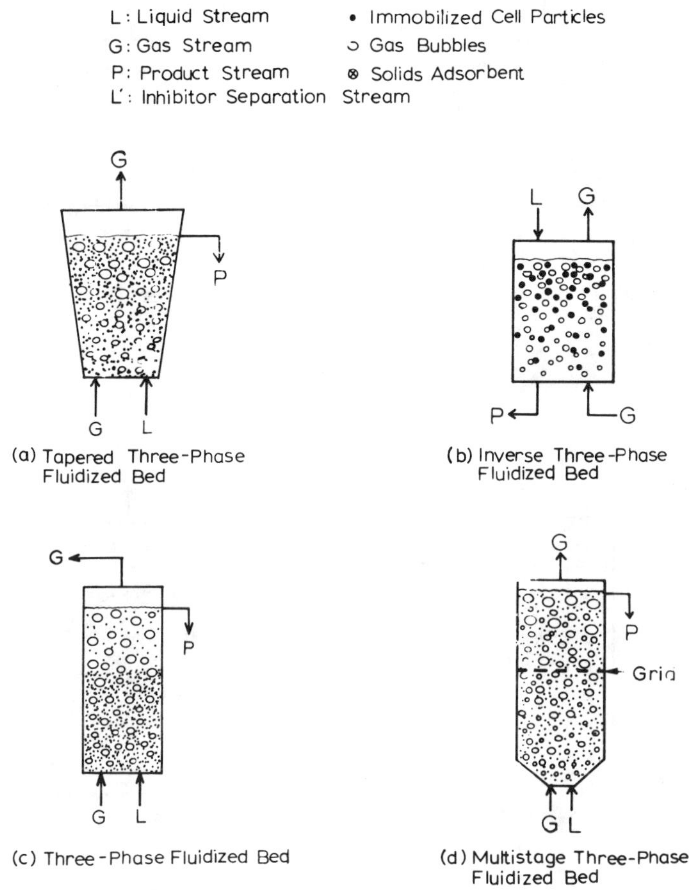

Figure 7.13 Configurations of gas-liquid-solid fluidized bed bioreactors.

may aggravate the problems of mass transfer and mixing and raises the need for a more delicate reactor design.

7.6.4 Bioreaction Kinetics Effects

There are various gas-liquid-solid fluidized bed bioreactor configurations which can be used for immobilized cell particle operation. Figure 7.13 shows schematic diagrams of some configurations. Selection of bioreactor configurations hinges on bioreaction kinetics involved. In the following, the effects of bioreaction kinetics on the selection of various bioreactor configurations given in the figure are described. Note that configurations given in Fig. 7.13-a and 7.13-b will be described

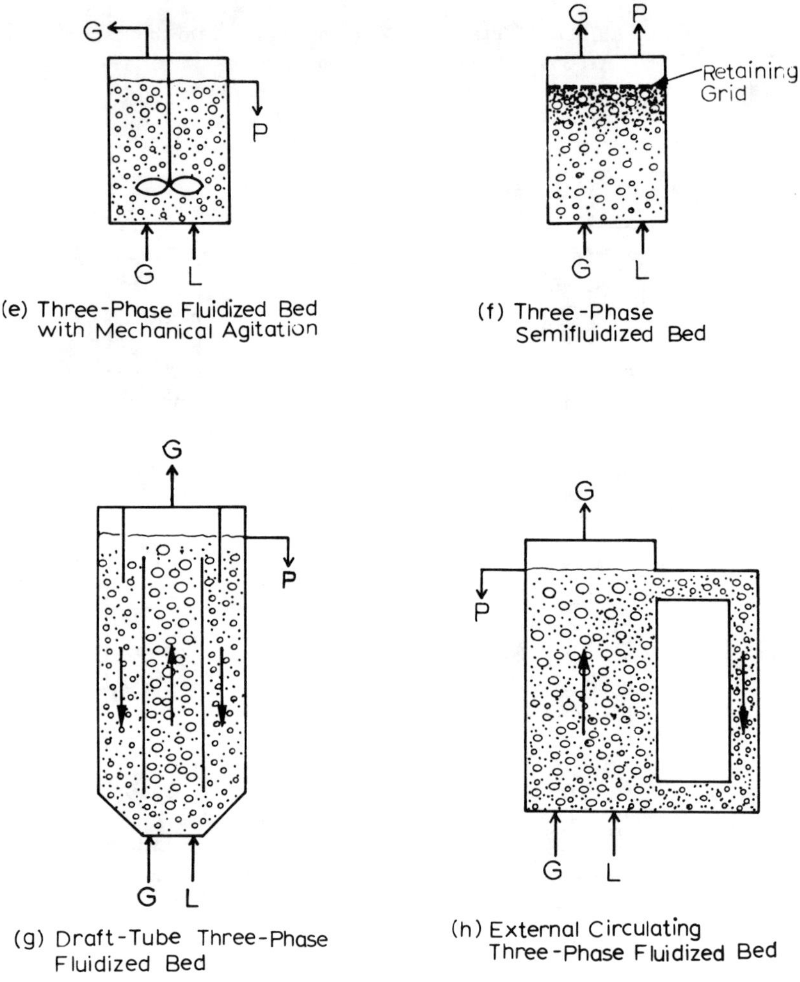

(Figure 7.13 continued)

in the context of wastewater treatment in Chapter 8 due to close relevance in their application.

7.6.4.1 Monod Kinetics

When substrates do not inhibit the reactions and reactions follow the Monod relationship, bioreactors can best be operated at high substrate concentration conditions when production of biomass or primary metabolites is desired. In continuous operation, this condition can be achieved with a plug flow pattern for the liquid flow. A large

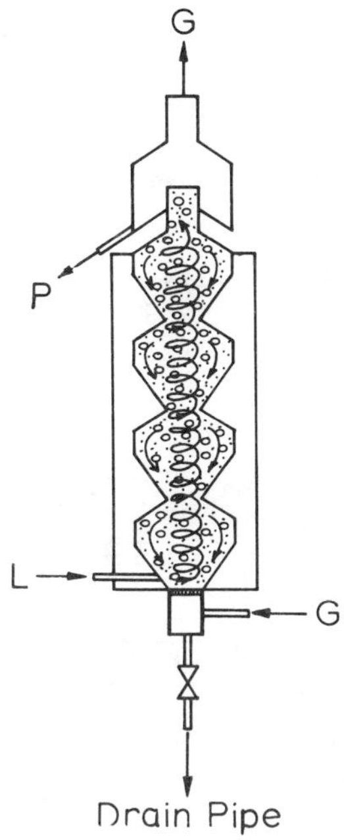

(i) Multiunit Three-Phase Fluidized Bed
(from Fukushima et al., 1982)

(Figure 7.13 continued)

length/diameter ratio is required for the gas-liquid-solid fluidized bed (Figure 7.13-c) to approximate this flow pattern. Note that because of the power consumption consideration in operating a fluidized bed bioreactor, the height of the bioreactor is normally limited. The use of a staging concept for bioreactor design with an equal or different volume for each stage of a multistage bioreactor is a viable operational consideration (Fig. 7.13-d). This staging strategy was applied by Fan (1985) in the design of a two-stage gas-liquid-solid draft tube fluidized bed bioreactor. In this bioreactor, the first stage served mainly for biofilm cultivation and partial biochemical reactions. By appropriate control of the gas and liquid flow rates, particles with biofilm growing to a desired thickness can be entrained to the second stage in which the main biochemical reactions occur. Variations of a multistage three-phase fluidized bed design include a configuration similar to Fig. 7.13-d but with an internal draft tube. The multistage design would provide simultaneously high oxygen transfer and solids mixing, and a high substrate consumption. Without

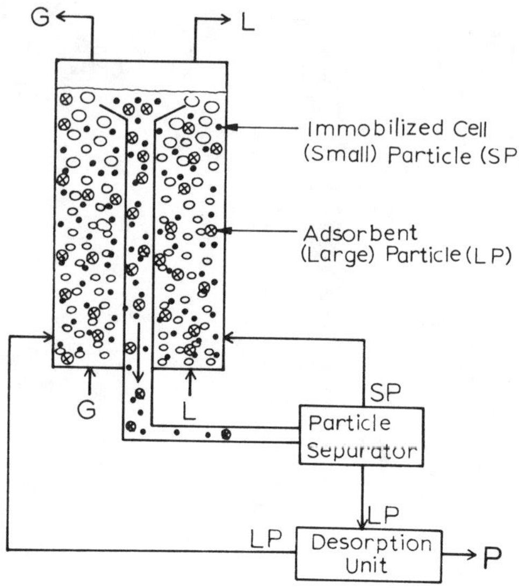

(j-1) Combined Three-Phase Fluidized Bed (Ebullated Bed) Bioreactor / Separator System - Internal Adsorption Separation

(Figure 7.13 continued)

resorting to internal or external means for liquid circulation, the liquid velocity in the bioreactor is normally very low. In most cases a plug flow pattern for the liquid cannot practically be approximated under such low liquid flow conditions. The use of a fixed bed to minimize the axial liquid mixing is not desirable due to inherent bed plugging problems when immobilized living cells are used.

7.6.4.2 Substrate Inhibiting Kinetics

When substrate inhibition occurs in a reaction, a fluidized bed with mechanical agitation (Fig. 7.13-e) can be used for the reaction. More appropriately, a continuous-flow stirred-tank reactor (CSTR) followed by a plug flow can provide an optimum reactor combination for the reaction. In practice, this combination can be achieved by the use of a large fluidized bed reactor followed by a small fluidized bed reactor or a series of small fluidized bed reactors. Alternatively, this combination can be accomplished by staging a bioreactor with a large

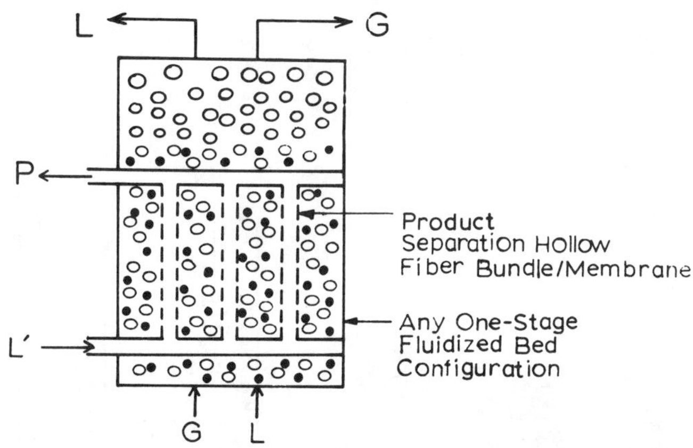

(j-2) Combined Three-Phase Fluidized Bed Bioreactor/Separator System - Internal Hollow Fiber Bundle/Membrane Separation

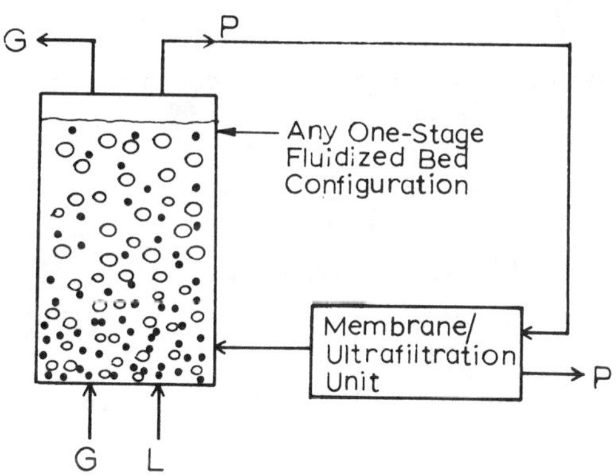

(j-3) Combined Three-Phase Fluidized Bioreactor/ Separator System - External Separation

(Figure 7.13 continued)

stage followed by a small, or a series of small stages in a bioreactor. The concept of semifluidization (L.T. Fan and Wen, 1959; Chern et al., 1984) can be applied to carry out this type of reactions. In a semifluidized bed (see Section 6.2), a fluidized bed and a packed bed are in series in a single reactor (Figure 7.13-f). The bed expansion is constrained by a retaining grid. In a semifluidized bed reactor, the fluidized portion of the semifluidized bed carries the main load of biological reaction while the packed portion acts as a polishing section. The liquid reaching the latter section will be lean in substrate, and

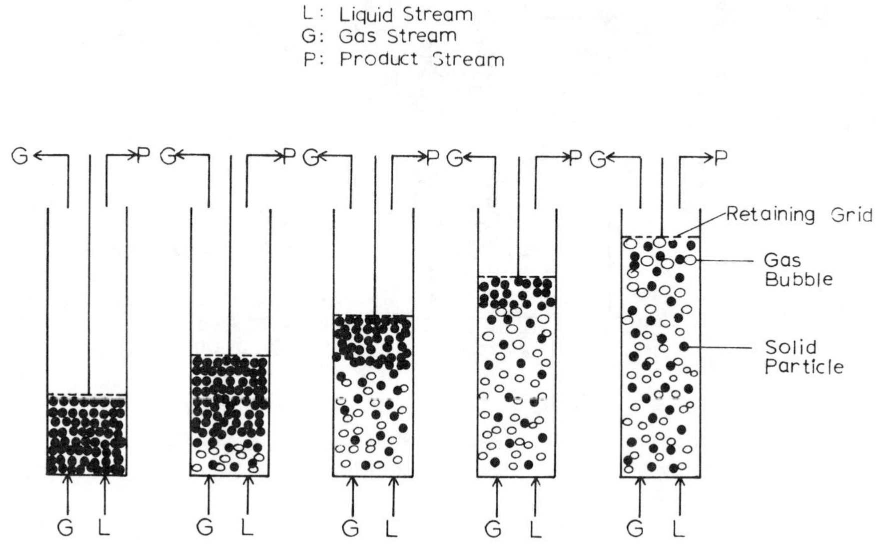

Figure 7.14 Operating cycle for a three-phase semifluidized bed (see Fig. 7.13-f) (from L.T. Fan, 1980).

thus, will not excessively contribute to increases in cell mass which can clog the bed. If the packed portion of the bed becomes plugged by excess cells or by suspended solids, the clogging can be cleared by raising the retaining grid and completely fluidizing the bed as shown in a operating cycle given in Fig. 7.14 (L.T. Fan, 1980). To control the microbial population, the entire bed should periodically be fully fluidized. During fluidization, excess cell mass is removed from the particles by shear stress.

The small length/diameter ratio for a gas-liquid-solid fluidized bed provides good mixing of the liquid phase. Independent of length/diameter ratio, however, excellent liquid mixing can be obtained when a draft tube is placed in the reactor (Fig. 7.13-g). An internal circulation of liquid and solids between the draft tube and annulus ensures excellent mixing. The circulation of the phases can also be achieved externally by the use of an external tube (Fig. 7.13-h). Note that the extent of axial dispersion for liquid is not crucial in determining the overall conversion when the biological reaction rate is extremely slow.

Good mixing of liquid phase in a three-phase fluidized bed bioreactor can also be achieved by modifying the reactor configuration. A multiunit gas-liquid-solid fluidized bed bioreactor with an irregular geometry was patented by Fukushima et al. (1982) for continuous production of bulk chemicals such as ethanol. This reactor is shown in Fig. 7.13-i. Each unit is constructed from a top cone connected to an inverted bottom cone by a short cylinder positioned slightly above the middle portion of the unit. The bottom of the lowest unit is connected to a gas disperser made of a sintered or porous plate. Fermentation broth is injected from the inlet pipe. The top of the fermentor is connected to a gas outlet and a liquid mash discharge pipe. A good mixing of liquid phase can be achieved in this bioreactor.

7.6.4.3 Product Inhibiting Kinetics

For systems with a product inhibition kinetics, a plug flow reactor is more favorable for high substrate conversion than a CSTR. However, the conversion can best be enhanced by removing any inhibitory components in situ: this consideration can be achieved by using solid particle adsorption or hollow-fiber separation methods. In the adsorption method (Fig. 7.13-(j-1)), the adsorbent, which is a second type particle in an ebullated bed with a size either considerably larger or smaller than that of the immobilized cell particles, can easily be separated from the immobilized cell particles. The sorbent should preferentially adsorb the inhibitors (or products) and resist attachment of free cells. Desorption of the adsorbed species can be performed in a separate desorption unit. The absorbent can also be another liquid phase immiscible with the fermentation broth. This in situ adsorption concept has been used by Allen and Kim (1986) in a multi-solid fluidized bed process design for production of bulk chemicals, such as acetone/butanol fermentation. In this process, cell-entrapped calcium alginate gel particles of about 1 ~ 2 mm diameter are considered to be the optimal particle size for biocatalytic activity. An immiscible organic (sorbent) liquid phase introduced either cocurrently or countercurrently to the fermentation broth flow provides an in situ adsorption of products (or intermediates) to alleviate product (or metabolite) feedback inhibition or repression activity.

In the hollow-fiber method (Fig. 7.13-(j-2)), a semipermeable hollow fiber bundle can be placed in the bed to provide optimum removal of inhibitors. This requires, however, that cells do not attach to the fiber and clog the liquid permeation path. The removal of inhibitors can also be achieved by the use of an integrated process scheme combining the external ultrafiltration separator with the bioreactor (Fig. 7.13-(j-3)).

7.6.5 Stability and Strength of the Immobilized Cell Particles

The durability (strength) and stability of the immobilized particle is one of the most important factors for successful operation of the gas-liquid-solid fluidized bioreactor. Aside from the physical properties of the immobilized cell particles, the maintenance of particle integrity is mainly governed by particle-particle collision and abrasion and liquid turbulent shear stresses generated by the gas flow. The disintegration of the particles would result in failure of the process. The effects of shear stress on the immobilized cells, noted earlier, depends also on the immobilization method used. For example, cells immobilized by adsorption are more susceptible to shear stress compared to those by entrapment. The shear stress, on the other hand, can be used as a measure for biofilm thickness control for adsorption immobilization. The use of a propeller, hydrocyclone or gas jetting has shown some success in maintaining the uniformity of biofilm thickness for adsorption immobilized cell particles (see Section 8.3.4).

7.6.6 Oxygen Transfer

Efficient oxygen transfer is crucial especially for systems containing high biomass loading. The O_2 transfer in the gas-liquid-solid fluidized bed depends on various operating parameters as indicated in Chapters 3 through 6. For both gas-liquid and liquid-solid O_2 transfer resistances, an optimum particle size exists for a given bioreaction kinetics which achieves maximum overall biological reaction rates (see Appendix A for a general approach). Without resorting to the use of pure oxygen or high air flow rates, O_2 transfer can be improved by either enhancing the mass transfer coefficient (k_ℓ) or increasing the interfacial area (a). Different designs or operational configurations such as those shown in Fig. 7.13 give rise to different k_ℓ and "a" and hence, different O_2 transfer rates. The O_2 transfer rate can be increased, for example, by placing a draft tube in the bed. The use of a fine bubble generator has shown to yield considerably higher O_2 transfer than does a coarse bubble generator. For highly viscous liquid media such as those involving filamentous organisms or biopolymer producing organisms, mechanical agitation is required to aid O_2 transfer.

7.7 ECONOMICS

There is relatively little development accomplished on three-phase fluidized beds bioreactors for fermentation in the production of food, chemicals, and pharmaceuticals compared to that for biological wastewater treatment. From a process viewpoint, fermentation is only one of several important steps, such as sterilization and separation, in a biological process. In fermentation, bioreactor design is complicated by implementation of sophisticated instrumentation and control. Thus, the capital cost of the fluidized bed bioreactor is expected to be high. However, due to a higher bioconversion efficiency in fluidized bed bioreactors as compared to the immobilized cell stirred tank bioreactor with mechanical agitation and the conventional suspended cell bioreactors, the operating cost for fluidized bed bioreactors is expected to be lower. In addition, the cost for downstream processing such as cell and product separation can be reduced. Therefore, the integrated fluidized bed bioreactor system offers attractive economic features to merit further development. The development of immobilized cell bioreactor technology is likely to have a far greater impact on fermentation processes involving large volumetric production, such as ethanol production, than on those involving small volumetric production such as antibiotics or enzyme production. Furthermore, considerable potential cost benefits prevail in the application of immobilized cell bioreactor technology to processes involving genetically engineered microorganisms due to their better containment and controllability (Venkatasubramanian and Karkare, 1983).

7.8 CONCLUDING REMARKS

Although commercial operation of a three-phase fluidized bed bioreactor using immobilized cells in fermentation has been demonstrated, its development is only in its infancy. Fundamentals such as the interactions among cellular physiology, production kinetics, immobilization techniques, and hydrodynamics, mass transfer, and mixing characteristics of a three-phase fluidized bed bioreactor have not been fully explored.

A variety of immobilization techniques have been developed. Each has its specific merits and weaknesses. When selecting an immobilization method, one must consider the retention of viability and/or activity, mass transfer limitations, mechanical stability and integrity of the carrier matrix, and ease in product recovery and separation.

An understanding of the physiology and the metabolic control mechanisms of cells in an immobilized state is required to illustrate the change in apparent kinetic behavior of cells upon immobilization. Independent evaluations of the intrinsic kinetics of immobilized cells, mass transfer characteristics, and spatial distribution of cells in immobilized cell particles are necessary for proper modeling and prediction of product synthesis of an immobilized cell system.

Three-phase fluidized bed bioreactors offer some unique advantages such as good mixing of solid and liquid phases, high gas-liquid and liquid-solid mass transfer, easy rejuvenation of biocatalyst, and low attrition of biocatalyst. Since attrition and shear effects counteract mixing and mass transfer in the performance of a fluidized bed bioreactor, the hydrodynamics, mass transport behavior, and mixing characteristics under various gas and liquid flow rates, and bioparticle properties and loadings must be further investigated such that optimization and scale-up can be achieved. More studies are also needed on the rheological properties of the viscous media in immobilized cell systems, to permit accurate assessment of O_2 transfer and power requirements.

7.9 NOMENCLATURE

A	: effective area through which diffusion occurs
a	: gas-liquid interfacial area
C_ℓ	: the limiting dissolved oxygen concentration on biofloc surface below which an anoxic core will develop inside the biofloc
C	: dissolved oxygen concentration
C_i	: dissolved oxygen concentration in vessel i
C^*	: dissolved oxygen concentration at the bioparticle surface

C_s : dissolved oxygen concentration saturated with oxygen gas or air

D_M : diffusivity of a substrate or oxygen within an immobilized cell particle, biofilm, biofloc, or microbial aggregate

D_m : diffusivity of a substrate or oxygen through membrane

D_t : overall diffusivity of a substrate or oxygen through the packed template

D_w : diffusivity of a substrate or oxygen in pure water

e : electrostatic charge

F : total substrate transferred through a calcium alginate gel slab

k_B : Boltzmann constant

L : thickness

N : number of particles

Q : volumetric liquid flow rate

r : radial distance

r_c : radius of the anoxic core

r_p : particle or bioparticle radius

$-r_{cv}$: oxygen uptake rate per unit volume of immobilized cell particles or biofilms attached to solid surface

$-r_{cv,max}$: maximum value of $-r_{cv}$ when C^* is greater than C_ℓ

$-r_{cw}$: oxygen uptake rate per unit dry weight of bioflocs

$-r_{cw,max}$: maximum value of $-r_{cw}$ when C^* is greater than C_ℓ

$-r_{sv}$: substrate consumption rate per unit volume of the immobilized cell particle or the biofilm attached to the solid particle

S : substrate concentration

S_b : bulk concentration of a substrate or product

$S_{b,i}$: bulk concentration of a substrate in chamber i

$S_b(0)$: initial bulk concentration of a substrate

$S_b(t)$: bulk concentration of a substrate at time t

S_l : local concentration of substrates or products

t : time

T : temperature

$\vec{v}$: convective velocity of fluid

V : volume of the liquid phase in a vessel

X : concentration of activated sludge

x_c : r_c/r_p

Z : number of charges

Greek Letters

α : $V/(N4\pi r_p^3/3)$

δ : effective thickness of a slab of gel, a biofilm, or a template assembly through which diffusion of solutes occurs

ψ : electrostatic potential

7.10 REFERENCES

Adamson, S.R., and B. Schmidli, Can. J. Chem. Eng., 64, 531 (1986).

Akao, T., and Y. Okamoto, in "Fluidization," ed. by D. Kunii and R. Toei, p. 631, Engineering Foundation Press (1984).

Allen, B.R., and B.C. Kim, 1st Quarterly Technical Progress Report to Jet Propulsion Laboratory, Contract No. 957241 (1986).

Atkinson, B., G.M. Black, P.J.S. Lewis, and A. Pinches, Biotech. Bioeng., 21, 193 (1979).

Baillod, R.C., and W.C. Boyle, J. Sanitary Engineering Div. ASCE, 96, 525 (1970).

Bandyopadhyay, K.K., and T.K. Ghose, Biotech. Bioeng., 24, 805 (1982).

Bauer, W.J., in "Fluidization," ed. by K. Ostergaard and A. Sorensen, p. 619, Engineering Foundation Press (1986).

Beddows, C.G., H.G. Gil, and J.T. Guthrie, Biotech. Bioeng., 27, 579 (1985).

Ben-Amotz, A., and M. Avron, TIBS, p. 297, Nov. (1981).

Berk, D., L.A. Behie, A. Jones, B.H. Lesser, and G.M. Gaucher, Can. J. Chem. Eng., 62, 112 (1984a).

Berk, D., L.A. Behie, A. Jones, B.H. Lesser, and G.M. Gaucher, Can. J. Chem. Eng., 62, 120 (1984b).

Birnbaum, S., R. Pendleton, P.-O. Larsson, and K. Mosbach, Biotech. Lett., 3, 393 (1981).

Briffaud, J., and J.M. Engasser, Biotech. Bioeng., 21, 2093 (1979).

Bronnenmeier, R., and H. Markl, Biotech. Bioeng., 24, 553 (1982).

Brown, D.E., and M.A. Zainudeen, Biotech. Bioeng., 19, 914 (1977).

Bryers, J.D., Biotech. Prog., 3, 57 (1987).

Bungay, H.R., W.J. Whalen, and W.M. Sanders, Biotech. Bioeng., 11, 765 (1969).

Burrill, H.N., L.E. Bell, P.F. Greenfield, and D.D. Do, Appl. Environ. Microbiol., 46, 716 (1983).

Cheetham, P.S.J., K.W. Blunt, and C. Bucke, Biotech. Bioeng., 21, 2155 (1979).

Chern, S.H., L.-S. Fan, and K. Muroyama, AIChE J., 30, 288 (1984).

Chibata, I., T. Tosa, and T. Sato, Appl. Microbiol., 27, 878 (1974).

Chibata, I., ACS Symposium Series, 106, 187 (1979).

Chibata, I., and T. Tosa, Ann. Rev. Biophys. Bioeng., 10, 197 (1981).

Chibata, I., T. Tosa, and T. Sato, in "Manual of Industrial Microbiology and Biotechnology," ed. by A.L. Demain and N.A. Soloman, Chap. 18, Am. Soc. Microbiology, Washington, DC (1986).

Constantinides, A., Biotech. Bioeng., 22, 119 (1980).

Davidson, B.H., and C.D. Scott, paper presented at the 9th Symposium on Biotechnology for Fuels and Chemicals, Boulder, Colorado, May 5-8 (1987).

Dinelli, D., W. Marconi, and F. Morisi, Meth. Enzymol., 44, 227 (1976).

Doran, P.M., and J.E. Bailey, Biotech. Bioeng., 28, 73 (1986).

Durand, G., and J.M. Navarro, Process Biochem., 13, 14 (1978).

Edgley, M., and A.D. Brown, J. Gen. Microbiol., 104, 343 (1978).

Eikmeier, H., F. Westmeier, and H.J. Rehm, Appl. Microbiol. Biotech., 19, 53 (1984).

El-Sayed, A.-H.M.M., and H.J. Rehm, Appl. Microbiol. Biotech., 24, 89 (1986).

Endo, I., T. Nagamune, N. Kato, T. Kobayashi, and M. Nishimura, paper presented at the AIChE Annual Meeting, Miami Beach, Nov. 2-7 (1986).

Endo, I., Nikkei New Material, p. 15, April 28 (1986).

Fan, L.-S., Progress Reports to U.S. Geological Survey, Contract No. 14-08-0001-G1150 (1985); paper presented at the AIChE Annual Meeting, Washington D.C., Nov. 27 - Dec. 2 (1988).

Fan, L.T., and C.Y. Wen, AICHE J., 5, 407 (1959).

Fan, L.T., Seminar Notes, Department of Chemical Engineering, Kansas State University (1980).

Freeman, A., and Y. Aharonowitz, Biotech. Bioeng., 23, 2747 (1981).

Fukuda, H., C. Webb, and B. Atkinson, in Proceedings of the 3rd European Congress on Biotechnology, Vol. I, p. 547 (1984).

Fukui, S., and A. Tanaka, Advances in Biochem. Eng., 29, 1 (1984).

Fukushima, S., H. Munenobu, and K. Yamada, U.S. Patent No. 4,337,315 (1982).

Galinski, E.A., and H.G. Truper, FEMS Microbiol. Lett., 13, 357 (1982).

Gainer, J.L., D.J. Kirwan, J.A. Foster, and E. Seylan, Biotech. Bioeng. Symposium Series, 10, 35 (1981).

Ghose, T.K., and K.K., Bandyopadhyay, Biotech. Bioeng., 22, 1489 (1980).

Hahn-Hagerdal, B., M. Larsson, and B. Mattiasson, Biotech. Bioeng. Symposium Series, 12, 199 (1982).

Hanai, S., Energy and Resources (in Japanese), 7, 168 (1986).

Hannoun, B.J.M., and G. Stephanopoulos, Biotech. Bioeng., 28, 829 (1986).

Hiemstra, H., L. Dykhuizen, and W. Harder, Eur. J. Appl. Microbiol. Biotech., 18, 189 (1983).

Jack, T.R., and J.E. Zajic, Advances in Biochem. Eng., 5, 126 (1977a).

Jack, T.R., and J.E. Zajic, Biotech. Bioeng., 19, 631 (1977b).

Jirku, V., J. Turkova, P. Kuchynkova, and V. Krumphanzl, Eur. J. Microbiol. Biotech., 6, 217 (1979).

Jones, A., D.N. Wood, T. Razniewska, G.M. Gaucher, and L.A. Behie, Can. J. Chem. Eng., 64, 547 (1986).

Kalogerakis, N., L.A. Behie, and G.M. Gaucher, in "Fluidization," ed. by K. Ostergaard and A. Sorensen, p. 611, Engineering Foundation Press (1986).

Kennedy, J.F., S.A. Barker, and J.D. Humphreys, Nature, 261, 242 (1976).

Kennedy, J.F., ACS Symposium Series, 106, 119 (1979).

Kennedy, J.F., and J.M.S. Cabral, in "Applied Biochemistry and Bioengineering, Vol. 4 - Immobilized Microbial Cells," ed. by I. Chibata and L.B. Wingard, Jr., p. 189, Academic Press, New York (1983).

Klein, J., and P. Schara, J. Solid-Phase Biochem., 5, 61 (1980).

Klein, J., and F. Wagner, in "DECHEMA Monograph 84-Characterization of Biocatalyst" ed. by K. Bucholz, p. 265, Verlag Chemie, Weinheim (1979).

Klein, J., and F. Wagner, in "Applied Biochemistry and Bioengineering, Vol. 4 - Immobilized Microbial Cells," ed. by I. Chibata and L.B. Wingard, Jr., p. 14, Academic Press, New York (1983).

Knorr, D., S.M. Miazga, and R.A. Teutonico, Food Technol., 39, 135 (1985).

Kolot, F.B., Devel. Indust. Microbiol., 21, 295 (1980).

Kopp, B., and H.J. Rehm, Appl. Microbiol. Biotech., 19, 141 (1984).

Kumakura, M., M. Yoshida, and I. Kaetsu, Eur. J. Microbiol. Biotech., 6, 13 (1978).

Kumakura, M., and I. Kaetsu, Eur. J. Appl. Microbiol. Biotech., 17, 197 (1983).

Lartigue, D. J., and H.H. Weetall, U.S. Patent 3,939,041 (1976).

Linko, Y.-Y., K. Poutannen, L. Weckstrom, and P. Linko, Enzyme Microb. Technol., 1, 26 (1979).

Mahmoud, W., and H.J. Rehm, Appl. Microbiol. Biotech., 23, 305 (1986).

Matson, J.V., and W.G. Characklis, Water Research, 10, 877 (1976).

Mattiasson, B., P.-O. Larsson, and K. Mosbach, Nature, 268, 519 (1977).

Mattiasson, B., and B. Hahn-Hagerdal, Eur. J. Appl. Microbiol. Biotech., 16, 52 (1982).

Mattiasson, B., in "Immobilized Cells and Organelles," Vol. 1, ed. by B. Mattiasson, p. 3, CRC Press, Boca Raton, Florida (1983).

Mattiasson, B., and B. Hahn-Hagerdal, in "Immobilized Cells and Organelles," Vol. 1, ed. by B. Mattiasson, p. 121, CRC Press, Boca Raton, FL (1983).

Measures, J.C., Nature, 257, 398 (1975).

Messing, R.A., and R.A. Oppermann, Biotech. Bioeng., 21, 49 (1979).

Messing, R.A., R.A. Oppermann, and F.B. Kolot, in "Immobilized Microbial Cells," ACS Symposium Series, 106, 13 (1979).

Michaels, A.S., Chem. Eng. Prog., 80(6), 19 (1984).

Middelhoven, W.J., and C.M. Bakker, Eur. J. Appl. Microbiol. Biotech., 15, 214 (1982).

Moebus, O., and M. Teuber, Eur. J. Appl. Microbiol. Biotech., 15, 194 (1982).

Morikawa, Y., I. Karube, and S. Suzuki, Biotech. Bioeng., 21, 261 (1979).

Morikawa, Y., I. Karube, and S. Suzuki, Biotech. Bioeng., 22, 1015 (1980).

Mosbach, K., and R. Mosbach, Acta Chem. Scand., 20, 2807 (1966).

Mosbach, K., ed., "Methods in Enzymology," vol. 44, Academic Press (1976).

Mosbach, K., ed., "Methods in Enzymology," vol. 135, Academic Press (1987a).

Mosbach, K., ed., "Methods in Enzymology," vol. 136, Academic Press (1987b).

Mosbach, K., ed., "Methods in Enzymoloby," vol. 137, Academic Press (1988).

Mueller, J.A., W.C. Boyle, and E.N. Lightfoot, Biotech. Bioeng., 5, 331 (1968).

Nagashima, M., M. Azuma, and S. Noguchi, Annals New York Academy of Sciences, 413, 457 (1983).

Navarro, J.M., and G. Durand, Eur. J. Appl. Microbiol. Biotech., 4, 243 (1977).

Nishida, Y., T. Sato, and I. Chibata, Enzyme Microb. Technol., 1, 95 (1979).

Okita, W.B., D.B. Bonham, and J.L. Gainer, Biotech. Bioeng., 27, 632 (1985).

Onuma, M., T. Omura, T. Umita, and J. Aizawa, Biotech. Bioeng., 27, 1533 (1985).

Petre, D., C. Noel, and D. Thomas, Biotech. Bioeng., 20, 127 (1978).

Poulsen, P., and L. Zittan, Meth. Enzymol., 44, 809 (1976).

Ryu, Y.W., J.M. Navarro, and G. Durand, Eur. J. Appl. Microbiol. Biotech., 15, 1 (1982).

Samejima, H., M. Nagashima, M. Azuma, S. Noguchi, and K. Inuzuka, Annals New York Academy of Sciences, 434, 394 (1984).

Sato, K., and K. Toda, J. Ferment. Technol., 61, 239 (1983).

Shinmyo, A., H. Kimura, and H. Okada, Eur. J. Appl. Microbiol. Biotech., 14, 7 (1982).

Sitton, O.C., and J.L. Gaddy, Biotech. Bioeng., 22, 1735 (1980).

Smith, P. G., and P. Coackley, Water Research, 18, 117 (1984).

Stathopoulos, N.A., and J.D. Hellums, Biotech. Bioeng., 27, 1021 (1985).

Takata, I., T. Tosa, and I. Chibata, J. Solid-Phase Biochem., 2, 225 (1978).

Tanaka, A., T.-N. Jin, S. Kawamoto, and S. Fukui, Eur. J. Appl. Microbiol. Biotech., 7, 351 (1979).

Tanaka, Y., T. Hayashi, K. Kawashima, T. Yokoyama, and T. Watanabe, Biotech. Bioeng., 24, 857 (1982).

Tanaka, H., M. Matsumura, and I.A. Veliky, Biotech. Bioeng., 26, 53 (1984).

Toda, K., and M. Shoda, Biotech. Bioeng., 17, 481 (1975).

Tomlinson, T.G., and D.M. Snaddon, Int. J. Air Water Pollut., 10, 865 (1966).

Van Haecht, J.L., M. Bolipombo, and P.G. Rouxhet, Biotech. Bioeng., 27, 217 (1985).

Veelken, M., and H. Pape, Appl. Microbiol. Biotech., 19, 146 (1984).

Venkatasubramanian, K., and S.B. Karkare, in "Immobilized Cells and Organelles," Vol. 2, ed. by B. Mattiasson, p. 133, CRC Press, FL (1983).

Vieth, W.R., and K. Venkatasubramanian, Meth. Enzymol., 44, 243 (1976).

Vieth, W.R., and K. Venkatasubramanian, ACS Symposium Series, 106, 1 (1979).

Vijayalakshmi, M., A. Marcipar, E. Segard, and G.B. Broun, Annals New York Academy of Sciences, 326, 249 (1979).

Vorlop, K.-D., and J. Klein, in "Enzyme Technology," ed. by R.M. Lafferty, p. 219, Springer-Verlag (1983).

Wada, M., T. Uchida, J. Kato, and I. Chibata, Biotech. Bioeng., 22, 1175 (1980).

Wang, D.I.C., paper presented at the 9th International Symposium on Chemical Reaction Engineering, Philadelphia, May 18-21 (1986).

Williamson, K.J., and P.L. McCarty, paper presented at the WPCF Meeting, Cincinnati, OH (1973).

Young, D.H., H. Kohle, and H. Kauss, Plant Physiol., 70, 1449 (1982).

Chapter 8

Aerobic Biological Wastewater Treatment

The fluidized bed has emerged in recent years as one of the most promising devices for biological wastewater treatment. In such a device, organic or inorganic pollutants in wastewater are removed by microbes immobilized on the surface of fluidized particles. The particles coated with biofilm are referred to here as bioparticles. The major advantages of fluidized bed bioreactors[1] over other types of bioreactors such as activated sludge aeration tanks, trickling filters, rotating biological contactors, and submerged filters for wastewater treatment come from (e.g., Jeris and Owens, 1975; Scott and Hancher, 1976; Klein and Lee, 1978; Sutton et al., 1979a,b; Barbara et al., 1980; Stathis, 1980; Atkinson, 1981a; Vanderborght and Gilliard, 1981; Kargi and Park, 1982; Rao et al., 1984; Konishi, 1985; Andrews, 1986):

1) a low wash out rate of microbes from the system;

2) a minimum sludge recycle and biomass recovery at a high substrate loading;

3) no clogging of the system due to biomass accumulation; and

4) a high and variable biomass concentration and large solid-liquid contacting area.

Consequently, fluidized bed bioreactors yield high overall pollutant removal rates and low generation rates of excess sludge per unit reactor volume.

This chapter presents factors underlying the optimal operation and design of fluidized bed processes for aerobic biological wastewater treatment. Merits and limitations of different configurations of fluidized bed bioreactors, desired particle properties, cultivation and thickness control of biofilms, biomass concentration in bioreactors, interaction between microbial reactions and substrate diffusion in biofilms, steady state multiplicity in biofilms and

1: As in Chapter 7, a general term "fluidized bed bioreactor" (FBB) is used throughout this chapter to represent various modes of fluidization system operation in aerobic wastewater treatment applications. More specific designations are presented as the various systems are described in the chapter.

bioreactors, and modeling of a fluidized bed bioreactor are examined in detail. Finally, power economy for fluidized bed bioreactor operation relative to other types of bioreactor operation is indicated.

8.1 EARLY WORK

The use of immobilized living cells on particle supports in a suspended bed, and an expanded bed or a fluidized bed for wastewater treatment was conceived as early as the 1930's (Miller, 1981). Engineering work utilizing living cells immobilized on particles in a mobile state, or more precisely a fluidized bed bioreactor (FBB), as a sole means for biodegradation in wastewater treatment was not attempted until the late 1960's. Among various early investigators, W.J. Weber Jr. and coworkers (Freidman et al., 1971; Weber et al., 1973; Weber, 1978) studied the physicochemical treatment of raw wastewater and/or primary effluent using granular activated carbon in a fluidized bed adsorber. They observed that bioactivity on the carbon surface contributed to the reduction of biochemical oxygen demand (BOD) and total organic carbon (TOC) in the adsorber. They demonstrated an important improved wastewater treatment process involving bacterial growth on activated carbon particles in a fluidized bed and reported that BOD and TOC removal using activated carbon supporting media is substantially higher than that using non-activated carbon supporting media. In the study of a nitrification process utilizing a fluidized bed of activated carbon particles and high-purity oxygen, an increase in treatment capacity was observed due to biological growth on the carbon particles (Biospherics Research Inc. 1969). The benefits of using an expanded or fluidized bed over a fixed bed for denitrification were recognized by Beer (1970) and Beer et al. (1972). Short (1973) treated ammonia wastewater using a fluidized sand bed and found that the ammonia removal efficiency depends on the total sand concentration in the bed. Besik (1973) has also noted an increase in the BOD removal efficiency in a fluidized bed of activated carbon over the conventional biological treatment process. Work on fluidized bed bioreactors for wastewater treatment has mushroomed since 1973. Brief accounts of historical development of fluidized bed bioreactor application in wastewater treatment were cited by Miller (1981), Opplet and Smith (1981), and Hosaka and Hirata (1986a,b).

Biological wastewater treatment using fluidized beds involve either aerobic or anaerobic operation. The former is conducted for removal of organic pollutants such as BOD (biochemical oxygen demand) and COD (chemical oxygen demand) or for nitrification in the presence of oxygen, while the latter is operated to remove organic pollutants of high strength in the absence of oxygen, such as methane fermentation or denitrification processes. The FBB has been used extensively in anaerobic applications due to operational simplicity and energy conservation requirements. Some technological assessments of anaerobic FBBs for wastewater treatment are available in Jewell (1982), U.S. EPA (1982) and in books edited by Stafford et al. (1979) and Cooper and Atkinson (1981). Relatively limited information concerning aerobic FBBs is available (e.g., Cooper, 1980, Cooper and Wheeldon, 1980, Cooper and Atkinson, 1981, Baker et al., 1981; Andrews, 1982; Atkinson and Mavituna, 1983; Andrews, 1986; Nakamura et al., 1986). Specifically, fundamental information accounting for complicated interactions among biological reactions, reactor hydrodynamics, and mass transfer for aerobic wastewater treatment using fluidized bed bioreactors is lacking.

8.2 PROCESS CONSIDERATIONS

The high gas-liquid mass transfer rates of three-phase fluidized beds can be used advantageously in aerobic wastewater treatment to supply necessary oxygen. Various reactor designs are available to allow tailoring of the process to a given application. Inherently related to the reactor operation are the properties of the particles necessary to support the biofilms. In this section, existing reactor configurations, characteristics of biofilm support particles, and typical operating conditions are discussed.

8.2.1 Reactor Types

Four types of fluidized bed bioreactor operations commonly used in aerobic wastewater treatment are shown in Fig. 8.1(Fujie and Fan, 1987). For Type I - III bioreactors, both pollutant removal and aeration occur simultaneously in the bioreactor. Type IV bioreactors, however, use a preaerator to enrich oxygen concentration in the feed wastewater and a liquid-solid fluidized bed bioreactor for pollutant removal.

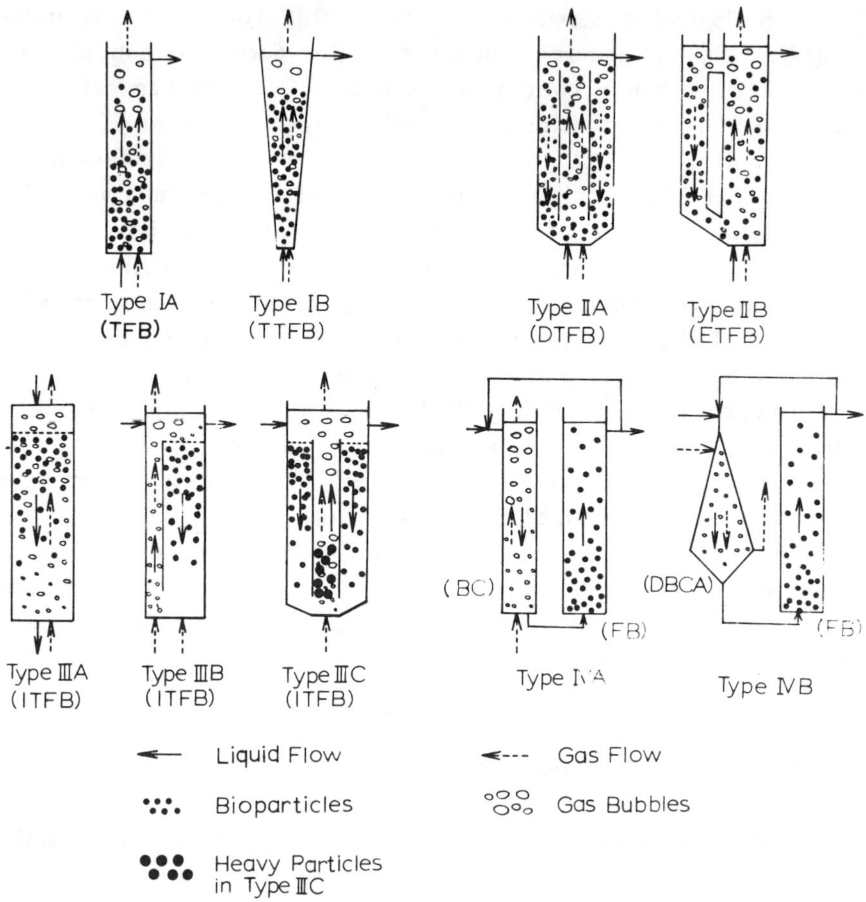

Figure 8.1 Various configurations of fluidized bed bioreactors for aerobic wastewater treatment (from Fujie and Fan, 1987).
BC: bubble column;
FB: liquid-solid fluidized bed;
TFB: three-phase fluidized bed;
DBCA: downward flow bubble contact aerator;
DTFB: draft tube three-phase fluidized bed;
ETFB: external circulating three-phase fluidized bed;
ITFB: inverse three-phase fluidized;
TTFB: tapered three-phase fluidized bed

8.2.1.1 Conventional (TFB) and Tapered (TTFB) Three-Phase Fluidized Beds

TFBs, denoted as Type IA in Fig. 8.1, consist of a straight

constant diameter column (Shen and Wang, 1980; Qian, 1980). TTFBs, denoted as Type IB (also see Fig. 7.13-a), on the other hand, consist of a tapered column with a small diameter at the bottom and large diameter at the top (Scott and Hancher, 1976; Holladay et al., 1978a,b and Lee et al., 1979). In both TFBs and TTFBs, bioparticles are supported by upward flows of gas and/or liquid (Mode E-I-a-1)[2]. For TTFBs, the bioparticles are stratified with low concentrations at the bottom and high concentrations at the top; whereas, for TFBs, the particle concentrations are relatively uniform. Transport properties described in Chapters 2 and 3 are valid for TFB systems as applied to aerobic wastewater treatment. For tapered fluidized beds, however, information on transport properties is limited to liquid-solid systems (Koloini and Farkas, 1973; Maruyama et al., 1984; Shi and Fan, 1984; Webster and Perona, 1988); little is known regarding gas-liquid-solid systems.

8.2.1.2 Draft Tube Fluidized Beds (DTFB)

DTFBs, denoted as Type IIA in Fig. 8.1, consist of a draft tube concentrically located inside the bed. DTFBs have been used on a commercial scale for wastewater treatment applications (Kataoka et al., 1979; Suzuki 1986; Suzuki and Tsunoda, 1986). Figure 8.2 is a view of a commercial size (4m (width) × 7m (length) × 9m (height) × 3 (cells)) DTFB located at The Osaka plant of the Takasaki Paper Manufacturing Co., Ltd., Japan. The draft tube is of 0.4m ID and 6.75m height. There are four draft tubes placed in each one of the three cells in the bioreactor. The bioreactor is licensed under the name Biolex by the Ebara Infilco, Co., Ltd. and has been in operation since 1981. It treats paperboard mill wastewater at a throughflow of 2,000 m^3/day (also see Section 1.4.1).

In the external circulating three-phase fluidized bed (ETFB), denoted as Type IIB, bioparticles carried by the continuous liquid phase are circulated via the air lift action between the main column and the external column located adjacent to the main column (Suzuki et al. 1983). The main column behaves as in Mode T-I-a-1 or T-I-a-2 (see Fig. 1.3 for mode designations). Transport properties described in Chapters 4 and 6 are valid for DTFB systems and the main column of ETFB systems, as applied to aerobic wastewater treatment.

2: The mode designation for tapered fluidized beds is based on the criterion of $U_t > 5$ cm/s.

Figure 8.2 A commercial size operation of Ebara Infilco's draft-tube fluidized bed bioreactor (Type IIA) for wastewater treatment (kindly supplied by K. Suzuki of Ebara Research Co., Japan).

8.2.1.3 Inverse Fluidized Beds (ITFB)

ITFBs, denoted as Types IIIA, IIIB, and IIIC in Fig. 8.1 (also see Fig. 7.13-b), were first applied to biological wastewater treatment by Edwards (1981), Shimodaira et al. (1981), and Nikolov et al. (1982). In Type IIIA bioreactors, liquid and gas flow countercurrently and gas (air) is introduced from the bottom (Mode E-II-a-1). In Type IIIB bioreactors, particles in the downcomer are fluidized by the downward circulative liquid flow created by the air lift action in the riser. In Type IIIC bioreactors, heavy and large particles in the draft tube are fluidized by an upward liquid flow, while light and small particles to which microbial films are attached are inversely fluidized by the downward liquid flow in the annulus.

Figure 8.3 is a view of a prototype size inverse fluidized bed bioreactor (Type IIIB) located at the Chiyoda Research Center in Yokohama, Japan. The bioreactor is of 1.33 m ID and 5 m height. It is used as a prototype unit to treat kitchen wastewater at a capacity of

Figure 8.3 A prototype of Chiyoda's inverse fluidized bed bioreactor (Type III-B) for wastewater treatment (kindly supplied by Y. Yushina of Chiyoda Chemical Engineering and Construction Co., Japan).

6 m^3/hr. For Type IIIC bioreactor operation, increasing biofilm thickness reduces the density difference between the bioparticles and the liquid, thus causing entrainment of the bioparticles into the draft tube. Inside the draft tube, extensive abrasion among the heavy particles and the entrained light bioparticles removes excess biofilm, providing a delicate means to control biofilm thickness (Chavarie et al., 1985; Chavarie and Karamanev, 1986; Karamanev and Nikolov, 1988). Since both the heavy particles and the light bioparticles have to be fluidized appropriately, the operating conditions used in Type IIIC bioreactors are more restricted than in other types. For example,

even for systems involving only low oxygen demand for biodegradation, excess air flow is required in a Type IIIC bioreactor just to fluidize both types of particles. Transport properties described in Chapter 5 are valid for Type IIIA (ITFB) systems as applied to the aerobic wastewater treatment.

8.2.1.4 Two-Phase Systems

Two-phase bioreactors, denoted as Types IVA and IVB, are the common bioreactors used in aerobic wastewater treatment and are included in Fig. 8.1 for comparison with the three-phase bioreactors. In Type IVA and IVB bioreactors, liquid is first oxygenated in a preaerator before being introduced into the liquid-solid fluidized bed. The extent of fluidization is moderate and creates only moderate shear stress in the bed. A conventional bubble column (BC) and a downward flow bubble contact aerator (DBCA), which have high oxygen transfer rates into the liquid phase, are used as preaerators for Type IVA and Type IVB FBBs, respectively. The dissolved oxygen (DO) concentration in the oxygenated liquid ranges from 8 to 9 mg/ℓ for air injection up to a maximum of 40 to 50 mg/ℓ for pure oxygen or oxygen enriched air injection at ambient pressure and temperature. The "Oxitron Process" developed by Dorr Oliver, Inc., (e.g., Sutton et al., 1979a, 1979b, 1981, 1984; Sutton, 1981) based on a patented process (the "Hy-Flo" process) licensed by Ecolotrol, Inc. (e.g., Jeris et al., 1977; Barbara et al., 1980; Stathis, 1980; Jewell, 1981) is representative of Type IV bioreactors.

8.2.1.5 Reactor Combination and Recycle Strategies

For nitrogen removal from domestic wastewater, Nakamura et al. (1986) effectively used an anaerobic liquid-solid fluidized bed followed by an aerobic three-phase fluidized bed (TFB). In this case, nitrates and nitrites produced in the aerobic stage are recycled to the anaerobic stage to be reduced to nitrogen gas. The anaerobic stage uses influent BOD (organic carbon) as the sole electron donor source; any effluent BOD from the first stage can be removed aerobically in the second stage (e.g., Melcer et al., 1984; Nakamura et al., 1986).

Note that in Type I, IIIA, and IV FBBs, the bioreactor effluent is often recycled to the bioreactor or to the preaerator to achieve a high pollutant conversion rate (e.g., Lee et al., 1979; Qian, 1980).

8.2.2 Particle Properties

In wastewater treatment, as in fermentation, cells are immobilized on support particles. Immobilization in wastewater treatment applications occurs mainly through adsorption (see Section 7.2.2). Matrix entrapment methods (see Section 7.2.3) have also been attempted, although no commercial scale wastewater treatment facilities have used matrix entrapped microbes to this date.

Desirable properties for particles (supports) used in a FBB wastewater treatment are (Atkinson et al., 1981b)

1) minimum size and shape variation;
2) appropriate specific gravity for fluidization;
3) appropriate surface conditions for cell immobilization;
4) abrasion resistant; and
5) low cost.

Particles with low terminal velocities experience problems of elutriation from the bioreactor and increase the loading for liquid-solid separation. Particles with high terminal velocities, on the other hand, have difficulty in cell immobilization (Fan et al., 1986) or in re-fluidization of a bed de-fluidized due to power failure during operation (Shimodaira and Yushina, 1983).

Activated carbon (A.C.), anthracite, and sand particles are the most widely used carriers for microbe attachment in wastewater treatment. Activated carbon is especially advantageous for microbe attachment. Its adsorptive properties enrich substrate concentrations at the carbon-liquid interface. The variety of functional groups on the surface enhance microbe attachment, and its craggy surface shelters attached microbes from fluid shear (Weber et al., 1978). Furthermore, the large adsorption capacity of activated carbon for organic pollutants greatly enhances the overall pollutant removal efficiency and the operational stability against pollutant shock loadings (Tang and Fan, 1985). Examples of application of granular or spherical activated carbon particles to an integrated biological-adsorption fluidized bed wastewater treatment process are abundant (e.g., Tsunoda et al., 1976; Tsunoda, 1977; Sawada, 1978; Tanemura et al., 1976; Tanemura and Morita, 1979; Ying and Weber, 1979; Tsezos and Benedek, 1980a,b; Andrews and Tien, 1981, 1982; Peel and Benedek, 1983; Suzuki et al., 1983; Wang and Tien, 1984a,b). Notably, Kureha Beads BAC, which are nearly spherical activated carbon particles with a remarkable abrasion resistance, are especially suitable for use as the immobilizing carrier in fluidized bed bioreactors and have been widely used (Hirata et al., 1982; Hosaka et al., 1983, 1985; Suzuki et al., 1983; Fan et al., 1984a,b; Fujie et al., 1985). Specifically, these particles can sustain the high shear stresses which promote the removal of excess biofilm, and can thus be used for bioparticle size control without a particle disintegration problem.

Sand particles of a small diameter were widely employed in pilot scale and commercial scale DTFBs (Kataoka et al., 1979; Suzuki et al., 1981; Sawai et al., 1982; Suzuki, 1986), solid-liquid FBBs of Types IVA and IVB (Jeris et al., 1977; Denac et al., 1983; Hoyland and Robinson, 1983; Melcer et al., 1984) and three-phase FBBs of Type IA (Shen and Wang, 1980; Qian, 1980; Takase et al., 1983). Sand particles also have a high abrasion resistance (Shich et al., 1981a,b; Cooper et al., 1981; Wheeldon and Bayley, 1981). Other particles such as anthracite (Scott and Hancher, 1976; Holladay et al., 1978a,b; Lee et al., 1979; Donaldson et al., 1984), porous alumina (Scott and Hancher, 1976), crushed brick (Chatib et al., 1981; Grasmick et al., 1981), and coke and fly ash (Zhang et al., 1981) have also been used.

In inverse three-phase fluidized beds, polypropylene cylinders with a specific gravity of 0.86 were used as immobilizing medium for industrial and domestic wastewater treatment (Shimodaira and Yushina, 1983; Yushina and Sato, 1983). The minimum downward fluidization velocity of these cylindrical particles (3 mm in diameter and 3.6 mm long) is 1 cm/s. Spherical polyethylene and polypropylene particles of sizes 4 to 10 mm and of densities 0.39 to 0.93 g/cm^3 were utilized by Edwards (1981) for phenol degradation. Granulated polyethylene particles of sizes 2 to 3 mm and of density 0.92 g/cm^3 were utilized by Chavarie et al. (1985) for ferrous sulfate degradation.

Tanemura and Morita (1979) and Walker and Austin (1981) assessed the feasibility of using plastic porous supports (e.g., polypropylene toroids and pads) and water adsorptive porous plastic (a copolymer of acrylonitrile and methylacrylate) for high rates of wastewater treatment. A large amount of anaerobic sludge was found inside the toroids, thus a low BOD loading is required to achieve reasonable BOD removal when using these supports. Atkinson and Lewis (1980) produced highly porous particles from knitted fine stainless steel wires via die-processing. These nearly spherical stainless steel wire particles, called Biomass Support Particles (BSP), have a porosity of 80% in which cells can attach and grow. The outgrowth of cells on the particle surface is largely prevented by interparticle abrasion.

Matrix entrapment immobilization methods can use a variety of support materials tailored to the application. κ-carrageenan has been used for denitrification (Shidara et al., 1983), polyacrylamide for BOD removal (Sumi et al., 1985) and nitrification (Mori et al., 1984), polyvinyl alcohol for BOD removal (Hashimoto et al., 1985, 1986), photo-cross-linkable resin for denitrification (Mishima and Watanabe, 1986), polyacrylamide-hydrazide for phenol biodegradation (Bettmann and Rehm, 1984), and calcium alginate for nitrification (Tramper et al., 1985). The support materials used in matrix entrapment, in addition to properties (1), (2), (4), and (5) mentioned previously, should have a high substrate diffusivity and no inhibitory effect on microbial activities, be nonbiodegradable, and allow for easy microbe entrapment. Sumi et al. (1985) suggested that polyacrylamide shows the best properties based on the criteria given above.

Additional information on the characteristics of solid supports and biomass support particles for use in fluidized beds can be found in Atkinson et al. (1979, 1981b) and Andrews and Trapasso (1985).

8.2.3 Operating Conditions

The application of different types of fluidized bed bioreactors to aerobic wastewater treatment are listed in Table 8.1. These reactors range from bench to commercial production scales with volumes ranging from 2.35 ℓ to 342 m^3. Organic chemicals or pollutants treated include domestic and industrial sewages, glucose, phenol, isobutanol, acetic acid, lactose and thiocyanate. This table also includes the operating conditions, the biomass concentration, and the performance of each reactor expressed in terms of the overall pollutant removal rate.

Table 8.1 summarizes the range of operating conditions and overall pollutant removal rates for each type of fluidized bed bioreactor listed in Table 8.1. As can be seen in these two tables, the volumetric BOD loading for either municipal sewage or industrial wastewater treatment in each type of FBB shown in Fig. 8.1 can reach about 15 kg-BOD/m^3•day. A biomass concentration in the bioreactor as high as 40 g/ℓ was obtained in a conventional three-phase fluidized bed (Takase et al., 1983). The highest biomass concentration (15 to 40 g/ℓ) and volumetric BOD loading in any type FBB shown in Fig. 8.1 are about 10 to 20 times higher than those in conventional activated sludge aeration tanks (Fujie et al., 1983; Fujie and Kubota, 1986).

In a Type IVB FBB, it was found that, under steady state BOD loading operation, the effluent BOD concentration achieved was

Table 8.1 Summary of Operating Conditions of Fluidized Bed Bioreactors for Aerobic Wastewater Treatment Employed by Various Investigators (from Fujie and Fan, 1987).

a) Three-Phase Fluidized Bed (Type IA) and Tapered Three-Phase Fluidized Bed (Type IB)

Author(s) (year)	Type of bio-reactor	Dimensions of the bioreactor	Solid particle	Wastewater (substrate)	Operational conditions	Loading and Removal Rate		Bioparticle	Comments
Qian (1980)	IA	D_c=14.5cm H=300cm V=49.5ℓ	Quartzite sand d_o=280~900μm	Slaughter house wastewater S_i=97.2~300.5 ppm-BOD S_i=226~576 ppm-COD	Q=131~148ℓ/hr $U_ℓ$=61~63.5m/hr γ=6.4~6.67	W_o=4.95~14.22 kg-BOD/m^3·day	x_f=60.6~82.8%BOD x_f=54.9~72.7%COD	X_m=6.2~7.7 g/ℓ	Liquid injection type air diffuser
Shen and Wang (1980)	IA	D_c=8cm H=295cm V=14.8ℓ	Quartzite sand $\bar{d}_o$=557μm Packed height: L_H=60cm	Municipal sewage S_i=82~199.6 ppm-BOD S_i=199.2~578.2 ppm-COD	Q=12.9~24ℓ/hr θ=32~51.6min	W_o=4~6 kg-BOD /m^3·day W_o=10~12kg-COD /m^3·day	x_f=70~86.2%BOD x_f=65~77.3%COD		Single nozzle air diffuser
Chatib et al. (1981) Grasmick et al. (1981)	IA	D_c=9.4cm H=200cm V=14ℓ	Crushed brick $\bar{d}_o$=830μm ρ_s=1.7g/cm^3 U_{mf}=14m/hr	Artificial wastewater (isobutanol + acetic acid) S_i=40ppm-COD	$U_ℓ$=12.7~31.8 m/hr U_g=12.7~38.2 m/hr γ=0	W_o=10kg-BOD/ m^3·day	x_f=6.8~30%		Y_o=0.714 g-O_2/g-COD Influent DO=6.6~8.9ppm
Hirata et al. (1982, 1986) Hosaka et al. (1985)	IA	D_c=5cm H=240cm V=8.1ℓ	Activated carbon particle d_o=272μm L_s=11.2vol% (74g-AC/ℓ-reactor)	Artificial wastewater (phenol) S_i=100~120ppm	Q=14.4~17.3ℓ/hr $U_ℓ$=7.3~8.8m/hr U_g=9m/hr γ=0	W_o=4~6kg/m^3·day	x_f=100%	δ=50~100μm (with mechanical stirring)	DO=2~4ppm T=30°C

Takase et al. (1983) (1987)	IA	D_c=5.4cm H=138.5cm V=3.1ℓ	Sand $\bar{d}_o$=390μm ρ_s=2.665 g/cm^3 L_s=9.15vol%	Artificial wastewater (glucose in BOD) S_i=80ppm	U_ℓ=0.22~0.5cm/s U_g=0~0.11cm/s (air bubbling) U_g=0~0.25cm/s (O_2 bubbling)	W_o=12kg-BOD/ m^3·day	x_f=87%BOD	δ=180~200μm X_m=30~40g/ℓ	Y_o=0.6g-O_2/g-BOD F/M=0.3g-BOD/ g-sludge·day
Scott and Hancher (1976)	IB	Bottom: D_c=2.54cm Top: D_c=7.6cm H=107cm V=2.35ℓ	Porous alumina $\bar{d}_o$=100μm (for lactose conversion) Anthracite coal d_o=148~177 μm (for denitrification)	Lactose S_i=50,000 ppm	Q=1.8~21.6ℓ/hr		Lactose conversion x_f=32.5~62.5%	Immobilized lactase on alumina particle	
Holladay et al. (1978a,b)	IB	Bottom: D_c=2.54cm Top: D_c=7.6cm H=107cm V=3.0ℓ and 10ℓ	Anthracite coal d_o=149~177 μm L_s=500~700g-dry solid/reactor	Artificial wastewater (phenol) S_i=45~2200ppm	Q=0.78~48ℓ/hr θ=0.21~3.85 hr G_v=3ℓ-air/min	Air bubbling W=4.04~16.66 kg-phenol/m^3·day O_2 bubbling W=1.90~21.17 kg-phenol/ m^3·day	x_f=56~99% x_f=87~99%		T=21°C Seeding: Bethlehem sludge or phenobac (mutant <u>Pseudomonas</u>) The minimum air flow: 38ℓ-air/g-phenol degraded Partly preaerator has been used.
Klein et al. (1978)	IB	Same as above	Coal Sand d_o=250~500μm	Coal conversion wastewater (artificial) Phenol S_i=200ppm	Q=0.5ℓ/min				
Klein et al. (1978)	IB	Bottom: D_c=10.2cm Top: D_c=15.2cm H=450cm	Coal d_o=250~500μm Alumina d_o=250μm	Same as above	Q=4ℓ/min	W =20kg-phenol/ m^3·day			

(Table 8.1 continued)

Author(s) (year)	Type of bio-reactor	Dimensions of the bioreactor	Solid particle	Wastewater (substrate)	Operational conditions	Loading and Removal Rate		Bioparticle	Comments
Lee et al. (1979)	IB	Bottom: D_c=1.54cm Top: D_c=7.6cm H=107cm V=3.0ℓ and 10ℓ (involving sand-liquid separator)	Anthracite coal d_o=149~177μm and d_o=417~701 μm L_s=1ℓ-solid/reactor	Artificial wastewater (phenol)	Q=17.1~42ℓ/hr γ>0 τ=5.6~15.0 min	Air bubbling in column W=5.3~15.3	S_i=14~140ppm x_f=29~99.6%		Porous metal frit gas diffuser Air or O_2 bubbling Seeding: Phenobac (mutant <u>Pseudomonas</u>) T=25±2°C pH=6.5~7.4
						Feed stream saturated with oxygen W=3.6~22.7 kg/m^3·day	S_i=9~63ppm x_f=68~100%		
						O_2 bubbling in column W=7.9~51.0 kg/m^3·day	S_i=48~260ppm x_f=85~100%		
				Artificial waste water (thiocyanate)	Q=1.8~4.98ℓ/hr τ= 11.7~13.6 min	W=0.71~1.54 kg-thiocyanate/m^3·day	S_i=50~170ppm x_f=30~96%		T=25±2°C Seeding: mutant <u>Pseudomonas</u>
				Coal hydrocar-bonization waste (phenol and thiocyanate)	Q=18±0.6ℓ/hr θ=15 min	Phenol: W=2.53~10.4 kg/m^3·day	S_i=35~135ppm x_f=76~99%		T=22±2°C pH=6~7.5 Seeding: mutant <u>Pseudomonas</u>
						Thiocyanate: W=0.03~0.8 kg/m^3·day	S_i=5~24ppm x_f=1.7~83%		
				Artificial waste water and coal hydrocarboniza-tion wastes	Q=24.3~35.5ℓ/hr τ= 7.6~11.1 min	Phenol: W=3.7~16.8 kg/m^3·day	S_i=19.4~184ppm x_f=49~99.9%		T=22±2°C pH=6~7.5 Seeding: Activated sludge
						Thiocyanate: W=0~1.2 kg/m^3·day	S_i=3.5~150ppm x_f=0~84%		

(Table 8.1 continued)

b) Draft-Tube Three-Phase Fluidized (Type IIA) and External Circulating Three-Phase Fluidized Bed (Type IIB)

Author(s) (year)	Type of bio-reactor	Dimensions of the bioreactor	Solid particle	Wastewater (substrate)	Operational conditions	Loading and Removal Rate		Bioparticle	Comments
Sawada (1978)	IIA	D_c=30cm D_T=10cm H=250cm V=180ℓ	Activated carbon particle 20~40 mesh L_s=20 vol%	Domestic waste (BOD) S_i=30~150ppm	θ=0.58~1.5hr	W_o=0.5~5.6 kg-BOD/m^3·day	x_f>90%BOD S_e= 10ppm	X_m=6~7g/ℓ δ<500μm	F/M=0.16~0.71kg-BOD/kg-sludge·day Y_x=0.6~0.65 g-cell/g-BOD
Kataoka et al. (1979)	IIA	D_c=40cm H=700cm V=870ℓ	Silica sand d_o=0.2~0.35mm L_s=20vol%	Domestic waste (BOD) S_i=40~250ppm	θ=0.5~2.17hr	W_o=1.4~12 kg-BOD/m^3·day	x_f>90% BOD for W_o<9kg/m^3·day	X_m=14~15 g/ℓ 35~40mg-VSS/g-sand	DO=2~3ppm 0.1~0.8kg-BOD/kg-cell·day T=14~15°C
Kataoka et al. (1979)	IIA	D_c=280cm H=750cm V=34x10^3ℓ	Silica sand	Night soil (effluent from anaerobic digester, BOD) S_i=1370 ppm	U_g=9.7~14.0m/hr	W_o=3.5kg-BOD/ m^3·day	S_e=27ppm-BOD		DO=0.8~1.0ppm
Suzuki et al. (1981)	IIA	D_c=50cm H=260cm V=510ℓ	Sand d_o=100~400μm L_s=5~15vol%	Municipal sewage Effluent of septic tank Photo processing wastes Food processing wastes Kraft pulp waste	θ=0.5hr θ=2 hr θ=6.3 hr θ=2.4 hr θ=45 hr	W_o=8kg-BOD/m^3·day W_o=2.5 W_o=4.5 W_o=10.4 W_o=7.8	S_i=250,S_e=12ppm-BOD S_i=140,S_e=6 S_i=1190,S_e=3.6 S_i=994,S_e=23 S_i=147,S_e=13.5	X_m=8.8g/ℓ X_m=6.3g/ℓ X_m=5.4g/ℓ X_m=10.4g/ℓ X_m=18.8g/ℓ	52.6~98.2mg-VSS/g-sand
Zhang et al. (1981)	IIA	D_c=18.5cm D_T=4.0cm H=30cm V=8.0ℓ	Coke	Municipal sewage (BOD) S_i=50~72ppm	Q(ℓ/hr)=3.6~5.4 θ(hr)=0.37~0.56	W_o= 2.2~3.1 kg-BOD/m^3· day	x_f=88~93%		T=12~14°C
			River sand	S_i=100~319ppm	Q=3.2~5.4 θ=0.37~0.62	6.0~14.0	x_f=89~95%		
			Fly ash	S_i=180~242ppm	Q=3.6 θ=0.56	9.0~10.4	x_f=86~92%		

(Table 8.1 b) *continued)*

Author(s) (year)	Type of bio-reactor	Dimensions of the bioreactor	Solid particle	Wastewater (substrate)	Operational conditions	Loading and Removal Rate		Bioparticle	Comments
Sawai et al. (1982)	IIA	D_c=50cm H=300cm V=500ℓ	Sand d_o=120~320μm L_s=10vol%	Fish processing wastewater (BOD) S_i=1870(790*)ppm *Filtrated sample	γ=0	W_o<10kg-BOD/m^3·day W_o=10~15kg-BOD/m^3·day	S_e <20* ppm-BOD S_e =20~50* ppm-BOD *Filtrated sample	X_m=6g/ℓ(L_s=3~4 vol%) 120mg-cell/g-sand	T=20~25°C Y_x=0.2~0.5 g-cell/g-BOD at 3~10kg-BOD/m^3·day
Fan et al. (1984b, 1987) Tang and Fan (1985) Tang et al. (1985)	IIA	D_c=7.6cm D_T=5.0cm V=5.5ℓ	Activated carbon particle $\bar{d}_o$=714μm L_s=3vol%	Artificial wastewater (phenol) S_i=100~300ppm	Q=3.1~9.8ℓ/hr U_g=18~65m/hr	W_o=0.5~12.8 kg-phenol/m^3·day	x_f=100%(W_o< 5kg-phenol/m^3·day)	δ<1000μm X_v=3.5~21 g/ℓ	DO=4~7ppm T=25±1°C
			$\bar{d}_o$=307μm L_s=7.2vol%		Q=14~22ℓ/hr U_g=65~130m/hr	W_o=6.5~21.1 kg-phenol/m^3·day	x_f=100%(W_o< 18kg-phenol/m^3·day)	δ<100μm X_v=20~180g/ℓ X_m=21 g/ℓ	DO=1.5~5ppm T=25±1°C
Suzuki (1986) Suzuki and Tsunoda (1986)	IIA	400(W)x700(L)x 900(H), in cm V=114m^3 x 3 (cells) = 342x10^3ℓ Draft tube: 40(φ)x675(H) x 4 (tubes)/reactor, in cm	Sand d_o=150~350μm L_s=10 vol%	Paper mill waste BOD=1000ppm COD=1000ppm SS=50ppm	Q=2000m^3/day θ=4.1hr	W_o=5.85 kg-BOD/m^3·day	x_f>80% BOD S_e<200ppm-BOD	X_m=10g/ℓ X_v=75~90 g/ℓ δ=130~250μm	F/M=0.56 kg-BOD/kg-MLVSS·day Biolex process
Suzuki et al. (1983)	IIB	D_c=8cm D_T=4cm* H=103.5cm V=6.3ℓ *external tube	Activated carbon particle $\bar{d}_o$=800μm* * estimated value	Artificial wastewater (phenol) S_i=147~202ppm	Q=6.3ℓ/hr θ=60min	W_o=3.5~4.85 kg-phenol/m^3·day		δ=620μm* *Estimated Value	Y_o=1.231g-O_2/g-phenol Y_x=0.704g-sludge/g-phenol F/M=1.44g-phenol/g-sludge·day

(Table 8.1 continued)

c) Inverse Three-Phase Fluidized Bed (Type IIIA and IIIB)

Author(s) (year)	Type of bio-reactor	Dimensions of the bioreactor	Solid particle	Wastewater (substrate)	Operational conditions	Loading and Removal Rate		Bioparticle	Comments
Shimo-daira and Yushina (1983)	IIIA and IIIB	Reactor: 160(W) ×180(L)×180(H), in cm $V=2.0\times10^3$ ℓ	Polypropylene 3mm(ϕ)×3.0mm(L) $\rho_s=0.86$ g/cm^3 $U_{mf}=39$m/hr $L_s=0.8$m^3 particles/reactor	Oil refinery wastewater (COD) $S_i=280$ppm	Q=500ℓ/hr	$W_o=1.9$kg-COD /m^3·day	$S_e=70$ppm-COD	$X_m=3\sim15$g/ℓ	T=26~32°C $Y_x=0.08\sim0.10$g-sludge/ g-COD
Hasegawa (1982)				Food processing wastewater (BOD) $S_i=810$ppm	Q=170ℓ/hr	$W_o=14.4$kg-BOD /m^3·day	$S_e=48$ppm	$X_m=10$g/ℓ	T=21~27°C $Y_x=0.65$g-sludge/g-BOD
		H=5.0m $V=6.9\times10^3$ ℓ	The same particle as above 3.8m^3 particle/ reactor	Kitchen waste-water (BOD) $S_i=160$ppm	Q=6m^3/hr θ=1.1hr $G_v=25$m^3(STP)/hr	$W_o=3.34$kg-BOD/ m^3·day	$S_e=30$ppm		T=20~24°C
Yushina (1985)	IIIB	500(L)×350(W)× 1400(H), in cm V=245 m^3	Cylindrical Polypropylene 3mm(Φ)×3mm(L)	Chemical plant wastewater $S_i=350$ ppm-BOD	Q=1000 m^3/day	$W_o=1.4$kg-BOD/ m^3·day	$S_e<20$ppm-BOD		Pilot plant study

(Table 8.1 continued)

d) Liquid-Solid Fluidized Bed Following Preaerator

Author(s) (year)	Type of bio-reactor	Dimensions of the bioreactor	Solid particle	Wastewater (substrate)	Operational condition	Loading and removal rate		Bioparticle	Comments
Tanemura et al. (1976)	IVA	D_c=2.0cm H=130cm V=0.4ℓ	Activated carbon particle $\bar{d}_o$=0.4mm	Fish processing wastewater (BOD) S_i=838~926ppm	U_ℓ=14~16m/hr θ=1.3~4.6hr γ=0.8~4.5	W_o=5.1~15.3 kg-BOD/m^3·day	x_f=58.9~81.2%BOD		T=20°C
Tsunoda et al. (1976)	IVA	D_c=10cm H=200cm V=15.71ℓ	Activated carbon particle d_o=0.6~1.2mm L_H=100cm	Nitrification (effluent of activated sludge process, NH_4-N) S_i=6.5ppm	U_ℓ=18.7~22.7 m/hr θ=6~6.5min γ=0	W=1kg-NH_4-N/m^3·day 5.6~6.9kg-NH_4-N/m^3-AC*·day *Activated carbon particle		δ=500μm	T=25~30.5°C Influent DO =27~35ppm
Tsunoda (1977)	IVA		Activated carbon particle	Artificial wastewater (Glucose in BOD) S_i=26~73ppm(BOD)	γ=0	W=10.2kg-BOD/m^3·day		X_m=20~30g/ℓ	O_2 bubbling in preaerator
Tanemura (1979)	IVA	V=1000ℓ	Activated carbon particle L_s=32 vol%	Fish processing wastewater (BOD) S_i=1200ppm	U_ℓ=4~5m/hr	W_o=3kg-BOD/m^3·day	x_f=98%BOD	δ=50μm	S_i=460ppm-COD
Tsezos and Benedek (1980a,b)	IVA	D_c=10.2cm	Activated carbon particle $\bar{d}_o$=557μm L_H= 60 cm	Artificial S_i=15~50ppm	U_ℓ=14.82m/hr γ=0	W=0.1224~0.263 g-TOC/m^2-bioparticle·hr	x_f=10.5~79% S_e=3.9~44.4 ppm-phenol	δ<1200μm	T=24°C DO influent=4.5~45ppm Y_o=1.7g-O_2/g-phenol
Muroyama et al. (1985)	IVA	D_c= 6 cm H = 1 m V=3.6~3.8ℓ	Spherical activated carbon particle (Kureha BAC) d_o=375~500μm	Artificial wastewater (Glucose) S_i=200~1400 ppm-COD	Q=34ℓ/hr τ=10~120min	W=5~20kg-COD/m^3· day	x_f=20~95%		T=25°C
Jeris et al. (1977)	IVB	D_c=46cm H=366cm V=1,680ℓ	Silica sand $\bar{d}_o$=0.6mm	Domestic sewage; Primary effluent (BOD) S_i=203ppm	Q=59ℓ/min U_ℓ=68.4m/hr θ=16min γ=2.22	W_o=10.3kg-BOD/m^3·day	x_f=93%BOD S_e=15ppm-BOD		T=24°C Y_x=0.5g-cell/g-BOD Y_o=1.1g-O_2/g-BOD

				Nitrification (NH_4-N) S_i=19.1 ppm	Q=52ℓ/min $U_ℓ$=62.6m/hr γ=2.35	W_o=0.815kg-NH_4-N /m^3·day	x_f=99% NH_4-N S_e=0.2ppm-NH_4-N	X_m=8.5 g/ℓ	Y_x=0.2g-cell/g-NH_4-N Y_o=4.3g-O_2/g-NH_4-N
Cooper and Wheeldon (1981)	IVB	D_c=48cm H=418cm V=512ℓ	Sand d_o=390μm (300~500μm) L_H=102cm	Sewage (Anoxic reactor effluent, BOD) S_i=60~190ppm	$U_ℓ$=15~35m/hr Q=840ℓ/hr θ=36.6min γ=1.59	W_o=5.39kg-BOD/ m^3·day W_o=3.71kg-SS/ m^3·day	W=2.95kg-BOD/m^3·day W=1.69kg-SS/m^3·day S_e=23~95ppm-BOD	Effluent suspended solids: 20~195ppm	DO Influent = ~8ppm T=15.9~17.3°C(influent) T=15.9~20.0°C(effluent)
Environment Canada (1981) Nutt et al.(1981)	IVB	D_c=28.9cm H=427cm	Quartzite sand d_o=480μm L_H=3.9m	Municipal sewage (BOD) S_i=102~124ppm	Q=0.1625m^3/hr $U_ℓ$=34.2~48.6m/hr γ=3 effluent DO= 5ppm(av.) τ=10 min	W_o=7.34~8.93 kg-BOD/m^3·day	S_e=28~39ppm x_f=68.5~72.5%	X_m=8.1~13.9g/ℓ	Sand-Biomass Separation Unit: Wemco Model 6 Spiral Classifer & Helical conveyor (Hapman Helix) T=16°C
Shieh et al. (1981b)	IVB	S_a=0.37m^2(A) S_a=2.22m^2(B) H=4.6m(A,B) V=1.7m^3(A) V=10.7m^3(B)	Quartzite sand L_H=3.05~3.66m A:d_o=547~562μm B:d_o=317~414μm	A:High strength carbohydrate wastewater S_i=4000ppm-BOD B:Municipal sewage S_i=80ppm-BOD	A:Q=0.23m^3/hr γ=23.7~56 $U_ℓ$=15.4~35m/hr B:Q(1+γ)=44.77~ 90.91m^3/hr	A:W_o=13.0kg-BOD/ m^3·day B: not reported		A:X_m=22.5~29.3 g/ℓ sand conc. =302~399g/ℓ B:X_m=6.7~14.7g/ℓ sand conc. =31.5~224g/ℓ	X_v / δ: A:30~72g/ℓ / 150~163μm; B:39~75g/ℓ / 89~312μm
Hovland and Robinson (1983)	IVB	183(L)x122(W) x400(H), in cm V=8.9x10^3ℓ Aerator:30(Φ)x 300 (H), in cm	Sand d_o=250~550μm	Domestic sewage (BOD) S_i=151~176 ppm	Q=150~400m^3/day $U_ℓ$=27~41m/hr θ=0.5~1.4hr γ=4.4~9.3	W_o=2.7~8.0 kg-BOD/m^3·day	x_f=62~99%	δ=40~160μm X_m=7.9~17g/ℓ	T=13~16°C Y_o=0.8~1.2g-O_2/g-BOD Y_x=0.5~0.9g-sludge/ g-BOD F/M=0.22~1.01g-BOD/ g-sludge·day
Melcer et al. (1984)	IVB	D_c=29cm H=390cm V=210ℓ	Quartzite sand $\bar{d}_o$=480μm	Blast furnace blowdown plus coke plant waste (NH_4-N)	$U_ℓ$=36~50.4m/hr θ=3.5~17.5hr γ=53~270	F/M=0.006~0.039 g-TKN*/g-sludge·day *Total kjeldahl nitrogen	x_f=90~99% S_e<1ppm-NH_4-N	X_m=16.6~24 g/ℓ	T=24~31°C Combined anoxic (denitrification + COD removal) and aerobic (nitrification) process
Li et al. (1987)	IVB	D_c=2.6cm H = 200cm	Activated carbon particle $\bar{d}_o$=0.6cm	Slaughter house wastewater	τ=8.8~30.8 min	W_o=8.5~98.5 kg-BOD/m^3·day	x_f=71-94%	x_m=8-16g/ℓ	T=25±2°C

Table 8.2 Ranges of Operating Conditions for Various Types of Fluidized Bed Bioreactors Described in Table 8.1.

Reactor type	Particles (diameter)	Operational conditions	Removal rate or loading		Cell mass concentration and/or remarks
			Domestic sewage or industrial wastes	Artificial wastewater	
IA & IB	Sand (280~900μm) Crushed brick (830μm) Spherical activated carbon (272μm) Anthracite (149~701μm)	U_ℓ=7.3~68.4m/hr U_g=4~38.2m/hr θ=5.6~52.7 min γ=0~6.67	W_o=4~15.3 kg-BOD/m^3·day	Air bubbling: W_o=4~14.22kg-phenol/m^3·day O_2 bubbling: W_o=7.9~ 51kg-phenol/m^3·day	X_m=6.2~40 g/ℓ
IIA & IIB	Spherical activated carbon (307~1200μm) Sand (100~1000μm) Coke (400~1000μm) Fly ash (100~1000μm)	U_g=9.7~130m/hr θ=0.25~6.3 hr	W_o=3.5~15 kg-BOD/m^3·day	W_o=6.5~21.1kg-phenol/m^3·day	X_m=5.4~21 g/ℓ D_T/D_C=0.22~0.5
IIIA & IIIB	Polypropylene cylinder 3mm(φ)x3.6mm(L),3mm(φ)x3.0mm(L)	θ = 1.1~4 hr G_V/Q = 4.2 m^3-air/m^3-liquid	W_o=2.7~15.3 kg-BOD/m^3·day		X_m=7.9~15 g/ℓ
IVA & IVB	Activated carbon (400~1800μm) Sand (280~600μm)	U_ℓ = 4~68.4 m/hr θ = 6~17.5* hr γ = 0~270* *for coke plant wastes	W_o=1.9~14.4kg-BOD/m^3·day		X_m=3~15 g/ℓ

approximately linearly proportional to the volumetric process loading (i.e., mass feed rate of BOD per unit bioreactor volume) (Environment Canada, 1981; Nutt et al., 1981). Under unsteady state BOD loading operation, the response of effluent dissolved organic carbon (DOC) was a function of the influent DOC concentration, DOC loading, and flow in the process. In a study of the effect of dissolved oxygen and glucose concentrations on the aerobic biooxidation rate of glucose in a Type IVA FBB, Muroyama et al. (1984) observed that an oxygen deficiency occurred around the exit of the fluidized bed. Such an oxygen deficiency is frequently observed for FBBs with a very high biomass holdup, such as in Types IA and IB (Holladay et al., 1978a,b; Donaldson et al., 1984). Note that for a Type IVA FBB, the oxygen supply rate in the preaerator determines the maximum biooxidation rate.

8.3 BIOREACTOR PERFORMANCE

In general, the performance of a FBB depends on the active biomass holdup, the overall biodegradation rate, the reactor configuration, and the feeding rates of the pollutants and dissolved oxygen. However, depending on the operating conditions, any of the above factors may be controlling. In this section, biofilm properties, including cell concentrations in the biofilm and bioreactor, are discussed. Furthermore, biological reaction kinetics in conjunction with transport behavior in the bioparticle and bioreactor, cultivation and control of biofilms, and bioreactor feasibility are described.

8.3.1 Cell Concentrations

The cell concentration within a bioreactor depends on the characteristics of the biofilm attached to the support particles (cell density, film thickness, etc.) and the bioparticle holdup. Factors affecting the bioparticle holdup include bioparticle shape, density, and size, bioparticle loading, gas and liquid velocities, and bioreactor configuration.

Bioparticles can be produced through attachment of a biofilm on the surface of non-porous or nearly non-porous particles inside which growth of biomass does not occur (see Section 7.2), or by confining cells within the porous structure of highly porous particles such as alginate gel particles (see Section 7.2) and Biomass Support

Particles (BSP, see Section 8.2.2). The apparent density of a bioparticle (ρ_p) can be given by

$$\rho_p = (\rho_s - \rho_{bw}) (\frac{d_o}{d_p})^3 + \rho_{bw} \tag{8.1a}$$

for biofilm-attached non-porous or nearly non-porous particles or

$$\rho_p = \rho_m(1-\epsilon_p) + \rho_W \epsilon_c \epsilon_p + \rho_\ell(1 - \epsilon_c)\epsilon_p \tag{8.1b}$$

for cell-entrapped porous particles where ρ_s is the density of the non-porous particle or the effective density (= $\rho_m(1-\epsilon_p) + \rho_\ell \epsilon_p$) of the nearly non-porous particle; d_o is the diameter of a biofilm-free solid particle; ρ_m is the density of the matrix material of the porous solid particle; ρ_W is the wet cell density which is comparable to the liquid density, ρ_ℓ; ϵ_p is the porosity of the porous particle; ϵ_c is the volume fraction of cells in the void of the porous particle; and ρ_{bW} is the biofilm wet density. The fluidization behavior of porous particles whose porosity is not sufficiently high to allow liquid through flow, is similar to that of non-porous particles. For highly porous particles, on the other hand, liquid through flow can occur during the initial stage of cell immobilization. The behavior of fluidization with particles under this condition significantly deviates from that with non-porous particles. Available information on fluidization characteristics involving particles with liquid through flow has been limited to two-phase (liquid-solid) systems (Webb et al. 1983); little is known regarding the three-phase systems. As noted, the majority of the wastewater treatment processes employ non-porous or nearly non-porous particles with attached biofilm, and thus, the following discussion deals mainly with these types of particles.

ρ_{bw} in Eqn. (8.1a) can be expressed by (Shieh et al., 1981a)

$$\rho_{bw} = X_v/(1-f_w) \tag{8.2}$$

where X_v is the dry weight of biomass per unit volume of wet biofilm, i.e., biofilm dry density, and f_w is the weight fraction of the moisture in the wet biofilm. Hence, the biomass concentration in a fluidized bed (X_m), the biomass dry weight per unit volume of the expanded bed, becomes

$$X_m = X_v(1-\epsilon)\{1 - (\frac{d_o}{d_p})^3\} \tag{8.3}$$

From Eqn. (8.3), the biomass concentration in the fluidized bed bioreactor can be predicted if the bed void fraction, the bioparticle size and the biofilm density are known; these properties are interrelated and depend on the flow conditions and reactor configuration. Edeline et al. (1969) reported that the classical equations used to calculate the terminal velocity of rigid particles were also valid for bioparticles. In the following, approaches to evaluate the void fraction (ϵ) for Type IV reactors are discussed. Note that for Type IV or liquid-solid fluidized bed reactors, ϵ is equal to ϵ_ℓ. As indicated earlier, the approaches to evaluating ϵ in Type I, II, and III reactors follow those given in Chapters 2, 4, 5 and 6.

8.3.1.1 Bioparticle Holdup in a Liquid-Solid Fluidized Bed Bioreactor (FB)

For a FB reactor, the minimum fluidization velocity can be given empirically by Eqn. (2.14). At given operating conditions, the bed void fraction can be obtained from (Wen and Yu, 1966)

$$\epsilon_\ell^{4.7} Ar = 18\ Re + 2.7\ Re^{1.687} \tag{8.4}$$

provided that Re < 1,000. Shieh et al. (1981a) compared the experimental biomass concentrations in a liquid-solid fluidized bed containing sand as support particles with those predicted based on Eqn. (8.4); the relative prediction error was within 18% for superficial liquid velocities from 0.42 to 1.14 cm/s.

The empirical equation proposed by Richardson and Zaki (1954), Eqn. (2.21), can also be used to estimate the bed expansion when the bioparticle size (d_p) is known (Ngian and Martin, 1980b; Hermanowicz and Ganczarczyk, 1983; Timmermans and Van Haute, 1984; Andrews and Trapasso, 1985); the Richardson-Zaki index n can be evaluated by equations given in Table 2.2. The Richardson and Zaki equation relates the bed voidage in a bioreactor, ϵ_ℓ, to that expected for a bed of biofilm-free particles, $\epsilon_{\ell o}$, as

$$\frac{\epsilon_\ell}{\epsilon_{\ell o}} = \left(\frac{U_{to}}{U_t}\right)^{1/n} \tag{8.5}$$

Andrews and Tien (1979) proposed

$$\frac{U_t}{U_{to}} = \frac{1+ Ax_p}{(1+x_p)^{1/3}} \tag{8.6}$$

as an expression for the bioparticle "settling" velocity in the bed, where x_p is defined by

$$x_p = \left(\frac{d_p}{d_o}\right)^3 - 1 \tag{8.7}$$

and A is

$$A = (\rho_{bw}-\rho_\ell)/(\rho_s-\rho_\ell) \tag{8.8}$$

Combining Eqns. (8.5) and (8.6) yields an expression for ϵ_ℓ in terms of x_p. In the above equations, a uniform biofilm thickness is assumed. In reality, there may exist a distribution of biofilm thickness and thus a distribution of x_p or ρ_s in the bioreactor. The variation in ρ_p tends to increase with the operating time and consequently may lead to a stratification and nonuniform expansion of the bioparticles in the reactor. Solids mixing behavior is inherently related to the distribution of biofilm thickness. For complete solids mixing ($x_p = \bar{x}_p$ everywhere), Andrews and Tien (1979) reported that the expanded bed height can be given by

$$\frac{L}{L_c} = \frac{(1-\epsilon_{\ell o})(1+\bar{x}_p)}{1-\epsilon_{\ell o}\left[\frac{(1+\bar{x}_p)^{1/3}}{1+A\bar{x}_p}\right]^{1/n}} \tag{8.9}$$

But, under a complete solids stratification condition, the expanded bed height could be given by

$$\frac{L}{L_c} = (1 - \epsilon_{\ell o}) \int_0^{x_{pmax}} \frac{(1+x_p)}{1 - \epsilon_\ell} g(x_p)dx_p \tag{8.10}$$

where $g(x_p)$ is a distribution function of biofilm volume, and x_{pmax} represents the value of x_p at the top of the fluidized bed. By comparing simulation results based on Eqns. (8.9) and (8.10), Andrews and Tien (1979) concluded that the bed height varied approximately linearly with the film coverage and was nearly unaffected by solids mixing conditions.

The analysis given above treats a bioparticle as if it were a rigid particle. More accurate prediction of the bed expansion, however, requires consideration of different drag effects on the particle due to inherently different surface characteristics between bioparticles and rigid particles. Mulcahy et al. (1981) reflected this point by modifying the Richardson-Zaki index to account for the fluidized bed porosity in terms of the settling velocity of bioparticles over a very narrow range of Re_t, i.e., $40 < Re_t < 90$. The modified Richardson and Zaki index (n′) can be correlated by

$$n' = 10.35\ Re_t^{-0.18} \tag{8.11}$$

For nonspherical particles, n can be modified to n" as given by (Cleasby and Fan, 1981)

$$n'' = n^{(-2.94\ \phi_s^{0.88}\ Re_t^{-0.36})} \tag{8.12}$$

It is important to recognize that while the overall relative differences among correlations proposed by Richardson and Zaki (1954), Wen and Yu (1966), Mulcahy et al. (1981), and Andrews and Tien (1979) in the prediction of bed expansion in bioparticle systems are not large (Wang and Tien, 1983), a general trend exists in that the relative differences for large bioparticles are greater than those for small particles (Ngian and Martin, 1980b). For biofilm thicknesses commonly encountered in the wastewater treatment operation, i.e., > 50μm, it is generally true that as the biofilm thickness increases, both the bioparticle density and settling velocity decrease.

8.3.1.2 Biomass Concentration and Biofilm Wet Density

To calculate the biomass concentration, either the biofilm wet density and moisture content or the biomass dry weight per unit volume of biofilm must be known; these parameters are interrelated by Eqn. (8.2). Generally, the biofilm contains mostly water. Suzuki et al. (1981) reported the moisture content reached as high as 94 to 96% in biofilms attached to 100 to 400 μm diameter sand particles cultivated in a DTFB bioreactor treating photo-processing wastewater; the corresponding apparent density of the wet biofilm (ρ_{bw}) was 1.005 to 1.006 g/cm^3. Ngian and Martin (1980a) reported ρ_{bw} was equal to 1.10 g/cm^3; Andrews (1982) assumed 1.03 g/cm^3; and Hermanowicz and

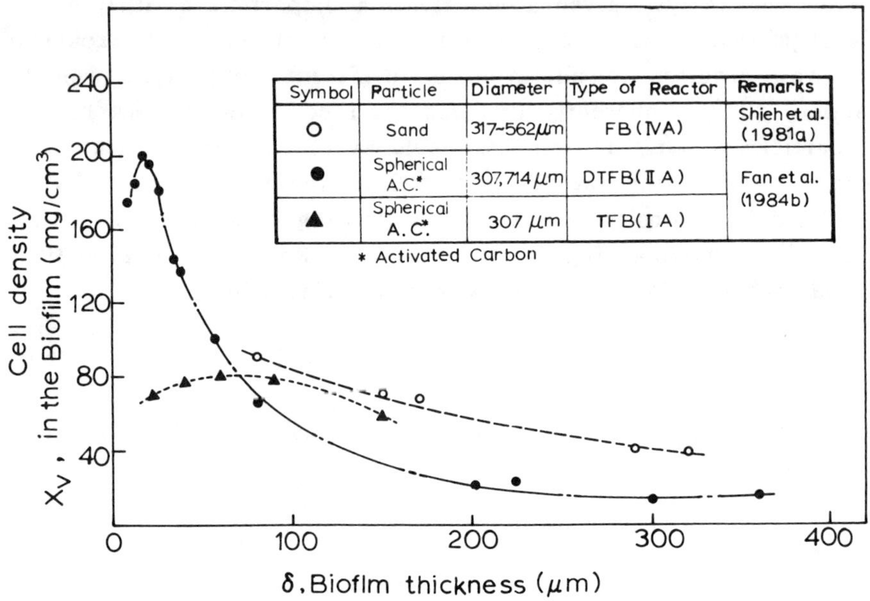

Figure 8.4 Variation of cell density in the biofilm with biofilm thickness for various fluidized bed bioreactors.

Ganczarczyk (1983) obtained ρ_{bw} equal to 1.14 g/cm^3 based on the observed terminal velocity of the bioparticle. Mueller et al. (1967) found that the dry and wet densities for the floc forming bacterium Zoogloea ramigera were 1.29 and 1.09 g/cm^3, respectively, and for activated sludge flocs were 1.52 and 1.07 g/cm^3, respectively.

Shieh et al. (1981a) evaluated the dry density (X_v) and the thickness of the biofilm (δ) attached to sand particles of 317 to 562 μm diameter. Their results are given in Fig. 8.4. In their experiments, the bioparticles were cultivated in the Oxitron System (see Section 8.2.1.4). In all data points but one, the biofilm dry densities are inversely proportional to biofilm thicknesses. Fan et al. (1986) determined the X_v and δ of the biofilm attached to spherical A.C. particles cultivated in a DTFB bioreactor. Their data, also plotted in Fig. 8.4, generally show the same inversely proportional relationship between δ and X_v. Note that very high cell densities (100 ~ 200 mg/cm^3) were obtained in biofilms less than 50 μm thick in a DTFB. The inherent relationship between X_v and δ may vary with substrate type, reactor design, and operating conditions such as organic loading, gas and liquid velocities, and DO concentration.

8.3.2 Single Particle Biodegradation

In aerobic wastewater treatment for a given bulk pollutant concentration, the overall or apparent pollutant removal rate by the biofilm of a bioparticle is governed by the biological activity, the active biomass concentration, and the diffusional flux of both pollutants and oxygen in the biofilm. Consequently, knowledge of the intrinsic microbial kinetics and of the physical and biological parameters of biofilms such as pollutant and oxygen diffusivities within the biofilm, biofilm thickness, and biomass concentration (e.g., biofilm dry density) are required to predict the overall pollutant degradation rate.

Depending upon the pollutant species (or substrate) of interest and the microbial species used, various intrinsic kinetic models for a single-substrate limiting biodegradation mechanism have been proposed. These kinetic models include rate expressions with respect to the limiting pollutant concentration of zero order (e.g., for denitrification and nitrification), first order (e.g., for valeric acid degradation), Monod type (e.g., for nitrification, denitrification, phenol degradation, BOD removal, and galactose digestion), Haldane type (e.g., for phenol degradation), or even of a very complex form (e.g., for phenol degradation). These kinetic expressions and their corresponding rate constants are summarized in Table 8.3. It is worth noting that Monod type kinetics can be approximated as a first order reaction at low substrate concentrations and a zero order reaction at high substrate concentrations (Fujie et al., 1979a, 1979c; Andrews and Trappaso, 1985). In addition, a simple approximation by a first order rate expression has been found to yield accurate results for a much wider range of bulk substrate concentrations than expected from inspection of the half velocity coefficient of the Monod equation for a given substrate (Jennings et al., 1976). Petersen (1965) suggested the following approximation for complex kinetic expressions of the substrate consumption rate (per unit biofilm volume):

$$-r_{SV} = -r_{SV}^{O} + K_V(S - S^O) \tag{8.13}$$

where

$$K_V = \{\partial(-r_{SV})/\partial S\}_{S=S^O} \tag{8.14}$$

Table 8.3 Intrinsic Kinetic Expressions and Parameters Applied to Evaluate the Aerobic Substrate Uptake Rate by Bioflocs.

Kinetic expression		Substrate	Parameters	Comments and References
Zero order	$-\nu_s = K_o$	phenol	$K_o = 0.18\sim0.20\ hr^{-1}$	Fan et al. (1984b)
	$-\nu_{N1} = K_{n1}$	ammonium	$K_{n1} = 0.05$g-N/kg-AC·hr*	Hano et al. (1983)
	$-\nu_{N2} = K_{n2}$	nitrite	$K_{n2} = 1.1$g-N/kg-AC·hr* *: nitrification rate per unit dry weight of carbon particles	Hano et al. (1983)
First order	$-r_{sv} = K_v S$ (aerobic region)	valeric acid	$K_v = 0.1\ s^{-1}$ $Y_x = 0.67$	Wang and Tien (1984b)
	$-r_{sv} = \alpha K_v S$ (anaerobic region)		$\alpha K_v = 0.079\ s^{-1}$ $Y_x = 0.56$	$\rho_c = 37 mg/cm^3$ (X_v* $= 69.7 mg/cm^3$) *estimated
Monod type	$\mu = \frac{\mu_m S}{K_s + S}$	valeric acid	$\mu_m = 0.144\ hr^{-1}$ $K_s = 50$ mg/ℓ	Andrews (1982)
	$-\nu_s = \frac{K_m S}{K_s + S}$	galactose	$K_m = 0.625\ hr^{-1}$ $K_s = 22.7$ mg/ℓ	Rittmann and Brunner (1984) $Y_x = 0.54$
		BOD* * solvent extracted coal gasification process wastewater	$K_m = 0.11\ day^{-1}$ $K_s = 2$ mg/ℓ	Cook et al. (1984)
	$-r_{N1} = -r_{m1}\left(\frac{N_1}{K_{N1}+N_1}\right)\cdot\left(\frac{C}{K_{c1}+C}\right)$	ammonia	$-r_{m1} = 20.8$mg-N/ℓ·min* $K_{N1} = 5.0$mg-N/ℓ $K_{c1} = 0.5$mg-O_2/ℓ	Denac et al. (1983)
	$-r_{N2} = -r_{m2}\left(\frac{N_2}{K_{N2}+N_2}\right)\cdot\left(\frac{C}{K_{c2}+C}\right)$	nitrite	$-r_{m2} = 12.6$mg-N/ℓ·min* $K_{N2} = 2.8$mg-N/ℓ $K_{c2} = 0.25$mg-O_2/ℓ * nitrification rate per unit biofilm volume	
Substrate inhibition type	$-\nu_s = \frac{K_m S}{K_s + S + S^2/K_i}\cdot\frac{C}{K_c + C}$	phenol	$K_m = 0.3158\ hr^{-1}$ $K_s = 44.54$ mg/ℓ $K_i = 202.9$ mg/ℓ $K_c = 0.042$ mg/ℓ	Suzuki et al. (1983) $Y_o = 1.23$ g-O_2/g-phenol $Y_x = 0.704$ g-cell/g-phenol
	$-\nu_s = \frac{K_m S}{K_s + S + S^2/K_i}$	phenol	$K_m = 0.308\ hr^{-1}$ $K_s = 3.14$ mg/ℓ $K_i = 65.06$ mg/ℓ	Fujie et al. (1985) $Y_x = 0.47\sim0.58$ g-cell/g-phenol $Y_o = 0.8\sim2.0$ g-O_2/g-phenol
	$\mu_s = \frac{\mu_m S}{K_s + S + S^2/K_i}\cdot\frac{C}{K_c + C}$	phenol	$\mu_m = 0.365\ hr^{-1}$ $K_s = 10.948$ mg/ℓ $K_i = 113.0$ mg/ℓ $K_c = 0.1$ mg/ℓ	Tang et al. (1985) $Y_x = 0.496$ g-cell/g-phenol $Y_x/Y_o = 0.354$ g-cell/g-O_2

In the above equations, $-r_{sv}$ is the substrate removal rate per unit biofilm volume and $-r^o_{sv}$ is the value of $-r_{sv}$ for a substrate concentration S of S^o. Note that Eqn. (8.13) was derived from a Taylor series expansion.

In aerobic wastewater treatment, the substrate consumption rate may depend on the dissolved oxygen concentration in the biofilm. Two different mechanisms for the oxygen dependency of the limiting substrate consumption rate are generally assumed. The first mechanism assumes that at dissolved oxygen concentrations higher than a critical value (C_c) the substrate consumption rate is independent of the oxygen concentration (depends only on the limiting substrate concentration); for DO concentrations lower than C_c no reaction of the limiting substrate occurs (Fujie et al., 1979a). The second mechanism assumes bi-substrate limiting kinetics in which the consumption rate of either the limiting substrate or dissolved oxygen is a function of the product of the kinetic expressions of both the limiting substrate and dissolved oxygen. The kinetic expression for oxygen is usually assumed a Monod type. Bi-substrate limiting kinetic models have been proposed for nitrification (N-source substrate and oxygen; e.g., Denac et al., 1983) and phenol degradation (C-source substrate and oxygen; e.g., Suzuki et al., 1983) in biofilms attached to particles.

Nitrification is a two-step biological process in which the ammonium ion is oxidized first to nitrite by Nitrosomonas and then to nitrate ion by Nitrobacter (Watson, 1974; Frobisher et al., 1974). Nitrification, followed by denitrification, occurs when the bioreactor effluent BOD concentration is lower than a critical value (Kubota and Miyajie, 1977; Kubota et al., 1978; Fujie and Kubota, 1980). Tanaka et al. (1981), Tanaka and Dunn (1982), Denac et al. (1983) and Dunn et al. (1983) reported that the intrinsic nitrification rate in a biofilm attached to sand particles can be given by the following equations: for the initial nitrification step

$$-r_{N1} = -r_{m1}\left(\frac{N_1}{K_{N1}+N_1}\right)\left(\frac{C}{K_{c1}+C}\right) \qquad (8.15)$$

and for the second nitrification step

$$-r_{N2} = -r_{m2}\left(\frac{N_2}{K_{N2}+N_2}\right)\left(\frac{C}{K_{c2}+C}\right) \qquad (8.16)$$

$-r_{N1}$ and $-r_{N2}$ are the rates of ammonium ion and nitrite ion (mass) oxidized, respectively, per unit biofilm volume; $-r_{m1}$ and $-r_{m2}$ are the maximum rates for the initial and the second steps, respectively, and were experimentally determined as summarized in Table 8.3. Note that when K_{N1} and K_{N2} in Eqns. (8.15) and (8.16) are less than 5 mg/ℓ, the nitrification rates can be considered to be independent of the ammonium and nitrite ion concentrations, which leads to the zero order reaction rate employed by Hano et al. (1983).

The bi-substrate (substrate and oxygen) limiting condition yielding the cell growth kinetic expression given below is exemplified by that for phenol and oxygen systems (Suzuki et al., 1983; Tang et al., 1985; Tang and Fan, 1985):

$$\mu = (-r_{sv}) \frac{Y_x}{X_v} = (-\nu_s)\, Y_x = \frac{\mu_m S}{K_s + S + S^2/K_i} \frac{C}{K_c + C} \tag{8.17}$$

where K_s and K_i are the saturation and inhibition constants, respectively, and K_c is the Monod constant. This equation uses a Monod type expression for oxygen depletion and a Haldane type expression, which considers a noncompetitive inhibition mechanism, for substrate consumption.

8.3.2.1 Steady (or Pseudo-Steady) State Behavior

The influence of diffusional transport on the overall reaction rate in a catalyst is traditionally represented by the effectiveness factor. Atkinson and coworkers (Atkinson et al., 1967, 1968; Atkinson and Daoud, 1968, 1970) applied the volume-based effectiveness factor concept to reactions in a biofilm attached to a flat surface for reactor operation at steady state. The volume-based effectiveness factor for a homogeneous biofilm is defined as

$$\eta_v = \frac{\text{[actual reaction rate in a biofilm]}}{\text{[reaction rate without diffusion hindrance in a biofilm]}} \tag{8.18}$$

For a spherical bioparticle, η_v for substrate S is

$$\eta_v = 3r_p^2 D_{sf}(dS/dr)_{r=r_p} / \{(r_p^{\,3} - r_o^{\,3})(-r_{sv}^*)\} \tag{8.19}$$

Solution of Eqn. (8.19) requires knowledge of the substrate concentration profile in the biofilm; the concentration profile can be described through a mass balance on S

$$\frac{D_{sf}}{r^2} \frac{d}{dr} \left(r^2 \frac{dS}{dr}\right) - (-r_{sv}) = 0 \tag{8.20}$$

In Eqn. (8.20), $-r_{sv}$, the intrinsic substrate consumption rate per unit volume of biofilm, depends upon both the limiting substrate and oxygen concentrations for aerobic biodegradation reactions. Equation (8.20), therefore, needs to be solved simultaneously with the oxygen mass balance within the biofilm:

$$\frac{D_{cf}}{r^2}\frac{d}{dr}\left(r^2\frac{dC}{dr}\right) - (-r_{cv}) = 0 \tag{8.21}$$

where $-r_{cv}$ is the intrinsic dissolved oxygen uptake rate per unit volume of biofilm. The boundary conditions for Eqns. (8.20) and (8.21) are

$$r = r_p\ ;\ C = C^* \text{ or } D_{cf}\frac{dC}{dr} = K_{\ell c}\,(C_b - C^*) \tag{8.22a}$$

$$S = S^* \text{ or } D_{sf}\frac{dS}{dr} = K_{\ell s}\,(S_b - S^*) \tag{8.22b}$$

$$r = r_o\ ;\ \frac{dS}{dr} = \frac{dC}{dr} = 0 \tag{8.22c}$$

The intrinsic consumption rates of both the limiting substrate and the dissolved oxygen can be related stoichiometrically through an observed yield factor (Y_o) as given by

$$-r_{cv} = Y_o\,(-r_{sv}) \tag{8.23}$$

From Eqns. (8.20), (8.21) and (8.23) a linear relation between the substrate concentration (S) and the dissolved oxygen concentration (C) in the biofilm can be obtained:

$$(S^* - S) = (D_{cf}/Y_oD_{sf})(C^* - C) \tag{8.24}$$

Due to the specific dependency of $(-r_{sv})$ on oxygen concentration, three distinct cases, i.e., Cases A, B and C, characterizing the concentration profiles of the substrate and oxygen inside the biofilm, can result (Fujie et al., 1979c). In Case A, the substrate and oxygen concentrations within the biofilm are sufficiently high such that all microbes in the biofilm are actively participating in the reaction. In Case B, the limiting substrate is completely consumed within a finite layer of the film leaving behind an inactive inner layer surrounding the particle. In Case C, the opposite situation from Case B, the dissolved oxygen is completely

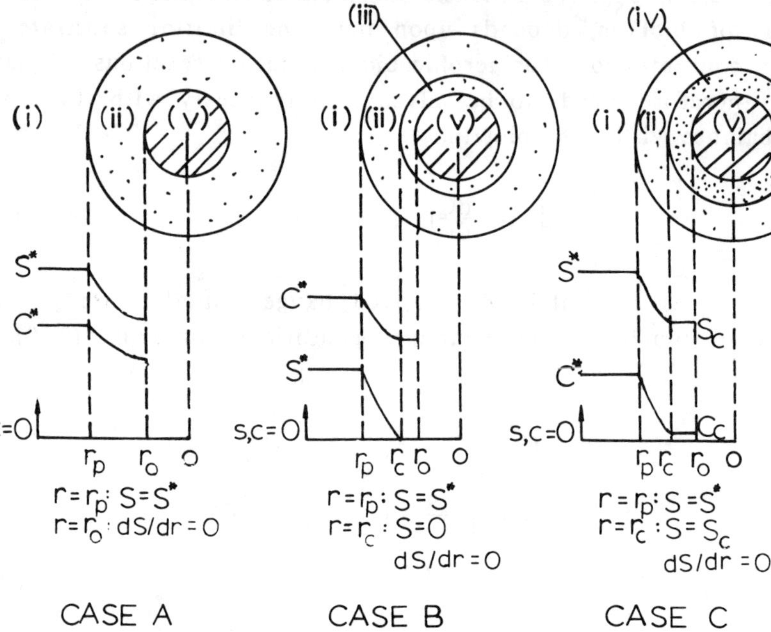

Figure 8.5 Concentration profile of organic substrate and dissolved oxygen in the bioparticle for three cases: (i) liquid phase; (ii) active zone; (iii) inactive aerobic zone; (iv) anaerobic zone; and (v) solid core (from Fujie et al., 1979a).

consumed within a finite layer of the film leaving behind an anaerobic inner layer surrounding the particle. These three cases are graphically shown in Fig. 8.5.

If, for DO concentrations larger than C_c, $-r_{sv}$ depends only on the limiting substrate concentration, the effectiveness factor can be solved analytically for zero and first order kinetics; otherwise a numerical solution can be obtained. Mathematical expressions for the volume-based effectiveness factors with zero order biodegradation kinetics and a generalized volume-based effectiveness factor based on linearized kinetic expression given by Eqn. (8.13) are given in Table 8.4 for each of the three cases mentioned above.

The volume-based effectiveness factor with $(-r_{sv})$ and $(-r_{cv})$ in Eqns. (8.20) and (8.21), respectively, represented by bi-substrate limiting kinetics, such as Eqns. (8.15) through (8.17), have to be solved numerically. One example for evaluating the effectiveness factor of phenol degradation by biofilms was detailed by Tong (1986).

Table 8.4 Volume-Based Effectiveness Factor of bioparticle with Solid Support for Aerobic Biodegradation (from Fujie et al., 1979c; and Fan et al., 1984b).

	Boundary conditions	Generalized volume-based effectiveness factor based on linearized kinetic expression given by Eqn. (8.13)	Volume-based effectiveness factor for zero-order kinetics	Notation for the table
Case A	$x=x_o;dy/dx=0$ $x=1;y=1$	$\eta_V=\frac{1}{1-x_o^3}\frac{3}{m}\left[\frac{mx_o\tanh\{m(1-x_o)\}+1}{\tanh\{m(1-x_o)\}+mx_o}-\frac{1}{m}\right]$ for $x_o=0$ $\eta_V=\frac{3}{m}\left[\frac{1}{\tanh m}-\frac{1}{m}\right]$	$\eta_V=1$ for $\Phi_s^2\le 6/(1-3x_o^2+2x_o^3)$ $\Phi_c^2\le 6(1-z_c)/(1-3x_o^2+2x_o^3)$	$x=r/r_p$ $x_o=r_o/r_p$ $x_c=r_c/r_p$ $y=S/S^*$
Case B	$x=x_c;y=0,dy/dx=0$ $x=1;y=1$	$\eta_V=\frac{1}{1-x_o^3}\frac{3}{m}\left[\left\{1-\frac{(x_c^2-1/m^2)\Phi^2}{(1+\Phi)^2}\right\}^{1/2}-\frac{1}{m}\right]$ x_c is obtained from $1+\Phi=\frac{\Phi}{m}[mx_c\cosh\{m(1-x_c)\}+\sinh\{m(1-x_c)\}]$	$\eta_V=\frac{(1-x_c^3)}{(1-x_o^3)}$ x_c is obtained from $2x_c^3-3x_c^2+(1-6/\Phi_s^2)=0$ for $1>x_c>x_o$ $\Phi_s^2\ge 6/(1-3x_o^2+2x_o^3)$ $\Phi_s^2\ge\Phi_c^2/(1-z_c)$	$y_c=S_c/S^*$ $z=C/C^*$ $z_c=C_c/C^*$ $S_c=S^*-D_{cf}(C^*-C_c)/Y_oD_{sf}$ $-r_{sv}=-r_{sv}^*+K_v(S-S^*)$ (8.13) $K_v=\{\partial(-r_{sv})/\partial S\}_{S=S^*}$ (8.14) $m^2=r_p^2K_v/D_{sf}$ $\Phi=-r_{sv}^*/K_vS^*-1$
Case C	$x=x_c;y=y_c,dy/dx=0$ $z=z_c,dz/dx=0$ $x=1\ ;\ y=1,\ z=1$	$\eta_V=\frac{1}{1-x_o^3}\frac{3}{m}\left[\left\{1-\frac{(x_c^2-1/m^2)(\Phi+y_c)^2}{(1+\Phi)^2}\right\}^{1/2}-\frac{1}{m}\right]$ x_c is obtained from $1+\Phi=\frac{\Phi+y_c}{m}[mx_c\cosh\{m(1-x_c)\}+\sinh\{m(1-x_c)\}]$	$\eta_V=\frac{(1-x_c^3)}{(1-x_o^3)}$ x_c is obtained from $2x_c^3-3x_c^2+1-(6/\Phi_c^2)(1-z_c)=0$ for $1>x_c>x_o$ $\Phi_c^2\ge 6(1-z_c)/(1-3x_o^2+2x_o^3)$ $\Phi_s^2\le\Phi_c^2/(1-z_c)$	$\Phi_s^2=r_p^2K_oX_v/D_{sf}S^*$ $\Phi_c^2=r_p^2K_oX_vY_o/D_{cf}C^*$ For Michaelis-Menten type $-r_{sv}=K_mX_vS/(K_s+S)$ $m^2=r_p^2K_mK_sX_v/D_{sf}(K_s+S^*)^2$ $\Phi=S^*/K_s$

Simultaneous reaction and diffusion of both the limiting substrate (pollutant) and oxygen in a biofilm may cause either the substrate or oxygen or both to be completely depleted inside the biofilm before reaching the biofilm-solid interface. Evidence of such a phenomenon is the existence of an optimal biofilm thickness above which the overall pollutant degradation rate remains the same. When oxygen is completely depleted before reaching the biofilm-particle interface, an anaerobic layer develops within the biofilm. If anaerobic microorganisms exist in the biofilm and the organic pollutant is not completely depleted, anaerobic digestion of the organic pollutant can occur. A typical anaerobic digestion of organic pollutant in wastewater treatment is the denitrification process. Denitrification results from the metabolic activity of certain facultative heterotrophic bacteria which utilize organic carbon both as a carbon and energy source, while using nitrate instead of oxygen as an electron acceptor in energy metabolism. Since the removal of nitrate to less than 10 mg/ℓ is required for drinking water, an aerobic microbial layer in which BOD is degraded followed by an anaerobic microbial layer in which degradation and removal of both BOD and nitrate occur is beneficial in some wastewater treatment systems. Such a bilayer treatment process in a liquid-solid fluidized bed bioreactor has been mathematically modeled by Wang and Tien (1983).

8.3.2.2 Dynamic Behavior

So far only steady state biodegradation has been discussed. In reality, reactor operation is usually in a transient condition during start-up to shut-down periods. Factors such as diurnal fluctuations and shock loading in the influent substrate concentration or flow rate produce transient conditions. In addition, biomass growth or decay in the biofilm and detachment of biomass due to fluid shear as well as abrasion and collision among bioparticles cause variations in biofilm thickness unless some type of biofilm control action is taken.

Tang et al. (1985) investigated the transient behavior in biofilm growth for phenol biodegradation subject to a step change in influent phenol concentration. Their analysis considered mass transfer from the bulk fluid to the biofilm, a biofilm of changing size, and a mass flux from the biofilm to the activated carbon particle limited by the adsorption characteristics of the activated carbon. Several basic assumptions were made including

i) The non-homogeneity of cell concentration in the biofilm is lumped into the cell density, which changes with radial location in the biofilm;
ii) The growth-limiting factors are both phenol and oxygen;
iii) There is no physiological change in the microbes when immobilized onto the activated carbon particles; and
iv) Effects of depletion of both phenol and oxygen due to adsorption in the biofilm are negligible, and adsorption of oxygen on activated carbon is negligible.

Tang et al.'s (1985) study provides a starting point for the analysis of the dynamics of aerobic biofilm degradation processes and will be briefly discussed below.

Variations in biofilm thickness result from changes in the cell mass attached to the bioparticle. Dynamic changes in cell mass are related to the rates of growth, decay, and detachment of the biofilm by

$$\frac{dm_f}{dt} = r_{growth} - r_{decay} - r_{detachment} \tag{8.25}$$

The biomass growth rate can be obtained by integrating the local specific growth rate over the biofilm volume:

$$r_{growth} = \int_{r_o}^{r_p} X_v(r) \frac{\mu_m S}{K_s + S + S^2/K_i} \frac{C}{K_c + C} 4\pi r^2 \, dr \tag{8.26}$$

Rittmann (1982) noted that the rate of cell mass lost per unit biofilm area was linearly proportional to the cell mass. Tang et al. (1985) found, for biofilms on activated carbon particles, that the decay and detachment rates were also linearly proportional to the biomass on a bioparticle and could be expressed as

$$r_{decay} + r_{detachment} = K_D \bar{X}_v V_f \tag{8.27}$$

where K_D is an overall decay coefficient, $\bar{X}_v$ is the average biofilm dry density and V_f is the biofilm volume on a single bioparticle. Combining Eqns. (8.25), (8.26), and (8.27) gives the variation in biofilm thickness during the transient state:

$$\frac{dr_p}{dt} = \frac{1}{\bar{X}_v r_p^2} \int_{r_o}^{r_p} \left\{ X_v(r) \frac{\mu_m S}{K_s + S + S^2/K_i} \frac{C}{K_c + C} r^2 \right\} dr - K_D r_p (1 - x_o^3)/3 \tag{8.28}$$

Within the biofilm, mass balances on the substrate (phenol) and oxygen which consider simultaneous transport and consumption can be expressed as

$$\frac{\partial S}{\partial t} = \frac{1}{\epsilon_f}\frac{1}{r^2}\frac{\partial}{\partial r}\left(r^2 D_{sf}\frac{\partial S}{\partial r}\right) - \frac{\mu_m X_v}{\epsilon_f Y_x}\frac{S}{K_s+S+S^2/K_i}\frac{C}{K_c+C} \tag{8.29}$$

$$\frac{\partial C}{\partial t} = \frac{1}{\epsilon_f}\frac{1}{r^2}\frac{\partial}{\partial r}\left(r^2 D_{cf}\frac{\partial C}{\partial r}\right) - \frac{\mu_m X_v Y_o}{\epsilon_f Y_x}\frac{S}{K_s+S+S^2/K_i}\frac{C}{K_c+C} \tag{8.30}$$

where ϵ_f is the porosity of the biofilm and can be estimated from $(1-X_v(r)/\rho_D)$. ρ_D is the dry density of microbes. The mass fluxes at both interfaces (bulk fluid and activated carbon particle) are considered in the boundary conditions.

Within the activated carbon particle, mass balances on phenol and oxygen which consider simultaneous diffusion of phenol and oxygen and adsorption of phenol can be expressed as

$$\epsilon_p\frac{\partial S}{\partial t} + \rho_p\frac{\partial q}{\partial t} = \frac{D_{sp}}{r^2}\left[\frac{\partial}{\partial r}\left(r^2\frac{\partial S}{\partial r}\right)\right] \tag{8.31}$$

$$\epsilon_p\frac{\partial C}{\partial t} = \frac{D_{cp}}{r^2}\left[\frac{\partial}{\partial r}\left(r^2\frac{\partial C}{\partial r}\right)\right] \tag{8.32}$$

In Eqn. (8.31), the adsorption characteristics of the activated carbon particle appear as $\partial q/\partial t$, where q is the amount of phenol adsorbed per unit weight of activated carbon. A Langmuir isotherm can be used to describe q:

$$q = \frac{b_1 S}{1 + b_2 S} \tag{8.33}$$

where b_1 and b_2 are isotherm constants.

The corresponding initial and boundary conditions for Eqns. (8.29) through (8.32) are

$$t=0\ ;\quad S = S_{io}\ ,\quad C = C_{io} \tag{8.34a}$$

$$r=0\ ;\quad \frac{\partial S}{\partial r} = \frac{\partial C}{\partial r} = 0 \tag{8.34b}$$

$$r=r_o;\ D_{sf}\left.\frac{\partial S}{\partial r}\right|_{r_o^+} = D_{sp}\left.\frac{\partial S}{\partial r}\right|_{r_o^-},\ D_{cf}\left.\frac{\partial C}{\partial r}\right|_{r_o^+} = D_{cp}\left.\frac{\partial C}{\partial r}\right|_{r_o^-} \tag{8.34c}$$

$$\left.S\right|_{r_o^+} = \left.S\right|_{r_o^-},\ \left.C\right|_{r_o^+} = \left.C\right|_{r_o^-} \tag{8.34d}$$

$$r=r_p\ ;\ N_s = D_{sf}\left.\frac{\partial S}{\partial r}\right|_{r_p} = K_{\ell s}(S_b-S^*) \tag{8.34e}$$

$$N_c = D_{cf}\left.\frac{\partial C}{\partial r}\right|_{r_p} = K_{\ell c}(C_b-C^*) \tag{8.34f}$$

At any given reactor time, the rates of phenol removal ($-R_{sv}$) and oxygen consumption ($-R_{cv}$) per unit volume of a DTFB are

$$-R_{sv} = \frac{3\epsilon_s}{r_p} N_s \tag{8.35}$$

$$-R_{cv} = \frac{3\epsilon_s}{r_p} N_c \tag{8.36}$$

To complete the analysis several important parameters must be evaluated.

Substrate and oxygen diffusivities are required for the solution of the biofilm mass balances. Numerous studies (e.g., Mulcahy et al., 1981) found lower diffusivities in the biofilm than in water; furthermore, diffusivities in the biofilm were inversely proportional to the cell density (see Fig. 8.6 (Fan et al., 1988); also see Section 7.4). Tang et al. (1985) obtained an empirical correlation equation relating phenol and oxygen diffusivities within the biofilm to the biofilm thickness on the activated carbon particles of

$$\frac{D_w}{D_f} - 1 = K_3 \exp[-K_4(r_p-r_o)] \tag{8.37}$$

where K_3 and K_4 are arbitrary constants.

The microbial specific growth rate ($-\nu_s$), as described by the growth kinetic models given in Table 8.3, is instantaneously affected by changes in the environmental substrate concentration. In reality, however, microbial cells require a finite time to synthesize or activate enzymes in response to variations in substrate levels. Chemostat transient studies have demonstrated that, macroscopically, the specific growth rates of microbial cultures lag the bulk substrate levels

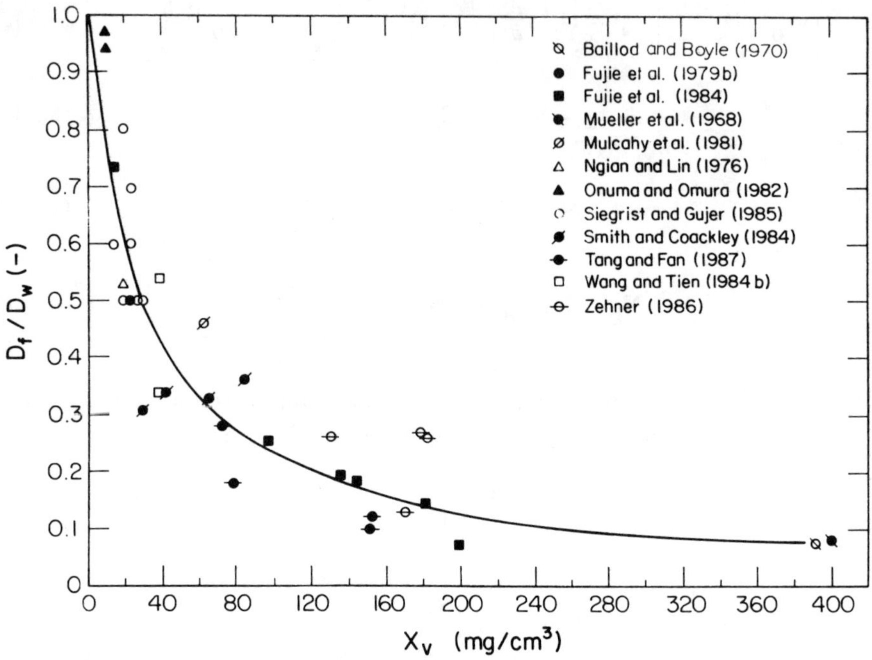

Figure 8.6 Relationship between the diffusivity and cell density in the biofilm attached to the solid supports (from Fan et al., 1988).

(Schaezler et al., 1971; Yang and Humphrey, 1975). Such a time delay phenomenon has been hypothesized to result from the limitation of the cellular RNA required to support a higher rate of protein synthesis after a step increase in substrate levels. Daigger and Grady (1982) showed that the RNA-limited hypothesis is inadequate for describing the transient responses of cultures grown over a wide range of specific growth rates, while Lievense and Lim (1985) modeled the transient response of a continuous culture based on the RNA-limited hypothesis and obtained a reasonable agreement between the experimental and predicted data. Some discussion regarding the role of RNA in the transient response of microbial cultures to changes in extracellular environments can be found in Koplove and Cooney (1979).

The simplest approach to model such a time lag in rate adjustment, however, is to incorporate an empirical expression for the time delay factor in the unstructured growth kinetic equation. Typical examples of this approach were given by Schaezler et al. (1971), Yang and Humphrey (1975), and Chiam and Harris (1983). The following linear function (Eqn. (8.38)) was proposed by Schaezler et al. (1971) and Yang and Humphrey (1975) to add to the growth kinetic model, and was used by Tang et al. (1985) to complement Eqns. (8.29) and (8.30) in the modeling of the transient response of phenol degradation by biofilms in a DTFB:

$$\mu(t) = (\mu_o + a_o) + a_1 t \tag{8.38}$$

where μ_o is the specific growth rate at prior steady state, a_1 is the time response constant, and a_o is the magnitude of immediate increase in μ at the time when the step change is introduced. It should be stressed that Eqn. (8.38) is strictly empirical and should be used with caution.

There have been some mechanistic models proposed to describe such unbalanced growth (e.g., Bremer and Dennis, 1975; Lievense and Lim, 1985). Most of these models require detailed information of cellular components such as RNA, DNA, or protein fractions. Note that mechanisms underlying these models may not apply well to low growth rate situations (Daigger and Grady, 1982).

Because the biofilm thickness changes with time during the transient state as given by Eqn. (8.28), the system of equations, Eqns. (8.28) through (8.38), constitute a moving boundary problem. By converting the moving coordinates to fixed coordinates using the methods presented by Park et al. (1984) and Tang et al. (1985), the dynamic behavior of phenol degradation by biofilms grown in a DTFB can then be solved numerically.

8.3.3 Bioreactor Models

The performance of a bioreactor in aerobic wastewater treatment depends upon several factors: 1) the pollutant biodegradation rate in the biofilm, which in turn depends on the intrinsic microbial degradation and growth kinetics, the biofilm characteristics (thickness, density, and spatial biomass distribution), and the substrate and oxygen diffusional characteristics in the biofilm; 2) the bioreactor hydrodynamic characteristics including the flow pattern, solids (bioparticles) holdup, and gas, liquid, and solid phase mixing states; and 3) the reactant interfacial gas-liquid and liquid-solid mass transfer. The general transport processes and kinetics of pollutant biodegradation in a single bioparticle have been discussed in Section 8.3.2. This section will illustrate the development of a comprehensive model to account for pollutant conversion in a FBB.

8.3.3.1 Steady (or Pseudo-Steady) State Conditions

Steady state bioreactor models have been developed by assuming that the particles are uniform in size and the biofilm is uniform in thickness and that the biofilm can be considered a homogeneous matrix through which oxygen and substrate diffuse and are consumed by the microbes. The liquid phase in the Type IA bioreactor substrate is considered to be axially dispersed, while the gas phase may be assumed to be in plug flow. Material balances on the substrate and dissolved oxygen in the bulk liquid phase yield (Wisecarver and Fan, 1987)

$$\epsilon_\ell E_{z\ell} \frac{d^2S_b}{dz^2} - U_\ell \frac{dS_b}{dz} - K_{\ell s}a_s(S_b - S^*) = 0 \tag{8.39}$$

$$\epsilon_\ell E_{z\ell} \frac{d^2C_b}{dz^2} - U_\ell \frac{dC_b}{dz} + K_\ell a(C_s - C_b) - K_{\ell c}a_s(C_b - C^*) = 0 \tag{8.40}$$

where

$$C^* = C|_{r=r_p} \text{ and } S^* = S|_{r=r_p}$$

A material balance on the gas phase yields

$$\left(\frac{U_g Z_i}{v_m}\right) \frac{d\eta}{dz} - K_\ell a(C_s - C_b) = 0 \tag{8.41}$$

where Z_i is the oxygen mole fraction in the feed air or oxygen enriched air, and v_m represents the specific volume of gaseous oxygen. C_s in Eqns. (8.40) and (8.41) is the saturated dissolved oxygen concentration with air or oxygen enriched air, which can be expressed by (Fujie and Fan, 1986)

$$C_s = C_o^* \{1 + \alpha(L-z)\} \frac{Z_i(1-\eta)}{1-Z_i\eta} \tag{8.42}$$

where α is the pressure conversion factor (= $\rho_a/(10{,}336\ P^*)$, in SI units) and η is the oxygen utilization factor. The boundary conditions for Eqns. (8.39), (8.40), and (8.41) are

$$z = 0;\ \eta = 0 \tag{8.43a}$$

$$\epsilon_\ell E_{z\ell} \frac{dC_b}{dz} + U_\ell C_{bi} = U_\ell C_b \tag{8.43b}$$

$$\epsilon_\ell E_{z\ell} \frac{dS_b}{dz} + U_\ell S_{bi} = U_\ell S_b \tag{8.43c}$$

$$z = L\ ;\ \frac{dC_b}{dz} = 0 \tag{8.43d}$$

$$z = L\ ;\ \frac{dS_b}{dz} = 0 \tag{8.43e}$$

The concentrations of substrate and dissolved oxygen at the surface of the bioparticle, S^* and C^*, will depend on the solid phase mixing (see Section 3.2.3). Conceptually, if the rate of solids intermixing is rapid compared with the rate of reaction, then solids mixing may occur before any significant reaction takes place, and the solid phase can be treated as being well-mixed. This is usually the case for the three-phase fluidized bed bioreactor, particularly when the liquid velocities are low and gas velocities are high. Thus the solids are assumed to be well-mixed in this analysis.

The oxygen concentrations in the gas phase and the bulk liquid phase may be found by solving Eqns. (8.40) and (8.41) with Eqns. (8.43a), (8.43b) and (8.43d) as the boundary conditions. For the well-mixed solids phase, the surface concentration is constant throughout the bioreactor. The solution under these conditions can be expressed similar to that for three-phase fluidized bed catalytic reactors (Wisecarver and Fan, 1988b), as given in Appendix A.

An overall material balance for oxygen in the reactor yields the following equation:

$$U_\ell(S_{bi} - S_{be})Y_o = U_\ell(C_{bi} - C_{be}) + U_g Z_i\ \eta_e/v_m \tag{8.44}$$

Note that in practice S_{bi} and C_{bi} are equivalent to S_i and C_i, respectively, representing the feed conditions indicated in Table 8.1.

The rate of reaction of substrate in the biofilm must equal the overall rate of disappearance of substrate in the bulk phase, i.e.,

$$D_{sf}a_sL \left.\frac{dS}{dr}\right|_{r=r_p} = U_\ell(S_{bi} - S_{be}) \tag{8.45}$$

The equations described above constitute a set of model equations for Type IA bioreactors which can be numerically solved. Based on phenol degradation kinetics, the model simulation (Wisecarver and Fan, 1987) shows that as the inlet phenol loading is increased, the rate of biodegradation increases until a break point is reached. As the inlet phenol loading is increased beyond this break point, the biodegradation rate levels off and the outlet phenol concentration begins to rise, apparently due to oxygen limitation.

When liquid-solid mass transfer resistance is negligible, as for bioreactors containing small bioparticles, Eqns. (8.39) and (8.40) can be simplified to

$$\epsilon_\ell E_{z\ell} \frac{d^2S_b}{dz^2} - U_\ell \frac{dS_b}{dz} - (-R_{sv}) = 0 \tag{8.46}$$

$$\epsilon_\ell E_{z\ell} \frac{d^2C_b}{dz^2} - U_\ell \frac{dC_b}{dz} + K_\ell a\,(C_s - C_b) - Y_o(-R_{sv}) = 0 \tag{8.47}$$

The boundary conditions for the above equations are given in Eqns. (8.43b) through (8.43e). In Eqns. (8.46) and (8.47), $(-R_{sv})$, the removal rate of substrate (pollutant) per unit bioreactor volume, can be expressed by

$$-R_{sv} = \eta_v(-r_{sv}^{\;*})\epsilon_s(1-x_o^{\;3}) \tag{8.48}$$

The volume based effectiveness factor for a bioparticle (η_v) can be obtained from Eqns. (8.18) and (8.19) or from Table 8.4 when applicable. ϵ_s can be estimated using the procedures mentioned in Section 8.3.1. $(-R_{sv})$ can also be related to the liquid-solid interfacial mass flux by

$$(-R_{sv}) = K_{\ell s}a_s\,(S_b - S^*) \tag{8.49}$$

For a liquid phase plug flow condition, Eqns. (8.46) and (8.47) reduce to

$$U_\ell \frac{dS_b}{dz} + (-R_{sv}) = 0 \tag{8.50}$$

$$U_\ell \frac{dC_b}{dz} - K_\ell a\,(C_s - C_b) + Y_o(-R_{sv}) = 0 \tag{8.51}$$

The plug flow model has been widely used for the liquid phase in Type IA FBBs, although a complete mixing model has also been used (Hosaka et al., 1983; Dunn et al., 1983). Based on plug flow for the liquid phase, Hirata et al. (1986) modeled phenol biodegradation with bi-substrate limiting kinetics with respect to phenol and oxygen, i.e., $[S/(K_s+S)][C/(K_c+C)]$. Based on complete mixing for the liquid phase, the mass balances on the pollutant and dissolved oxygen under steady state operation give

$$(S_{bi}-S_{be})/\tau = (-R_{sv}) = K_\ell a\,(C_s-C_{be})/Y_o + (C_{bi} - C_{be})/(\tau Y_o) \tag{8.52}$$

where S_{bi} and S_{be} are the influent and effluent pollutant concentrations and τ is the wastewater space time in the bioreactor.

The mass balance of oxygen in the gas phase for a complete mixing condition is given by

$$\left(\frac{U_g Z_i}{L\ v_m}\right)\eta_e = K_\ell a\,(C_s-C_{be}) \tag{8.53}$$

Where L is the fluidized bed height. C_s in Eqn. (8.53) can be expressed by (Fujie and Fan, 1986)

$$C_s = C_o^{*}\,(1 + \alpha L/2)\ \frac{Z_i(1-\eta_e)}{1-Z_i\,\eta_e} \tag{8.54}$$

When U_g, K_ℓ and C_b are known, η_e and C_s can be evaluated based on Eqns. (8.41) and (8.42) or (8.53) and (8.54) (Kubota et al., 1980; Sekizawa et al., 1985). The value of η_e calculated based on gas phase plug flow is slightly larger than that based on complete mixing flow. The difference between them, however, is not substantial (Fujie and Fan, 1986) mainly due to the low solubility of oxygen in the liquid phase. Complex flow models, such as the dispersion model, are not generally required for the gas phase. A complete mixing flow model may be applicable for FBBs with a low ratio of bed height to bed diameter and plug flow for those with high ratio.

In Type IV FBBs, the gas-liquid oxygen transfer in preaerators, such as a conventional bubble column or a downflow bubble contact aerator, has been extensively studied (Shah et al., 1982; Ade Bello et al., 1985). A plug flow model is commonly employed to describe the liquid phase behavior (e.g., Stathis, 1980; Shieh et al., 1981b). Mass balances on the pollutant and dissolved oxygen in the liquid-solid fluidized bed following the preaerator can be generally described by Eqns. (8.39) and (8.40), with $K_{\ell}a\ (C_s - C_b)$ in Eqn. (8.40) being omitted. The dissolved oxygen concentration is usually the limiting factor for BOD removal in this type of bioreactor and effluent recycling is usually required to achieve the desired effluent BOD level. For recycling operation, the pollutant concentration in the influent stream already mixed with the recycle stream (S_{bi}) can be related to the effluent concentration (S_{be}) by

$$S_{bi} = (S_o + \gamma S_{be})/(1 + \gamma) \tag{8.55}$$

where S_o is the pollutant concentration in the feed prior to mixing with the recycle stream and γ is the recycle ratio (Stathis, 1980). By solving the mass balance equations for the pollutant and dissolved oxygen in the liquid phase and for oxygen in the gas phase simultaneously, relationships among the operational parameters, such as the effluent pollutant concentration, DO in the liquid, required air feed rate, etc., can be evaluated. A similar procedure was developed to evaluate BOD removal performance and power economy in an activated sludge process (Fujie and Kubota, 1986). Modeling and simulation of the process dynamics for Type IV FBBs can also be found in Stathis (1980), Ngian and Martin (1980a), and Kleinstreuer and Poweigha (1984).

As mentioned earlier, kinetic expressions of biodegradation encountered most often in aerobic wastewater treatment are of zero order, first order, Monod type, Haldane type, and bi-substrate limiting equations. If the biodegradation is of zero or first order, the bioreactor models for plug and complete mixing liquid flow patterns can be solved analytically. Model equations under different flow conditions for various types of fluidized bed systems are summarized in Table 8.5.

8.3.3.2 Dynamic Conditions

Tang et al. (1985) investigated the dynamic behavior of phenol degradation by biofilm operated in a DTFB (Type IIA FBB) using activated carbon particles as the cells immobilizing carrier. The mass balance equations for the substrate and oxygen in the completely mixed bulk liquid phase in such a system are given by

Table 8.5 Model Equations under Different Flow Conditions to Account for Various Types of Fluidized Bed Systems.

Phases	Components \ Types of FBBs	IA, IB	IIA, IIB, IIIB*, IIIC: Riser	IIA, IIB, IIIB*, IIIC: Downcomer	IIIA, IIIB**	IVA, IVB: Preaerator	IVA, IVB: Fluidized Bed
Gas phase	Oxygen	Plug flow: Eqn. (8.41)	Plug flow: Eqn. (8.41)	No significant oxygen transfer	Plug flow: Eqn. (8.41)	Plug Flow in bubble column: Eqn. (8.41) Complete mixing for DBCA in Fig. 8.1: Eqn. (8.53)	———
			Complete mixing: Eqn. (8.53)				
Liquid phase	Oxygen	Dispersed flow: Eqn. (8.40) or (8.47) Plug flow: Eqn. (8.51)	Plug flow: Eqn. (8.51)	Plug flow: Eqn. (8.51)****	Dispersed flow: Eqn. (8.40) or (8.47) Complete mixing: Eqn. (8.52)	Dispersed flow: Eqn. (8.40)*** or (8.47)*** Plug flow: Eqn. (8.51)***	Plug flow: Eqn. (8.51)**** Dispersed blow: Eqn. (8.40)**** or (8.47)****
			Complete mixing: Eqn. (8.52)				
	Substrate	Dispersed flow: Eqn. (8.39) or (8.46) Plug flow: Eqn. (8.50)	Plug flow: Eqn. (8.50)	Plug flow: Eqn. (8.50)	Dispersed flow: Eqn. (8.39) or (8.46) Complete mixing: Eqn. (8.52)	Dispersed flow: Eqn. (8.39)*** or (8.46)*** Plug flow: Eqn. (8.50)***	Plug flow: Eqn. (8.50) Dispersed flow: Eqn. (8.39) or (8.46)
			Complete mixing: Eqn. (8.52)				

* - Riser
** - Downcomer
*** - $(-R_{sv})$ is omitted
**** - $K_\ell a(C_s - C_b)$ is omitted

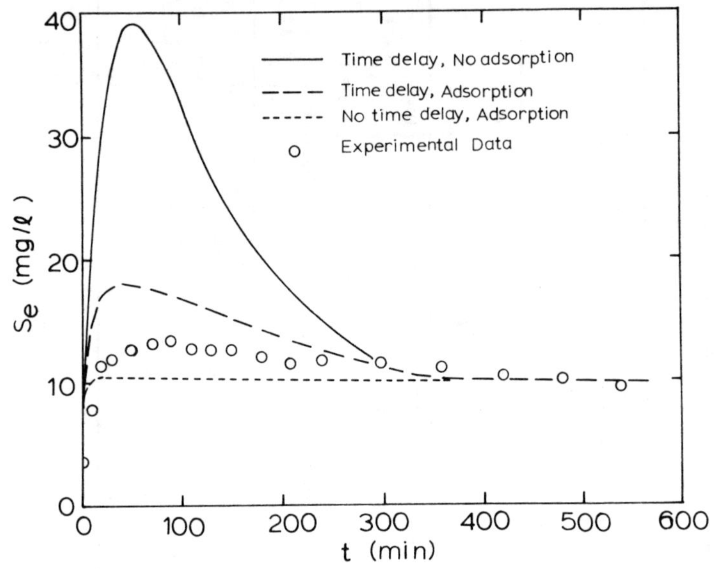

Figure 8.7 Transient response of a DTFB to step increase in inlet phenol concentration from 49.17 to 96.20 mg/ℓ (from Tang et al., 1985).

$$V \epsilon_\ell \frac{dS_{be}}{dt} = Q (S_{bi} - S_{be}) - VK_{\ell s}\, a_s (S_{be} - S^*) \tag{8.56}$$

$$V \epsilon_\ell \frac{dC_{be}}{dt} = Q (C_{bi} - C_{be}) + VK_{\ell}a (C_s - C_{be}) - VK_{\ell c}\, a_s(C_{be} - C^*) \tag{8.57}$$

Tang et al. (1985) coupled Eqns. (8.56) and (8.57) with Eqns. (8.25) through (8.38) to establish a simulation model and compared the model simulation with experimental results. Figure 8.7 shows the transient response in a DTFB to a step increase in inlet phenol concentration from 48.17 to 96.20 mg/ℓ reported by Tang et al. (1985). The model, without incorporating any time delay effect through Eqn. (8.38), predicts a rapid rise in the bulk phenol concentration to the final value without going through a maximum. Incorporating the time delay effect in the model, on the other hand, successfully predicts the transient response behavior qualitatively, but the predicted phenol concentration values are slightly higher than the experimental values. The predicted bulk phenol concentration goes through a maximum after about one hour, in agreement with experimental data. In addition, results simulated from the model excluding adsorption effects of the activated carbon show a very large transient change, especially immediately after the introduction of the step change. This result signifies the importance of the adsorption effect of activated carbon particles when used as solid supports and its effect on rapid response to the substrate shock loading for steady state operation.

During the operation of a DTFB bioreactor, the slow growth of biofilm counter-balanced by the removal of biomass from the biofilm surface by fluid shear and particle attrition results in an approximately constant bioparticle size within a 1- to 3-day period. Pseudo-steady state operation, during which the bulk substrate concentrations remain constant and the biofilm thickness is approximately constant, can thus be realized for each time frame of 1 to 3 days.

8.3.3.3 Concentration Multiplicity

The existence of multiplicity in a reactor is of major concern since it may invoke operational instability when the reactor is subjected to environmental disturbances. Multiplicity of steady states can be due to the nonlinear nature of the substrate inhibition kinetic expression. Concentration multiplicity in a completely mixed reactor containing a dissolved enzyme system with substrate inhibiting kinetics has been reported and investigated by Bruns et al. (1973). Detailed analyses of steady state multiplicity and stability of a suspended cell system were given by Lim et al. (1982).

For an immobilized cell system, concentration multiplicity may exist in either the Type IA bioreactor or the Type IIA bioreactor. For Type IA bioreactors, based on the axial dispersion model described in Section 8.3.3.1, Wisecarver and Fan (1987) reported that for phenol degradation in aeration with pure oxygen, which follows bi-substrate limiting cell growth kinetics described by Eqn. (8.17), steady state multiplicity in the bioreactor can occur at low liquid velocities. The possible existence of steady state multiplicity in a Type IIA bioreactor or a DTFB bioreactor with bi-substrate limiting cell growth kinetics dictated by Eqn. (8.17) was mathematically analyzed by Tong and Fan (1988), and will be briefly discussed in the following.

In a DTFB reactor, mass balances on the limiting substrate S (S.S) and substrate C (S.C) in bulk liquid phase give

$$QS_{bi} - QS_b - Va_sD_{sf}\left.\frac{dS}{dr}\right|_{r=r_p} = 0 \qquad (8.58)$$

$$QC_{bi} - QC_b - Va_sD_{cf}\left.\frac{dC}{dr}\right|_{r=r_p} = 0 \tag{8.59}$$

Note that in Eqns. (8.58) and (8.59) and subsequent equations, S_b and C_b are equivalent to S_{be} and C_{be} due to the complete mixing condition assumed for the DTFB. Equations (8.58) and (8.59) apply to both a liquid-solid fluidized bed and a gas-liquid-solid DTFB where the gas phase is inert. For a three-phase DTFB where the gas phase is a source for S.C, Eqn. (8.59) is replaced by

$$QC_{bi} - QC_b - Va_sD_{cf}\left.\frac{dC}{dr}\right|_{r=r_p} + VK_\ell a(C_s - C_b) = 0 \tag{8.60}$$

The fluxes of both S.S and S.C through the biofilm surface requires information on the concentration profiles of S.S and S.C within the biofilm. This information can be obtained by solving the diffusion-reaction equations as given by Eqns. (8.20) and (8.21) and the corresponding boundary conditions, Eqns. (8.22a) through (8.22c).

Multiplicity in the Biofilm

Before analyzing the steady state multiplicity of the DTFB bioreactor, the steady state multiplicity of substrate concentration profiles within the biofilm will first be examined. Concentrations of S and C within the biofilm at the same locus are related by Eqn. (8.24) as given by

$$C = C^* + \frac{Y_o\ D_{sf}}{D_{cf}}(S - S^*) = C^* + K_1(S - S^*) \tag{8.61}$$

where

$$K_1 = \frac{Y_o\ D_{sf}}{D_{cf}}$$

Combining Eqns. (8.17), (8.20), and (8.61) yields

$$\frac{1}{r^2}\frac{d}{dr}(r^2\frac{dS}{dr}) - \lambda f(S) = 0 \tag{8.62}$$

where

$$\lambda = X_v\mu_m/(D_{sf}Y_x) \tag{8.63a}$$

and

$$f(S) = \frac{S}{S + K_s + S^2/K_i} \frac{K_1(S - S^*) + C^*}{K_1(S - S^*) + C^* + K_c} \tag{8.63b}$$

For a thin biofilm attached to a relatively large spherical particle, the equation in spherical coordinates can be approximated using rectangular coordinates by

$$\frac{d^2S}{dr^2} - \lambda f(S) = 0 \tag{8.64}$$

Since C is non-negative, the following physical constraint based on Eqn. (8.61) is required:

$$S \geq S^* - C^*/K_1 \tag{8.65}$$

If G(S) is defined as

$$G(S) = \int_0^S f(t)dt \tag{8.66}$$

Equation (8.64) can be integrated to

$$(S')^2 - 2\lambda G(S) = M \tag{8.67}$$

where $S' = dS/dr$. From the boundary condition given in Eqn. (8.22c), this gives rise to

$$M = -2\lambda G(\omega) \tag{8.68}$$

where ω is the substrate concentration at $r = r_0$. Since $dS/dr \geq 0$, Eqns. (8.67) and (8.68) yield

$$\frac{dS}{dr} = (2\lambda)^{1/2}[G(S)-G(\omega)]^{1/2} \tag{8.69}$$

Integrating Eqn. (8.69) with Eqn. (8.22b) as the boundary condition results in

$$\Omega = h(\omega) \tag{8.70}$$

where

$$h(\omega) = \int_{\omega}^{S^*} \frac{d\xi}{[G(\xi) - G(\omega)]^{1/2}} \tag{8.71a}$$

and

$$\Omega = (2\lambda)^{1/2}(r_p - r_o) \tag{8.71b}$$

The derivative of h with respect to ω is

$$h'(\omega,S^*) = -\frac{1}{2}\int_{\omega}^{S^*} \frac{f(\xi)-f(\omega)}{[G(\xi)-G(\omega)]^{3/2}}\,d\xi - \frac{1}{[G(S^*)-G(\omega)]^{1/2}} \tag{8.72}$$

where $h' = dh/d\omega$. As can be seen, ω can be any positive value determined by Ω if C^* is sufficiently large. Under this condition, the second term on the right-hand side of Eqn. (8.72) approaches zero as S^* approaches infinity. The first term is positive if ω has a value such that $f'(\omega)(= df/d\omega) < 0$. Thus, Eqn. (8.72) yields a positive h'. From Eqn. (8.71a), when S^* is infinity, h approaches infinity as ω approaches 0, and h approaches 0 as ω approaches S^*. Consequently, there will be an even number of extrema on the h versus ω plot. If the value of Ω falls between the two extrema of h, multiple solutions are possible for Eqn. (8.70) for a given value of S^*. If C^* is small, the minimum ω is close to S^* and h' will not be a positive value. Thus, one may conclude the necessary and sufficient conditions for the existence of multiplicity in the biofilm as follows:

$$h'(\omega,S^*) > 0 \quad \text{for some values of } \omega \tag{8.73}$$

and

$$\max h(\omega,S^*) \geq \Omega \geq \min h(\omega,S^*) \tag{8.74}$$

Figure 8.8 shows the variation in $\bar{h}(\bar{\omega})$ with $\bar{\omega}$ as a function of C_b where $\bar{\omega}$ and $\bar{h}(\bar{\omega})$ are the dimensionless forms of ω and $h(\omega)$ and equal to ω/S^* and $h(\omega)/(S^*)^{1/2}$, respectively. For a small C_b, it is seen that $\bar{h}(\bar{\omega})$ monotonically decreases with $\bar{\omega}$ and the minimum value of $\bar{\omega}$ that can exist is large. For a large C_b, $\bar{h}(\bar{\omega})$ decreases, increases and decreases with $\bar{\omega}$ and the minimum value of $\bar{\omega}$ is extended to zero. Thus, two extrema exist for a large C_b, which allows multiplicity to exist,

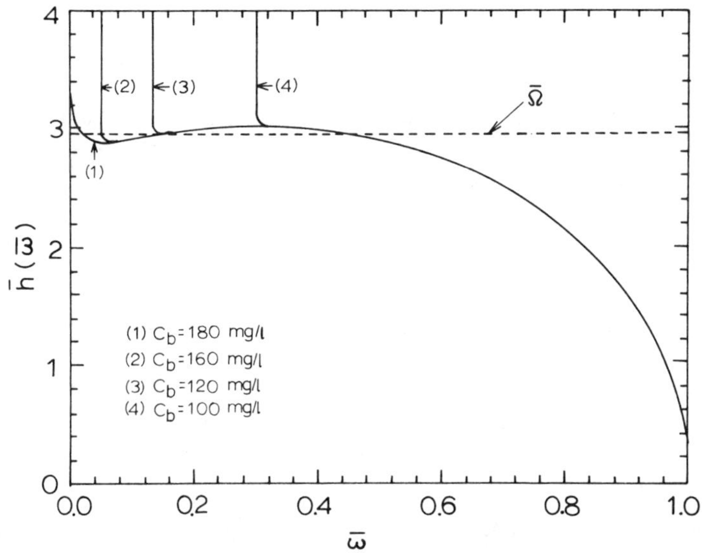

Figure 8.8 Variation of $\bar{h}(\bar{\omega})$ with $\bar{\omega}$ as a function of C_b reflecting the behavior of concentration multiplicity in a biofilm (from Tong and Fan, 1988).

provided that $\bar{\Omega}$ (= Ω/r_p) is located between the two values of $\bar{h}(\bar{\omega})$ corresponding to the two extrema for $\bar{\omega}$. Consequently, steady state multiplicity may exist in the biofilm provided that Ω, which is inherently related to the physical and biological parameters (Eqns. (8.63a) and (8.71b)) of the system of interest, has appropriate values.

Multiplicity in the Bioreactor

The multiplicity of steady state bulk substrate concentrations in a DTFB bioreactor is subjected to the following analysis. Note that the multiplicity analysis of Witmer et al. (1986) for a reaction with Langmuir-Hinshelwood kinetics in an isothermal porous catalyst slab can also be extended to the present system. The volume-based effectiveness factor of the biofilm (η_v) can be derived from Eqns. (8.19) and (8.69) as

$$\eta_v = \frac{6[G(S^*)-G(\omega)]^{1/2}}{(r_p-r_o^3/r_p^2)(2\lambda)^{1/2}g(S^*,C^*)} \tag{8.75}$$

where

$$g(S^*,C^*) = \frac{S^*}{S^* + K_s + S^{*2}/K_i} \frac{C^*}{C^* + K_c} \tag{8.76}$$

For the case in which the external mass transfer resistance at the biofilm surface is negligible, $S^* = S_b$ and $C^* = C_b$, then combining Eqns. (8.24), (8.58), and (8.59) or (8.60) yields

$$C_b = C_{bi} - Y_o(S_{bi} - S_b) \tag{8.77}$$

for a two-phase DTFB and a three-phase DTFB in which the gas phase is inert; or

$$C_b = \frac{QC_{bi} - Y_oQ(S_{bi} - S_b) + VK_\ell aC_s}{Q + VK_\ell a} \tag{8.78}$$

for a three-phase DTFB in which the gas phase is a source for S.C. Substituting Eqns. (8.19), (8.77), or (8.78) into (8.58) along with $S^* = S_b$ and $C^* = C_b$ gives

$$QS_{bi} - QS_b - \theta_b \eta_v H(S_b) = 0 \tag{8.79}$$

where

$$\theta_b = \frac{\mu_m X_v 4\pi N_p (r_p^3 - r_o^3)}{3\, Y_x} \tag{8.80a}$$

$$H(S_b) = \frac{S_b}{S_b + K_s + S_b^2/K_i} \frac{C_1 - Y_o(S_{bi} - S_b)}{C_2 - Y_o(S_{bi} - S_b)} \tag{8.80b}$$

and for a three-phase DTFB

$$C_1 = C_{bi} + VK_\ell aC_s/Q \quad ; \quad C_2 = C_1 + K_c(1 + VK_\ell a/Q)$$

or for a two-phase DTFB

$$C_1 = C_{bi} \quad ; \quad C_2 = C_1 + K_c$$

Rearranging Eqn. (8.79) gives

$$\theta_b = \frac{Q(S_{bi} - S_b)}{\eta_v H(S_b)} \tag{8.81}$$

Let

$$F(S_b) = \frac{Q(S_{bi}-S_b)}{\eta_v H(S_b)} \tag{8.82}$$

The value of this function will be zero as S_b approaches S_{bi} and will be infinity as S_b approaches 0 or $S_{bi}-C_1/Y_o$. Taking the derivative of $F(S_b)$ with respect to S_b yields $F'(S_b)$ as given below

$$F'(S_b) = -\{(QS_{bi}-QS_b)[\eta_v \left(\frac{K_s-S_b^2/K_i}{\Lambda^2} \frac{C_1-Y_o(S_{bi}-S_b)}{C_2-Y_o(S_{bi}-S_b)} + \frac{S_b}{\Lambda} \cdot \frac{Y_o(C_2-C_1)}{[C_2-Y_o(S_{bi}-S_b)]^2} \right) + H(S_b)\frac{d\eta_v}{dS_b}] + Q\eta_v H(S_b)\}/[\eta_v H(S_b)]^2 \tag{8.83}$$

where

$$\Lambda = K_s+S_b+S_b^2/K_i \tag{8.84}$$

The derivative $d\eta_v/dS_b$ can be negative if S_b becomes large. Since C_b must be non-negative, from Eqn. (8.77) or (8.78), it yields

$$C_1-Y_o(S_{bi}-S_b) \geq 0 \tag{8.85}$$

Based on the same argument given for multiplicity in the biofilm, if there exists some values of S_b such that $F'(S_b)$ is positive, there may be multiple solutions for S_b for a given S_i and C_{bi}. Thus, the necessary and sufficient conditions for multiplicity in a bioreactor are

$$F'(S_b) > 0 \quad \text{for some values of } S_b \tag{8.86}$$

and

$$\max F(S_b) \geq \theta_b \geq \min F(S_b) \tag{8.87}$$

Figure 8.9 shows the variation of $\bar{F}(\bar{S}_b)$ with $\bar{S}_b$ as a function of S_{bi}, where $\bar{S}_b$ and $\bar{F}(\bar{S}_b)$ are dimensionless forms of S_b and $F(S_b)$ and are equivalent to S_b/S_{bi} and $F(S_b)/(QS_{bi})$, respectively. The existence of two extrema in the curves indicates the possibility of three steady-states in the bioreactor for given inlet substrate concentrations. Since

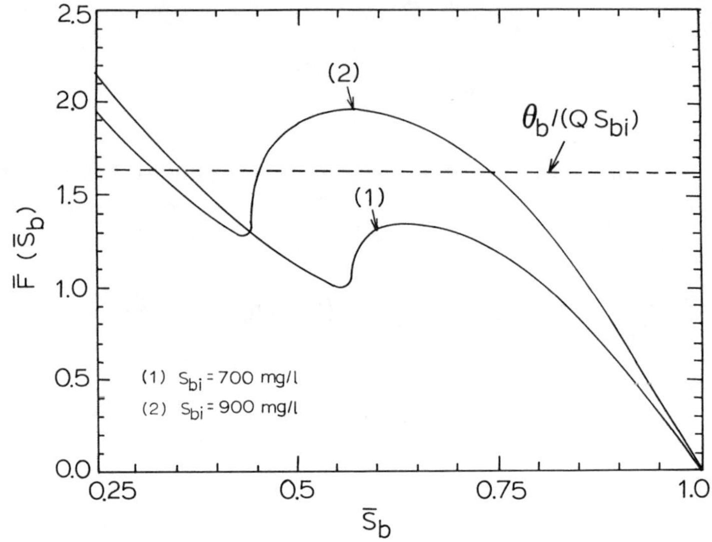

Figure 8.9 Variation of $\bar{F}(\bar{S}_b)$ with $\bar{S}_b$ as a function of S_{bi} reflecting the behavior of concentration multiplicity in a DTFB (from Tong and Fan, 1988).

the solutions of Eqn. (8.81) in dimensional form are given by the intersection of the $F(S_b)$ curve and the straight line for θ_b, multiplicity occurs when the value of θ_b is located between the two extrema of the $F(S_b)$ curve. Alternatively, in dimensionless form, multiplicity occurs when the value of $\theta_b/(QS_{bi})$ is located between the two extrema of the $\bar{F}(\bar{S}_b)$ curve.

Tong and Fan (1988) evaluated the possible existence of steady-state multiplicity for phenol biodegradation in a DTFB with all the parameters in the above equations being experimentally determined. They concluded that when a DTFB bioreactor is operated with air at low phenol concentrations (less than 50 mg/ℓ), as is commonly encountered in practical situations, a steady-state multiplicity does not exist. For systems with more than two limiting substrates, the analysis given above is directly extendable as the variation of substrates can be interrelated via the yield factors.

8.3.4 Cultivation and Biofilm Control

During the start-up of a FBB, microbes are introduced into the reactor to establish attachment to support particles. Appropriate substrates are supplied to allow the attached microbes to develop into visible biofilms. Completion of the cultivation of biofilms is indicated by their capability for a preset degradation rate. The FBB then assumes normal operation, and usually biofilm thickness control action has to be taken periodically to prolong the operating lifespan of the FBB.

The capability of microbes to attach to particles under applied shear stress due to liquid flow is critical to the initial microbial seeding and the subsequent biofilm development. The critical surface shear stress above which microbial adhesion can not be achieved can be evaluated by use of the LH-Flower cell adhesion measurement module (Atkinson and Mavituna, 1983). In this module, a microbial suspension is pumped at a constant volumetric flow rate through the center of a radial flow chamber consisting of two parallel discs with only a narrow separation between them. The suspension then flows out radially between the discs with a continuous reduction in surface shear stress conditions from the center to the periphery under which the attachment of the microbial cells can be studied.

In the initial seeding and the subsequent cultivation of biofilms in FBBs, the culture of seed microbes or activated sludges is added to the reactor containing particles and an appropriate substrate concentration in the liquid phase. The reactor may then be operated in batch mode without any liquid circulation for 12 to 24 hours (Hirata et al., 1982; Fan et al., 1984a) or with the liquid suspension circulated for 8 to 12 hours (Holladay et al., 1978a) to allow sufficient initial attachment of the microbes. The FBBs are then placed under continuous operation for approximately one to two weeks to reach the desired biofilm thickness and biodegradation rate. The additional shear stress resulting from the direct injection of gas into the reactor should be considered, especially during the initial biofilm cultivation. In fact, difficulty in biofilm cultivation has been observed in a 7.6 cm diameter three-phase fluidized bed at a superficial gas velocity beyond 2.5 cm/s (Fan et al., 1986).

During the transient operation of the FBBs, the biofilm will continue to grow, though at a low rate due to the fluid shear and abrasion among bioparticles. One useful parameter for characterizing the growth rate of the biofilm relative to the substrate (pollutant) loading rate is the apparent biomass yield factor of biofilm, which is defined as the amount of cell mass yielded and retained in the biofilm per unit amount of substrate consumed. For example, Fan et al. (1986) reported that the overall yield factor for biological phenol degradation in the DTFB bioreactor was 0.469, while that value in the TFB, operated under the same condition as the DTFB, was 0.579. Significantly higher biofilm dry density was obtained in the DTFB than in the TFB at a given biofilm thickness. Apparently, the higher fluid shear resulting from the extensive internal circulation of liquid and solid phases in the DTFB account for the lower biofilm growth rate in the DTFB.

The increase in biofilm thickness reduces the buoyant density of the bioparticles, resulting in elutriation of bioparticles out of the reactor and shortening of the operating life-span of the reactor. Biofilm control is therefore required to maintain a long and active FBB operation. Excess biofilm thickness in a FBB is frequently removed by air scouring and back-washing methods. Such methods, however, interrupt normal operation of the FBBs. Mechanical devices situated externally have been used for biofilm control by providing mechanical shear stress to remove excess biomass from the biofilm. Mechanical stirrers placed at the top of the fluidized bed have also been applied for in-situ biofilm control via mechanical shear stress for steady state operation (Jeris et al., 1977, Hirata et al., 1982, Takase et al., 1983). Jeris et al. (1977) employed a sand separation tank, called a swirl concentrator, which creates centrifugal force to achieve biofilm removal together with sludge separation. Holladay et al. (1978b) and Hoyland and Robinson (1983) employed the sludge pump and the centrifugal pump to remove the biofilm, and the peeled particles were recycled to the fluidized bed. Sawada (1978) and Sawai et al. (1982), however, reported that a constant biofilm thickness could be maintained in the steady state commercial operation using a Type IIA bioreactor containing sand or activated carbon particles without any external shear control device. The amount of microbial cells attached to the particles in this system was 120 mg-VSS/g-particle, where VSS denotes volatile suspended solids. This steady state operation is achieved possibly due to the internally generated high shear stress coupled with relatively low BOD loadings. Hirata et al. (1982) reported that without biofilm removal, the film thickness can increase at a rate of 16 μm per day when the phenol loading is 3 to 4 kg/m^3•day in a TFB bioreactor. Even though the apparent cell mass yield factor of biofilm in the DTFB bioreactor is lower than that in other types of bioreactors (Fujie et al., 1983; Fan et al., 1986), the DTFB bioreactor requires biofilm control for stable operation. Fan et al. (1986) satisfactorily controlled the biofilm thickness attached to the spherical activated carbon particles by a positive displacement liquid pump coupled with a liquid cyclone. As noted earlier, the in-situ biofilm control can also be achieved through the bombardment action of the dense bed particles with bioparticles, as demonstrated in the Type IIIC bioreactor (Nikolov et al., 1982).

It should be noted, however, that the induced shear stress generated by mechanical means produces rough surfaces on the biofilms, which promote abrupt expansion of the biofilm in the bioreactor (Hirata et al., 1982; Fan et al., 1986).

8.3.5 Power Economy

In aerobic wastewater treatment, energy is mainly consumed by supplying oxygen to the bioreactor. The power economy is one of the most important factors in evaluating the feasibility of biological wastewater treatment processes (Hosono et al., 1980). The power economy of organic removal can be defined as the amount of pollutant, e.g., BOD or COD, removed per unit power consumption (P_w) or expressed by $(-R_{sv})V/P_w$ ($= Q(S_{bi}-S_{be})/P_w$). The P_w required for feeding air into the bioreactor can be given by (in SI units) (Kubota et al., 1980)

$$P_w = 0.027\ G_v\ \{(P_e/P_i)^{0.286} - 1\}/\epsilon_b \tag{8.88}$$

where G_v is the volumetric air flow rate evaluated at 0°C and 1 atm, P_i and P_e are the inlet and outlet pressures of the air blower, and ϵ_b is the overall efficiency of the blower and the electric motor used.

Fujie et al. (1985) reported that the power economy of Types I and II FBBs strongly depends on the relationship between $K_\ell a$ and U_g in the bioreactors. For a given U_g a higher $K_\ell a$ results in a higher removal rate of organic pollutant. Less power is required to remove a given amount of pollutant when $K_\ell a$ is higher, yielding a higher power economy. The power economy of the FBBs of Types I and II is lower than that of the conventional bubble column bioreactor, e.g., the aeration tank used in a conventional activated sludge process, and the power economy of the DTFB is higher than that of the TFB. For Type IVB FBBs, oxygenation using oxygen enriched air requires 80% or higher oxygen utilization to give a power economy comparable to that of Type IIA FBBs (Fujie et al., 1980). Thus, the power economy of Type IVB depends strongly on the cost of oxygen enriched air. Under the optimum operating conditions of this type of bioreactor, the power consumption by the liquid pump for fluidization of bioparticles amounts to one-third of the total power consumption. Fujie (1987)

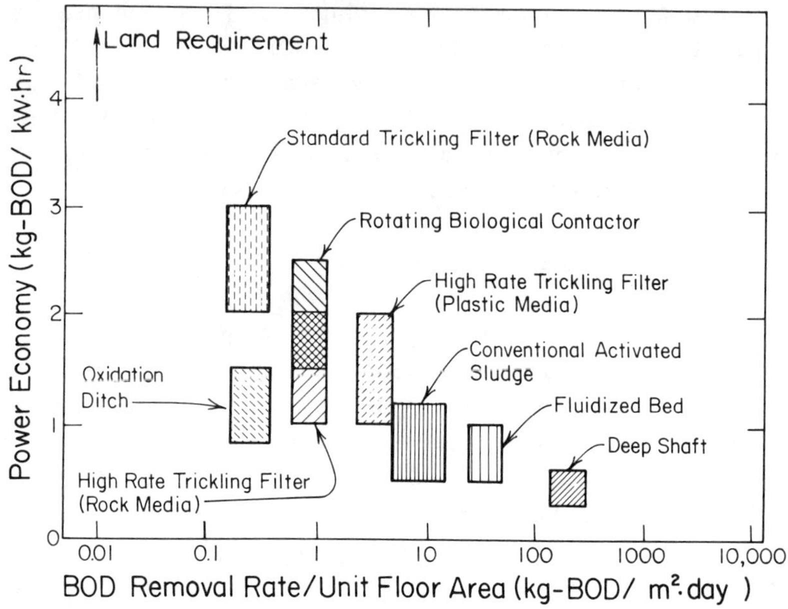

Figure 8.10 Power economy of various biological wastewater treatment processes under typical operating conditions for domestic sewage treatment (from Fujie, 1987).

also reported that the power economy of various wastewater treatment processes under typical operating conditions for domestic wastewater treatment is inversely proportional to the organic pollutant removal rate per unit floor area of the process. Such a relationship is shown in Fig. 8.10. The power economy of FBBs ranges from 0.4 to 0.8 kg-BOD/kW•hr. Note that the power economy of the FBB is the second lowest next to the deep shaft process.

Regarding investment costs, Ishihara and Sawada (1978) indicated that the capital investment and the floor area for the FBB process are 70% and 46%, respectively, of those for a conventional activated sludge process with the same treatment capacity. Wheeldon and Bayley (1981) studied the economics of Type IV FBBs and also concluded that the capital cost of a FBB is significantly lower than that of a conventional activated sludge aeration tank having the same conversion for wastewater treatment because of the high biomass concentration in a FBB. However, excessive costs of oxygen enriched air and liquid pumping for fluidization of bioflocs may amount to as high as 57% of the total operating cost for this type of bioreactor.

Hoyland and Robinson (1983) obtained the same conclusion as Wheeldon and Bayley (1981), i.e., the aerobic treatment of sewage in a FBB process is more expensive than treatment in an activated sludge process. They also concluded that this high operating cost is due to aeration and pumping of the liquid.

Although the FBB is regarded as an alternative bioreactor to the trickling filter in wastewater treatment, FBBs do not outperform the trickling filter from a power economy viewpoint. Furthermore, particulate organic substances, which account for about one-half of the total organic substances in domestic sewage, cannot be removed by the fluidized bed bioreactor. For the treatment of such wastewater, additional equipment to remove suspended solids from the effluent is needed (Suzuki, 1986). It should be noted, however, that the cell mass concentration in FBBs which is equivalent to the MLSS of 15 to 40 g/ℓ in an activated sludge aeration tank and the volumetric BOD loading which is greater than 10 kg-BOD/m^3•day are about 5 to 20 times higher than those in a conventional activated sludge process as described earlier (e.g., Goto, 1986). As pointed out by Hoyland and Robinson (1983), a FBB may best be applied to the wastewater treatment where the small ground area requirement can be turned into an economic advantages. In such a case, a FBB plant would be economically competitive with an activated sludge process (Barbara et al., 1980).

8.4 CONCLUDING REMARKS

In gas-liquid-solid fluidized bed bioreactors including Types I, II and IIIA FBBs, there are complex interactive activities taking place between the transport properties such as gas-liquid oxygen transfer, liquid dispersion, and biological reactions. The concept of volume-based effectiveness factor can be conveniently utilized to account for diffusional effects on the aerobic biological reactions in biofilms. The overall cell growth rate or apparent cell yield factor in biofilms in the aerobic FBBs is lower than that in a conventional activated sludge process for a given removal rate of BOD.

The volumetric BOD loading in any type of FBBs shown in Fig. 8.1 could reach to 15 kg-BOD/m^3•day or even higher. Proper biofilm control is normally needed for long term and stable operation of a FBB under high organic loading in the bioreactor. New biofilms should be cultivated on the completely peeled particles to maintain high microbe activity. Though the capital investment of a FBB is low, the operational cost of the FBB is higher than the conventional activated sludge process. It can be stated that FBBs presently merit use over conventional activated sludge processes in wastewater treatment with high BOD loadings. The FBB has particular advantages in populated areas where the land value is high.

Further studies on the relationship between the biofilm growth rate and the organic carbon loading for various substrates, e.g., organic pollutants, and the internal structure of biofilm and its effect on substrate diffusivity are needed. The microbial behavior in the immobilized state should be explored in comparison with that in the suspended states. Methods for effective biofilm thickness control are also required to ensure long term operation of the FBB with a high biodegradative activity. In this connection, distribution of shear stress in the flow field and shear stress effects on the biofilm thickness requires attention. Furthermore, dynamic behavior of the FBB in response to the unsteady state loading needs to be explored. Moreover, overall feasibility studies involving capital investment and operating cost beyond power economy studies for FBBs in comparison with other biological wastewater treatment processes should be further examined under various operating conditions.

8.5 NOMENCLATURE

a	: surface area of gas bubbles per unit bed volume
a_s	: surface area of particles or bioparticles per unit bed volume
A	: density ratio defined by Eqn. (8.8)
Ar	: Archimedes number, $d_p{}^3\rho_\ell(\rho_p-\rho_\ell)g/\mu_\ell^2$
C	: dissolved oxygen concentration; C_i in the feed; C_e at the exit
C^*	: dissolved oxygen concentration on the surface of biofloc particle
C_b	: value of C in the bulk liquid; C_{bi} at the inlet; C_{be} at the exit
C_c	: critical value of C below which aerobic reaction ceases
C_{io}	: initial dissolved oxygen concentration
C_o^*	: value of C saturated with pure oxygen at atmospheric pressure

C_s	: value of C saturated with air
d_p	: bioparticle diameter
d_o	: diameter of biofilm-free solid particle; $\overline{d}_o$, the mean value
D_c	: column diameter
D_{cf}	: diffusivity of oxygen in the biofilm
D_{cp}	: diffusivity of oxygen in the particle
D_f	: diffusivity in the biofilm
D_{sf}	: diffusivity of substrate in the biofilm
D_{sp}	: diffusivity of substrate in the particle
D_T	: draft tube diameter
D_w	: diffusivity in water
$E_{z\ell}$	: axial dispersion coefficient of the liquid based on the interstitial liquid velocity
f_w	: weight fraction of moisture content in the biofilm
g	: gravitational acceleration
$g(x_p)$	: distribution function of biofilm thickness
G_v	: volumetric gas flow rate
H	: reactor height
K_1	: parameter defined as Y_oD_{sf}/D_{cf}
K_c	: Monod constant for oxygen uptake; K_{c1} for initial nitrification; K_{c2} for second nitrification
K_D	: overall decay coefficient of biofilm
K_i	: inhibition constant
K_ℓ	: gas-liquid mass transfer coefficient for oxygen
$K_{\ell c}$	: liquid-solid mass transfer coefficient for oxygen
$K_{\ell s}$	: liquid-solid mass transfer coefficient for substrate
K_m	: reaction rate constant, μ_m/Y_x
K_n	: zero-order nitrification rate constant; K_{n1} for initial nitrification; K_{n2} for second nitrification
K_N	: Monod constant of nitrification; K_{N1} for initial nitrification; K_{N2} for second nitrification
K_o	: zero-order reaction rate constant
K_s	: Monod constant for substrate

K_v	: first-order reaction rate constant
L	: fluidized bed height
L_c	: fluidized bed height of clean particle
L_H	: fixed bed height
L_s	: volumetric solids loading
m_f	: cell mass on a single bioparticle
n	: Richardson-Zaki index
n'	: modified Richardson-Zaki index given by Eqn. (8.11)
n''	: modified Richardson-Zaki index given by Eqn. (8.12)
N_1	: concentration of ammonia nitrogen
N_2	: concentration of nitrite nitrogen
N_c	: mass flux of oxygen through biofilm surface
N_s	: mass flux of organic substrate through biofilm surface
N_P	: number of bioparticles
P^*	: atmospheric pressure
P_e	: pressure at blower outlet
P_i	: pressure at blower inlet
P_w	: power required for aeration
q	: amount of substrate adsorbed per unit amount of activated carbon particle
Q	: volumetric liquid flow rate
r	: radial coordinate
r_c	: critical radius of r where the reaction ceases
$-r_{cv}$	: oxygen consumption rate per unit biofilm volume
$-r_m$	: maximum nitrification rate; $-r_{m1}$ for initial nitrification; $-r_{m2}$ for second nitrification
$-r_{N1}$	: rate of initial nitrification reaction
$-r_{N2}$	: rate of second nitrification reaction
r_o	: radius of the biofilm-free solid particle
r_p	: radius of a bioparticle
$-r_{sv}$	: substrate removal rate per unit biofilm volume
$-r_{sv}^o$	: value of $-r_{sv}$ for a substrate concentration of S^o

$-r_{sv}^*$	: value of $-r_{sv}$ at the surface of biofloc particle
$-R_{cv}$	: oxygen consumption rate per unit reactor volume
$-R_{sv}$	: substrate removal rate per unit reactor volume
Re	: Reynolds number, $d_p U_\ell \rho_\ell / \mu_\ell$
Re_t	: Reynolds number, $d_p U_t \rho_\ell / \mu_\ell$
S	: substrate concentration; S_i in the feed; S_e at the exit
S^o	: reference substrate concentration
S^*	: substrate concentration on the surface of bioparticle
S_a	: cross-sectional area of the bioreactor
S_b	: value of S in bulk liquid; S_{bi} at the inlet; S_{be} at the exit
S.C	: substrate C
S_{io}	: initial substrate concentration
S.S	: substrate S
t	: time
T	: temperature
U_g	: superficial gas velocity
U_ℓ	: superficial liquid velocity
U_{mf}	: minimum fluidization velocity
U_t	: terminal velocity of a single particle or bioparticle
U_{to}	: terminal velocity of a clean particle without biofilm
v_m	: specific volume of oxygen
V	: reactor volume
V_f	: biofilm volume on a single bioparticle
W	: removal rate of pollutant per unit reactor volume in the FBB bioreactor
W_o	: volumetric loading of pollutant per unit reactor volume in the FBB bioreactor
x	: r/r_p
x_c	: r_c/r_p
x_f	: fractional conversion of pollutant
x_o	: r_o/r_p

x_p : biofilm volume/clean particle volume, $(d_p/d_o)^3$-1; $\bar{x}_p$, average value over the axial distance; x_{pmax} at the top of the bed

X_m : cell concentration in the bioreactor

X_v : dry weight of biomass per unit volume of wet biofilm or biofilm dry density; $\bar{X}_v$, average biofilm dry density

Y_o : overall oxygen yield factor defined as the amount of oxygen consumed per unit amount of substrate consumed

Y_x : overall cell mass yield factor, defined as the amount of cell mass produced per unit amount of substrate consumed

z : axial distance

Z_i : oxygen mole fraction in the feed gas

Greek Letters

α : pressure conversion factor

γ : recycle ratio of effluent

δ : biofilm thickness

ϵ : void fraction of bed

ϵ_b : overall efficiency of the air blower

ϵ_c : volume fraction of cells in the void of the highly porous particle

ϵ_f : porosity of biofilm

ϵ_ℓ : liquid holdup

$\epsilon_{\ell o}$: liquid holdup in the bed of clean particles

ϵ_p : porosity of particle

ϵ_s : solid holdup in the bed

η : fraction of oxygen utilized in the bioreactor; η_e at the exit

η_v : volume-based effectiveness factor

μ : specific cell growth rate; μ_o at prior steady state; μ_m, the maximum value

μ_ℓ : viscosity of liquid

$-\nu_N$: specific rate of nitrification; $-\nu_{N1}$ for initial nitrification; $-\nu_{N2}$ for second nitrification

$-\nu_s$: specific substrate removal rate defined in Eqn. (8.17)

ρ_a : apparent bed density

ρ_{bw} : wet density of biofilm

ρ_D : dry density of microbes

ρ_ℓ : liquid density

ρ_m : density of matrix material for the highly porous particle

ρ_p : density of the bioparticle

ρ_s : density of the biofilm-free non-porous particle or effective density of the biofilm-free nearly non-porous particle

ρ_W : wet density of microbes

τ : wastewater space time in the bioreactor

ϕ_s : particle sphericity

ω : concentration of substrate S at $r=r_o$

θ : wastewater hydraulic retention time in the bioreactor

θ_b : parameter defined by Eqn. (8.80a)

8.6 REFERENCES

Ade Bello, R., C.W. Robinson, and M. Moo-Young, Biotech. Bioeng., 27, 369 (1985).

Andrews, G.F., and C. Tien, AIChE J., 25, 720 (1979).

Andrews, G.F., and C. Tien, AIChE J., 27, 396 (1981).

Andrews, G.F., and C. Tien, AIChE J., 28, 182 (1982).

Andrews, G.F., Biotech. Bioeng., 24, 2013 (1982).

Andrews, G.F., and R. Trapasso, J. Water Pollution Control Fed., 57, 143 (1985).

Andrews, G.F., Biotech. Prog., 2, 16 (1986).

Atkinson, B., E.L. Swilley, A.W. Busch, and D.A. Williams, Trans. Inst. Chem. Engrs., 45, T257 (1967).

Atkinson, B., and I.S. Daoud, Trans. Inst. Chem. Engrs., 46, T19 (1968).

Atkinson, B., I.S. Daoud, and D.A. Williams, Trans. Inst. Chem. Engrs., 46, T245 (1968).

Atkinson, B., and I.S. Daoud, Trans. Inst. Chem. Engrs., 48, T245 (1970).

Atkinson, B., G.M. Black, P.J.S. Lewis, and A. Pinches, Biotech. Bioeng., 21, 193 (1979).

Atkinson, B., and P.J.S. Lewis, in "Fungal Biotechnology," ed. by J.E. Smith, D.R. Berry, and B. Kristiansen, p. 153, Academic Press, London (1980).

Atkinson, B., in "Biological Fluidized Bed Treatment of Water and Wastewater," ed. by P.F. Cooper and B. Atkinson, p. 22, Ellis Horwood, Chichester, England (1981a).

Atkinson, B., G.M. Black, and A. Pinches, in "Biological Fluidized Bed Treatment of Water and Wastewater," ed. by P.F. Cooper and B. Atkinson, p. 75, Ellis Horwood, Chichester, England (1981b).

Atkinson, B., and F. Mavituna, "Biochemical Engineering and Biotechnology Handbook," Nature Press (1983).

Baillod, C.R., and W.C. Boyle, J. Sanitary Engineering Div. ASCE, 96, 525 (1970).

Baker, C.G.J., A. Margaritis, and M.A. Bergougnou, in Proceedings of the 6th International Fermentation Symposium, p. 635, Pergamon Press, New York (1981).

Barbara, M., F. Flood, and J. Jeris, Water and Wastes Engineering, 17, 35 (1980).

Beer, C., Discussion, J. Sanitary Engineering Div. ASCE, 96, 1452 (1970).

Beer, C., J.S. Jeris, and J.A. Mueller, New York State Department of Environmental Conservation, Environmental Quality Technical Paper No. 11 (1972).

Besik, F., Water and Sewage Works, 120, 68 (1973).

Bettmann, H., and H.J. Rehm, Appl. Microbiol. Biotech., 20, 285 (1984).

Biospherics Research Inc., Final Report to U.S. Department of Interior, Project No. PB-184-445, May (1969).

Bremer, H., and P.P. Dennis, J. Theor. Biol., 52, 365 (1975).

Bruns, D.D., J.E. Bailey, and D. Luss, Biotech. Bioeng., 15, 1131 (1973).

Chatib, B., A. Grasmick, S. Elmaleh, and R. Ben Aim, in "Biological Fluidized Bed Treatment of Water and Wastewater," ed. by P.F. Cooper and B. Atkinson, p. 192, Ellis Horwood, Chichester, England (1981).

Chavarie, C., D. Karamanev, L.N. Nikolov, and J. Champagne, paper presented at the 4th European Congress on Fluidization, Toulouse, France, Sep. 18-20 (1985).

Chavarie, C., and D. Karamanev, paper presented at the International Conference on Bioreactor Fluid Dynamics, Cambridge, England, April 15-17 (1986).

Chiam, H.F., and I.J. Harris, Biotech. Bioeng., 25, 1613 (1983).

Chu, J.C., J. Kalil, and W.A. Wetteroth, Chem. Eng. Prog., 49, 142 (1953).

Cleasby, J.L., and K. Fan, J. Environmental Engineering Div. ASCE, 107, 454 (1981).

Cook, J.A., C.J. Drummond, R.P. Noceti, and J.G. Diskin, paper presented at ASCE National Conference on Environmental Engineering, Los Angeles, CA, June 25-27 (1984).

Cooper, P.F., Waster Services, 84, 475, Aug. (1980).

Cooper, P.F., and D.H.V. Wheeldon, Water Pollution Control, 79, 286 (1980).

Cooper, P.F., and B. Atkinson, eds., "Biological Fluidized Bed Treatment of Water and Wastewater," Ellis Horwood, Chichester, England (1981).

Cooper, P.F., and D.H.V. Wheeldon, in "Biological Fluidized Bed Treatment of Water and Wastewater," ed. by P.F. Cooper and B. Atkinson, p. 121, Ellis Horwood, Chichester, England (1981).

Cooper, P.F., D.H.V. Wheeldon, P.E. Ingram-Tedd, and D.W. Harrington, in "Biological Fluidized Bed Treatment of Water and Wastewater," ed. by P.F. Cooper and B. Atkinson, p. 361, Ellis Horwood, Chichester, England (1981).

Daigger, G.T., and C.P.L. Grady, Jr., Biotech. Bioeng., 24, 1427 (1982).

Denac, M., S. Uzman, H. Tanaka, and I.J. Dunn, Biotech. Bioeng., 25, 1841 (1983).

Donaldson, T.L., G.W. Strandberg, J.D. Hewitt, and G.S. Shields, Environ. Prog., 3, 248 (1984).

Dunn, I.J., H. Tanaka, S. Uzman, and M. Denac, Annals New York Academy of Sciences, 413, 168 (1983).

Edeline, F., I. Tesarik, and J. Vostrcil, in "Advances in Water Pollution Research," ed. by S.H. Jenkins, p. 523, Pergamon Press (1969).

Edwards, A.M., M.S. Thesis, The Ohio State University (1981).

Environment Canada, Sewage Collection and Treatment Report SCAT-11, Dec. (1981).

Fan, L.-S., T.-R. Long, and K. Fujie, paper presented at the AIChE Annual Meeting, San Francisco, Nov. 25-30 (1984a).

Fan, L.-S., K. Fujie, and T.-R. Long, AIChE Symposium Series, 80, No. 241, 102 (1984b).

Fan, L.-S., K. Fujie, T.-R. Long, and W.-T. Tang, in Proceedings of World Congress III of Chemical Engineering, III, p. 560 (1986).

Fan, L.-S., K. Fujie, T.-R. Long, and W.-T. Tang, Biotech. Bioeng., 30, 498 (1987).

Fan, L.-S., R. Leyva-Ramos, B.J. Zehner, and K.D. Wisecarver, Biotech. Bioeng. (in press, 1988).

Freidman, L.D., W.J. Weber, R. Bloom, and C.B. Hopkins, Water Pollution Control Research Series, No. 17020 GDN 07/71, EPA (1971).

Frobisher, M., R.D. Hinsdill, K.T. Grabtree, and G.R. Goodheart, "Fundamentals of Microbiology," 9th ed., Saundres, Philadelphia (1974).

Fujie, K., T. Furuya, and H. Kubota, J. Ferment. Technol., 57, 99 (1979a).

Fujie, K., T. Tsukamoto, and H. Kubota, J. Ferment. Technol., 57, 539 (1979b).

Fujie, K., M. Takaine, and H. Kubota, J. Ferment. Technol., 57, 574 (1979c).

Fujie, K., and H. Kubota, J. Japan Water Pollution Research, 3, 53 (1980).

Fujie, K., M. Takaine, and H. Kubota, Hakko Kogaku, Japan, 50, 179 (1980).

Fujie, K., T. Tsubone, H. Kubota, and S. Shibuya, J. Japan Sewage Works Assoc., 20, 25 (1983).

Fujie, K., W.-T. Tang, T.-R. Long, and L.-S. Fan, unpublished results (1984).

Fujie, K., L.-S. Fan, and H. Kubota, paper SB307 presented at 19th Autumn Meeting of Soc. Chem. Engrs. Japan, Nagoya, Japan, Oct. 17-19 (1985).

Fujie, K., and H. Kubota, Water Research, 20, 1051 (1986).

Fujie, K., and L.-S. Fan, in "Fluidization," ed. by K. Ostergaard and A. Sorensen, p. 369, Engineering Foundation Press (1986).

Fujie, K., Kagaku to Kogyo, Japan, 40, 168 (1987).

Fujie, K., and L.-S. Fan, unpublished results (1987).

Goto, C., Kagaku Sochi, Japan, 28, 87 (1986).

Grasmick, A., B. Chatib, S. Elmaleh, and R. Ben Aim, Water Research, 15, 719 (1981).

Hano, T., S. Kokubu, Y. Sato, M. Morishita, and K. Takagi, in Proceedings of the Symposium on Biological Wastewater Treatments, p. 29, Soc. Chem. Engrs. Japan, Sendai, Japan (1983).

Hasegawa, J., Hakko Kogyo Kyokai, Japan, Oct. 15 (1982).

Hashimoto, S., K. Furukawa, and H. Hama, J. Japan Sewage Works Assoc., 22, 41 (1985).

Hashimoto, S., K. Furukawa, and H. Hama, J. Japan Sewage Works Assoc., 23, 16 (1986).

Hermanowicz, S.W., and J.J. Ganczarczyk, Biotech. Bioeng., 25, 1321 (1983).

Hirata, A., Y. Hosaka, H. Mukai, and M. Kaihou, Water Purification and Liquid Wastes Treatment, Japan, 23, 15 (1982).

Hirata, A., Y. Hosaka, and H. Umezawa, in Proceedings of World Congress III of Chemical Engineering, III, p. 556 (1986).

Holladay, D.W., C.W. Hancher, C.D. Scott, and D.D. Chilcote, J. Water Pollution Control Fed., 50, 2573 (1978a).

Holladay, D.W., C.W. Hancher, D.D. Chilcote, and C.D. Scott, AIChE Symposium Series, 74, No.172, 241 (1978b).

Hosaka, Y., M. Kaihou, H. Onishi, and A. Hirata, in Proceedings of the Symposium on Biological Wastewater Treatments, p. 22, Soc. Chem. Engrs. Japan, Sendai, Japan (1983).

Hosaka, Y., M. Kaihou, and A. Hirata, Water Sci. and Technol., 17, 1437 (1985).

Hosaka, Y., and A. Hirata, Kagaku Sochi, Japan, 28, 40 (1986a).

Hosaka, Y., and A. Hirata, Kagaku Sochi, Japan, 28, 102 (1986b).

Hosono, Y., H. Kubota, and Y. Miyaji, Water Research, 14, 81 (1980).

Hoyland, G., and P.J. Robinson, Water Pollution Control, 82, 479 (1983).

Ishihara, M., and S. Sawada, Kagaku Sochi, Japan, 20, 100 (1978).

Jennings, P.A., V.L. Snoeyink, and E.S.K. Chian, Biotech. Bioeng., 18, 1249 (1976).

Jeris, J.S., and R.G. Owens, J. Water Pollution Control Fed., 47, 2043 (1975).

Jeris, J.S., R.W. Owens, R. Hickey, and F. Flood, J. Water Pollution Control Fed., 49, 816 (1977).

Jewell, W.J., in "Biological Fluidized Bed Treatment of Water and Wastewater," ed. by P.F. Cooper and B. Atkinson, p. 251, Ellis Horwood, Chichester, England (1981).

Jewell, W.J., "Anaerobic Attached Film Expanded Bed Process - A Collection of Papers," Cornell University, Ithaca, NY (1982).

Karamanev, D.G., and L.N. Nikolov, Biotech. Bioeng., 31, 295 (1988).

Kargi, F., and J.K. Park, J. Chem. Technol. Biotech., 32, 744 (1982).

Kataoka, K., S. Ito, and S. Tsunoda, Kankyo Souzou, p. 59, Japan, Sep. (1979).

Klein, J.A., and D.D. Lee, Biotech. Bioeng. Symposium Series, 20, 379 (1978).

Kleinstreuer, C., and T. Poweigha, Advances in Biochem. Eng./ Biotech., 30, 91 (1984).

Koloini, T., and E.J. Farkas, Can. J. Chem. Eng., 51, 499 (1973).

Konishi, K., Kagaku Sochi, Japan, 27, 75 (1985).

Koplove, H.M., and C.L. Cooney, Advances in Biochem. Eng., 12, 1 (1979).

Kubota, H., and Y. Miyaji, J. Japan Sewage Works Assoc., 14, 16 (1977).

Kubota, H., S. Odai, and Y. Miyaji, J. Japan Sewage Works Assoc., 15, 1 (1978).

Kubota, H., K. Fujie, T. Hasumoto, T. Kasakura, and A. Mizuno, J. Japan Sewage Works Assoc., 17, 31 (1980).

Lee, D.D., C.F. Scott, and C.W. Hancher, J. Water Pollution Control Fed., 51, 974 (1979).

Li, C.T., W.K. Shieh, C.S. Wu, and J.S. Huang, Water Sci. and Technol., 19, 1 (1987).

Lievense, J.C., and H.C. Lim, paper presented at the ACS National Meeting, Chicago, Sep. 8-13 (1985).

Lim, H.C., P. Agrawal, C. Lee, and D. Ramkrishna, Chem. Eng. Sci., 37, 453, (1982).

Maruyama, T., H. Maeda, and T. Mizushina, J. Chem. Eng. Japan, 17, 132 (1984).

Melcer, H., S. Nutt, I. Marvan, and P. Sutton, J. Water Pollution Control Fed., 56, 192 (1984).

Miller, D.G., in "Biological Fluidized Bed Treatment of Water and Wastewater," ed. by P.F. Cooper and B. Atkinson, p. 35, Ellis Horwood, Chichester, England (1981).

Mishima, K., and A. Watanabe, Ebara Infilco Report, No. 95, p. 1 (1986).

Mori, N., H. Nakamura, and Y. Otake, paper presented at the 21st Annual Meeting of Japan Sewage Works Assoc., p. 347 (1984).

Mueller, J.A., J. Morand, and W.C. Boyle, Appl. Microbiol., 15, 125 (1967).

Mueller, J.A., W.C. Boyle, and E.N. Lightfoot, Biotech. Bioeng., 10, 331 (1968).

Mulcahy, L.T., W.K. Shieh, and E.J. Lamotta, AIChE Symposium Series, 77, No.209, 273 (1981).

Muroyama, K., T. Fujii, R. Tsujimoto, and A. Yasunishi, paper E203 presented at the Annual Meeting, Soc. Chem. Engrs. Japan (1984).

Muroyama, K., T. Fujii, and A. Yasunishi, Water Purification and Liquid Wastes Treatment, 26, 149 (1985).

Nakamura, H., Y. Otake, N. Mori, and I. Nakajima, Kogai to Taisaku, Japan, 22, 135 (1986).

Ngian, K.F., and S.H. Lin, Biotech. Bioeng., 18, 1623 (1976).

Ngian, K.F., and W.R.B. Martin, Biotech. Bioeng., 22, 1007 (1980a).

Ngian, K.F., and W.R.B. Martin, Biotech. Bioeng., 22, 1843 (1980b).

Nikolov, L.N., D. Karamanev, and D.G. Elenkov, Bulgarian Patent No. 53,798 (1982).

Nutt, S.G., J.P. Stephenson, and J.H. Pries, in "Biological Fluidized Bed Treatment of Water and Wastewater," ed. by P.F. Cooper and B. Atkinson, p. 145, Ellis Horwood, Chichester, England (1981).

Onuma, M., and T. Omura, Water Sci. and Technol., 14, 553 (1982).

Oppelt, E.T., and J.M. Smith, in "Biological Fluidized Bed Treatment of Water and Wastewater," ed. by P.F. Cooper and B. Atkinson, p. 165, Ellis Horwood, Chichester, England (1981).

Park, Y., M.E. Davis, and D.A. Wallis, Biotech. Bioeng., 26, 468 (1984).

Peel, R.G., and A. Benedek, J. Water Pollution Control Fed., 55, 1168 (1983).

Petersen, E.E., "Chemical Reaction Analysis," Prentice-Hall (1965).

Qian, W.X., Research Report, No. 5, Wuhan Institute of Municipal Engineering, China, Jan. (1980).

Rao, B.S.M., P.F. Greenfield, and P.R.F. Bell, in Proceedings of the 12th Australian Chemical Engineering Conference, p. 35, Melbourne, Aug. 26-29 (1984).

Richardson, J.F. and W.N. Zaki, Trans. Inst. Chem. Engrs., 32, 35 (1954).

Rittmann, B.E., Biotech. Bioeng., 24, 501 (1982).

Rittmann, B.E., and C.W. Brunner, J. Water Pollution Control Fed., 56, 874 (1984).

Sawada, S., Yosui to Haisui, Japan, 20, 1051 (1978).

Sawai, T., K. Sugada, and T. Miki, The Mitsubishi Heavy Industry Technical Report, 19(5), 510, Sep. (1982).

Schaezler, D.J., W.H. Mcharg, and A.W. Busch, Biotech. Bioeng. Symposium Series, 2, 107 (1971).

Scott, C.D., and C.W. Hancher, Biotech. Bioeng., 18, 1393 (1976).

Sekizawa, T., K. Fujie, H. Kubota, T. Kasakura, and A Mizuno, J. Water Pollution Control Fed., 57, 53 (1985).

Shah, Y.T., B.G. Kelkar, S.P. Godbole, and W.-D. Deckwer, AIChE J., 28, 353 (1982).

Shen, Y.M., and B.L. Wang, Research Report, No. 2, Chengdu Institute of Municipal Engineering, China, Nov. (1980).

Shi, Y.F., and L.T. Fan, paper presented at the AIChE Annual Meeting, San Francisco, Nov. 25-30 (1984).

Shidara, A., Watanabe, and K. Suzuki, J. Japan Sewage Works Assoc., 20, 234 (1983).

Shieh, W.K., P.M. Sutton, and P. Kos, J. Water Pollution Control Fed., 53, 1574 (1981a).

Shieh, W.K., L.T. Mulcahy, and E.J. Lamotta, paper presented at the 2nd World Congress of Chemical Engineering and World Chemical Exposition, Montreal, Canada, Oct. 4-9 (1981b).

Shimodaira, C., Y. Yushina, H. Kamata, H. Komatsu, A. Kurima, O. Mabu, and Y. Tanaka, U.S. Patent 4,256,573 (1981).

Shimodaira, C., and Y. Yushina, in Proceedings of the 3rd Pacific Chemical Engineering Congress, IV, p. 239, Seoul (1983).

Short, C.S., Technical Paper, TP101, Water Research Association, Medmenham, England (1973).

Siegrist, H., and W. Gujer, Water Research, 19, 1369 1985).

Smith, P.G., and P. Coackley, Water Research, 18, 117 (1984).

Stafford, D.A., B.I. Wheatley, and D.E. Hughes, eds., "Anaerobic Digestion," Elsevier Biomedical (1979).

Stathis, T.C., J. Environmental Engineering Div. ASCE, 106(EE1), 227 (1980).

Sumi, T., M. Kon, N. Mori, and K. Nakajima, J. Water and Wastewater (Japanese), 27(10), 1024 (1985).

Sutton, P.M., W.K. Shieh, C.P. Woodcock, and R.W. Morton, paper presented at the Joint Annual Conference of the Air Pollution Control Association and Water Pollution Control Association, Toronto, Canada, April (1979a).

Sutton, P.M., D.F. Langley, K.E. Warner, and K. Park, paper presented at the 34th Industrial Waste Conference, Purdue University, May (1979b).

Sutton, P.M., W.K. Shieh, P. Kos, and P.R. Dunning, in "Biological Fluidized Bed Treatment of Water and Wastewater," ed. by P.F. Cooper and B. Atkinson, p. 285, Ellis Horwood, Chichester, England (1981).

Sutton, P.M., Pollution Engineering, 13, 38 (1981).

Sutton, P.M., R.R. Evans, and P.N. Mishra, paper presented at the Am. Soc. Lubrication Engrs. Annual Meeting, Chicago, IL, May (1984).

Suzuki, K., K. Kataoka, and S. Tsunoda, Ebara Infilco Report, No. 85 (1981).

Suzuki, M., S. Momonoi, and H. Harada, in Proceedings of the Symposium on Biological Wastewater Treatment, p. 34, Soc. Chem. Engrs. Japan, Sendai, Japan (1983).

Suzuki, K., Chem. Eng., 31, 556 (1986).

Suzuki, K., and S. Tsunoda, in Proceedings of World Contress III of Chemical Engineering, III, p. 791 (1986).

Takase, H., D. Ishibashi, and M. Sugimoto, in Proceedings of the Symposium on Biological Wastewater Treatment, p. 17, Soc. Chem. Engrs. Japan, Sendai, Japan (1983).

Takase, H., N. Naeta, and M. Sugimoto, J. Japan Sewage Works Assoc., 24, 54 (1987).

Tanaka, H., S. Uzman, and I.J. Dunn, Biotech. Bioeng., 23, 1683 (1981).

Tanaka, H., and I.J. Dunn, Biotech. Bioeng., 24, 669 (1982).

Tanemura, T., T. Uno, K. Yamazaki, S. Sasaki, I. Hori, T. Saito, and A. Akamatsu, Hokkaidoritsu Kogyoshikenjo Houkoku, Japan, No. 256, p. 30 (1976).

Tanemura, T., and Y. Morita, PPM (Jan.), p. 90 (1979).

Tang, W.-T., and L.-S. Fan, paper presented at the ACS National Meeting, Chicago, Sep. 8-14 (1985); AIChE J., 33, 239 (1987).

Tang, W.-T., K.D. Wisecarver, and L.-S. Fan, paper presented at the AIChE Annual Meeting, Chicago, Nov. 10-14 (1985); Chem. Eng. Sci., 42, 2133 (1987).

Timmermans, P., and A. Van Haute, Appl. Microbial Biotech., 19, 36 (1984).

Tong, C.C., M.S. Thesis, The Ohio State University (1986).

Tong, C.C., and L.-S. Fan, Biotech. Bioeng., 31(1), 24 (1988).

Tramper, J., G. Suwinska-Borowiec, and A. Klapwijk, Enzyme Microb. Technol., 7, 155 (1985).

Tsezos, M., and A. Benedek, J. Water Pollution Control Fed., 52, 578 (1980a).

Tsezos, M., and A. Benedek, Water Research, 14, 689 (1980b).

Tsunoda, S., K. Shimada, and Y. Aoyanagi, Kagaku Kogaku, 40, 32 (1976).

Tsunoda, S., Kankyo Souzou, 49 (1977).

U.S. EPA, "Technology Assessment of Anaerobic Systems for Municipal Wastewater Treatment," EPA-600/2-82-004, Feb. (1982).

Vanderborght, J.P., and Ch. Gilliard, in "Biological Fluidized Bed Treatment of Water and Wastewater," ed. by P.F. Cooper and B. Atkinson, p. 383, Ellis Horwood, Chichester, England (1981).

Walker, I., and E.P. Austin, in "Biological Fluidized Bed Treatment of Water and Wastewater," ed. by P.F. Cooper and B. Atkinson, p. 272, Ellis Horwood, Chichester, England (1981).

Wang, S.C.P., and C. Tien, Can. J. Chem. Eng., 61, 64 (1983).

Wang, S.C.P., and C. Tien, AIChE J., 30, 786 (1984a).

Wang, S.C.P., and C. Tien, AIChE J., 30, 794 (1984b).

Watson, S.W., in "Bergey's Manual of Determinative Bacteriology," ed. by R.E. Buchanan and W.E. Gibbons, 8th ed., Chap. 12, Williams and Wilkens, Baltimore (1974).

Webb, C., G.M. Black, and B. Atkinson, Chem. Eng. Res. Des., 61, 125 (1983).

Weber, W.J., Jr., L.D. Friedman, and R. Bloom, Jr., in Proceedings of the 6th Conference of the International Association Water Pollution Research, Pergamon Press, Oxford, U.K. (1973).

Weber, W.J., Jr., J. Water Pollution Control Fed., 50, 781 (1978).

Weber, W.J. Jr., M. Pirbazari, and G.L. Melson, Environ. Sci. and Technol., 12, 817 (1978).

Webster, G.H., and J.J. Perona, AIChE J., 34, 1398 (1988).

Wen, C.Y., and Y.H. Yu, Chem. Eng. Prog. Symposium Series, 62, 100 (1966).

Wheeldon, D.H.V., and R.W. Bayley, in "Biological Fluidized Bed Treatment of Water and Wastewater," ed. by P.F. Cooper and B. Atkinson, p. 306, Ellis Horwood, Chichester, England (1981).

Wisecarver, K.D., and L.-S. Fan, paper presented at the AIChE Annual Meeting, New York, Nov. 15-20 (1987); Biotech. Bioeng. (in press, 1988a).

Wisecarver, K.D., and L.-S. Fan, Chem. Eng. Sci., 43, 2741 (1988b).

Witmer, G.S., V. Balakotaiah, and D. Luss, Chem. Eng. Sci., 41, 179 (1986).

Yang, R.D., and A.E. Humphrey, Biotech. Bioeng., 17, 1211 (1975).

Ying, W.C., and W.J. Weber, Jr., J. Water Pollution Control Fed., 51, 2661 (1979).

Yushina, Y., and H. Sato, in Proceedings of the Muroran Meeting, p. 131, Soc. Chem. Engrs. Japan (1983).

Yushina, Y., Zousui Gijutsu, Japan, 11, 50 (1985).

Zehner, B.J., M.S. Thesis, The Ohio State University (1986).

Zhang, Z.J., B.N. Lu, and X. Gao, Research Report, No. 4, Harbin Institute of Architecture and Engineering, China, April (1981).

Chapter 9

Flue Gas Desulfurization and Particulate Removal

Both sulfur dioxide and particulates are commonly found in flue gas of processes such as heavy oil and coal combustion, smelting operations, sulfuric acid manufacture, and metallurgical processes. Consequently, sulfur dioxide and/or particulate removal is of considerable environmental importance. Wet scrubbers provide a relatively simple, yet efficient, method of removal.

Several types of wet scrubbers operated in gas-liquid-solid fluidization modes have been considered for sulfur dioxide (SO_2) removal from a flue gas including venturi (Mode T-III-b[1]) (Wen and Uchida, 1973), spray (Mode T-II-b[1]) (Epstein, 1973), jet bubbling (Mode T-II-a[1]) (Idemura et al., 1978; Sawahata et al., 1987), turbulent bed contactor (TBC) (Modes E-II-a-2 and E-II-b[2]) (McMichael et al., 1976), and hydro-filter (marble bed) (Mode E-II-a-2[2]) (Epstein, 1973). See Fig. 1.3 for mode designations. For particulate-laden flue gas containing SO_2, some wet scrubbers, such as a TBC, have been utilized to remove particulates and SO_2 simultaneously from a flue gas. Others, such as spray towers, have been utilized primarily to remove SO_2 alone, with particulates removal performed by a separate device prior to it, such as an electrostatic precipitator.

In SO_2 scrubbing and/or particulate removal, the TBC has been demonstrated as an effective scrubbing device; however, operational problems have impeded widespread commercial use of the TBC in flue gas cleaning (see Section 9.1.3). Despite these problems, the TBC remains a viable device for applications in mass transfer with or without chemical reactions. Thus, it is of sufficient merit to examine the fundamental features of a TBC for such applications as flue gas desulfurization and particulate removal.

1: Designation of the mode of operation is based on the behavior of the slurry particles.
2: Designation of the mode of operation is based on the behavior of the low density large fluidized particles rather than the slurry particles.

9.1 SO_2 SCRUBBING

SO_2 scrubbing processes basically consist of a scrubber, a hold tank, and a liquid-solid separator. Typical absorbents used in the wet scrubbing process include lime, limestone, or dolomite based slurries with/without additives such as magnesium oxide and carboxylic acids. The scrubber provides intimate contact between the flue gas and the scrubbing medium. The hold tank(s) (one or two in series) provides additional residence time for the reactions of absorbed SO_2 with dissolved absorbents in the presence or absence of oxygen and for the crystallization of the products $CaSO_3$ and $CaSO_4$. Fresh make-up absorbents are added to the hold tank to bring up the recycle slurry pH. The liquid-solid separator removes the reaction products as wet sludges.

Liquid phase chemical reactions are complex as there are numerous ionic and neutral species present in the solution. Examples of the major components present are HSO_3^-, $SO_3^=$, $CaSO_3$, HSO_4^-, $MgSO_3$, H_2SO_3, $SO_4^=$, Ca^{++}, $CaHCO_3^+$, HCO_3^-, $MgHCO_3^+$, $CO_3^=$, H_2CO_3, $CaOH^+$, OH^-, H^+, Mg^{++}, and $CaCO_3$. It is recognized that the strategy underlying the reaction schemes in the slurry phase is the key factor to the successful operation of a wet scrubbing process. Improved strategies continue to evolve. In the following, examples of process considerations are given. Furthermore, operational problems and scrubber analysis are illustrated.

9.1.1 Simple Scrubbing Processes

The simplest wet scrubbing process for SO_2 removal, shown in Fig 9.1, is known as the throwaway scrubbing process (Slack and Hollinden, 1975; Rochelle and King, 1978; Henzel et al., 1981). The major consideration +20Xoperation of this process lies in alleviation of scaling and plugging problems in the scrubbers due to solid solutions containing $CaSO_3$ and gypsum ($CaSO_4 \cdot 2H_2O$) (see Section 9.1.3). In this process, the pH in the hold tank with lime/limestone makeup is maintained at about 5.5 ~ 7.5. Only a small fraction of the slurry from the hold tank, which contains about 10 ~ 15% suspended solids, is discharged to the liquid-solid separator. A sufficient liquid circulation rate is required to avoid excessive saturation or super-

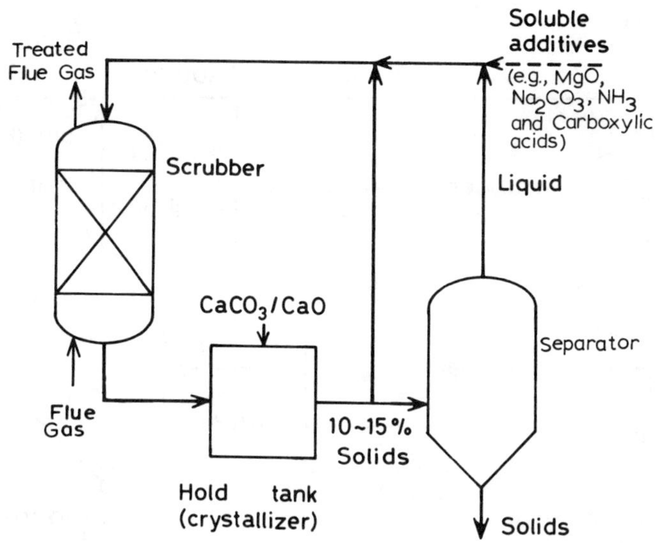

Figure 9.1 A simple throwaway scrubbing process (from Rochelle and King, 1978).

saturation[3] of $CaSO_4$ in the scrubber. Oxidation in the scrubber was also reported to improve SO_2 absorption efficiency (Borgwardt, 1978).

The double alkali process provides an alternative method using lime/limestone slurry in the scrubbing process. Figure 9.2 shows the schematic diagram of the process. In the scrubber of this process, SO_2 is absorbed in a clear solution of soluble sulfite, e.g., Na_2SO_3 or $(NH_4)_2SO_3$ by the following reaction:

$$SO_3^{=} + SO_2 + H_2O \rightleftharpoons 2HSO_3^{-}$$

Lime/limestone is then used in the hold tank to neutralize the scrubbing solution containing HSO_3^{-} by the following reaction:

$$2HSO_3^{-} + CaO \rightleftharpoons CaSO_3\downarrow + SO_3^{=} + H_2O$$

In this process, effluent from the hold tank containing about 2 ~ 5% suspended solid product $CaSO_3$ is then discharged to the liquid-solid separator. The separated liquid is replenished with fresh solution of

---

3: Supersaturation of a compound refers to a condition when the relative saturation (R.S.), defined as the ratio of the product of the activities of all the ions of that compound to the solubility product constant of that compound, is greater than 1.

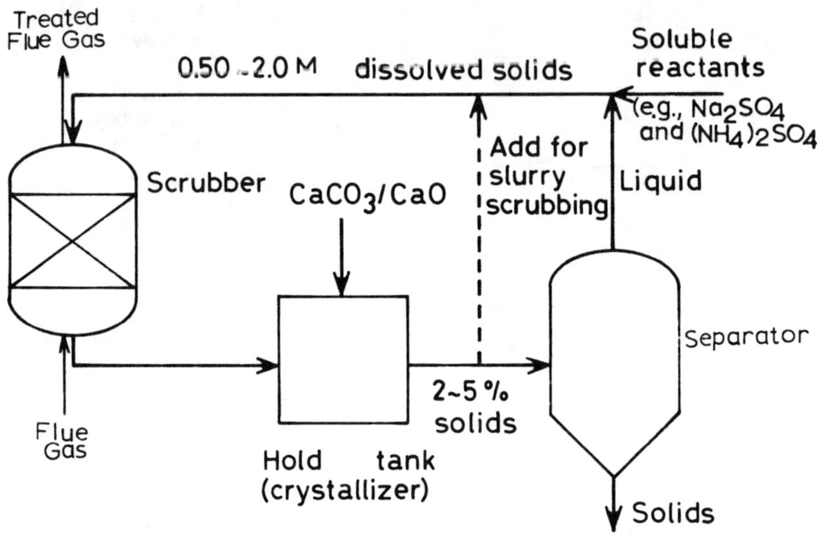

Figure 9.2 Double alkali scrubbing process (from Kaplan, 1978).

soluble reagent, e.g., Na_2CO_3 or NaOH, before returning to the scrubber.

Instead of soluble sulfite solutions, a dilute sulfuric acid solution has also been used as the scrubbing medium (Noguchi, 1974). In this case, SO_2 reacts with a sulfuric acid solution in the scrubber. The scrubber effluent is then oxidized with air before discharge to the hold tank. When SO_2 is absent in the inlet scrubbing medium, SO_2 scrubbing is possible even at a very low pH (1 ~ 2). However, scrubbing at a very low pH requires very large liquid circulation rates. Addition of a soluble alkali, such as NH_3, can produce a pH as high as 10 ~ 13 in the hold tank, providing an effective process for sulfate crystallization (Kaplan, 1973).

SO_2 scrubbing in either a simple throwaway process or a double alkali process can be improved via a hybrid approach. In the improved simple throwaway process, SO_2 scrubbing is enhanced by addition of soluble additives (see Section 9.1.2) to the lime/limestone slurries as indicated by the dotted lines in Fig. 9.1. In the improved double alkali process, lime/limestone/$CaSO_3$ or $CaSO_4$ slurries, in part, are recycled from the hold tank to the scrubber directly as indicated by the dotted lines in Fig. 9.2. The use of slurry recycle in the improved double alkali process yields lower soluble sulfite concentration for SO_2 scrubbing and lower operating loads in the liquid-solid separator. However, operational problems may occur in the scrubber due to the presence of suspended solids.

9.1.2 Soluble Additives and Forced Oxidation

One limiting factor in the throwaway process is known to be the liquid-side gas-liquid mass transfer. Liquid-side mass transfer in a simple slurry system is probably accomplished through the buffering couples HCO_3^-/CO_2 and $SO_3^=/HSO_3^-$ (Chang, 1979). At the gas-liquid interface, the alkali HCO_3^- and $SO_3^=$ react with dissolved SO_2 as given by (Fan, 1975; Chang, 1979)

$$SO_3^= + SO_2 + H_2O \rightleftharpoons 2HSO_3^-$$

$$HCO_3^- + SO_2 \rightleftharpoons CO_2 + HSO_3^-$$

HCO_3^- and $SO_3^=$ are regenerated at the liquid-solid interface through dissolution of $CaCO_3$ and $CaSO_3$. The reactions are postulated as (Chang, 1979)

$$CaCO_3 + CO_2 + H_2O \rightleftharpoons Ca^{++} + 2HCO_3^-$$

$$CaCO_3 + 2HSO_3^- \rightleftharpoons Ca^{++} + 2SO_3^= + CO_2 + H_2O$$

$$CaSO_3 \rightleftharpoons Ca^{++} + SO_3^=$$

For systems where these reactions are sufficiently rapid, the rate of SO_2 transfer through the liquid-side at the interface strongly depends on the dissolved alkalinity[4] of the bulk liquid.

Soluble alkali additives such as MgO, Na_2CO_3, and NH_3 accumulate primarily as dissolved sulfates and enhance the effect of dissolved alkalinity of HCO_3^- and $SO_3^=$. Carboxylic acid additives such as adipic, formic, acetic, maleic, sulfopropionic and sulfosuccinic acids, which have buffer effects yielding a pH in the bulk liquid phase of 5 ~ 6, generate almost one equivalent of dissolved alkalinity per equivalent of these additives, compared to about 0.008 to 0.03 equivalent of dissolved alkalinity per equivalent soluble alkali

4: Dissolved alkalinity in the liquid is the sum of activity of ions which are more alkaline than HSO_3^-, such as $CO_3^=$, HCO_3^-, $SO_3^=$ and OH^-.

additives (Rochelle and King, 1978). 10 to 20 mM of adipic acid buffer additive, for example, is adequate to produce substantial gas-liquid mass transfer enhancement for a typical lime/limestone slurry scrubber for desulfurization of flue gas derived from high sulfur coal (Chang, 1979). Buffer additives thus provide a higher lime/limestone utilization, allow fewer scrubber stages, reduce the pressure drop, and require a lower liquid flow rate and/or a simpler scrubber design (Rochelle and King, 1978). The addition of magnesium oxide to the scrubbing slurry can allow the scrubbing system to operate in the unsaturated[5] mode thus preventing calcium sulfate scale formation (Borgwardt, 1974). Furthermore, magnesium oxide improves scrubbing efficiency of the system.

As noted above, adipic acid effectively buffers the pH in the lime/limestone scrubber slurry and improves the SO_2 scrubbing efficiency (Rochelle and King, 1977; Mobley and Chang, 1981; Chang and Mobley, 1983). In fact, any acid which is intermediate in acidic strength between a carboxylic acid and a sulfurous acid with a reasonably soluble calcium salt can be used. However, since adipic acid is one of the most cost effective organic acids commercially available, it is attractive as a slurry additive. The buffering action of adipic acid results in higher gas-liquid mass transfer due to permission of higher concentrations of SO_2 at the interface. The presence of calcium adipate in the bulk liquor also increases the capability of the bulk liquor to react with SO_2. Thus, the rate of SO_2 absorption becomes less dependent on the dissolution rate of lime/limestone or calcium sulfite in the scrubber to provide the required alkalinity (Wang and Burbank, 1982).

Forced oxidation in the scrubber to convert $CaSO_3$ to $CaSO_4$, accompanied by a buffer to resist changes in pH, was suggested by Rochelle and King (1978) as an economical means of using additives in the slurries. Adipic acid decomposition, however, may take place during the course of SO_2 scrubbing. The decomposition products include valeric acid, glutaric acid, butynic acid, succinic acid, butane, methane and carbon dioxide (Meserole et al., 1979). Furthermore, the rate of adipic acid decomposition appears to be proportional to the rate of sulfite oxidation (Meserole et al., 1979) and increases with increasing pH (Wang and Burbank, 1982). In the following, performance of three-stage TBC scrubbers with and without fluidized

5: Unsaturation or subsaturation of a compound refers to the situation when the relative saturation (R.S.) of that compound is less than 1.

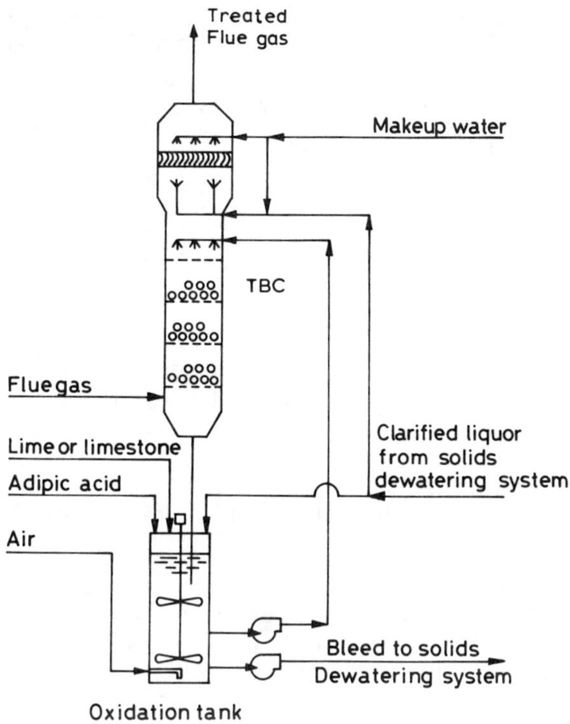

Figure 9.3 Flow diagram for adipic acid-enhanced lime/limestone scrubbing process using a TBC with one tank and forced oxidation (from Wang and Burbank, 1982).

particles is illustrated based on two different process configurations using a pilot scale TVA (Tennessee Valley Authority) TBC (1.7m (width) × 1.7m (length) × 11.7m (height)). Nitrile foam spheres with a density of 0.2 g/cm^3 are used as fluidized particles, yielding fluidization without flooding (Type I TBC, see Chapter 5), in each of the three stages. Note that operation of a TBC without spheres is essentially that of a spray tower.

One Hold Tank System with/without Forced Oxidation (Wang and Burbank, 1982)

Figure 9.3 shows a single-loop scrubbing configuration with adipic acid additives. The scrubber inlet pH ranges from 5.6 to 6.1. The hold tank both with and without forced oxidation[6] is considered.

6: Forced oxidation of the scrubber effluent in the hold tank assists formation of easily dewatered waste solids for sludge landfill purposes since gypsum crystallizes in much larger, more easily dewatered crystals than does $CaSO_3$.

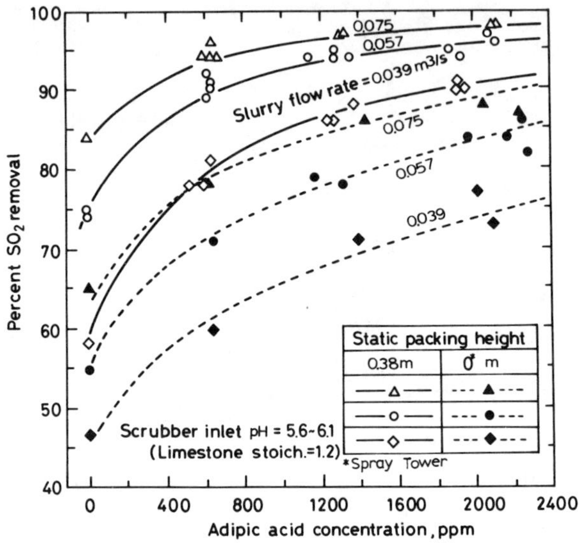

Figure 9.4 Effect of adipic acid concentration and slurry flow rate on SO_2 removal in the TBC with and without packing spheres (with one tank and with or without forced oxidation) (from Wang and Burbank, 1982).

In the hold tank, high liquor sulfite concentrations may lead to limestone blinding[7] problems. The limestone blinding problem occurs under in-loop forced oxidation conditions when the recirculated slurry is deficient in solid calcium sulfite crystal seeds and thus calcium sulfite in the liquor preferentially precipitated on the surface of the limestone particles[8]. Limestone blinding has detrimental effects on limestone utilization.

Figure 9.4 shows the variation of SO_2 removal versus adipic acid concentration as a function of slurry flow rate for TBCs with and without spheres and with and without forced oxidation. Note that the mechanism by which adipic acid promotes SO_2 removal is unaffected by forced oxidation. When adipic acid is used with lime, calcium adipate becomes a buffer species and both good SO_2 removal

7: When the relative saturation (R.S.) is greater than 1, crystal growth phenomenon occurs where solids precipitate slowly, mainly on the surface of existing particles of the same compound. When R.S. is significantly greater than 1, nucleation phenomenon occurs where ion clusters form and grow into solid particles. When crystal growth or nucleation occurs on the surface of lime/limestone surface, it is known as blinding phenomena (EPRI, 1987).

8: The limestone blinding problem can be alleviated, however, as is commonly done in industry, by controlling the ratio of the O_2 molar flow rate in the hold tank to the molar absorption rate of SO_2 in the scrubber in a one hold tank system with forced in-loop oxidation.

and sulfite forced oxidation can be achieved using the scrubber-tank forced oxidation system. With adipic acid addition at scrubber inlet pHs of 5.6 ~ 6.1[9], the difference in SO_2 removal with and without forced oxidation is not apparent.

As shown in Fig. 9.4, for TBCs both with and without spheres, SO_2 removal increases with an increase in adipic acid concentration and with an increase in the slurry flow rate. For example, at a slurry flow rate of 0.057 m^3/s, as the adipic acid concentration increases from 0 to 600 ppm, the SO_2 removal rate increases from 75% to 90% for a TBC with spheres and from 55% to 72% for a TBC without spheres. The results suggest the TBC with spheres achieves a far higher SO_2 removal efficiency than a TBC without spheres or a spray tower for a given operation.

Two Hold Tank System with Forced Oxidation

The role of the hold tank given in Fig. 9.3 can be improved by the use of two tanks in series. The schematic diagram of a TBC-two hold tank system is given in Fig. 9.5. In the figure, adipic acid is added to the first hold tank which is also used for oxidation. Limestone is added to the second hold tank. Relocation of the limestone feed to the second tank eliminates the high liquor sulfite concentrations and limestone blinding problems experienced in the one tank system. Figure 9.6 shows the variation in SO_2 removal with adipic acid concentration as a function of TBC inlet liquor pH. As shown, SO_2 removal has a higher dependency on adipic acid concentration at a low pH than at a high pH for adipic acid concentrations above 600 ppm. The lower dependency at a high pH is consistent with the data given in Fig. 9.4. At a low pH, a significant amount of adipic acid does not dissociate and thus, is only partially effective. For example, it is seen in Fig. 9.4 that at the slurry rate of 0.057 m^3/s, 90% SO_2 removal can be achieved with an adipic acid concentration of 600 ppm and a slurry inlet pH of 5.6 ~ 6.1. Achieving the same SO_2 removal efficiency, however, requires an adipic acid concentration of 1400 ppm at a slurry inlet pH of 5.3 or 1900 ppm at a slurry inlet pH of 5.0, as shown in Fig. 9.6. The higher dependency at a lower pH suggests that SO_2 removal is more sensitive to pH and adipic acid concentration fluctuations at low pH than at high pH.

9: In current industrial practice, the slurry pH at the scrubber inlet is controlled at a moderate level between 5.2 and 6. This is because a pH too high yields low limestone utilization and causes scrubber scaling problems, whereas a pH too low yields a low SO_2 scrubbing efficiency and causes scrubber corrosion problems.

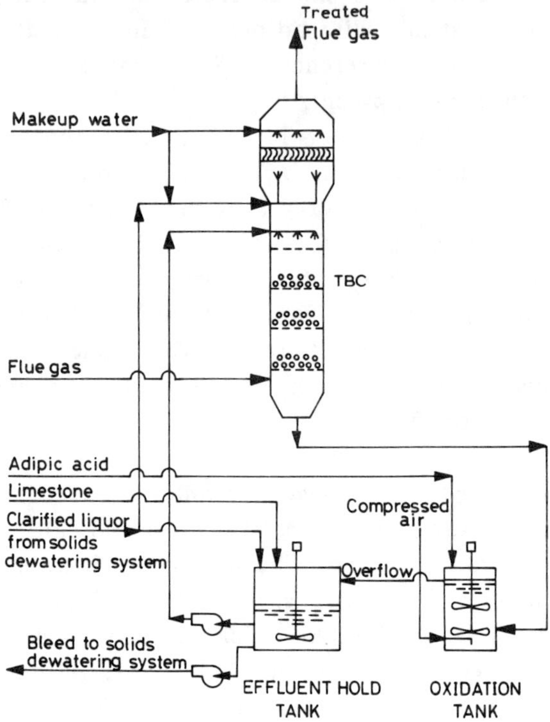

Figure 9.5 Flow diagram for adipic acid-enhanced lime/limestone scrubbing process using a TBC with two tanks and forced oxidation (from Wang and Burbank, 1982).

9.1.3 Operational Problems

A typical problem of wet scrubbers using lime/limestone slurries is internal plugging due to scale formation, as noted earlier. The reaction products, calcium sulfite and gypsum, precipitate on the scrubber wall, gas and liquid distributors, grid and demister blade, and eventually plug the system. Scale formation in the scrubber can be minimized. If the relative solubility (R.S.) of calcium sulfite and gypsum in the solution can be properly controlled (see Section 9.1.2), their precipitation occurs in the hold tank instead of the scrubber. Note that R.S. is a measure of the tendency for solids to precipitate or dissolve in the solution.

Other operating problems of TBCs are associated with the properties and behavior of the low density particles used in the scrubber. These problems were evident from the operational

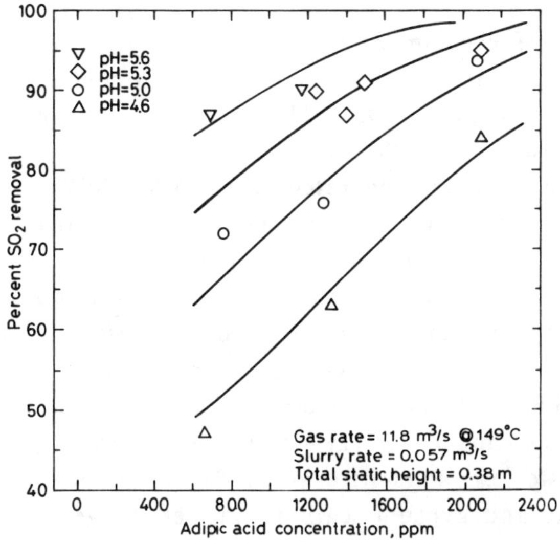

Figure 9.6 Effect of adipic acid concentration and scrubber inlet pH on SO_2 removal in the TBC (with two-tank forced oxidation) (from Wang and Burbank, 1982).

experience with TBC scrubbers at pilot and commercial scales (Wang and Burbank, 1982; Zemo, 1988). Particles used have included polyethylene ping-pong balls and nitrile foam balls of 1.3 ~ 5.1 cm ID. For ping-pong balls, the vigorous tumbling action the particles undergo leads to surface thinning, erosion, and ultimately to breakage. Ping-pong balls filled with slurry sink to the bottom of the bed and therefore may plug the distributor plate. Nitrile foam particles have low thermal resistance and can be broken into pieces by thermal stresses inherent in scrubbing operations. Broken pieces of nitrile foam balls or ping-pong balls may be discharged from the bed and plug the valve or pump in a closed or open loop scrubbing process. Maldistribution of the low density particles also occurs due, in part, to scale formation on the grid and gas and liquid distributors, and also due to the effect of broken particles mentioned above. Apparently, irregular particle migration coupled with gas channelling can establish a considerably nonuniform gas-liquid flow field which affects the stability of the scrubber performance. Thus, the long term reliability of TBC operation for flue gas desulfurization using lime/limestone slurries needs to be demonstrated further. The problems described above have prompted the current use of a relatively low pressure drop, simple design slurry spray tower for SO_2 scrubbing of coal combustion flue gas.

A view of a commercial slurry spray tower scrubber located at Conesville, Ohio, currently operated by the Columbus Southern Power Company is shown in Fig. 9.7. The scrubber measures of 6.1 m wide, 12.2 m deep, and 38.1 m high. There are four scrubbers situated in parallel to treat flue gas generated from an 800 MW (two 400 MW coal combustors) utility plant. The towers were operated in the TBC mode from the mid-1970's until 1981 when they were converted to the spray tower mode[10]. Note that in the more recent SO_2 scrubbing processes, a "dry" scrubbing system has been favorably used in a spray tower where flue gas containing SO_2 is contacted with a fine spray of aqueous solution or slurry of reactive alkali in such a manner that SO_2 removal and solution or slurry drying occur simultaneously yielding dry solids waste (e.g., Parsons et al., 1981). This process alleviates the calcium sulfite and gypsum scaling and wet sludge handling problems occurring in the wet scrubbing processes.

9.1.4 Scrubber Analysis (Fan, 1975; McMichael et al., 1976)

SO_2 scrubbing by lime/limestone slurries with/without additives is principally a mass transfer operation with chemical reactions, and as such, the analysis and design of contacting devices used for SO_2 scrubbing must evaluate the SO_2 flux across the gas-liquid interface. The most difficult problem in analyzing the SO_2 scrubbing is associated with the chemical reactions occurring in the liquid phase and their effects on the mass transfer rates. Furthermore, as indicated in Chapter 5, information on liquid-side gas-liquid mass transfer (for physical absorption) is very limited, which adds additional difficulty to the evaluation of the SO_2 flux. In the present analysis of SO_2 scrubbing, gas-side and liquid-side mass transfer and enhancement of liquid-side mass transfer due to chemical absorption are, for practical purposes, treated empirically.

9.1.4.1 Scrubber Model

The molar flux of SO_2 across the gas-liquid interface can be written in terms of the gas-side (gas-film) resistance as

---

10: Conversion was done mainly by removing the spheres and grids and adding spray nozzles.

Figure 9.7 Commercial lime slurry spray tower scrubbers for flue gas desulfurization, located at Conesville, Ohio (elevation looking east typical of one of the four scrubber models, lime preparation building shown at lower left, one of the four hold tanks pictured at lower right). These towers had been operated in the TBC mode from the mid-1970's until 1981 when they were converted to the spray tower mode (kindly supplied by B.R. Zemo of Columbus Southern Power Company, Ohio).

$$N_{SO_2} = k_g(P_{SO_2} - P^i_{SO_2}) \tag{9.1}$$

where (k_g) is the gas-side mass transfer coefficient. The molar flux can also be expressed in terms of the liquid-side (liquid-film) resistance as

$$N_{SO_2} = k_\ell \phi (C^i_A - C_A) \tag{9.2}$$

where the liquid-side mass transfer coefficient has been subdivided into a physical part, i.e., the liquid-side mass transfer coefficient for physical absorption (k_ℓ), and a chemical part, i.e., the enhancement factor (ϕ) (see Section 6.1.2.1). ϕ in Eqn. (9.2) takes into account the reaction of the diffusing H_2SO_3 with various ions contributing to the dissolved alkalinity in the liquid phase.

By defining the Henry's Law constant, He, as P_{SO_2}/C_A and combining Eqns. (9.1) and (9.2), the molar flux of SO_2 can be rewritten as

$$N_{SO_2} = \left(\frac{1}{k_g} + \frac{He}{\phi k_\ell}\right)^{-1} (P_{SO_2} - P^*_{SO_2}) \tag{9.3}$$

Note that a large ϕ corresponds to the situation in which the gas-side resistance dominates the SO_2 transfer. "He" has been given as a function of temperature by Vivian (1973) as (in SI units)

$$\ln He = 15.07 - \frac{3175.2}{T} \tag{9.4}$$

In Eqn. (9.3) both ϕ and $P^*_{SO_2}$ are functions of position and depend on the mechanism of SO_2 absorption into the limestone slurry. Nevertheless, the equilibrium SO_2 partial pressure of an aqueous system will usually be much less than the SO_2 partial pressure found in a flue gas (i.e., $P_{SO_2} >> P^*_{SO_2}$ for pH > 4.7 in a calcium carbonate, calcium sulfite, carbon dioxide, sulfur dioxide, and water system at equilibrium).

Ignoring $P^*_{SO_2}$ relative to P_{SO_2} in Eqn. (9.3), and assuming a plug flow for the gas phase, the rate of SO_2 absorption in a differential height of the scrubber (dz) can be written as

$$\frac{-F}{P_T}\frac{dP_{SO_2}}{dz} = \left(\frac{1}{k_g a} + \frac{He}{\phi k_\ell a}\right)^{-1} P_{SO_2} \tag{9.5}$$

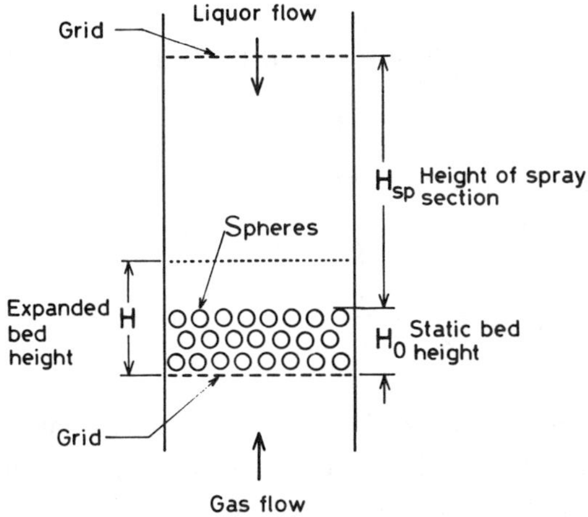

Figure 9.8 An idealized model system for a TBC (from McMichael et al., 1976).

Equation (9.5) can be integrated over the height of the column section (from Z_1 to Z_2) to give

$$\frac{F}{\Delta Z P_T} \ln \frac{P_{SO_2}\big|_{z=Z_1}}{P_{SO_2}\big|_{z=Z_2}} = \left(\frac{1}{k_g a} + \frac{He}{\bar{\phi}\, k_\ell a}\right)^{-1} \tag{9.6}$$

where $\bar{\phi}$ is an average enhancement factor and $\Delta Z = Z_2 - Z_1$.

A simple TBC or a single stage of a multi-stage TBC can be subdivided axially into two sections; one contains the spheres and the other contains no spheres and behaves much like a spray device. This situation is illustrated in Fig. 9.8. The packed and spray sections have significantly different contacting mechanisms and consequently, different mass transfer properties. Also, this differentiation allows for a better comparison between columns of different packing heights.

In the development to follow it will be assumed that the mass transfer behavior of the spray sections is constant regardless of their position in the column. Furthermore, the overall mass transfer behavior can be conveniently characterized by the mass transfer in the packed (static) section in series with that in the spray section. The integrated mass balance on SO_2 in the TBC scrubber can be written as

$$\frac{F}{P_T} \ln \frac{P_{SO_2}^{in}}{P_{SO_2}^{out}} = K_g{}^p a H_o + K_g{}^s a H_{sp} \tag{9.7}$$

where $K_g a$, the overall gas-side mass transfer coefficient for each section, is defined by

$$K_g a = \left(\frac{1}{k_g a} + \frac{He}{\bar{\phi}\, k_\ell a}\right)^{-1} = k_g a \left(\frac{R}{R+1}\right) \tag{9.8}$$

where $k_g a$ and R, the ratio of the volumetric mass transfer resistance in the gas-side to that in the liquid-side (or $\bar{\phi} k_\ell a/(He\ k_g a)$), are specific to the respective sections (denoted by superscript p in the packed section and superscript s in the spray section). Equation (9.7) can be rewritten in terms of R and $k_g a$ as

$$\frac{F}{P_T} \ln \frac{P_{SO_2}^{in}}{P_{SO_2}^{out}} = k_g{}^p a \left(\frac{R_p}{R_p + 1}\right) H_o + k_g{}^s a \left(\frac{R_s}{R_s + 1}\right) H_{sp} \tag{9.9}$$

From Eqn. (9.9) and values of $k_g{}^p a$, $\left(\frac{R_p}{R_p + 1}\right)$, $k_g{}^s a$, and $\left(\frac{R_s}{R_s + 1}\right)$ the amount of SO_2 absorbed per stage of a scrubber can be calculated.

9.1.4.2 Model Simulation

In this section, data obtained from two TBC scrubbers of widely different scales are used to illustrate flue gas desulfurization behavior in TBC scrubbers. One scrubber, operated by the Environmental Protection Agency in Research Triangle Park (EPA/RTP), is of a bench scale with 23cm ID and 3m height. The other scrubber, operated at the TVA Shawnee plant, is of a pilot scale and is already shown schematically in Figs. 9.3 and 9.5. The TVA facility is capable of treating approximately 10 MW equivalent (14.2 m^3/s at 149°C) of flue gas containing SO_2. As shown in Fig. 9.3 or 9.5, the scrubbing slurry is sprayed from a single arrangement of nozzles at the top of the column and falls through a series of grids and sections filled with low density spheres. Flue gas containing SO_2 in concentrations up to 4500 ppm flows countercurrent to the limestone slurry. A large scale TVA spray column (2.44 m ID, circular cross section), as shown in Fig. 9.9, is also used to illustrate the limiting case

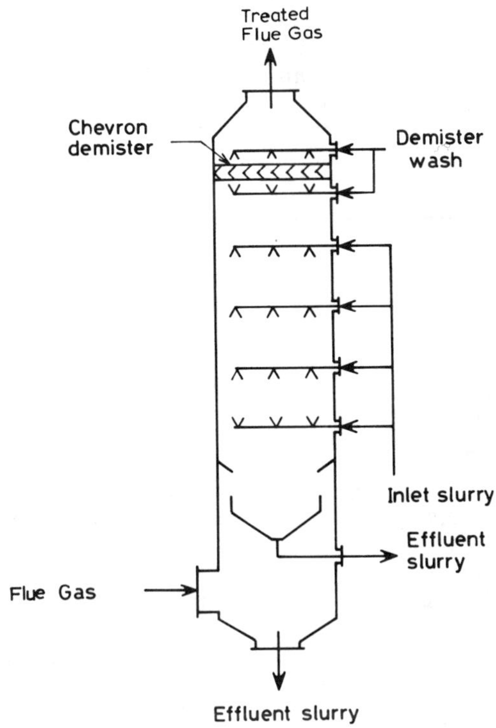

Figure 9.9 Schematic diagram of the pilot scale TVA Shawnee Spray Tower (from Epstein, 1973).

of a TBC scrubber operation in the absence of spheres. The TVA spray column has the same flue gas treatment capacity as the TVA TBC. The spray column with four spray headers is slightly more complicated than the TBC without spheres in that the scrubbing slurry is sprayed at four levels.

For a TBC operating without spheres, the first term on the right hand side of Eqn. (9.9) drops out and the gas-side mass transfer coefficient ($k_g^s a$) can be calculated based on experiments on SO_2 absorption into sodium carbonate solutions with pH greater than 8.5 in which the SO_2 transfer is essentially gas-side controlled (i.e., $R_s/(R_s + 1) \approx 1$). Data for slurries of low pH values for both the gas-side and liquid-side mass transfer are important can then be used to calculate the quantity $R_s/(R_s + 1)$. With $k_g^s a$ and R_s known, the gas-side mass transfer coefficient ($k_g^p a$) can be calculated from Eqn. (9.9) using the TBC data for slurries of high pH (i.e., $R_p/(R_p + 1) \approx 1$), and $R_p/(R_p + 1)$ can then be determined from the TBC data for slurries of low pH.

Table 9.1 Summary of Equations (in SI Units) Necessary for Simulating the Performance of the Pilot Scale TVA Shawnee TBC and Spray Column and the bench scale EPA/RTP TBC (from Fan, 1975)

Spray section	Packed section
$k_g{}^s a = 3.32 \times 10^{-6} F^{0.8} G_\ell{}^{0.4}$	$k_g{}^p a = 8.83 \times 10^{-5} F^{0.47} G_\ell{}^{0.51}$
$R_s = (A_s/\Delta_s) e^{-3.26 \times 10^{-3} P_{SO_2}^{in}}$	$R_p = (A_p/\Delta_p) e^{-3.26 \times 10^{-3} P_{SO_2}^{in}}$
$A_s^{-1} = \exp(-1.35\ \mathrm{pH} + 7.82) - 0.15$; for $\mathrm{pH} < 7.2$	$A_p^{-1} = 0.308$; for $\mathrm{pH} \leq 6.0$
	$A_p^{-1} = -0.517\ \mathrm{pH} + 3.41$; for $6.0 < \mathrm{pH} \leq 6.6$
$A_s^{-1} = 0$; for $\mathrm{pH} > 7.2$	$A_p^{-1} = 0$; for $\mathrm{pH} > 6.6$
$\Delta_s = 1.0$; for $(10^6 \delta_{Mg}) < 350$ (ppm)	$\Delta_p = 1.0$; for $(10^6 \delta_{Mg}) < 3600$ (ppm)
$\Delta_s = 50.1\,(10^6 \delta_{Mg})^{-0.668}$; for $(10^6 \delta_{Mg}) \geq 350$ (ppm)	$\Delta_p = 2.2 \times 10^7 (10^6 \delta_{Mg})^{-2.065}$, for $(10^6 \delta_{Mg}) \geq 3600$ (ppm)
$K_g{}^s a = k_g{}^s a R_s/(1 + R_s)$	$K_g{}^p a = k_g{}^p a R_p/(1 + R_p)$

$$\frac{F}{P_T} \ln \frac{P_{SO_2}^{in}}{P_{SO_2}^{out}} = K_g{}^s a H_{sp} + K_g{}^p a H_o \tag{9.7}$$

Table 9.1 summarizes the correlations for $k_g{}^p a$, R_p, $k_g{}^s a$, and R_s which are obtained based on the procedure described above. It is noted that R_s and R_p vary strongly with the inlet slurry pH; the inlet slurry pH reflects the slurry pH in the scrubber since the amount of SO_2 absorbed per unit volume (liter) of slurry is relatively small and the slurry residence time fairly short. The range of operating conditions covered in developing these correlations is as follows:

G_ℓ: 6.52 ~ 33.2 kg/m^2•s
F: 54.8 ~ 138 mol/m^2•s
inlet slurry temp: 14 ~ 53°C; inlet pH: 5.2 ~ 6.8

inlet SO_2 concentration: 1750 ~ 4500 ppm
total static height of packing: 0 ~ 0.76 m
magnesium concentration: 0 ~ 16,700 ppm

Based on these correlations, the SO_2 removal efficiencies of the large scale TVA Shawnee TBC, TVA spray column, and the EPA/RTP TBC can be computed fairly accurately (Fan, 1975). Note that in the simulation of the Shawnee spray tower with four spray headers, the tower is divided into four sections; the top section has a liquid flow rate of 1/4 G_ℓ, the next a liquid flow rate of 1/2 G_ℓ and so on until in the bottom (fourth section) the liquid flow rate is equal to G_ℓ (Fan, 1975). Furthermore, the same gas and liquid flow rate dependencies (i.e., $F^{0.8} G_\ell^{0.4}$; see Table 9.1) in the Shawnee spray tower as in the Shawnee TBC without spheres are employed to account for the gas-side mass transfer coefficient. Consistency in SO_2 scrubbing performance between the pilot scale TVA TBC scrubber and the commercial scale TBC (100 MW) scrubber located at Springfield City Utilities Southwest Station near Springfield, Missouri was noted by Wang and Burbank (1982). This fact gives added confidence that the correlations reported here can be used over a wide range of scrubber sizes including possible commercial TBC scrubbers.

When alkali additives or organic acid buffer additives other than magnesium oxide are used, Δ_s and Δ_p in Table 9.1 may vary to reflect their effects on the overall mass transfer effects.

9.2 PARTICULATE COLLECTION

A TBC also exhibits excellent particulate collection ability. The TBC scrubber has been used to clean effluents in granulating or drying processes and to control dust emissions in smelting and combustion processes (Kielback, 1961; Calvert, 1977). Very limited literature, however, is available concerning particulate collection phenomena in a TBC (Uysal, 1978; Zadlo and Kupiec, 1986).

9.2.1 Collection Mechanism and Efficiency

In a TBC, particulates are collected by liquid films, rivulets, or droplets. There are two basic classes of mechanisms for particulate collection: the aerodynamic and induced electrostatic precipitation or phoretic mechanisms. The aerodynamic mechanism includes gravi-

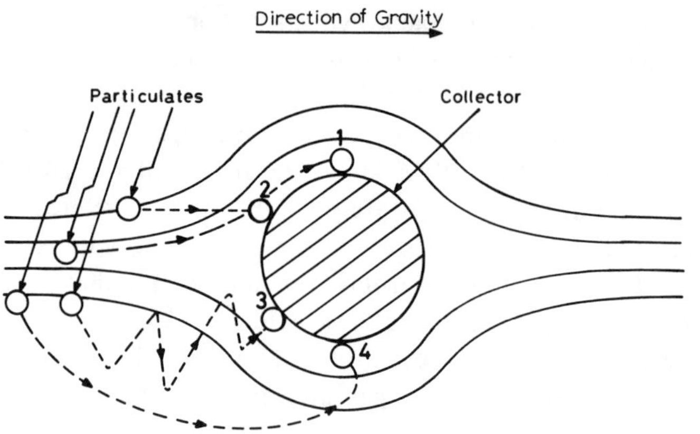

Figure 9.10 Schematic diagram of collection mechanisms for a single collector (after Peters et al., 1982).

tational sedimentation, interception, inertial impaction, and Brownian diffusion effects. The induced electrostatic precipitation and phoretic mechanism includes thermophoresis, diffusiophoresis, electrophoresis and magnetophoresis effects (Calvert, 1977; Uysal, 1978; Peters et al., 1982). In industrial scrubbers, gravitational effects and centrifugal deposition can usually be neglected since the particulates to be collected are usually smaller than about 5 μm in diameter. Calvert (1977) pointed out that Brownian diffusion effects prevail for particulates below 0.3 μm in diameter. A schematic diagram for particulate collection based on several aerodynamic mechanisms is shown in Fig. 9.10. In a TBC scrubber, phoretic effects are insignificant and, in most cases (with particulate sizes 0.3 ~ 10μm), aerodynamic capture processes are the dominant mechanisms as indicated by Uysal (1978).

Because of the complexity of particulate collection phenomena, an overall mass balance approach has been adopted to evaluate the particulate collection performance in a TBC. The particulate penetration is defined as

$$Pt = \frac{\text{number of particulates leaving the scrubber}}{\text{number of particulates entering the scrubber}} \tag{9.10}$$

The collection efficiency is then

$$\eta = 1 - Pt \tag{9.11}$$

For isolated single sphere collection efficiencies, Peters et al. (1982) provided a summary of various particulate capture mechanisms. An average penetration or efficiency can also be defined for particulates of nonuniform sizes.

9.2.2 Scrubber Model

The particulate penetration or collection efficiency in a TBC has been correlated either empirically or semi-empirically in terms of particulate size, gas velocity, liquid velocity, fluidized particle size, and other operating parameters. Early work on the particulate collection efficiency in a TBC was reviewed by Uysal (1978). Epstein et al. (1971) obtained an empirical correlation for the particulate collection efficiency based on limited data reported by Pollack et al.(1966) for fly ash recovery from a coal burning power plant flue gas. Calvert (1977) recommended using the aerodynamic particle diameter, d_{pa}, instead of the physical diameter in characterizing particulate collection phenomena. The aerodynamic diameter is defined as $d_p'(C' \rho_p')^{1/2}$ where d_p' is the particulate diameter and ρ_p' is the particulate density. The Cunningham slip factor, C', accounts for an increased particle velocity when the particle size approaches the mean free path of the gas molecules.

Uysal (1978) systematically and comprehensively studied particulate collection characteristics in a single-stage TBC column 0.29 m in diameter and 1.5 m high. Four types of particulates were studied: uniform sized ferrous sulphate particulates (average diameter varying from 0.35 to 5.5 μm), 2.5 μm silica particulates, 2.0 μm polyvinyl toluene latex particles and 2.5 μm dioctyl phthalate particulates. The contact angles (see Section 2.1.6) of these particulates vary from zero (ferrous sulphate and silica) to 63.5° (dioctyl phthalate). Uysal (1978) used fluidized particles with a density of 157 kg/m^3 and diameters ranging from 19 to 38 mm. Uysal developed a particulate transfer model which defines a volumetric particle transfer coefficient analogous to a volumetric mass transfer coefficient to account for the particulate transfer from the gas phase to the liquid phase. Based on the model, the particulate penetration through the fluidized bed section of the TBC, Pt_f, can be expressed as

$$Pt_f = \exp\left(-k_f \frac{H}{V_g}\right) \tag{9.12}$$

where k_f is the volumetric particulate transfer coefficient, V_g is the interstitial gas velocity in the fluidized bed section, and H is the height of the fluidized bed section. The particulate penetration through the spray section of the TBC, Pt_s, can also be expressed as

$$Pt_s = \exp\left(-k_s \frac{H_s}{V'_g}\right) \tag{9.13}$$

where k_s is the volumetric particulate transfer coefficient, V'_g is the interstitial gas velocity in the spray section, and H_s is the height of the spray section. The total penetration between the column inlet and outlet is the product of the penetration in the fluidized bed section and that in the spray section, i.e.,

$$Pt = Pt_f \cdot Pt_s \tag{9.14}$$

Extensive experiments using ferrous sulphate particulates showed that Pt_f decreases with increasing liquid flow rate, fluidized particle size, and static bed height. On the other hand, Pt_f is typically either almost independent of gas flow rate or increases only slightly with increasing gas flow rate. A correlation for k_f is given by

$$k_f = 3765\, St^{0.8}\, Re_g^{-2.64}\, Re_\ell^{2.1} \left(\frac{H_o}{d_p}\right)^{1.52} \tag{9.15}$$

where

$$St = \frac{C' d_p'^2 \rho_p' V_r}{9 \mu_g d_p}, \quad Re_g = \frac{V_g \rho_g d_p}{\mu_g}, \quad Re_\ell = \frac{V_\ell \rho_\ell d_p}{\mu_\ell}, \quad V_r = V_g + V_\ell$$

Equation (9.15) applies for $0.29 < H_o < 0.58$ m; $0.5 < G_g < 5.5$ kg/m^2·s; $4.7 < G_\ell < 33$ kg/m^2·s; d_p = 19, 25, 38 mm; ρ_p = 157 kg/m^3; and $0.35 < d_p'$ < 5.5 μm. k_f is in 1/s.

A correlation for k_s is given by (in SI units)

$$k_s = 0.0165\, G_\ell^{1.508}\, G_g^{0.745}\, St_d^{-0.413} \tag{9.16}$$

where $V'_r = V'_g + V'_\ell$; $St_d = C' d'^2_p \rho'_p V'_r /(9\mu_g d_c)$. k_s is in 1/s, and d_c is the volume mean diameter for the liquid drop, which can be estimated by (in SI units) (Merrington and Richardson, 1947)

$$d_c = \frac{0.315}{V'_r} \left[\frac{\mu_\ell}{\rho_\ell}\right]^{0.2} \tag{9.17}$$

Equation (9.16) is valid under the same conditions as those for Eqn. (9.15).

9.2.3 Scrubber Performance

Rowbottom (1973) reported the results of pilot scale experiments on particulate removal in a multi-stage TBC containing 38 mm diameter spheres. Mlodzinski and Warych (1975) studied the collection efficiency in a small TBC for zinc sulfide (hydrophobic) and superphosphate (hydrophilic) particulates. They found that there is little effect of hydrophobicity on the collection efficiency. The collection efficiency for both kinds of particulates increases with increasing liquid flow rate and increases to a less extent with increasing gas flow rate. The experimental results of Rowbottom (1973) led to the observation by Uysal (1978) that varying the inlet particulate concentration over the range 1.8 ~ 3.3 g/m^3 has no effect on the exit concentration for liquid flow rates greater than 14 $kg/m^2 \cdot s$. Some quantitative findings by Uysal (1978) are presented in Figs. 9.11 through 9.14 and are given below:

Figure 9.11 shows the variation in particulate (ferrous sulfate) penetration with liquid flux as a function of particulate diameter for a constant gas flux. As shown, an increase in G_ℓ decreases Pt and, at a constant G_ℓ, an increase in d'_p decreases Pt; this behavior occurs since an increase in G_ℓ increases liquid holdup and, consequently, increases the particulate collection. As noted earlier, the dominant mechanism for particulate collection over the particulate size range and gas and liquid flow rates considered is inertial impaction. Thus, larger particulate sizes show greater extents of inertial impaction and hence, yield higher particulate collection efficiencies. Figure 9.12 shows the variation in Pt with G_g as a function of d'_p for a given G_ℓ; as shown, at a low G_g, Pt is independent of G_g whereas Pt slightly increases with G_g at a high G_g. This behavior represents an overall effect resulting from two counteracting factors. That is, an increase in G_g increases the efficiency of inertial impaction, thus promoting a

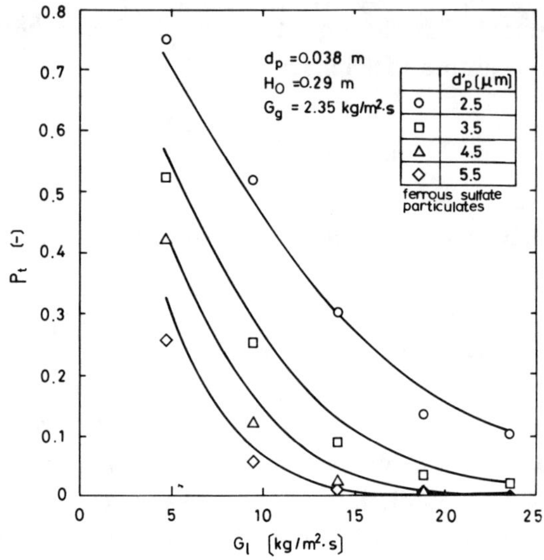

Figure 9.11 Variations of particulate penetration with liquid mass flux as a function of particulate sizes for a TBC scrubber (from Uysal, 1978).

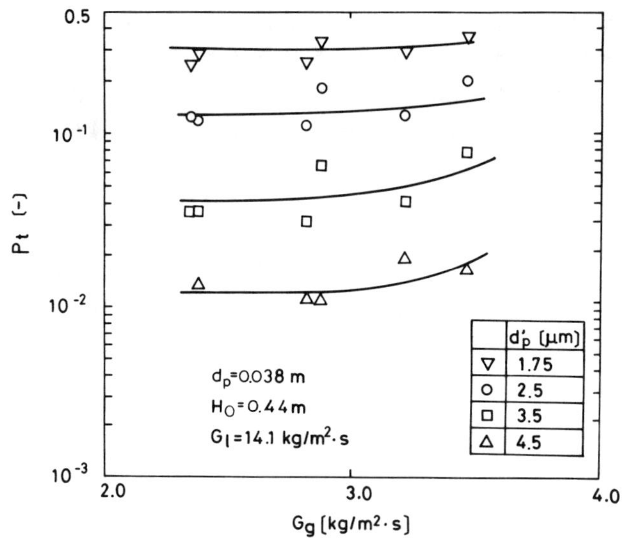

Figure 9.12 Variations of particulate penetration with gas mass flux as a function of particulate sizes for a TBC scrubber (from Uysal, 1978).

higher particulate collection efficiency, but reducing the particulate residence time. The lower residence time which, in turn, reduces the probability of particulate-collector contact, thus lowering particulate collection efficiency. Figure 9.13 shows the variation in Pt with the static bed height as a function of the fluidized particle diameter; as shown, an increase in H_o decreases Pt and, at a given H_o, an increase in d_p increases Pt. This behavior occurs since an increase in H_o increases the particulate residence time which, in turn, increases the probability of their contact with collectors. An increase in d_p decreases the intensity of the mixing among phases and the liquid holdup, yielding a lower particulate collection.

Figure 9.14 shows the variation in Pt with G_ℓ for polyvinyl toluene latex, dioctyl phthalate and silica particulates. These particulates have different diameters and/or densities and different contact angles, i.e., 0°, 46°, and 64°, respectively, and are all insoluble in water. The particulate penetration behavior for these particles is, however, predictable based on the model equations developed for ferrous sulfate particulate collection despite the obvious differences in particulate properties. Note that ferrous sulfate has a contact angle of 0° and is water soluble. Thus, it can be stated that the particulate collection efficiency in a TBC is not affected by such particulate properties as solubility, hydrophobicity, or θ_c when particulates are under the conditions of $\theta_c < 90°$ and $0.35 < d_p' < 5.5$ μm.

When water vapor condensation from a hot, humid particulate-laden gas stream becomes important, effects of diffusiophoresis on particle penetration need to be considered. Thus, the overall particulate penetration will be the product of the penetration due to aerodynamic collection given in Eqn. (9.14) and that due to diffusiophoresis. Based on different diffusiophoretic velocities of the particle, the penetration due to diffusiophoresis can be expressed based on the outlet and inlet ratio of the Schmitt-Waldmann velocity (Waldmann and Schmitt, 1966), gas mean mass velocity or gas molar velocity for the moving particle. Due to small differences in these three velocities, Uysal (1978) indicated that the one based on the outlet and inlet ratio of the mean mass velocity can provide a slightly more conservative estimate for particulate collection. Uysal (1978) also reported that for an identical total static packing height, a two-stage TBC yields a higher particulate collection efficiency than a single stage TBC for small spheres, i.e., 19 and 25 mm. However, the two TBCs yield essentially the same particulate collection efficiency for large spheres, i.e., 38 mm.

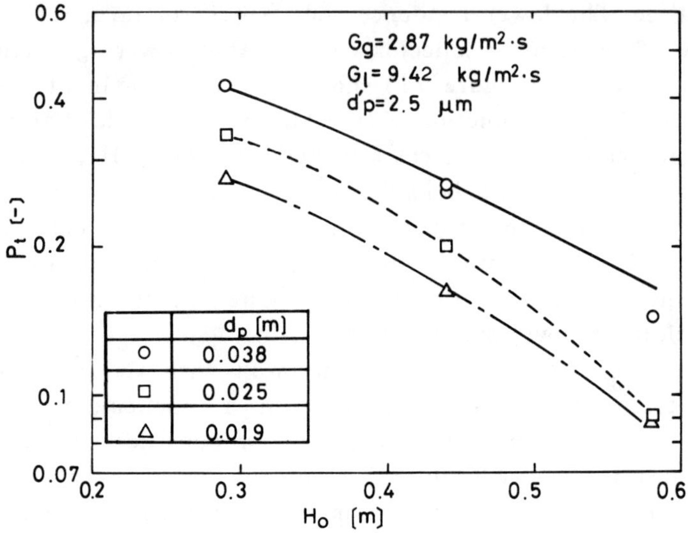

Figure 9.13 Variations of particulate penetration with static packing height as a function of sphere sizes for a TBC scrubber (from Uysal, 1978).

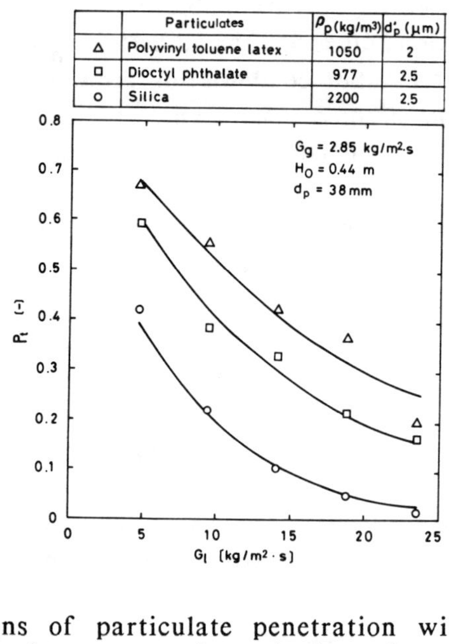

Figure 9.14 Variations of particulate penetration with liquid mass flux for three types of particulates for a TBC scrubber (from Uysal, 1978).

Recently, Tabei and Kuroda (1982) studied particulate collection efficiency for various TBC operating conditions. They reported that the collection efficiency increases with increasing liquid holdup and hence, with increasing pressure drop. The particulate collection efficiency correlated by them is a function of the pressure drop and particulate size. Wang and Burbank (1982) reported the particulate size distribution between inlet and outlet streams of the flue gas using the TBC shown in Fig. 9.3 or 9.5, operated for simultaneous flue gas desulfurization and particulate removal. The collection efficiency in a TBC was found to be generally higher than that in a spray tower operated at similar conditions. A semi-theoretical equation was proposed by Zadlo and Kupiec (1986) to describe the particulate collection efficiency in a TBC based on a flow model considering an idealized curvature channel. The model was verified with limited data.

Concerning the choice of a scrubber, functions and characteristics of various industrial scrubbers currently available were discussed by Calvert (1977) and Calvert et al. (1980); they compared the performance of the scrubbers for particulate removal using the cut diameter, defined as the diameter of particulates with 50% collection efficiency. The cut diameters were correlated in terms of the gas-phase pressure drop for the TBC scrubber as well as for other types of scrubbers. Calvert (1977) showed that at low gas pressure drops, the cut diameter for a TBC scrubber is higher than for rings or saddles, fixed bed, fibrous fixed bed, or gas-atomized spray column scrubbers. At high gas pressure drops, on the other hand, compared to these scrubbers the cut diameter for a TBC is substantially improved.

9.3 CONCLUDING REMARKS

In the application of a TBC scrubber to flue gas desulfurization, optimal operation requires a comprehensive understanding of the role played by the hold tank and the process chemistry in both the scrubber and hold tank in the system. The mathematical model introduced here takes into consideration the gas-side and liquid-side mass transfer and chemical reactions in the liquid phase in both the packed and spray sections of a TBC scrubber. The model can be utilized to describe the SO_2 scrubbing behavior over a wide range of scrubber sizes.

Buffer additives, such as adipic acid, can improve SO_2 scrubbing efficiency by permitting a higher SO_2 concentration in the gas-liquid interface and enhancing the bulk liquor reactions with SO_2 in the presence of calcium salt. Thus, the SO_2 scrubbing rate becomes less dependent on the rate of dissolution of lime/limestone or calcium sulfite in the scrubber to provide the required dissolved alkalinity. The slurry pHs at the scrubber inlet can be controlled at 5.2 ~ 6.0 to avoid scaling problems while maintaining significant limestone utilization. Proper balance of the ratio of the molar flow rate of O_2 in the hold tank and the molar absorption rate of SO_2 in the scrubber in a one hold tank system with in-loop forced oxidation can alleviate limestone blinding problems.

In the application of a TBC scrubber for particulate collection, the collection efficiency increases with increasing liquid holdup and hence with increasing pressure drop. The particulate penetration through both the fluidized bed section and spray section varies inversely with the ratio of the section height to the linear gas velocity in the section. Inertial impaction is the dominant mechanism for particulate collection. The overall particulate penetration is a product of the particulate penetration in the fluidized bed section and that in the spray section. The particulate collection efficiency in the fluidized bed section is significantly higher than that in the spray tower section. The diffusiophoresis mechanism becomes important when water vapor condensation from a hot, humid particulate-laden gas stream is encountered.

9.4 NOMENCLATURE

a	: specific interfacial area available to mass transfer
A_p	: pre-exponential factor in the expression for R for the packed section
A_s	: pre-exponential factor in the expression for R for the spray section
C_A	: H_2SO_3 concentration in the bulk liquid phase
$C_A{}^i$	: interfacial H_2SO_3 concentration
C'	: Cunningham's slip factor
d_c	: volume mean diameter of liquid droplet
d_p	: diameter of sphere or low density large particle

d'_p : diameter of aerosol particle in flue gas

d_{pa} : aerodynamic diameter of particulate collected

F : molar flux of gas based on cross-sectional area of the scrubber

G : mass flux of fluid based on cross-sectional area of the scrubber; G_ℓ for liquid; G_g for gas

H : height of fluidized bed in the TBC

H_o : height of static bed in the TBC

H_s : H_T - H

H_{sp} : H_T - H_o

H_T : height of the TBC

He : Henry's law constant

k_g : gas-side mass transfer coefficient

k_ℓ : liquid-side mass transfer coefficient

$k_g{}^pa$: volumetric gas-side mass transfer coefficient for the packed section

$k_g{}^sa$: volumetric gas-side mass transfer coefficient for the spray section

k_f : volumetric particulate transfer coefficient in the fluidized bed section of a TBC

k_s : volumetric particulate transfer coefficient in the spray section of a TBC or in a spray tower

K_ga : overall volumetric gas-side mass transfer coefficient

$K_g{}^pa$: overall volumetric gas-side mass transfer coefficient for the packed section

$K_g{}^sa$: overall volumetric gas-side mass transfer coefficient for the spray section

N_{SO_2} : molar flux of SO_2

P_{SO_2} : partial pressure of SO_2 in the bulk gas phase

$P_{SO_2}{}^i$: interfacial partial pressure of SO_2

$P_{SO_2}{}^{in}$: inlet partial pressure of SO_2 in the bulk gas phase

$P_{SO_2}{}^{out}$: outlet partial pressure of SO_2 in the bulk gas phase

$P^*_{SO_2}$: partial pressure of SO_2 in equilibrium with the bulk liquid phase concentration

P_T : total pressure

Pt	: particulate penetration in a scrubber
Pt_f	: particulate penetration in the fluidized bed section
Pt_s	: particulate penetration in the spray section of a TBC or in a spray tower
R	: ratio of the gas-side to liquid-side mass transfer resistances
R_p	: R for the packed section
R_s	: R for the spray section
Re_g	: gas Reynolds number, $V_g \rho_g d_p / \mu_g$
Re_ℓ	: liquid Reynolds number, $V_\ell \rho_\ell d_p / \mu_\ell$
St	: Stokes number for a TBC, $C' d_p'^2 \rho_p' V_r / 9 \mu_g d_p$
St_d	: Stokes number for a spray tower, $C' d_p'^2 \rho_p' V_r' / 9 \mu_g d_c$
T	: temperature
V_g	: interstitial gas velocity in the fluidized section
V_g'	: interstitial gas velocity in the spray section
V_ℓ	: interstitial liquid velocity in the fluidized section
V_ℓ'	: interstitial liquid velocity in the spray section
V_r	: relative velocity between the liquid and gas in the fluidized section, $V_g + V_\ell$
V_r'	: relative velocity between the liquid and gas in the spray section, $V_g' + V_\ell'$
z	: axial distance from gas inlet
Z_i	: height of the i-th position

Greek Letters

δ_{Mg}	: magnesium concentration (in weight fraction) in the bulk liquid phase
Δ_s	: magnesium correction factor for A_s
Δ_p	: magnesium correction factor for A_p
η	: particulate collection efficiency
θ_c	: contact angle
μ_g	: gas viscosity
μ_ℓ	: liquid viscosity
ρ_g	: gas density
ρ_ℓ	: liquid density

ρ_p : density of sphere or low density large particle

ρ_p' : density of aerosol particle

ϕ : enhancement factor for mass transfer in the liquid film due to chemical reactions

$\bar{\phi}$: average enhancement factor

9.5 REFERENCES

Borgwardt, R.H., paper presented at the EPA Flue Gas Desulfurization Symposium, Atlanta, GA, Nov. 4-7 (1974).

Borgwardt, R.H., in Proceedings of Symposium on Flue Gas Desulfurization, Hollywood, FL, EPA-600/7-78-058a (1978).

Calvert, S., Chem. Eng., 84(23), 133 (1977).

Calvert, S., S.-C. Yung, H. Barbarika, G. Monahan, L.E. Sparks, and D.L. Harmon, AIChE Symposium Series, 76, No. 192, 74 (1980).

Chang, J.C.S., Ph.D. Dissertation, University of Texas at Austin (1979).

Chang, J.C.S., and J.D. Mobley, APCA J., 33, 955 (1983).

EPRI (Electric Power Research Institute), "Introduction to Limestone Flue Gas Desulfurization," Tech Tape CC87-01 (1987).

Epstein, M., C.C. Leivo, and C.H. Rowland, in Proceedings of the 2nd International Lime/Limestone Wet Scrubbing Symposium, 1, 45 (1971).

Epstein, M., EPA Series, EPA-650/2-73/013 (1973).

Fan, L.-S., Ph.D. Dissertation, West Virginia University (1975).

Henzel, D.S., B.A. Laseke, E.O. Smith, and D.O. Swenson, EPA Series, EPA-600/8-81-017 (1981).

Idemura, H., T. Kanai, and H. Yanagioka, Chem. Eng. Prog., 74(2), 46 (1978).

Kaplan, N., EPA Series, EPA-650/2-73-038 (1973).

Kielback, A.W., Chem. Eng. Prog. Symposium Series, 57, No. 35, 51 (1961).

McMichael, W.J., L.-S. Fan, and C.Y. Wen, I&EC Process Des. & Dev., 15, 459 (1976).

Merrington, A.C., and E.G. Richardson, Proc. Phys. Soc., 59, No. 331, 1 (1947).

Meserole, F.B., D.L. Lewis, A.W. Nichols, and G.T. Rochelle, EPA Series, EPA-600/7-79-224 (1979).

Mlodzinski, B., and J. Warych, Chem.-Ing.-Tech., 47, 813 (1975).

Mobley, J.D., and J.C.S. Chang, APCA J., 31, 1249 (1981).

Noguchi, M., EPA Series, EPA-650/2-74-1266 (1974).

Parsons, E.L. Jr., L.F. Hemenway, G.T. Kragh, T.G. Brna, and R.L. Ostop, in Proceedings of Symposium on Flue Gas Desulfurization, Vol. II, EPA-600/9-81-019b, p. 801 (1981).

Peters, M.H., L.-S. Fan, and T.L. Sweeney, AIChE J., 28(1), 39 (1982).

Pollack, W.A., J.P. Tomany, and G. Frieling, ASME Publication, 66-WA/CD-4 (1966).

Rochelle, G.T., and C.J. King, I&EC Fundam., 16, 67 (1977).

Rochelle, G.T., and C.J. King, Chem. Eng. Prog., 74(2), 65 (1978).

Rowbottom, R.S., Domtar Ltd. Research Center, CPAR Project No. 211, Dec. 31 (1973).

Sawahata, M., Y. Kogawa, A. Kumagai, and B.R. Gilbert, paper presented at the American Power Conference, Chicago, IL, April 27-29 (1987).

Slack, A.V., and G.A. Hollinden, "Sulfur Dioxide Removal from Waste Gases," Noyes Data Corp., Park Ridge, NJ (1975).

Tabei, T., and M. Kuroda, paper presented at the 16th Autumn Meeting of Soc. Chem. Engrs. Japan, Toyoda, Japan, Oct. 4 - 6 (1982).

Uysal, B.Z., Ph.D. Thesis, McGill University (1978).

Vivian, J.E., Report to HEW, Sep. (1973).

Waldmann, L., and K.M. Schmitt, in "Aerosol Science," ed. by C.N. Davis, p. 137, Academic Press, London (1966).

Wang, S.C., and D. Burbank, EPA Series, EPA-600/7-82-010a (1982).

Wen, C.Y., and S. Uchida, EPA Series, EPA-650/2-73-003 (1973).

Zadlo, J., and K. Kupiec, in Proceedings of World Congress III of Chemical Engineering, III, p. 643 (1986).

Zemo, B.R., Columbus Southern Power Company, personal communication (1988).

Chapter 10

Hydrotreating and Conversion of Resids

One of the most important commercial applications of the three-phase fluidized bed is in heavy oil, petroleum resid, and synthetic crude processing. The technology has been successfully demonstrated for petroleum resid processing at the Kuwait Shuaiba, Pemex Salamanca, Texaco Convent, and Amoco Texas City refineries. The processing capacity for petroleum resids in these refineries ranges from 18,500 barrels per day (bbl/day) to 60,000 bbl/day. Resid processing technology has grown in importance recently since about 70% of world oil reserves consist of heavy crude and bitumen (Parkinson, 1985).

Feedstocks of petroleum resids, heavy oils, and synthetic crudes from coal and tar sands show some similarities in properties. Table 10.1 lists various properties of these heavy feedstocks. Despite the obvious differences in metal and heteroatom contents between petroleum resids and coal liquids or coal-solvent slurries, reactor engineering for hydrotreating of these heavy feedstocks is similar. More specifically, similarities exist in the processing of petroleum heavy feedstocks and coal-solvent slurries in coal liquefaction (Li and Liu, 1981).

In this chapter resid processing alternatives are discussed with a specific focus on the alternative using ebullated bed reactors. An analysis of ebullated bed reactors is given based on the viewpoint of transport properties, reactor design, and reactor operation. Furthermore, catalysis and kinetics of various reactions taking place in the reactor such as thermal cracking, hydrocracking, hydrodesulfurization, hydrodemetallization, and Ramsbottom carbon conversion are described. The chapter is described in the context of hydrotreating of petroleum resids as feedstocks. However, wherever appropriate comments on operational differences in processing other feedstocks including heavy oils and coal-solvent slurries are included.

Table 10.1 Properties of Heavy Petroleum and Synthetic Crude from Coal, Oil Shale and Tar Sands (from Thakur and Thomas, 1985)

Properties	Petroleum crude	Synthetic crude					
	Atm.resid Khafji	Coal (Bituminous)	SRC II (Bituminous)	H-Coal (Bituminous)	EDS Lignite	Oil Shale	Tar Sands
Gravity/°API	14.7	-	18.6	25.8	29	20.2	11.6
Carbon/wt%	83.9	82.9	84.6	86.9	86.4	84.3	82.6
Hydrogen/wt%	11.0	7.3	10.5	11.4	11.2	11.3	10.3
Sulfur/wt%	4.3	2.3	0.3	0.3	0.1	0.7	4.8
Oxygen/wt%	-	5.9	3.8	1.8	1.9	1.2	1.8
Nitrogen/wt%	0.3	0.9	0.8	0.5	0.3	2.2	0.5
H/C (atom)	1.58	1.06	1.47	1.56	1.56	1.60	1.48
Iron/ppm			7.5	22.0	-	70.0	-
Nickel/ppm	32.0						112.0
Vanadium/ppm	93.0						180.0
Arsenic/ppm						28.0	

Table 10.2 Resid Processing Alternatives (after Dautzenberg and de Deken, 1984)

C-Rejection (Concentration into a residual by-product)	C-Rejection (Indirect conversion into low-Btu gas or steam)	H-Addition (Direct conversion into oil products)
• Visbreaking • Delayed Coking • Deasphalting	• Fluid Coking • Flexicoking • Dynacracking • Heavy oil catalytic cracking • ART (Asphalt Resid Treating)	• Shell-HDM/HDS • Gulf-HDS • H-Oil • LC-Fining • Aurabon • M-Coke • Canmet • Veba-Combi-Cracking

10.1 PROCESSING ALTERNATIVES

Table 10.2 shows the available industrial processing alternatives for petroleum resids. Two basic approaches to resid processing are well recognized in industry: Carbon(C)-rejection and Hydrogen(H)-addition. C-rejection processes can be further subdivided into processes either those which reject substantial parts of the fuel as a residual by-product or coke or those which convert major portions of the fuel via indirect conversion to low Btu gas, steam, and heat. Examples of the first type of C-rejection process include delayed coking and deasphalting. Examples of the second type of C-rejection process include fluid coking, flexicoking, dynacracking, heavy oil catalytic cracking, and ART (Asphalt Resid Treating). H-addition processes directly convert resids into light products such as naphthas, middle distillates, and gas oils at a relatively high yield compared to C-rejection processes. Examples of this type process include Shell-HDM/HDS, Gulf-HDS, H-Oil, LC-Fining, Aurabon, M-Coke, Canmet and Veba-Combi/Cracking. Economic factors dictate the selection between H-addition and C-rejection processes. For example, when the price difference between resid and light oils is large, economics favor the H-addition process. When the price difference is small, on the other hand, economics favor the C-rejection process. Note that the cost difference in processing heavy oils over conventional lighter oils is about \$3 ~ 7 per barrel (Parkinson, 1985).

In H-addition (hydrotreating) processes, hydrogen gas, liquid resids, and a solid catalyst are contacted in the reactor and many complex reactions take place including thermal cracking, hydrocracking, hydrodesulfurization, hydrodeoxygenation, hydrodenitrogenation, hydrodemetallization, and carbonaceous solid (sludge) formation. The actual reactor can encompass three different operating modes: the fixed, moving or trickle bed (Shell-HDM/HDS and Gulf-HDS), the fluidized or ebullated bed (H-Oil and LC-Fining), and the slurry bubble column (UOP-Aurabon, Exxon-M-Coke, Canadian Government-Canmet, and Veba-Combi/Cracking). Of these, the trickle bed and ebullated bed (H-Oil and LC-Fining process) are commercially proven technologies and the slurry bubble column (Canmet process) has been successfully demonstrated on a mini-commercial scale[1]. Typically, hydrotreating of heavy crudes is conducted at temperatures between 300 and 425°C and at pressures between 5.5 and 21 MPa.

The ebullated bed reactor is normally operated at a higher temperature than the trickle bed reactor. The merits and demerits of the ebullated bed reactor over the trickle bed reactor are given below (Thakur and Thomas, 1985 and Beaton et al., 1986).

(1) The well-mixed characteristics of the ebullated bed reactor permit a rather uniform temperature distribution in the reactor. During steady state operation, the temperature difference between any two locations in the reactor is less than 6°C unless the liquid recycle ratio is low. Thus, internal quenching is not necessary in the ebullated bed reactor and the reactor quenching is usually accomplished at the reactor feed inlet. Careful monitoring of the

---

1: In this case, the mini-commercial scale is about 1/4 to 1/5 of the full-commercial scale.

temperature in the ebullated bed reactor is, however, required to prevent temperature run away due to formation of stagnant regions or maldistribution of gas and liquid flows. The trickle bed reactor implements temperature control via multiple internal quenching points using cold hydrogen injection. Due to the multiple internal quench points trickle bed reactors can avoid run-away situations more effectively than ebullated bed reactors based on normal quenching methods.

(2) On-stream catalyst replacement can be easily accomplished in an ebullated bed reactor. The trickle bed reactor, however, must employ special designs to achieve on-stream catalyst replacement. The moving bed and bunker bed types of the trickle bed reactor allow on-stream catalyst replacement, but the fixed bed type does not.

(3) Plugging problems exist in the trickle bed reactor due to solids formation and the presence of attrited catalysts, clay, and sand in the influent stream of the reactor. Such problems do not occur in the ebullated bed reactor. Dirt tolerance of the ebullated bed reactor is better than the trickle bed reactor. Particle attrition and agglomeration occur to some extent in the ebullated bed reactor.

A photograph of the LC-Fining unit using the ebullated bed process at Amoco's Texas City refinery is shown in Fig. 10.1. The reactor section of the process normally has two or three independent parallel trains with two or three reactors in series per train. The reactor, operated in Mode E-I-a-1, is about 3.9 m in diameter and approximately 25 m high (MacArthur et al., 1983) (see Fig. 1.3 for mode designation). Figure 10.2 shows a simplified flow diagram for one train of the process with three reactors in series. Resid oil and hydrogen are heated separately, then mixed and introduced into the ebullated bed reactor. The reactor contains a bed of catalyst expanded by the upflow of feed hydrogen and oil and internal recycle of liquid products. The internal recycling establishes a complete mixing of the liquid phase in the reactor. The solid phase is also in a completely mixed state due to the bubbling action of the gas phase.

The intermediate products from the first stage are quenched with hydrogen gas or oil before being introduced to the second reactor. The intermediate products from the second reactor are also quenched before being introduced to the third reactor. The first reactor is designed for demetallization while the second and third reactors are designed for desulfurization (Beaton et al., 1986). The products then go through a series of separators to yield unconverted residual oil, light oils, and fuel gases. Typical resid conversion is 60 ~

Figure 10.1 LC-fining resid hydrocracking unit at Amoco's Texas City Refinery. The unit, with its nine expanded bed reactors shown to the right, converts 60,000 bbl/day of residual oil into lighter transportation fuels. Lummus Crest, Cities Service R&D Co., and Amoco Co. are co-licensors of the process (kindly supplied by R.D. Buttke of Amoco Oil Company).

65%. However, conversion can be increased to 95% or higher by recycling unconverted resids if coke precursors in the intermediate product stream can be removed continuously prior to recycling (Van Driesen et al., 1986; Boening et al., 1987).

The economic outlook for ebullated-bed processing depends strongly on the technology available for reducing carbonaceous solid

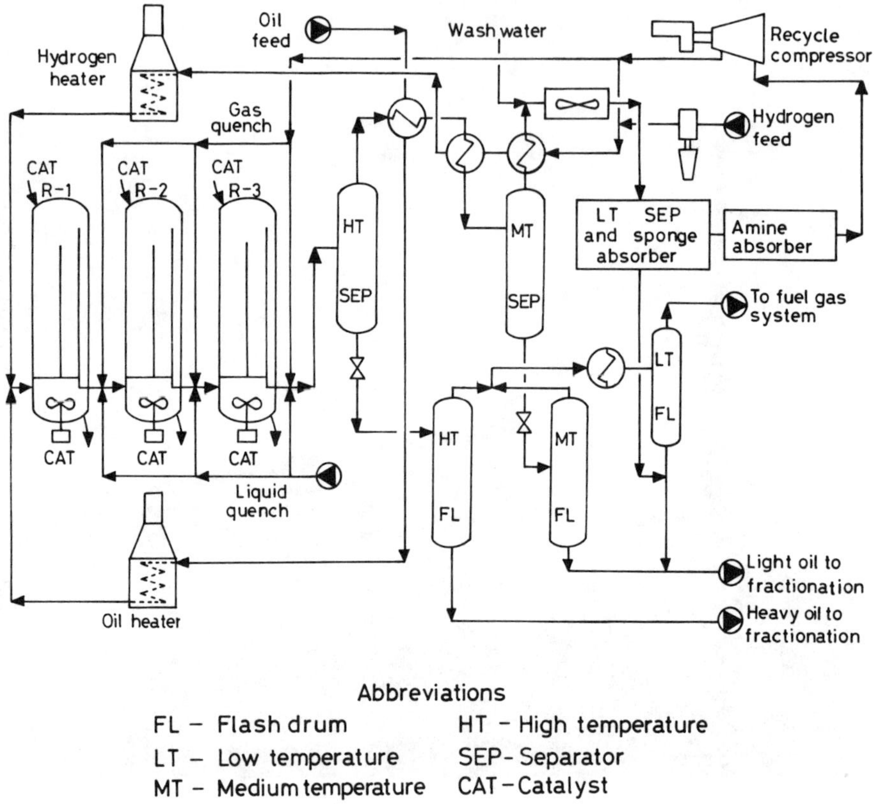

Figure 10.2 A simplified flow scheme for three-stage ebullated bed reactors in the LC-fining process (from Beaton et al., 1986).

formation and prolonging catalyst activities or reducing catalysis associated costs.

The slurry bubble column (Mode T-1-a-2) as demonstrated by that in the Veba-Combi-Cracking process and the Canmet process has recently emerged as an important reactor in the resid hydrotreating technology. Notably, slurry bubble column reactors have several advantages over ebullated bed reactors (Dautzenberg and de Deken, 1984; also see Appendix A): the small catalyst particle sizes used in the slurry bubble column reactor have low intraparticle and liquid-solid mass transfer resistances and a low solids holdup requirement to achieve a large particle surface area, and thus, enhancing overall reaction rates while suppressing liquid-phase free radical polymerization reactions. Moreover, no special mechanical strength is needed for the catalyst particles. The noted disadvantages of the slurry bubble column reactor compared to the ebullated bed reactor are the difficulty in the downstream catalyst-liquid separation and the inflexibility in on-stream catalyst replacement.

Relatively more is known regarding the Canmet process than the Veba-Combi-Cracking process. The Veba-Combi-Cracking process was developed based on the Bergius-Pier technology (see Section 1.4.1) for coal liquefaction and is being demonstrated at a mini-commercial scale with a throughput of 3,500 bbl/day (Wenzel et al., 1986). To avoid catalyst deactivation problems, the Canmet process uses an additive[2] to promote hydrotreating reactions. The process achieves a high conversion of the 524+°C material (pitch) to lighter distillate products in a single pass and at a relatively low pressure (Waugh and Silva, 1984). A pitch conversion of 86% ~ 90% has been achieved and asphaltene conversions have exceeded 70% for heavy Canadian and Venezuelan crudes. The main emphasis of this process is the conversion of the heavy material to lighter products, while desulfurization is only a secondary goal. Desulfurization of the products is moderate and downstream use of conventional desulfurization technology is necessary. Currently, this process is licensed to Petro-Canada and treats about 5000 bbl/day of 560+°C vacuum bottoms. Figure 10.3 shows a photograph of the Canmet process located at the Petro-Canada Products Montreal refinery.

Figure 10.4 shows a simplified flow diagram of the Canmet process. The reactor, a slurry bubble column operated in Mode T-I-a-2, is 2 m in diameter and 22 m high. The low cost additive, a coal particle-iron compound mixture, is added to the vacuum tower bottoms prior to the feed preheater; this combined stream is then passed through the reactor cocurrently with the recycle and makeup hydrogen. Downstream from the reactor, the product is separated into the light ends, naphtha, gas oil, and pitch fractions. The additive preparation passes through the reactor only once and the residue exits the system in the pitch fraction. The pitch contains nearly all metals entering with the feed.

2: Disposable catalysts used in a once-through process are commonly known in industry as an "additive".

Figure 10.3 Canmet resid hydrocracking mini-commercial plant at the Petro-Canada Products Montreal Refinery. This photograph is an overall view of the reactor section, showing liquid and gas furnaces, reactor structure and some heat exchangers. The fractionation section is in the background (kindly supplied by B. B. Pruden of Petro-Canada Products).

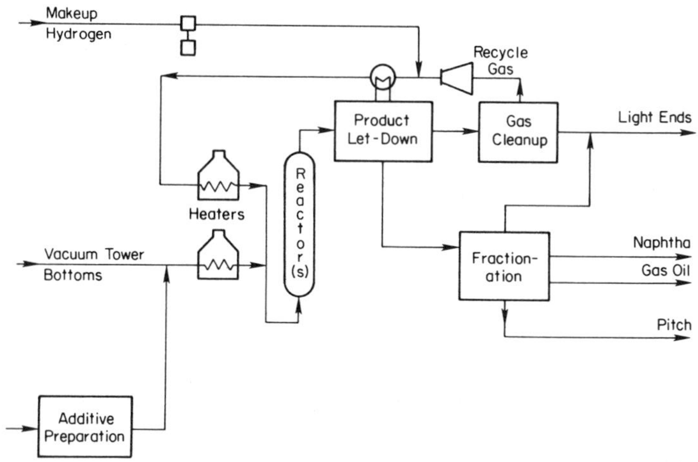

Figure 10.4 A simplified flow scheme for a slurry bubble column reactor in the Canmet process (from Lunin et al., 1985).

10.2 FEEDSTOCK CHARACTERISTICS

The different hydrotreating feedstocks from heavy oils, petroleum resids, and synthetic crudes are complex mixtures of organic compounds and consequently are difficult to characterize. Petroleum resids, depending on the extent of fractionation, can be either atmospheric or vacuum resids. Atmospheric resids typically boil above 343 °C, while vacuum resids boil above 538 °C. Heavy oils generally refer to crude oils with an API (American Petroleum Institute) gravity of less than 20° and usually have higher sulfur contents than lighter oils.

Petroleum resids and heavy oils contain a large number of compounds which have high molecular weights, high metals content, and numerous multi-functional groups. Feedstocks of coal liquids and other synthetic crude oils have different amounts and types of functional groups and contain different metals. The most important heteroatoms present in hydrotreating feedstocks are sulfur, nitrogen, and oxygen. The complex organic compounds in petroleum resids and heavy oils can be grouped into three basic subclasses based on differences in solubilities: oils, resins, and asphaltenes. The asphaltene fraction is precipitated out from the other fractions using a solvent such as n-pentane (or in some cases n-heptane or n-hexane). The oils and resins can then be separated by differences in solubility in liquid propane or adsorption behavior on fuller's earth (Speight, 1981).

Typically, oils, resins, and asphaltenes account for 35 ~ 55 wt%, 25 ~ 40 wt%, and 10 ~ 25 wt% of the petroleum resids, respectively.

The aromatic content of petroleum feedstocks differs from that of synthetic crude oils. Coal liquids contain about 60 wt% aromatics, predominantly with condensed two to five ring compounds, of which about half are saturated. Petroleum crude oils contain only about 20 wt% aromatics of which most are alkyl benzenes (Tsonopoulos et al., 1986). Resids have an aromatic content in the range of 35 ~ 45 wt%.

10.2.1 Atomic H/C Ratio

Petroleum resids and synthetic crude oils have an atomic hydrogen to carbon (H/C) ratio ranging from 1.6 to 1.3. The atomic H/C ratio for coal liquids reflects the large amount of hydrogen required in the liquefaction process (H/C ratio of coal is 0.5 ~ 1.0). The asphaltene fraction of petroleum resids has a H/C ratio of about 1.2.

10.2.2 Molecular Weights and Sizes

Takeuchi et al. (1985) gave molecular weight distributions for a petroleum resid and several characteristic compound subclasses from that resid as shown in Fig. 10.5; they reported the average molecular weight of the resid to be 1100. Sulfur containing compounds show nearly the same molecular weight distribution as that of the entire resid. Metal containing compounds, however, show a molecular weight distribution closer to that for the asphaltene fraction.

Molecular size plays an important role in the hydrodesulfurization process. For residual oils of Kuwait and West Coast crude oils, molecules range in size from 10 to 10^3 Å with averages about 25 and 40 Å, respectively (Ohtsuka, 1977).

10.2.3 Metal Compounds

Petroleum resids contain significant amounts of nickel and vanadium while coal liquids contain iron and tungsten. In resids, nickel and vanadium contents range from 10 to 120 ppm and 30 to 160 ppm, respectively. The vanadium content can reach higher than 1400 ppm in the asphaltene fraction. A distribution of metals among the various fractions of a resid is given in Table 10.3. Most of the metals are concentrated in the resin and asphaltene fractions.

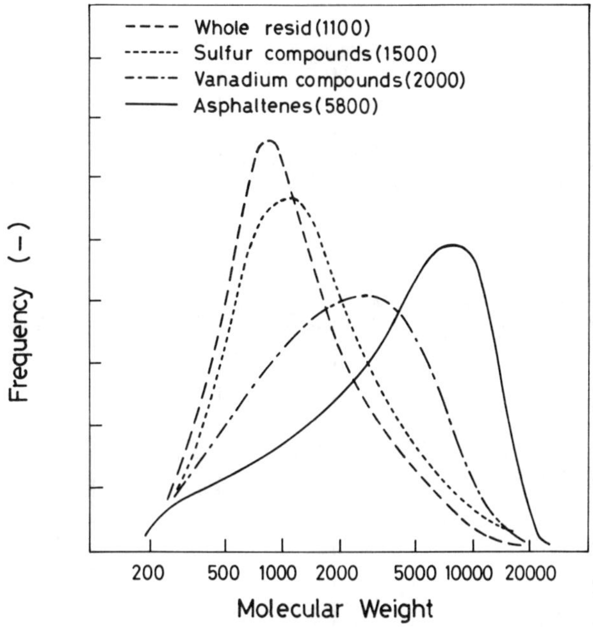

Figure 10.5 Molecular weight distributions in a petroluem resid (from Takeuchi et al., 1985). Values in parentheses are average molecular weights.

Metal compounds can be classified as porphyrinic or non-porphyrinic with most metal compounds in petroleum resids being porphyrinic. Porphyrinic compounds have the basic structure given below.

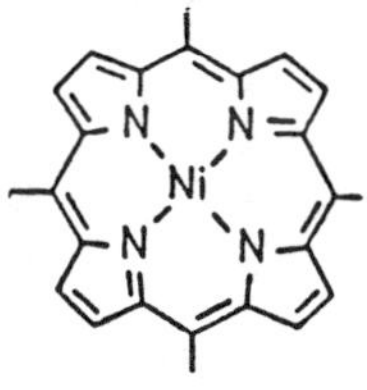

The metal atoms, V and Ni for resids, occupy the center location of the molecule. The structures of non-porphyrinic compounds are not well understood. The reactivity of non-porphyrinic metals exceeds that of porphyrinic metals. (Galiasso et al., 1985).

Table 10.3 Property Distributions Across the Various Fractions of a Resid

Resids		Properties	Saturates	Aromatics	Resins	Asphaltenes
Atmospheric Resids	Khafji[a]	Composition (wt%)	26.6	38.5	21.2	13.8
		Sulfur[f] (wt%)	-	2.06	1.21	1.10
		Vanadium (ppm)[b]	-	-	139	786
		Nickel (ppm)[b]	-	-	36	250
	Ceueta[a]	Composition (wt%)	36.4	45.9	13.7	4.0
		Sulfur[f] (wt%)	-	1.32	0.58	0.17
		Vanadium (ppm)	-	-	na	na
		Nickel (ppm)	-	-	na	na
	Mixed Iranian[a]	Composition (wt%)	32.7	40.0	22.6	4.7
		Sulfur[f] (wt%)	-	1.47	0.77	0.23
		Vanadium (ppm)	-	-	na	na
		Nickel (ppm)	-	-	na	na
	Romashkino[c]	Composition (wt%)	na	na	na	5.5
		Sulfur[f] (wt%)	-	na	na	0.26
		Vanadium (ppm)	-	-	na	1750[e]
		Nickel (ppm)	-	-	na	
Vacuum Resids	Gach Saran[b]	Composition (wt%)	28		62[d]	10
		Sulfur[f] (wt%)	-	na	na	na
		Vanadium (ppm)	-	-	282	1455
		Nickel (ppm)	-	-	100	530

a: Data from Ohtsuka (1977)
b: Data from Oxenreiter et al. (1972)
c: Data from Kubicka et al. (1968)
d: Combination of aromatics and resins
e: V + Ni (ppm)
f: Defined relative to total resid

10.2.4 Sulfur Compounds

Sulfur is the major heteroatom contained in petroleum crude oils and exits mostly in non-polar or weakly aromatic compounds (Tsonopoulos et al., 1986). In resids, most sulfur exists in the aromatic fraction with progressively smaller amounts in the resin and asphaltene fractions. The distributions of sulfur among the aromatic, resin, and asphaltene fractions for several resids are also included in Table 10.3.

Coal liquids contain only small amounts of sulfur. The low sulfur content results from sulfur removal by hydrogenation in the liquefaction process.

10.2.5 Nitrogen and Oxygen Compounds

Both nitrogen and oxygen concentrations are low in petroleum resids. Oxygen concentrations are insignificant while nitrogen concentrations range from about 2000 to 6000 ppm (Beaton, 1977).

Feedstocks of synthetic crude oils (coal liquids, tar sands, oil shale, etc...), however, show significantly higher contents of oxygen and nitrogen. 25 wt% of the compounds in direct coal liquefaction liquids contain heteroatoms (Tsonopoulos et al., 1986) while in light crude oils only 10 wt% of the compounds contain heteroatoms (S included). Oxygen and nitrogen contents may reach as high as 3.8 and 0.8 wt% in coal liquids, respectively.

10.2.6 Asphaltenes

Asphaltenes are aggregates of compounds which contain polycyclic aromatics and various functional groups. In addition, part of the aggregate shows a close packing tendency which imparts a crystalline like character to the aggregate. A hypothetical structure and reaction scheme for an asphaltene particle is given in Fig. 10.6 (Asaoka et al., 1983). The proposed structure shows a section of closely packed aromatic disks containing several heteroatoms. The proposed reaction scheme initially breaks the asphaltene into smaller compounds with the removal of the metal followed by heteroatom removal from the smaller compounds.

The ratio of sulfur to nitrogen atoms typically ranges from 3 or 4 to 1. Asphaltenic sulfur makes up about 0 to 1 wt% of the total feed sulfur or from 4 to 7 wt% of the asphaltene fraction and nickel and

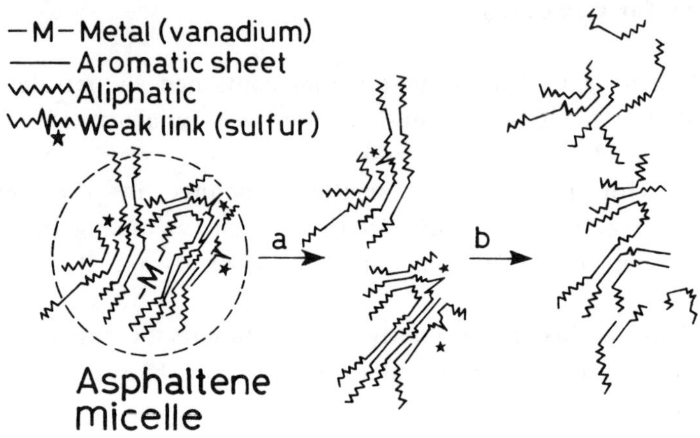

Figure 10.6 Proposed structure and reaction scheme for an asphaltene aggregate (from Asaoka et al., 1983).

vanadium contents are high (see Table 10.3). The molecular weights of asphaltenes range from several hundred to nearly 5,000; however, agglomerates of asphaltenes may have molecular weights as high as 100,000. The aggregate size ranges from about 20 to 1000 Å. West coast crude and Kuwait crude asphaltenes show bimodal distributions with modes located at 70 and 100 Å.

10.2.7 Ramsbottom Carbon

Ramsbottom carbon residue is quantified by carbonaceous residue formed during evaporation and pyrolysis of an oil under given conditions (ASTM D524-81). Information on the quantity of Ramsbottom carbon residue, or a similar property known as Conradson carbon, provides some indication of relative coke-forming propensity for each feedstock. In a sense, measurement of Ramsbottom carbon residue acts as a measurement of coke-precursor in the hydrotreating process. Beaton (1977) reported Ramsbottom carbon varied from about 8 wt% for atmospheric resids to near 17 wt% for a vacuum resid; the asphaltene content for the same resids ranged from only 2 to 8 wt%.

10.3 CHEMICAL REACTIONS

The main objective in hydrotreating is product upgrading of heavy feedstocks through the removal of heteroatoms and metals.

Because of the complex nature of the chemical processes involved, hydrotreating processes include several types of reactions which compete for available hydrogen. These reactions include hydrodesulfurization (HDS), hydrodemetallization (HDM), hydrodenitrogenation (HDN), hydrodeoxygenation (HDO), hydrocracking, coking and thermal cracking.

High hydrogen partial pressures in HDS reactors increase the importance of hydrogenation reactions which, consequently, become significant contributors to the overall hydrogen consumption. Hydrogenation steps are often required for heteroatom removal in the more stable compounds. HDM and coking reactions cause catalyst deactivation through the formation of deposits on the catalyst surface. Nitrogen compounds contribute to catalyst deactivation through their tendency to strongly adsorb on active sites and their subsequent slow rate of reaction.

10.3.1 Hydrodesulfurization

Hydrodesulfurization reactions can be represented by the following general formula

$$\begin{matrix}\text{Organic}\\\text{Sulfur Compound}\end{matrix} + H_2 \rightarrow H_2S + \begin{matrix}\text{Desulfurized}\\\text{Organic Compound}\end{matrix}$$

Hydrodesulfurization may occur through a combination of hydrogenation and hydrogenolysis reactions or directly through hydrogenolysis reactions. HDS occurs catalytically with both the sulfur compound and hydrogen adsorbed onto the catalyst surface. Sulfur adsorption appears to control the reaction sequence. Adsorption through the sulfur atom leads directly to sulfur removal; however, adsorption of the sulfur compound with an orientation parallel to the catalyst surface allows initial hydrogenation steps to occur (Vrinat, 1983). Several general reviews exist on hydrodesulfurization: Schuman and Shalit (1970), Schuit and Gates (1973), Beaton (1977), Gates et al. (1979), Grange (1980), Speight (1981), Vrinat (1983), and Topsoe et al. (1986).

HDS reactions are exothermic with heat of reactions from 445 to 890 kcal/m^3(STP)-H_2 consumed (Schuit and Gates, 1973). Those reactions which involve the most hydrogenation show the highest heat of reaction.

Desulfurization rates vary according to the type of organic sulfur compound being reacted. Aromatic fraction sulfur compounds

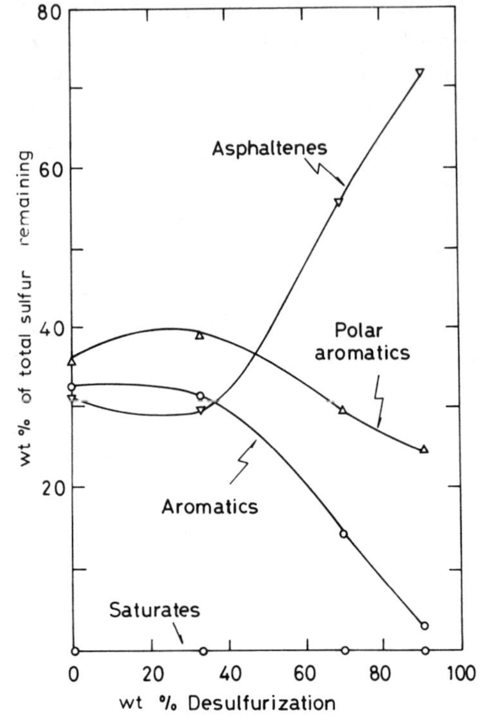

Figure 10.7 Relative sulfur contents of resid fractions as a function of desulfurization (from Drushel, 1972).

show the highest reactivity and those in the asphaltenes the lowest. However, as the extent of desulfurization increases, the sulfur compounds remaining are increasingly less reactive. As shown in Fig. 10.7, at high extents of desulfurization over 70 % of the remaining sulfur in the resid exists in the asphaltene fraction (Drushel, 1972).

Figure 10.8 shows the relative HDS reaction rates for asphaltenic and non-asphaltenic sulfur compounds for an Arabian light atmospheric resid through a plot of sulfur content versus contact time. Non-asphaltenic sulfur compounds (aromatic and polar aromatic combined) react faster than asphaltenic sulfur compounds. During the initial period, the total desulfurization rate reflects the high reactivity of non-asphaltenic sulfur compounds. At later times the total desulfurization rate reflects the lower reactivity of the asphaltenic sulfur. Removal of the asphaltene fraction from the reactor feedstock enhances the apparent desulfurization rate and leads to a higher conversion for a given reactor configuration.

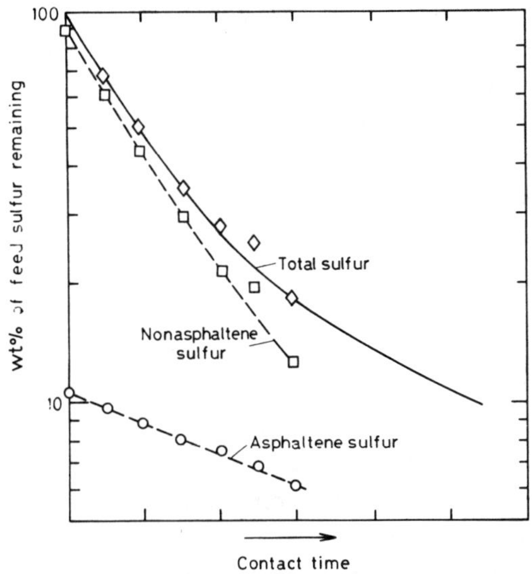

Figure 10.8 HDS reaction rates of asphaltenic and non-asphaltenic sulfur compounds from an Arabian light atmospheric resid (from Scott et al., 1970).

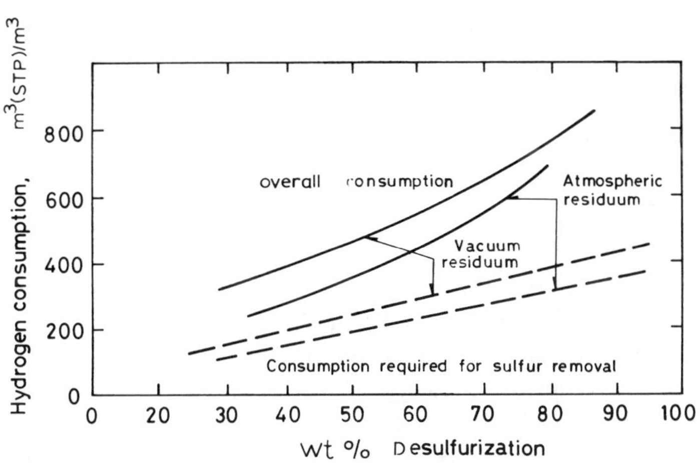

Figure 10.9 Dependence of hydrogen consumption on desulfurization for a Kuwait resid (from Beuther and Schmid, 1963). Note the increase in overall consumption is sharper than the increase in consumption required for sulfur removal alone.

The lower reactivity of the asphaltene fraction towards desulfurization leads to a lower degree of hydrogen selectivity for desulfurization over hydrogenation than is experienced in removing the non-asphaltenic sulfur. As shown in Fig. 10.9, the overall hydrogen consumption increases more rapidly than the hydrogen consumption required for desulfurization; this is especially true at high extents of desulfurization. The increased level of hydrogenation may result from the need for C-S bond activation through ring saturation or the "cracking" of asphaltene aggregates to smaller more readily desulfurized molecules.

Laboratory studies of HDS reactions focusing on model compounds have established a relative order of reactivity for various sulfur compounds (Schuman and Shalit, 1970). The low reactivity and aromatic nature of thiophene and dibenzothiophene made these compounds ideal as model compounds for study of petroleum hydrodesulfurization processes.

Thiophene HDS has been suggested (Vrinat, 1983) to follow either of two pathways. One path involves a parallel adsorption of thiophene leading to hydrogenation of the ring followed by ring scission through sulfur removal. The other path involves thiophene adsorbing on the catalyst through the sulfur atom followed by its eventual loss. The C-S bond rupture is reported to occur first, however (Schuman and Shalit, 1970). Subsequently, the desulfurized compound is hydrogenated to butadiene. Dibenzothiophene, a larger, slower reacting molecule, also may proceed either through a direct desulfurization step followed by hydrogenation steps or through an initial hydrogenation step followed by desulfurization steps as shown in Fig. 10.10. The initial desulfurization path proceeded at a significantly higher rate than the initial hydrogenation path, in fact, all hydrogenation steps in the reaction network occurred at lower rates than did the desulfurization steps.

The desulfurization and hydrogenation steps occurring in the thiophene reaction network have different thermodynamic behavior. The hydrogenation step is favorable only under low temperatures (below about 340°C) and high H_2 pressures while the desulfurization step is favorable under most industrial conditions.

Several kinetic expressions have been presented for thiophene HDS. Lee and Butt (1977) suggested the reaction to have a single-point thiophene adsorption in competition with H_2S adsorption and a separate, noncompeting, adsorption of hydrogen. The Langmuir-Hinshelwood rate expression for this model is

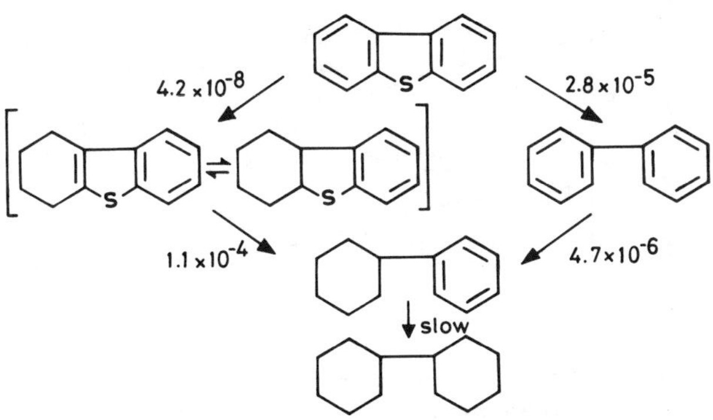

Figure 10.10 Proposed HDS reaction network for dibenzothiophene (from Houalla et al., 1978). Note the rate constants for hydrogenolysis (sulfur removal) reactions is higher than for hydrogenation reactions.

$$r_{HDS} = k \frac{K_T P_T}{(1 + K_T P_T + K_S P_S)} \left[\frac{K_H P_H}{(1 + K_H P_H)} + k_a K_H P_H \right] \quad (10.1)$$

where k_a reflects changes in the hydrogen adsorption behavior with temperature. Hydrogen adsorption changes from a weakly adsorbed state at low temperatures to a much stronger adsorbed state at high temperatures.

The most recent kinetic expressions for HDS of Dibenzothiophene are Langmuir-Hinshelwood expressions based on a single-point adsorption of DBT with accompanying, but noncompetitive adsorption of hydrogen. Vrinat and de Mourgues (1982) suggested the following expression based on a vapor phase kinetic study using a Co-Mo/Al_2O_3 industrial catalyst:

$$r_{HDS} = k \frac{K_D P_D}{(1 + K_D P_D + K_S P_S)} \frac{K_H P_H}{(1 + K_H P_H)} \quad (10.2)$$

Singhal et al. (1981) found similar results with the exception that they accounted for adsorption of products in addition to H_2S. Vrinat (1983) lists proposed kinetic expressions for HDS of Dibenzothiophene available in the literature.

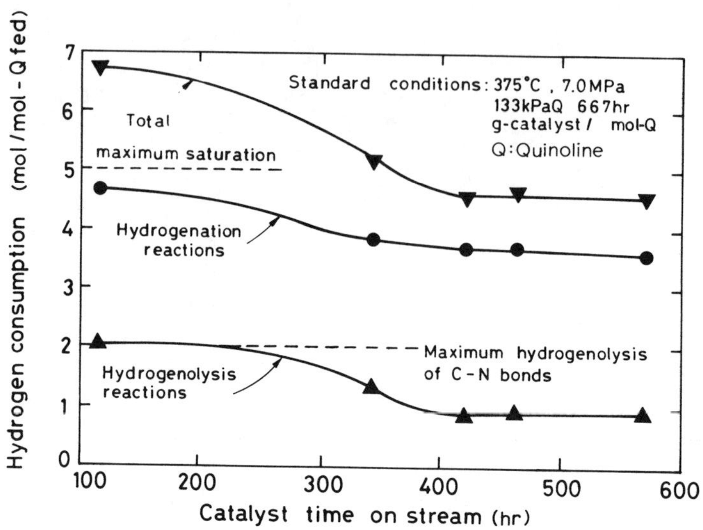

Figure 10.11 Hydrogen consumption for HDN of quinoline over a Co-Mo/Al_2O_3 catalyst (from Satterfield and Cocchetto, 1981).

10.3.2 Hydrodenitrogenation

Organic nitrogen compounds present in the resids, like organic sulfur compounds, are subject to reaction with hydrogen. HDN reactions have been suggested to occur through the adsorption of the nitrogen compounds on one site and noncompetitive adsorption of hydrogen on a second site. The difficult step in HDN reactions is heterocyclic C-N bond scission to ring opened products (Laine, 1983). In HDN reactions, hydrogenation steps are required for activation of the C-N bond.

Hydrogen consumption in HDN reactions reflects both the hydrogen required to remove nitrogen directly and that used in saturation and C-N bond activation. Figure 10.11 shows hydrogen consumption for HDN of quinoline over a Co-Mo/Al_2O_3 catalyst as a function of catalyst life. The relative amount of hydrogen consumed for saturation as compared to that for denitrogenation increases with catalyst age. This behavior suggests a preferential poisoning of the hydrogenolysis sites. H_2S acts to promote denitrogenation reactions while simultaneously inhibiting the hydrogenation steps (Satterfield and Gultekin, 1981). Laine (1983) suggested that the H_2S promotes

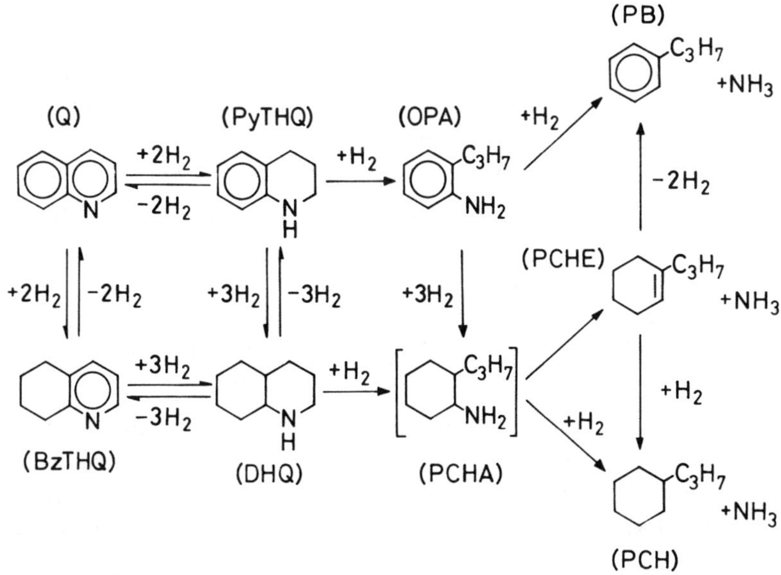

Q	quinoline
PyTHQ	Py(or1,2,3,4)-tetrahydroquinoline
BzTHQ	Bz(or 5,6,7,8)-tetrahydroquinoline
DHQ	decahydroquinoline(cis and trans isomers)
OPA	o-propylaniline
PCHA	propylcyclohexylamine
PB	propylbenzene
PCHE	propylcyclohexene
PCH	propylcyclohexane

Figure 10.12 Proposed HDN reaction network for quinoline (from Cocchetto and Satterfield, 1981).

denitrogenation through nucleophillic attack on the metal complexed (adsorbed) ring structure.

The relative reactivities for nitrogen compounds depend on both their relative susceptibility to nitrogen elimination and their relative affinity for adsorption on the active sites. Koros et al. (1967) found that quinoline-type compounds showed a higher rate of reaction than indole-type compounds despite their lower reactivity in an isolated environment. Basic nitrogen compounds strongly adsorb on acidic sites on the catalyst surface, while nonbasic compounds only adsorb mildly.

The reaction network for HDN of quinoline proposed by Cocchetto and Satterfield (1981) is shown in Fig. 10.12. The two main products are propylcyclohexane and propylbenzene. The initial series of reactions are equilibrium hydrogenation steps to form decahydroquinoline (DHQ). These hydrogenation reactions are exothermic and have unfavorable equilibrium constants at industrial HDN conditions. Hydrogenolysis reactions, however, are less exothermic and essentially irreversible (Cocchetto and Satterfield, 1981). Propylcyclohexane, the thermodynamically favored product, requires nearly twice as much hydrogen as the preferred route to propylbenzene (propylbenzene has a higher octane rating). Satterfield and Gultekin (1981) indicated that in the presence of H_2S the two initial, parallel hydrogenation steps of quinoline became irreversible.

Kinetic modeling of this reaction by Satterfield and Cocchetto (1981) followed a Langmuir-Hinshelwood relationship assuming competitive adsorption of nitrogen compounds on one site and noncompetitive adsorption of hydrogen on a second site. First order or pseudo-first order reaction constants for HDN reactions decrease with an increase in initial nitrogen content due to inhibition by the strongly adsorbed nitrogen compounds (Satterfield and Gultekin, 1981).

10.3.3 Hydrodeoxygenation

For coal liquids and other synthetic crude oils oxygen removal contributes significantly to hydrogen consumption in a hydrotreator. As in the case of HDN reactions a combination of hydrogenation and hydrogenolysis steps is required. Rollmann (1977) and Badilla-Ohlbaum et al. (1979) observed significantly higher hydrogen consumption required for oxygen removal compared to that of sulfur removal for corresponding heterocyclic compounds. This behavior suggests a greater need for hydrogenation steps in HDO (Furimsky, 1983). Similarities in the electron withdrawing effects and unpaired electrons of sulfur and oxygen atoms suggest similar behavior on the catalyst surface; however, the significant difference in atom size may lead to a greater degree of steric hindrance by the hydrocarbon portion of the molecule in oxygen compounds (Furimsky, 1983).

The least reactive oxygen compounds are furans, phenols, and quinones (Furimsky, 1983). In the HDO of phenols, it has been suggested that ring saturation must occur prior to the OH group removal (Furimsky, 1978). Studies of furan and dibenzofuran model compounds show that the expected reaction networks depend on the availability of surface hydrogen. HDO of dibenzofuran can follow three pathways depending on the availability of surface hydrogen as

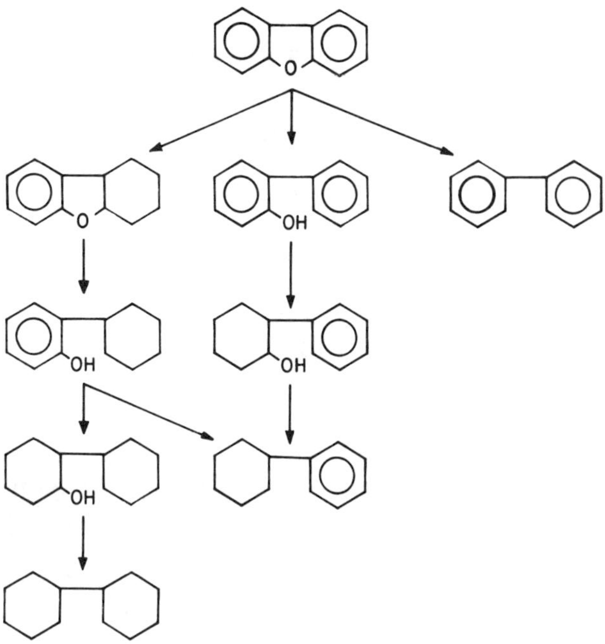

Figure 10.13 Proposed HDO reaction network for dibenzofuran (from Furimsky, 1983).

shown in Fig. 10.13. The initial hydrogenation pathway will be favored under high hydrogen partial pressures, but as the availability of surface hydrogen decreases the extent of initial hydrogenation will decrease and the ring scission or oxygen removal pathways will become more prominent.

10.3.4 Hydrodemetallization

The removal of vanadium and nickel from complex organometallic compounds can occur through the coordination of the metal with sulfur (Dautzenberg and de Deken, 1985) or through a combination of hydrogenation and ring scission steps (Ware and Wei, 1985a,b) as shown in Fig. 10.14. The reaction sequence proposed by Ware and Wei shows two initial, reversible hydrogenation steps followed by ring scission and metal deposition. Metal deposition occurs either directly or following an additional hydrogenation step. The reactions occur on or near the catalyst surface.

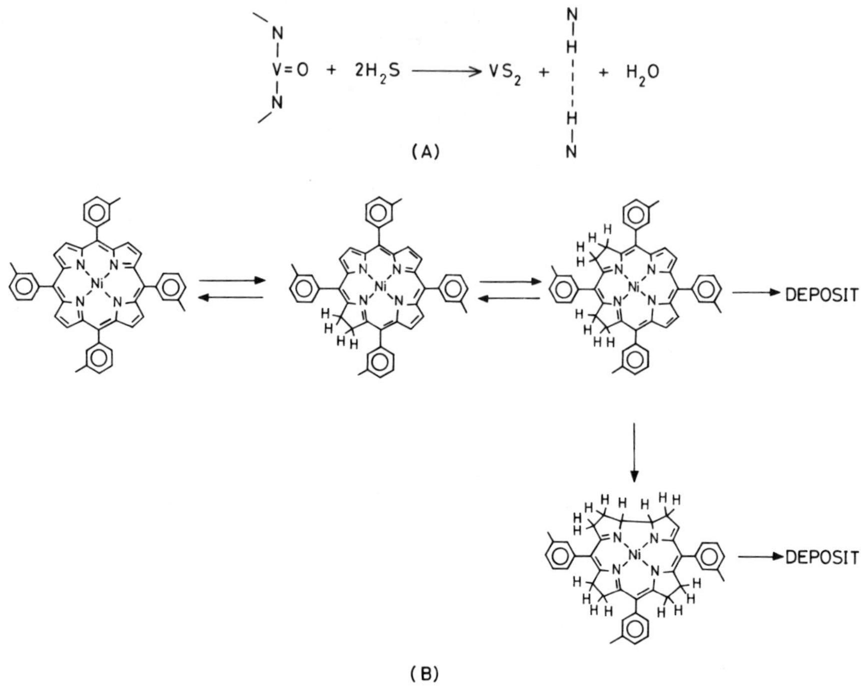

Figure 10.14 Types of demetallization reactions: (A) demetallization through sulfur coordination with vanadium (from Dautzenberg and de Deken, 1985); (B) proposed nickel hydrodemetallization reaction network (from Ware and Wei, 1985a,b).

The metal deposited on the catalyst surface promotes further demetallization activity. Figure 10.15 shows the relative activity of a metal/carrier catalyst and the carrier alone as a function of wt% vanadium deposited for demetallization and desulfurization reactions. For demetallization, the activity of the carrier increased as vanadium was deposited and ultimately attained the activity of the metal/carrier catalyst (Takeuchi et al., 1985). The effect on desulfurization rate, however, was not as significant.

Galiasso et al. (1985) reported HDM reactions follow first order kinetics with respect to metals content for both the resin and asphaltene fractions and different orders with respect to hydrogen partial pressure.

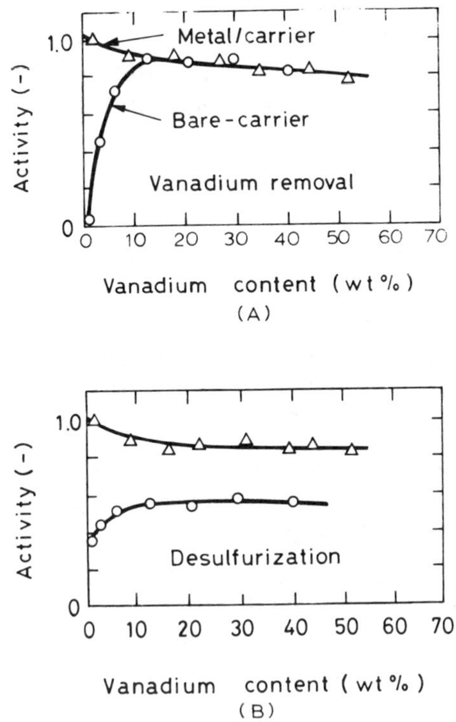

Figure 10.15 Catalytic activities of deposited metals on carrier surface (from Takeuchi et al., 1985): (A) activity towards demetallization as a function of deposited vanadium; (B) activity towards desulfurization as a function of vanadium deposited. Vanadium concentration is defined based on the weight of fresh catalyst.

10.3.5 Hydrogenation and Hydrocracking

Hydrogenation reactions are an integral part of both HDN and HDO reactions in addition to being an important secondary reaction on its own. Hydrogenation of aromatic compounds proceeds slowly under industrial conditions, yet ring saturation is required for nitrogen removal from ring compounds. Some typical hydrogenation reactions are shown in the various HDN and HDO reaction schemes previously presented. The reactions occur catalytically. It is generally believed that the active sites on the catalyst surface are anion vacancies which allow interaction with the π electrons of the aromatic ring structure. The adsorbed species may rearrange to give a σ bonded species that is activated for hydrogen addition (Sapre and Gates, 1981). Hydrogenation reactions are exothermic and are thermodynamically favored only at low temperatures (usually below about 340°C). At the high temperatures used in industrial hydrotreators the low equilibrium constants are offset by high hydrogen partial pressures. Hydrogen consumption due to hydrogenation reactions increases as the hydrogen partial pressure increases.

The properties of the catalyst support material influence other possible side reactions. Acidic sites common to the alumina supports used for industrial HDS catalysts provide sites for hydrocracking. Both high operation temperatures and the presence of acidic sites on the catalyst surface provide the right setting for the primary reactions of catalytic cracking. However, under conditions of high hydrogen partial pressures the secondary reactions prevalent in catalytic crackers are suppressed. For example (Speight, 1980), in the cracking of a paraffin the initial step is the formation of a carbonium ion through interaction with an acidic site. Subsequently, the carbonium ion may migrate or undergo beta-scission into a smaller olefin and another carbonium ion, and the resulting olefin is quickly hydrogenated to a paraffin.

$$C_{16}H_{34} + \text{Cat} \rightarrow C_{16}H_{33}^{+} + \text{Cat H}^{-}$$

$$C_5H_{11}C^{+}H\text{-}CH_2\text{-}CH_2C_8H_{17} \rightarrow C_5H_{11}CH{=}CH_2 + C^{+}H_2C_8H_{17}$$

$$C_5H_{11}CH{=}CH_2 + H_2 \rightarrow C_7H_{16}$$

For cracking of ring structures it is believed that the ring is broken followed by subsequent saturation of the chain ends. Polynuclear aromatic structures react more readily than single ring structures by stepwise ring saturation followed by ring scission (Speight, 1980).

The amount of hydrocracking reactions occurring in HDS reactors is limited by the susceptibility of acidic sites to deactivation through both adsorption of nitrogen compounds and coke formation.

10.3.6 Coking

Unstable molecules, cracking reaction intermediates, and asphaltenes contribute to the formation of coke on the catalyst surface (Ohtsuka, 1977). Large molecules, already existing as asphaltenes or formed through condensation reactions of unstable molecules or cracking reaction intermediates, may lack mobility in catalyst pores or may adsorb strongly on the catalyst surface preventing their access to hydrogenation sites. Consequently, these molecules may continue to increase in size by further condensation reactions and eventually form a coke deposit.

Coke deposits increase as the reaction temperature rises and decrease as the hydrogen partial pressure increases. Removal of the asphaltene fraction slightly reduces the amount of coke deposited on the catalyst surface (Beuther and Schmid, 1963).

The acidic sites associated with the γ-Al_2O_3 carrier are susceptible to compounds adsorbing strongly and forming coke. Hydrocracking rates have been observed to decline more rapidly through coke formation than hydrogenation rates (Weisz and Prater, 1957). It has been postulated (Beuther and Larson, 1965) that during hydrocracking only those acidic sites in close proximity to hydrogenation sites remain free of coke. This postulate suggests that initially the catalyst activity is high due to the total number of acidic sites; subsequently the activity falls to a steady-state level. This steady-state level represents only those acidic sites in proximity close enough to hydrogenation sites to prevent coke formation.

10.3.7 Thermal Cracking

Thermal cracking is the decomposition of organic molecules at high temperatures (>350°C) by the formation of active free radicals. The pathways for free radical chain reactions are complex with the product distribution dictated by the reaction sequences that are favored thermodynamically. The coal liquefaction process also involves extensive free radical reactions; details of these liquefaction reactions were given by Shah (1981).

Free radical chain reactions involve three steps, namely, initiation, propagation, and termination. The initiation step is the formation of free radicals consisting of an atom or a group of atoms possessing an unpaired electron via carbon-carbon bond rupture of a hydrocarbon molecule. After initiation, propagation steps occur in which a free radical reacts to form a hydrocarbon compound and another free radical. The reaction may take place either with a hydrocarbon molecule by abstracting a hydrogen atom yielding another hydrocarbon molecule and a new free radical or the free radical may undergo internal rearrangement to give a stable compound and a free radical (beta-scission). Reactions of free radicals can lead to both higher and lower molecular weight compounds than the original material. The termination step occurs when two free radicals react to form a stable molecule. Examples of these reactions are

(i) Initiation

$$CH_3\text{-}(CH_2)_n\text{-}(CH_2)_m\text{-}CH_3 \rightarrow CH_3\text{-}(CH_2)_n\cdot + \cdot\,(CH_2)_m\text{-}CH_3$$

(ii) Propagation

$$CH_3\cdot + CH_3\text{-}(CH_2)_n\text{-}(CH_2)_m\text{-}CH_3 \rightarrow CH_4 + CH_3\text{-}\dot{C}H\text{-}(CH_2)_{n-1}\text{-}(CH_2)_m\text{-}CH_3$$

$$CH_3\text{-}\dot{C}H\text{-}(CH_2)_{n-1}\text{-}(CH_2)_m\text{-}CH_3 \rightarrow CH_2{=}CH_2 + \cdot(CH_2)_{n-1}\text{-}(CH_2)_m\text{-}CH_3$$

$$\cdot(CH_2)_mCH_3 \rightarrow CH_2{=}\,CH_2\text{-}(CH_2)_{m-2} + \cdot CH_3$$

(iii) Termination

$$CH_3\cdot + CH_3(CH_2)_n\cdot \rightarrow CH_3\text{-}(CH_2)_n\text{-}CH_3$$

It is generally recognized that hydrocarbon free radicals resist isomerization, the aromatic ring is fairly stable for temperatures between 350 ~ 500°C, and alkylated aromatics are more likely to undergo dealkylation than ring destruction (Speight, 1980). The following are several examples of thermal decomposition offered by Speight (1980):

$$CH_3\text{-}(CH_2)_n\text{-}CH_3 \rightarrow CH_3\text{-}(CH_2)_x\text{-}CH_3 + CH_3\text{-}(CH_2)_y\text{-}CH = CH_2$$

where n > x + y

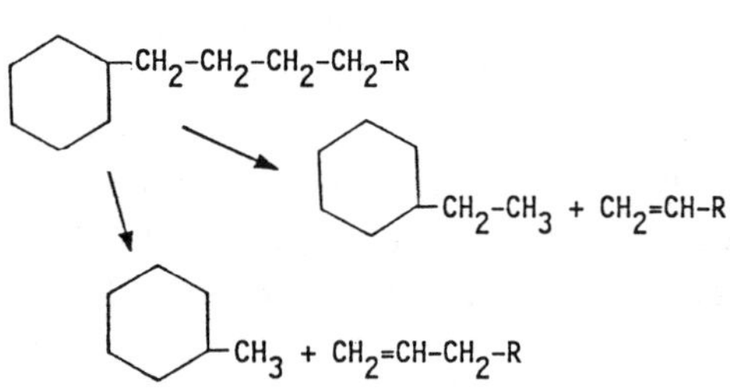

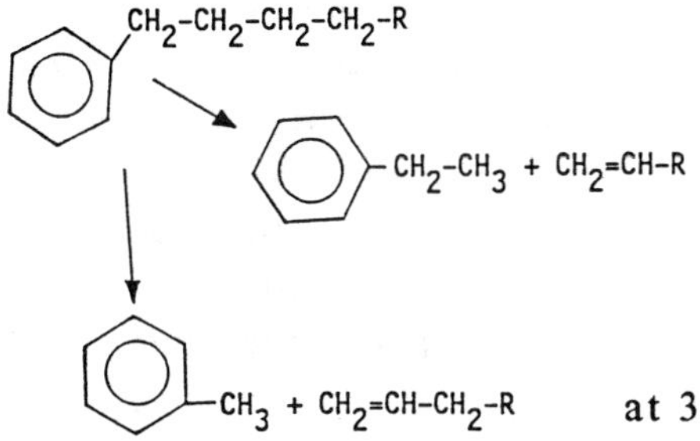

at 350 ~ 500°C

Information on thermal decomposition of pure compounds provides a basis for estimation of product types and distributions in the thermal cracking of various feedstocks.

10.3.8 Industrial Kinetics

Kinetic information for industrial feedstocks is limited to approximate expressions. The desulfurization rate behavior illustrated in Fig. 10.8 suggests that a linear combination of two first order reactions may be appropriate (Gates et al., 1979):

$$r_{HDS} = \alpha_1 k_1 C_S + \alpha_2 k_2 C_S \tag{10.3}$$

This expression is valid only under conditions of constant H_2 and H_2S partial pressures. At high conversions, where the more reactive fraction has been desulfurized, only an expression for the unreactive sulfur compounds need be considered. In this situation a Langmuir-Hinshelwood type kinetic expression can be used to approximate the basic kinetic behavior of the system (Gates et al., 1979):

$$r_{HDS} = \frac{\alpha_2 k_2 P_H C_S}{(1 + K_H P_H + K_S P_S + K_{AR} P_{AR})} \tag{10.4}$$

where the term $K_{AR}P_{AR}$ represents the aromatic compounds including sulfur compounds. This type of expression requires the experimental determination of the parameters for each feedstock.

Le Page (1978) suggested a power law expression in terms of sulfur conversion:

$$r_{HDS} = k[C_{S0}(1-X_S)]^a \, f(P_H) \exp(-E/RT) \tag{10.5}$$

where C_{S0} is the initial sulfur concentration. The term $f(P_H)$ in Eqn. (10.5) expresses the rate dependence on hydrogen partial pressure; one possible form for $f(P_H)$ would be the H_2 adsorption term of Vrinat and de Mourgues (1982) kinetic expression for HDS of DBT (Eqn. (10.2)).

A kinetic treatment of the entire conversion process can be done based on a lumped parameter approach. Mosby et al. (1986) developed a model based on experimental evidence for predicting the product yields in the treating of heavy resids. According to available evidence the cracking reactions of the petroleum feedstock to lighter products are independent of the catalytic reactions for removal of feed contaminants; consequently, cracking reactions can be treated separately from catalytic reactions. It is further recognized from experimental findings that the thermal cracking reactions of heavy resids can not be represented by a single lumped first-order reaction, the reactions are irreversible, and the product yield structure (lumps) is independent of the type of feedstock. Furthermore, materials with boiling temperatures both above and below 538°C are cracked to lighter products. To incorporate these known effects, Mosby et al. (1986) proposed the following:

Feed is subdivided into "hard"-to-react feed resid (538+°C), "easy"-to-react feed resid (538+°C), and feed gas oil (343 ~ 538°C), while products are subdivided into product gas oil (343 ~ 538°C), product gases (C_1 ~ C_4), naphtha (C_5 ~ 182°C), and distillate (182 ~ 343°C) fractions. Figure 10.16 shows a matrix indicating the interrelationship among these feed and product fractions for cracking of a heavy resid. Reactions are irreversible and of first order for each component reaction in the matrix. Relative reaction rate constants for each matrix component are provided in the figure. Note that due to the nonuniformity in resid feed reactivity, the resids are also divided into an "easy-to-remove" lump with a large reaction rate constant and a "hard-to-remove" lump with a small constant in addition to the various boiling ranges. Figure 10.17 shows that the calculated product and feed distributions based on the model vary with reaction temperature. Gases, naphtha, and distillate products monotonically increase with an increase in temperature. The gas oil product increases and then decreases with an increase in temperature. In addition, the resid conversion monotonically increases with an increase in temperature.

A similar approach can be taken for the catalytic reactions. Feed and product matrices can be set up for each catalytic reaction in terms of appropriate compound subclasses.

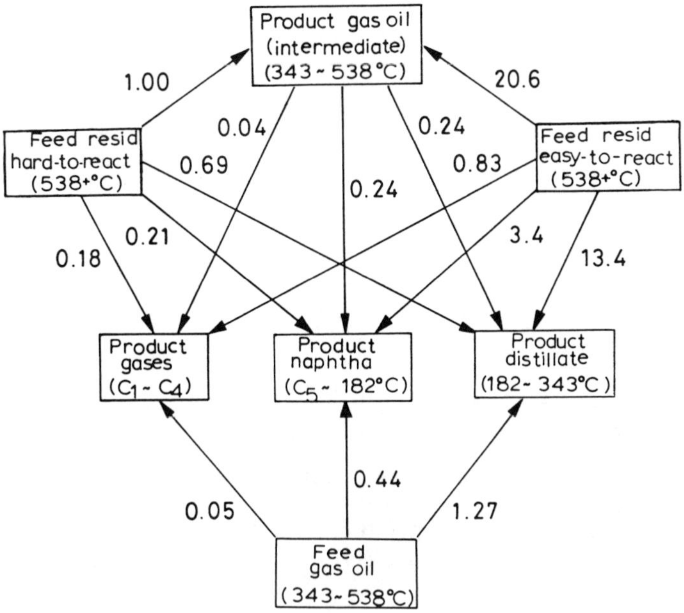

Figure 10.16 Lumped parameter reaction scheme for the thermal cracking of petroleum resids. Numbers indicate relative first-order reaction rate constants based on the hard-to-react fraction (from Mosby et al., 1986).

The rate of metal deposition has been observed to be first order with respect to metals composition. Mosby et al. (1986) presented the following rate equation for metals deposition, M, in terms of V and Ni content:

$$\frac{dM}{dt} = \alpha \frac{\rho_o}{\rho_c} A_m (k_v C_V + k_n C_{Ni}) \tag{10.6}$$

where α is a constant, A_m is the demetallization activity, k_v is a kinetic constant for vanadium removal, and k_n is a kinetic constant for nickel removal. Note that the catalyst demetallization activity included in this equation is a unique function of the amount of metals deposited.

Evidence indicates that the apparent reaction order of both desulfurization and demetallization reactions with respect to reactant concentration varies with reactor type. The demetallization reaction has been observed to be first order in a well-stirred reactor

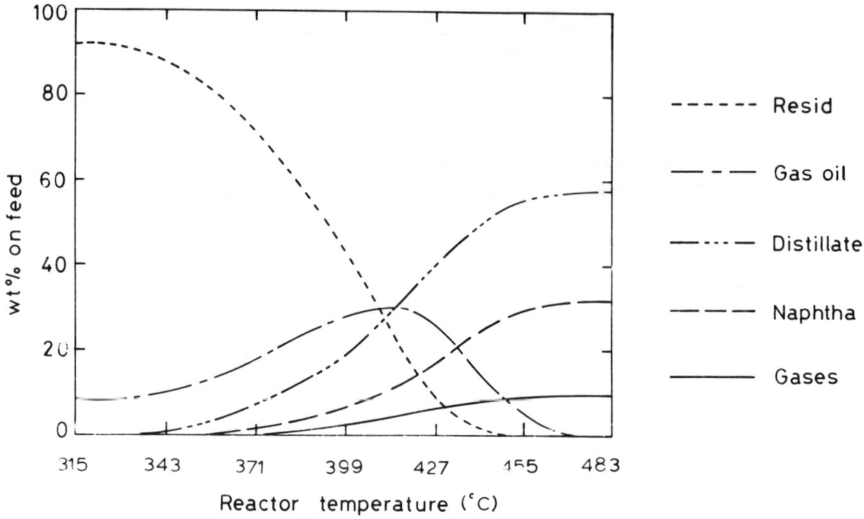

Figure 10.17 Calculated product and feed distributions as a function of temperature (from Mosby et al., 1986).

(Dautzenberg et al., 1978; Van Zijll Langhout et al., 1980; Dautzenberg et al., 1982) and 1.5 order in a tubular flow reactor (Van Zijll Langhout et al., 1980; Dautzenberg et al., 1982). The desulfurization reaction has been observed to be of 1.5 order in a well-stirred reactor (Dautzenberg et al., 1978; Van Zijll Langhout et al., 1980) and second order in a tubular flow reactor (Beuther and Schmid, 1963; Massagutov et al., 1967; de Brujin, 1976; Van Deemter, 1984).

10.4 CATALYSTS

Typically, hydrotreating catalysts for petroleum resids consist of Mo on γ-Al_2O_3 with Co or Ni promoters. Tungsten also shows good catalytic behavior but its use is limited commercially due to its high cost. The specific surface area of hydrotreating catalysts ranges from 100 to over 300 m^2/g and pore sizes regularly show a distribution between 40 and 200 Å. Table 10.4 lists for several commercial catalysts the preferred metals and other important properties.

Catalyst sites active in hydrotreating result both from any acidic nature of the support material (alumina) and Co-Mo interactions. The acidic sites show activity for cracking reactions and are subject to strong adsorption by basic compounds (nitrogen

Table 10.4 Composition and Properties of Industrial Hydrotreating Catalysts

Company	Preferred Metals	Catalyst Bulk Density (g/cm^3)	Average Pore Volume (cm^3/g)	Pore Diameter (Å)	Surface Area (m^2/g)
Esso[a]	Ni/Mo Co/Mo	<0.70	>0.25	30~70	>150
UOP[a]	Ni/Mo	0.4~0.9	<0.15	40~150	100~300
Standard[a] Oil of Indiana	Co/Mo	—	>0.5	100~200	150~500
Gulf[a]	Ni/Co/Mo	—	0.46	0~240[d]	165~220
Texaco[a]	Ni/W Ni or Co/Mo	0.5~0.7	0.1~0.6	—	300~800
Shell[a]	Ni/Co & Wo/Mo	0.71	—	90	292
Chevron[a]	Group VI + VII	—	—	>60	>100
Nippon[a] Oil Co.	Co/Mo or Ni/W	—	—	1000~ 50,000	—
Hydrocarbon[a] Research	Co/Mo +Ni	—	0.45~0.5	60~70 100~1000[e] 1000+[e]	260~355
Girdler[b] G-35B	Co/Mo	0.96	0.22	<200	270
Pro Catalyse[b] IFD HR 304	Co/Mo	0.5	0.78	<5000[d]	283
Davison[b] Chemical	Ni/Co/Mo	0.64	0.39	—	222

a: Data appears in Gates et al. (1979).
b: Data appears in Ohtsuka et al. (1968).
c: Condensed form of a Table appearing in Ohtsuka (1977)
d: Pore sizes show a regular distribution.
e: These pores are considered channels.

compounds). A lack of understanding exists about the interactions between Mo and Co or Ni and the resulting active sites they form. Discussions about these interactions and the resulting active sites can be found in Schuman and Shalit (1970), Schuit and Gates (1973), Gates et al. (1979), Chianelli (1984), Topsoe and Clausen (1984), and Topsoe et al. (1986). It is generally believed that the sulfur compounds adsorb at anionic vacancies on the catalyst surface. Interactions with nearby

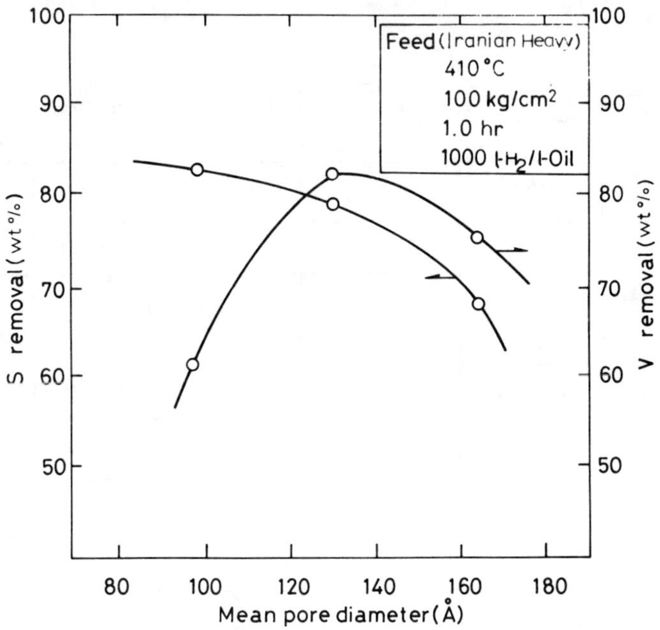

Figure 10.18 Effect of mean pore diameter on sulfur and vanadium removal (from Inoguchi, 1976).

hydrogen adsorption sites through proton and electron transfers make up the HDS reactions.

10.4.1 Catalyst Infrastructure

The mean pore size, the pore size distribution, the surface area, and the pore volume all are important to the effectiveness of the catalyst towards desulfurization and demetallization. Increasing the surface area, pore size, and pore volume increases the desulfurization rate (Beuther and Schmid, 1963; Ohtsuka, 1977). The pore size affects the desulfurization and demetallization rates through diffusional and surface area effects. Pore sizes less than about 40 Å are ineffective for desulfurization due to molecular size considerations. As shown in Fig. 10.18, for pore sizes above 80 Å the extent of desulfurization declines slightly, while demetallization increases above 100 Å and reaches a maximum around 130 Å (Inoguchi, 1976). The low effectiveness towards demetallization below 100 Å results from the exclusion of asphaltenes from the catalyst pores and the decrease in

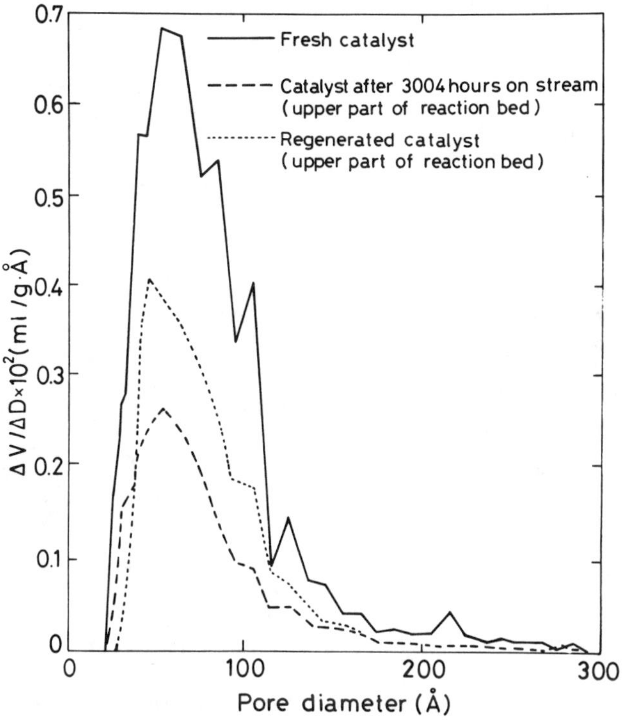

Figure 10.19 Pore size distribution for a typical HDS catalyst before use, after use, and after partial regeneration (from Ohtsuka, 1977).

the extent of desulfurization may be due to a decrease in overall surface area with increasing pore size.

The effect of particle size on desulfurization also suggests mass transfer limitations exist. Ohtsuka et al. (1968), using a fixed capacity layer of catalyst, found the extent of sulfur removal increased as catalyst size decreased. In addition, for a constant bed volume, Kato et al. (1971) revealed that for catalyst sizes smaller than 1 mm a first order rate constant varied proportionally to the reciprocal of the catalyst size, but above 1 mm the rate constant was nearly independent of catalyst size. A limit exists on the size of an individual catalyst above which mass transfer limits the observed reaction rate and below which intraparticle diffusion becomes important; the exact value of the limit depends on the internal pore structure.

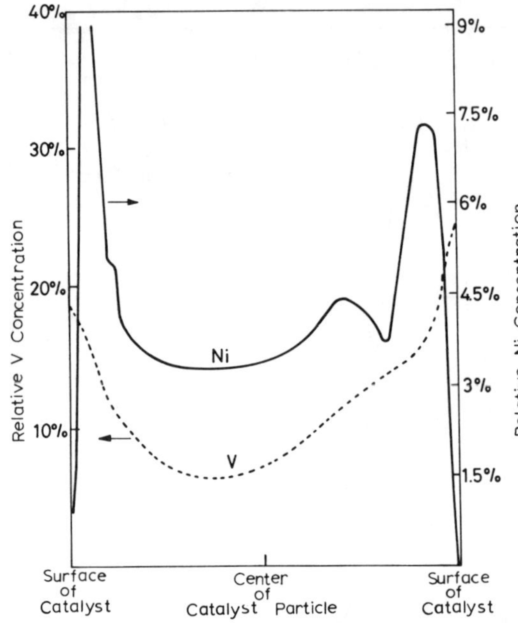

Figure 10.20 Nickel and vanadium distribution in used catalyst particles. The curves represent typical electron microprobe scans of catalyst particles from the top of the bed (from Oxenreiter et al., 1972).

10.4.2 Catalyst Deactivation

Deactivation of HDS catalysts occurs through surface area loss and active site poisoning. Products from both coking and HDM reactions form deposits on the catalyst surface. These deposits, consequently, may prevent reactant molecules access to the internal active sites by blocking pores. Typically, the deposits of HDM reactions block the internal pore structure and those of coking reactions block the pore mouths. Figure 10.19 shows a pore size distribution for a typical HDS catalyst. The differences between the distributions of the fresh and used catalyst express the loss in pore volume due to deposits. Only a small portion of that lost surface area may be regenerated by burning off the coke. The permanent loss is due to irreversible metal deposition.

Metals deposit profiles depend on the diffusional limitations of the organometallic compounds from which they originate. For the demetallization of resins the resulting metal deposit profile proves to be homogeneous while for asphaltenes the profile shows large deposits in the external surface regions (Galiasso et al., 1985). Figure 10.20 shows a typical profile of Ni and V deposits. Vanadium deposits mainly near the external surface while Ni deposits more uniformly. In high hydrogen partial pressure environments, essentially coke suppressing environments, smaller catalyst sizes show longer lives than larger sizes because of a more uniform distribution of metal deposits (Thakur and Thomas, 1985).

The pore size distribution significantly affects the deposit distribution of V. A bimodal pore size distribution allows greater access to the internal pore structure and consequently a more complete penetration by the metal complexes. However, changes in pore structure towards a bimodal distribution also result in reduced overall surface area.

Hydrogen partial pressure affects catalyst deactivation through the relative rates of metal or coke deposition. High hydrogen pressures suppress coke formation and increase HDM reaction rates. Under conditions where the deactivation rate is controlled by coke formation, high hydrogen partial pressures will reduce the deactivation rate. However, when HDM reactions control, an increase in hydrogen partial pressure will increase the demetallization rate (Speight, 1981).

Catalyst demetallization activity varies as a unique function of the weight fraction, based on fresh catalyst, of metals deposited. Mosby et al. (1986) presented the catalyst deactivation relative to metal deposition on the catalyst obtained for an ebullated bed reactor as given in Fig. 10.21. Catalyst deactivation behavior shows an initial rapid decline in activity followed by a relatively long period of nearly linear decay. The sharp decrease during the very early stage of the catalyst life is attributed to coke deposition (Mosby et al., 1986). Metals deposition is a slow process relative to coke deposition and contributes to deactivation during the bulk of the catalyst life. Thus, catalyst deactivation kinetics can be expressed solely in terms of the extent of metal deposition on catalyst. Dautzenberg et al. (1978 and 1982) proposed a two parameter pore mouth plugging model to describe catalyst deactivation behavior due to metal deposition. The model assumes zero order kinetics with respect to metal content:

$$A_m/A_{m0} = 1 - t^* \tag{10.7}$$

$$C_m/C_{m,max} = 2t^* - t^{*2} \tag{10.8}$$

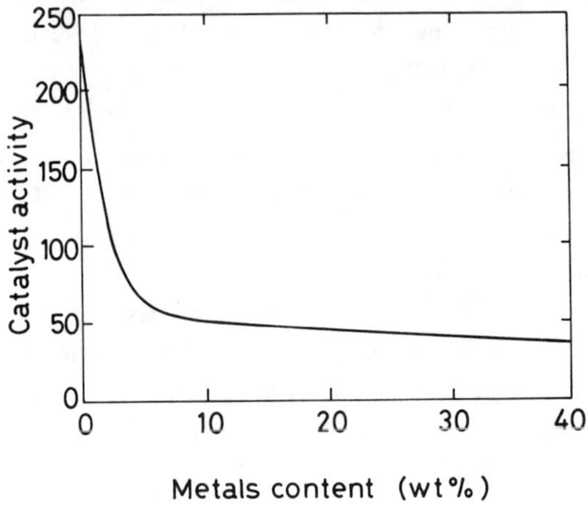

Figure 10.21 Catalyst deactivation behavior in an ebullated bed reactor (from Mosby et al., 1986).

In these equations, t^* is a dimensionless time (0 for fresh catalyst and 1 for spent catalyst); A_m and A_{m0} are the activity and initial activity, respectively; and C_m and $C_{m,max}$ are the metal content and metal uptake capacity, respectively. The two parameters for this model A_m and $C_{m,max}$ can be expressed in terms of catalyst dimensions and physicochemical constants of the reaction system (Dautzenberg and de Deken, 1984).

10.4.3 Catalyst Selectivity

An important issue in the design of catalysts for hydrotreating stems from the conflict between HDS activity and metal tolerance. HDS reactions occur within the catalyst infrastructure and rates show an increase with an increase in total surface area. Metal deposition poses the most serious problem in terms of deactivation. For a long catalyst life, large pores, and consequently smaller surface areas, are required for adequate use of the entire catalyst for metal deposition. Thus the two optimum cases, a high surface area for desulfurization and large pores for demetallization, are not compatible. The catalyst may be tailored for either HDS or HDM reactions but not for both as mentioned. A high extent of metal removal can be obtained at a sacrifice in extent of HDS.

In recent years the trend has been to carry out the HDS process in stages. The initial stage is designed for demetallization with a catalyst optimized accordingly; the succeeding stages are designed for desulfurization. Toulhoat et al. (1985) suggested several guidelines for the development of HDM catalysts. First, since metals are complexed in large molecules a proper pore size distribution is necessary to allow the metals to reach all active sites. Second, the catalyst should use efficiently the available porosity for metal accumulation by having the HDM reaction as rate limiting. And third, coke deposits have to be minimized during the early operational age of the catalyst.

10.5 TRANSPORT PROPERTIES, DESIGN, AND OPERATION OF EBULLATED BED REACTORS

In the operation of an ebullated bed for either hydrotreating of petroleum resids (H-Oil process or LC-Fining process) or direct liquefaction of coal (H-Coal process) in the presence of catalysts, the catalyst bed is mainly supported by product liquid recycled from the reactor freeboard through the inner column back to the reactor plenum chamber. Since catalyst expansion takes place only in the annular region of the fluidized bed, the operation can be more precisely termed "annular three-phase fluidization" (see Section 6.3), as opposed to "conventional" three-phase fluidization (see Chapters 2 and 3). In an air-water system at atmospheric conditions the behavior of annular three-phase fluidized beds is different from the behavior of a conventional three-phase fluidized bed with equal cross-sectional area. This altered behavior is due to the imposed physical constraints on the bed hydrodynamics which are absent in a conventional fluidized bed of equal cross-sectional area (Fan et al., 1985). Note, however, that a commercial ebullated bed reactor is designed with a small inner column to outer column diameter ratio and that hydrogen-hydrocarbon systems at industrial hydrotreating or liquefaction conditions have extremely high gas holdup with small dispersed bubbles; consequently, the physical constraints on the hydrodynamics due to the presence of the inner column under these conditions can be negligibly small.

10.5.1 Transport Properties

Little has been reported regarding the transport properties of the ebullated bed reactor for resid hydrotreating applications. However, under resid hydrotreating conditions or coal liquefaction conditions, the gas phase behavior in an ebullated bed reactor and that

Table 10.5 Unit Characteristics, Operating Conditions and Phase Properties for the Study of Exxon Donor Solvent Reactor for Coal Liquefaction (from Tarmy et al., 1984)

		PILOT UNITS			COLD FLOW UNITS	
		RCLU	CLPP	ECLP	SMALL	LARGE
	Coal Throughput kg/day	34	1000	250,000	-	-
Unit Characteristics	Diameter m	0.024	0.066	0.61	0.15	0.61
	Height m	1.2	7.3	18	2.5	8.5
	No. of Columns	10	3	4	1	1
	Distributor	Single Nozzle	Single Nozzle	Bubble Cap or Single Nozzle	Bubble Cap or Multiple Nozzles	Bubble Cap
Conditions	Temperature °C	450	450	450	25	25
	Pressure kPa	17,101	17,101	17,101	241~621	122
	Gas Velocity cm/s	0.6~2.5	1.0~4.0	4.6~6.5	4.5~20	6.0~15
	Liquid Velocity cm/s	0.15~0.22	0.4~0.6	1.6	0.45~.15	0.6~1.2
Phase Properties	Liquid density kg/m^3	830	850	810	685	685
	Viscosity mPa·s	0.3	0.4	0.3	0.4	0.4
	Surface tension N/m	0.005	0.006	0.005	0.02	0.02
	Gas density kg/m^3	70	70	75	1.6	3~7.3
	Particle Density kg/m^3	2020	2600	2200	1270	-
	Particle Size maximum (95 wt%)/ weight mean mm	.26/.11	.45/.24	1.70/.70	1.70/.60 0.20/0.12	-

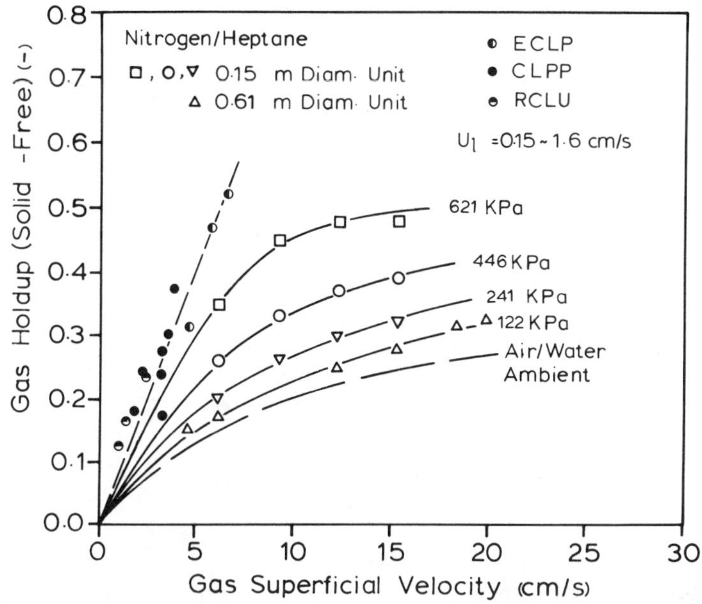

Figure 10.22 Effect of gas velocity on gas holdup (solid-free) at ambient and reaction conditions for EDS coal liquefaction (from Tarmy et al., 1984).

in a slurry bubble column reactor bear certain similarities. The transport properties reported for the slurry bubble column reactor (Mode E-I-a-2) characterizing the Exxon Donor Solvent (EDS) coal liquefaction process by Tarmy et al. (1984) may thus provide invaluable information for the understanding of the transport behavior and scaleup for the ebullated bed reactor in hydrotreating applications.

Table 10.5 shows the unit characteristics, operating conditions, and phase properties for EDS coal liquefaction experimentation reported by Tarmy et al. (1984). The experimental work covered three scales of hot units i.e., recycle coal liquefaction unit (RCLU), coal liquefaction pilot plant (CLPP) and EDS coal liquefaction pilot plant (ECLP) and two scales of cold units, i.e., a small high pressure unit and a large ambient pressure unit. In an attempt to simulate the liquid properties of hot units, a liquid with very low surface tension and low viscosity was used for cold unit operation.

Reactor scaleup from cold unit experimental data to pilot or commercial hot units is difficult; however, far less difficulty occurs in scaleup from bench hot units to pilot or commercial hot units. This is

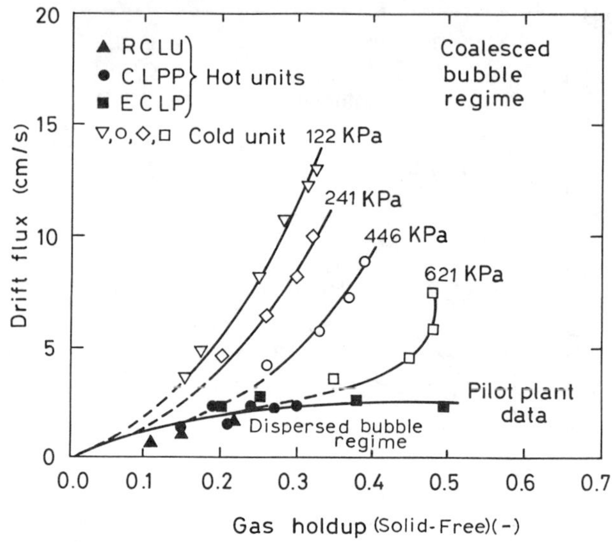

Figure 10.23 Relationship between gas holdup (solid-free) and the drift flux in the cold and hot units for EDS coal liquefaction reactors (from Tarmy et al., 1984).

evident in the hydrodynamic (gas holdup) similarity between different scale hot units for coal liquefaction as shown in Fig. 10.22. The solid-free gas holdup data follows the same line for each scale pilot (hot) unit. Such behavior indicates only minimal scale-up effects occur in going from bench scale to large scale resid hydrotreating or coal liquefaction systems. Also, as shown, the similarity does not extend to the cold unit operating conditions. Cold unit studies, which are essential for system design, provide qualitative information regarding the interrelationship among design parameters. An accurate evaluation of scaleup behavior requires that the flow regime in different scale units, either cold or hot, be consistent.

The drift flux model discussed in Section 2.3.6 was employed successfully by Tarmy et al. (1984) in describing phase holdup behavior in coal liquefaction reactors. Figure 10.23 shows the experimental drift flux versus gas holdup for cold and hot units reported by Tarmy et al. (1984). The drift flux, $j'_{g\ell}$, here is defined as $U_g - \delta_g(U_g + U_\ell)$ (also see Eqn. (2.28)), which can be correlated by $U_{b\infty}$ $\delta_g (1 - \delta_g)^m$ (also see Eqn. (2.128)) where δ_g is the solid-free based gas holdup. As indicated, several pressures were used in cold unit operations. A pressure increase in the cold flow units increases the

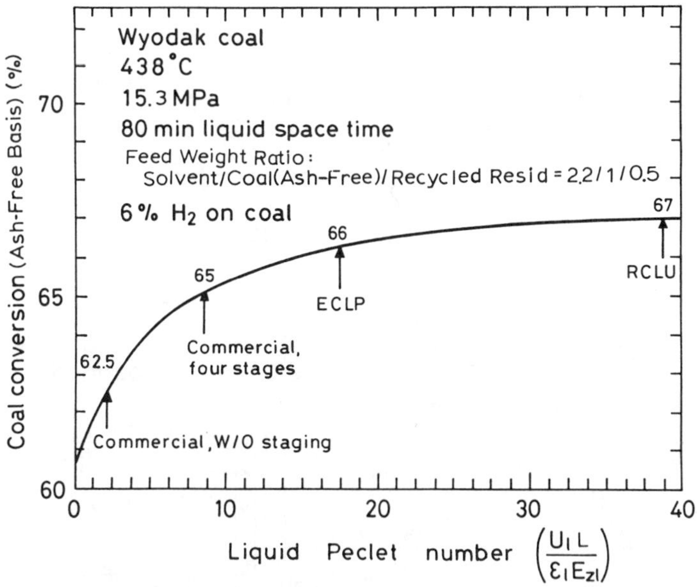

Figure 10.24 Predicted effect of liquid dispersion on coal conversion in EDS coal liquefaction reactors (from Tarmy et al., 1984).

gas holdup at a given drift flux. This behavior signifies that the hydrodynamic compatibility between cold and hot units increases with an increase in cold unit pressure. In addition, the cold unit (621 kPa) and hot unit show similar drift flux and gas holdup behavior. The hot unit is believed to be operated in the dispersed bubble regime. Thus, the 621 kPa cold unit was operated in the dispersed bubble regime over a wide range of gas flow rates. Based on 621 kPa cold unit data, $U_{b\infty}$ is determined by Eqn. (2.130) as 10 cm/s.

Liquid phase mixing considerably affects coal conversion in the reactor. The liquid mixing in the cold unit considerably exceeds liquid mixing in the hot unit, especially when the cold unit uses air and water at atmospheric pressure. Figure 10.24 shows the predicted scaleup effects of liquid backmixing on coal conversion for hot units including that at the commercial scale of 3.7 m ID as reported by Tarmy et al. (1984): an increase in the reactor scale decreases the liquid Peclet number, $U_\ell L/E_{z\ell}\epsilon_\ell$, and hence the extent of coal conversion; as noted, reactor staging enhances conversion.

In essence, catalyst particles are completely mixed in an ebullated bed reactor operated under large gas holdup conditions.

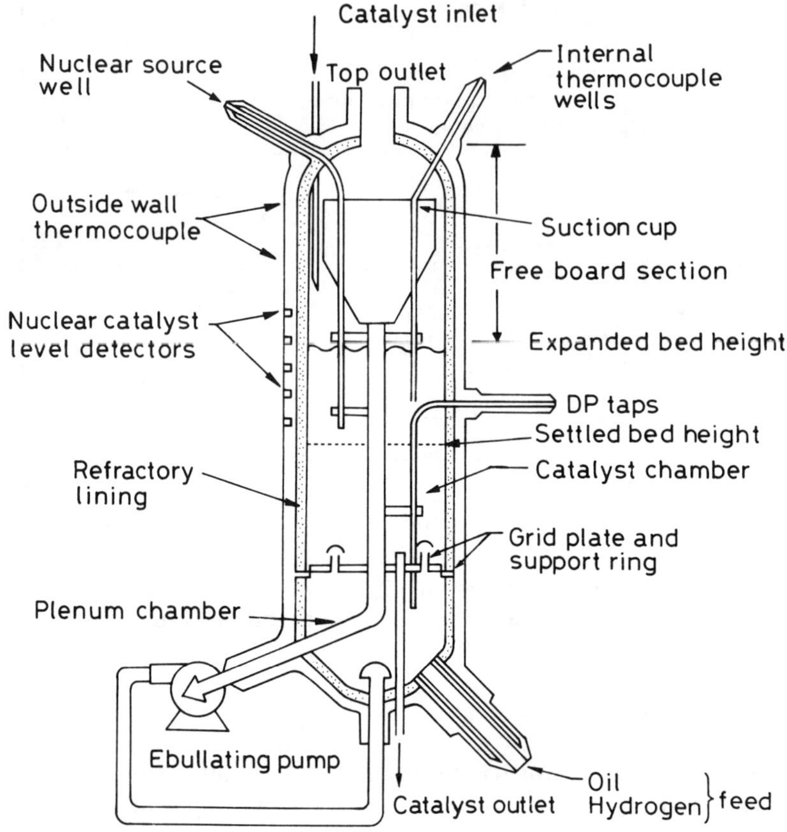

Figure 10.25 Design scheme for an ebullated bed reactor (from Li and Liu, 1981 and Ashland Synthetic Fuels Inc., 1984).

Although the density of spent catalyst can be twice as high as that of fresh catalyst, bed segregation is suppressed through bubble motion. Bickel and Thomas (1982) measured catalyst particle residence times during pilot plant operations of the H-Coal process using radioactive tracer methods and observed a catalyst residence time distribution characteristic of a CSTR reactor.

10.5.2 Design and Operation

Figure 10.25 shows the basic design configuration of an ebullated bed for hydrotreating of petroleum resid and coal liquefaction reported by Li and Liu (1981) and Ashland Synthetic

Fuels, Inc. (1984). Cylindrical catalysts with an average diameter of 0.8 mm and length 3 mm are fed into the top of the bed and withdrawn from the bottom of the bed. Feeding and withdrawing the catalyst can be conducted either continuously or periodically. Fresh feed (petroleum resid or coal slurry) from the preheater is introduced into the reactor plenum chamber along with internally recycled products. As noted earlier for ebullating purposes, the gas-liquid mixture (for resid hydrotreating) or gas-slurry mixture (for coal liquefaction) is then introduced into the bed, ideally at least, uniformly through the distributor. The extent of bed expansion ranges from near the minimum fluidization condition to a moderate expansion. The bed height is monitored by nuclear catalyst level detectors. Radiation sources are positioned within the vessel and detectors and accompanying electronics are mounted externally. Based on the effect of density on radiation flux, bed density or circulating fluid density is continuously monitored.

Pilot scale operations at Ashland Synthetic Fuels Inc. for coal liquefaction were carried out in a unit with a 1.8 m diameter reactor shell. The reactor was constructed of SA-387 Gr 22 Class 2 low alloy steel, a nominal 2.25% chrome - 1% molybdenum carbon steel, due to its hydrogen embrittlement resistance and high tensile strength at temperatures up to 482°C. This material is, however, subject to temper embrittlement. That is, the material may become brittle upon cooling after operation at high temperatures. Thus, warming the material above the brittle temperature, which increases over continued operation, must be done at low stress, i.e., low pressure (Ashland Synthetic Fuels, Inc., 1984).

Several problems have been reported in the operation of ebullated beds which cause processing outages. These problems include recycle pump electrical failures, recycle gas compressor seal failures, seal leakage between the grid plate and its support ring, and catalyst carry over (Ashland Synthetic Fuels, Inc., 1984; Beaton, et al., 1986). Among these, catalyst carryover presents a major concern because it induces long downtimes. Catalyst carryover is normally caused by excessive expansion of the catalyst bed by the flow of gas bubbles and liquid during refluidization. Catalysts can be carried over into the ebullating pump suction line or into the reactor effluent separator. Only a limited amount of elutriated catalyst can be handled by the ebullating pump system since the catalyst tends to collect in the piping and reactor plenum chamber. Uniform distribution of gas and liquid reactants across the reactor distributor

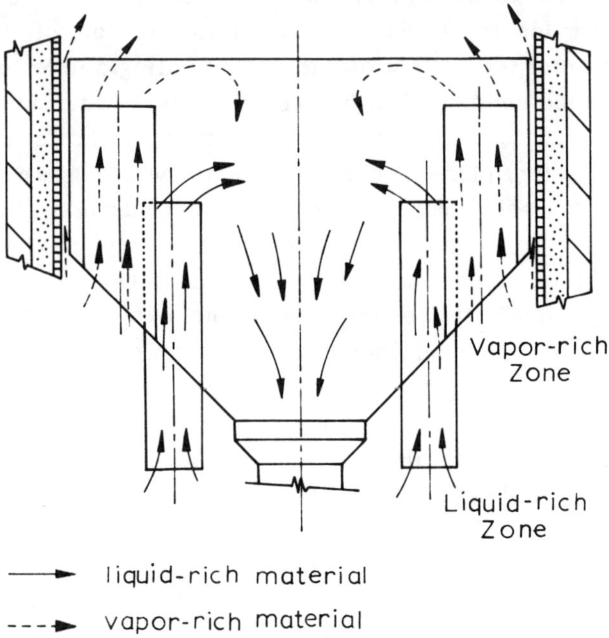

Figure 10.26 Design of the tubular recycle cup for an ebullated bed reactor (from Li and Liu, 1981).

is essential to the establishment of temperature uniformity in the bed. Maldistribution of gas and liquid reactants leads to significant local temperature excursions. The consequences of these temperature excursions are increased coke formation, possible catalyst sintering, or, at worst, temperature runaway and explosion. Temperature excursions may also be caused by sudden loss of reactor feed or utilities. Under such transient operating conditions, the temperature can be controlled by using a reactor emergency quench oil pump to cool the reactor feed temperature by pumping material to the reactor feed heater inlet or to the piping between the heater and the reactor.

A major concern in the design and operation of an ebullated bed reactor is the efficiency of the product recycle. The product internal recycle loop consists of a suction cup, inner column, and an ebullating pump. The detailed design of a tubular ebullating cup is shown in Fig. 10.26. As shown, the cup consists of a conical shape separation-collection design with risers flush mounted for receiving the gas-rich mixture and risers for receiving the liquid-rich mixture protrude through the cup. The tubular ebullating cup is an improved version of the original "H-Oil cup" which consists of a conical shape

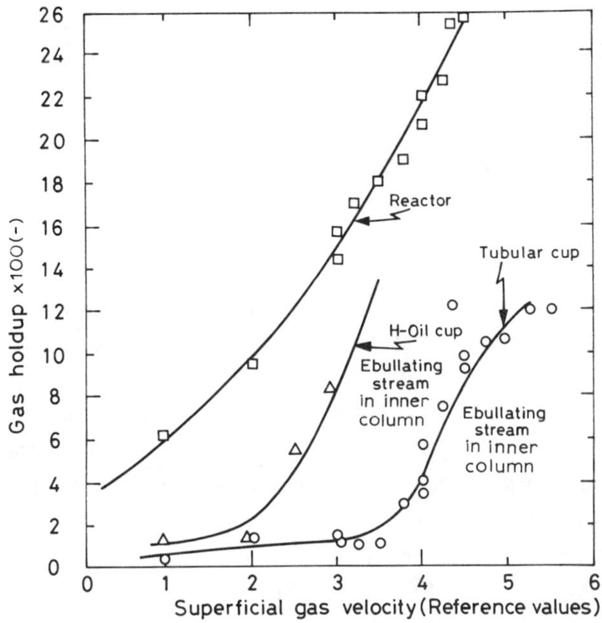

Figure 10.27 Effect of suction cup design on gas bubble entrainment in the recycle system (from Li and Liu, 1981). System operated using kerosene and nitrogen to simulate operating conditions.

separation-collection design without any risers. The cup design greatly affects the vapor-liquid separation required at the mouth of the inner tube. Figure 10.27 shows the gas holdup in the ebullated bed and in the inner tube for both the "H-Oil cup" and the tubular ebullating cup. Note that in arriving at Fig. 10.27 kerosene and nitrogen at a temperature of 38°C were used to simulate actual hydrotreating conditions. At a given gas velocity, the gas holdup in the inner tube was significantly lower when using a tubular cup than when using the H-Oil cup. The holdup for both cups, however, was considerably lower than in the ebullated bed itself. Based on a value of 10 cm/s for the isolated bubble rise velocity for the EDS reactor described earlier, the bubble size at industrial conditions for resid hydrotreating or coal liquefaction would be 0.68 mm with a Re, defined as $U_{b\infty} d_p \rho_\ell / \mu_\ell$, of 176; this calculated value takes into consideration the physical properties of the liquid and gas given by Tarmy et al. (1984). Such small bubbles require only very low shear stress or low ebullating liquid velocity in the inner tube to reach a full bubble recycle.

In normal operation of ebullated bed reactors, the reactor feed temperature is the control variable. The desired reactor feed temperature depends on both the feed rate and feed composition. The feed temperature is chosen such that the overall heat generation in the reactor is used to elevate the low temperature feed material to the bed temperature during the adiabatic operation of the reactor. Additional temperature control can also be provided by immediately shutting off the hydrogen supply.

Ideally the ebullated bed reactor operates without any internal nonuniformities; however, in practice such nonuniformities may occur. Adequate pressure drop across the distributor plate is needed to provide uniform gas and liquid distribution in the reactor. During steady state operation, the superficial gas velocity decreases axially due to hydrogen consumption, yet, for a given axial location, the velocity is nearly uniform with respect to radial and angular position. Pilot operation of an ebullated bed has shown that some radial temperature differences do exist within the first 2 to 2.5 meters above the grid plate (Ashland Synthetic Fuels, Inc., 1984); such behavior indicates preferential flow patterns occur near the distributor. Figure 10.28 shows the axial temperature variation in a commercial ebullated bed reactor for hydrotreating of heavy residual oils (Robson et al., 1980). The temperature variation between the bottom and top of the bed is about 1 °C.

10.5.3 Analysis of Ebullated Bed Reactors

The volume required to achieve a given extent of desulfurization is an important factor in the selection of a reactor technology. Dautzenberg and de Deken (1985) compared reactor volume requirements of trickle bed and ebullated bed reactors at various degrees of conversion. Inherent in their development is the assumption that an averaged reactivity for contaminant removal provides an adequate description of reactor volume requirements. They presented the following:

The required volume of an ebullated bed reactor, V_B (m^3), for a given extent of demetallization, based on CSTR governing equations for reactors in series and assuming a first order reaction rate, is

$$V_B = \frac{N}{(1-\varsigma)}\frac{F}{k_{mb}}\left[\frac{1-(1-X_M)^{1/N}}{(1-X_M)^{1/N}}\right] \tag{10.9}$$

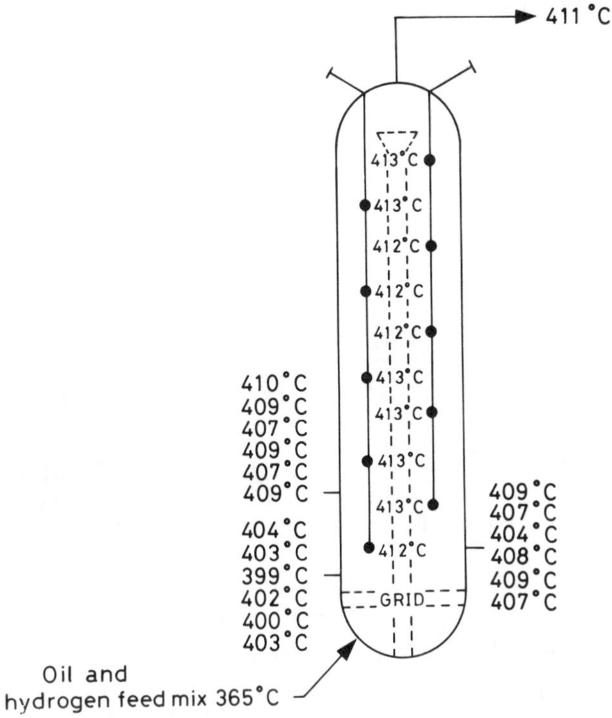

Figure 10.28 Axial temperature variation in a commercial ebullated bed reactor for hydrotreating of resids (from Robson et al., 1980).

where N is the number of the reactors in series, F is the volumetric feed rate, k_{mb} is the demetallization rate constant, ς is the extent of bed expansion relative to the height of the expanded bed, and X_M is the degree of demetallization. The required volume of a trickle bed reactor for a given extent of demetallization, based on plug flow governing equations and assuming a 1.5 order reaction rate, is

$$V_P = \frac{2F}{k_{mp}} \left[\frac{1}{(1 - X_M)^{0.5}} - 1 \right] \qquad (10.10)$$

where k_{mp} is the demetallization rate constant.

The ratio of these two equations determines the relative reactor volume requirements in terms of the ratio of average rate constants for each type reactor (k_{mb}/k_{mp}); this parameter is referred to as the apparent catalyst activity ratio. Figure 10.29 shows the relative

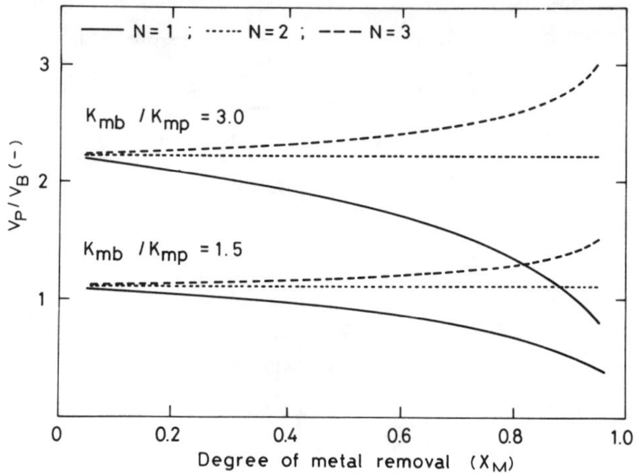

Figure 10.29 Relative reactor volume requirements of ebullated bed and trickle bed reactors as a function of demetallization (from Dautzenberg and de Deken, 1985).

volume requirements (V_P/V_B) as a function of desulfurization for one, two, or three ebullated bed reactors in series and (k_{mb}/k_{mp}) values of 1.5 and 3.0. The V_P/V_B versus conversion relationship appears quite sensitive to the apparent catalyst activity ratio. For most degrees of conversion the ebullated bed reactor requires the least volume.

A relationship between the extents of desulfurization and demetallization can be derived by first expressing the volume of a trickle bed reactor in terms of the extent of desulfurization

$$V_P = \frac{F}{k_{sp}} \frac{X_S}{(1 - X_S)} \tag{10.11}$$

where k_{sp} is the desulfurization rate constant and the desulfurization reaction is assumed to be second order. And second, combining Eqns. (10.10) and (10.11) and solving for X_M

$$X_M = 1 - \left[\frac{1}{(1 + \frac{k_{mp}}{2k_{sp}} (\frac{X_S}{(1 - X_S)}))} \right]^2 \tag{10.12}$$

Equation (10.12) relates X_M and X_S through (k_{mp}/k_{sp}); this ratio is a measure of catalyst selectivity.

The above relationships for the trickle bed reactor can be extended to estimate information on catalyst life for the ebullated bed reactor. By defining a catalyst metal uptake capacity as the overall metal load on the catalyst and making a material balance on the metal components a relationship can be developed between the metal content in the feed, the catalyst life, and the metal uptake capacity (Dautzenberg and de Deken, 1985).

The catalyst loading per reactor volume for an ebullated bed reactor is smaller than in a trickle bed reactor due to bed expansion. Since the average reaction temperature for an ebullated bed reactor (~425°C) is higher than that for a trickle bed reactor, an apparent catalyst activity in an ebullated bed reactor can easily be a factor of 1.5 to 2.0 higher than that in a trickle bed reactor (Dautzenberg and de Deken, 1984).

Li and Liu (1981) analyzed the temperature stability in an ebullated bed reactor and presented the following:

The temperature change through an ebullated bed reactor can be estimated by

$$T_{out} - T_{pl} = \left(\frac{-\Delta H_r}{\rho_m C_{pm}}\right)\left(\frac{1}{1+\gamma}\right)(C_{A,feed} - C_{A,out}) \qquad (10.13)$$

where T_{out} is the reactor outlet temperature, T_{pl} is the reactor plenum chamber temperature, $C_{A,feed}$ is the reactant feed concentration, $C_{A,out}$ is the reactant effluent concentration, ΔH_r is the heat of reaction per mole of reactant consumed, γ is a dimensionless constant related to recycle liquid/feed ratio and internal liquid backmixing (γ ranges from 5 to 15), and ρ_m and c_{pm} are the average density and heat capacity, respectively, for the gas-liquid mixture.

The temperature change through a trickle bed reactor can be estimated by

$$T_{out} - T_{feed} = \left(\frac{-\Delta H_r}{\rho_m C_{pm}}\right)(C_{A,feed} - C_{A,out}) \qquad (10.14)$$

where T_{feed} denotes the temperature of the reactor feed stream after the preheater. Calculations based on Eqns. (10.13) and (10.14) yield a temperature change in a trickle bed reactor about one order of magnitude higher than that in an ebullated bed reactor (less than 6°C). Theoretically, however, an ebullated bed reactor is more prone to thermal runaway than a trickle bed reactor for which inter-bed quenching can be used to control the exothermic nature of the reactions (Van Zijll Langhout et al., 1980). This point will be elaborated upon later. In fact, the possibility of temperature runaway exists particularly for ebullated bed operation at high feed conversions where the H_2 consumption rate is high.

For series reactor operation, inter-reactor quenching normally controls temperature effectively in a reactor operated under adiabatic conditions. The heat generated in an upstream reactor exceeds that in a downstream reactor due to higher reaction rates in the upstream reactor. Thus, the difference between reactor effluent and feed temperatures for an upstream reactor exceeds that for a downstream reactor. A process simulation model developed by Mosby et al. (1986) calculated the inter-reactor quenching necessary to maintain the reactor in heat balance. Furthermore, the model, which incorporates the reactor kinetics for each reaction matrix mentioned earlier, a heat balance, and catalyst deactivation, has been used to guide commercial operation of ebullated bed reactors for resid hydrotreating. Although the details of this model are not fully available in the literature, the model appears to be the most relevant one to predict the behavior of the commercial ebullated bed reactor for resid hydrotreating. Major features of this model are given below:

The energy balance of the reactor can be given by

$$Q_{gen} = H_{out} - H_{in} + Q_{\ell} \tag{10.15}$$

where Q_{gen} is the heat released by chemical reaction and H_{out} and H_{in} are the enthalpy of the effluent and influent gas-liquid mixtures. For adiabatic operation, we have

$$Q_{\ell} = 0 \tag{10.16}$$

Thus, Eqn. (10.15) can be simplified to

$$H_{out}(T_r) - H_{in}(T_{in}) = Q_{gen}(T_r) \tag{10.17}$$

Q_{gen} can be related to the hydrogen consumption by

$$Q_{gen} = \sum_i (-\Delta H_r)_i \, (\Delta C_{H_2})_i \tag{10.18}$$

where $(\Delta H_r)_i$ is the heat of reaction i per mole hydrogen consumed and $(\Delta C_{H_2})_i$ is the hydrogen consumption for reaction i. The reaction types considered include both thermal cracking reactions and catalytic reactions of such feed contaminants as metals and sulfur. Reaction models for all reaction types are based on the approach by Mosby et al. (1986) outlined earlier.

Catalyst activity can be described as a unique function of the amount of metal deposited on its surface. Knowledge of catalyst metal content, and consequently its activity, is important to any model of the reactor system. A fundamental approach to acquiring this knowledge is to determine the catalyst age distribution for a given reactor and then determine an average catalyst activity through a relationship between catalyst age, metal deposition, and activity. Thus, knowledge of the catalyst age distribution in the reactor is essential to the evaluation of catalyst activity. Assuming completely mixed catalyst in the reactor, a difference equation for the fraction of catalyst on stream day "i" with a particular reactor age "j", $F_{i,j}$, in one reactor of a multiple reactor system is given by

$$F_{i,j} = F_{i-1,j-1} + A_{i,j}\frac{C_c}{C_I} - F_{i-1,j-1}\frac{C_w}{C_I}, \quad i \geq j \qquad (10.19)$$

where $F_{i-1,j-1}$ is the fraction of catalyst which was "j-1" reactor days old on stream day "i-1"; $A_{i,j}$ is the fraction of catalyst in batch added which is "j" days old on stream day "i"; C_c is the quantity of catalyst in batch added; C_I is the reactor catalyst inventory; and C_w is the quantity of catalyst in batch withdrawn. Note that $A_{i,j}$ in Eqn. (10.19) is zero for j > 1 when fresh catalyst is added to the reactor. i = 1 normally refers to the time of startup. Similar equations to Eqn. (10.19) can be written for each reactor in a multiple reactor system. Thus, for a given stream day "i", the average catalyst reactivity, $\overline{A}_i$, in the reactor can be given by

$$\overline{A}_i = \sum_{j=1}^{n} A_i|_{M = M_j} F_{i,j} \qquad (10.20)$$

where M_j is the extent of metal deposition on catalyst "j" reactor days old. As n increases beyond 999, the catalyst activity is essentially zero. With the above considerations, a calculation scheme can be developed based on the model of Mosby et al. (1986) to account for the interrelationship among the enthalpy, $H_{in}(T_{in})$ or $[H_{out}(T_r) - Q_{gen}(T_r)]$, T_{in}, and T_r. Such a relationship for the first reactor in a series of reactors is shown in Fig. 10.30. As shown, a plot of $H_{in}(T_{in})$ versus T_r shows a slightly convex shape indicating that, for a given T_{in}, a

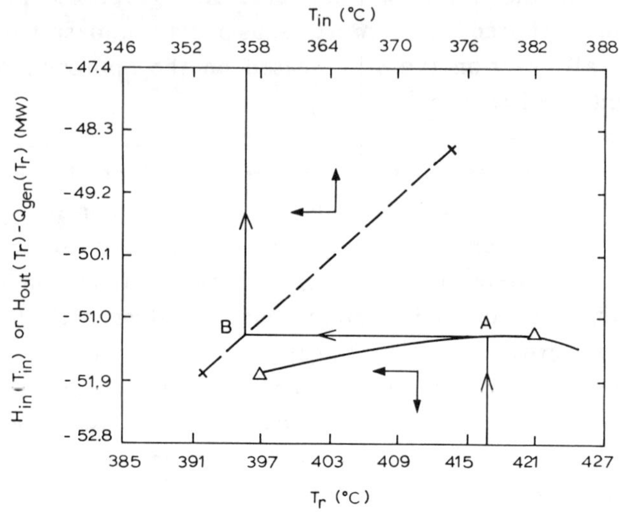

Figure 10.30 Calculation of reactor inlet temperature required to obtain a desired temperature for the 1st reactor in a reactor train (after Mosby et al., 1986).

possibility of temperature multiplicity exists in the reactor which involves a low and very high stable temperatures and an intermediate metastable temperature. It is also noted that the reactor temperature, T_r, is very sensitive to T_{in}. Thus proper control of T_{in} is necessary to avoid temperature runaway in the reactor.

10.6 CONCLUDING REMARKS

The hydroprocessing of heavy crude oils is one of the most important commercial applications of three-phase fluidized beds. Feedstocks of heavy petroleum crudes and synthetic crudes from coal, tar sands, and oil shales present important property similarities and consequently, the processing of these different materials also presents similarities.

Hydrotreating processes upgrade petroleum resids by removing unwanted heteroatoms, such as sulfur, and through Ramsbottom carbon conversion. The process contacts the feedstock with hydrogen over a bed of catalysts. A complex montage of both catalytic and noncatalytic reactions occurs, which includes hydrodesulfurization, hydrodenitrogenation, hydrodeoxygenation, hydrodemetallization, hydrogenation, hydrocracking, thermal cracking, and coking. In practice the complex chemistry of hydrotreating can be effectively described by subdividing both the feedstock and products into lumps (Mosby et al., 1986). Feedstock lumps are based on boiling range and reactivity while those of the products are based on boiling range. A matrix of irreversible first order reactions can be set up between each feedstock lump and each product lump for all reaction types considered.

A major concern of hydrotreating processes is the catalyst deactivation resulting from metal and coke deposits on the catalyst surface. Optimum demetallization catalysts have pore structures designed specifically to prolong catalyst life by a uniform deposition of metals throughout the particle. However, the larger pores required for long catalyst life reduce the catalytic activity towards desulfurization.

Industrial hydrotreating units which employ ebullated bed reactor technology consist of a train of reactors operated in series to enhance product yield and accommodate the complex reactions. Gas holdup in the ebullated bed can be as high as 0.50 (Tarmy et al., 1984) while the bubble size may be as small as 0.68 mm. The suction cup design, liquid velocity, and gas velocity all significantly affect catalyst and gas bubble entrainment in the internal recycle stream of the ebullated bed reactor. Steady-state temperature multiplicities could occur, particularly under conditions of high resid conversions where a high hydrogen consumption is required.

10.7 NOMENCLATURE

A_i : catalyst activity on stream day "i"; $\bar{A}_i$, average activity

$A_{i,j}$: fraction catalyst in batch added to the reactor which is "j" reactor days old on stream day "i"

A_m : demetallization activity

A_{m0} : initial demetallization activity

$C_{A,feed}$: reactant concentration in the feed

$C_{A,out}$: reactant concentration at the reactor outlet

C_c : quantity of catalyst in batch added to the reactor

C_I : reactor catalyst inventory

C_m : metals content in catalyst

$C_{m,max}$: metals uptake capacity of catalyst

C_{Ni} : nickel concentration in the bulk liquid

C_{pm} : average heat capacity of gas-liquid mixture

C_S : sulfur concentration in the bulk liquid phase

C_{S0} : initial sulfur concentration

C_V : vanadium concentration in the bulk liquid phase

C_w : quantity of catalyst in batch withdrawn from the reactor

$(\Delta C_{H_2})_i$: hydrogen consumption for reaction i

d_p : particle diameter

ΔD : increment of pore diameter variation

E : activation energy of overall HDS reaction

$E_{z\ell}$: liquid phase dispersion coefficient

$f(P_H)$: function describing overall HDS reaction dependency on hydrogen partial pressure

F : volumetric feed rate

$F_{i,j}$: fraction catalyst which is "j" reactor days old on stream day "i"

ΔH_r : heat of reaction per mole hydrogen consumed; $(\Delta H_r)_i$ for reaction i

H_{in} : enthalpy of the influent gas-liquid mixture

H_{out} : enthalpy of the effluent gas-liquid mixture

$j'_{g\ell}$: drift flux

k : general rate constant in Langmuir-Hinshelwood expression

k_1 : reaction rate constant for reactive sulfur fraction

k_2 : reaction rate constant for unreactive sulfur fraction

k_a : hydrogen adsorption constant

k_{mb} : first-order demetallization rate constant in an ebullated bed reactor

k_{mp} : 1.5-order demetallization rate constant in a trickle bed reactor

k_n : reaction rate constant for nickel removal

k_{sp} : second-order desulfurization rate constant in a trickle bed reactor

k_v : reaction rate constant for vanadium removal

K_{AR} : aromatic compounds adsorption equilibrium constant (sulfur compounds included)

K_D : dibenzothiophene adsorption equilibrium constant

K_H : hydrogen adsorption equilibrium constant

K_S : hydrogen sulfide adsorption equilibrium constant

K_T : thiophene adsorption equilibrium constant

L : reactor length

m : correlation constant

M : extent of metal deposition on catalyst; M_j for catalyst reactor day "j"

N : number of ebullated bed reactors in series

P_{AR} : aromatic compounds partial pressure

P_D : dibenzothiophene partial pressure

P_H : hydrogen partial pressure

P_S : hydrogen sulfide partial pressure

P_T : thiophene partial pressure

Q_{gen} : heat released by reactions

Q_ℓ : heat exchanged or loss to atmosphere

r_{HDS} : rate of hydrodesulfurization

R : ideal gas constant

Re : Reynolds number, $U_{b\infty} d_p \rho_\ell / \mu_\ell$

t : time

t^* : dimensionless time

T : temperature

T_{feed} : reactor feed temperature (after the preheater)

T_{in} : reactor inlet temperature

T_{out} : reactor outlet temperature

T_{pl} : reactor plenum chamber temperature

T_r : reactor temperature

$U_{b\infty}$: isolated bubble rise velocity

U_g : superficial gas velocity

U_ℓ : superficial liquid velocity

V_B : reactor volume of an ebullated bed reactor

V_P : reactor volume of a trickle bed reactor

ΔV : increment of pore volume variation

X_S : degree of desulfurization

X_M : degree of metal removal

Greek Letters

α, α_1, α_2 : empirical constants

γ : dimensionless constant related to recycle liquid/feed ratio and internal liquid backmixing

ζ : extent of bed expansion relative to the height of the expanded bed

ρ_c : bulk density of catalyst

ρ_ℓ : liquid density

ρ_m : average density of gas-liquid mixture

ρ_o : feed specific gravity

δ_g : solid-free gas holdup, $\epsilon_g/(\epsilon_g + \epsilon_\ell)$

ϵ_g : gas holdup

ϵ_ℓ : liquid holdup

μ_ℓ : liquid viscosity

10.8 REFERENCES

Asaoka, S., S. Nakata, Y. Shiroto, and C. Takeuchi, I&EC Process Des. & Dev., 22, 242 (1983).

Ashland Synthetic Fuels, Inc., Final Report to U.S. Department of Energy, Contract No. DE-AC05-76ET10143, April (1984).

Badilla-Ohlbaum, R., C.K. Pratt, and D.L. Trimm, Fuel, 58(4), 309 (1979).

Beaton, W. I., in "Chemical Reactions as a Means of Separation-Sulfur Removal," ed. by B.L. Crynes, p. 1, Marcel Dekker, New York (1977).

Beaton, W.I., N.K. McDaniel, W.E. McWhirter, R.D. Petersen, and R.P. Van Driesen, paper presented at the National Petroleum Refiners Association Annual Meeting, March 23-25 (1986).

Beuther, H., and B.K. Schmid, in Proceedings of the 6th World Petroleum Congress, 3, 297 (1963).

Beuther, H., and O.A. Larson, I&EC Process Des. & Dev., 4, 177 (1965).

Bickel, T. C., and M.G. Thomas, I&EC Process Des. & Dev., 21, 377 (1982).

Boening, R.E., N.K. McDaniel, R.D. Petersen, and R.P. Van Driesen, Hydrocarbon Processing, 66, 59, Sep. (1987).

de Brujin, A., in Proceedings of the International Congress on Catalysis, paper B34, London (1976).

Chianelli, R.R., Catal. Rev. - Sci. Eng., 26, 361 (1984).

Cocchetto, J.F., and C. N. Satterfield, I&EC Process Des. & Dev., 20, 49 (1981).

Dautzenberg, F.M., J. Van Klinen, K.M.A. Pronk, S.T. Sie, and J.-B. Wijffels, ACS Symposium Series, 65, 254 (1978).

Dautzenberg, F.M., S.E. George, C. Ouwerkere, and S.T. Sie, paper presented at the Advances in Catalytic Chemistry Symposium, Salt Lake City, Utah, May (1982).

Dautzenberg, F.M., and J.C. de Deken, Catal. Rev. - Sci. Eng., 26, 421 (1984).

Dautzenberg, F.M., and J.C. de Deken, Symposium on Developments in Hydrodemetallization Catalysts, ACS Div. of Petrol. Chem., Preprints, p. 8 (1985).

Drushel, H.V., ACS Div. of Petrol. Chem., Preprints, 17(4), F92 (1972).

Fan, L.-S., K. Kitano, and B.E. Kreischer, paper presented at the AIChE Annual Meeting, Chicago, Nov. 10-14 (1985); AIChE J., 33, 177 (1987).

Furimsky, E., Fuel, 57(8), 494 (1978).

Furimsky, E., Catal. Rev. Sci. - Eng., 25(3), 421 (1983).

Galiasso, R., J. Garcia, L. Caprioli, J.M. Pazos, and A. Soto, Symposium on Developments in Hydrodemetallization Catalysts, ACS Div. of Petrol. Chem., Preprints, p. 50 (1985).

Gates, B.C., J.R. Katzer, and G.C.A. Schuit, "Chemistry of Catalytic Processes," McGraw-Hill (1979).

Grange, P., Catal. Rev. - Sci. Eng., 21(1), 135 (1980).

Houalla, M., N.K. Nag, A.V. Sapre, D.H. Broderick, and B. C. Gates, AIChE J., 24, 1015 (1978).

Inoguchi, M., Shokubai, Japan, 18(3), 78 (1976).

Kato, J., K. Shimada, M. Suzuki, H. Ose, and S. Ohshima, Kogyo Kagakushi, Japan, 74(6), 1156 (1971).

Koros, R.M., S. Bank, J.E. Hofmann, and M.I. Kay, ACS Div. Petrol. Chem., Preprints, 12(4), B-165 (1967).

Kubicka, R., J. Cir, V. Novak, and J. Veprek, Brennst.-Chem., 49(10), 308 (1968).

Laine, R.M., Catal. Rev. Sci. - Eng., 25(3), 459 (1983).

Lee, H.C., and J.B. Butt, J. Catal., 49, 320 (1977).

Le Page, J.F., "Catalyse de Contact," Technip, p. 457, Paris (1978).

Li, A., and D. Liu, in Proceedings of the 2nd World Congress of Chemical Engineering, IV, p. 170 (1981).

Lunin, G., J. Lamarre, and K. Whitham, paper presented at the UNITAR 3rd International Conference on Heavy Crude and Tar Sands, Long Beach, CA, July 22-31 (1985).

MacArthur, J.B., A. Li, and A.S. Ambegaonkar, paper presented at the 186th ACS National Meeting, Washington, DC, Aug. (1983).

Massagutov, R.M., G.A. Berg, G.M. Kulinich, and T.S. Kirilov, in Proceedings of the 7th World Petroleum Congress, 4, 177, Mexico (1967).

Mosby, J.F., R.D. Buttke, J.A. Cox, and C. Nikolaides, Chem. Eng. Sci., 41, 989 (1986).

Ohtsuka, T., Y. Hasegawa, and N. Takanari, Bull. Japan Petrol. Inst., 10, 14 (1968).

Ohtsuka, T., Catal. Rev. - Sci. Eng., 16(2), 291 (1977).

Oxenreiter, M.F., C.G. Frye, G.B. Hoekstra, and J.M. Sroka, Japan Petrol. Inst., Fuel Oil Desulfurization Symposium, Tokyo, Japan, Nov. 30 (1972).

Parkinson, G., Chem. Eng., p. 17, Sep. 16 (1985).

Robson, B., K. Chinnery, C. Herster, C. Viens, and J. Caspers, in Proceedings of the C.E. Lummus Montreal Area Refiner's Seminar, May 21 (1980).

Rollmann, L.D., J. Catal., 46, 243 (1977).

Sapre, A.V., and B.C. Gates, I&EC Process Des. & Dev., 20, 68 (1981).

Satterfield, C.N., and J.F. Cocchetto, I&EC Process Des. & Dev., 20, 53 (1981).

Satterfield, C.N., and S. Gultekin, I&EC Process Des. & Dev., 20, 62(1981).

Schuit, G.C.A., and B.C. Gates, AIChE J., 19(3), 417 (1973).

Schuman, S.C., and H. Shalit, Catal. Rev., 4(2), 245 (1970).

Scott, J.W., A.G. Bridge, R.I. Christensen, and G.D. Gould, Japan Petrol. Inst., Fuel Oil Desulfurization Symposium, Tokyo, Japan, March 10-11 (1970).

Shah, Y.T., "Reaction Engineering in Direct Coal Liquefaction," Addison-Wesley (1981).

Singhal, G.P., R.L. Espino, J.E. Sobel, and G. A. Huff, J. Catal., 67, 457 (1981).

Speight, J.G., "The Chemistry and Technology of Petroleum," Marcel Dekker (1980).

Speight, J.G., "The Desulfurization of Heavy Oils and Residual," Marcel Dekker (1981).

Takeuchi, C., S. Asaoka, S. Nakata, and Y. Shiroto, Symposium on Developments of Hydrodemetallization Catalysts, ACS Div. of Petrol. Chem., Preprints, p. 96 (1985).

Tarmy, B.L., M. Chang, C.A. Coulaloglou, and P.R. Ponzi, paper presented at the 8th International Symposium on Chemical Reaction Engineering, Edinburgh, Sep. 10-13 (1984).

Thakur, D.S., and M.G. Thomas, Appl. Catal., 15, 197 (1985).

Topsoe, H., and B.S. Clausen, Catal. Rev. - Sci. Eng., 26, 395 (1984).

Topsoe, H., B.S. Clausen, N. Topsoe, and E. Pederesen, I&EC Fundam., 25, 25 (1986).

Toulhoat, H., J.C. Plumail, G. Martino, and Y. Jacquin, Symposium on Developments in Hydrodemetallization Catalysts, ACS Div. of Petrol. Chem., Preprints, p. 85 (1985).

Tsonopoulos, C., J.L. Heidman, and S.-C. Hwang, "Thermodynamic and Transport Properties of Coal Liquids," John Wiley and Sons, New York (1986).

Van Deemter, J.J., in Proceedings of the 3rd European Symposium on Chemical Reaction Engineering, p. 215 (1984).

Van Driesen, R.P., V.A. Strangio, A. Rhoe, and J.J. Kolstad, paper presented at the AIChE National Meeting, New Orleans, April 6-10 (1986).

Van Zijll Langhout, W.E., C. Ouwerkerk, and K.M.A. Pronk, paper presented at the AIChE National Meeting, Philadelphia, PA, June 8-12 (1980).

Vrinat, M.L., and L. de Mourgues, J. Chem. Phys., 79, 45 (1982).

Vrinat, M.L., Appl. Catal., 6, 137 (1983).

Waugh, R.J., and A.E. Silva, paper presented at the ASME/CSME Energy Sources Technology Conference and Exhibition, New Orleans, Feb. 12-17 (1984).

Ware, R.A., and J. Wei, Symposium on Developments in Hydrodemetallization Catalysts, ACS Div. of Petrol. Chem., Preprints, p. 62 (1985a).

Ware, R.A., and J. Wei, J. Catal., 93, 100 (1985b).

Weisz, P.B., and C.D. Prater, Advances in Catalysis, 9, 575 (1957).

Wenzel, F.W., U. Graeser, and K. Niemann, paper presented at the Japan Petrol. Inst., Petroleum Refining Conference, Tokyo, Japan, Oct. 27-28 (1986).

Chapter 11

Miscellaneous Systems

This chapter describes additional applications of three-phase fluidization systems involving either reactive or nonreactive solid particles. These applications include electrodes, granulation, adiponitrile production, calcium bisulfite acid production, sand filter cleaning, crystallization, flotation, polymerization, coal gasification, methanation, Fischer-Tropsch synthesis and methanol synthesis. The modes of operation covered include Modes E-I-a-1, E-I-a-2, E-I-b, E-II-a-2, E-II-b, E-III-a, T-I-a-1, T-I-a-2, T-II-a, and T-III-a (see Fig. 1.3 for mode designations).

11.1 ELECTRODES

Since the advent of fluidized bed electrodes in the 1960's, there has been considerable research on both theory and experiments conducted in this area. Work performed is mainly on two-phase (liquid-solid) fluidized bed electrodes (e.g., Wilkinson and Haines, 1972; Goodridge and Wright, 1980; Walker and Wragg, 1980; Masterson and Evans, 1982). Relatively little, however, has been reported on three-phase (gas-liquid-solid) fluidized bed electrodes (Mode E-I-a-1) (e.g., Oloman and Watkinson, 1975; Kusakabe et al., 1981).

11.1.1 General Principles and Applications

Basically, the fluidized bed electrode is a diaphragm or membrane-divided electrochemical cell in which a bed of conducting particles is added to a half cell (usually a cathode half cell). The bed expansion is maintained by an upflowing electrolyte. The overwhelming industrial application using fluidized bed electrodes has centered around metal winning (e.g., Kreysa, 1978; Bareau and Coeuret, 1979; Sabacky and Evans, 1979; Morooka et al., 1980) either in trace metal removal in effluent stream treatment or metal recovery

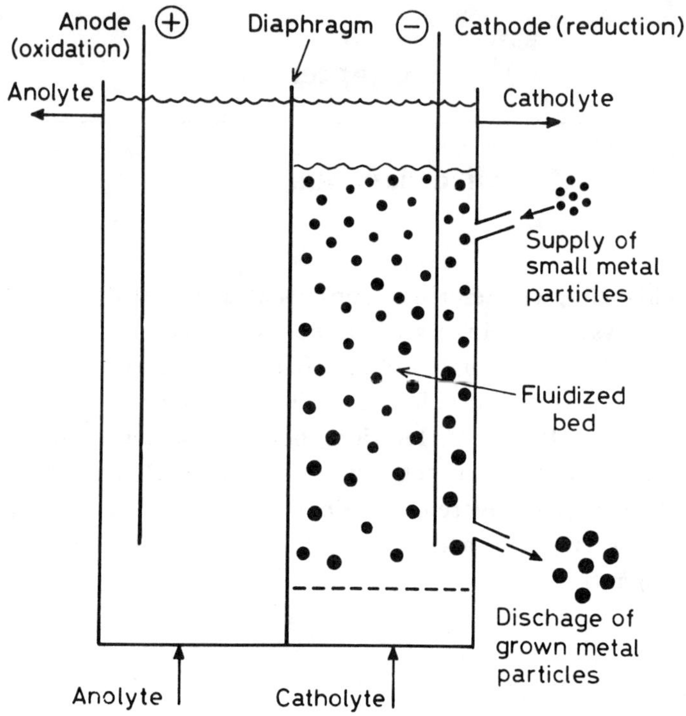

Figure 11.1 Continuous particle growth and recovery in a fluidized bed electrode (from Van der Heiden et al., 1978).

from leaching solution in hydrometallurgy applications. These applications depend on environmental legislation and economics relating to prices of metals and electric energy (Kelsall, 1983). For information regarding the fundamentals of electrochemistry, readers are referred to Adams (1969) and Bard and Faulkner (1980).

Figure 11.1 shows a schematic diagram of a two-phase fluidized bed electrode for metal deposition applications. In this figure, the fluidized particles are conductive metal particles which act as moving electrodes (cathode). These particles adsorb cations in the electrolyte and charges from a stationary, or feeder, electrode, and grow in size before exiting from the bed. The electrode geometry greatly affects the prevailing current and energy efficiencies. Figure 11.2 shows a commercial design of a cylindrical fluidized bed electrode (1.2 m in height) with a hexagonal anode arrangement (Van der Heiden et. al., 1978). This fluidized bed electrode was developed by Akzo Zout Chemie to treat 1 ~ 5 ppm mercury containing brine effluents from

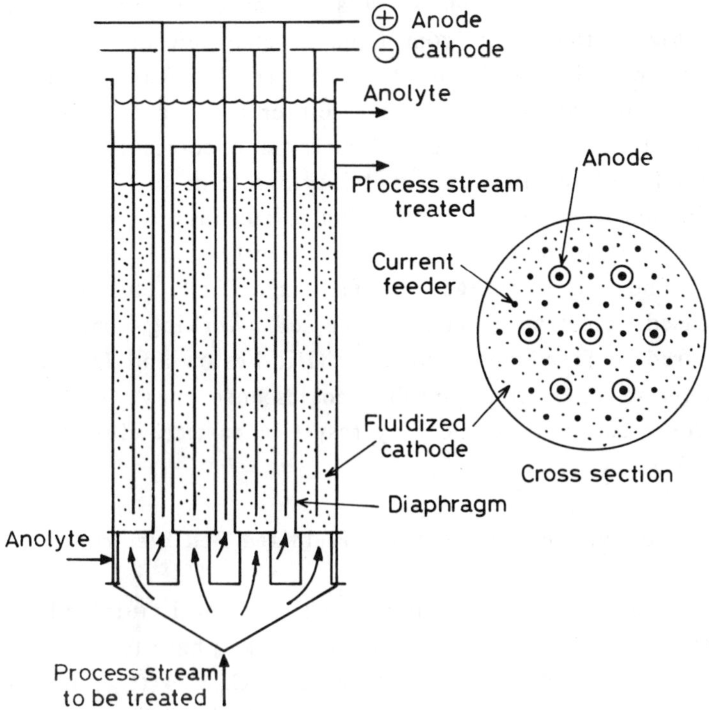

Figure 11.2 Cylindrical fluidized bed electrode with hexagonally spaced membranes and anode: developed by Akzo Zout Chemie, a sublicensee of Constructors John Brown Ltd. (from Van der Heiden et al., 1978).

mercury chlor-alkali cells. A similar unit was also used for the recovery of copper, nickel, and cobalt. Fluidized bed electrodes have also been applied to the reaction of oxygen and hydrogen in fuel cells (Bockris and Srinivasan, 1969), sulfide mineral oxidation/electro-dissolution (Coeuret, 1980), electro-synthesis of organic and inorganic chemicals (Fry, 1972), electro-refining of metals (Coeuret and Storck, 1984), reduction of m-nitrobenzene sulfonic acid (Backhurst et al., 1969), and oxidation of methanol (Berent et al., 1971).

The use of moving particles as working electrodes presents several advantages (Kelsall, 1983). First, moving particle electrodes have a high specific liquid-solid contacting area (e.g., 3500 m^{-1} compared to 15 m^{-1} for a conventional planar electrode) and low mass transfer resistance, thus yielding a high rate of mass transfer of ionic species to the fluidized particles, a large space-time, and a small cell volume requirement. By applying sufficient voltage, the fluidized bed electrode results in a mass transfer limited current several hundred times higher than that from a planar electrode at the same flow rate. Furthermore, it can be used for electrochemical reactions involving low real current densities due to either low reactant concentration or low reaction rates. A second advantage comes from more uniform potential and reaction rate distributions than in fixed bed electrodes. In addition, a desirable application to continuous processes for growth and recovery of metals can be conveniently made.

The disadvantage of fluidized bed electrodes resides in decreased electronic conduction of working particle electrodes compared to fixed bed electrodes (Hiddleston and Douglas, 1968). In three-phase operations, bubble coalescence may further reduce the electronic conduction and may result in particle elutriation (Berent et al., 1971).

11.1.2 Behavior of Three-Phase Fluidized Bed Electrodes

The addition of a gas into a liquid-solid fluidized bed electrode serves either of two purposes: an inert gas phase (e.g., N_2) can be used to enhance liquid-solid mass transfer - Case A (e.g., Kusakabe et al., 1981) or a gas phase reactant may be required for an electrochemical reaction - Case B (e.g., H_2 and O_2 for fuel cells (Bockris and Srinivasan, 1969) and O_2 for H_2O_2 production (Oloman and Watkinson, 1975)). Examples of each of these applications are given in the following.

Case A - Copper deposition from acid aqueous solution containing copper ion (Kusakabe et al., 1981).

Using a D.C. power supply, the electroreduction of copper in an acidic aqueous solution can be expressed as

$$Cu^{2+} + 2e^- \rightarrow Cu \text{ (Cathode)}$$

$$H_2O \rightarrow 2H^+ + 1/2\ O_2 + 2e^- \text{ (Anode)}$$

Figure 11.3 shows the experimental apparatus used by Kusakabe et al. (1981) to carry out the above electrochemical reactions. The dimensions of the cathode chamber of the electrode are given in Table 11.1. As shown in Fig. 11.3, a Luggin capillary probe (connected to a saturated silver/silver chloride electrode) and a copper wire probe were used for the measurement of solution phase potential (ϕ_ℓ) and

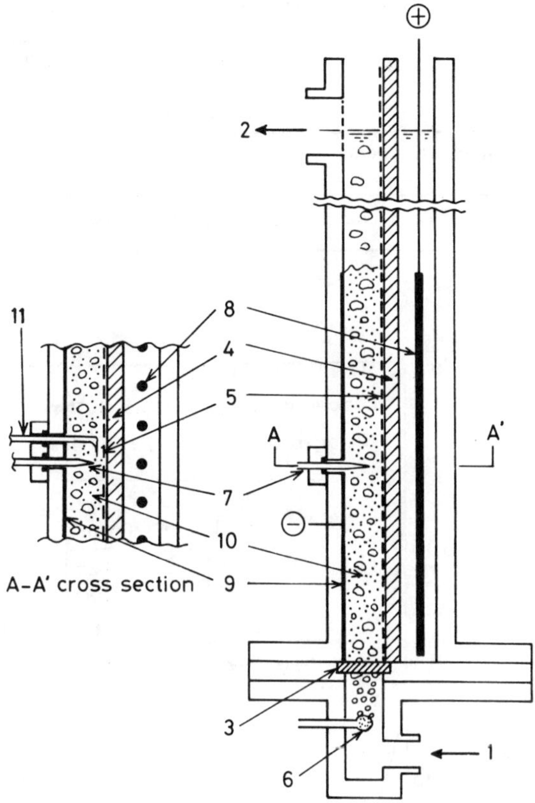

Figure 11.3 Schematic diagram of an experimental three-phase rectangular fluidized bed electrode used by Kusakabe et al. (1981).

1: Electrolyte inlet
2: Electrolyte outlet
3: Liquid distributor
4: Porous plate
5: Cation-exchange membrane
6: Gas distributor
7: Luggin probe (for measurement of solution phase potential)
8: Carbon counter electrode
9: Copper plate electrode
10: Gas-liquid-solid fluidized bed
11: Copper wire probe (for measurement of particle phase potential)

particle phase potential (ϕ_p), respectively. Figure 11.4 shows the variation in potential difference between the particle and solution phases (ϕ_p - ϕ_ℓ) with the dimensionless distance x/L, where L is the distance between the cathode and membrane, and x is the distance from the flat cathode in the cathode chamber of the three-phase

Table 11.1 Dimensions of the Cathode Chamber of the Three-Phase Fluidized Bed Electrodes Used by Kusakabe et al. (1981).

Dimensions	65 cm (H) (above distributor) × 15 cm (L) × 2cm (W) (in direction of current flow)
Area of feeder	90, 150, 300 cm^2
Fluidized bed height	6, 10, 20 cm

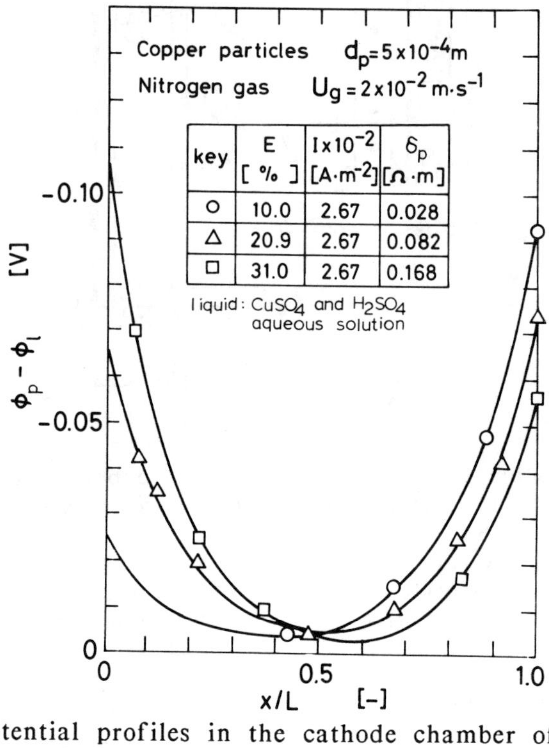

Figure 11.4 Potential profiles in the cathode chamber of the three-phase fluidized bed electrode (from Kusakabe et al., 1981).

fluidized bed electrode. Define a fluidized bed expansion ratio (E), as (expended bed height)/(static bed height) - 1. It is seen in the figure that for an E of 31%, the most electrochemically active region is near the feeder, whereas for an E of 10% it is near the membrane. The total current density (I) is the same for values of E of 10%, 20.9% and 31%. The effective specific charge transfer resistance in the particle phase (δ_p), which is a result of the specific resistances due to the charge conductive and the particle collision mechanisms, increases with an increase in E. Note that the copper reduction rate increases with an increase in (ϕ_p - ϕ_ℓ).

Case B - Electroreduction of oxygen to hydrogen peroxide on graphite particles (Oloman and Watkinson, 1975).

Electroreduction of oxygen in alkaline solution involves the following reactions (Latimer, 1952)

$$O_2 + H_2O + 2e^- \rightleftharpoons OH^- + HO_2^- \quad , \quad E_o = 0.076V$$

$$HO_2^- + H_2O + 2e^- \rightleftharpoons 3OH^- \quad , \quad E_o = -0.88V$$

$$H_2O_2 \rightleftharpoons H^+ + HO_2^- \quad , \quad K = 2.4 \times 10^{-12} \text{ M at } 25°C$$

From electrode kinetics for reduction of oxygen on carbon, it was found that production of peroxide is favored in an alkaline solution (Yeager et al., 1964). In these reactions, oxygen can be introduced into the electrolyte either by presaturation of the electrolyte with oxygen and admitting into the cell or by direct sparging of air into the cell with graphite particles as cathodes in either cell. Thus, the cell for the former is a liquid-solid (two-phase) fluidized bed electrode and the latter is a gas-liquid-solid (three-phase) electrode.

Table 11.2 gives the dimensions of the fluidized bed electrode used by Oloman and Watkinson (1975). They found that in a two-phase fluidized bed electrode, the maximum H_2O_2 production rate increases with bed expansion up to E of 5% and then decreases with a further increase in bed expansion. The decrease in product formation with increasing bed expansion is common for fluidized bed electrodes (Backhurst et al., 1969; Janssen, 1971; Flett, 1971). It is likely that increasing bed expansion creates an unfavorable potential distribution in the bed for product formation. Particularly, at high bed expansion, the potential will become high in the cathode electrode as noted earlier (see Fig. 11.4). The resulting high current density would lead to low peroxide product yield due to an increase in the extent of the undesirable secondary reaction (Oloman and Watkinson, 1975). In both two-phase and three-phase fluidized bed electrodes, the current efficiency for peroxide formation, defined as the ratio of the rate of peroxide produced by the cell to the electrochemical equivalent of current in the first reaction, decreases with an increase in current

Table 11.2 Dimensions of the three-phase Fluidized Bed Electrodes Used by Oloman and Watkinson (1975).

Cathode chamber	42 cm (H) (above distributor) x 5 cm (L) x 1 cm (W) (in direction of current flow)
Total area of cathode feeder plate	20 cm^2
Active area of diaphragm with full bed	210 cm^2
Catholyte distributor	30, 0.04 cm holes in a 0.63 cm thick block of Plexiglas
Oxygen distributor	12 needle point (~0.01 cm) holes in an 80 cm long, 0.16 cm O.D. teflon tube
Anode chamber	45 cm (H) x 5 cm (L) x 1 cm (W) (in direction of current flow)
Membrane support	100 mesh nylon backed by 0.31 cm thick Plexiglas with 1.3 cm holes drilled on about 1.9 cm centres

density. The product peroxide concentration increases and then decreases with increasing current density.

Figure 11.5 shows the effect of oxygen flow rate on current efficiency and product H_2O_2 concentration for a three-phase fluidized bed electrode. An increase in the oxygen flow rate increases both product H_2O_2 concentration and current efficiency for a given current density. Note that the introduction of oxygen into the liquid-solid fluidized bed cell at a high electrolyte flow rate produces a large bed contraction and high oxygen concentration in the catholyte. Consequently, the superficial current density increases yielding a considerably higher rate of peroxide production than that from a two-phase fluidized bed electrode under comparable operating conditions.

Figure 11.6 shows the effect of particle sizes between 0.42 mm and 0.59 mm and 0.71 mm and 0.99 mm on the maximum H_2O_2 production rate as a function of oxygen flow rate over a moderate range of catholyte flow rates or bed expansion ratios. As shown, the two particle sizes have a similar impact on the H_2O_2 production rate due to two counteracting effects: an increase in particle size increases the gas-liquid mass transfer, but decreases the surface area for liquid-solid mass transfer. The maximum H_2O_2 production rate generally

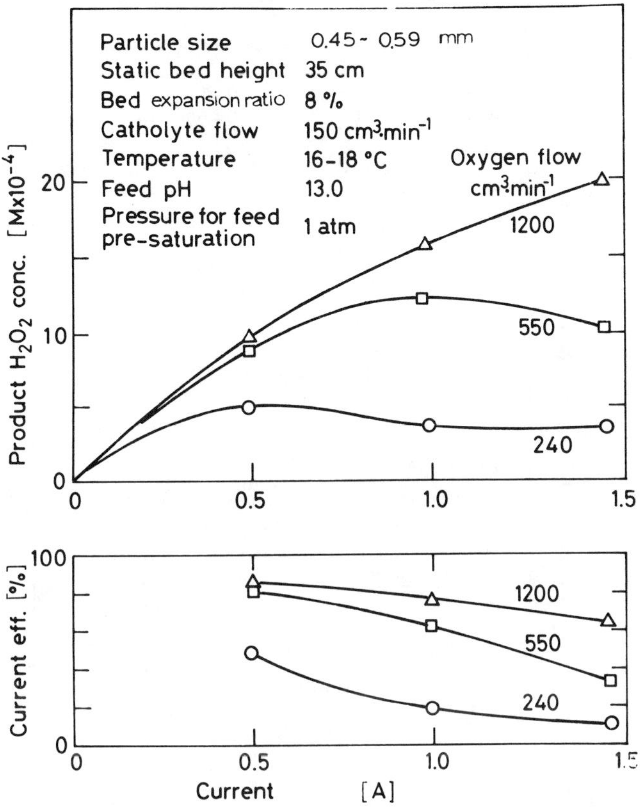

Figure 11.5 Effect of oxygen flow on peroxide concentration and current efficiency in the three-phase fluidized bed cathode (from Oloman and Watkinson, 1975).

increases with increasing catholyte flow rate for a given oxygen flow rate.

Three-phase fluidized beds with the gas phase consisting solely of products have also been considered for use in photoelectrolysis or photocatalysis in which radiant energy is converted to chemical energy (Augugliaro et al., 1982). The particles used are normally small in size due to large surface area requirements for photon and/or electron transfer. The energy conversion for photoelectrolysis is based on the intramolecular boost of electrons upon absorbing visible or UV lights to high energy levels and intermolecular transfer of electrons from these excited molecules. Separation of the photogenerated electron-hole pairs before recombination is essential for efficient utilization of radiant energy. This separation, which requires the light energy to be

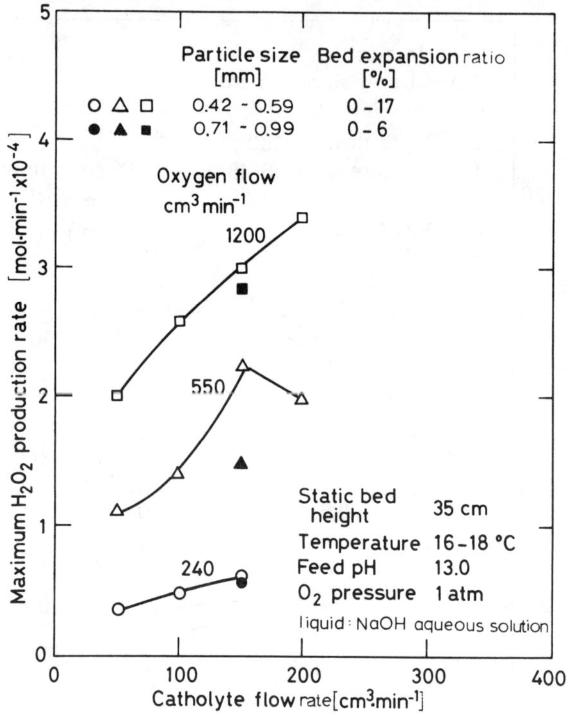

Figure 11.6 Effect of catholyte and oxygen flows on peroxide production in the three-phase fluidized bed cathode (from Oloman and Watkinson, 1975).

greater than the band gap energy, has been found to be efficient at the interface between a semiconductor and an electrolyte solution (Fujishima and Honda, 1972; Bard, 1979). An example of such an application is the photosynthesis of ethane and hydrogen gases in a particulate system consisting of particles of a single crystal TiO_2 as an anode with Pt which is deposited on it as a cathode. The electrolyte is an aqueous solution of CH_3COOH. The overall reactions are:

$$2H^+ + 2e^- \rightarrow H_2 \text{ (cathode)}$$

$$2CH_3COO^- \rightarrow C_2H_6 + 2CO_2 + 2e^- \text{ (anode)}$$

Considerable research is indeed needed to warrant commercial application of three-phase fluidized bed electrodes.

11.2 GRANULATION AND ADIPONITRILE PRODUCTION

Three-phase fluidization systems under Mode E-I-b or E-II-b, with gas as the continuous phase and solids in either batch or a continuous flow have been applied to granulation of powdery material into pellets in the pharmaceutical industry (Schaefer and Worts, 1978; Waldie et al., 1987) and in the fertilizer industry (Kono, 1981). In granulation, granules (0.5 ~ 3 mm in diameter) and powder (or seed, 10 ~ 100 μm in diameter) as feedstocks constitute the solid phase whereas water or a binder solution and hot or cool air constitute the liquid and gas phases, respectively. In the following, an example of a granulation application based on Mode E-I-b is illustrated. Extension of Mode E-I-b operation to other applications such as adiponitrile production is briefly described.

11.2.1 Granulation

Figure 11.7 shows a design of a reactor, or more specifically, a spouted bed reactor for powder granulation (Kono, 1980). The figure shows a double tube injector used for liquid feed injection. The liquid feed is introduced in the inner tube with gas flow in the outer tube for liquid atomization. The liquid feed injector is concentrically situated in a gas jet nozzle which generates a gas spout in the bed. The remaining gas is introduced into the bed through a conical perforated or porous plate distributor. This gas injector-distributor arrangement yields a circulation flow pattern of solid particles in the bed.

In the reactor, the powdery material grows in size according to two postulated mechanisms which take place simultaneously: a layering mechanism and a collision mechanism (Kono and Huang, 1985). In the layering mechanism, dry powder directly adheres to the wet surface of granules. The adhered granule then becomes wet again for further adherence of dry powder. In the collision mechanism, (Rumpf, 1958; Ouchiyama and Tanaka, 1975; Kapur et al., 1981) feed powder becomes wet first and then attaches to dry granules by collision. The granulation process is achieved via capillary forces of liquid and powdery material which are present on the surface of granules. The capillary effects are noted to strongly depend on the packing structure in the gas-liquid-solid system. The same processing application can also be extended to granule coating operations.

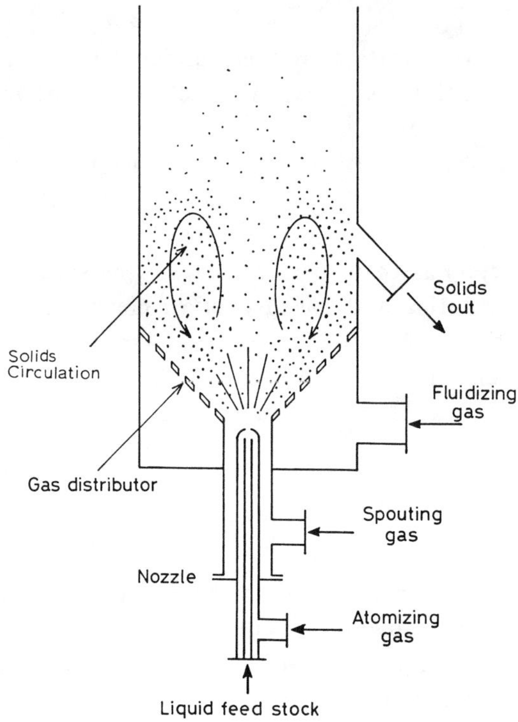

Figure 11.7 Schematic diagram of a three-phase spouted bed for granulation (from Kono, 1980).

The liquid phase required for granulation can also be produced through sintering reactions of powdery material at high temperatures. An example of such a granulation process is the sintering of mullite mineral, i.e., Al_2O_3- SiO_2-MgO, at 1350 through 1500°C. The quantity of molten mineral which constitutes the liquid phase is determined by phase equilibrium dictated by the temperature and chemical compositions of the mineral. Figure 11.8 shows the equilibrium amount of molten material in an Al_2O_3-SiO_2-MgO mixture in the bed as a function of MgO composition at various temperatures. It is seen that the amount of molten material increases sharply with an increase in MgO composition. The operable range of the spouted bed for granulation is approximately when MgO composition is from 3 to 4%.

For simultaneous drying and granulation of suspension melts or solutions in a three-phase fluidized bed, a simplified mechanistic model was proposed by Morl and Kunne (1986) to describe the initial

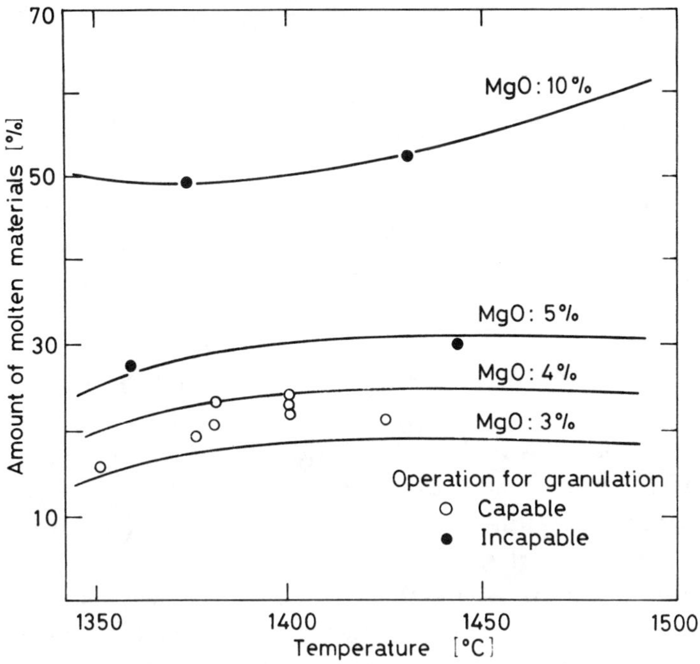

Figure 11.8 Percent of equilibrium amount of molten material versus temperature for Al_2O_3-Si_2O_3-MgO mixtures (from Kono, 1980).

unsteady granulation and drying processes and transition to steady-state operation. The model explained the experimental data well.

11.2.2 Adiponitrile Production

Three-phase spouted bed reactors under Mode E-I-b can also be applied to the production of adiponitrile, where adipic acid as feed stock in the form of a hot aqueous solution is fed into a bed of B_2O_3 catalyst fluidized by ammonia gas (Kono, 1980). A complete vaporization of adipic acid in the bed would reduce the product yield and hence, a three-phase system is preferred over a two-phase (gas-solid) system for the reaction. A yield of adiponitrile as high as 95% was reported when using a three-phase spouted bed reactor.

The spouted bed reactor shown in Fig. 11.7 can also be applied in cracking of heavy hydrocarbons and gasification of residual oil. It is noted that, in these applications, vaporization of the liquid feed is almost complete at the lower part of the bed, which leaves behind a two-phase region at the upper part of the bed.

11.3 CALCIUM BISULFITE ACID PRODUCTION

Calcium bisulfite acid solution used as a leaching agent for pulp production in the paper industry can be produced by absorbing sulfur dioxide in water in the presence of limestone. The overall chemical reaction is

$$CaCO_3 + 2SO_2 + H_2O \rightarrow Ca(HSO_3)_2 + CO_2$$

The use of gas-liquid-solid fluidized bed reactors provides an alternative over conventional processes such as the Jenssen tower (Stephenson, 1952) in which SO_2 is absorbed in water over a fixed bed of limestone particles. For calcium bisulfite acid production, the gas-liquid-solid fluidization system with cocurrent gas-liquid upward flow can be operated with either liquid or gas as the continuous phase. Limestone particles serve as fluidized particles with diameters ranging from 500 μm to 10 mm. In the following, the two fluidization modes represented by the gas and the liquid as a continuous phase for the production of calcium bisulfite acid are discussed (Volpicelli and Massimilla, 1965, 1970).

In the operation with liquid as a continuous phase (Mode E-I-a-1), the superficial gas velocity, U_g, is limited to 10 cm/s with a high (75%) SO_2 concentration in the gas phase. Achieving a desirable concentration of SO_2 in the liquid phase requires U_g/U_ℓ to be in the range of 100 ~ 150. That is, for U_g of 10 cm/s, a desired U_ℓ would be 0.07 ~ 0.1 cm/s, which yields a throughput capacity for the liquid phase of 0.067 ~ 0.1 $cm^3/cm^2 \cdot s$. This throughput capacity is substantially lower than that of practical significance.

In the operation with gas as a continuous phase (Mode E-I-b), on the other hand, high U_g, e.g., 150 cm/s is used with a low (1 ~ 2%) SO_2 concentration in the gas phase (Volpicelli and Massimilla, 1970). The gas phase holdup is about 0.3. Superficial liquid velocities of 0.09 ~ 0.6 cm/s are used, yielding a high liquid throughput rate of 0.083 ~ 0.56 $cm^3/cm^2 \cdot s$. Thus, to achieve high liquid throughput for the reaction, Mode E-I-b appears to be a more viable alternative than Mode E-I-a-1 for the production of $Ca(HSO_3)_2$.

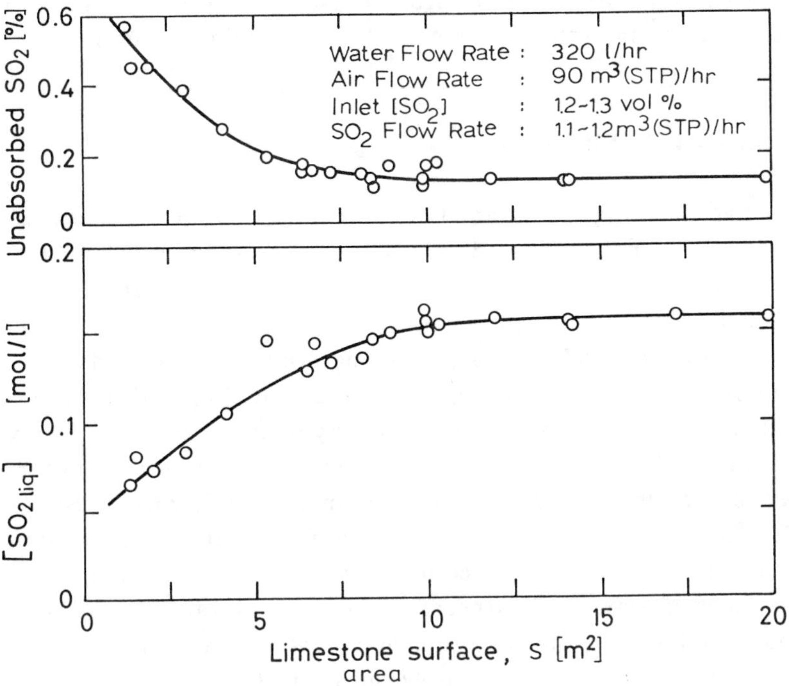

Figure 11.9 Effect of limestone surface area in the reactor on the calcium dissolution rate (from Volpicelli and Massimilla, 1970).

Figure 11.9 shows the variation in total SO_2 concentration in the liquid phase ($[SO_{2\ liq}]$) with the total surface area of limestone in the reactor (S). The reactor is of 14.3 cm ID and 4 m height and is operated under Mode E-I-b. Also note that $[SO_{2\ liq}]$ reflects the concentration of HSO_3^- ($[HSO_3^-]$). It is seen from the figure that $[SO_{2\ liq}]$ increases with an increase in S up to 10 m^2 beyond which $[SO_{2\ liq}]$ levels off. The figure also shows the corresponding decrease in the SO_2 concentration in the gas phase to a low level. It is noted that the behavior is not significantly affected by the particle size ranging from 1 to 6 mm. Furthermore, the SO_2 and $CaCO_3$ surface reaction controls the overall reaction rate for S less than 10 m^2; beyond this value thermodynamic equilibrium is reached for HSO_3^- formation. When limestone surface areas are exceedingly high in the reactor, however, the concentrations of Ca^{++} and $SO_3^=$ may significantly increase to form insoluble calcium sulfite which would lead to scaling problems in the reactor (also see Section 9.1). Thus optimal $CaCO_3$ surface areas are critical for application of a three-phase fluidization system in calcium bisulfite acid production.

11.4 SAND FILTER CLEANING

Backwashing of sand filters employs what is essentially a fluidized bed operation utilizing two or three-phases (Mode E-I-a-1). Batch gravity water filters are widely used in industrial and municipal water treatment plants (Brownell, 1951). These filters consist of tanks filled with a porous material, such as sand, supported by layers of progressively coarser gravel packing and broken stone placed on a perforated bottom. The hydrostatic head of the water above the bed forces the fluid through the sand, which filters out suspended solids, silt, bacteria, etc., mainly via adsorption (Halvorson, 1955). During continuous operation, a fine sediment and microbial flocs accumulate at the surface of the sand layer increasing resistance to the flow of water. This necessitates the removal of these solids by periodic backwashing to maintain or renew the capacity of the filter. Sand filters operate with a 3 to 4.6 m head of water and may require backwashing every 10 ~ 24 hours. During backwashing, filtered water is pumped back through the sand from the underdrains under sufficient pressure to bring about the fluidization of the sand layer. The sludge entrapped in the sand is removed by the agitation and separation of the sand achieved by fluidization. Air is sometimes blown into the bed, thus creating a three-phase fluidization system. Mechanical agitation may be introduced, if necessary, to generate additional turbulence to enhance the cleaning of the sand (Halvorson, 1955). The backwash water containing the removed sludge is run into gutters placed on the tank above the sand and drained into a waste line. After a few minutes of fluidization the backwash water comes through clean and the bed may be put back into service.

11.5 CRYSTALLIZATION

The process of crystallization may be defined as the precipitation of solid material from solution (Thompson, 1955). From a practical standpoint crystallization may be used to obtain pure crystals from a relatively impure solution, provided mixed crystals do not form. Design and operating considerations include desired

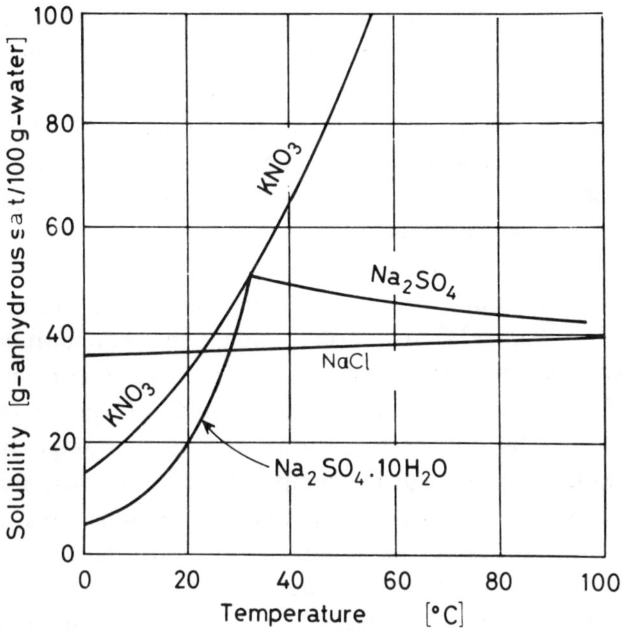

Figure 11.10 Solubility of some salts in water as a function of temperature (from Thompson, 1955).

crystal shape, size, uniformity, and state of agglomeration, as well as purity. Crystallizer designs may be batch or continuous, with internal or external heat exchange, and with mechanical or nonmechanical agitation. Fluidized bed crystallizers are well known, particularly two-phase solid-liquid beds. However, three-phase designs (Modes E-II-a-2 and E-III-a) such as the Robinson continuous evaporative crystallization process (described below) and similar three-phase crystallizers are also used.

In order for crystallization to occur, it is necessary that supersaturation of the solution takes place. This is normally achieved by either adjusting the temperature of the solution so as to decrease the solubility or removing solvent by evaporation. Supersaturation brought on by cooling a concentrated salt solution is the most common method of producing crystals in the case of materials that show a rapid increase in solubility with temperature (e.g., potassium nitrate; see Fig. 11.10). Supersaturation via evaporation may be used successfully for materials which undergo little change in solubility with temperature (e.g., sodium chloride). Supersaturation produced by adiabatic evaporation is a third method of crystallization employed for substances with an intermediate change in solubility with temperature (e.g., potassium chloride). Regardless of the method employed, heat must be either supplied to or removed from the crystallizing solution. This heat effect may involve the sensible heat of actual changes in temperature of the solution, the latent heat of evaporation of the solvent, and the heat of crystallization itself. The Robinson continuous evaporative crystallization process, in which an inert gas, such as air, is blown through a system of aqueous crystallizing solution, combines evaporation and heat exchange between gas and liquid in a unique crystallizer operation (Chemical Engineering, 1953). Its application to ammonium sulphate crystallization, as an example system, is given below (Robinson, 1952).

Feed to the crystallizer is a saturated solution between 1 and 15°C below the "boiling point saturation limit", that is, the point of maximum solubility and boiling temperature of the solution system at atmospheric pressure. Feed flow enters the apparatus (see Figure 11.11) at a point above the gas inlet, and is controlled by a liquid-level controller on the crystallizer. The gas stream is blown into the crystallizer at ambient temperature near the bottom of the conical section and passes through the solution, creating evaporation and cooling of the salt solution. The cooling and evaporation bring about supersaturation of the solution and subsequent crystal formation. The air flow rate is adjusted to maintain the crystallizer temperature at the desired level. This optimum temperature is based on the amount of humidity removed by the gas stream, and thus the degree of supersaturation achieved relative to the cost of blowing the gas. Lower solution temperatures result in increased blower costs, higher temperatures in lower crystal yield.

A second action of the gas stream is the creation of turbulence and a net upward force resulting in the fluidization of the crystals. Fluidization brings about an automatic separation of the crystals based on size: smaller particles being entrained upward where they can continue to grow in the solution, larger crystals settling to the bottom of the crystallizer where they are removed via a diaphragm pump. The gas flow rate also controls the size of crystals settling to the bottom of the crystallizer. High gas flow rates will allow only crystals which have achieved a large size to settle out, while lower gas rates allow the collection of relatively smaller crystals. In either case, a uniform size of crystal may be collected, thus reducing the need for further classification equipment downstream. Undesirable aggregation of the growing crystals is also eliminated by the agitation and

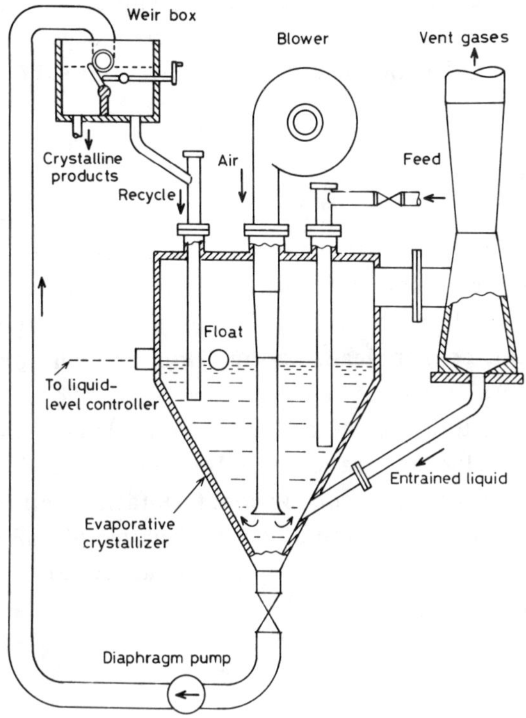

Figure 11.11 Robinson continuous evaporative crystallizer (from Chemical Engineering, 1953).

separation of the fluidized crystals. Also, since introduction of the gas stream does not involve contacting heat exchange equipment or mixing impellers with the crystallizing solution, there is no reduced efficiency of cooling or agitation due to crystal buildup on equipment surfaces.

The crystals contained in mother liquor removed from the bottom of the crystallizer are sent to a weir box which splits the stream into product and recycle streams. The recycle stream enters the column below the liquid level, but above the gas inlet. The ratio of recycle to product is in the range from 1:1 to 5:1, adjusted to control the rate of crystallization; excessive crystallization may occur unless a high liquid velocity is maintained. The product crystals are separated from the mother liquor of the product stream in external crystal recovery equipment.

Recent Japanese patents (Okochi et al., 1978; Kagawa, 1982) involving a novel three-phase fluidized bed evaporative crystallizer design indicate a continued interest in the subject. Much of the new design is similar to the Robinson evaporative crystallizer, with the exception that the gas is preheated before entry into the crystallizer.

11.6 FLOTATION

Flotation of minerals has long been used as a method of recovering valuable minerals from a mining slurry. This gas-liquid-solid process is used in the separation of such valuable minerals as molybdenum, coal, copper, lead, kaolin, zinc, and phosphate, from the undesirable mineral residues, known as gangue. Several reviews are available in the literature describing conventional fine particle or froth flotation, including Trahar and Warren (1976), Chander (1978), and Fuerstenau (1980). The conventional flotation cell is a tank with mechanical impellers for mixing and maintenance of small bubble size, but vertical columns are coming into use as an alternative process (Parkinson, 1986).

11.6.1 General Principles

Air flotation phenomena are complex and involve physicochemical-mechanical interaction between the gas bubbles and the liquid and solid phases. For small mineral particle flotation, the interaction between bubbles and particles is usually established by the surface forces, as compared to ordinary particle fluidization, where inertial and drag forces dominate. Once the particle adheres to a bubble, the particle is carried to the top of the separation vessel. The mineral particle size is usually less than 200 μm.

In the flotation process, the wettability of mineral surfaces is selectively controlled by addition of surfactants, classified as collectors (promoters), modifiers and frothing agents (Aplan, 1985). The interfacial forces between the phases are modified to develop an affinity between the bubbles and the mineral particles, and reduce the affinity of the mineral particles for the liquid phase.

The surfactant molecules, having an amphipatic character, orient themselves so that the polar end of the molecule adsorbs onto the polar surface of the mineral while the other, hydrophobic, end makes the mineral surface more non-polar increasing the affinity between the desired mineral and the gas phase. The adsorption of the surfactant onto the mineral surface increases the interfacial tension between the mineral (S) and the water solution, γ_{SL}, which results in a decrease in the spreading coefficient, $S_{L/S}$, a measure of wettability, where (Somasundaran, 1973)

$$S_{L/S} = \gamma_{SA} - \gamma_{LA} - \gamma_{SL} \tag{11.1}$$

Subscripts L, S, and A denote the liquid, solid and air phases, and γ_{ij} is the interfacial tension between the phases i and j. The hydrophobic coating "acts as a bridge" (Aplan, 1985) between the solid and the air, keeping the desired solid "out of the water".

The surface electrical effects can be important, and affect the selectivity in flotation (Fuerstenau, 1980). An electrical double layer is formed at the solid surface because of the presence of ions in an impure aqueous solution. The solid matter builds up an electrical charge, and counter ions in the solution are drawn to the surface of the solid. The effective thickness, often called the Debye length, of this layer is given by the expression (Rosen, 1978):

$$\text{Debye length} = \frac{(\delta_r \, \delta_o RT)^{1/2}}{4\pi F^2 \sum_i C_i \, Z_i^2} \tag{11.2}$$

where δ_r (= δ/δ_o) is the relative static permittivity or dielectric constant of the solution, δ is the static permittivity of the solution, δ_o is the permittivity of a vacuum, R is the ideal gas constant, T is the absolute temperature, F is the Faraday constant, C_i is the molar concentration of ion i in solution phase, and Z_i is the valence of ion i. The ionic surfactants can form this double layer, or replace existing ions if one is already present (Rosen, 1978). If the double layer is thick, the particles will remain dispersed, whereas if it is thin or absent, the attractive forces will dominate and the particles will coagulate (Fuerstenau, 1980). Heteropolar surfactants can also be used to change the polarity of a mineral in solution (Aplan, 1985).

The value of contact angle θ_c (see Section 2.1.6 and Fig. 2.13) can be evaluated by the Young equation (Young, 1805):

$$\cos \theta_c = \frac{(\gamma_{SA} - \gamma_{SL})}{\gamma_{LA}} \tag{11.3}$$

It is desired to have the contact angle as large as possible, so that the solid surface is wetted as little as possible and will adhere to the air bubble. The contact angle will also be increased by decreasing γ_{LA}.

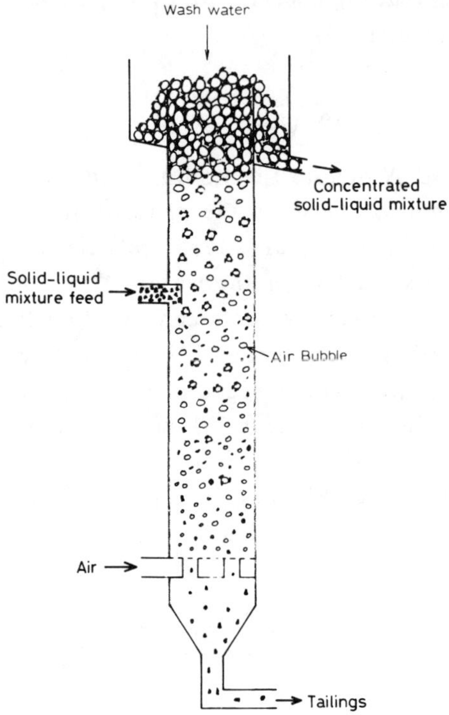

Figure 11.12 Schematic representation of a flotation column (after Fuerstenau, 1980).

Addition of a frothing agent, a second surfactant which acts at the liquid-air interface, reduces γ_{LA} and also creates a "locking in" effect, where the solid and bubble surface films merge, helping to keep the bubble-solid system stable (Adamson, 1982). The particle is carried to the top of the separation column when the net buoyant force on the particle is greater than the gravitational forces.

11.6.2 Column Flotation

Figure 11.12 shows a countercurrent gas-liquid-solid flotation column. Bubbles flow upward while the solid (gangue)-liquid mixture flows down. A dense bed is formed at the bottom of the column. A large freeboard region above the dense bed permits the desired minerals to be separated from the gangue. A wash water flow is usually added at the top of the column, creating a cleaning zone in the freeboard region. The undesired particles, the gangue, are continuously discharged from the bottom of the bed.

The bulk part of the column can be described as operating under three-phase fluidization of Mode T-II-a. In this case the mode designation is based on the flows of the bulk of the liquid and the undesired gangue counter to the gas phase. There is some flow of liquids and desired minerals, cocurrent with the gas phase, as the desired minerals are floated to the top of the column.

The three-phase flotation column has the following advantages over its conventional counterpart (Foot et al.,1986; Parkinson, 1986):

1) It is less turbulent, hence disengagement of the desired mineral is less likely;
2) It requires fewer units to perform the same separation, and hence less floor space;
3) It has a higher energy efficiency and requires less maintenance, since there is no mechanical agitation;
4) It allows for the use of wash water from overhead, hence a cleaning zone is created within the column. Less of the raw slurry feed water is floated off; and
5) It is usually more selective over less time.

The three-phase flotation column has the following disadvantages compared with its conventional counterpart:

1) Tighter control of bubble size and gas and liquid flow rates is required;
2) Being more selective, it is not as good for the initial, "roughing" stages, where the material is less finely ground;
3) It requires initial generation of small bubbles which usually requires an increase in frother dosage; and
4) It is subject to air sparger plugging by the fines.

There is some industrial use of three-phase flotation columns. Industrial flotation columns are generally 5 to 13 m in height, and range from 0.3 to 2.5 m in diameter (Dobby and Finch, 1985; Foot et al. 1986; Parkinson, 1986; Deister Co., 1987). The Chinese adopted columns in the late 1960's, but most sites stopped using them due to control and operating difficulties (Parkinson, 1986). The Deister Concentrator Company (Ft. Wayne, Indiana) installed its first commercial flotation column, 19.1 m^3, in 1979, in a Florida phosphoric acid plant (Zipperian and Christophersen, 1985; Parkinson, 1986), and have installed over 40 units worldwide since then (Deister Co. 1987). In comparisons between a 0.2 m diameter pilot scale column and conventional flotation cells, recovery of copper, coal, and phosphate were all increased from 10 ~ 30% when using the column system (Zipperian and Christophersen, 1985). Foot et al. (1986) found that column flotation was faster and more efficient than a standard laboratory flotation cell in recovery of chromite ores. Column Flotation Co. of Canada (CFCC) also has several commercial installations, with its first installation in 1980-81. Three of its columns were used to replace thirteen conventional stages in a molybdenum recovery process at Noranda Inc.'s Mines Gaspe Div. (Quebec). Those columns were reported to yield a 70% recovery, compared to 40% with the conventional cells (Parkinson, 1986).

In column flotation, different techniques are used to control bubble size, an important flotation parameter. Large bubbles will carry undesired particles in their wake, thus reducing column efficiency. Small bubble size is maintained by mechanical impellers in a conventional cell, whereas bubbles are externally generated in column flotation. Foot et al. (1986) investigated three methods for fine bubble generation, electrolytic generation (energy intensive), mechanical air induction, and pressure dissolution. They found that a combination of the latter two methods required less frothing agent. The Deister Company uses a venturi to aspirate the air into the column, producing 40 ~ 50 μm bubbles (Parkinson, 1986), thus reducing any plugging problem, whereas CFCC uses a perforated rubber compound as a sparger. Yianatos et al. (1986) studied gas holdup and bubble size in flotation column froths. They found that the wash water increased the froth depth and stability. Models were developed for gas holdup (ϵ_g) in both the packed bubble ($\epsilon_g > 0.74$) and expanded bubble ($0.2 < \epsilon_g < 0.74$) regions. The bubble size distribution was found to be very narrow in the flotation column, due to the surfactant effects. Bubbles were spherical in shape. This was also found in a conventional three-phase annular fluidized bed system using glass beads in a 0.5 wt% t-pentanol solution by Fan et al. (1987), even though no flotation effects were observed (see Section 6.3).

Flotation effects in a three-phase column can also occur unintentionally, when there is an affinity between the solid and gas phases. In three-phase fluidized bed operation using large benzoic acid particles (2 ~ 4mm ID), air, and water, Arters and Fan (1986) reported the attachment of small air bubbles to benzoic acid particles when a turbulent fluidized bed is suddenly made quiescent. The attachment of bubbles would alter the particle density and interfacial area. It is possible that at zero or low liquid flow rates the bubble attachment is large enough to lower the apparent density of the "pseudo-particle", leading to flotation phenomena. However, in an active fluidized bed, at high gas and liquid flow rates, the turbulent action serves to rapidly disengage attached bubbles. Thus, the effect on hydrodynamic properties in an active fluidized bed due to bubble attachment is not significant.

11.7 POLYMERIZATION

Three-phase fluidization systems have been successfully applied to polymerization reactions for the production of polyolefins. Such an application can best be exemplified by the production of polyethylene including both high density polyethylene (HDPE) and low density polyethylene (LDPE) as given below.

For the production of HDPE, there are two major processes employed in industry, namely, the three-phase (gas-liquid-solid) slurry process and the vapor phase (gas-solid) process. The three-phase slurry process is represented by the Solvay process using Ziegler-Natta catalysts (Modes T-I-a-1 and T-III-a). The process is conducted at 80 ~ 90°C and less than 11 atm. The vapor phase process is exemplified by the Unipol process using a gas-solid fluidized bed reactor developed by Union Carbide Corporation (Union Carbide Literature, 1986). The vapor phase HDPE reactor is operated at 85 ~ 105°C and less than 21 atm. The advantages of the vapor phase process over the three-phase slurry process are that no solvent recovery is necessary and control of polymer density is easy. The advantages of three-phase slurry process over the vapor phase process are that no ethylene recycle and monomer recovery are necessary. In general, these two processes present comparable operational features and are both widely used for the production of HDPE.

The conventional production of LDPE employs high pressure slurry processes such as the Gulf process (Hydrocarbon Processing, 1979). The process is operated at 100 ~ 200°C and 1360 ~ 2040 atm. Due to the severity in operating conditions, LDPE production has been gradually shifted to the vapor phase represented by the Unipol process using a gas-solid fluidized bed reactor. The vapor phase LDPE reactor is conducted at less than 100°C and 8 ~ 21 atm. Besides the appreciably reduced temperature and pressure for the reaction, the advantages of the vapor phase process over the high pressure process include significantly reduced capital investment and energy consumption, no risk of explosion, and ease in catalyst handling (Oil and Gas J., 1977).

The tremendous advantages of the vapor phase (gas-solid) process over conventional high pressure processes for LDPE production have been further enhanced recently by incorporating the unique properties of three-phase fluidization operation. Specifically, the increased ability of a gas-liquid-solid mixture to remove heat from the exothermic polymerization reaction via vaporization of the liquid over that of a gas-solid mixture results in increased space-time yields for the process (Jenkins et al., 1985).

In the two-phase Unipol process for LDPE production, gaseous ethylene and a catalyst are introduced continuously into the reactor. The gaseous mixture exiting from the top of the reactor is compressed, cooled, and mixed with fresh feed to supply monomer for consumption in the reactor and to remove the heat of polymerization within the bed. Product polyethylene is removed directly from the reactor as a granular solid and requires separation from only minimal amounts of gaseous monomer. This monomer is combined with the gas exiting from the top of the reactor and recycled back into the reactor. The highly active catalyst used plays a key role in the success of the process. Catalyst productivity is so high, and consequently catalyst residue concentrations in the polyethylene are so low, that removal of catalyst residue is not necessary. Likewise, no special steps for removal of low molecular weight oligomers are required. No diluent is used, eliminating the concerns associated with its presence. The granular product may be taken directly to market or storage.

While the use of the gas-solid fluidized bed reactor is a significant improvement over the conventional high pressure process for the production of LDPE, it too suffers from certain limitations. Polymerization reactions are, in general, highly exothermic reactions and removal of heat from the fluidized bed reactor is the limiting factor in increasing the rate of production. The most common method of heat removal is that used by the Unipol process, in which, as noted earlier, the recycle gas stream is compressed and cooled at a point external to the reactor. Cooling of the stream has traditionally been limited to lowering the temperature to a point slightly above the dew point of the recycle gas stream. This limit was imposed based on the belief that introduction of liquid into the gas phase system would result in a variety of problems, including plugging of recycle lines, heat exchangers, and distributor, nonuniformity of monomer concentrations within the fluidized bed, and, in the extreme case, collapse of the fluidized bed and sintering of the solid particles into a solid mass (Mangin et al., 1974; Caumartin and Havas, 1976; Fraser et al., 1979). This operational limit severely restricts the production rate for a reactor of a given size, however, since the amount of fluid required for support of the fluidized bed is greatly exceeded by the amount of fluid which must be circulated to remove the heat of polymerization. Fluid velocity within the reactor must be limited so that excessive entrainment of solids into the recycle line does not occur. This conflicts with the need for greater fluid flow rates to increase cooling.

Recently, however, it has been discovered (Jenkins et al., 1985, 1986) that the recycle stream may be cooled below the dew point and partially condensed, within certain constraints, without plugging or other problems associated with liquid entrained in the gas stream. By operating in the so-called "condensing mode" characterized by three-phase fluidization Mode E-I-b, the cooling capacity of the recycle stream is increased both via the vaporization of the condensed liquids entrained in the recycle stream and via the greater temperature differential between the entering recycle stream and the reactor. The volume of recycle stream required to meet reactor cooling needs is thus reduced. It is noted, however, that due to vaporization, the upper part of the reactor is still operated under the two-phase (gas-solid) condition. In condensing mode operation adequate cooling of the reactor is achieved for high production rates with neither oversizing the reactor to meet cooling demands nor operating with excessive reactor gas velocities and subsequent solid carryover. Examples within the patent literature claim two- to three-fold space-time yield increases of three-phase condensing mode fluidized reactor systems over conventional two-phase fluidized processes. One such example is summarized in Table 11.3 (Jenkins et al., 1986) in which a fluidized bed continuous reaction was operated with and without a partially condensed recycle stream to produce a hexene-1/propylene/ethylene copolymer. In this example a 4.6% liquid recycle stream increased the reactor space-time yield by 155%.

Figure 11.13 (Jenkins et al., 1986) is a schematic diagram of a continuous, fluidized bed polymerization process utilizing the condensing mode operation. The reactor itself is composed of a reaction region and a gas-solid disengaging region. The reaction region

Table 11.3 Comparison of Noncondensing (Two-Phase) and Condensing (Three-Phase) Modes of Operation in Unipol process: $C_2/C_3/C_6$ Copolymer Example (after Jenkins et al., 1986).

Recycle stream composition (mol%)	Two-phase	Three-phase
Nitrogen	36.5	34.7
Ethylene	38.2	38.1
Ethane	11.2	8.9
Propylene	5.0	5.1
Hexene-1	4.1	4.8
Hydrogen	4.8	7.3
Methane	0.0	0.1
Recycle stream dew point (°C)	53.2	57.8
Reactor inlet temperature (°C)	60.6	48.0
Liquid in recycle (wt%)	0.0	4.6
Reactor temperature (°C)	87.1	88.0
Reactor pressure (atm)	18.4	18.4
Superficial gas velocity (cm/s)	61.0	61.0
Bed height (cm)	1130	1130
Space-time-yield $(g/cm^3 \cdot s) \times 10^6$	8.9	22.7
Melt index of resin (dg/min)	1.0	1.0
Density of resin (g/cm^3)	0.918	0.918
Activator	triethyl	aluminum

contains the gas-continuous fluidized bed of solid catalyst, growing polymer granules, gaseous monomer, and vaporizing liquid recycle. A three-phase gas-liquid-solid fluidization condition exists in the lower portion of the reaction region, near the distributor. At some point above the distributor in the reaction region the recycled liquid is completely vaporized by the heat of polymerization and the fluidized bed operates at a two-phase gas-solid condition as noted earlier. The ratio of the length of the reaction region to the bed diameter ranges

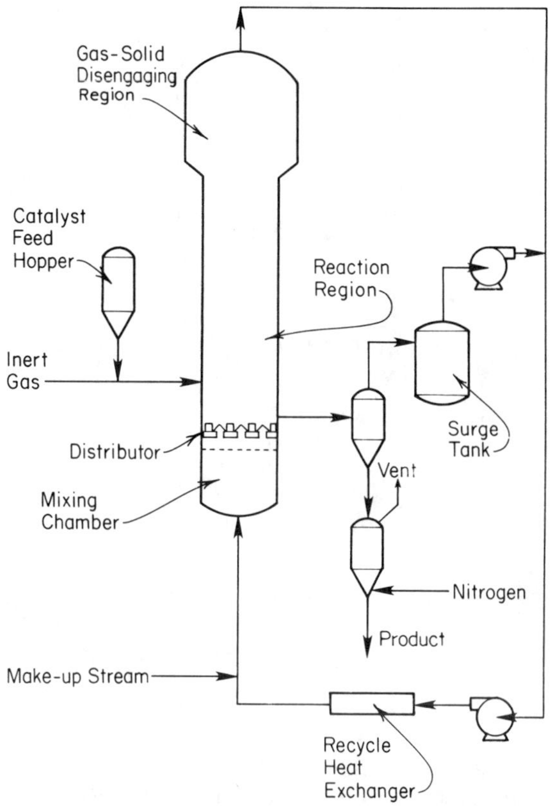

Figure 11.13 Condensing mode polymerization process (from Jenkins et al., 1986).

from about 2.7:1 to about 5:1 and the cross-sectional area of the disengaging region is about 2.5 to 2.9 times that of the reaction region. The superficial gas velocity within the reaction region is less than 1.5 m/s. Makeup catalyst is introduced into the reaction region at a point 20 to 30% of the reactor diameter away from the wall and above the distributor at a distance of about 5 to 30% of the total bed height. Particulate polymer product and some reactor fluid are withdrawn directly from the fluidized bed portion of the reactor and introduced into the product discharge tank, where the gas is separated, recompressed, and combined with the recycle stream. Product granules are subsequently discharged to the product surge tank and then to storage or distribution.

The recycle stream of unreacted monomer leaves the reactor from the top of the disengaging region and is compressed, cooled to

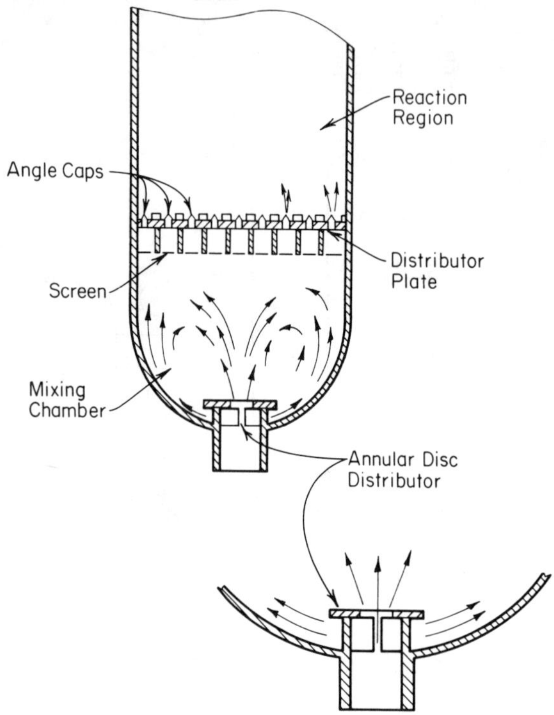

Figure 11.14 Reactor inlet and distributor of the condensing mode polymerization process (from Jenkins et al., 1986).

the desired level, and returned to the bottom of the reactor, where it is combined with the makeup feed at a point prior to the reactor inlet. Introduction of the stream into the fluidized bed is achieved via a diffuser mixing chamber and gas distributor plate, detailed in Fig. 11.14. An annular disc supported by spacers divides the entering recycle stream into a central upward flow, which aids in entraining the liquid phase, and an upward peripheral annular flow along the lower side walls of the reactor, which keeps solid particles from falling down the inner surface of the reactor walls into the bottom head. The flow streams mix and pass through a protective screen, the holes of the distributor plate and the angle caps, and finally into the fluidized bed.

Application of the three-phase condensing mode operation has been found to be most advantageous in polymerization of such monomers as homopolymers of ethylene, propylene, or butene, or copolymers of a major mole percent of ethylene, propylene, or butene and a minor mole percent of one or more C_2 to C_8 α-olefins. More specifically, the applicability of the invention to the production of any given polymer can be determined based on the operation benefit value, X, expressed by

$$X = \frac{F\ (-\Delta H_r)}{G_m\ C_{pm}\ (T_r - T_i)} \tag{11.4}$$

where F is the polymer production rate, constrained to rates giving X < 1 for noncondensing mode operation; ΔH_r, heat of polymerization for the specific polymer; G_m, mass flow rate of the recycle stream; C_{pm}, heat capacity of the recycle stream; T_r, temperature of the reaction region; and T_i, minimum temperature of the recycle stream entering the reaction region as limited by noncondensing operation (i.e., slightly above the dew point of the recycle, or the cooling limit of the heat exchanger, whichever is higher). If the value of X is larger than 1.0, the condensing mode operation will afford a benefit. The larger the value of X, the greater the benefit.

11.8 COAL GASIFICATION AND METHANATION

During the 1970's, research and development for coal gasification or synthesis gas technology such as methanation were accelerated to augment the dwindling supply of petroleum. A number of gasification processes were conceived and conducted in various stages of development from bench scale to demonstration scale during this period. Liquid medium gasification (Modes E-I-a-2 and T-I-a-2) is one unique process for low Btu gas production among these coal gasification processes. The M. W. Kellogg Company (Cover et al., 1973) and Rockwell International (Kohl et al., 1978; Yosim and Barclay, 1981) developed such gasifiers using molten salt as a reaction medium. Liquid medium methanation (Mode E-I-a-1) was also conceived for high Btu gas production. They are described below.

11.8.1 Coal Gasification

The concept of the molten salt gasification process is shown in Fig. 11.15. Coal is gasified at a temperature of about 980°C and at pressures up to about 30 atm by using either air or oxygen (with steam) in a melt of sodium carbonate. Coal is injected beneath the

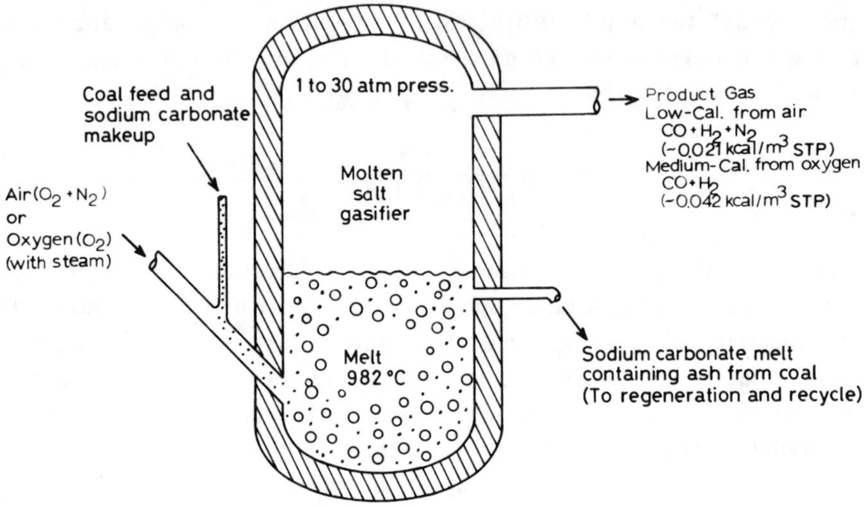

Figure 11.15 Concept of molten salt gasification (from Yosim and Barclay, 1981).

surface of the melt together with required sodium carbonate makeup. The ash in the coal is retained in the melt. A small stream of melt is continuously removed from the gasifier and replaced by the addition of sodium carbonate with the coal in order to prevent excessive buildup of ash. The sodium carbonate recycle rate is adjusted to maintain a steady-state ash concentration of 20% in the gasifier melt.

The molten salt gasification of coal offers several important advantages. First, the molten bath serves to disperse coal throughout the reactor, which permits direct gasification of all coals, including highly-caking coals. Second, the molten salt bath operates at a uniform temperature in the gasification range and, combined with the catalytic effect of sodium carbonate on steam-carbon gasification, yields a raw gas that is free of heavy hydrocarbons and tars. Third, the NO_x concentration in the product gas is low since the gasifier is operated at a relatively low temperature. Fourth, sulfur entering with the coal accumulates in the bath as sodium sulfide.

Note that the molten salt coal gasification reactor is not entirely operated in an expanded bed regime. Large coal particles upon introduction to the reactor would disintegrate into small particles during the course of the reactions which results in a portion of the bed operated in the transport regime.

11.8.2 Methanation

The methanation of gas streams containing carbon oxides and hydrogen is an extremely exothermic reaction. Synthesis gas derived from coal contains a high percentage of carbon oxides, and cannot be methanated with a high conversion in a single-pass adiabatic reactor. Methanation of high carbon oxide gas streams is thus usually accomplished by gas recycle and cold feed quench systems or packed multitubular reactors (Blum and Toman, 1977). These systems tend to be expensive, and there are potential safety problems involved because of rapid heat removal requirements in the reactor due to a large heat of reaction.

A process was developed for the methanation of high carbon oxide gas streams using a three-phase fluidized bed reactor (Blum and Toman, 1977). In this process, the synthesis gas is reacted on a NiO/SiO_2 type catalyst of 3.5 to 5.4 mm in equivalent diameter in the presence of an inert circulating gas oil. The oil serves to fluidize the catalyst bed and acts as a heat transfer medium, removing the large exothermic heat of reaction as both sensible and latent heat. The reaction is carried out at about 350°C and 68 atm pressure. This process was claimed to be more efficient and safer than conventional gas phase methanation of high carbon oxide gas streams.

11.9 FISCHER-TROPSCH SYNTHESIS

One of the most important, and perhaps the most and longest studied, application of three-phase fluidized systems is that concerning the hydrogenation of carbon monoxide by the Fischer-Tropsch (F-T) process in the liquid phase. Here, synthesis gas of relatively low hydrogen to carbon monoxide ratio (0.6 ~ 0.7) is bubbled through a slurry of precipitated catalyst suspended in a heavy oil medium in a slurry bubble column (most often Mode E-III-a). The F-T synthesis forms saturated and unsaturated hydrocarbon compounds ranging from methane to high-melting paraffin waxes (MW>20,000). Selectivity to desired products including light hydrocarbons, gasoline, or diesel fuel depends upon the catalyst employed, the reactor temperature, and type of process employed. By-products formed on a lesser scale include alcohols, ketones, acids, esters, and aromatics.

Transport properties and reaction engineering aspects of F-T synthesis have been the subject of comprehensive reviews in recent years, e.g., Frohning et al. (1977), Deckwer (1980), Kolbel and Ralek (1980), and Anderson (1984). Readers are referred to these references for details on the F-T synthesis and the extensive literature listings contained in them. In the following, an historical perspective, reactions of the F-T synthesis and the catalyst used, and application of the synthesis in slurry bubble columns are briefly described.

11.9.1 Historical Perspective

As discussed in Section 1.4.1, pioneering efforts in the area of the liquid phase reaction were performed by Kolbel and coworkers in Germany during the period 1938-44 and 1951-53 (Kolbel et al., 1951a,b, 1956a,b,c; Kolbel and Ackermann, 1956), culminating in the construction and operation of the semi-industrial scale plant by the companies Rheinpreussen and Koppers for the production of liquid fuels and chemical feedstocks. Other early investigators of pilot plant or larger scale included an American effort by the Pittsburgh Coke and Chemical Company and the U.S. Bureau of Mines at Bruceton (Schlesinger et al., 1951, 1954); British investigations at the Fuel Research Station at Greenwich (Hall and Taylor, 1955; Farley and Ray, 1964) and by Calderbank et al. (1964); and work at the Indian Institute of Technology at Kharagpur (Mitra and Roy, 1963).

Development on F-T synthesis effectively ceased in the 1960's with the availability of inexpensive crude oil supplies but revived in the mid 1970's with the advent of unstable and rapidly rising crude oil prices. Recent pilot and bench scale studies of liquid phase F-T synthesis include the work of Deckwer and coworkers (Zaidi et al., 1979; Deckwer et al., 1982; Sanders et al., 1986), Mobil's two-stage system (Haag, 1980; Kuo, 1983, 1985); and a somewhat older effort at the Department of Synthetic Chemistry of Tokyo University (Sakai and Kunugi, 1974).

11.9.2 Reactions and Catalyst

The highly exothermic F-T synthesis may be reduced to the hydrogenation of carbon monoxide:

$$CO + 2H_2 \rightarrow (\text{-}CH_2\text{-}) + H_2O \qquad \Delta H_r(227\,°C) = -165 \text{ kJ/mol}$$

and the consecutive conversion of water:

$$CO + H_2O \rightarrow H_2 + CO_2 \qquad \Delta H_r(227°C) = -39.8 \text{ kJ/mol}$$

The second reaction, which occurs preferentially on the iron catalysts most commonly employed in F-T synthesis, may be depressed by using lower reactor temperatures, shorter residence times, and recirculation of synthesis gas from which the reaction water has been removed (Kolbel and Ralek, 1980). The overall reaction of the desired hydrocarbon synthesis is thus

$$2CO + H_2 \rightarrow (-CH_2-) + CO_2 \qquad \Delta H_r(227°C) = -204.8 \text{ kJ/mol}$$

Products of the F-T synthesis are suitable for further chemical processing because of their predominantly straight chain structure and the position of the double bond at the end of the chain.

The Kolbel-Engelhardt (K-E) synthesis is a modification of the F-T synthesis, particularly suitable for conversion of industrial gases such as producer gas and blast-furnace gas, which forms hydrocarbons from carbon monoxide and water (Kolbel and Engelhardt, 1952; Kolbel, 1957; Frohning et al., 1977; Anderson, 1984) according to the overall reaction:

$$3CO + H_2O \rightarrow (-CH_2-) + 2\ CO_2 \qquad \Delta H_r(227°C) = -244.2 \text{ kJ/mol}$$

Development of catalysts of high conversion activity and proper selectivity remains the primary requirement for the successful development of the F-T synthesis. At the same time, the catalyst must display convenient preparation, activation, and regeneration procedures, and long term stability under the synthesis environment.

Almost all the catalysts developed to date are produced from iron precipitation. Varying proportions of minor constituents, such as copper, potassium, and cobalt are contained in classical, broad spectrum product catalysts (Kolbel et al., 1951a,b; Schlesinger et al., 1951; Farley and Ray, 1964; etc.). Modern efforts have concentrated on the formation of specific, short-chain hydrocarbons by utilizing catalysts containing oxides of titanium, vanadium, molybdenum, manganese, and cobalt (Kolbel and Ralek, 1980) and high alkali content (Kuo, 1983, 1985). Support-free catalysts have been found to yield superior slurry characteristics and provide the highest activity (Kolbel et al., 1951a,b). Catalyst concentration within the reactor has been optimized at around 10 wt%, below which reactor efficiency decreases and above which slurry viscosity increases, decreasing the interfacial gas-liquid area due to larger bubble formation. Regeneration of catalyst is required due to poisoning, recrystallization, and carbonization and may be accomplished through a simple oxidation.

11.9.3 Synthesis in Slurry Bubble Columns

The highly exothermic nature of the desired synthesis is the cause of two of the main technical problems associated with the F-T synthesis: that of removal of the heat of reaction, amounting to about 25% of the heat of combustion; and avoidance of local overheating of the catalyst, which favors methane formation and catalyst deactivation (Kolbel and Ralek, 1980). The use of the slurry bubble column reactor effectively eliminates these problems by virtue of its excellent heat transfer characteristics and temperature uniformity. Other advantages offered by the slurry bubble column over fixed beds or moving beds in the absence of liquid media include high catalyst efficiency; favorable catalyst regeneration conditions; flexibility in establishing operating conditions, allowing economical control of the synthesis; ease of addition or removal of catalyst; and simplicity of construction and therefore lower capital costs. One disadvantage of the slurry bubble column in this case is the degree of backmixing of the liquid and, to a lesser extent, the gas phase, limiting the conversion and specific product selectivity attainable. The backmixing problems, however, can be reduced by using baffles.

Figure 11.16 provides a simplified diagram of the reactor for F-T synthesis at the Rheinpreussen-Koppers demonstration plant which contains many features common to F-T process designs. F-T liquid-phase reactors require cooling, achieved in large reactors, such as in Fig. 11.16, by immersed heat exchangers and in smaller reactors by jacketing. Reactor diameters range from 1.55 m (Fig. 11.16) to 4 ~ 5 cm in bench scale units. Suspension height is controlled by adding or removing high-boiling synthesis fractions, depending on whether the process is being geared towards the production of low or high boiling products. Catalyst is added or removed with these fractions as well.

Flexibility of product composition can be controlled by changing reaction parameters such as temperature, pressure, CO/H_2 ratio in the synthesis gas, and space velocity. Slurry bubble columns are particularly adaptable to operation over a wide range of conditions. Fixed bed processes, for instance, are susceptible to reactor clogging under conditions of synthesis gas feed rich in CO, due to carbon formation by the Boudouard reaction (Calderbank et al.,

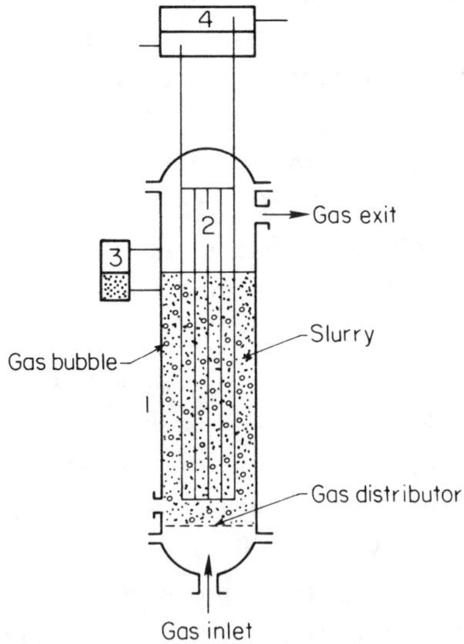

Figure 11.16 Reactor for liquid-phase synthesis (Rheinpreussen Process) (from Kolbel and Ralek, 1980).
1: Reactor cylinder. 2: Cooling pipe register.
3: Liquid level regulator. 4: Steam drum.

1964). Entrained bed processes, on the other hand, are limited to production of low-boiling products due to the high flow velocities required to fluidize or transport the bed; short residence times require higher reactor temperatures for reasonable conversion, and are limited by the undesirable reactions to methane and carbon and catalyst disintegration which occur above 300 °C.

Temperature may be used to control both synthesis rate and product composition (Kolbel and Ackermann, 1950; Schlesinger et al., 1951, 1954; Kuo, 1985; Sanders et al., 1986). As mentioned, elevated temperatures (274 ~ 282 °C) promote the formation of relatively low-boiling, olefin-rich hydrocarbon. The olefin content of the hydrocarbon product boiling below 200 °C can be as high as 80%, and n-α-olefins may make up as much as 85% of the total n-olefins. Low temperatures (240 ~ 260 °C), in combination with a high alkali content catalyst and a chain extending molecular buildup of the hydrocarbons which are suspended or recycled, promote the formation of higher molecular weight products (Sakai and Kunugi, 1974).

The effect of pressure is complicated by changes in reactant solubility, reaction rates, bubble characteristics, and transport phenomena. Schlesinger et al. (1951), Kuo (1983), and Sanders et al. (1986) found increases in conversion with pressure over the range 0.76 ~ 4 MPa, due mainly to increases in reactant solubility and promotion of the formation of larger molecules. The same authors also found a decrease in conversion with increasing gas velocity over the range 1 ~ 5 cm/s. Effects of the CO/H_2 feed ratio are also complex and interrelated with such parameters as the space velocity, catalyst employed, and reactor operating conditions. Schlesinger et al. (1954), Deckwer et al. (1982), Kuo (1983), and Sanders et al. (1986) have discussed the effect of CO/H_2 on both conversion and the nature of the synthesis products.

The successful application of liquid-phase F-T synthesis to commercial scale units in the future will require the development of catalysts of high activity and proper selectivity and improved understanding of the reaction mechanisms involved in order to compete with the still plentiful supplies of crude oil as a source of fuel and chemical feedstocks. Given the enormous potential and supply of the primary raw material (coal), it seems inevitable that this technology will one day become an important part of the chemical processing industry.

11.10 METHANOL SYNTHESIS

Synthesis gas can be reacted or reformed to make methanol through highly exothermic reactions:

$$CO + 2H_2 \rightleftharpoons CH_3OH\ (\ell) \qquad \Delta H_r(255\ ^\circ C) = -91.9\ kJ/mol$$

$$CO_2 + 3H_2 \rightleftharpoons CH_3OH\ (\ell) + H_2O \qquad \Delta H_r(255\ ^\circ C) = -51.2\ kJ/mol$$

The first reaction produces methanol with a low hydrogen consumption, but evolves significantly greater amounts of heat. The second reaction evolves less heat, but consumes more hydrogen and produces the by-product steam. Thermodynamically, low temperatures and high pressures favor methanol formation. The process is carried out catalytically to ensure high selectivity. Typically, copper-containing catalyst formulations are used and the reactor is operated near 260°C and 5 MPa (Probstein and Hicks, 1982). The overall reaction rate increases with an increase in hydrogen from stoichiometrically deficient to balanced conditions and with the presence of a small percentage of carbon dioxide (Sherwin and Frank, 1976).

Reactors used for methanol production must effectively contact the gaseous phase reactants with the solid phase catalysts and allow for removal of large amounts of heat due to the high heats of reactions. Several reactor types have been used for methanol formation. Conventional reactors involve contacting the reactant gases over a fixed bed of catalysts (Lurgi or ICI reactors). To provide the necessary heat removal either a series of adiabatic reactors separated by cold gas injection (quench) steps or high heat transfer area reactors such as multi-tubular reactors are used. But, with the introduction of an inert liquid phase to the system with catalyst particle suspension, more efficient heat transfer can be achieved because the liquid phase acts as a heat carrier. For methanol synthesis, Chem Systems Inc. (Sherwin and Frank, 1976) developed such a three-phase reactor system where an inert hydrocarbon serves as the liquid phase.

In the three-phase system, the inert liquid phase circulates between the reactor and an external heat exchanger providing the necessary heat removal. Operation can be either fluidization Mode E-I-a-1 or Mode T-I-a. The increased complexity inherent with a three-phase reactor system can be offset by several advantages. For example, in the three-phase system, the methanol concentration in the effluent stream can reach 15 ~ 20% by volume compared to 5 ~ 6% in a conventional process; eliminating the need for redistribution and quench along the reactor length simplifies reactor design; the inherent temperature uniformity increases catalyst life, insures optimum selectivity and reaction rate, and reduces NO_x formation; and catalyst can be added or withdrawn without system shutdown. Comparing the two types of processes, Sherwin and Frank (1976) predicted that, based on a capacity of 1800 tons per day of methanol produced, a three-phase fluidized bed system would require less total power (9.7×10^5 versus 4.9×10^6 watts) and would be more energy-efficient (97.9 versus 86.3%) than a conventional multi-fixed bed adiabatic quench system. The use of a slurry methanol process also offers potential for producing co-generation products for utility industry when a gasifier will be used as a source of feed for methanol process and to meet peak shaving requirements of the electric utility.

A process demonstration unit (PDU), with a capacity of 8 tons of methanol per day, was designed and constructed by Air Products (Studer et al., 1987) and a view of their three-phase reactor system is shown in Fig. 11.17. The PDU operates at temperatures between 225

Figure 11.17 The liquid phase methanol process development unit (PDU) owned by the U.S. Department of Energy and operated by Air Products and Chemicals, Inc. at the company's LaPorte, Texas facility (kindly supplied by B.J. Halper of Air Products and Chemicals, Inc.).

and 270°C and pressures between 3.5 and 6.3 MPa. Their technology can be applied for indirect coal liquefaction because it is particularly suited to synthesis gas rich in CO, such as that derived from coal. In Mode E-I-a-1 operation, catalyst particles between 3 and 6 mm in size are used and the superficial gas and liquid velocities are about 12 and 61 cm/s, respectively. In Mode T-I-a operation, a slurry with catalyst particles about 50 μm in size circulates with a velocity of 6 cm/s in the reactor and contains from 10 to 33 % solids by weight. In the PDU, the reactor also contains an internal heat exchanger; it occupies only 3.5% of the reactor cross-sectional area and is not expected to interfere with the reactor hydrodynamics. Further, the three-phase flow over the heat exchanger tubes will enhance the heat transfer (Studer et al., 1987).

Catalyst problems play an important role in reactor design. The catalysts are subject to deactivation by adsorption of trace poisons and consequently, can show a loss in activity over time. However, catalyst powders can be activated in-situ in the liquid medium with results equivalent to vapor phase reduction (Air Products DOE report, 1987). Mode T-I-a operation is presently the preferred route because catalyst attrition is not a problem as it is in Mode E-I-a-1. With the catalyst loading at 25% by weight, sustained reactor performance is achieved. However, when the loading is increased to 45%, a decrease in reactor productivity is observed due to limitations in mass transfer and mixing (Air Products DOE report, 1988). Slurry rheology is an important factor in the three-phase methanol synthesis process.

In the original PDU design, the reactor, primary vapor-liquid separator, slurry heat exchanger and slurry pump were separate units. However, the liquid circulation and turbulence induced by the feed gas is found to be sufficient to homogeneously suspend the catalyst particles within the slurry; the internal heat exchanger provides adequate heat removal without altering the hydrodynamics; and adequate vapor-liquid disengagement can be achieved by incorporating a reactor freeboard section. With these findings, a process simplification was proposed to perform the vapor-liquid separation and the heat exchange inside the reactor, thereby combining these units into a single unit (Studer et al., 1987). Research efforts in catalyst activation procedures, choice of liquid medium, reactor hydrodynamics, and reactor modeling are necessary for improving the process efficiency, design, and evaluations (Air Products DOE report, 1987).

11.11 CONCLUDING REMARKS

This chapter describes additional examples of applications using fluidization systems operated mostly at a bench or pilot scale. The success of the application of these systems strongly relies on the comprehensive understanding of the principles underlying the physicochemical-mechanical interaction among gas, liquid and solid phases of the system under which the application is intended. More specifically, extensive knowledge is needed in electrochemistry for electrodes application, in powder properties and capillary forces of liquid and powder for granulation and adiponitrile production, in kinetics and equilibrium of calcium carbonate and sulfur dioxide reactions for calcium bisulfite acid production, in physical properties of sand filter for sand filter cleaning, in solubility and boiling point of solutions for crystallization, in interfacial phenomena for flotation, in polymer properties and reaction kinetics for polymerization application, in catalytic reaction kinetics for liquid phase reactions between carbon and steam and carbon monoxide and hydrogen for coal gasification and methanation, respectively, and in reaction kinetics, catalysis, and equilibrium of carbon monoxide and hydrogen reactions for Fischer-Tropsch and methanol syntheses.

In some cases, the complex nature of the process system requires a three-phase fluidization system operated under more than one unique flow condition. For instance, in coal gasification and methanation, the reactors are operated under the expanded bed regime at the lower part of the reactor and transport regime at the upper part of the reactor. In granulation and adiponitrile production and polymerization, the reactors are operated with three-phase (gas-liquid-solid) fluidization at the lower part of the reactor and two-phase (gas-solid) fluidization at the upper part of the reactor. These types of complex multiphase reactor operation have gained growing importance in three-phase fluidization applications.

11.12 NOMENCLATURE

Electrodes

d_p : particle diameter

E : fluidized bed expansion ratio

E_o : standard electrode potential

I : current density

K : chemical equilibrium constant

L : distance between flat electrode and membrane

U_g : superficial gas velocity

x : distance from flat electrode

Greek Letters

δ_p : effective specific charge transfer resistance of particle phase

ϕ_ℓ : potential of solution phase

ϕ_p : potential of particle phase

Calcium Bisulfate Acid Production

S : limestone surface area

U_g : superficial gas velocity

U_ℓ : superficial liquid velocity

Flotation

C_i : molar concentration of ion i in solution phase

F : Faraday's constant

R : the gas constant

$S_{L/S}$: spreading coefficient between solid and liquid phases

T : absolute temperature

Z_i : valence of ion i

Greek Letters

γ_{LA} : interfacial tension between liquid and air phases

γ_{SA} : interfacial tension between solid and air phases

γ_{SL} : interfacial tension between solid and liquid phases

δ : static permittivity of the solution

δ_o : permittivity of a vacuum

δ_r : relative static permittivity or dielectric constant, δ/δ_o

θ_c : contact angle between the solid and liquid phases

ϵ_g : gas holdup

Polymerization

C_{pm} : heat capacity of recycle stream

F : polymer production rate

G_m : mass flow rate of recycle stream

ΔH_r : heat of polymerization

T_i : minimum temperature of recycle stream entering reaction region

T_r : temperature of reaction region

X : condensing mode operation benefit value

Fischer-Tropsch Synthesis

ΔH_r : heat of reaction

11.13 REFERENCES

Electrodes

Adams, R.N., "Electrochemistry at Solid Electrodes," Marcel Dekker (1969).

Augugliaro, V., F.D'Alba, L. Rizzuti, M. Schiavello, and A. Sclafani, Int. J. Hydrogen Energy, 7(11), 851 (1982).

Backhurst, J.R., J.M. Coulson, F. Goodridge, and R.E. Plimley, J. Electrochem. Soc., 116, 1600 (1969).

Bard, A.J., J. Photochemistry, 10, 59 (1979).

Bard, A.J., and L.R. Faulkner, "Electrochemical Methods - Fundamentals and Applications," John Wiley, New York (1980).

Bareau, J.Y., and F. Coeuret, J. Appl. Electrochem., 9(6), 737 (1979).

Berent, T., R. Mason, and I. Fells, J. Appl. Chem. Biotech., 21, 71 (1971).

Bockris, J. O'M, and S. Srinivasan, "Fuel Cells: Their Electrochemistry," McGraw-Hill (1969).

Coeuret, F., J. Appl. Electrochem. 10(6), 687 (1980).

Coeuret, F., and A. Storck, "Elements de Genie Electrochimique," Chap. 7, Technique et Documentation, Paris (1984).

Flett, D.S., Chemistry and Industry (London), No. 11, p. 300, March 13 (1971).

Fry, A.J., "Synthetic Organic Electrochemistry," Harper and Row, New York (1972).

Fujishima, A., and K. Honda, Nature, 238, 37 (1972).

Goodridge, F., and A.R. Wright, in "Comprehensive Treatise of Electrochemistry," ed. by O'M Brockris and E.B. Yeager, Chap. 6, Plenum Press (1980).

Hiddleston, J.N, and A.F. Douglas, Nature, 218, 601 (1968).

Janssen, L.J.J., Electrochim. Acta, 16, 151 (1971).

Kelsall, G.H., Chem. Eng., 90, 63, Feb. 21 (1983).

Kreysa, G., Electrochim. Acta, 23, 1351 (1978).

Kusakabe, K., S. Morooka, and Y. Kato, J. Chem. Eng. Japan, 14(3), 208 (1981).

Latimer, W.M., "The Oxidation States of the Elements and Their Potentials in Aqueous Solutions, " 2nd ed., Prentice-Hall, New York (1952).

Masterson, I.F., and J.W. Evans, Metallurgical Trans. B, 13B, 3, March (1982).

Morooka, S., K. Kusakabe, T. Watari, and Y. Kato, Kagaku Kogaku Ronbunshu, 6(1), 40 (1980).

Oloman, C., and A.P. Watkinson, Can. J. Chem. Eng., 53, 268 (1975).

Sabacky, B.J., and J.W. Evans, J. Electrochem. Soc., 126, 1180 (1979).

Van der Heiden, G., C.M.S. Raats, and H.F. Boon, Chemistry and Industry (London), No. 13, p. 465, July 1 (1978).

Walker, A.T.S., and A.A. Wragg, Chem. Eng. Sci., 35, 405 (1980).

Wilkinson, J.A.E., and K.P. Haines, Trans. Inst. Mining and Metallurgy, 81, C157 (1972).

Yeager, E., P. Krouse, and K.V. Rao, Electrochim. Acta, 9, 1057 (1964).

Granulation and Adiponitrile Production

Kapur, P.C., K.V.S. Sastry, and D.W. Fuerstenau, I&EC Process Des. & Dev., 20, 519 (1981).

Kono, H., Hydrocarbon Processing, 59, 123, Jan. (1980).

Kono, H., paper presented at the 3rd International Symposium on Agglomeration, G16, Nurnberg, May 6-8 (1981).

Kono, H., and C.C. Huang, paper presented at the 4th International Symposium on Agglomeration, Toronto, June (1985).

Morl, L., and H.J. Kunne, Int. Chem. Eng., 26(3), 423 (1986).

Ouchiyama, N., and T. Tanaka, I&EC Process Des. & Dev., 14(3), 286 (1975).

Rumpf, H., Chem.-Ing.-Tech., 30, 144 (1958).

Schaefer, T., and O. Worts, Arch. Pharm. Chem. Sci. Ed., 6, 14 (1978).

Waldie, B., D. Wilkinson, and L. Zachra, Chem. Eng. Sci., 42, 653 (1987).

Calcium Bisulfate Acid Production

Stephenson, J.N., "Pulp and Paper Manufacture," Vol. I, Chap. IV, McGraw-Hill, New York (1952).

Volpicelli, G., and L. Massimilla, Pulp Paper Mag. Canada, 66(10), T512 (1965).

Volpicelli, G., and L. Massimilla, Chem. Eng. Sci., 25, 1361 (1970).

Sand Filter Cleaning

Brownell, L.E., in "Encyclopedia of Chemical Technology," ed. by R.E. Kirk and D.F. Othmer, Vol. 6, p. 516, Interscience Encyclopedia, New York (1951).

Halvorson, H.O., in "Encyclopedia of Chemical Technology," ed. by R.E. Kirk and D.F. Othmer, Vol. 14, p. 956, Interscience Encyclopedia, New York (1955).

Crystallization

Kagawa, S., U.S. Patent 4,331,826, May 25 (1982).

Okochi, I., H. Yamazaki, K. Izumi, and S. Takahashi, Japan Kokai Tokkyo Koho 78,119,782, Oct. 19 (1978).

Robinson, S.P., U.S. Patent 2,614,035, Oct. 14 (1952).

Thompson, A.R., in "Encyclopedia of Chemical Technology," ed. by R.E., Kirk and D.F. Othmer, Vol. 4, p. 619, Interscience Encyclopedia, New York (1955).

——, Chem. Eng., p. 338, Jan. (1953).

Flotation

Adamson, A.W., "Physical Chemistry of Surfaces," Wiley-Interscience, New York (1982).

Aplan, F.F., in "Kirk-Othmer Concise Encyclopedia of Chemical Technology," ed. by E. Graber, A. Klingsberg, and P.M. Siegel, p. 494, Wiley-Interscience (1985).

Arters, D., and L.-S. Fan, Chem. Eng. Sci., 41, 107 (1986).

Chander, S., Trans. Indian Inst. Metals, 31, 12 (1978).

Deister Concentrator Co. Literature, "Fast Relief from Mineral Recovery Headaches is Really Quite Simple" (1987).

Dobby, G.S., and J.A. Finch, Chem. Eng. Sci., 40(7), 1061 (1985).

Fan, L.-S., F. Bavarian, R.L. Gorowara, B.E. Kreischer, R.D. Buttke, and L.B. Peck, Powder Technol., 53(3), 285 (1987).

Foot, D.G., Jr., J.D. McKay, and J.L. Huiatt, Can. Metal. Q., 25(1), 15 (1986).

Fuerstenau, D.W., in "Fine Particle Processing," ed. by P. Somasundaran, Vol. 1, p. 669, Am. Inst. Mining, Metallurgical and Petroleum Engrs., New York (1980).

Parkinson, G., Chem. Eng., p. 27, March 31 (1986).

Rosen, M.J., "Surfactants and Interfacial Phenomena," Wiley-Interscience, New York (1978).

Somasundaran, P., in "Separation and Purification Methods," ed. by E.S. Perry and C.J. Van Oss, Vol. 1, p. 117, Marcel Dekker, New York (1973).

Trahar, W.J., and L.J. Warren, Int. J. Mineral Processing, 3, 103 (1976).

Yianatos, J.B., Finch, J.A., and Laplante, A.R., Can. Metal Q., 25(1), 23 (1986).

Young, T., Phil. Trans. Roy. Soc., London, 95, 65 (1805).

Zipperian, D.E., and J.A. Christophersen, paper presented at the Annual Meeting of Soc. Mining Engrs. of AIME, New York, Feb. 24-28 (1985).

Polymerization

Caumartin, F., and L. Havas (Naphtachimie S. A.), Ger. Offen. 2,623,693 (1976).

Fraser, W.A., N.J. Maraschin, F.J. Karol, and A.J. Makai, U.S. Patent 4,359,561 (1979).

Jenkins, III, J.M., R.L. Jones, and T.M. Jones, U.S. Patent 4,543,399 (1985).

Jenkins, III, J.M., R.L. Jones, T.M. Jones, and S. Beret, U.S. Patent 4,588,790 (1986).

Mangin, P., L. Havas, and R. Dormenval (Naphtachimie S. A.), Ger. Offen. 2,357,120 (1974).

Union Carbide Literature, "Unipol Process" (1986).

——, Hydrocarbon Processing, 58(11), 222 (1979).

——, Oil and Gas J., p. 73, Nov. 21 (1977).

Coal Gasification and Methanation

Blum, D.B., and J.J. Toman, AIChE Symposium Series, 73, No. 161, 115 (1977).

Cover, A.E., W.C. Schreiner, and G.T. Skaperdas, Chem. Eng. Prog., 69(3), 31 (1973).

Kohl, A.L., R.B. Harty, and J.G. Johanson, Chem. Eng. Prog., 74(8), 73 (1978).

Yosim, S.J., and K.M. Barclay, in Proceedings of the 2nd International Symposium on Molten Salts, ed. by J.R. Selman and J. Braunstein, p. 313, Electrochemical Society (1981).

Fischer-Tropsch Synthesis

Anderson, R.B., "The Fischer-Tropsch Synthesis," Academic Press (1984).

Calderbank, P.H., F. Evans, R. Farley, G. Jepson, and A. Poll, "Catalysis in Practice," p. 66, Inst. Chem. Engrs., London (1964).

Deckwer, W.-D., Oil and Gas J., 78(45), 198 (1980).

Deckwer, W.-D., Y. Serpmen, M. Ralek, and B. Schmidt, I&EC Process Des. & Dev., 21, 222 (1982).

Farley, R., and D.J. Ray, J. Inst. Petrol., 50, 27 (1964).

Frohning, C.D., H. Kolbel, M. Ralek, W. Rottig, F. Schnur, and H. Schulz, in "Chemierohstoffe aus Kohle," ed. by J. Falbe, Thieme, p. 219, Stuttgart (1977).

Haag, W.O., Synergy, 2(1), 21 (1980).

Hall, C.C., and A.H. Taylor, J. Inst. Petrol., 41, 101 (1955).

Kolbel, H., and P. Ackermann, Brennst.-Chem., 31, 10 (1950).

Kolbel, H., P. Ackermann, E. Ruschenburg, R. Langheim, and F. Engelhardt, Chem.-Ing.-Tech., 8, 153 (1951a).

Kolbel, H., P. Ackermann, E. Ruschenburg, R. Langheim, and F. Engelhardt, Chem.-Ing.-Tech., 8, 183 (1951b).

Kolbel, H., and F. Engelhardt, Erdoel Kohle, 5, 1 (1952).

Kolbel, H., and P. Ackermann, Chem.-Ing.-Tech., 28, 381 (1956).

Kolbel, H., P. Ackermann, and F. Engelhardt, Erdoel Kohle, 9, 153 (1956a).

Kolbel, H., P. Ackermann, and F. Engelhardt, Erdoel Kohle, 9, 225 (1956b).

Kolbel, H., P. Ackermann, and F. Engelhardt, Erdoel Kohle, 9, 303 (1956c).

Kolbel, H., Chem.-Ing.-Tech., 29, 505 (1957).

Kolbel, H., and M. Ralek, Catal. Rev. - Sci. Eng., 21(2), 225 (1980).

Kuo, J.C.W., Final Report to U.S. Department of Energy, Contract No. DE-AC22-80PC30022, June (1983).

Kuo, J.C.W., Final Report to U.S. Department of Energy, Contract No. DE-AC22-83PC60019, Oct. (1985).

Mitra, A.K., and A.N. Roy, Indian Chem. Eng., 127 (1963).

Sakai, T., and T. Kunugi, Sekyu Gakkai Shi, Japan, 17, 863 (1974).

Sanders, E., S. Ledakowicz, and W.-D. Deckwer, Can. J. Chem. Eng., 64, 133 (1986).

Schlesinger, M.D., J.H. Crowell, M. Leva, and H.H. Storch, I&EC, 43(6), 1474 (1951).

Schlesinger, M.D., H.E. Benson, W.M. Murphy, and H.H. Storch, I&EC, 46(6), 1322 (1954).

Zaidi, A., Y. Louisi, M. Ralek, and W.-D. Deckwer, Ger. Chem. Eng., 2, 94 (1979).

Methanol Synthesis

Air Products and Chemicals Inc., Final Report to U.S. Department of Energy, Contract No. DE-AC22-81PC30019, Aug. (1987).

Air Products and Chemicals Inc., Final Report to U.S. Department of Energy, Contract No. DE-AC22-85PC80007, Feb. (1988).

Probstein, R.F., and R.E. Hicks, "Synthetic Fuels," McGraw-Hill, New York (1982).

Sherwin, M.B., and M.E. Frank, Hydrocarbon Processing, 55(11), 122 (1976).

Studer D.W., E.P. Holley, T.H. Hsiung, and R.L. Mednick, paper presented at the DOE/FE Indirect Liquefaction Contractors Review Meeting, Pittsburgh, Dec. 8-9 (1987).

Appendices

Appendix A

Optimum Particle Size in Catalytic Reactors – An Example of Reactor Analyses for Modes E-I-a-1, E-I-a-2 and T-I-a-2 (Figure 1.3)

In industrial catalytic operation involving gas-liquid-solid fluidization systems with liquid as the continuous phase and a cocurrent upward flow of gas and liquid, selection of catalyst particle size and density is closely associated with the selection of reactor types. That is, for large/heavy catalyst particles, the reactions would be more likely to be conducted in fluidized bed reactors (Mode E-I-a-1) whereas for small/light catalyst particles, use of slurry bubble column reactors (Mode E-I-a-2 or T-I-a-2) is more likely. Determination of the optimum catalyst particle size and density requires consideration of both reaction kinetics and bed transport properties (given in Chapters 2, 3, and 4).

A.1 A BRIEF REVIEW OF REACTOR MODELS

Considerable studies have appeared in the literature regarding the modeling of slurry bubble column reactors with batch liquid (Mode E-III-a). However, relatively little work has been done in modeling slurry bubble columns with cocurrent gas-liquid flow (Mode E-I-a-2 or T-I-a-2).

Parulekar and Shah (1980) modeled a cocurrent flow slurry bubble column reactor for hydrocracking. They found that all product yields undergo a maximum as feed catalyst concentration is varied, due to the competing effects of catalyst concentration and liquid-phase holdup. Considering a first-order rate expression for hydrogen consumption and axial dispersion for both the gas and liquid phases, Deckwer et al. (1982) developed a model to account for reactant conversion for Fischer-Tropsch reactions in a slurry bubble column. The model was verified with data from a demonstration reactor. The model simulation indicated that liquid-solid mass transfer resistance and catalyst settling can be neglected and gas-liquid mass transfer limitations are small in large reactors.

Govindarao and Chidambaram (1983) simulated a catalytic reaction in a cocurrent flow slurry bubble column reactor using the model differential equations presented by Govindarao (1975), which assume axially dispersed flow of the gas and liquid phases and a sedimentation-dispersion model for the solid phase. They found that the conversion reaches a minimum with respect to particle size. However, they only considered catalytic reactions occurring on the particle surface; intraparticle diffusion and reaction was not considered in their simulations. Serpemen and Deckwer (1983) modeled the hydrogenation of butynediol in a cocurrent flow slurry bubble column. They concluded that the nonuniformity of the catalyst distribution has a negligible effect on reactor performance for large column diameters (greater than 30 cm) but the effect can be considerable for small column diameters.

Relatively few papers have appeared in the literature regarding catalytic reactions in gas-liquid-solid fluidized beds. Lee et al. (1974) analyzed a gas-liquid-solid fluidized bed catalytic reaction by considering a global reaction rate based on mass transfer resistances. They calculated an overall resistance based on gas-liquid mass transfer, liquid-solid mass transfer, and catalytic reaction in the solid. They noted that, over a wide range of particle diameters, the gas-liquid mass transfer resistance decreases with increasing particle diameter due to decreasing bubble size, while the reaction resistance increases with particle diameter due to increasing intraparticle diffusion effect. Lee et al. (1974) showed that this combined effect can lead to the existence of an optimum particle diameter for gas-liquid-solid fluidized beds. However, their analysis was based only on reaction resistances. They did not consider the bed hydrodynamics or phase holdups and did not quantitatively examine the relationship between the optimum particle size and process variables.

Sastri et al. (1983) presented a noncatalytic gas-liquid-solid fluidized bed model for the production of zinc hydrosulfite from sulfur dioxide and particulate zinc in aqueous solution. The model considered the cases of complete mixing, axially dispersed flow, and plug flow for the liquid phase. Since the gas phase consisted of pure SO_2, with constant gas composition throughout the reactor, the gas phase hydrodynamics was not considered. Phase holdups were determined using the generalized wake model of Bhatia and Epstein

(1974). The bubble size was assumed to remain constant while the bubble frequency decreased with bed height due to absorption of SO_2. They found that the measured concentration profiles of dissolved SO_2 and reacted zinc were best predicted by the axially dispersed flow model for the liquid phase.

For a three-phase catalytic reaction with a specified catalyst particle density, the particle size used as well as the flow mode for the reactor can have a strong effect on the overall conversion. Such effects are illustrated in this appendix using a model system which involves a pseudo-first order reaction with hydrodynamic properties of the systems comparable to those of an air-water-particle system at ambient conditions. A general modeling approach to predict optimum particle size and operating regimes is presented for both the fluidized bed and the slurry bubble column in the following (Wisecarver and Fan, 1988).

A.2 A CATALYTIC REACTOR MODEL

A three-phase catalytic reaction may involve a gas phase component reacting with a liquid phase component, such as the hydrogenation reactions; it may be a gas phase component reacting in the presence of a liquid solvent, such as an isomerization reaction; or it may involve the catalytic reaction of two or more gas phase components. In general, reaction in a three-phase reactor involves mass transfer of the reactant from the gas phase to the liquid phase, diffusion of reactants in the liquid phase to the particle surface, and diffusion and reaction within the porous catalyst particles. For simplicity, the catalytic reaction is assumed here to be pseudo-first order with respect to the gas component; all other reactants are assumed to be present in excess. The reactor is also assumed to be isothermal.

The liquid phase in a gas-liquid-solid fluidized bed is axially dispersed. A material balance for the reactant in a liquid phase with axially dispersed flow yields the following equation:

$$\epsilon_\ell E_{z\ell}\frac{d^2C_\ell}{dz^2} - U_\ell \frac{dC_\ell}{dz} + K_\ell a\,(C_g/He - C_\ell) - k_s a_p \epsilon_s (C_\ell - C_s) = 0 \tag{A.1}$$

If the gas used is not an inert component, the material balance of the gas phase must also be considered. If the extent of absorption of the gas phase is high, then the gas velocity can vary considerably through the length of the reactor. There would also be a considerable axial variation in $K_\ell a$ due to the decrease in gas-liquid interfacial area as the gas is depleted in the reactor. It is assumed here that the gas is sparingly soluble in the liquid so that an average gas velocity and an average gas-liquid mass transfer coefficient for the reactor can be used.

Since the degree of axial dispersion for the gas phase in a gas-liquid-solid fluidized bed is low, plug flow can be assumed for the gas phase. A material balance on the gas phase yields

$$U_g \frac{dC_g}{dz} + K_\ell a\,(C_g/He - C_\ell) = 0 \tag{A.2}$$

The concentration of reactant at the solid surface, C_s, will depend on the solids phase mixing. As indicated in Chapter 3, at high liquid and low gas velocities, the solids mixing time, i.e., the time required to reach complete solids mixing in the bed, is long, while at low liquid and high gas velocities, the solids mixing time is short. The effect of solids mixing on the conversion depends on the solids mixing time in relation to the rate of reaction. Conceptually, when the reaction rate is much higher than the rate of solids mixing, then reaction will occur before any significant solids movement can take place and the rate will be the same as it would have been without any solids mixing. For a reaction rate which is very slow compared to the rate of solids mixing, complete mixing of the solids may occur before any significant reaction takes place and the solid phase can be treated as well mixed. It is therefore desirable to examine the effect of solids mixing characterized by these two limiting cases, i.e., complete solids mixing and no solids mixing, on the reactant conversion in a three-phase fluidized bed reactor.

If there is no solids mixing, the surface concentration, C_s, can be found from

$$k_s a_p (C_\ell - C_s) = k_r \eta C_s \tag{A.3}$$

where η is the effectiveness factor for a spherical particle which, for a first-order reaction with a spherical catalyst, is

$$\eta = \frac{1}{\Phi}\,[\coth(3\Phi) - 1/(3\Phi)] \tag{A.4}$$

where $\Phi\,[\,= (d_p/6)\sqrt{(k_r/D_{eff})}\,]$ is the Thiele modulus.

For the case of complete solids mixing, the surface concentration is constant throughout the reactor. C_s can therefore be found by making an overall material balance on the reactor:

$$C_s = \frac{U_g\,(C_{g0} - C_g|_{z=H}) + U_\ell\,(C_{\ell 0} - C_\ell|_{z=H})}{k_r\,\eta\epsilon_s\,H} \tag{A.5}$$

The boundary conditions for Eqns. (A.1) and (A.2) are

$$z=0\,,\quad C_g = C_{g0} \tag{A.6a}$$

$$z=0\,,\quad \epsilon_\ell E_{z\ell}\frac{dC_\ell}{dz} + U_\ell C_{\ell 0} = U_\ell C_\ell \tag{A.6b}$$

$$z=H\,,\quad \frac{dC_\ell}{dz} = 0 \tag{A.6c}$$

For the case of complete solids mixing, Eqns. (A.1), (A.2), and (A.5) with Eqns. (A.6a) to (A.6c) as the boundary conditions can be solved analytically. The solution, for a physically meaningful range of Pe and St, is as follows:

$$C_g/C_{g0} = \alpha_1\exp(\beta_1\xi) + \alpha_2\exp(\beta_2\xi) + \alpha_3\exp(\beta_3\xi) + C_s He/C_{g0} \tag{A.7}$$

$$\frac{He\,St\;U\,C_\ell}{C_{g0}} = \alpha_1(He\,\beta_1 + St\,U)\exp(\beta_1\xi) + \alpha_2(He\,\beta_2 + St\,U)\exp(\beta_2\xi)$$

$$+\,\alpha_3(He\,\beta_3 + St\,U)\exp(\beta_3\xi) + He\,St\,U\,C_s/C_{g0} \tag{A.8}$$

and

$$C_s He/C_{g0} = \frac{(\lambda St\;\,U + Pe\;St\;U\,R_1)C_{\ell 0}\,He/C_{g0} + He\,\lambda\,St\;\; - R_2}{St^2\;U\,\kappa_1\lambda/\kappa_2 + St\,(He + U)\,\lambda + Pe\;St\;\,U\,R_1 - R_2} \tag{A.9}$$

where

$$\beta_1 = \frac{2}{3He}\sqrt{p}\,\cos(\theta/3) + \frac{A}{3He}$$

$$\beta_2 = \frac{2}{3He}\sqrt{p}\,\cos[(\theta + 2\pi)/3] + \frac{A}{3He}$$

$$\beta_3 = \frac{2}{3\text{He}} \sqrt{p} \cos[(\theta + 4\pi)/3] + \frac{A}{3\text{He}}$$

$$\theta = \cos^{-1}(q\ p^{-3/2}/2)$$

$$p = 3\text{He Pe St}(\text{He} + \text{He}\ \kappa_1 + U) + A^2$$

$$q = 2A^3 + 9\text{He Pe St}\ A(\text{He} + \text{He}\ \kappa_1 + U) + 27\text{He}^2\ \text{Pe St}^2\ U\kappa_1$$

$$\alpha_1 = \lambda_1/\lambda,\ \alpha_2 = \lambda_2/\lambda,\ \alpha_3 = \lambda_3/\lambda$$

$$\lambda = b_{12} + b_{23} + b_{31}$$

$$\lambda_1 = \text{Pe St}\ U(E_2 - E_3)C_{\ell 0}\text{He}/C_{g0} + b_{23} - [\text{Pe St}\ U(E_2 - E_3) + b_{23}]C_s\text{He}/C_{g0}$$

$$\lambda_2 = \text{Pe St}\ U(E_3 - E_1)C_{\ell 0}\text{He}/C_{g0} + b_{31} - [\text{Pe St}\ U(E_3 - E_1) + b_{31}]C_s\text{He}/C_{g0}$$

$$\lambda_3 = \text{Pe St}\ U(E_1 - E_2)C_{\ell 0}\text{He}/C_{g0} + b_{12} - [\text{Pe St}\ U(E_1 - E_2) + b_{12}]C_s\text{He}/C_{g0}$$

$$A = \text{He Pe-St}\ U$$

$$b_{12} = E_1F_2 - E_2F_1$$

$$b_{23} = E_2F_3 - E_3F_2$$

$$b_{31} = E_3F_1 - E_1F_3$$

$$E_1 = \beta_1(\text{He}\ \beta_1 + \text{St}\ U)\exp(\beta_1)$$

$$E_2 = \beta_2(\text{He}\ \beta_2 + \text{St}\ U)\exp(\beta_2)$$

$$E_3 = \beta_3(\text{He}\ \beta_3 + \text{St}\ U)\exp(\beta_3)$$

$$F_1 = \beta_1(\beta_1\text{He} - A) - \text{Pe St}\ U$$

$$F_2 = \beta_2(\beta_2 He - A) - Pe\ St\ U$$

$$F_3 = \beta_3(\beta_3 He - A) - Pe\ St\ U$$

$$U = U_\ell/U_g$$

$$\kappa_1 = \frac{k_s\ a_p \epsilon_s}{K_\ell a}$$

$$R_1 = G_1(E_3-E_2)+G_2(E_1-E_3)+G_3(E_2-E_1)$$

$$R_2 = b_{23}G_1+b_{31}G_2 + b_{12}G_3$$

$$G_1 = [St(He+U)+He\beta_1]\ \exp(\beta_1)$$

$$G_2 = [St(He+U)+He\beta_2]\ \exp(\beta_2)$$

$$G_3 = [St(He+U)+He\beta_3]\ \exp(\beta_3)$$

and

$$\kappa_2 = \frac{k_s a_p}{k_r \eta}$$

For the case of no solids mixing, the solution to Eqns. (A.1), (A.2) and (A.3) with Eqns. (A.6a) to (A.6c) as the boundary conditions is

$$C_g/C_{g0} = \gamma_1\ \exp(\beta_1'\xi) + \gamma_2\ \exp(\beta_2'\xi) + \gamma_3\ \exp(\beta_3'\xi) \tag{A.10}$$

$$\frac{He\ C_\ell}{C_{g0}} = \gamma_1[He\beta_1'/(St\ U) + 1]\ \exp(\beta_1'\xi) + \gamma_2[He\beta_2'/(St\ U) + 1]\ \exp(\beta_2'\xi)$$
$$+ \gamma_3[He\beta_3'/(St\ U) + 1]\ \exp(\beta_3'\xi) \tag{A.11}$$

and

$$C_s = \frac{k_s\ a_p\ C_\ell}{k_s\ a_p + k_r \eta} \tag{A.12}$$

where

$$\beta_1' = \frac{2}{3\mathrm{He}}\sqrt{p'}\,\cos(\theta'/3) + \frac{A}{3\mathrm{He}}$$

$$\beta_2' = \frac{2}{3\mathrm{He}}\sqrt{p'}\,\cos[(\theta' + 2\pi)/3] + \frac{A}{3\mathrm{He}}$$

$$\beta_3' = \frac{2}{3\mathrm{He}}\sqrt{p'}\,\cos[(\theta' + 4\pi)/3] + \frac{A}{3\mathrm{He}}$$

$$\theta' = \cos^{-1}(q' p'^{-3/2}/2)$$

$$p' = 3\mathrm{He}\ \mathrm{Pe}\ \mathrm{St}\ [\mathrm{He} + \mathrm{He}\ \kappa_1/(1 + \kappa_2) + U] + A^2$$

$$q' = 2A^3 + 9\mathrm{He}\ \mathrm{Pe}\ \mathrm{St}\ A\ [\mathrm{He} + \mathrm{He}\ \kappa_1/(1 + \kappa_2) + U] + 27\mathrm{He}^2\ \mathrm{Pe}\ \mathrm{St}^2\ U\kappa_1/(1 + \kappa_2)$$

$$\gamma_1 = \frac{C_{\ell 0}\mathrm{He}/C_{g0}(E_2 - E_3) + b_{23}}{b_{23} + b_{12} + b_{31}}$$

$$\gamma_2 = \frac{C_{\ell 0}\mathrm{He}/C_{g0}(E_3 - E_1) + b_{31}}{b_{23} + b_{12} + b_{31}}$$

and

$$\gamma_3 = \frac{C_{\ell 0}\mathrm{He}/C_{g0}(E_1 - E_2) + b_{12}}{b_{23} + b_{12} + b_{31}}$$

Note that E_1, E_2, E_3, b_{12}, b_{23}, b_{31}, F_1, F_2, F_3, A, κ_1, κ_2 and U are defined as before.

For the special case of constant gas composition and surface reaction on the solid particles, the model equations for the liquid phase with no solids mixing reduce to Eqns. (A.1) and (A.3) with the effectiveness factor equal to unity. Sastri et al. (1983) solved this special case for the zinc hydrosulfite reaction in which $K_\ell a$ decreased with bed height as the volume of gas in the bed was depleted; the model showed good agreement with their data.

For comparison, conversion for a slurry bubble column reactor with a continuous solids flow is also calculated. The sedimentation-dispersion model (given in Section 4.6.3) has been shown to accurately describe the axial distribution of solids for the slurry bubble column (Kato et al., 1972; Smith and Ruether, 1985), while the liquid phase has been shown to be axially dispersed (see Section 4.5.1). It is assumed here that the gas is in plug flow. Equations (A.1) and (A.2) can therefore be used for the gas and liquid phases, respectively. The sedimentation-dispersion model yields an axial solids concentration profiles given by Eqn. (4.39). A material balance over a control element for the solids phase yields the following equation:

$$E^*_{zs} \frac{d}{dz} (\eta C_s \epsilon_p \frac{d\epsilon_s}{dz}) - V_p \frac{d(\eta C_s \epsilon_p \epsilon_s)}{dz} + k_s a_p \epsilon_s (C_\ell - C_s) - k_r \eta C_s \epsilon_s = 0 \tag{A.13}$$

where V_p is given by Eqn. (4.41). Combining Eqns. (4.39) and (A.13) results in the following equation for the solid phase:

$$\epsilon_p \eta [E^*_{zs} \frac{d\epsilon_s}{dz} - V_p \epsilon_s] \frac{dC_s}{dz} + k_s a_p \epsilon_s (C_\ell - C_s) - k_r \eta C_s \epsilon_s = 0 \tag{A.14}$$

The boundary condition for Eqn. (A.14) is

$$z=0, \qquad C_s = 0 \tag{A.15}$$

Equations (A.1), (A.2), and (A.14) can be solved numerically with Eqns. (A.6a) to (A.6c) and (A.15) as boundary conditions (with $d\epsilon_s/dz$ evaluated by the sedimentation-dispersion model, see Eqn. (4.39)).

A.3 MODEL PARAMETERS USED IN SIMULATIONS

Transport properties described in Chapters 2, 3 and 4 are used to predict the properties in fluidized bed and slurry bubble column reactor. For the gas-liquid-solid fluidized bed reactor model, the correlation of Ohashi et al. (1981) for the liquid-solid mass transfer coefficient given in Eqn. (3.18) is used. The axial dispersion coefficient, $E_{z\ell}$, is determined from the correlation of Kim and Kim (1983) (see Eqn. (T-3.3.1)). The empirical correlations of Begovich and Watson (1978) are used for calculation of phase holdups (see Eqns. (T-2.1.1) and (T-2.1.2)). The volumetric gas-liquid mass transfer coefficient can be represented by the liquid-side volumetric gas-liquid mass transfer coefficients; the correlation of Wisecarver and Fan (1988) given by Eqn. (T-3.1.2b) is used for it.

For the slurry bubble column reactor, the equations of Smith and Ruether (1985) and O'Dowd et al. (1987) given as Eqns. (T-4.6.1) and (T-4.6.2) are used, respectively, to determine the solids axial dispersion coefficient (E_{zs}^*) and the slip velocity between the liquid and solid phases (U_p) (for calculation of V_p). The liquid phase axial dispersion coefficient is determined from the correlation of Kato et al. (1972) (Eqn. (T-4.4.2)). The gas-liquid mass transfer coefficient is calculated using the correlation of Nguyen-Tien et al. (1985) (Eqn. (T-3.1.2a). Sanger and Deckwer's correlation (1981) is used to calculate the liquid-solid mass transfer coefficient (Eqn. (T-4.8.5)). Gas holdup is determined from Hughmark's correlation (1967) given in Eqn. (T-4.1.1).

Since most of the above correlations are based on an aqueous liquid phase, the physical properties of water are used for the liquid phase in the simulations. The gas properties used correspond to the physical properties of air at atmospheric pressure. The parameters used in the model simulations are summarized in Table A.1.

A.4 SIMULATION

Particle size directly impacts the prevailing hydrodynamic behavior in a three-phase fluidized bed including liquid and solids mixing, individual phase holdups, and mass transfer; consequently, particle size directly affects the chemical reaction rate. Figure A.1 shows the overall reaction rate in a gas-liquid-solid fluidized bed catalytic reactor as a function of particle diameter for two different combinations of reaction rate constant and fluid velocities. Curves are given for both complete mixing and no mixing of the solid phase. It can be seen that the reaction rate goes through a maximum with respect to particle size, as was predicted by Lee et al. (1974). At a low particle diameter the terminal velocity of the particle is approached. The solids holdup is thus extremely low, resulting in a low overall reaction rate. As the particle diameter is increased, the reaction rate increases due to both the increase in solids holdup and the increase in the gas-liquid mass transfer coefficient. At large particle diameters, this increase in $K_{\ell}a$ is offset by the decrease in effectiveness factor due to the intraparticle mass transfer resistance. These effects result in a maximum reaction rate at an intermediate particle size. It can be seen from Fig. A.1, however, that the value of this optimum particle size can differ markedly for different values of the process variables.

Table A.1 Parameters Used in the Simulation

Parameter	Value
Pressure, P	1 atm
Temperature, T	293 K
Particle density, ρ_s	1.5 g/cm^3
Liquid density, ρ_ℓ	1.0 g/cm^3
Gas density, ρ_g	1.2×10^{-3} g/cm^3
Reactor diameter, D_c	15 cm
Bed or column height, H	200 cm
Liquid viscosity, μ_ℓ	0.01 g/cm·s
Gas viscosity, μ_g	1.8×10^{-4} g/cm·s
Henry's Law constant, He	30
Inlet gas concentration, C_{g0}	8.6×10^{-6} mol/cm^3
Reactant diffusivity in liquid, D_M	2.4×10^{-5} cm^2/s
Surface tension of liquid, σ	72 dyne/cm

It can be seen from Fig. A.1 that there is very little effect of solids mixing on the amount of reactant consumed for the conditions shown. For a low reaction rate constant (0.1 s^{-1}) there is a greater relative difference in overall reaction rate for the two extremes of solids mixing, but the difference is still less than 10%. It is noted that complete mixing of the solids results in a slightly higher reaction rate than no solids mixing. If the inlet liquid stream contains no dissolved gas, then the gas must diffuse into the liquid phase at the bottom of the reactor and then into the catalyst particles before any reaction can take place; therefore, the particles at the very bottom of the bed do not contribute to the reaction if there is no solids mixing. If the solids are mixed, then the bottom of the bed contains reactant-rich particles which contribute to the reaction, thereby increasing the rate at the bottom of the bed. However, the particles at the top of the bed have a lower reactant concentration, and therefore a lower reaction rate, than in the case of no solids mixing. Apparently the effect of the increased reaction rate at the bottom of the bed outweighs the effect of the decreased reaction rate at the top of the bed.

It is found that when the liquid feed is saturated with the gas phase reactant ($C_{\ell 0} = C_{g0}/He$), there is no discernible difference in

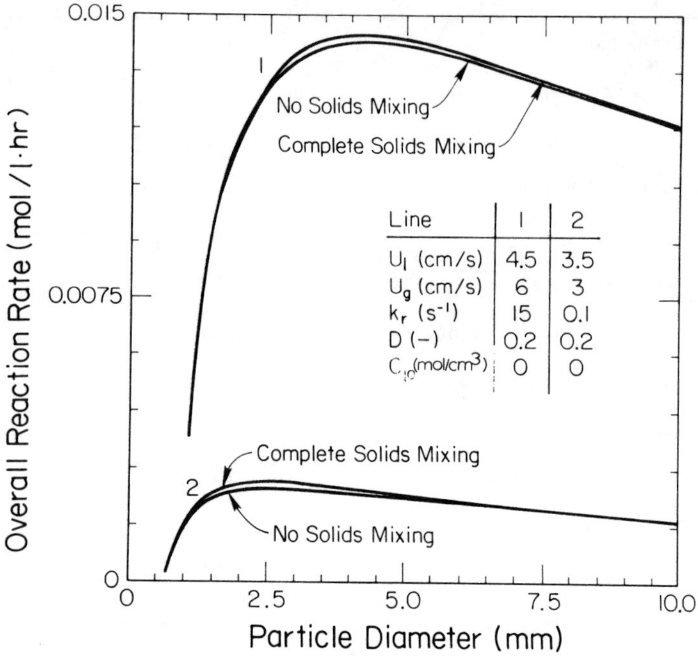

Figure A.1 Overall reaction rate versus particle diameter for complete solids mixing and no solids mixing for fluidized bed reactors (from Wisecarver and Fan, 1988).

reactor performance for the two extremes of solids mixing. The effect of solids mixing is relatively more prominent for lower values of reaction rate constant and for lower values of intraparticle diffusivity because the concentration at the surface of the particles, C_S, is higher under these conditions, as can be seen from Eqn. (A.5). The amount of reactant brought to the bottom of the bed by solids mixing is therefore greater. If the reaction rate constant is large enough that the reaction can be considered instantaneous, it can be seen that solids mixing would have no effect at all on reactor performance, since the surface concentration C_S would then go to zero and the solid phase would no longer contain any reactant. It is also noted that the effect of solids mixing is influenced by the extent of mixing in the liquid phase. If the liquid were completely mixed, then every catalyst particle would be exposed to the same liquid concentration regardless of the extent of solids mixing, thus resulting in the same reaction rate. The closer the liquid phase is to the plug flow state, the greater should be the importance of solid phase mixing. It was found that the difference in

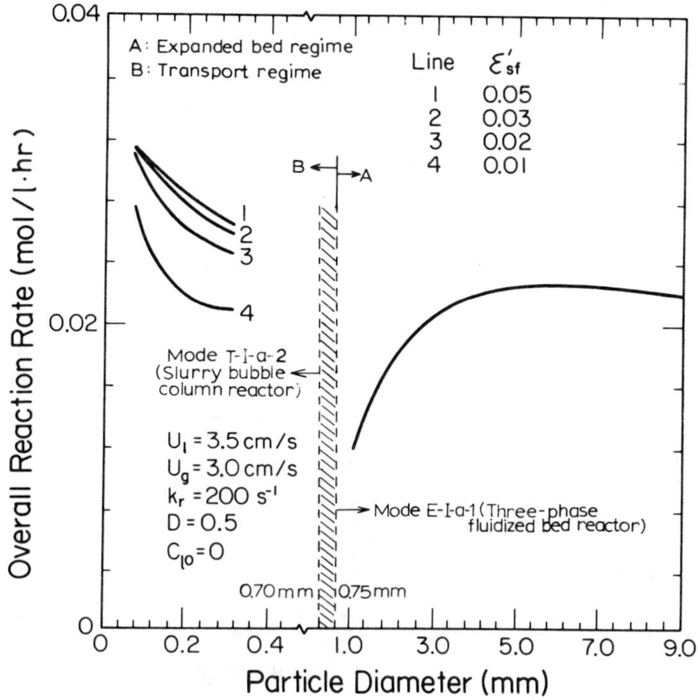

Figure A.2 Comparison of the overall reaction rate in the reactor between the slurry bubble column and gas-liquid-solid fluidized bed for fast reactions (after Wisecarver and Fan, 1988).

overall reaction rate for the two extremes of solids mixing decreases as U_ℓ and U_g are increased, corresponding to a higher axial dispersion coefficient for the liquid.

A comparison is made between the overall rate of reaction for the gas-liquid-solid fluidized bed and for the slurry bubble column in Figs. A.2 and A.3. The figures also indicate the range of particle diameters under which Mode E-I-a-1 (in the expanded bed regime) for three-phase fluidized beds and Mode T-I-a-2 (in the transport regime) for slurry bubble columns (see Sections 1.2 and 2.3.5 for the criterion or model regarding mode or regime designation) prevail. It is noted that under the operating conditions specified in both figures, the slurry bubble column is not operated under Mode E-I-a-2. Define a parameter ϵ'_{sf} as the relative volumetric solids flow rate in the slurry flow for the slurry bubble column, i.e.,

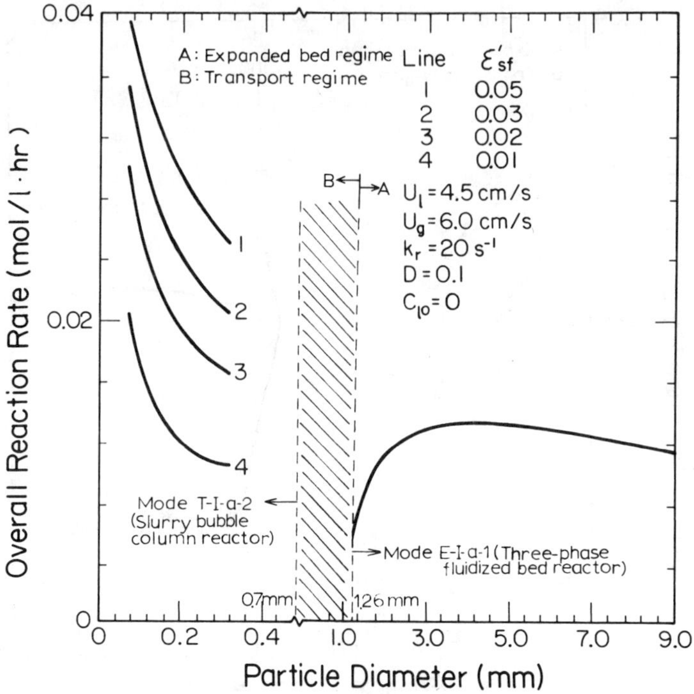

Figure A.3 Comparison of the overall reaction rate in the reactor between the slurry bubble column and gas-liquid-solid fluidized bed for moderate rate reactions (after Wisecarver and Fan, 1988).

$$\epsilon'_{sf} = U_s/(U_\ell + U_s) \tag{A.16}$$

Four values for ϵ'_{sf}, 0.01, 0.02, 0.03, and 0.05, are considered in the figure. For the conditions shown in Fig. A.2, it can be seen that the slurry bubble column gives a somewhat higher rate of reaction than the gas-liquid-solid fluidized bed, especially at very low particle diameters. The reaction rate for the slurry bubble column decreases monotonically with particle diameter and increases with increasing ϵ'_{sf}.

For a lower reaction rate constant, a lower intraparticle diffusivity, and a higher gas velocity, as shown in Fig. A.3, the overall rate of reaction for a slurry bubble column is up to three times higher than for a gas-liquid-solid fluidized bed when ϵ'_{sf} is high. The higher reaction rate in the slurry bubble column is due to the larger specific surface area of solids, resulting in a higher liquid-solid mass transfer rate, and the much higher effectiveness factor for small particles. These advantages are partially offset by the increased gas-liquid mass transfer in the gas-liquid-solid fluidized bed resulting from bubble breakup with large particles. In general, the performance of the gas-liquid-solid fluidized bed as compared with the slurry bubble column is improved with increasing reaction rate constant, increasing intraparticle diffusivity, and decreasing gas velocity, because the relative importance of gas-liquid mass transfer to the overall reaction rate is higher. The gas-liquid-solid fluidized bed also fares better when ϵ'_{sf} in the slurry bubble column is low.

Under conditions of high pressure and high temperature, as are often encountered in industrial gas-liquid-solid reactions, the bubble size is considerably smaller and the gas holdup is considerably larger (as noted in Section 10.5). The difference in reaction rate between the transport reactor and the expanded bed reactor would therefore be even greater, since there would be less effect of bubble size reduction with increasing particle size. It should be noted, however, that the increased reaction rate in slurry bubble columns as compared with gas-liquid-solid fluidized beds is only one among many factors to be considered in deciding which reactor type or flow regime to use for a given catalytic reaction. Many process variables enter into such a decision. A slurry bubble column system, for example, requires low power consumption, possesses a high slurry phase heat capacity and liquid components can be easily vaporized or condensed to work as a heat source or sink, resulting in a thermally stable operation. The settling of particles, creating a dead space; erosion of the reactor inner wall, the piping or the operating pump; difficulty in separation of catalyst particles from effluent liquid stream; and rapid catalyst deactivation, however, are operational disadvantages of slurry bubble column systems. Catalyst replacement in the reactor is a primary operational advantage and catalyst attrition is the primary disadvantage (also see Chapter 10) for fluidized bed systems.

A.5 CONCLUDING REMARKS

The catalytic reactor model presented here indicates the importance of selecting the proper catalyst size when designing or operating a catalytic three-phase reactor using either the fluidized bed or slurry bubble column. The model is developed by considering both reaction kinetics and the inherent transport properties of the system.

The model simulation reflects the relative merits of the slurry reactor and the three-phase fluidized bed reactor from the viewpoint of reactant conversion.

A.6 NOMENCLATURE

a_p	: solids surface area per unit volume of solids
C_g	: reactant concentration in gas phase
C_{g0}	: inlet gas concentration
C_ℓ	: reactant concentration in liquid phase
$C_{\ell 0}$	: reactant concentration in liquid feed
C_s	: reactant concentration at particle surface
d_p	: particle diameter
D	: D_{eff}/D_M
D_c	: column diameter
D_{eff}	: effective intraparticle diffusivity
D_M	: molecular diffusivity of gas component in liquid
$E_{z\ell}$	: liquid phase axial dispersion coefficient
E_{zs}^*	: solid phase axial dispersion coefficient
H	: bed height (for fluidized bed) or column height (for slurry bubble column)
He	: Henry's law constant
$K_\ell a$	: overall volumetric gas-liquid mass transfer coefficient
k_r	: pseudo-first-order reaction rate constant
k_s	: liquid-solid mass transfer coefficient
P	: pressure
Pe	: Peclet number, $U_\ell H/\epsilon_\ell E_{z\ell}$
St	: Stanton number, $K_\ell aH/U_\ell$
T	: temperature
U_g	: superficial gas velocity
U_ℓ	: superficial liquid velocity
U_p	: slip velocity between the liquid and solid phases

U_s : superficial solids velocity

V_p : velocity of solid particle

z : axial distance from the bottom of bed

Greek Letters

ϵ_ℓ : holdup of liquid phase

ϵ_p : void fraction of catalyst particle

ϵ_s : holdup of solid phase

ϵ'_{sf} : relative volumetric solids flow rate in the slurry flow, defined by Eqn. (A.16)

μ_g : viscosity of gas

μ_ℓ : viscosity of liquid

ρ_g : density of gas

ρ_ℓ : density of liquid

ρ_s : density of particle

σ : surface tension

ξ : z/H

η : effectiveness factor

Φ : Thiele modulus

A.7 REFERENCES

Begovich, J.M., and J.S. Watson, in "Fluidization," ed. by J.F. Davidson and D.L. Keairns, p. 190, Cambridge University Press (1978).

Bhatia, V.K., and N. Epstein, in "Fluidization and Its Applications," ed. by H. Angelino, J.P. Couderc, H. Gibert, and C. Laguerie, p. 380, Cepadues-Editions, Toulouse (1974).

Deckwer, W.-D., Y. Serpemen, M. Ralek, and B. Schmidt, I&EC Process Des. & Dev., 21(2), 231 (1982).

Govindarao, V.M.H., Chem. Eng. J., 9, 229 (1975).

Govindarao, V.M.H., and M. Chidambaram, Chem. Eng. J., 27, 29 (1983).

Hughmark, G.A., I&EC Process Des. & Dev., 6(2), 218 (1967).

Kato, Y., A. Nishiwaki, T. Fukuda, and S. Tanaka, J. Chem. Eng. Japan, 5, 112 (1972).

Kim, S.D., and C.H. Kim, J. Chem. Eng. Japan, 16(3), 172 (1983).

Lee, J.C., A.J. Sherrard, and P.S. Buckley, in "Fluidization and Its Applications," ed. by H. Angelino, J.P. Couderc, H. Gibert, and C. Laguerie, p. 407, Cepadues-Editions, Toulouse (1974).

Nguyen-Tien, K., A.N. Patwari, A. Schumpe, and W.-D. Deckwer, AIChE J., 31, 194 (1985).

O'Dowd, W., D.N. Smith, J.A. Ruether, and S.C. Saxena, AIChE J., 33, 1959 (1987).

Ohashi, H., T. Sugawara, K. Kikuchi, and H. Konno, J. Chem. Eng. Japan, 14, 433 (1981b).

Parulekar, S.J., and Y.T. Shah, Chem. Eng. J., 20, 21 (1980).

Sanger,P., and W.-D. Deckwer, Chem. Eng. J., 22, 179 (1981).

Sastri, N.V.S., N. Epstein, A. Hirata, I. Koshijima, and M. Izumi, Can. J. Chem. Eng., 61, 635 (1983).

Serpemen, Y., and W.-D. Deckwer, in "Mass Transfer with Chemical Reaction in Multiphase Systems, Volume II: Three-Phase Systems," ed. by E. Alper, p. 239, NATO ASI Series E: Applied Sciences-No. 73, Martinus Nijhoff, The Hague (1983).

Smith, D.N., and J.A. Ruether, Chem. Eng. Sci., 40, 741 (1985).

Wisecarver, K.D., and L.-S. Fan, Chem. Eng. Sci., 43, 2741 (1988).

Appendix B

SI Units Used for the Empirical Correlation Equations in This Book

Quantity	Name for SI Units	Symbol for SI Units
Apparent Viscosity (Dynamic)*	newton-second to the n-th power per square meter	$N \cdot s^n/m^2$
Bubble Chord Length	meter	m
Bubble Diameter	meter	m
Cell Concentration	kilogram per cubic meter	kg/m^3
Column Diameter	meter	m
Density	kilogram per cubic meter	kg/m^3
Dispersion Coefficient	square meter per second	m^2/s
Diffusivity	square meter per second	m^2/s
Distance	meter	m
Energy Dissipation Rate per Unit Mass of the System	watt per kilogram	W/kg
Gas-Liquid Interfacial Area per Unit Volume of the System	square meter per cubic meter	m^2/m^3
Gas-Side Gas-Liquid Mass Transfer Coefficient	mole per pascal-square meter -second	$mol/Pa \cdot m^2 \cdot s$
Generalized Viscosity Constant*	newton-second to the n-th power per square meter	$N \cdot s^n/m^2$
Gravitational Acceleration	meter per square second	m/s^2
Heat Transfer Coefficient	watt per square meter-kelvin	$W/m^2 \cdot K$

* for a power law fluid with rheological index n

Quantity	Name for SI Units	Symbol for SI Units
Henry's Law Constant	pascal-cubic meter per mole	$Pa \cdot m^3/mol$
Gas Law Constant	pascal-cubic meter per mole-kelvin	$Pa \cdot m^3/mol \cdot K$
Liquid-Side Gas-Liquid Mass Transfer Coefficient	meter per second	m/s
Mass Flow Rate Per Unit Cross-Sectional Area (or Mass Flux)	kilogram per square meter-second	$kg/m^2 \cdot s$
Molar Flow Rate Per Unit Cross-Sectional Area (or Molar Flux)	mole per square meter-second	$mol/m^2 \cdot s$
Particle Diameter	meter	m
Pressure	pascal	Pa
Specific Heat	joule per kilogram-kelvin	$J/kg \cdot K$
Surface Tension	newton per meter	N/m
Temperature	kelvin	K
Thermal Conductivity	watt per meter-kelvin	$W/m \cdot K$
Velocity	meter per second	m/s
Viscosity (Dynamic)	newton-second per square meter	$N \cdot s/m^2$
Viscosity (Kinematic)	square meter per second	m^2/s
Volume	cubic meter	m^3
Volumetric Flow Rate	cubic meter per second	m^3/s
Volumetric Gas-Side Gas-Liquid Mass Transfer Coefficient	mole per pascal-cubic meter -second	$mol/Pa \cdot m^3 \cdot s$
Volumetric Liquid-Side Gas-Liquid Mass Transfer Coefficient	reciprocal second	1/s

Subject Index